Fundamentals of Translational Tissue Engineering

Fundamentals of Translational Tissue Engineering

Edited by

Yogesh Kumar Verma
Regenerative Biology Research lab, Division of Clinical Research and Medical Management, Institute of Nuclear Medicine and Allied Sciences, Defence Research and Development Organisation (DRDO), New Delhi, India

Monalisa Mukherjee
Amity Institute of Click Chemistry Research and Studies (AICCRS) Professor, Amity Institute of Biotechnology (AIB), Amity University Uttar Pradesh Sector 125, Noida, Uttar Pradesh India

Subodh Kumar
Department of Biomedical Sciences, School of Biosciences and Technology, Galgotias University, Greater Noida, Uttar Pradesh, India; Institute of Nuclear Medicine and Allied Sciences, Defence Research and Development Organisation, Delhi, India

Sachin Kumar
Regenerative Biology Research lab, Division of Clinical Research and Medical Management, Institute of Nuclear Medicine and Allied Sciences, Defence Research and Development Organisation (DRDO), New Delhi, India

ELSEVIER

ACADEMIC PRESS
An imprint of Elsevier

Academic Press is an imprint of Elsevier
125 London Wall, London EC2Y 5AS, United Kingdom
50 Hampshire Street, 5th Floor, Cambridge, MA 02139, United States

ISBN: 978-0-443-33455-9

For Information on all Academic Press publications
visit our website at https://www.elsevier.com/books-and-journals

Publisher: Megan Ball
Acquisitions Editor: Elizabeth Brown
Editorial Project Manager: Vaishali Thapa
Production Project Manager: Vijayaraj Purushothaman
Cover Designer: Miles Hitchen

Typeset by MPS Limited, Chennai, India

Contents

16. Animals as preclinical models for tissue engineering in retina

Shikha Upreti, Neilanjan Maitra and Madhumita P. Ghosh

17. Circadian mechanisms and the importance of biological timing in tissue engineering

Amlan Chakraborty, Kamal Dua and Hannah J. Durrington

18. Tissue engineering strategies in retina

Shikha Upreti, Neilanjan Maitra and Madhumita P. Ghosh

Section III
Biosensors in tissue engineering

19. Biomaterial-based biosensors

Sabina Khan, Ruma Rani, Parth Malik and Tapan Kumar Mukherjee

20. Functional Biomaterials for Biosensing Technologies

Ramesh Kumar, Muskan, Teena Gakhar and Suman Nagpal

List of contributors

Ijaz Ahmed Department of Biology, Medgar Evers College, The City University of New York, New York, NY, United States

Shilpa Anupurba Department of Pathology, IMS and Sum Hospital, Bhubaneswar, Odisha, India

Ashima Bagaria Department of Physics, Manipal University Jaipur, Jaipur, Rajasthan, India

Biswarup Basu Department of Neuroendocrinology and Experimental Hematology, Chittaranjan National Cancer Institute, Kolkata, West Bengal, India

Bhupendra Singh Butola Department of Textile and Fibre Engineering, Indian Institute of Technology Delhi, New Delhi, India

Amlan Chakraborty Division of Immunology, Immunity to Infection and Respiratory Medicine, University of Manchester, Manchester, United Kingdom

Nidhi Chauhan School of Health Sciences and Technology (SoHST), UPES, Dehradun, Uttarakhand, India

Anita Chopra Lab Oncology, Dr. BRA Institute-Rotary Cancer Hospital, All India Institute of Medical Sciences, New Delhi, India

Peter R. Corridon Department of Biomedical Engineering and Biotechnology, Khalifa University of Science and Technology, Abu Dhabi, United Arab Emirates; Healthcare Engineering Innovation Group, Khalifa University, Abu Dhabi, United Arab Emirates; Center for Biotechnology, Khalifa University of Science and Technology, Al Zafranah Area, Abu Dhabi, United Arab Emirates

Devansh Dahiya Applied Chemistry and Nanoscience Department, Institute of Nuclear Medicine and Allied Sciences, DRDO, Brig S.K. Mazumdar Marg, New Delhi, Delhi, India

Eva C. Das Department of Biomaterial Science and Technology, Biomedical Technology Wing, Sree Chitra Tirunal Institute for Medical Sciences and Technology, Thiruvananthapuram, Kerala, India

Souvik Das Department of Neuroendocrinology and Experimental Hematology, Chittaranjan National Cancer Institute, Kolkata, West Bengal, India

Saumya Dash Stem Cell Facility, DBT-Centre of Excellence for Stem Cell Research, All India Institute of Medical Sciences, New Delhi, India

Vishal Kumar Deb School of Health Sciences and Technology (SoHST), UPES, Dehradun, Uttarakhand, India

Kamal Dua NICM Health Research Institute & School of Science, Western Sydney University, Sydney, NSW, Australia

Harshita Dubey Lab Oncology, Dr. BRA Institute-Rotary Cancer Hospital, All India Institute of Medical Sciences, New Delhi, India

Akansha Durgapal Department of Applied Physiology, Defence Research and Development Organisation (DRDO), Defence Institute of Physiology and Allied Sciences (DIPAS), New Delhi, India

Hannah J. Durrington Division of Immunology, Immunity to Infection and Respiratory Medicine, University of Manchester, Manchester, United Kingdom; Wythenshawe Hospital, Manchester University NHS Foundation Trust, Manchester, United Kingdom

Makayla J. Fahmy Department of Biomedical Engineering and Biotechnology, Khalifa University of Science and Technology, Abu Dhabi, United Arab Emirates

Teena Gakhar University School of Information and Communication Technology, GGSIPU, New Delhi, India

Gauri Gaur Department of Biotechnology, School of Health Sciences and Technology, UPES, Dehradun, Uttarakhand, India

Madhumita P. Ghosh Centre for Medical Biotechnology, Amity Institute of Biotechnology, Amity University, Noida, Uttar Pradesh, India

Shantanu Ghosh Amity Centre for Cognitive Science, Amity University Uttar Pradesh, Noida, Uttar Pradesh, India

Tanushri Ghosh Translational and Clinical Trials Unit, Kolkata Gynecological Oncology Trials and Translational Research Group (KolGOTrg), Kolkata, West Bengal, India

Anthony Guiseppi-Elie Department of Biomedical Engineering, Bioelectronics, Biosensors and Biochips (C3B®), Texas A&M University, College Station, TX, United States; ABTECH Scientific, Inc., Biotechnology Research Park, Richmond, VA, United States; Department of Cardiovascular Sciences, Houston Methodist Research Institute, Houston, TX, United States

Apoorv Gupta Reviana Innovations Private Limited, Delhi, India

Asheesh Gupta Department of Pharmacology, Defence Institute of Physiology and Allied Sciences (DIPAS), DRDO, Delhi, India

Shivangi Gupta Eicher Shroff Center for Stem Cell Research, Delhi, India

Zayd Hashem Department of Biomedical Engineering and Biotechnology, Khalifa University of Science and Technology, Abu Dhabi, United Arab Emirates

Utkarsh Jain School of Health Sciences and Technology (SoHST), UPES, Dehradun, Uttarakhand, India

Jabeen Kamal Department of Biology, Medgar Evers College, The City University of New York, New York, NY, United States

Devanshi Kapoor School of Biotechnology, Gautam Buddha University, Greater Noida, Uttar Pradesh, India

Sabina Khan Kadi Sarva Vishwavidyalaya, Gandhinagar, Gujarat, India

Anoushka Khanna Department of Textile and Fibre Engineering, Indian Institute of Technology Delhi, New Delhi, India

Manoj Komath Department of Biomaterial Science and Technology, Biomedical Technology Wing, Sree Chitra Tirunal Institute for Medical Sciences and Technology, Thiruvananthapuram, Kerala, India

Dhruv Kumar Department of Biotechnology, School of Health Sciences and Technology, UPES, Dehradun, Uttarakhand, India

Krishnanandan Kumar Department of Surgery, BMG Hospital, New Delhi, India

Ramesh Kumar Department of Biotechnology, Indira Gandhi University Meerpur, Rewari, Haryana, India

Sachin Kumar Regenerative Biology Research Lab, Division of Clinical Research and Medical Management, Institute of Nuclear Medicine and Allied Sciences, Defence Research and Development Organisation (DRDO), New Delhi, India

Sushmita Kushwaha Nanomaterial Toxicology Laboratory, Food, Drug & Chemical, Environment and Systems Toxicology (FEST) Division, Drug and Chemical Toxicology Group, CSIR-Indian Institute of Toxicology Research (CSIR-IITR), Lucknow, Uttar Pradesh, India; Academy of Scientific and Innovative Research (AcSIR), Ghaziabad, Uttar Pradesh, India

Maya Lakshmanan Manipal Institute of Regenerative Medicine, Manipal Academy of Higher Education, Manipal, Karnataka, India

Ashitha Mariam Leslie Manipal Institute of Regenerative Medicine, Manipal Academy of Higher Education, Manipal, Karnataka, India

Neilanjan Maitra Centre for Medical Biotechnology, Amity Institute of Biotechnology, Amity University, Noida, Uttar Pradesh, India

Misba Majood Department of Physiology, Johns Hopkins School of Medicine, Baltimore, MD, United States

Krishnajit Malakar Department of Pharmaceutical Regulatory Affairs, Amity Institute of Pharmacy, Amity University Noida, Noida, Uttar Pradesh, India

Parth Malik Swarrnim School of Sciences, Swarrnim Startup and Innovation University (SSIU), Gandhinagar, Gujarat, India

Priyam Mangal Applied Chemistry and Nanoscience Department, Institute of Nuclear Medicine and Allied Sciences, DRDO, Brig S.K. Mazumdar Marg, New Delhi, Delhi, India

Rashi Mathur Applied Chemistry and Nanoscience Department, Institute of Nuclear Medicine and Allied Sciences, DRDO, Brig S.K. Mazumdar Marg, New Delhi, Delhi, India

Robabeh Mazhabi West China School of Basic Medical Sciences & Forensic Medicine, Sichuan University, Chengdu, P.R. China

Sathya Veera Merla All India Institute of Medical Sciences, New Delhi, India

Ayman Mobin Department of Biomedical Engineering and Biotechnology, Khalifa University of Science and Technology, Abu Dhabi, United Arab Emirates

Mansi Modi Department of Biochemistry, BHR Hospital, Delhi, India

Sujata Mohanty Stem Cell Facility, DBT-Centre of Excellence for Stem Cell Research, All India Institute of Medical Sciences, New Delhi, India

Tapan Kumar Mukherjee Department of Biotechnology, Amity University, Kolkata, West Bengal, India

Asima Mukhopadhay Translational and Clinical Trials Unit, Kolkata Gynecological Oncology Trials and Translational Research Group (KolGOTrg), Kolkata, West Bengal, India

Muskan Department of Biotechnology, Indira Gandhi University Meerpur, Rewari, Haryana, India

Suman Nagpal Environmental Sciences, Indira Gandhi University, Rewari, Haryana, India

Ashrit Nair Department of Textile and Fibre Engineering, Indian Institute of Technology, Delhi, India

Dyer Narinesingh Department of Chemistry, University of the West indies, St. Augustine, Trinidad and Tobago

Shakra Nazli Nanomaterial Toxicology Laboratory, Food, Drug & Chemical, Environment and Systems Toxicology (FEST) Division, Drug and Chemical Toxicology Group, CSIR-Indian Institute of Toxicology Research (CSIR-IITR), Lucknow, Uttar Pradesh, India; Academy of Scientific and Innovative Research (AcSIR), Ghaziabad, Uttar Pradesh, India

Manasa Nune Manipal Institute of Regenerative Medicine, Manipal Academy of Higher Education, Manipal, Karnataka, India

Alam Nur-e-kamal Department of Biology, Medgar Evers College, The City University of New York, New York, NY, United States

Surya Kiran Panga Department of Forensic Medicine, All India Institute of Medical Sciences, New Delhi, India

Shefali Pareek Department of Physics, Manipal University Jaipur, Jaipur, Rajasthan, India

Virinder Parmar Department of Chemistry and Environmental Science, Medgar Evers College, The City University of New York, Brooklyn, NY, United States

Shiv Dutt Purohit Department of Biomedical Engineering and Biotechnology, Khalifa University of Science and Technology, Abu Dhabi, United Arab Emirates

Ekta Rahul Department of Pathology, AIIMS-CAPFIMS Campus, New Delhi, India

Ruma Rani Indian Council of Agricultural Research (ICAR)-National Research Centre on Equines, Hisar, Haryana, India

Amar Ranjan Lab Oncology, Dr. BRA Institute-Rotary Cancer Hospital, All India Institute of Medical Sciences, New Delhi, India

E. Pranshu Rao Stem Cell Facility, All India Institute of Medical Sciences, New Delhi, India

Raja Gopal Rayavarapu Nanomaterial Toxicology Laboratory, Food, Drug & Chemical, Environment and Systems Toxicology (FEST) Division, Drug and Chemical Toxicology Group, CSIR-Indian Institute of Toxicology Research (CSIR-IITR), Lucknow, Uttar Pradesh, India; Academy of Scientific and Innovative Research (AcSIR), Ghaziabad, Uttar Pradesh, India

Khushboo Razdan Amity Institute of Biotechnology, Amity University, Noida, Uttar Pradesh, India

Sumantra Dutta Roy Department of Electrical Engineering, Indian Institute of Technology Delhi, New Delhi, India

Mukesh Saran Directorate of Research, Manipal University Jaipur, Jaipur, Rajasthan, India

Somoshree Sengupta Translational and Clinical Trials Unit, Kolkata Gynecological Oncology Trials and Translational Research Group (KolGOTrg), Kolkata, West Bengal, India

Anil Sharma Department of Pathology, SKS Medical College, Mathura, Uttar Pradesh, India

Navneet Sharma Department of Pharmaceutical Regulatory Affairs, Amity Institute of Pharmacy, Amity University Noida, Noida, Uttar Pradesh, India

Vikesh Kumar Shukla Department of Pharmaceutical Regulatory Affairs, Amity Institute of Pharmacy, Amity University Noida, Noida, Uttar Pradesh, India

Sweta Singh Clinical Research and Medical Management Department, Institute of Nuclear Medicine and Allied Sciences, DRDO, Brig S.K. Mazumdar Marg, New Delhi, Delhi, India

Ruchi Sinha Department of Pathology, All India Institute of Medical Sciences, New Delhi, India

Sachin Sony Applied Chemistry and Nanoscience Department, Institute of Nuclear Medicine and Allied Sciences, DRDO, Brig S.K. Mazumdar Marg, New Delhi, Delhi, India

Tushar Srivastava The Krinsky Company, San Francisco, CA, United States

Sthuthi Jose M. Department of Chemistry, School of Advanced Sciences, Vellore Institute of Technology, Vellore, Tamil Nadu, India

S. Sumathi Department of Chemistry, School of Advanced Sciences, Vellore Institute of Technology, Vellore, Tamil Nadu, India

Pranay Tanwar Lab Oncology, Dr. BRA Institute-Rotary Cancer Hospital, All India Institute of Medical Sciences, New Delhi, India

Nishant Tyagi Voxvess Biotech Lab Private Limited, New Delhi, India

Shikha Upreti Centre for Medical Biotechnology, Amity Institute of Biotechnology, Amity University, Noida, Uttar Pradesh, India

Raunak Varshney Applied Chemistry and Nanoscience Department, Institute of Nuclear Medicine and Allied Sciences, DRDO, Brig S.K. Mazumdar Marg, New Delhi, Delhi, India

Yogesh Kumar Verma Regenerative Biology Research lab, Division of Clinical Research and Medical Management, Institute of Nuclear Medicine and Allied Sciences, Defence Research and Development Organisation (DRDO), New Delhi, India

Divya Yadav Department of Pharmacology, Defence Institute of Physiology and Allied Sciences (DIPAS), DRDO, Delhi, India

Simran Yadav Department of Pharmaceutical Regulatory Affairs, Amity Institute of Pharmacy, Amity University Noida, Noida, Uttar Pradesh, India

About the editors

Dr. Yogesh Kumar Verma, PhD, is a senior research scientist in the Institute of Nuclear Medicine and Allied Sciences (INMAS), Defence Research and Development Organisation (DRDO), situated in Delhi, India. He has obtained his basic qualification in biomedical sciences and bioinformatics. He has over 15 years of experience in the fields of molecular biology, stem cell research, regenerative medicine, tissue engineering, and bioinformatics. He has more than 60 publications to his credit, including patents, book chapters, and copyrights. Previously, he edited a book entitled *Stem Cells: An Alternative Therapy for COVID-19 and Cytokine Storm*, an Elsevier publication (eBook ISBN: 9780323955461), in the year 2023. Dr. Verma has been honored with many awards and fellowships for his exemplary work in the field of regenerative medicine and tissue engineering. He is a member of several professional bodies and committees and is engaged as a reviewer and editor for numerous journals.

Prof. (Dr.) Monalisa Mukherjee, PhD, FRSC is an accomplished academic and researcher in chemical sciences and materiobiology, serving as director at Amity Institute of Click Chemistry Research and Studies (AICCRS), Amity University Noida. A recipient of the National Scholarship Award (1998), Young Scientist Award (2010), and UK-India Distinguished Visiting Scientist Award (2011), she was elected Fellow of the Royal Society of Chemistry (FRSC, London) in 2021. Her pioneering work includes the discovery of origami-inspired ice helix structures within hydrogel nanotubes, which was featured in *Materials Horizons* and widely acclaimed. With over nine completed research projects and three ongoing initiatives, her innovations span biomaterials, nano-biomedicine, and sustainable chemistry. Prof. Mukherjee holds multiple patents and leads a transformative industry-academia collaboration on diabetic wound healing therapeutics. An accomplished mentor (9 PhD and 20 MTech scholars) and editor (Scientific Reports; Biocatalysis and Biotransformation), she has authored significant publications in premier journals, including *JACS*, *ACS Nano*, *ACS Applied Materials and Interfaces,* and *Green Chemistry*. Her edited volume *Bioinspired and Green Synthesis of Nanostructures* (Wiley) underscores her commitment to sustainable science.

Dr. Subodh Kumar is an Indian-born researcher and currently works as an assistant professor at the School of Biological and Biomedical Sciences, Galgotias University, Greater Noida, India. He completed his PhD from the INMAS, DRDO, in collaboration with Bharathiar University, Coimbatore, Tamil Nadu, India. His areas of specialization are stem cell research, tissue engineering, regenerative medicine, materials science, molecular biology, and bioinformatics. He has earned his bachelor's degree from Hansraj College, University of Delhi, and his master's degree (medical biotechnology) from Maharshi Dayanand University, Rohtak, Haryana, sponsored by the Department of Biotechnology, Government of India (PG research fellow). He has published more than 20 peer-reviewed articles in international journals and edited one book entitled *Stem Cells: An Alternative Therapy for COVID-19 and Cytokine Storm*. Dr. Kumar is endeavoring to put his intellect to the research community through his writing, scientific research, and ideas.

Sachin Kumar, MSc, is a researcher currently working at the INMAS, DRDO, Timarpur, New Delhi. His research focuses on developing innovative solutions for tissue healing, neuronal repair, and biomedical applications, contributing significantly to regenerative biology. Born and raised in India, Sachin Kumar earned his master's degree in genetics from Maharshi Dayanand University (Indian University). Sachin is passionate about contributing to the scientific community through innovative research and collaborative academic efforts. He is committed to translating laboratory discoveries into real-world clinical applications to improve patient outcomes and advance healthcare practices.

Foreword

This book offers a comprehensive description and scope of the translational tissue engineering concepts and real-world clinical applications. It guides learners through a spectrum of topics, from selection of biomaterials to design of scaffolds and critical regulatory considerations essential for bringing tissue engineering products to market.

Tissue engineering thrives through the collaboration and integration of diverse perspectives. The creativity of engineers, clinical insight of physicians and surgeons, and the scientific expertise of biologists and chemists propel this book to an excellent basis for the students and researchers in the field. This multidisciplinary constructive collaboration is what drives innovation forward, giving hope for advancing biomedical engineering in helping organ treatments and transplants.

I commend the authors and editors for providing a text that is scientifically and innovatively sound and accessible, making it not only an excellent introduction for researchers but also a valuable resource for experienced scientists and clinicians. The inclusion of critical thinking exercises further encourages active engagement with the material, fostering the next generation of innovators equipped to tackle the technological and scientific challenges ahead.

As you begin this journey into tissue engineering, remember that your curiosity and perseverance are as essential as any technique or material described within this book. The future of medicine relies on those willing to push boundaries. Hence, this book serves as your guide and inspiration in that pursuit.

Sincerely yours,

Prabir Patra, PhD

Jerome A. Gilbert, Chair Professor and Chair, Department of Biomedical and Electrical Engineering, College of Engineering and Computer Sciences, Marshall University, Huntington, WV, United States

Preface

Tissue engineering (TE) combines the fields of biology, engineering, and medicine to develop biological substitutes for the repair and regeneration of tissues and organs. However, there remains a significant gap between laboratory validation and clinical application, which requires a thorough understanding of scientific, technical, regulatory, and ethical domains. *Fundamentals of Translational Tissue Engineering* seeks to equip readers with comprehensive foundational knowledge as well as an applied perspective to translate scientific discoveries into clinical realities.

As researchers in the field of TE, we remain so engrossed in scientific experiments that the final goal of product development remains obscured by the publication and patenting. Horizontal level of experimentation on the vertical ladder of product development never increases the technology readiness level. In addition, lab scientists are, in most cases, unaware of the complex route of product translation from the lab-level proof-of-concept studies. This book title was envisaged to reconcile the diverse information available in TE and bridge the "valley of death" between research and translation. With this thought, we proposed the book title "Fundamentals of Translational Tissue Engineering" to Ms. Brown, senior acquisitions editor, Elsevier. The proposal was kindly accepted by Elsevier. The editorial team for this book comprises experts in the field of TE and regenerative medicine, who have contributed to shaping this book and brought it to fruition.

This book is meticulously divided into four sections.

I. Materials in Tissue Engineering with 09 chapters
II. Biological Aspects of Tissue Engineering with 09 chapters
III. Biosensors in Tissue Engineering with 04 chapters
IV. Clinical Aspect in Tissue Engineering with 08 chapters

The first section, "Material Sciences," introduces tissue engineering, mechanisms for cellular interaction with extracellular matrix or scaffolds, preparation techniques for scaffolds, specific interaction of biomaterials with different types of stem cells, and tumor microenvironment. A chapter is dedicated to the nonmedical applications of biomaterials.

The second section, "Biological Aspects of Tissue Engineering," discusses the assays and models available for in vitro and in vivo evaluation of the scaffolds and TE strategies for retina and reproductive tissue. This section also covers the importance of biological timing in TE and detection modeling for Parkinson's disease tremors.

The third section, "Biosensors in Tissue Engineering," deliberates on the application of biosensors in TE. This section covers how biomaterials-based biosensors enable real-time monitoring and analysis of biological processes within the engineered tissues. Moreover, how do biosensors facilitate the assessment of cell behavior, tissue maturation, and the response to stimuli, leading to more effective and predictable tissue regeneration?

The fourth section, "Clinical Aspects of Tissue Engineering," explores the guidelines for patenting the proof-of-concept outcomes, its third-party validation under GLP conditions, and regulatory guidelines for the translation of the prototype in clinical trials. This section also includes the ethical issues in TE and organ donation.

Each section is written by experts in their respective fields, providing an in-depth discussion of the complex pathway of translational TE. This book is intended for graduate students, researchers, clinicians, and professionals in the biomedical industry who aspire to contribute to the progression of TE. Whether readers are interested in biomaterial design and tissue cultivation or clinical translation, the text offers a comprehensive and pragmatic guide, supplemented with current challenges and case studies.

We extend our appreciation to the contributors, collaborators, and mentors who have shaped this book. The editorial team is especially thankful to Elsevier and its team, who guided us and the authors throughout the two-year journey with some hiccups and delayed deadlines. We express our gratitude to all the authors who have taken the time to share their hard work with readers, benefiting from their deliberations. The authors are highly appreciated for their effort in dealing with the new ELSA platform for writing and uploading the chapters, who never got bogged down by "technology" to share their experiences in the form of book chapters. We also appreciate the endeavors of all those who indirectly contributed to this project.

We hope that this volume will aid and inspire readers to effect positive changes within their professional sphere and ultimately in the lives of those who benefit from the advances of TE.

Welcome to the field of translational tissue engineering.

The views expressed in this book are solely by editors and authors and not that of Government of India.

Section I

Materials in tissue engineering

Chapter 1

Introduction to translational tissue engineering

Sachin Kumar[1], Akansha Durgapal[2], and Yogesh Kumar Verma[1]

[1]*Regenerative Biology Research Lab, Division of Clinical Research and Medical Management, Institute of Nuclear Medicine and Allied Sciences, Defence Research and Development Organisation (DRDO), New Delhi, India,* [2]*Department of Applied Physiology, Defence Research and Development Organisation (DRDO), Defence Institute of Physiology and Allied Sciences (DIPAS), New Delhi, India*

1.1 Introduction

Tissue translation engineering and regenerative medicine represent a transformative approach in biomedical science, merging engineering principles with biological insights to restore or replace damaged tissues and organs. This chapter focuses on the principles, technologies, and clinical applications that drive the field, emphasizing its potential to address medical gaps through regenerative techniques. Tissue translation engineering is an interdisciplinary process that involves designing, fabricating, and validating tissue constructs, typically through the use of scaffolds, cells, and bioactive molecules, for clinical applications. Unlike conventional tissue engineering (TE), which focuses on in vitro or preclinical development, translation engineering crosses the gap to human use by emphasizing scalability, regulatory compliance, and therapeutic efficacy [1]. It includes biomaterials (natural or synthetic), cellular therapies (e.g., stem cells), and advanced fabrication techniques (such as 3D bioprinting and electrospinning), all tailored to regenerate specific tissues, including muscle, bone, and reproductive organs. The scope is vast, targeting conditions from traumatic injuries like volumetric muscle loss (VML) to degenerative diseases and congenital defects. To succeed in vivo, TE integrates with regenerative medicine to restore physiological function, often requiring vascularization and immune compatibility [2].

1.2 Ancient history of tissue engineering

Regenerative medicine and TE have a long history. The Sushruta Samhita (6th century BC) mentions several TE techniques. One of these techniques is the forehead flap, a rhinoplasty method that involves removing a skin flap from the forehead to repair damage to the nose. Sushruta utilized splints made of bamboo, bark, or cloth to treat bone fractures and ensure that the bones were aligned adequately at the fracture site. A technique for repairing injured earlobes using cheek tissue is described in the Charaka Samhita (4th century BCE) [3].

An early depiction of regenerative medicine's vision is the well-known painting "Healing of Justinian," by St. Cosmas and St. Damien (278 AD), which depicts the transplanting of a homograft limb onto the patient [4].

1.2.1 History of tissue translational engineering

This field extends beyond static implants to dynamic systems, including innovative biomaterials that respond to environmental cues, reflecting a shift toward personalized, adaptive therapies. Its ultimate aim is to transition laboratory innovations into clinical practice, addressing both technical and translational challenges. Biomaterials, such as polyglycolic acid, were first investigated as cell growth scaffolds in the 1980s, marking the establishment of TE as a distinct field [5]. The 1990s saw advancements in TE, including the FDA clearance of Apligraf, a bilayered skin replacement with both dermal and epidermal components, in 1997 [6]. Thomson et al. isolated human embryonic stem cells, which expanded cellular options in the regenerative field [7]. These milestones laid the foundation for integrating engineering with biology, shifting focus from repair to regeneration. In the 2000s the development of vascular tissue

Fundamentals of Translational Tissue Engineering. DOI: https://doi.org/10.1016/B978-0-443-33455-9.00021-3

marked a significant advancement in scaffold design and 3D bioprinting [8]. The 2010s have shown promise in clinical research with synthetic tissues, such as decellularized extracellular matrix (ECM) scaffolds for TE [9]. Nowadays, advances in smart biomaterials and in silico modeling have accelerated translation, enabling real-time optimization and personalized design. This improvement represents a transition from basic to complex scaffolds and bioresponsive systems, driven by technological advancements and a greater understanding of tissue biology. Regenerative medicine, aided by tissue translation engineering, fills essential gaps in modern healthcare where traditional therapies (surgery, drugs, and prostheses) fall short. Chronic illnesses, including heart failure, osteoarthritis, and infertility, as well as severe injuries like VML or burns, frequently result in permanent tissue loss or malfunction [10]. Regenerative techniques have the potential to restore tissue structure and function, eliminating the need for organ donation, which is currently in limited availability.

1.3 Fundamentals of tissue engineering

TE integrates engineering and biological sciences to produce functioning tissue constructs that repair, maintain, or improve damaged tissues. TE is built on three crucial components: biomaterials, cellular elements, and scaffolds or matrices [11]. These basics provide the foundations for designing and developing regenerative therapies, allowing engineered tissues to be translated into therapeutic medicines. This section examines the diverse types and characteristics of biomaterials, the role of biological components, and the design and production of scaffolds, thereby establishing the framework for understanding future translational methods.

1.3.1 Biomaterials: types and properties

Biomaterials serve as the structural component of tissue-engineered constructs, providing a supportive environment for cell growth, differentiation, and tissue formation. They are broadly categorized into natural and synthetic types, each with unique characteristics suited for specific regeneration needs (Table 1.1) [2]. Natural biomaterials, such as collagen and fibrin, are derived from biological sources and offer inherent biocompatibility and bioactivity, promoting cell adhesion and signaling through motifs like RGD peptides [2]. However, their mechanical strength and degradation rates can be varied, limiting their use in load-bearing applications. Synthetic biomaterials, including polycaprolactone (PCL) and polylactic-co-glycolic acid (PLGA), provide precise control over mechanical properties (e.g., stiffness and elasticity) and degradation rate, making them ideal for customizable scaffolds [12]. Their lack of intrinsic bioactivity often necessitates surface modifications or the incorporation of growth factors. Hybrid biomaterials, such as collagen-PCL blends, combine the bioactivity of natural materials with the tunability of synthetics, optimizing both cellular response and structural integrity [13]. The properties that influence biomaterial selection include porosity for nutrient diffusion, biodegradability to match tissue regeneration, and biocompatibility to minimize immune rejection, all of which are crucial for successful TE outcomes.

1.3.2 Green biomaterials for bone and tissue engineering

Green biomaterials are sustainable, biodegradable, and often derived from natural sources, offering an environmentally friendly approach to bone and TE (BTE), which reduces reliance on synthetic polymers while supporting regeneration. Materials such as chitosan (derived from crustacean shells), cellulose (extracted from plant fibers), and silk fibroin (obtained from silkworms) offer biocompatible scaffolds with tailored properties for bone repair and soft tissue applications [2]. For bone regeneration, chitosan-hydroxyapatite composites emulate the mineralized matrix, promoting osteoblast activity and achieving a 45% increase in bone density in rat femoral defects within 6 weeks [14]. These scaffolds degrade into nontoxic byproducts, aligning with sustainability goals. In TE, cellulose-based hydrogels foster cartilage regeneration, with porcine models demonstrating a 30% enhancement in chondrocyte viability compared to synthetic alternatives [15]. Silk fibroin scaffolds, fabricated via electrospinning, support skin repair by accelerating vascularization, with preclinical studies showing a 50% faster wound closure rate [16].

1.4 Scaffold-based approaches

Scaffold-based approaches utilize three-dimensional (3D) constructs to support cell attachment, proliferation, and differentiation, mimicking the ECM. Scaffolds fabricated from natural or synthetic biomaterials are designed to have specific porosity, stiffness, and degradation rates that match the regeneration of tissues [2]

TABLE 1.1 Applications of natural, synthetic, and hybrid biomaterials in tissue engineering.

Natural biomaterial	Applications	Advantages	Disadvantages
Natural rubber	Cartilage repair, wound healing, and bone regeneration	High elasticity, biocompatibility, and durability	Low biodegradability, potential allergenicity
Collagen	Skin, bone, cartilage regeneration	Supports cell growth, natural ECM	Batch variability, risk of immunogenicity
Cellulose	Wound healing and scaffold development	Abundant, mechanically strong, biocompatible	Poor cell adhesion, insoluble in water
Gelatin	Wound healing, skin scaffolds	Biodegradable	Low mechanical strength
Lignin	Create scaffolds, hydrogels, bone regeneration, and wound dressing	Antioxidant, UV resistance, bio-based	Brittle, hydrophobic, low cell affinity
Pectin	Create scaffolds, hydrogels, and wound healing	Biocompatible, gel-forming, biodegradable	Weak mechanical properties, pH-sensitive
Alginate	Create a scaffold, form hydrogels, and promote wound healing	Biocompatible, easy gelation with Ca^{2+}, nontoxic	Lacks cell adhesion motifs, is mechanically weak
Fibrin	Create a scaffold, wound healing, biofunctionalization, and act as a sustained release system.	Natural ECM component promotes cell growth	Rapid degradation, low mechanical strength
Chitosan	Bone scaffold, cartilage repair, wound healing, nerve guidance	Biodegradable, antimicrobial, promotes cell adhesion	Limited solubility at neutral pH, brittle
Alginate	Create scaffold, wound healing, 3d bioprinting	Biodegradable, inexpensive, easy to process	Poor mechanical properties, high swelling
Keratin	Scaffold fabrication, wound healing, tissue regeneration (bone, skin, etc.)	Supports cell attachment, biocompatible	Brittle, difficult to process uniformly
Synthetic biomaterial	**Applications**	**Advantages**	**Disadvantages**
PLA	Bone, cartilage, and nerve scaffold	Biodegradable, good mechanical strength, easy to process	Brittle, acidic degradation byproducts, and slow degradation
PGA	Sutures, bone tissue scaffolds	Fast degradation, high porosity	Poor mechanical strength over time, acidic degradation
PLGA	Drug delivery, bone, and cartilage scaffolds	Tunable degradation rate, biocompatible, FDA-approved	Acidic byproducts, inflammation risk
PCL	Long-term implants, bone/cartilage scaffolds	Slow degradation, good flexibility, easy fabrication	Very slow degradation, limited cell adhesion
PEG	Hydrogels, drug delivery, 3D cell encapsulation	Hydrophilic, nontoxic, resistant to protein adsorption	Nonbiodegradable (unless modified), poor mechanical strength
PU	Vascular grafts, skin, heart valves	Elastic, durable, customizable properties	Potentially toxic degradation products, limited bioactivity
PEEK	Bone implants, spinal cages	High mechanical strength, bio-inert, radiolucent	Nonbiodegradable, expensive, and low bioactivity
PMMA	Bone cement, cranial implants	Rigid, durable, and easy to shape	Nonbiodegradable, heat generation during curing

(Continued)

TABLE 1.1 (Continued)

Natural biomaterial	Applications	Advantages	Disadvantages
PVA	Hydrogels for cartilage/ocular tissue	Biocompatible, flexible, and forms stable hydrogels	Poor biodegradability, low bioactivity
Polyanhydrides	Drug delivery systems, temporary scaffolds	Surface-eroding, tunable degradation, good for controlled release	Brittle, complex synthesis, limited mechanical strength
Hybrid biomaterial	**Applications**	**Advantages**	**Disadvantages**
SBGS-gelatin	Bone tissue engineering, implants	Enhanced compressive strength, improved osteoblastic activity, biocompatible	Complex fabrication, potential for uneven silica distribution
Electrospun polyurethane-gelatin/silk fibroin	Muscle regeneration, tissue scaffolds	Improved myotube formation, electroactive for cell stimulation	Requires precise electrospinning, potential cytotoxicity of synthetic polymers
Silicate bioactive glass-gelatin blend	Bone tissue engineering, implants	Enhanced compressive strength, improved osteoblastic activity, biocompatible	Complex fabrication, potential for uneven silica distribution
Collagen-PCL blend	Urethral tissue regeneration, soft tissue scaffolds	High biocompatibility, reduced complications (e.g., stone formation), mimics ECM	Limited mechanical strength for hard tissues, potential immunogenicity
Polyurethane-gelatin/silk fibroin blend	Muscle regeneration, tissue scaffolds	Improved myotube formation, electroactive for cell stimulation	Requires precise electrospinning, potential cytotoxicity of synthetic polymers

ECM, Extracellular matrix; *PCL*, polycaprolactone; *PEEK*, polyether ether ketone; *PEG*, polyethylene glycol; *PGA*, polyglycolic acid; *PLA*, polylactic acid; *PLGA*, polylactic-co-glycolic acid; *PMMA*, polymethyl methacrylate; *PU*, polyurethane; *PVA*, polyvinyl alcohol; *SBGS*, silica-based bioactive glass sol.

1.4.1 Scaffolds: design and fabrication

Scaffolds are three-dimensional structures that mimic the native ECM, guide cell behavior, and tissue organization. Design considerations include porosity, cell infiltration and gaseous exchange, mechanical strength to match the target tissue, and degradation rate to match the rate of regeneration [17]. Fabrication techniques range from conventional techniques like solvent casting and salt leaching, which create porous polymer scaffolds, to advanced approaches like electrospinning and 3D bioprinting (Fig. 1.1). Electrospinning produces nanofibrous scaffolds with a high surface area, thereby enhancing cell attachment, while bioprinting enables the precise deposition of cells and biomaterials to form complex, vascularized structures [8]. Hydrogels provide a hydrated environment for cell encapsulation, making them ideal for soft tissue regeneration, such as cartilage. Recent innovations incorporate innovative features such as stimulus-responsive degradation or drug release, enhancing scaffold adaptability to physiological cues [18].

1.4.2 Cell sheet engineering

Cell sheet engineering creates scaffold-free tissue constructs by culturing cells (e.g., fibroblasts and endothelial cells) on temperature-responsive surfaces (e.g., poly(N-isopropylacrylamide), PNIPAAm) that release intact cell layers upon cooling [19]. These sheets are rich in ECM and are stacked to form 3D tissues without the need for synthetic supports, which preserves cell-cell junctions and bioactivity [20]. This method is utilized in applications such as corneal repair (e.g., Holoclar) and skin regeneration, providing high cellularity and minimal immunogenicity. However, its reliance on cell-intensive processes limits thickness (typically <100 μm per sheet) and mechanical strength, restricting use in load-bearing tissues like bone; therefore hybrid approaches with scaffolds are emerging.

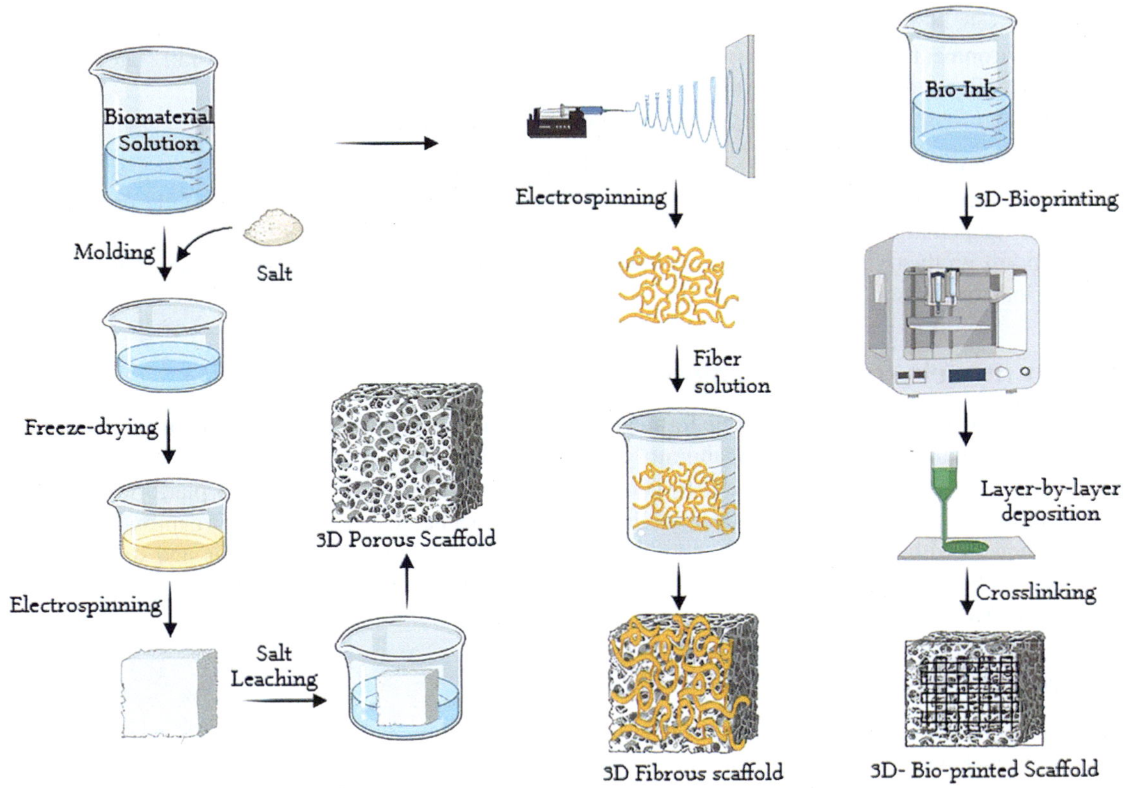

FIGURE 1.1 **Scaffold fabrication technique.** Overview of scaffold fabrication techniques: from conventional methods to advanced 3D bioprinting for tissue engineering applications.

1.4.3 Decellularization and recellularization

Decellularization and recellularization involve removing cellular components from donor tissues to create acellular ECM scaffolds, which are then repopulated with patient-specific cells. Detergents (e.g., SDS) or enzymatic treatments strip immunogenic antigens, which preserve the ECM's architecture and bioactivity. Recellularization with stem cells (e.g., induced pluripotent stem cells, iPSCs) or progenitors restores functionality [21]. These techniques are advantageous in terms of biomimetic cues and reduced rejection risk. Still, challenges such as incomplete antigen removal and cell repopulation efficiency hinder their large-scale adoption, driving research into optimized protocols.

1.4.4 Bioreactor cultivation

Bioreactors are a valuable tool for rapidly and efficiently growing large numbers of cells. They create the optimal environment, including the appropriate temperature and pH. Compared to standard techniques, such as growing on agar, bioreactors with a membrane layer generate more cells and facilitate their easier collection. Bioreactors also enable the development of 3D cell cultures, which are superior to static cultures. This is particularly useful in TE, where maintaining constant conditions is essential for cell development on scaffolds [21]. Bioreactor cultivation enhances tissue maturation by providing dynamic environments that simulate physiological conditions (e.g., shear stress and oxygenation). Many types of bioreactors are being utilized in the culturing of cells.

1. Stirred-tank bioreactors (STRs): STRs are versatile and suitable for both E. coli and mammalian cells. Scalable but may produce shear stress.
2. Airlift bioreactors: Low shear, suitable for plant/animal cells and vaccines, but less effective for high oxygen needs.
3. Packed-bed bioreactors: Ideal for immobilized systems and continuous usage.
4. Perfusion bioreactors maintain a continuous flow of nutrients and oxygen to cells, while simultaneously eliminating waste products and producing desired bioproducts.

5. Rotating wall vessels bioreactor: Cylindrical bioreactors that revolve along a horizontal axis to provide a low-shear, microgravity environment. The rotation suspends cells or tissue structures in a constant free-fall state, which replicates space conditions, allowing for the formation of three-dimensional (3D) tissue [22].
6. Photobioreactors are best suited for algae and photosynthetic activities [23].

Systems like perfusion bioreactors deliver nutrients to thick constructs (e.g., cartilage and bone), while rotating wall vessels promote uniform cell distribution in microgravity-like conditions [24]. Bioreactors provide precise control over conditions such as temperature, pH, and nutrients, ensuring consistent and scalable cell growth. They support high cell densities, which boost productivity, for instance, in the production of monoclonal antibodies. Their versatility allows the cultivation of various organisms for applications in healthcare, agriculture, and energy.

1.5 In vitro models for tissue engineering

In vitro models for TE offer controlled platforms for developing, testing, and refining tissue constructs before their in vivo application, thereby bridging the gap between laboratory research and preclinical studies. These models range from simple 3D cell cultures (e.g., spheroids and organoids) to complex systems, such as organ-on-chip platforms, which integrate microfluidics to mimic physiological conditions (e.g., fluid flow and mechanical stress) [25]. Organ-on-chip systems, such as liver-on-chip models with hepatocytes and endothelial cells, replicate metabolic functions and drug responses [26]. Neural models using ESC-derived neurones on 3D scaffolds simulate synaptic activity, providing a platform for drug screening for neurodegenerative diseases.

1.5.1 In vitro assay methods

In vitro assay methods are critical for evaluating tissue-engineered constructs before animal studies, which provide insights into cellular behavior, toxicity, biomaterial performance, and construct functionality under controlled conditions. These assays include viability tests (e.g., MTT and Live/Dead staining), proliferation assays (e.g., BrdU incorporation), and differentiation assessments (e.g., qPCR for gene expression and immunohistochemistry for protein markers) [27].

Advanced techniques, such as 3D imaging (e.g., confocal microscopy) and mechanical testing (e.g., tensile strength analysis), assess scaffold-cell interactions and structural integrity, respectively. Microfluidic platforms enable high-throughput screening of drug responses in engineered tissues. However, there are some limitations to in vitro assays, including the inability to fully replicate in vivo complexity (e.g., immune responses) and variability in assay sensitivity [28].

1.5.2 Stem cells in regenerative scaffolds

Cells are the functional units of tissue-engineered constructs, which drive regeneration through proliferation, differentiation, and ECM production. Stem cells are multipotent, undifferentiated cells that can replicate themselves to differentiate into specialized cells. Stem cells in regenerative scaffolds enhance tissue repair by providing a renewable source of progenitor cells that can differentiate into target tissues and secrete bioactive factors [29]. Stem cells, particularly mesenchymal stem cells (MSCs) and induced pluripotent stem cells (iPSCs), are widely utilized due to their self-renewal and multilineage differentiation potential [30]. MSCs derived from the bone marrow or adipose tissue differentiate into osteoblasts, chondrocytes, or myocytes, making them versatile for the repair of musculoskeletal or reproductive tissues. iPSCs (induced pluripotent stem cells) are reprogrammed from adult cells to provide patient-specific options, as these cells are derived from patient fibroblasts, which offer personalized options while reducing immunogenicity [31]. Preclinical studies have demonstrated that iPSC-derived cardiomyocytes improve cardiac patch function in infarcted rats; however, their clinical application is limited due to concerns about tumorigenicity [32]. Progenitor cells are lineage-committed cells that specialize in particular functions, having differentiated from stem cells and retaining the potential to self-renew, for example, endothelial progenitor cells (EPCs) [33]. Differentiated cells are specialized or mature cells that perform unique functions and are more specific in terms of function, but they lack proliferative capacity. They are frequently cocultured with stem cells to enhance construct maturity, such as fibroblasts and endothelial cells. The selection of a cellular component depends on the target tissue's needs for repair or regeneration and delivery methods, which involve seeding onto scaffolds or injecting directly, emphasizing the need for specific approaches in regenerative medicine [34].

Embryonic stem cells (ESCs) on 3D surfaces provide a powerful platform for regenerative medicine, owing to their pluripotency and ability to differentiate into any cell type under controlled conditions. Unlike 2D cultures, 3D scaffolds (e.g., hydrogels and nanofibrous matrices) mimic the native ECM, providing spatial cues that enhance ESC proliferation, self-renewal, and lineage-specific differentiation [35].

1.6 Translational strategies in regenerative medicine

Translational strategies in regenerative medicine aim to convert TE innovations from laboratory concepts into clinically viable therapies, addressing the complex journey from bench to bedside. This process involves rigorous preclinical validation, optimization of biocompatibility and immune responses, and scalable manufacturing with stringent quality control. These strategies are crucial for ensuring that engineered tissues, whether for musculoskeletal repair or cardiovascular regeneration, meet the safety, efficacy, and practical standards required for human use.

1.6.1 Preclinical models: from bench to animal studies

Preclinical models serve as the critical link between in vitro experimentation and human application, testing the feasibility, safety, and efficacy of tissue-engineered constructs. Bench studies start with material synthesis and cell culture, evaluating scaffold degradation, cellular viability, and bioactivity in controlled settings (e.g., bioreactors) [24]. These experiments often utilize human cell lines or stem cells to establish proof of concept, such as VEGF-loaded scaffolds promoting angiogenesis in vitro before animal studies.

Small animal models, such as mice or rats, are used for initial in vivo testing due to their cost-effectiveness and genetic manipulability [36]. Larger animals, such as pigs or sheep, more closely mimic human physiology, particularly in load-bearing tissues or vascular integration [37]. Preclinical studies assess outcomes such as tissue integration, immune response, and long-term stability, providing data to refine constructs before conducting clinical trials. The choice of model depends on the target tissue and therapeutic goal.

1.6.2 Biocompatibility and immune response

Biocompatibility is a construct's ability to function without eliciting adverse effects, and it is critical for translational success. Biomaterials should promote cellular activity without causing inflammation, fibrosis, or rejection in the host [38]. Natural materials, such as collagen, have excellent biocompatibility due to their resemblance to the original ECM. In contrast, synthetic polymers such as PCL may require surface modifications (e.g., peptide coatings) to minimize immunological response [12]. Immune responses vary depending on the substance and host; for example, xenogeneic ECM scaffolds can activate macrophages, polarizing them toward pro-regenerative (M2) or pro-inflammatory (M1) phenotypes, depending on the processing [39].

To improve biocompatibility, strategies include decellularization to remove immunogenic antigens and the use of immunosuppressive coatings, as well as the utilization of autologous cells to avoid rejection [21]. In regenerative medicine, vascularization is an important biocompatibility parameter; without it, structures fail owing to hypoxia. Preclinical testing assesses these reactions using histology (e.g., inflammatory cell infiltration) and cytokine profiling, verifying that constructs integrate with host tissues. These factors are crucial for overcoming translational problems observed in early trials, where immune responses reduced efficacy.

1.6.3 Manufacturing and quality control

To ensure uniformity, safety, and reproducibility when scaling up tissue-engineered constructions from lab prototypes to clinical applications, strong manufacturing techniques and quality control (QC) are required (Fig. 1.2). Manufacturing entails transitioning from manual scaffold development (e.g., electrospinning) to automated technologies, such as 3D bioprinting, which can produce massive, homogeneous structures [8]. Bioreactors play a crucial role in providing dynamic settings to cultivate cell-seeded scaffolds at scale, thereby imitating physiological conditions (e.g., shear stress for vascular tissues). [24].

Quality control ensures product integrity by adhering to Good Manufacturing Practices (GMP). The parameters include sterility (e.g., absence of microbiological contamination), mechanical consistency (e.g., tensile strength within 5–15 kPa for muscle tissue), and bioactivity (e.g., drug or growth factor release rates) [1,17].

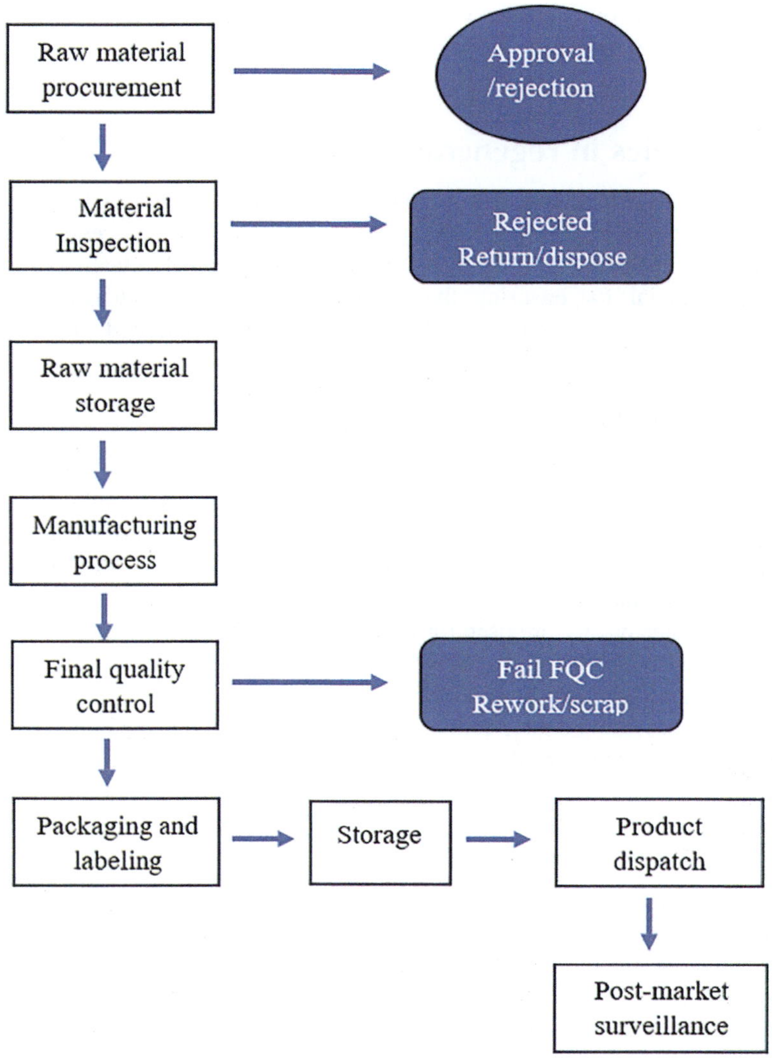

FIGURE 1.2 **Manufacturing and quality control.** Integrated manufacturing and quality control processes in tissue engineering: from biomaterial processing and scaffold fabrication to sterility assurance, mechanical testing, and regulatory compliance.

1.7 Advanced technologies in tissue translation

The translation of TE into regenerative medicine is driven by advanced technologies that address limitations in traditional approaches, such as poor vascularization, inconsistent bioactivity, and scalability. Technologies such as 3D bioprinting, controlled drug delivery systems, innovative biomaterials, and in silico modeling enable the precise fabrication, dynamic functionality, and optimization of tissue constructs. These technologies are reshaping the regenerative strategies by integrating engineering precision with biological complexity.

1.7.1 3D bioprinting and biofabrication

3D bioprinting revolutionizes TE, which enables the layer-by-layer deposition of cells, biomaterials, and bioactive molecules to create complex, structurally accurate constructs. Techniques such as extrusion, inkjet, and laser-assisted bioprinting utilize bioinks, typically hydrogels loaded with cells, to fabricate constructs with predefined structures [24]. These approaches can overcome the diffusion limitations of conventional scaffolds by incorporating vascular channels, which are essential for supplying nutrients to thick structures, such as muscle [40].

Biofabrication extends beyond printing to include postprocessing, such as bioreactor maturation, ensuring cellular viability and tissue functionality [24].

1.7.2 Controlled drug delivery systems

Controlled drug delivery systems enhance tissue translation by providing precise and sustained release of bioactive agents, such as growth factors or antiinflammatory drugs, within engineered constructs. These systems, including microspheres, hydrogel coatings, and electrospun fibers, regulate release kinetics through diffusion, degradation, or external cues (e.g., pH and temperature) [41]. This technology ensures localized therapy, minimizing systemic side effects and optimizing cellular responses, such as angiogenesis or differentiation.

1.7.3 Smart biomaterials and responsive systems

Innovative biomaterials respond dynamically to physiological cues such as pH, temperature, or hypoxia, enhancing the adaptability of tissue constructs. Examples include thermo-responsive hydrogels (e.g., PNIPAAm), which gel at body temperature for injectable delivery, and hypoxia-sensitive PEG scaffolds that release VEGF under low oxygen conditions [18]. Shape-memory polymers (e.g., polyurethane) adapt to defect shapes postimplantation, thereby improving fit and function in applications such as amputation stump repair [42]. These systems mimic the responsiveness of native tissue, promoting regeneration by aligning with environmental needs.

1.7.4 In silico modeling and computational approaches

In silico modeling utilizes computational tools to simulate TE processes, predict outcomes, and optimize designs before experimental validation (Fig. 1.3). Agent-based models (ABMs) replicate the individual behaviors of cells (e.g., endothelial cell movement). In contrast, continuum models utilize partial differential equations (PDEs) to represent tissue-level processes (e.g., oxygen diffusion, hypoxia, and mechanical stress) [43]. Finite element analysis (FEA) assesses scaffold mechanical properties, while molecular dynamics (MD) simulations explore protein-ligand interactions. These

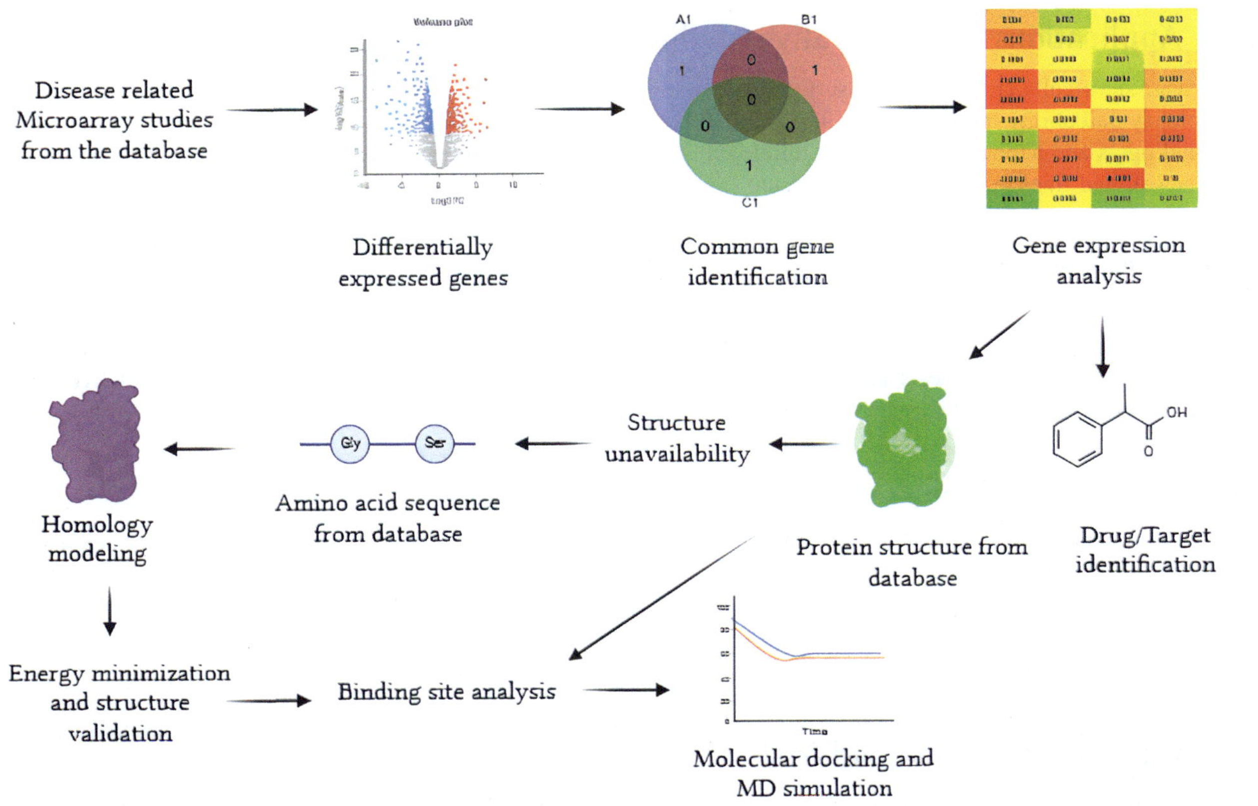

FIGURE 1.3 Computational approaches in tissue engineering. Computational approaches in tissue engineering: modeling, simulation, and AI-driven design for optimizing scaffold architecture.

models integrate multiscale data from molecular signaling to tissue construct, providing insights into how vascular networks form and support muscle regeneration in complex injury scenarios [44]. Machine learning enhances these models by analyzing large datasets (e.g., from GEO, BioModels) to predict biocompatibility or regeneration rates, thereby accelerating translation.

Various databases for in silico studies

Several databases provide critical data for building and validating in silico models:

- **Gene Expression Omnibus (GEO)**: Provides the microarray data.
- **Protein Data Bank (PDB)**: 3D structures of proteins for molecular docking and MD simulations, aiding in drug or peptide design.
- **Reactome**: Maps signaling pathways.
- **BioModels Database**: Provides prebuilt computational models.
- **Human Cell Atlas**: Offers single-cell RNA-seq data.

These repositories supply experimental data (e.g., gene expression and protein interactions) to parameterize and validate simulations, ensuring biological relevance.

1.7.5 Biomaterial-based biosensors and electroconductive hydrogels

Biomaterial-based biosensors use bioactive scaffolds and responsive materials to sense biological signals (such as pH, glucose, and cytokines) in tissue-engineered constructs or host environments to enhance control and responses. These sensors combine biomaterials, such as hydrogels or conductive polymers, with enzymes and antibodies, which allows for real-time analysis [45]. Electroconductive hydrogels combine the hydration and biocompatibility of hydrogels with electrical conductivity, enhancing TE by mimicking the bioelectric properties of native tissues, such as muscle, nerve, and heart. These materials, often composed of polymers like polypyrrole, polyaniline, or graphene blended with hydrogels (e.g., alginate and PEG), facilitate cell signaling and differentiation through electrical stimulation [46].

1.8 Applications in regenerative medicine

Embryonic stem cells (ESCs) on 3D surfaces provide a powerful platform for regenerative medicine, owing to their pluripotency and ability to differentiate into any cell type under controlled conditions. Unlike 2D cultures, 3D scaffolds (e.g., hydrogels and nanofibrous matrices) mimic the native ECM to provide spatial cues that enhance ESC proliferation, self-renewal, and lineage-specific differentiation [35].

1.8.1 Musculoskeletal regeneration

Musculoskeletal regeneration targets conditions such as VML, where significant muscle tissue is lost due to trauma or surgery, exceeding the body's natural repair capacity. TE approaches use decellularized ECM scaffolds, often seeded with stem cells, to promote myogenesis and angiogenesis [36]. Synthetic materials-based scaffolds with VEGF delivery further enhance vascular support, which is critical for significant defects [17].

1.8.2 Cardiovascular tissue engineering

Cardiovascular TE addresses heart failure, myocardial infarction, and vascular deficits by regenerating cardiac muscle and blood arteries. In preclinical models, acellular scaffolds, such as decellularized pig hearts, serve as a template for recellularization with cardiomyocytes or induced pluripotent stem cells (iPSCs) to restore partial pump function [47]. Bioprinted vascular patches that incorporate endothelial cells and smooth muscle cells show promise in healing infarcted hearts, with preclinical experiments finding increased ejection fraction [8].

1.9 Clinical translation and trials

Clinical translation and trials are the pinnacle of TE and regenerative medicine in transforming laboratory innovations into therapies that improve patient outcomes. This process involves navigating complex regulatory frameworks, conducting rigorous clinical studies, and learning from both successes and setbacks.

1.9.1 Regulatory frameworks and approval processes

Regulatory frameworks ensure the safety, efficacy, and quality of regenerative medicine products, striking a balance between innovation and patient protection. In the United States, the Food and Drug Administration (FDA) oversees tissue-engineered products under the 21st Century Cures Act (2016). The Regenerative Medicine Advanced Therapy (RMAT) designation accelerates approval for therapies addressing unmet medical needs. RMAT products, like cell therapies and tissue constructs, undergo phased clinical trials (I–III) but benefit from expedited review, requiring robust preclinical data on biocompatibility and functionality [1]. The FDA's Center for Biologics Evaluation and Research (CBER) enforces Good Manufacturing Practices (GMP) and postmarket surveillance to ensure consistency.

In the European Union, the European Medicines Agency (EMA) regulates Advanced Therapy Medicinal Products (ATMPs) under Regulation (EC) No 1394/2007 to categorize them as gene therapies, somatic cell therapies, or tissue-engineered products (EMA, 2007). The Committee for Advanced Therapies (CAT) evaluates ATMPs, emphasizing donor traceability and risk-based assessments with a centralized approval streamlining market access across member states. Japan's Pharmaceuticals and Medical Devices Agency (PMDA) offers a dual-track system, as outlined in the Regenerative Medicine Promotion Act 2014, which allows for conditional approval after early phase trials, provided safety is demonstrated. Full approval is contingent upon postmarket efficacy data [48]. These frameworks, while sharing goals of rigorous evidence generation and adaptability to rapid scientific advancements, shape global translation strategies.

1.9.2 Current clinical trials in regenerative medicine

Clinical trials in regenerative medicine are expanding globally, with over 5000 studies registered on ClinicalTrials.gov that involve stem cells, tissue scaffolds, and gene therapies [49]. Musculoskeletal applications dominate, with trials like NCT01159899 (Table 1.2) testing bone marrow stem cells seeded ECM scaffolds for the articular cartilage of the knee defects, reporting improved restoration of the osteochondral defects in Phase I. Cardiovascular trials, such as NCT05068674, explore cardiac patches with human embryonic stem cells, aiming to enhance survival and cardiac function in patients with chronic left ventricular dysfunction secondary to MI (Myocardial Infarction), with interim Phase I results showing safety.

Neural regeneration trials, such as NCT03105882, a pilot study on the clinical safety and Feasibility of the Neuro-Spinal Scaffold for the Treatment of Complete (AIS A) Traumatic Acute Spinal Cord Injury at the C5-T1 Neurological Levels. Reproductive health trials are emerging, notably NCT00001306, which tests the hypothesis that a lower risk therapy (alternate-day, lower dose, shorter term prednisone) will cause a remission of autoimmune ovarian failure. These trials focus on safety, effectiveness, and scalability, with many utilizing accelerated methods (e.g., RMAT, Japan's conditional approval) to expedite clinical adoption. At the same time, larger cohorts are required for more robust results.

1.9.3 Animals as preclinical models

Animals serve as indispensable preclinical models for evaluating tissue-engineered constructs before human trials, providing valuable insights into biocompatibility, functionality, and host responses under physiological conditions. Animal models for TE involve the implantation of engineered constructs into living organisms to assess integration, functionality, and host responses under physiological conditions. Common models include rodents (e.g., rats and mice), larger animals (e.g., rabbits and pigs), and immunocompromised strains (e.g., nude mice) to minimize rejection of human cell-based constructs [50]. Larger models, such as porcine coronary artery grafts, assess mechanical performance and patency, which are critical for cardiovascular applications. The in vivo studies provide preclinical validation of bone, cartilage, and neural constructs, providing insights into immune responses, remodeling, and long-term efficacy. However, there are some limitations, including species-specific differences, ethical concerns, and high costs, which challenge scalability. Specialized models, such as those for bone implants in sheep or neural repair in dogs, address load-bearing and complex neural outcomes, respectively. Humanized mouse models with engrafted human immune systems enhance relevance for immunotherapy studies [51]. These models provide important insight before human trials, which include safety profiling, efficacy testing, and regulatory data generation, bridging in vitro findings to clinical trials.

1.9.4 GLP testing for tissue engineering

Good laboratory practice (GLP) testing for TE ensures that preclinical research adheres to high-quality standards established by regulatory agencies (e.g., FDA and EMA) to evaluate the safety, effectiveness, and repeatability of tissue-

TABLE 1.2 Clinical trials in tissue engineering.

Clinical trial no.	Title	Status	Disease/Treatment
NCT01301664	Osteoarthritic cartilage fragments can promote chondrogenesis of MSCs	Phase I	Total knee/hip replacement
NCT04878445	Tissue Engineering Approaches to Treat COPD	Phase I	Improve lung function and clinical outcome in patients with COPD
NCT02668042	Skin substitutes safety and efficacy studies for the treatment of difficult-to-heal wounds	Phase I	Liveness tissue skin safety and effectiveness for the treatment of difficult-to-heal wounds
NCT05721261	Regeneration of Acutely Injured Nerves With Temporary Electrical Stimulation	Phase I	A novel temporary peripheral nerve stimulation system that delivers a single dose of electrical stimulation therapy for 1 hour
NCT04001842	Mandibular Reconstruction With Axially Vascularised Bone Substitutes	Phase I	Axial **vascularization** of constructs using a microvascular arteriovenous fistula/loop (AV loop) aims at providing the construct with blood supply through a defined and dedicated vascular axis
NCT05068674	Human Embryonic Stem Cell-Derived Cardiomyocyte Therapy for Chronic Ischemic Left Ventricular Dysfunction (HECTOR)	Phase II	Cell therapy approach (Human embryonic stem cell-derived cardiomyocytes or hESC-CMs) to improve survival and cardiac function in patients with chronic left ventricular dysfunction secondary to MI (Myocardial Infarction)
NCT01159899	Transplantation of Bone Marrow Stem Cells Stimulated by Proteins Scaffold to Heal Defects Articular Cartilage of the Knee	Phase I	To investigate the efficacy and safety of autologous transplantation, under arthroscopy, of bone marrow mesenchymal stem cells, using, with a cell separator, a fresh nonculture expanded autologous bone marrow derived mesenchymal stem, mixed and activated with proteins scaffold in patient with Knee cartilage defects and osteoarthritis
NCT03105882	Pilot Study of the Neuro-Spinal Scaffold for the Treatment of AIS A Cervical Acute SCI	Completed	Neuro-spinal scaffoldTM for the treatment of complete (AIS A) traumatic acute spinal cord injury at the C5-T1 neurological levels
NCT00001306	Steroid Therapy in Autoimmune Premature Ovarian Failure	Completed	Autoimmune premature ovarian failure: a controlled trial of alternate-day prednisone therapy

engineered constructs before clinical trials. This method includes standardized protocols for study design, documentation, and quality control for biomaterials, cells, and structures that adhere to OECD GLP guidelines (OECD, 1998). It supports safety profiling (cytotoxicity, genotoxicity), efficacy validation (functional restoration), and batch consistency for scaled manufacturing.

1.9.5 Clinical trials guidelines for tissue engineering

Clinical trial guidelines for TE provide a systematic framework for designing, performing, and reporting studies, ensuring safety, effectiveness, and regulatory compliance for TE products, including scaffolds, cell treatments, and bioengineered organs. Key guidelines, such as ICH E6 (R3) Good Clinical Practice (GCP), emphasize participant protection, data integrity, and trial robustness, adapted for TE's unique challenges (e.g., biocompatibility and long-term integration). For example, a Phase I trial of a bioprinted skin transplant must assess adverse events (e.g., rejection) by GCP, with 90% of patients demonstrating no severe reactions judged acceptable for progression. Guidelines also require risk-based monitoring, such as imaging every 12 weeks, to track scaffold degradation in implants and address any abnormalities as soon as possible.

1.9.6 Clinical trials documents in nanomaterials and scaffolds

Clinical trial documents for nanomaterials and scaffolds in TE are crucial for regulatory filing, trial execution, and posttrial analysis to ensure the safety, effectiveness, and reproducibility of TE products. Key documents include the Investigator's Drug (IND), detailing nanomaterial properties and preclinical safety data (e.g., cell viability in vitro); the Clinical Trial Protocol (CTP), outlining trial phases, endpoints (e.g., 50% bone regeneration in 12 weeks), and nanomaterial-specific risks (e.g., toxicity); and Case Report Forms (CRFs), which capture patient outcomes (e.g., scaffold integration rates). These documents comply with ICH E6(R3) and GCP, as well as FDA 21 CFR Part 11, ensuring the auditable integrity of electronic records.

1.9.7 Regulatory aspects of tissue engineering

Regulatory aspects of TE encompass specific considerations and challenges that distinguish TE products like scaffolds, cell therapies, and bioengineered organs from traditional drugs or devices and require regulations by agencies like the FDA, EMA, and PMDA. TE products are frequently classified as combined products (e.g., biologic-device under the FDA's CBER/CDRH), which require compliance with standards such as those for human cells/tissues and biocompatibility (21 CFR Part 1271) (ISO 10993).

Key regulatory challenges include standardization, as TE products vary widely (e.g., hydrogel vs 3D-printed bone), which complicates the development of universal guidelines; long-term safety, with risks such as tumorigenesis from stem cells requiring 5–10-year follow-ups; and global harmonization, where differing standards (e.g., FDA vs China's NMPA) delay approvals. Now, regulatory bodies are adapting to TE-specific pathways, such as the FDA's Regenerative Medicine Advanced Therapy (RMAT). These aspects ensure patient safety and market entry but require ongoing refinement to keep pace with TE's rapid evolution.

1.10 Challenges and future directions

Tissue translation engineering and regenerative medicine have made significant advancements, but their widespread clinical adoption remains hindered by technical, ethical, and economic challenges. Overcoming these barriers is essential to realizing the field's potential to transform healthcare. Simultaneously, emerging trends such as personalized medicine and the integration of artificial intelligence (AI) offer promising avenues to enhance efficacy and accessibility. This section examines the primary obstacles to vascularization, scalability, and integration, alongside ethical and economic considerations, and highlights future directions that could redefine regenerative therapies.

1.10.1 Technical barriers: vascularization, scalability, and integration

Technical barriers hinder the translation of tissue-engineered structures into functional therapies. Vascularization remains a significant hurdle, as constructs thicker than 200 μm require blood vessels to supply oxygen and nutrients while preventing necrosis [40]. Current approaches, such as bioprinted vascular channels or VEGF delivery, improve perfusion but often result in immature or leaky arteries, limiting long-term survival [8].

Scalability challenges the transition from lab to clinic because traditional fabrication (such as electrospinning) cannot fulfill population demand. Automated technologies, such as 3D bioprinting and bioreactors, address this issue, but high costs and variability in cell sources (e.g., donor ECM) hinder large-scale production [1]. Integration with host tissues is another obstacle, as mismatched mechanical properties or immune responses (e.g., fibrosis) disrupt functionality. For example, cardiac patches frequently fail to electrically synchronize with the original myocardium and require advancements in biomaterials and cellular alignment [47]. These barriers demand innovative engineering solutions to ensure clinical success.

1.10.2 Ethical and economic considerations

Ethical and economic factors significantly influence the trajectory of regenerative medicine. Ethical concerns arise from cell sourcing; ESCs face moral debates, while iPSCs raise safety issues, such as teratoma risk [31]. Xenogeneic scaffolds pose zoonotic risks, necessitating stringent decellularization; however, regulatory approval varies globally, complicating standardization [21]. Another ethical challenge is equity of access, as high-cost treatments may exclude low-income groups, leading to healthcare disparities.

Economic considerations are equally pressing, with development costs for a single regenerative product often exceeding $1 billion, driven by R&D, trials, and GMP compliance [17]. Patient-specific therapies, such as bioprinted tissues, increase expenses, while reimbursement models lag, deterring investment (Grand View Research, 2023). In India, where trauma-related amputations burden the economy, affordable solutions are critical yet elusive. Balancing innovation with cost-effectiveness is essential for sustainable translation.

1.11 Emerging trends: personalized medicine and artificial intelligence integration

Personalized medicine and AI integration are shaping future directions in regenerative medicine, with promises of targeted and efficient therapy. Personalized techniques utilize patient-derived cells and leverage genomic data to create constructs tailored to individual tissue characteristics, thereby minimizing rejection and enhancing outcomes [52]. CRISPR-edited cells for precise functional restoration are an advancement in TE, but scaling remains a challenge. AI integration enhances design and translation by refining scaffold structure, predicting cellular responses, and facilitating trials. Machine learning techniques examine large datasets to predict biocompatibility or drug release kinetics, decreasing development time by up to 30% [53]. In silico models in TE aid in estimating scaffold porosity, mechanical strength, and degradation rate, hence accelerating preclinical validation. AI-driven automation may reduce costs in the future, making regenerative therapies more accessible; nevertheless, continuous improvement is needed in data quality and addressing technology limitations.

1.12 Limitations of tissue engineering

TE has inherent limitations that restrict its current applications and scalability, despite technological advancements. A fundamental restriction is the complexity of construction, as reproducing the delicate architecture of organs like the liver or kidney remains challenging. In preclinical models, 3D bioprinted constructs achieve only 20–30% of the performance of natural tissue. Cell sourcing poses another barrier, like the limited availability of autologous cells and ethical concerns over allogeneic or stem cells [31]. Material constraints also limit progress; biomaterial constructs often lack the mechanical strength required for load-bearing tissues, such as bone, and degradation rates may not match tissue regeneration [2]. Vascularization deficits restrict construct thickness without engineered vessels, which leads to necrosis in thicker tissues [54]. Finally, the cost and accessibility of production are high for complex scaffolds, which limit clinical adoption, especially in resource-poor settings.

References

[1] A. Khademhosseini, R. Langer, A decade of progress in tissue engineering, Nature Protocols 11 (10) (2016) 1775–1781.

[2] F.J. O'brien, Biomaterials & scaffolds for tissue engineering, Materials Today 14 (3) (2011) 88–95.

[3] M.A. Gandhi, B.K. Patil, Sushruta: the father of surgery and ancient medical innovations, Cureus 16 (9) (2024) .

[4] U. Meyer, The history of tissue engineering and regenerative medicine in perspective, Fundamentals of Tissue Engineering and Regenerative Medicine, Springer, Berlin Heidelberg, (2009), pp. 5–12 Berlin, Heidelberg.

[5] J.P. Vacanti, M.A. Morse, W.M. Saltzman, A.J. Domb, A. Perez-Atayde, R. Langer, Selective cell transplantation using bioabsorbable artificial polymers as matrices, Journal of Pediatric Surgery 23 (1) (1988) 3–9.

[6] V. Falanga, D. Margolis, O. Alvarez, M. Auletta, F. Maggiacomo, M. Altman, ... Human Skin Equivalent Investigators Group, Rapid healing of venous ulcers and lack of clinical rejection with an allogeneic cultured human skin equivalent, Archives of Dermatology 134 (3) (1998) 293–300.

[7] J.A. Thomson, J. Itskovitz-Eldor, S.S. Shapiro, M.A. Waknitz, J.J. Swiergiel, V.S. Marshall, et al., Embryonic stem cell lines derived from human blastocysts, Science 282 (5391) (1998) 1145–1147.

[8] S.V. Murphy, A. Atala, 3D bioprinting of tissues and organs, Nature Biotechnology 32 (8) (2014) 773–785.

[9] J. Dziki, S. Badylak, M. Yabroudi, B. Sicari, F. Ambrosio, K. Stearns, ... J.P. Rubin, An acellular biologic scaffold treatment for volumetric muscle loss: results of a 13-patient cohort study, NPJ Regenerative Medicine 1 (1) (2016) 1–12.

[10] A.S. Mao, D.J. Mooney, Regenerative medicine: current therapies and future directions, Proceedings of the National Academy of Sciences 112 (47) (2015) 14452–14459.

[11] P. Sharma, P. Kumar, R. Sharma, V.D. Bhatt, P.S. Dhot, Tissue engineering; current status & futuristic scope, Journal of Medicine and Life 12 (3) (2019) 225.

[12] R. Langer, D.A. Tirrell, Designing materials for biology and medicine, Nature 428 (6982) (2004) 487–492.

[13] B.P. Chan, K.W. Leong, Scaffolding in tissue engineering: general approaches and tissue-specific considerations, European Spine Journal 17 (Suppl 4) (2008) 467–479.

[14] J. Venkatesan, S.K. Kim, Chitosan composites for bone tissue engineering—an overview, Marine drugs 8 (8) (2010) 2252–2266.
[15] C. Chang, L. Zhang, Cellulose-based hydrogels: present status and application prospects, Carbohydrate Polymers 84 (1) (2011) 40–53.
[16] D.N. Rockwood, R.C. Preda, T. Yücel, X. Wang, M.L. Lovett, D.L. Kaplan, Materials fabrication from Bombyx mori silk fibroin, Nature Protocols 6 (10) (2011) 1612–1631.
[17] S.J. Hollister, Porous scaffold design for tissue engineering, Nature Materials 4 (7) (2005) 518–524.
[18] A.S. Caldwell, G.T. Campbell, K.M. Shekiro, K.S. Anseth, Clickable microgel scaffolds as platforms for 3D cell encapsulation, Advanced Healthcare Materials 6 (15) (2017) 1700254.
[19] M. Yamato, T. Okano, Cell sheet engineering, Materials Today 7 (5) (2004) 42–47.
[20] T. Shimizu, H. Sekine, M. Yamato, T. Okano, Cell sheet-based myocardial tissue engineering: new hope for damaged heart rescue, Current Pharmaceutical Design 15 (24) (2009) 2807–2814.
[21] S.F. Badylak, D. Taylor, K. Uygun, Whole-organ tissue engineering: decellularization and recellularization of three-dimensional matrix scaffolds, Annual Review of Biomedical Engineering 13 (1) (2011) 27–53.
[22] R.P. Schwarz, T.J. Goodwin, D.A. Wolf, Cell culture for three-dimensional modeling in rotating-wall vessels: an application of simulated microgravity, Journal of Tissue Culture Methods 14 (2) (1992) 51–57.
[23] C. Posten, Design principles of photo-bioreactors for cultivation of microalgae, Engineering in Life Sciences 9 (3) (2009) 165–177.
[24] I. Martin, D. Wendt, M. Heberer, The role of bioreactors in tissue engineering, Trends in Biotechnology 22 (2) (2004) 80–86.
[25] D. Huh, G.A. Hamilton, D.E. Ingber, From 3D cell culture to organs-on-chips, Trends in Cell Biology 21 (12) (2011) 745–754.
[26] S.N. Bhatia, D.E. Ingber, Microfluidic organs-on-chips, Nature Biotechnology 32 (8) (2014) 760–772.
[27] R.I. Freshney, Culture of Animal Cells: A Manual of Basic Technique and Specialized Applications, John Wiley & Sons, (2015).
[28] S. Caddeo, M. Boffito, S. Sartori, Tissue engineering approaches in the design of healthy and pathological in vitro tissue models, Frontiers in Bioengineering and Biotechnology 5 (2017) 40.
[29] N. Ono, D.H. Balani, H.M. Kronenberg, Stem and progenitor cells in skeletal development, Current Topics in Developmental Biology 133 (2019) 1–24.
[30] M.F. Pittenger, A.M. Mackay, S.C. Beck, R.K. Jaiswal, R. Douglas, J.D. Mosca, ... D.R. Marshak, Multilineage potential of adult human mesenchymal stem cells, Science 284 (5411) (1999) 143–147.
[31] K. Takahashi, S. Yamanaka, Induction of pluripotent stem cells from mouse embryonic and adult fibroblast cultures by defined factors, Cell 126 (4) (2006) 663–676.
[32] J. Zhang, W. Zhu, M. Radisic, G. Vunjak-Novakovic, Can we engineer a human cardiac patch for therapy? Circulation Research 123 (2) (2018) 244–265.
[33] T. Asahara, T. Murohara, A. Sullivan, M. Silver, R. van der Zee, T. Li, ... J.M. Isner, Isolation of putative progenitor endothelial cells for angiogenesis, Science 275 (5302) (1997) 964–966.
[34] A.I. Caplan, Adult mesenchymal stem cells for tissue engineering versus regenerative medicine, Journal of Cellular Physiology 213 (2) (2007) 341–347.
[35] T.P. Kraehenbuehl, R. Langer, L.S. Ferreira, Three-dimensional biomaterials for the study of human pluripotent stem cells, Nature Methods 8 (9) (2011) 731–736.
[36] B.M. Sicari, J.P. Rubin, C.L. Dearth, M.T. Wolf, F. Ambrosio, M. Boninger, ... S.F. Badylak, An acellular biologic scaffold promotes skeletal muscle formation in mice and humans with volumetric muscle loss, Science Translational Medicine 6 (234) (2014) 234ra58-234ra58.
[37] I. Ribitsch, P.M. Baptista, A. Lange-Consiglio, L. Melotti, M. Patruno, F. Jenner, ... L.C. Penning, Large animal models in regenerative medicine and tissue engineering: to do or not to do, Frontiers in Bioengineering and Biotechnology 8 (2020) 972.
[38] E. Ambrożewicz, G. Tokajuk, M. Muszyńska, I. Zaręba, E. Skrzydlewska, Cross talk between redox signalling and metabolic activity of osteoblasts and fibroblasts in the presence of hydroxyapatite-based biomaterials influences bone regeneration, Journal of Applied Biomedicine 17 (2) (2019) .
[39] B.N. Brown, J.E. Valentin, A.M. Stewart-Akers, G.P. McCabe, S.F. Badylak, Macrophage phenotype and remodeling outcomes in response to biologic scaffolds with and without a cellular component, Biomaterials 30 (8) (2009) 1482–1491.
[40] R.K. Jain, P. Au, J. Tam, D.G. Duda, D. Fukumura, Engineering vascularized tissue, Nature Biotechnology 23 (7) (2005) 821–823.
[41] N. Kamaly, B. Yameen, J. Wu, O.C. Farokhzad, Degradable controlled-release polymers and polymeric nanoparticles: mechanisms of controlling drug release, Chemical Reviews 116 (4) (2016) 2602–2663.
[42] J.D. Kingsley, S. Ranjan, N. Dasgupta, P. Saha, Nanotechnology for tissue engineering: need, techniques and applications, Journal of pharmacy research 7 (2) (2013) 200–204.
[43] J.T. Nardini, R.E. Baker, M.J. Simpson, K.B. Flores, Learning differential equation models from stochastic agent-based model simulations, Journal of the Royal Society Interface 18 (176) (2021) .
[44] S.M. Peirce, Computational and mathematical modeling of angiogenesis, Microcirculation 15 (8) (2008) 739–751.
[45] A.P. Turner, Biosensors: sense and sensibility, Chemical Society reviews 42 (8) (2013) 3184–3196.
[46] A. Guiseppi-Elie, Electroconductive hydrogels: synthesis, characterization and biomedical applications, Biomaterials 31 (10) (2010) 2701–2716.
[47] H.C. Ott, T.S. Matthiesen, S.K. Goh, L.D. Black, S.M. Kren, T.I. Netoff, et al., Perfusion-decellularized matrix: using nature's platform to engineer a bioartificial heart, Nature Medicine 14 (2) (2008) 213–221.
[48] K. Azuma, Regulatory landscape of regenerative medicine in Japan, Current Stem Cell Reports 1 (2) (2015) 118–128.
[49] A. Petrosyan, P.N. Martins, K. Solez, B.E. Uygun, V.S. Gorantla, G. Orlando, Regenerative medicine applications: an overview of clinical trials, Frontiers in Bioengineering and Biotechnology 10 (2022) 942750.

[50] J.I. Dawson, D.A. Wahl, S.A. Lanham, J.M. Kanczler, J.T. Czernuszka, R.O. Oreffo, Development of specific collagen scaffolds to support the osteogenic and chondrogenic differentiation of human bone marrow stromal cells, Biomaterials 29 (21) (2008) 3105–3116.

[51] J. Haier, F. Schmidt, In vivo animal models in tissue engineering, Fundamentals of Tissue Engineering and Regenerative Medicine, Springer, Berlin Heidelberg, (2009), pp. 773–779 Berlin, Heidelberg.

[52] M.M. Laronda, A.L. Rutz, S. Xiao, K.A. Whelan, F.E. Duncan, E.W. Roth, ... R.N. Shah, A bioprosthetic ovary created using 3D printed microporous scaffolds restores ovarian function in sterilized mice, Nature Communications 8 (1) (2017) 15261.

[53] W. Sun, B. Starly, J. Nam, A. Darling, Bio-CAD modeling and its applications in computer-aided tissue engineering, Computer-aided Design 37 (11) (2005) 1097–1114.

[54] J. Rouwkema, A. Khademhosseini, Vascularization and angiogenesis in tissue engineering: beyond creating static networks, Trends in Biotechnology 34 (9) (2016) 733–745.

Chapter 2

Scaffold integration: extracellular matrices and cellular interfaces

Shiv Dutt Purohit[1], Makayla J. Fahmy[1], Ayman Mobin[1], Zayd Hashem[1], and Peter R. Corridon[1,2,3]

[1]*Department of Biomedical Engineering and Biotechnology, Khalifa University of Science and Technology, Abu Dhabi, United Arab Emirates,* [2]*Healthcare Engineering Innovation Group, Khalifa University, Abu Dhabi, United Arab Emirates,* [3]*Center for Biotechnology, Khalifa University of Science and Technology, Al Zafranah Area, Abu Dhabi, United Arab Emirates*

2.1 Introduction

Within tissue engineering, scaffolds are designed from natural and synthetic materials to facilitate cellular attachment, structural support, and tissue development. Accordingly, interactions between the extracellular matrix (ECM) and resident or reseeded cellular populations play an essential role in achieving this goal. These interactions comprise dynamic events that facilitate cellular integration, proliferation, differentiation, and ECM remodeling for tissue functionality.

The structure and make-up of these scaffolds can play a crucial role in modulating cellular responses, thus determining the success of tissue engineering applications. An ideal scaffold would mimic the native ECM characteristics (biochemical and biophysical) so that the cellular activities could considerably be improved. Scaffolds based on biocompatible materials (e.g., collagen, gelatin, and silk fibroin) to enhance cell adhesion form a suitable microenvironment, resulting in ECM deposition [1–3]. This further improves the functionality of scaffolds by including ECM-derived components. That is, the decellularized matrix is an abundant source of matric proteins and bioactive factors [4–6].

Previous studies have shown that mesenchymal stem cells (MSCs) cultured on ECM-loaded scaffolds exhibit enhanced proliferation and differentiation compared to conventional scaffold culture [7,8]. This improved efficacy is primarily attributed to abundant bioactive molecules within the ECM, which provide indispensable signaling cues for numerous cellular processes [7]. The mechanical properties of scaffolds are also critical for controlling cell behavior. Scaffolds must have proper mechanical properties, that is, sufficient strength and elasticity, to support tissue regeneration [2,9,10]. The structural properties of the biomaterial scaffold, such as pore size and alignment, are essential determinants of cellular behavior. For example, scaffolds with regularly arranged pores exhibit superior mechanical properties and cell infiltration compared to those with random pore structures [11,12]. This enables directional cell migration and improves tissue organization, crucial in bone, cartilage, ocular, and vascular regenerative applications [5,6,13–18].

Cellular migration within scaffolds involves the activation of signaling pathways triggered by interactions with the ECM. For example, integrins are cell surface receptors that mediate cellular adhesion and chemo-/mechano-transduction within the ECM [7]. These dynamic interactions influence numerous cellular functions, including gene expression, migration, and differentiation [7]. Accordingly, such interactions can also directly influence cellular behavior in response to the chemical and physical characteristics of the ECM, which can be controlled more efficiently during the scaffold design process to achieve optimal conditions for tissue regeneration [19]. Moreover, the immunomodulatory characteristics of scaffolds are essential in promoting cytocompatibility and cell integration into scaffolds, in vitro and within the posttransplantation environment [20,21]. To illustrate, this is a critical issue for allogeneic cell therapies, given the high risk for immune rejection. Accordingly, ECM-derived scaffolds hold great promise in overcoming this challenge by facilitating cell engraftment and functional integration, thereby advancing effective tissue engineering strategies.

This chapter provides a systematic overview of scaffold designing approaches by outlining different materials used in scaffold fabrication and describing innovative strategies that seek to enhance cell-ECM interactions. Additionally, the chapter briefly discusses testing methods applied to evaluate interactions between cells and the ECM, which ultimately

Fundamentals of Translational Tissue Engineering. DOI: https://doi.org/10.1016/B978-0-443-33455-9.00011-0

inform scaffold fabrication. Lastly, challenges associated with scaffold design and future research and development trends in tissue engineering are highlighted.

2.2 Basic concepts of scaffold design

In tissue engineering, the term "scaffold" refers to a structure miming the natural ECM of the tissue in which it will be implanted [22]. As the biochemical composition and mechanical structure of the ECM vary widely between tissue types, the ideal scaffold would effectively replicate that of the target tissue. However, this presents a significant challenge as the complexity and dynamic nature of native ECM make it difficult to replicate precisely. Thus current approaches in this field aim to partially replicate the ECM's functions in a way that aligns with the target tissue's biological, structural, and chemical properties [23,24]. Although the key characteristics of the ideal scaffold will vary depending on the tissue, generally, all scaffolds should promote cell adhesion, proliferation, and differentiation, as these factors are essential for tissue growth and regeneration [22,25]. Scaffolds should also enable cellular interactions by providing the conditions necessary for cell survival, such as oxygen and nutrient delivery, waste removal, and mechanical stability [26,27]. These key features fall into two broader categories: biological properties (encompassing biocompatibility, biodegradability, and porosity) and mechanical properties [28,29].

2.2.1 Biological characteristics of scaffolds

The primary requirement for any tissue engineering scaffold is biocompatibility. The chosen scaffold must support cell adhesion, migration, and proliferation without triggering a foreign body immune response that could lead to rejection [30]. Another essential feature of a scaffold in tissue engineering is biodegradability. Cells within the scaffold should eventually form their own ECM, remodeling the environment within the implanted scaffold [26,30], ultimately resulting in tissue regeneration without adversely disrupting the surroundings [27].

Porosity (defined as the total number of pores per unit volume) and pore diameter are other factors that significantly affect the adhesion and proliferation of the cells cultured on the scaffold. Porosity and pore size influence many other scaffold characteristics, such as biocompatibility, mechanical strength, and biodegradability. Highly porous biomaterials with an interconnected pore structure allow for cellular penetration, diffusion of nutrients to the implant site, and exchange of waste products from the scaffold [30,31]. High porosity also provides optimal conditions for vascularization, a process essential for regenerating highly metabolic organs [31]. However, this characteristic, which varies across tissue types, needs to be balanced to support sufficient adhesion and proliferation.

The pore diameter is a factor that can be altered easily; thus its effect on scaffold efficacy is widely researched [32]. The ideal pore size for a scaffold varies depending on the diameter of the suspended cells cultured on it [31]. It is commonly thought that scaffold pore size should exceed the size of the cell, as smaller pore sizes would inhibit cell migration and infiltration [29].

Porosity is another parameter to consider when designing a scaffold, particularly for ensuring effective cell infiltration, which can be a limitation of thicker scaffolds [33]. While low porosity is correlated with larger surface areas that enhance initial cell adhesion, high porosity contributes to greater diffusion of nutrients and water permeability, facilitating cell penetration. However, increased porosity compromises mechanical strength due to the loss of Young's modulus and is also linked to lower cell density, which can delay cell proliferation [32].

Sussman et al. observed that macrophages within 34 μM porous implants exhibited a pronounced M1 phenotype, with up to 63% higher M1 marker expression and an 85% reduction in M2 marker expression compared to macrophages external to the implants, which predominantly displayed an M2 phenotype [34]. Conversely, Madden et al. found that acellular scaffolds with 30–40μm pore diameters promoted angiogenesis and reduced fibrosis when implanted in cardiac tissue. This corresponded with a shift in macrophage polarity to an M2 phenotype [35]. In another study, Matsiko et al. demonstrated that collagen-hyaluronic acid scaffolds with larger pore diameters (300 μm) promoted cell proliferation, cartilage-like matrix deposition, and compressive modulus in comparison to scaffolds with smaller pore sizes of 94 μm and 130 μm [36]. These findings highlight how pore size and porosity can significantly affect the scaffold's efficacy, particularly in enhancing cell adhesion and inhibiting adverse immune responses.

2.2.2 Mechanical properties of scaffolds

Mechanical properties are an essential consideration when choosing a scaffold for use in tissue engineering. Compile strength, elasticity modulus, and stiffness must be considered when designing a scaffold to ensure structural integrity and durability [29]. Ideally, the scaffold should have mechanical properties consistent with the site in which it is being inserted while also being strong enough to withstand the surgical procedure necessary to implant it [37]. The mechanical strength of a scaffold can be assessed using compressive and tensile strength tests to ensure that the scaffold's strength is compliant with that of the tissue in which it is being implanted [29].

The mechanical properties of scaffolds can also affect cell behavior, such as proliferation or the ability to infiltrate the scaffold [32]. This applies particularly to mechanosensitive cells, such as mesenchymal cells, which respond to scaffold stiffness by differentiating into tissue-specific lineages [23,26].

2.3 Materials and methods for enhancing scaffold-extracellular matrix interactions

Several material forms can be utilized to create scaffolds, each having numerous desirable qualities and potential complications. Researchers must determine the optimal material after comprehensively reviewing advantages and possible logistical and financial restraints. This section will explore using natural, synthetic, and composite materials for scaffolds and their benefits and limitations.

2.3.1 Use of natural materials for scaffold construction

Natural materials, often sourced from biological matter that closely resembles the ECM found in humans, are commonly used to create scaffolds. These materials are biologically akin to human tissue, which makes them more desirable as scaffolds and less likely to be rejected by hosts [38]. The primary natural materials comprise proteins and polysaccharides, which are sugar multimers. Their innate architecture promotes cell attachment and proliferation and aids in the appropriate differentiation of cells. Common natural materials are collagen, chitosan, alginate, and silk fibroin [39].

In the human body, collagen is an essential structural protein. Hence, it is commonly used in tissue engineering as it is biocompatible with preexisting human tissue. Creating collagen into a scaffold involves freeze-drying or electrospinning the sample to create a structure filled with pores to allow cell growth [40]. The following material, chitosan, is a natural material with antibacterial benefits. It is made from the exoskeletons of crustaceans, such as lobsters and crabs. It has the advantage of being biodegradable.

Additionally, it is less likely to be rejected as the recorded interactions with human cells have been favorable [41]. The next biomaterial, alginate, is derived from brown seaweed. It is predominantly used in soft tissue engineering because it forms hydrogels with tunable mechanical properties [42]. Finally, silk fibers are created by silkworks and used for skin, bone, or cartilage. It's known for its mechanical strength; hence, it is most useful in tensile tissues [43].

Scaffolds derived from decellularized tissues and organs (dECM) are becoming increasingly popular natural alternatives for tissue engineering due to their MHC-free nature and interspecies ECM conservation, which enhance biocompatibility and reduce immunogenicity and rejection [44]. The ECM, typically from a xenogeneic source, has been used as a scaffold for the replacement or reconstruction of many different tissues, including studies utilizing ECM derived from organs such as the small intestine [45]. Furthermore, as native ECM structures can effectively support cell division, differentiation, and maturation, they provide an ideal environment for tissue development [25]. However, challenges remain, as variations in decellularization and postprocessing can impact dECM immunogenicity, limiting their clinical application [44]. To address this, researchers are developing advanced decellularization techniques to create ECM-based materials that more closely resemble human physiology. Tissue-derived materials are well-suited for encapsulating new cellular and biological components, supporting in vivo delivery for controlled degradation and tissue maturation [25].

The most significant advantage of natural materials is their high biocompatibility, which prevents rejection from the host and promotes cellular interactions for the booming scaffold implantation [38]. Along with benefits, natural materials also have several limitations. The most considerable challenge in creating such scaffolds is the intra and interbatch variability. Since these materials are derived and sourced from biological materials, there are natural differences in properties from one batch to another [46]. Furthermore, most natural materials, except silk fibroin, do not have the mechanical strength to be used in weight-bearing tissues such as bone or cartilage. Finally, they also have the potential to be contaminated with inappropriate cells if they are not purified in the correct manner, which leads to a higher risk of rejection [44].

2.3.2 Use of synthetic materials for scaffold construction

Another medium commonly used as a scaffold is synthetic materials. They are created in controlled environments, allowing them to be very carefully crafted to ensure the presence of particular desired properties [47]. Selecting a synthetic material requires engineers to analyze the tissue in question, identify the needed specific criteria of the scaffold, and then create it. Common examples of synthetic materials used in tissue engineering are polylactic acid (PLA), polyglycolic acid

(PGA), polycaprolactone (PCL), and polyethylene glycol (PEG) [48,49]. PLA is created from renewable sources such as corn starch. It provides mechanical strength and is slow to degrade, making it a desirable option for bone and cartilage scaffolds [50].

On the other hand, PGA degrades faster and is combined with PLA to create a copolymer of polylactic-glycolic acid (PLGA). This copolymer is very useful as it can be modified to account for when the scaffold will be used [51]. PCL can also support bone tissue as a long-term scaffolding option as it degrades much slower than PLA and PGA [52]. Finally, PEG is used for soft tissue scaffolds, where hydrogels retain water and support the ECM's structure [53].

The main advantage of using synthetic material for scaffolds is the mechanical strength they can provide for weight-bearing tissues such as bones and cartilage. They can be created with careful manipulation of their degradation rate by selecting specific polymer types [53]. Compared to natural materials, they have consistency in their batches, which means a standardized approach to creating scaffolds. However, they do not have the innate bioactivity of natural materials like collagen, so engineers must create bioactive molecules to mimic cellular interactions [54]. A potential risk these materials pose is the toxic degradation byproduct that can disrupt tissue regeneration [55]. Synthetic scaffolds can also trigger a foreign body response (FBR), marked by neutrophil and macrophage migration and cytokine production, leading to inflammation, degradation, and encapsulation by fibrotic tissue, which persists until the scaffold degrades or is removed, risking chronic inflammation [44]. Despite these challenges, synthetic materials remain popular due to their customizability and consistency.

2.3.3 Use of composite materials for scaffold construction

Composite materials are combinations of natural and synthetic materials. They can be used to create scaffolds for both mechanical strength and biocompatibility. These materials encompass numerous advantages of both natural and synthetic materials as they can be designed to be bioactive and strong. Additionally, their degradation rates can be meticulously controlled [56]. Composites of collagen with PLA are often used to create bone and cartilage to allow for cellular interactions in a structurally supported environment [57]. Bone regeneration is also promoted through hydroxyapatite and PCL composite materials, which increase the scaffold's osteoconductivity [58]. Alginate-PEG hydrogels can provide mechanical power similar to bioactivity combined [59]for soft tissue scaffolds. Finally, chitosan and polyvinyl alcohol (PVA) are often combined as scaffolds in wound healing [60].

Composite materials are the most multifunctional as they can implement the bioactive composition of natural materials in the structural strength of synthetic materials. These properties can be controlled and adjusted based on the tissue's needs, allowing for more flexibility [60]. However, the increased flexibility and customization come with increased complexity and cost [61]. Creating these scaffolds is more problematic as it requires specialized techniques to ensure that each scaffold exhibits consistent properties and has the correct material ratio [62,63]. However, the complications and potential for uneven degradation in these scaffolds [64,65]come with the additional challenges of creating such a material.

2.3.4 Conclusion

In conclusion, selecting the correct material in scaffold construction is vital to ensure successful implantation and function. Each material provides its advantages. Natural materials offer biocompatibility, which enhances their bioactivity. At the same time, synthetic materials provide mechanical strength and consistency. Composite materials are a balance of natural and artificial. However, they are more complex and have a greater risk of malfunction. Engineers must carefully analyze each material's properties while considering the scaffold's goal to optimize the design.

2.4 Techniques to promote scaffold-cell adhesion and integration

Improving scaffold-cell adhesion and integration is crucial for any tissue engineering approach that uses scaffolds for regenerative therapy. To achieve this, various strategies exist, including biochemical modifications and the incorporation of adhesion molecules in the scaffold design or by understanding cell signaling mechanisms influenced by scaffold interactions.

2.4.1 Biochemical modifications for cell adhesion and proliferation

Biochemical modifications of scaffolds could also dramatically improve cell attachment and growth on the scaffold. It can be achieved by the inclusion of bioactive molecules, which encourage cellular responses. A study depicted that enamel aids in improving the initial cell adhesion of osteoblast-like cells on polylactic-co-glycolic acid (PLGA) films [66]. The biofunctionalization of PLGA surfaces with Arg-Gly-Asp (RGD) peptides and hydroxyapatite nanoparticles has also been shown to promote cell adhesion and proliferation, suggesting surface chemistry's importance in scaffold design [67]. Similarly, gelatin coatings on scaffolds are advantageous by increasing hydrophilicity and, thus, cells' ability to adhere, thereby improving tissue colonization [68].

2.4.2 Adhesion molecules in scaffold design

Adhesion molecules mediate cell adhesion to scaffolds. As a study involving integrins [69]. The introduction of integrin-binding sites at specific locations on scaffold surfaces might have a substantial impact on cell activity. Scaffolds with higher collagen content provide increased integrin-binding sites to improve osteoblast differentiation and dispersion. Meanwhile, hydroxyapatite incorporation in scaffold materials has delineated stronger osteoblast adhesion and proliferation emanating from the augmented cell-matrix interactions [70]. Including adhesion molecules in scaffolds to mimic the natural ECM may improve MSCs' adhesion, promoting better integration and functionality within tissue engineering applications. Similarly, as shown in Fig. 2.1 hydroxyapatite and graphene oxide nanocomposite with a gelatin–alginate blend depicted increased alkaline phosphatase activity [71].

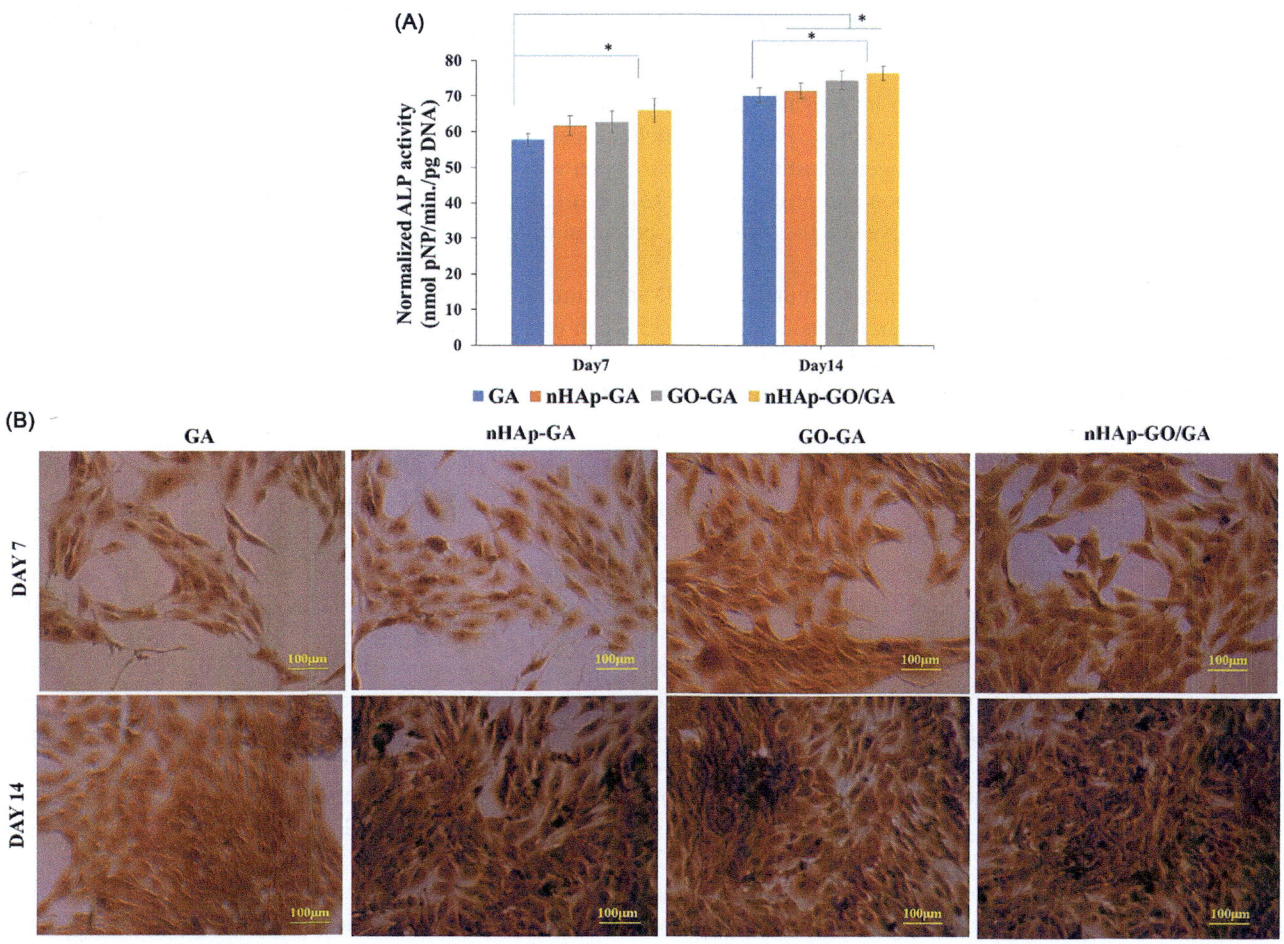

FIGURE 2.1 Quantitative analysis of ALP activity (A), visual depiction of alkaline phosphatase activity over GA, nHAp-GA, GO-GA, and nHAp-GO/GA nanocomposite scaffolds by ALP assay kit (B).
Reproduce with permission from S.D. Purohit et al., Fabrication of graphene oxide and nanohydroxyapatite reinforced gelatin–alginate nanocomposite scaffold for bone tissue regeneration, Frontiers in Materials 7 (2020).

2.4.3 Overview of cell signaling mechanisms influenced by scaffold interactions

The engagement of scaffolds with cells results in the activation of multiple signaling pathways that control critical cellular events, including migration, proliferation, and differentiation. For example, cyclic AMP (cAMP) enhances osteoblastic adhesion and scaffold differentiation. This suggests that the cellular response could be modulated by scaffold-induced signaling, as in the cAMP signaling pathway [72]. Beyond the physical properties of scaffolds, like surface topography and porosity, have also played a significant role in cell behavior by allowing their ecological mode of communication between proximity, affecting nutrient presence and removal, and assisting cell survival [73,74].

Therefore scaffold-cell adhesion and integration can be promoted by one or a combination of many strategies ranging from biochemical modifications to the incorporation of adhesion molecules and an understanding of the signaling mechanisms influenced by scaffold interactions. This is critical for tissue regenerative and repair scaffolds.

2.5 Methods of characterization and testing

Scaffold-cell interactions are critical to the success of any tissue-engineered product and are represented by characterization and testing. This includes several in vitro and in vivo techniques that examine the interaction of scaffolds with ECM components, cells, imaging, and biomechanical testing measures to quantify these interactions. These interactions can be scored based on set criteria, which are critical in guiding scaffold design optimizations.

2.5.1 In vitro methods for the evaluation of scaffold-extracellular matrix and scaffold-cell interactions

In vitro techniques have been used to investigate scaffold and cell interactions. This includes conventional methods for measuring and estimating possible cellular responses, such as cell adhesion, proliferation, and differentiation. For instance, scaffolds composed of collagen, elastin, and fibrin proteins have been documented to improve adipose tissue regeneration by enhancing cell adhesion and proliferation [75]. Similarly, gelatin–alginate-based scaffolds with graphene oxide or cerium oxide nanoparticles revealed the osteogenic potential of MSCs within the scaffolds by evaluating RUNX2 and Osteocalcin expressions [10,71].

Further, electrospun nanofibers can also imitate the 3D structure of the ECM, which might be beneficial for cell attachment and growth [76,77]. Research conducted by Vatankhah et al. fabricated a scaffold composed of electrospun tecophilic/gelatin nanofibers. Applying the as-fabricated scaffolds for small-diameter blood vessel tissue engineering depicted better cell migration, cell proliferation, and reduced protein adsorption related to thrombosis [78].

2.5.2 Imaging and biomechanical testing methods

X-ray microtomography (μCT) and high-resolution magnetic resonance imaging (MRI) are used to image scaffold architecture and monitor in vivo degradation. For example, the degradation of collagen scaffolds in vivo was monitored by μCT, where a substantial volume loss occurred over time that must be considered for an optimal synchronization of scaffold degradation with neo-tissue formation [75]. MRI has been used for the non-invasive evaluation of scaffold integration with surrounding tissues [79].

Another crucial property of the scaffolds is their mechanical properties. Thus it is necessary to evaluate the mechanical properties of scaffolds using biomechanical testing strategies such as tensile and compressive strength tests. These tests assist in verifying that the scaffolds can withstand normal physiological loads and provide a medium for cell attachment and growth [71]. For instance, scaffolds with hierarchical porous structures have exhibited augmented mechanical properties and enhanced cell attachment to elucidate pronounced cell-scaffold integration [80].

In scaffolds, regularly arranged pores are very important for osteochondral, ocular, and vascular regeneration due to the mechanical cues that promote cell migration and improve tissue integration. The ideal pore size and interconnectivity enable efficient nutrient and oxygen transport, which is crucial for cellular proliferation and vascularization. Moreover, aligned pore architectures direct cellular assembly and matrix deposition (which mimics tissue architectures). On one hand, this design facilitates cell attachment, and on the other hand, it modulates cellular activities to enhance tissue regeneration and integration. In conclusion, the patterned organization of pores in scaffolds could be a key factor in enhancing tissue regeneration, particularly in osteochondral, ocular, and vascular applications.

Various criteria can be used to assess scaffold-cell interactions, including cell adhesion, proliferation rates, differentiation potential, and support for ECM deposition. For a scaffold design to succeed, it should either modify or enhance such interactions, conferring appropriate mechanical properties and biodegradability. As such, it has been demonstrated that PLGA-collagen hybrid meshes stimulate adipogenic differentiation of human MSCs (making strong cell-ECM interactions) compared to other polymers [81].

A few scaffold design methods are mentioned below to provide a basic understanding of scaffold fabrication.

A study conducted by Xu and coworkers developed hierarchical porous nagelschmidtite bioceramic-silk scaffolds. Bioceramics significantly improved cell adhesion and vascularization, which can overcome poor cellular integration in larger-size scaffolds [82]. Such scaffolds show ECM-like structure and benefit better cell behavior and tissue regeneration. Similarly, electrospun gelatin scaffolds combined with controlled nano-architectures have been shown to promote angiogenesis and oriented tissue formation, demonstrating the significant influence of the scaffold architecture on cell behavior and tissue regeneration [83]. The fibers that make up these scaffolds are aligned to better guide cell growth and improve vascularization.

Further, cryogenic 3D printing can fabricate hierarchical porous and rhBMP-2-loaded Ca-P/PLLA nanocomposite scaffolds for bone tissue engineering using a similar method. Also, these scaffolds mimic the ECM and can contribute to osteogenic biochemical signals [80].

The scaffolds that replicate the biochemical and mechanical properties of the ECM lead to enhanced cell adhesion and proliferation. For example, incorporating fibronectin into ECM-deposited scaffolds improves cell adhesion and differentiation [81].

Moreover, incorporating bioactive chemical moieties, such as growth factors or adhesion peptides, can significantly augment the biological capacity of scaffolds. For instance, it has been found that RGD peptide-modified scaffolds possess higher cell adhesion and proliferation capacities, thus promoting tissue incorporation [84].

Overall, the characterization and evaluation of scaffold-cell interactions must be approached via a multifaceted strategy that integrates in vitro and in vivo techniques, advanced imaging modalities, biomechanical testing methods, and specific assessment criteria. We showed that these techniques have utility in scaffold design for mediating interactions with ECM and cells to improve the overall outcome of tissue engineering.

2.6 Challenges, opportunities, and future directions

The current challenges, recent improvements, and future perspectives in developing new scaffold strategies are described briefly in this section.

A primary obstacle in scaffold design is optimizing porosity and mechanical properties. The scaffold must be porous enough to permit the diffusional influx of nutrients and infiltration of cells but tough enough to withstand natural loads [85,86]. Furthermore, the scaffold materials' hydrophilicity could enhance cell adhesion and proliferation. Cell attachment was greatly restricted on hydrophobic surfaces, causing bio-tectonic materials to lack bonds with surrounding tissues [87]. Moreover, the absence of bioactive molecules that support the signaling of cells and cell differentiation limits their effectiveness for tissue regeneration [88,89].

Another challenge is mimicking the complex structure of native ECM, requiring scaffold designs. Conventional scaffold designs cannot reproduce natural tissues' hierarchical organization and biochemical cues, which can hamper the cellular responses necessary for successful tissue regeneration [90]. More recently, improvements in scaffold design have concentrated on developing material at the fabrication level to alter scaffold properties. As an example, the incorporation of electrospinning allows the obtaining of nanofibrous scaffolds that resemble the ECM architecture and thus favor cell adhesion and proliferation [68,91].

Recently, scaffold design has been facilitated by emerging technologies like 3D printing and bioprinting, which offer unprecedented control over scaffold architecture and composition [92]. For instance, 3D-printed scaffolds can be customed to have pore size and geometries that may boost the process of cell migration and nutrient transport in these ways [21,93]. Moreover, incorporating innovative materials that can respond to environmental cues (e.g., temperature and pH) may enhance the scaffolding performance by allowing dynamic cell interactions [21].

Researchers are currently exploring different nanomaterials, like carbon-based nanomaterials and metallic nanoparticles, which possess advantageous properties for deliberate cell adhesion and proliferation [94,95]. The results elsewhere showed that graphene oxide is suitable in providing a good environment for the cells to be grown and includes signals of biochemical cues from ECM [10,96]—also, the incorporation of mussel-derived coatings onto the scaffold enhanced cell adhesion and osteogenic potential [97].

In the future, scaffold design may shift towards an integral part of personalized medicine, where scaffolds are designed to accommodate the tissue characteristics and healing regiments needed explicitly by each patient. In this regard, biomaterial innovations such as decellularized ECMs are expected to contribute significantly towards developing scaffolds that mimic, more precisely, the native tissue environment [88,89]. In addition, the combination of gene therapy and stem cell technology with scaffold design is anticipated to improve tissue regeneration results. Gene therapies may benefit from these scaffolds that deliver them with stem cells and enhance tissue repair by directly regulating cellular behavior [90].

2.7 Conclusion

In conclusion, despite the challenges in scaffold design, such innovative new strategies and technologies provide encouraging ways to improve the integration of scaffolds with ECM and cells. Future scaffolds for tissue engineering can be designed with potential fabrication and therapeutic significance in mind.

AI disclosure

During the preparation of this work, the author(s) used Grammarly and ChatGPT to check grammar and spelling. After using this tool/service, the author(s) reviewed and edited the content as needed and take(s) full responsibility for the content of the publication.

References

[1] Z. Ding, et al., Simulation of ECM with silk and chitosan nanocomposite materials, Journal of Materials Chemistry B (2017).

[2] T. Long, et al., Fabrication of three-dimensional porous scaffold based on collagen fiber and bioglass for bone tissue engineering, Journal of Biomedical Materials Research. Part B, Applied Biomaterials (2014).

[3] S.D. Purohit, et al., Gelatin—alginate—cerium oxide nanocomposite scaffold for bone regeneration, Materials Science and Engineering C 116 (2020).

[4] J. Li, et al., The application of composite scaffold materials based on decellularized vascular matrix in tissue engineering: a review, Biomedical Engineering Online 22 (1) (2023) 62.

[5] X. Wang, V. Chan, P.R. Corridon, Decellularized blood vessel development: current state-of-the-art and future directions, Front Bioeng Biotechnology 10 (2022) 951644.

[6] X. Wang, V. Chan, P.R. Corridon, Acellular tissue-engineered vascular grafts from polymers: methods, achievements, characterization, and challenges, Polymers (Basel) 14 (22) (2022).

[7] L.S. Chua, H.W. Kim, J.H. Lee, Signaling of extracellular matrices for tissue regeneration and therapeutics, Tissue Engineering and Regenerative Medicine (2016).

[8] S. Thakkar, et al., Mesenchymal stromal cell-derived extracellular matrix influences gene expression of chondrocytes, Biofabrication (2013).

[9] N.E. Muzzio, S. Moya, G. Romero, Multifunctional scaffolds and synergistic strategies in tissue engineering and regenerative medicine, Pharmaceutics (2021).

[10] S.D. Purohit, et al., Development of a nanocomposite scaffold of gelatin–alginate–graphene oxide for bone tissue engineering, International Journal of Biological Macromolecules 133 (2019) 592–602.

[11] S.E. Dunphy, et al., Mechanical characterisation of directionally frozen polycaprolactone scaffolds using 1,4-dioxane and glacial acetic acid for articular cartilage tissue engineering, Biomedical Physics & Engineering Express (2018).

[12] Y. Qin, et al., Design and characterization of porous poly(glycerol-dodecanedioate) scaffolds for cartilage repair (2023).

[13] W. Zhang, et al., Cell-free biomimetic scaffold with cartilage extracellular matrix-like architectures for *in situ* inductive regeneration of osteochondral defects, Acs Biomaterials Science & Engineering (2020).

[14] Z.M. Ali, et al., A sustainable approach to derive sheep corneal scaffolds from stored slaughterhouse waste, Regenerative Medicine 19 (6) (2024) 303–315.

[15] I.V. Pantic, et al., Computational approaches for evaluating morphological changes in the corneal stroma associated with decellularization, Frontiers in Bioengineering and Biotechnology 11 (2023) 1105377.

[16] I.V. Pantic, et al., Analysis of vascular architecture and parenchymal damage generated by reduced blood perfusion in decellularized porcine kidneys using a gray level co-occurrence matrix, Frontiers in Cardiovascular Medicine 9 (2022) 797283.

[17] X. Wang, et al., A proposed model of xeno-keratoplasty using 3D printing and decellularization, Frontiers in Pharmacology 14 (2023) 1193606.

[18] X. Wang, et al., A scalable corneal xenograft platform: simultaneous opportunities for tissue engineering and circular economic sustainability by repurposing slaughterhouse waste, Frontiers in Bioengineering and Biotechnology 11 (2023) 1133122.

[19] B. Dhandayuthapani, et al., Polymeric scaffolds in tissue engineering application: a review, International Journal of Polymer Science (2011).

[20] X. Qiu, et al., Mesenchymal stem cells and extracellular matrix scaffold promote muscle regeneration by synergistically regulating macrophage polarization toward the M2 phenotype, Stem Cell Research & Therapy (2018).

[21] H. Yuan, K. Xing, H.-Y. Hsu, Trinity of three-dimensional (3D) scaffold, vibration, and 3D printing on cell culture application: a systematic review and indicating future direction, Bioengineering (2018).

[22] S. Sachdeva, et al., Scaffolds in tissue engineering, Journal of Cellular Biotechnology 8 (2) (2022) 95–103.

[23] B.P. Chan, K.W. Leong, Scaffolding in tissue engineering: general approaches and tissue-specific considerations, European Spine Journal Official Publication of the European Spine Society, the European Spinal Deformity Society, and the European Section of the Cervical Spine Research Society 17 (Suppl 4) (2008) 467–479.

[24] I.M. Adel, M.F. ElMeligy, N.A. Elkasabgy, Conventional and recent trends of scaffolds fabrication: a superior mode for tissue engineering, Pharmaceutics 14 (2) (2022).
[25] Y.-H. Kim, S. Vijayavenkataraman, G. Cidonio, Biomaterials and scaffolds for tissue engineering and regenerative medicine, BMC Methods 1 (1) (2024) 2.
[26] M. Mabrouk, H.H. Beherei, D.B. Das, Recent progress in the fabrication techniques of 3D scaffolds for tissue engineering, Materials Science and Engineering: C 110 (2020) 110716.
[27] A. Eltom, G. Zhong, A. Muhammad, Scaffold techniques and designs in tissue engineering functions and purposes: a review, Advances in Materials Science and Engineering 2019 (1) (2019) 3429527.
[28] S. Kona, A.S. Wadajkar, K.T. Nguyen, 6 - Tissue engineering applications of injectable biomaterials, in: B. Vernon (Ed.), Injectable Biomaterials, Woodhead Publishing, (2011), pp. 142–182.
[29] L. Suamte, et al., Various manufacturing methods and ideal properties of scaffolds for tissue engineering applications, Smart Materials in Manufacturing 1 (2023) 100011.
[30] R. Mohanraj, Chapter 2 - Plant-derived resorbable polymers in tissue engineering, in: V. Grumezescu, A.M. Grumezescu (Eds.), Materials for Biomedical Engineering, Elsevier, (2019), pp. 19–40.
[31] C.W. Li, G.J. Wang, 8 - MEMS manufacturing techniques for tissue scaffolding devices, in: S. Bhansali, A. Vasudev (Eds.), MEMS for Biomedical Applications, Woodhead Publishing, (2012), pp. 192–217.
[32] G. Lutzweiler, A. Ndreu Halili, N. Engin Vrana, The overview of porous, bioactive scaffolds as instructive biomaterials for tissue regeneration and their clinical translation, Pharmaceutics 12 (7) (2020).
[33] K.N. Bitar, E. Zakhem, Design strategies of biodegradable scaffolds for tissue regeneration, Biomedical Engineering and Computational Biology 6 (2014) 13–20.
[34] E.M. Sussman, et al., Porous implants modulate healing and induce shifts in local macrophage polarization in the foreign body reaction, Annals of Biomedical Engineering 42 (7) (2014) 1508–1516.
[35] L.R. Madden, et al., Proangiogenic scaffolds as functional templates for cardiac tissue engineering, Proceedings of the National Academy of Sciences 107 (34) (2010) 15211–15216.
[36] A. Matsiko, J.P. Gleeson, F.J. O'Brien, Scaffold mean pore size influences mesenchymal stem cell chondrogenic differentiation and matrix deposition, Tissue Engineering. Part A 21 (3–4) (2014) 486–497.
[37] N. Sultana, Y. Deng, J. Kuiper (Eds.), 1 - Mechanical and Biological Properties of Scaffold Materials, in Functional 3D Tissue Engineering Scaffolds, Woodhead Publishing, (2018), pp. 1–21.
[38] T.G. Kim, H. Shin, D.W. Lim, Biomimetic scaffolds for tissue engineering, Advanced Functional Materials 22 (12) (2012) 2446–2468.
[39] S. Ullah, X. Chen, Fabrication, applications and challenges of natural biomaterials in tissue engineering, Applied Materials Today 20 (2020) 100656.
[40] S.K. Nune, et al., Chapter 11 - Electrospinning of collagen nanofiber scaffolds for tissue repair and regeneration, in: D. Ficai, A.M. Grumezescu (Eds.), Nanostructures for Novel Therapy, Elsevier, (2017), pp. 281–311.
[41] S.L. Levengood, M. Zhang, Chitosan-based scaffolds for bone tissue engineering, Journal Materials Chemistry B 2 (21) (2014) 3161–3184.
[42] P. Rastogi, B. Kandasubramanian, Review of alginate-based hydrogel bioprinting for application in tissue engineering, Biofabrication 11 (4) (2019) 042001.
[43] L.-D. Koh, et al., Structures, mechanical properties and applications of silk fibroin materials, Progress in Polymer Science 46 (2015) 86–110.
[44] M. Kasravi, et al., Immunogenicity of decellularized extracellular matrix scaffolds: a bottleneck in tissue engineering and regenerative medicine, Biomaterials Research 27 (1) (2023).
[45] S.F. Badylak, Xenogeneic extracellular matrix as a scaffold for tissue reconstruction, Transplant Immunology 12 (3) (2004) 367–377.
[46] M. Abbasian, et al., Scaffolding polymeric biomaterials: are naturally occurring biological macromolecules more appropriate for tissue engineering? International Journal of Biological Macromolecules 134 (2019) 673–694.
[47] J.-F. Lutz, et al., From precision polymers to complex materials and systems, Nature Reviews Materials (2016) 16024.
[48] M.S. Lopes, A.L. Jardini, R.M. Filho, Poly (Lactic Acid) production for tissue engineering applications, Procedia Engineering 42 (2012) 1402–1413.
[49] O. Janoušková, Synthetic polymer scaffolds for soft tissue engineering, Physiological Research / Academia Scientiarum Bohemoslovaca 67 (Suppl 2) (2018) S335–s348.
[50] N. Khouri, et al., Polylactic Acid (PLA): Properties, synthesis, and biomedical applications - a review of the literature, Journal of Molecular Structure 1309 (2024) 138243.
[51] B. Prabhu, et al., Bioresorbable materials for orthopedic applications (lactide and glycolide based), in: B. Li, T. Webster (Eds.), Orthopedic Biomaterials: Progress in Biology, Manufacturing, and Industry Perspectives, Springer International Publishing, Cham, (2018), pp. 287–344.
[52] E. Malikmammadov, et al., PCL and PCL-based materials in biomedical applications, Journal of Biomaterials Science, Polymer Edition 29 (7–9) (2018) 863–893.
[53] J.K. Tessmar, A.M. Göpferich, Customized PEG-derived copolymers for tissue-engineering applications, Macromolecular Bioscience 7 (1) (2007) 23–39.
[54] P.X. Ma, Scaffolds for tissue fabrication, Materials Today 7 (5) (2004) 30–40.
[55] F.M. Michael, et al., Effect of nanofillers on the physico-mechanical properties of load bearing bone implants, Materials Science and Engineering: C 67 (2016) 792–806.

[56] J.M. Shapiro, M.L. Oyen, Hydrogel composite materials for tissue engineering scaffolds, JOM 65 (4) (2013) 505–516.

[57] A.-M. Haaparanta, et al., Preparation and characterization of collagen/PLA, chitosan/PLA, and collagen/chitosan/PLA hybrid scaffolds for cartilage tissue engineering, Journal of Materials Science: Materials in Medicine 25 (4) (2014) 1129–1136.

[58] P.X. Ma, R. Langer, Degradation, structure and properties of fibrous nonwoven poly(glycolic acid) scaffolds for tissue engineering, MRS Online Proceedings Library 394 (1) (1995) 99–104.

[59] N. Ramesh, S.C. Moratti, G.J. Dias, Hydroxyapatite-polymer biocomposites for bone regeneration: a review of current trends, Journal of Biomedical Materials Research. Part B, Applied Biomaterials 106 (5) (2018) 2046–2057.

[60] M. Hasany, et al., Synthesis, properties, and biomedical applications of alginate methacrylate (ALMA)-based hydrogels: current advances and challenges, Applied Materials Today 24 (2021) 101150.

[61] A. Rafique, et al., Chitosan functionalized poly(vinyl alcohol) for prospects biomedical and industrial applications: a review, International Journal of Biological Macromolecules 87 (2016) 141–154.

[62] A.R. Boccaccini, J.J. Blaker, Bioactive composite materials for tissue engineering scaffolds, Expert Review of Medical Devices 2 (3) (2005) 303–317.

[63] D.W. Hutmacher, et al., State of the art and future directions of scaffold-based bone engineering from a biomaterials perspective, Journal of Tissue Engineering and Regenerative Medicine 1 (4) (2007) 245–260.

[64] B. Thavornyutikarn, et al., Bone tissue engineering scaffolding: computer-aided scaffolding techniques, Progress in Biomaterials 3 (2) (2014) 61–102.

[65] N. Grossiord, et al., Degradation mechanisms in organic photovoltaic devices, Organic Electronics 13 (2012).

[66] K.W.H. Lo, et al., Evaluating the feasibility of utilizing the small molecule phenamil as a novel biofactor for bone regenerative engineering, Journal of Tissue Engineering and Regenerative Medicine (2012).

[67] H. Pan, et al., Effects of functionalization of PLGA-[Asp-PEG]ncopolymer surfaces with Arg-Gly-Asp peptides, hydroxyapatite nanoparticles, and BMP-2-derived peptides on cell behavior<i>in Vitro</I>, Journal of Biomedical Materials Research. Part A (2014).

[68] S. Bertoldi, et al., Functionalization of PU Foams via Inorganic and Organic Coatings to Improve Cell and Tissue Interactions, 2019.

[69] F.D.O. Gonçalves, et al., Hybrid membranes of PLLA/collagen for bone tissue engineering: a comparative study of scaffold production techniques for optimal mechanical properties and osteoinduction ability, Materials 8 (2) (2015) 408–423.

[70] B. Yin, et al., Hybrid macro-porous titanium ornamented by degradable 3D Gel/nHA micro-scaffolds for bone tissue regeneration, International Journal of Molecular Sciences (2016).

[71] S.D. Purohit, et al., Fabrication of graphene oxide and nanohydroxyapatite reinforced gelatin–alginate nanocomposite scaffold for bone tissue regeneration, Frontiers in Materials 7 (2020).

[72] K.W.H. Lo, et al., Activation of cyclic amp/protein kinase: a signaling pathway enhances osteoblast cell adhesion on biomaterials for regenerative engineering, Journal of Orthopaedic Research® (2010).

[73] F. Dini, et al., Tailored star poly (E-caprolactone) wet-spun scaffolds for in vivo regeneration of long bone critical size defects, Journal of Bioactive and Compatible Polymers (2015).

[74] J. Li, et al., Microporosity mediated proliferation of preosteoblast cells on 3D printed bone scaffolds, Nano Select (2021).

[75] P. Sawadkar, et al., Three dimensional porous scaffolds derived from collagen, elastin and fibrin proteins orchestrate adipose tissue regeneration, Journal of Tissue Engineering (2021).

[76] S. Gautam, et al., Gelatin-polycaprolactone-nanohydroxyapatite electrospun nanocomposite scaffold for bone tissue engineering, Materials Science and Engineering: C 119 (2021) 111588.

[77] S.D. Purohit, et al., Chitosan-based electrospun fibers for bone-tissue engineering: recent research advancements, International Journal of Biological Macromolecules 281 (2024) p. 136530-136530.

[78] E. Vatankhah, et al., Electrospun tecophilic/gelatin nanofibers with potential for small diameter blood vessel tissue engineering, Biopolymers (2014).

[79] Poirier, M. Quinot, et al., High-resolution 1.5-tesla magnetic resonance imaging for tissue-engineered constructs: a noninvasive tool to assess three-dimensional scaffold architecture and cell seeding, Tissue Engineering. Part C, Methods (2010).

[80] C. Wang, Q. Zhao, M. Wang, Cryogenic 3D printing for producing hierarchical porous and rhBMP-2-loaded Ca-P/Plla Nanocomposite scaffolds for bone tissue engineering, Biofabrication (2017).

[81] Y. Chen, et al., Preparation of stepwise adipogenesis-mimicking ECM-deposited PLGA–collagen hybrid meshes and their influence on adipogenic differentiation of hMSCs, Acs Biomaterials Science & Engineering (2019).

[82] M. Xu, et al., Hierarchically porous nagelschmidtite bioceramic–silk scaffolds for bone tissue engineering, Journal of Materials Chemistry B (2015).

[83] R.B. Montero, et al., bFGF-containing electrospun gelatin scaffolds with controlled nano-architectural features for directed angiogenesis, Acta Biomaterialia (2012).

[84] A. Koroleva, et al., Osteogenic differentiation of human mesenchymal stem cells in 3-D Zr-Si organic-inorganic scaffolds produced by two-photon polymerization technique, PLoS One (2015).

[85] Y. Ando, et al., Chitosan scaffolds as microcarriers for dynamic culture of human neural stem cells, Pharmaceutics (2023).

[86] C. Gandhimathi, et al., Osteogenic differentiation of mesenchymal stem cells with silica-coated gold nanoparticles for bone tissue engineering, International Journal of Molecular Sciences (2019).

[87] Q. Xü, et al., Recombinant human BMP-7 grafted poly(lactide-<i>co</I>-glycolide)/hydroxyapatite scaffolds <i>via</I> polydopamine for enhanced calvarial repair, RSC Advances (2018).

[88] Á. Paim, et al., Human dental pulp stem cell adhesion and detachment in polycaprolactone electrospun scaffolds under direct perfusion, Brazilian Journal of Medical and Biological Research (2018).

[89] Y. Zhao, et al., Cartilage extracellular matrix scaffold with kartogenin encapsulated PLGA microspheres for cartilage regeneration, Frontiers in Bioengineering and Biotechnology (2020).

[90] M.-X. Yao, Cartilage tissue healing and regeneration based on biocompatible materials: a systematic review and bibliometric analysis From 1993 to 2022, Frontiers in Pharmacology (2024).

[91] B. Sun, et al., Use of electrospun phenylalanine/poly-ε-caprolactone chiral hybrid scaffolds to promote endothelial remodeling, Frontiers in Bioengineering and Biotechnology (2021).

[92] M. Govindharaj, et al., 3D bioprinting of human mesenchymal stem cells in a novel tunic decellularized ecm bioink for cartilage tissue engineering, Materialia 23 (2022) 101457.

[93] Z. Xu, et al., In vitro and in vivo analysis of the effects of 3d-printed porous titanium alloy scaffold structure on osteogenic activity, Biomed Research International (2022).

[94] A. Madhusudhan, et al., Carbon based polymeric nanocomposite hydrogel bioink: a review, Polymers 16 (23) (2024).

[95] T.T. Truong, et al., Precision-engineered metal and metal-oxide nanoparticles for biomedical imaging and healthcare applications, Advances in Colloid and Interface Science 332 (2024) p. 103263-103263.

[96] S. Thampi, et al., Accelerated outgrowth of neurites on graphene oxide-based hybrid electrospun fibro-porous polymeric substrates, Acs Applied Bio Materials (2020).

[97] M. Wu, et al., Mussel-inspired multifunctional surface through promoting osteogenesis and inhibiting osteoclastogenesis to facilitate bone regeneration, NPJ Regenerative Medicine (2022).

Chapter 3

Tissue engineering: methods for scaffolds to extracellular matrix and scaffold to cell interaction

Harshita Dubey[1], Mansi Modi[2], Anil Sharma[3], Ekta Rahul[4], Tushar Srivastava[5], Sathya Veera Merla[6], Surya Kiran Panga[7], Ruchi Sinha[8], Anita Chopra[1], Pranay Tanwar[1], Krishnanandan Kumar[9], Shilpa Anupurba[10], and Amar Ranjan[1]

[1]*Lab Oncology, Dr. BRA Institute-Rotary Cancer Hospital, All India Institute of Medical Sciences, New Delhi, India,* [2]*Department of Biochemistry, BHR Hospital, Delhi, India,* [3]*Department of Pathology, SKS Medical College, Mathura, Uttar Pradesh, India,* [4]*Department of Pathology, AIIMS-CAPFIMS Campus, New Delhi, India,* [5]*The Krinsky Company, San Francisco, CA, United States,* [6]*All India Institute of Medical Sciences, New Delhi, India,* [7]*Department of Forensic Medicine, All India Institute of Medical Sciences, New Delhi, India,* [8]*Department of Pathology, All India Institute of Medical Sciences, New Delhi, India,* [9]*Department of Surgery, BMG Hospital, New Delhi, India,* [10]*Department of Pathology, IMS and Sum Hospital, Bhubaneswar, Odisha, India*

3.1 Introduction

The subject Tissue engineering (TE) aims to produce biological material for clinical use [1]. For example, bioengineered cartilage has been developed [2]. The implementation of tissue engineering is seen in organ transplant and disease modeling. It has also been used in clinical trials of drugs and personalized medicine. It has also a role in organ donation [3]. tissue engineering involves two techniques: scaffold-based or scaffold-free [3]. Scaffold-based technique is required in solid organs that promotes a replica of the existing organ. It uses the extracellular matrix (ECM) and proteinaceous materials of tissues. ECM gives strength to cell-to-cell interaction [4] and a firm structure [5] to the tissues. It helps in the signaling mechanisms [6] for proper functioning. ECM involves a number of proteins such as collagens, elastin, laminin, fibronectin, and various glycans [7].

3.2 Translational tissue engineering

Translational TE is a complex and multidisciplinary field (Fig. 3.1) that integrates the knowledge from various scientific disciplines to develop functional biological products for repairing or replacing damaged tissues in the body [8]. This innovative approach leverages the body's healing capabilities and exploits the remarkable reproductive potential of cells. The process involves utilizing specialized cells that produces a diverse array of growth factors and cytokines that serve as essential coordinators for tissue regeneration and repair. These cells are typically cultivated in advanced bioreactors or on precisely designed three-dimensional scaffolds, potentially leading to the regeneration of complex tissues and even entire organs [9].

Scaffolds, a fundamental component of tissue engineering , are fabricated either through synthetic means or by decellularizing existing tissues to produce collagen-rich matrices. These scaffolds are designed to gradually degrade upon implantation, being replaced over time by extracellular matrix protein secreted by infiltrating cells, thus replicating the natural environment [10].

Stem cells have emerged as particularly valuable assets in tissue engineering due to their unique self-renewal capabilities and remarkable ability to differentiate into various specialized cell types. This versatility renders them ideal candidates for regenerating diverse tissue types, ranging from cardiac muscle to neural tissue. Mesenchymal stem cells, for instance, have demonstrated potential in regenerating bone and cartilage, while induced pluripotent stem cells offer possibilities for patient-specific tissue engineering applications [11].

Fundamentals of Translational Tissue Engineering. DOI: https://doi.org/10.1016/B978-0-443-33455-9.00003-1

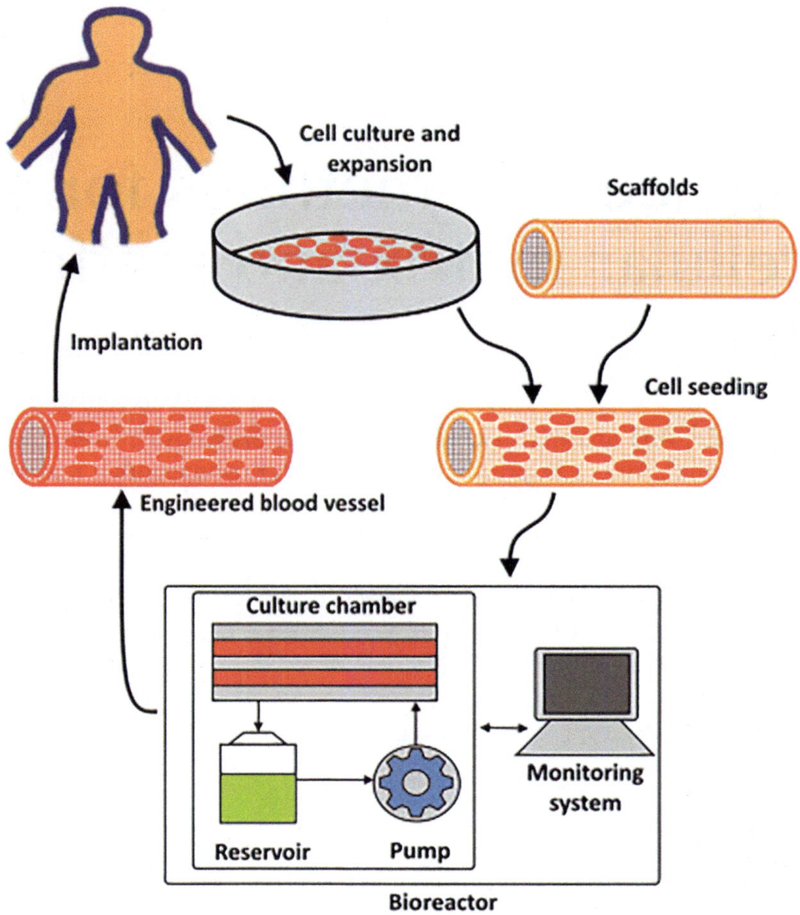

FIGURE 3.1 An overview of tissue engineering demonstrating the whole process.

However, the transition from promising research to successful product development in tissue engineering faces numerous challenges, often referred to as the "valley of death" in the biomedical industry. This critical phase necessitates extensive collaboration among engineers, biologists and clinicians to overcome significant hurdles in design, manufacturing, preclinical studies and clinical trials. Issues such as scalability, vascularization of engineered tissues, and long-term functionality must be addressed to ensure the viability and efficacy of tissue-engineered products [12].

As the field continues to advance, emerging technologies are expected to play increasingly crucial roles in shaping the future of personalized regenerative medicine. 3D-bioprinting, for example offers the potential to create complex, multicellular tissues with precise spatial organization, while gene editing techniques such as CRISPR-Cas9 could enable cell modification to enhance therapeutic effects or reduce immunogenicity. The regulatory landscape for tissue-engineered products also presents unique challenges, as these novel therapies often fall into the gray area between traditional pharmaceuticals and medical devices. Regulatory bodies worldwide are working to develop appropriate frameworks to ensure the safety and efficacy of these innovative treatments while not impeding innovation in the field. Continued multidisciplinary efforts and substantial investments will be essential to realize the full clinical potential of translational tissue engineering. As the researchers make progress in understanding the complex interplay between cells, scaffolds and the host environment, the prospect of on-demand, personalized tissue and organ replacements becomes increasingly feasible. The successful integration of tissue engineering into clinical practice has the potential to revolutionize healthcare, offering hope for patients with previously untreatable conditions and reducing the burden of organ shortages [13].

3.3 Tissue engineering scaffolds

Tissue engineering uses scaffolds (Fig. 3.2) are temporary frameworks that in conjunction with biological elements like cells and growth factors try to heal or regenerate damaged tissues [14]. Though it has limited use in patient care currently,

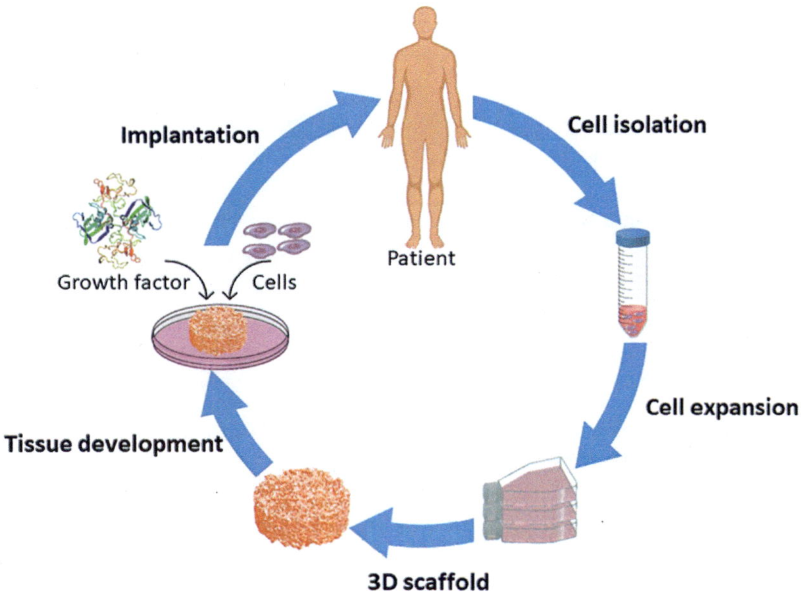

FIGURE 3.2 Illustration of tissue regeneration. This image demonstrates how scaffolds and cellular components contribute to the healing and rebuilding of tissue.

it holds great promise for treating a wide range of conditions from artificial cartilage and skin grafts to testing of the toxicity of experimental medicine.

Scaffolds are essential in tissue engineering process that offer a three-dimensional framework that facilitates cell adhesion, growth and differentiation. Acting as a structural guide, scaffolds support tissue development and regeneration with the goal of being gradually replaced by the body's natural ECM. Additionally, scaffolds can deliver growth factors and other bioactive agents that enhance the healing and repair processes [15].

3.4 Scaffold material, design and fabrication techniques

Scaffolds produced must have a range of properties like biocompatibility, biodegradability and mechanical properties that are consistent with the anatomic site, interconnected porous architecture and cost-effective scalable manufacturing processes to be suitable for use [15]. Natural materials from collagen, alginate and chitosan to highly complex decellularized whole-organ scaffolds have been used in scaffold design due to their biocompatibility and ability to mimic natural ECM (Fig. 3.3). However, they may lack the mechanical strength required and may degrade too quickly. There is also the possibility of potential impurities from the proteins or polysaccharides before implantation, which can be a source of immunogenicity [16]. Polyethylene glycol, polylactic acid and polyglycolic acid are examples of synthetic materials that provide more control over mechanical characteristics and rates of degradation. Given their lack of intrinsic bioactivity, these materials might need to have their surfaces modified in order to enhance cell adhesion and encourage the creation of ECM [17]. In bone tissue engineering, ceramics like hydroxyapatite and calcium phosphate are commonly used due to their similarity to bone mineral content. These materials are highly bioactive and support bone cell differentiation but can be brittle and challenging to shape [18].

There are various conventional fabrication techniques used to develop scaffolds, like electrospinning, solvent casting/particle leaching, freeze-drying, thermal induced phase separation (TIPS), gas foaming (GF), powder foaming and sol–gel, which play a noteworthy role in producing features of the architecture responsible for the success of the tissue engineering process [19].

The ability of the scaffold to replicate the native ECM in terms of both biochemical activity and physical structure is a crucial component of scaffold design. Scaffolds must facilitate cell attachment as well as promote the production of natural ECM constituents like collagen and elastin [15]. Scaffolds need to mimic the mechanical characteristics of the tissue they aim to regenerate. For instance, softer scaffolds are typically better suited for regenerating soft tissues like skin, whereas more rigid scaffolds are required for harder tissues such as bone [20]. To enhance ECM deposition, scaffolds can be functionalized with bioactive molecules such as RGD peptides, growth factors (e.g., VEGF for angiogenesis) or cytokines. These molecules help signal cells to secrete their own ECM, accelerating tissue regeneration [21]. The concept

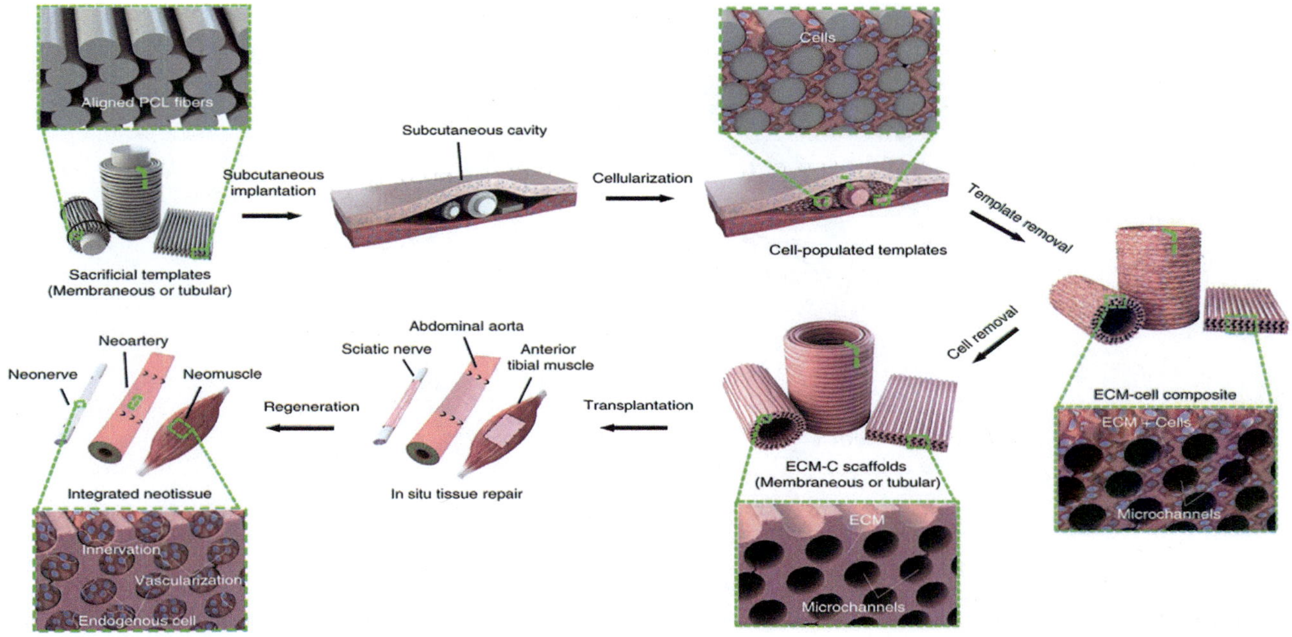

FIGURE 3.3 The interaction scaffold with ECM. This image depicts how engineered scaffolds integrate with the ECM to support cellular functions and tissue formation. *ECM*, Extracellular matrix.

of tissue engineering has evolved since 1980 and has been used in the diversified fields to date. It is being used for the regeneration and grafting of biological materials [22,23]. Tissue engineeringconsists of three parts: cells, scaffolds and growth-stimulating signals. Scaffold proteins are a group of proteins that bring together other proteins for Signal transduction events in cells. Scaffold stimulates the interaction of two proteins for a specified activity [24]. The Signal transduction is required for metabolism in the human body [25].

Scaffolds have a significant role in the repair of inflamed tissue. It protects cells involved in repair pathways, especially fibroblasts for their optimal functioning. In the repair of hernias, fistulas or burns, extracellular membranes from animal tissues are used. The dermis of pig and human is a common example [26]. Scaffolds are made up of a polymer that creates a framework for attachment of cells to convert it into tissue for repair [27].

3.5 Summary

Scaffolds are an important component of tissue regeneration that involves the extracellular matrix. ECM establishes a number of structural and biochemical mechanisms to produce a tissue. Tissue generation has numerous clinical roles in human beings. It is involved in the repair of damaged tissue, organ transplants, clinical trials of drugs and the utilization of various drugs in the form of personalized medicine.

3.6 Future perspectives

Tissue engineering has come a long way from its early conceptual stages and its future looks incredibly promising. As technologies evolve, we are moving closer to truly personalized regenerative therapies that could change the way we approach organ failure, chronic injuries and degenerative diseases.

3D bioprinting is a technique that allows scientists to layer different types of cells and biomaterials with precision, creating tissues that closely mimic the structure and function of native organs. With continued progress, printing fully functional and vascularized tissues for transplantation may become a clinical reality.

Another development is the use of *gene-editing tools*, like CRISPR-Cas9, to enhance cell properties. These tools can help correct genetic mutations or make cells less likely to trigger immune reactions, making engineered tissues safer and more effective, especially when combined with patient-derived stem cells. Also, the next-generation biomaterials based on scaffolds could play a vital role in improving integration with the host tissue and supporting long-term functionality.

In the coming years, tissue engineering is likely to shift from experimental treatments to *mainstream medical solutions*. With continued research and development, the dream of growing tailor-made tissues or organs for each patient, once science fiction, may soon become standard clinical practice. However, the road ahead still includes some challenges. Questions about *long-term safety, regulatory approvals, cost-effectiveness* and *ethical use* of emerging tools like gene editing remain important.

AI disclosure

The authors acknowledge the use of generative AI assistance exclusively to improve language readability, streamline phrasing, and check grammatical clarity during the drafting of the chapter. The conceptual framework, scientific analysis, and final manuscript verification were conducted entirely by the authors.

References

[1] C.C. Bell, D.F.G. Hendriks, S.M.L. Moro, E. Ellis, J. Walsh, A. Renblom, et al., Characterization of primary human hepatocyte spheroids as a model system for drug-induced liver injury, liver function and disease, Scientific Reports 6 (2016) Available from https://doi.org/10.1038/srep25187, www.nature.com/srep/index.html.

[2] M. Younesi, V.M. Goldberg, O. Akkus, A micro-architecturally biomimetic collagen template for mesenchymal condensation based cartilage regeneration, Acta Biomaterialia 30 (2016) 212–221. Available from https://doi.org/10.1016/j.actbio.2015.11.024, http://www.journals.elsevier.com/acta-biomaterialia.

[3] A.I. Caplan, Tissue engineering: then, now, and the future, Tissue Engineering - Part A 25 (7-8) (2019) 515–517. Available from https://doi.org/10.1089/ten.tea.2019.0011, http://www.liebertonline.com/tea.

[4] H.J. Kuo, C.L. Maslen, D.R. Keene, R.W. Glanville, Type VI collagen anchors endothelial basement membranes by interacting with type IV collagen, Journal of Biological Chemistry 272 (42) (1997) 26522–26529. Available from https://doi.org/10.1074/jbc.272.42.26522.

[5] S. Wang, R. Sekiguchi, W.P. Daley, K.M. Yamada, Patterned cell and matrix dynamics in branching morphogenesis, The Journal of Cell Biology 216 (3) (2017) 559–570. Available from https://doi.org/10.1083/jcb.201610048.

[6] Extracellular matrix: not just pretty fibrils - PMC. Available from: https://pmc.ncbi.nlm.nih.gov/articles/PMC3536535/.

[7] The extracellular matrix at a glance - PMC. Available from: https://pmc.ncbi.nlm.nih.gov/articles/PMC2995612/.

[8] R.G.L. da Silva, L. Au, A. Blasimme, Organizational aspects of tissue engineering clinical translation: insights from a qualitative case study, Translational Medicine Communications 9 (1) (2024). Available from https://doi.org/10.1186/s41231-024-00179-7.

[9] J.J. Vranckx, M.D. Hondt, Tissue engineering and surgery: from translational studies to human trials, Innovative Surgical Sciences 2 (4) (2017) 189–202. Available from https://doi.org/10.1515/iss-2017-0011, www.degruyter.com/view/j/iss.

[10] J.L. Olson, A. Atala, J.J. Yoo, Tissue engineering: current strategies and future directions, Chonnam Medical Journal 47 (1) (2011) 1. Available from https://doi.org/10.4068/cmj.2011.47.1.1.

[11] B. Bakhshandeh, P. Zarrintaj, M.O. Oftadeh, F. Keramati, H. Fouladiha, S. Sohrabi-jahromi, et al., Tissue engineering: strategies, tissues, and biomaterials, Biotechnology and Genetic Engineering Reviews 33 (2) (2017) 144–172. Available from https://doi.org/10.1080/02648725.2018.1430464, https://www.tandfonline.com/action/journalInformation?journalCode=tbgr20.

[12] C.E. Dlaska, G. Andersson, M. Brittberg, N.P. Suedkamp, M.J. Raschke, M.A. Schuetz, Clinical translation in tissue engineering—the surgeon's view, Current Molecular Biology Reports 1 (2) (2015) 61–70. Available from https://doi.org/10.1007/s40610-015-0013-3.

[13] E.S. Bishop, S. Mostafa, M. Pakvasa, H.H. Luu, M.J. Lee, J.M. Wolf, et al., 3-D bioprinting technologies in tissue engineering and regenerative medicine: current and future trends, Genes and Diseases 4 (4) (2017) 185–195. Available from https://doi.org/10.1016/j.gendis.2017.10.002, https://www.keaipublishing.com/en/journals/genes-and-diseases/.

[14] Tissue engineering and regenerative medicine. Available from: https://www.nibib.nih.gov/science-education/science-topics/tissue-engineering-and-regenerative-medicine.

[15] F.J. O'Brien, Biomaterials & scaffolds for tissue engineering, Materials Today 14 (3) (2011) 88–95. Available from https://doi.org/10.1016/S1369-7021(11)70058-X, http://www.journals.elsevier.com/materials-today/.

[16] M. Brovold, J.I. Almeida, I. Pla-Palacín, P. Sainz-Arnal, N. Sánchez-Romero, J.J. Rivas, et al., Naturally-derived biomaterials for tissue engineering applications, Advances in Experimental Medicine and Biology 1077 (2018) 421–449. Available from https://doi.org/10.1007/978-981-13-0947-2_23, http://www.springer.com/series/5584.

[17] X. Liu, P.X. Ma, Polymeric scaffolds for bone tissue engineering.

[18] K. Rezwan, Q.Z. Chen, J.J. Blaker, A.R. Boccaccini, Biodegradable and bioactive porous polymer/inorganic composite scaffolds for bone tissue engineering, Biomaterials 27 (18) (2006) 3413–3431. Available from https://doi.org/10.1016/j.biomaterials.2006.01.039.

[19] L. Suamte, A. Tirkey, J. Barman, P.J. Babu, Various manufacturing methods and ideal properties of scaffolds for tissue engineering applications, Smart Materials in Manufacturing 1 (2023) 100011. Available from https://doi.org/10.1016/j.smmf.2022.100011.

[20] A.J. Engler, S. Sen, H.L. Sweeney, D.E. Discher, Matrix elasticity directs stem cell lineage specification, Cell 126 (4) (2006) 677–689. Available from https://doi.org/10.1016/j.cell.2006.06.044.

[21] V. Kesireddy, F.K. Kasper, Approaches for building bioactive elements into synthetic scaffolds for bone tissue engineering, Journal of Materials Chemistry B 4 (42) (2016) 6773–6786. Available from https://doi.org/10.1039/c6tb00783j, http://pubs.rsc.org/en/journals/journalissues/tb.

[22] J.M. Karp, R. Langer, Development and therapeutic applications of advanced biomaterials, Current Opinion in Biotechnology 18 (5) (2007) 454–459. Available from https://doi.org/10.1016/j.copbio.2007.09.008.

[23] R. Langer, D.A. Tirrell, Designing materials for biology and medicine, Nature 428 (6982) (2004) 487–492. Available from https://doi.org/10.1038/nature02388.

[24] Scaffold Protein - an overview | ScienceDirect Topics. Available from: https://www.sciencedirect.com/topics/biochemistry-genetics-and-molecular-biology/scaffold-protein.

[25] Signal Transduction - an overview | ScienceDirect Topics. Available from: https://www.sciencedirect.com/topics/biochemistry-genetics-and-molecular-biology/signal-transduction.

[26] Regenerative medicine - Tissue Scaffolds, Repair Factors | Britannica. (2025), Available from: https://www.britannica.com/science/regenerative-medicine/Tissue-scaffolds-and-soluble-repair-factors.

[27] B.P. Chan, K.W. Leong, Scaffolding in tissue engineering: general approaches and tissue-specific considerations, European Spine Journal 17 (4) (2008) S467–S479. Available from https://doi.org/10.1007/s00586-008-0745-3.

Chapter 4

Advances in biomanufacturing techniques, biomaterials, and stem cells for tissue engineering

Vishal Kumar Deb, Nidhi Chauhan, and Utkarsh Jain
School of Health Sciences and Technology (SoHST), UPES, Dehradun, Uttarakhand, India

4.1 Introduction

The development of novel material-based medicinal solutions for biological systems and their industrialization is the only focus of the developing field of biomanufacturing. This technology links the process of producing the products to biological elements. Scientific techniques for biomanufacturing are emerging every 10–15 years to address healthcare-related difficulties, thanks to the advancement of several fields of biology such as cell biology, proteomics, genomics, 3D bioprinting, process engineering, and regenerative medicine. As a result, these cutting-edge technologies are helping us develop new, sophisticated biomanufacturing techniques.

The field of advanced biomanufacturing comprises numerous subtechnologies, including structural DNA nanotechnology, polymer engineering-based cellular approach, and synthetic biology as molecular building blocks. Biological building blocks from nature, such as modified sugars and non-native amino acids, can be synthesized to create new metabolic pathways and design and implement a cellular level of rewiring using molecular building blocks, an advanced biomanufacturing tool [1–4]. Conversely, structural DNA nanotechnology facilitates the creation of digital self-assembly using biomolecules, such as 2D and 3D nanoscale structures and DNA bricks. Nevertheless, biomanufacturing tools to assist biological systems could also be made possible by polymer engineering–based technologies; for instance, polymers mediated by elastin can help biological tissues retain their elasticity [5–8].

Tissue engineering (TE), however, is also an interdisciplinary field that directly links to biomanufacturing to produce various regenerative medications or devices to enhance biological tissues' functionality. It has become clear that this is a groundbreaking strategy for solving the organ scarcity problems [9]. TE is making a significant contribution to the field of regenerative medicine research by creating intelligent biomaterials targeted at organ regeneration. Porous polymeric scaffolds, for example, are employed in TE to provide an environment for tissue restoration where cells and other cellular components interact and potentially serve as a favorable environment for tissue regeneration [10].

4.2 Current methods of biomanufacturing for tissue engineering

TE is a fascinating biological method that combines engineering principles with the use of various biomaterials and bioactive substances for replacement or scaffolding. Even if there are numerous ways to heal the wounds, there may be obstacles to these treatments. In this situation, TE is helpful in resolving these problems. TE is a novel approach that repairs defects at the site of injury by combining bioactive substances, cells, and biomaterials within a scaffold. Among the various biomanufacturing techniques, 3D scaffolds are useful for directly customizing wounded tissues since they are made with precise tissue sizes and orientations [11–15].

Several biomanufacturing methods have been introduced in the prospect of TE. These methods are also described in the following sections.

Fundamentals of Translational Tissue Engineering. DOI: https://doi.org/10.1016/B978-0-443-33455-9.00020-1

4.2.1 Conventional technologies

4.2.1.1 Foaming technology

To achieve gaseous decomposition, this approach uses CO_2 gas in addition to a blowing agent, which is a chemical. This results in elevated pressure. Pore structures resembling the fundamental structure of skin are created by this process.

4.2.1.1.1 Gas foaming technique

This method uses natural cytostatic solvents at high temperatures along with inert gas-foaming agents such as hydrogen, carbon monoxide, nitrogen, and methane. Furthermore, until the gas bubbles stop forming, this approach uses biodegradable polymers constantly [16]. Remarkably, this method also helps generate micropores around 10 μm, which can be used in the 3D procedure. By cutting down on processing time, the fused deposition approach combines macro porosity through attached channels that have evolved saturation capacities. The key feature of this method is the production of porosity, which can reach 85%, and sizes ranging from 30 to 700 μm, which, combined, create a structure like a sponge [13].

4.2.1.2 Salt leaching technology

This approach creates a polymer membrane with salt particles by mixing sieved NaCl particles with polymer liquids. Following this procedure, the salt particles are removed using water, creating a permeable membrane. The weight percentage and size of the salt particles determine the pore size.

4.2.1.3 Phase separation technology

This method involves dissolving the polymer in an appropriate solvent, pouring the mixture into a mold, and then freezing it. Following the solvent's removal by the freeze-drier, a foam of polymer with pores ranging in diameter from 1 to 20 nm is created.

4.2.1.3.1 Freeze-drying

This popular production method, known as lyophilization, is typically used to freeze synthetic polymers that have been dissolved in a solvent. The procedure requires a temperature between -20°C and -80°C, which could help solidify the result. Evaporate the frozen sample with a lyophilizer to create a solid scaffold with linked pores [17]. Soumya et al. [18] used this same idea to create an osteoinductive herbal scaffold by combining plant extracts with alginate and O-carboxymethyl chitosan (CS). This manufactured scaffold was then utilized for the TE application, taking into account the results of the biocompatibility investigation.

The technologies mentioned above have certain limits when it comes to controlling pore size, distribution, and geometry. In addition to these difficulties, producing these technologies entails longer lead times and the use of hazardous organic solvents. As a result, a new class of additive biomanufacturing processes, such as extrusion-based bioprinting, laser sintering, inkjet bioprinting, stereolithographic processes, and 3D bioprinting technologies, has been introduced as nonconventional biomanufacturing approaches.

4.2.2 Stereolithographic processes

A key biomanufacturing approach is stereolithography, which uses a multilayer procedure to create 3-dimensional solid objects by a polymer's selective photo-mediated response [19]. Three distinct irradiation mechanisms are used in this procedure. The first type of irradiation transfers the picture into a liquid polymer and solidifies it using pattern mask–based irradiation. Conversely, the second approach concentrates on producing polymer structures using a UV beam [9]. The final one is inkjet technology, which creates a planned architecture via sprayed photopolymer materials onto a tray.

4.2.3 Inkjet bioprinting

Inkjet technology is one of the most popular bioprinting techniques for depositing tiny droplets to create a multilayer structure during printing. Using a thermal actuator, this technique regulates the size of the printed droplets [20]. When the temperature of the heating components reaches about 300°C, thermal inkjet-based methods are employed to generate vapor bubbles [21]. The ink droplets are forced to eject onto the substrate by the vapor bubbles that are created [22]. Nevertheless, there are disadvantages to this kind of printing technique as well. First, it is not very good at processing bioinks with a high cell density. As a result, blockage may occur in the print head, leading to an uneven distribution of cells [23].

4.2.4 Laser sintering

Typically, the laser sintering technique uses a thin coating of sintering powder and infrared radiation that the laser emits to trace the contour of each cross-section of the model. Afterwards, additional energy is added to fuse the nearby powder particles and create a connection between the freshly sintered layers. Following this stage, each layer is allowed to set before the model is purged and a fresh layer of powder is applied once more. Using this method, Wiria et al. created a polycaprolactone (PCL) and HA biocomposite in which HA was added to increase the scaffold's overall activity. Different HA ratios, such as 10, 20, and 30 wt.%, were blended in this method. This same study was also carried out to make another biocomposite with PVA for bone replacement application [24,25].

4.2.5 3D bioprinting technology

In an effort to pave the way for regenerative medicine, scientists have been working on creating artificial organ systems or tissues over the past 30 years. After much work, a combination of physics, developmental biology, stem cell biology, and clinical translation research led to the development of regeneration medicine, which includes 3D bioprinting.

One of the best technologies for creating a biological construct that resembles the target organ in hierarchical design is 3D bioprinting. Using a layer-by-layer approach, this technology uses additive biomanufacturing to produce materials with biological functionality [26]. The development of scaffold-free or scaffold-based tissue as well as organ-on-a-chip model systems is aided by 3D bioprinting technologies [27–33]. But there are also a lot of difficulties with this technique, such as finding printable biomimetic materials, intricately patterning various tissues, and maintaining cell viability over an extended period of time.

4.2.5.1 3D bioprinting in cardiac tissue

As 3D bioprinting became more significant, cardiac bioengineering became more appealing as a means of addressing the growing problems associated with heart failure and heart attacks [34,35]. There are many reasons that have already been addressed by several scientific reports, not only the single nucleotide polymorphism in several inflammatory molecules such as IL-1, IL-6, IL-10, TNF-α, TGF-β, and CRP is also one of the major reasons for cardiovascular diseases [36]. The heart tissue was 3D bio-printed with two connected ventricles, a rectangular sheet, and an inner diameter of 1 cm (3 × 0.8 × 0.5 cm). By printing one layer at a time, this printing might work. When comparing this kind of cardiac printing to hand-made scaffolds, the degree of elastic modules is lower. Gelatin/alginate that has been connected to calcium chloride is printed on the materials needed to create the heart cells.

4.2.5.2 3D bioprinting for bone tissue

Bone bioprinting has emerged as a substitute method for replacing damaged bone tissue using allografts and autografts [37]. In this case, layer-by-layer PCL deposition via 3D printing has been utilized to create a scaffold with nano-structured pores. Pigs have also been used to evaluate these artificially created nano-structured pores with PCL scaffolds for calvarial deficiencies.

4.2.6 Extrusion-based bioprinting

Because it is simpler and more inexpensive, extrusion-based bioprinting is a crucial fabrication approach in TE biomanufacturing methods. This method introduces various mechanical forces to the cells. Shear stress is the most dangerous type of mechanical force because it can harm cells. Nonetheless, due to its improved structural integrity, this bioprinting method is considered practical. Rather than using liquid droplets, it relies on the precise deposition of filaments. In contrast to other bioprinting methods like inkjet, which have a resolution of 50 μm, this one has 200 μm [38–41]. The loading of bioinks into a syringe barrel that is extruded by a micro-nozzle tip is the foundation of the operation of extrusion-based bioprinters [42]. To preserve the construction's structural integrity, supporting materials must also be filled on the print heads. Furthermore, based on the fluid dispensing system [43], this technology is divided into three categories: a. mechanical, focused on screw-driven; b. pneumatic, focused on paper; and c. solenoid [42] (Table 4.1).

TABLE 4.1 Different types of biomanufacturing techniques and materials in distinct tissue engineering prospects.

Different techniques	Materials used	Distinct tissue engineering prospect	References
Freeze-drying	CS polyelectrolyte complex	Cartilage tissue engineering	[44]
Thermally induced phase separation	Hydrogel-filled polylactic acid	Cartilage and bone tissue engineering	[45]
3D bioprinting (extrusion)	Nano-cellulose and alginate	Cartilage tissue engineering	[46]
Inkjet printing	Poly(ethylene glycol) dimethacrylate	Cartilage tissue formation	[47]
Electrospinning	Polycaprolactone-gelatin	Tissue engineering	[48]
Stereolithography	Poly(ethylene glycol) dimethacrylate	Vascular endothelial growth factor secretion	[49]

4.3 Materials used in biomanufacturing for tissue engineering

4.3.1 Collagen

Collagen is one of the many different kinds of polymeric proteins that are commonly found as an element of the extracellular matrix (ECM). It is interesting to note that it makes up about 20%–30% of the total weight of all proteins found in mammals [50]. Its distinctive structure is a triple helix composed of several procollagen polypeptide chains [51]. Osteoblasts create collagen type I, which regulates the development of hydroxyapatite, which is deposited as calcium and phosphate ions, in the setting of bone. When the ECM breaks down, exposed collagen fibers come into contact with cells. To further explore this context, consider how integrin-binding motifs RGD allow cells to interact with collagen [52]. Interestingly, because collagen type I is an organic component of the ECM of bones, it has been used to construct a scaffold material for bone tissue creation. Collagen type I, however, has several drawbacks, including high biodegradability and limited mechanical strength. It could be necessary to address this issue [53].

In a similar setting, a study was carried out to treat ligament and tendon injuries. To create a unique electrospun tissue-engineered device, collagen type I and poly(D, L-lactide) (PDLLA) were dissolved in DMSO. This created product may offer strong biomechanical stability because of its good porosity and cell infiltration capabilities. Remarkably, this new substance, when implanted in a rabbit Achilles tendon damage model, opens the way for tissue regeneration and maturation [54].

4.3.2 Gelatin

A biocompatible polymer made up of water, salts, and 85–92% proteins is called gelatin [55,56]. As a molecular derivative of collagen type I, it is employed to create culinary and pharmaceutical goods [57,58]. Gelatin exhibits remarkable properties akin to collagen, facilitating the creation of products at the cellular level. It is noteworthy that it does not cause any toxicity or antigenicity in human cells [59].

Since 3D bioprinting is a promising new method for creating various complex scaffold materials, gelatin has been employed in a study to produce a complex structure using gelatin methacrylate, which has been photocrosslinked, and dynamic hydrazone-crosslinked hyaluronic acid. This new complex structure functions as a printable hydrogel bioink for extrusion-based 3D printing. Additionally, it produces a double network hydrogel that enhances mechanical strength and promotes self-healing. Furthermore, it has been demonstrated to be a supporter of both the proliferation and survival of bone marrow mesenchymal stem cells (MSCs) based on the in vitro cytocompatibility test [60].

4.3.3 Chitosan

After being treated with alkali, chitin is deacetylated and becomes chitosan [61,62]. It is made up of many components that are soluble in acidic aqueous solutions, such as N-acetyl-D-glucosamine and deacetylated D-glucosamine [63]. Its mechanical characteristics and molecular structure are comparable to those of glycosaminoglycans [64].

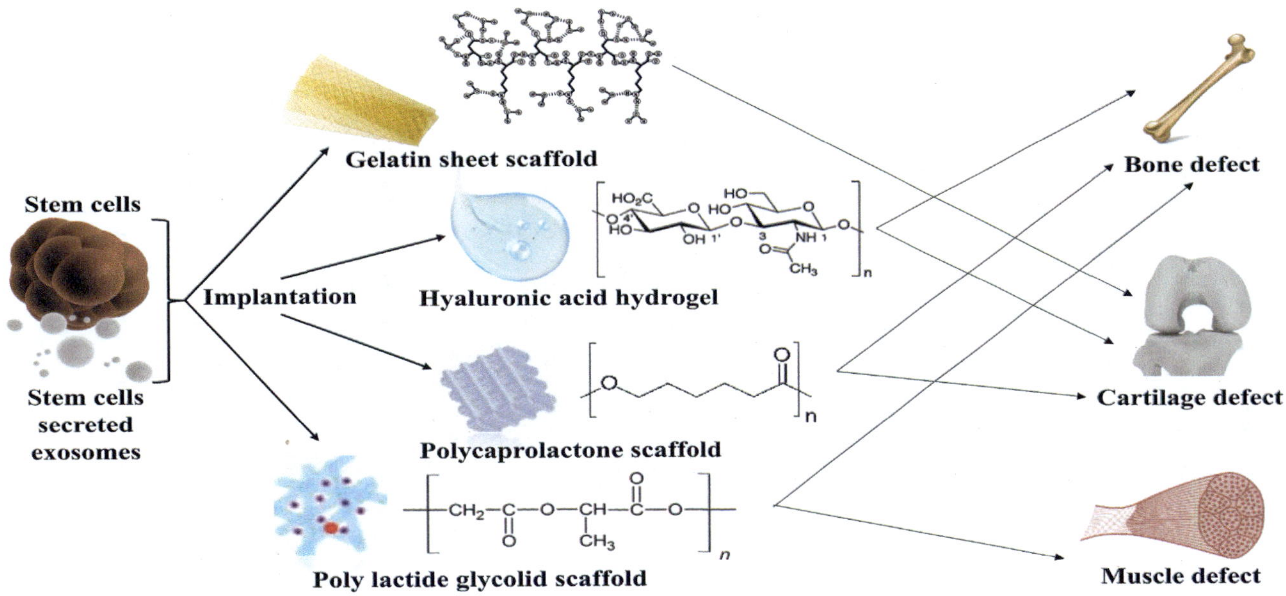

FIGURE 4.1 Biomaterials in tissue engineering. Role of stem cells secreted exosomes combined with different types of scaffolds in biomaterials manufacturing for tissue engineering.

These properties of chitosan, however, rendered it more advantageous for applications involving the engineering of bone and cartilage [65]. It was found that applying a chitosan coating to PCL scaffolds improved the scaffold's capacity for cell adhesion. A chitosan/hydroxyapatite scaffold was created in this situation by Ang et al. to improve the scaffold's mechanical strength and osteoconductivity. Strong attachment qualities occur as a result of the produced product's chitosan layer [66] (Fig. 4.1).

4.3.4 Polylactic acid

The polymer form of lactic acid present in living things is called poly(lactic acid) (PLA). Metabolism produces it in creatures such as plants, animals, and microbes. Lactic acid's chemical structure is made up of hydroxyl and carbonyl groups. According to recent research, PLA is spun using a rotary jet to create polymeric fibers that can be used to wound dressings as a skin TE application [67]. Conversely, musculoskeletal TE also makes use of creating products based on PGA-PLA copolymer. In this sense, PGA functions as an excellent biomaterial for cell growth; additionally, combining it with PLA increases the effectiveness of cell growth [68–73].

A 3D framework composed of PLA and PLA modified with hydroxyapatite (PLA-HAp) was described in another study. However, only scaffolds made of PLA show poor cell adhesion. The PLA demonstrated improved cell attachment characteristics after being manufactured using nano Hap [74].

4.3.5 Polycaprolactone

A well-researched polymer called PCL is utilized as a scaffold in a number of TE applications. In addition to its poor mechanical qualities, this polymer has poor cell adhesion properties. Using a melt-blending technique, PCL was blended with a natural polymer, specifically Bombyx mori silk microparticles, to enhance these properties. High interconnectedness between printed fibers with planned pore topologies has been seen in this fabrication product. As a result, this compliant study produced a stronger scaffold that is suitable for bone tissue creation and has high mechanical reinforcement and increased cell proliferation [75].

Within the same framework, an additional investigation has demonstrated the use of PCL in bone TE through the use of AM (additive manufacturing) techniques to prepare the scaffold with ECM. In this work, mesenchymal stem/stromal cells were cultured on three-dimensional porous PCL scaffolds, which promoted MSC proliferation and increased attachment. The results of the overall investigation indicated that MSC osteogenic differentiation may be influenced in a way that mimics the natural bone niche [76] (Table 4.2).

TABLE 4.2 Different types of materials used in biomanufacturing for distinct tissue-engineering materials.

S. No.	Types of materials	Biomanufacturing method (Bioprinting)	In vitro model used	Printed tissue	Observations	References
1.	Collagen	Laser-assisted bioprinting	NIH 3T3 and HaCaT	Skin tissue	Printed tissue model is analogous to the native skin archetype.	[77]
2.		Extrusion-based bioprinting	hMSCs	Bone tissue	Anisotropic soft collagen-rich substrates supported osteogenesis.	[78]
3.		Droplet ejector-based bioprinting	SMCs	-	Cell viability over long-term culture.	[79]
4.	Gelatin	Extrusion-based bioprinting	VIC and SMCs	Aortic valve tissue	Cell viability for VIC (~91% on day 1 and ~86% on 7th day). Cell viability for SMCs (~90% on day 1 and ~84% on 7th day).	[80]
5.		Extrusion-based bioprinting	BMSCs, ACPCs, and chondrocytes	Cartilage tissue	Cell viability ~75% on day 1 and ~90% on 14th day.	[81]
6.	Agarose	Bioprinting with a valve ejector	NIH 3T3	-	Porosity and interconnectivity of scaffolds can be controlled by incorporating microorganisms in hydrogels.	[82]
7.		Extrusion-based bioprinting	L929 and MG-63	-	Cell viability was 96% on 1st day, 99% on 4th day and 97% on 7th day.	[83]
8.	Hyaluronic acid	Dual syringe deposition tool	Int-407, HepG2 C3A, and NIH 3T3	-	Cytocompatible and supported cell attachment as well as cell proliferation.	[84]
9.		Multimaterial bioprinter	HAVIC	Heart valve tissue	Manufactured hydrogel decreased the stiffness of the hydrogel and increased the cell adhesion property.	[85]
10.	PEG	Modified inkjet	hMSCs	Bone tissue	Effective towards osteogenic differentiation and osteogenic ECM production.	[86]
11.		Extrusion-based bioprinting	Int-407, HepG2 C3A, and NIH 3T3	Vascular tissue	Easy to prepare and biocompatible for printing a tubular tissue construct.	[87]
12.	PCL	Pneumatic dispenser	C20A4	-	Enhanced Cell viability due to PCL in the construct rather than alone alginate-based scaffolds after printing.	[88]
13.		Extrusion-based bioprinting	Chondrocytes and osteoblasts	Osteochondral tissue	Higher amounts of osteogenic markers and chondrocyte markers were seen with Col-1 and HA hydrogels.	[89]
14.	PVP	Inkjet	HFF-1	-	Improved the homogeneity and viability of printed cells.	[90]

PCL, Polycaprolactone; *PEG*, polyethylene glycol; *PVP*, polyvinylpyrrolidone; *SMC*, aortic root sinus smooth muscle cells; *VIC*, aortic valve leaflet interstitial cells.

4.4 Stem cells used in tissue engineering

4.4.1 Embryonic stem cells

The ability of embryonic stem cells to develop into several lineages makes them a renewable source of cells. It's interesting to note that ESCs are employed alone or in conjunction with other materials to replace or transplant cells when they take the shape of scaffolds. It does, however, face certain difficulties in these situations, such as suitable and homogenous differentiation into desirable cells under regulated conditions. ESCs are clarifying or immobilizing in cues for the signaling systems in TE at this point.

4.4.1.1 Genetic reprogramming

Additionally, ESCs are employed in genetic reprogramming, a process in which particular genes are inserted into hESCs to generate particular cell types. Nonetheless, methods for introducing genes based on knock-out or knock-in have already been devised in hESCs to cause them to differentiate into certain cell fates. To transfect plasmids into undifferentiated hESCs that may be further beneficial for homologous recombination events, chemical reagents or electroporation procedures have also been applied [91].

4.4.1.2 Microenvironmental cues

Microenvironmental regulation of ESC destiny decision-making has also been implemented. Using this method, the ESC's inner cell mass is exposed to the carefully controlled signals from its environment. At the moment, embryonic stem cells are cultured as clumped colonies of cells, a rudimentary form [92].

4.4.2 Induced pluripotent stem cells

After studying induced pluripotent stem cells (iPSCs), working expectations with stem cells are starting to raise certain concerns. This particular facet of stem cell biology embodies the exceptional framework of cell therapy. Sox4, cMyc, oct3/4, and klf4 are four transcription factors that are crucial in the process of creating iPSCs from somatic committed cells [93].

When using iPSCs as a cell source, a number of concerns must be resolved, such as the differences between different iPSC lines and their degree of development for particular TE applications. An investigation into the production of NCSCs using iPSCs as a model for peripheral nerve generation and evaluation of therapeutic potential was carried out.

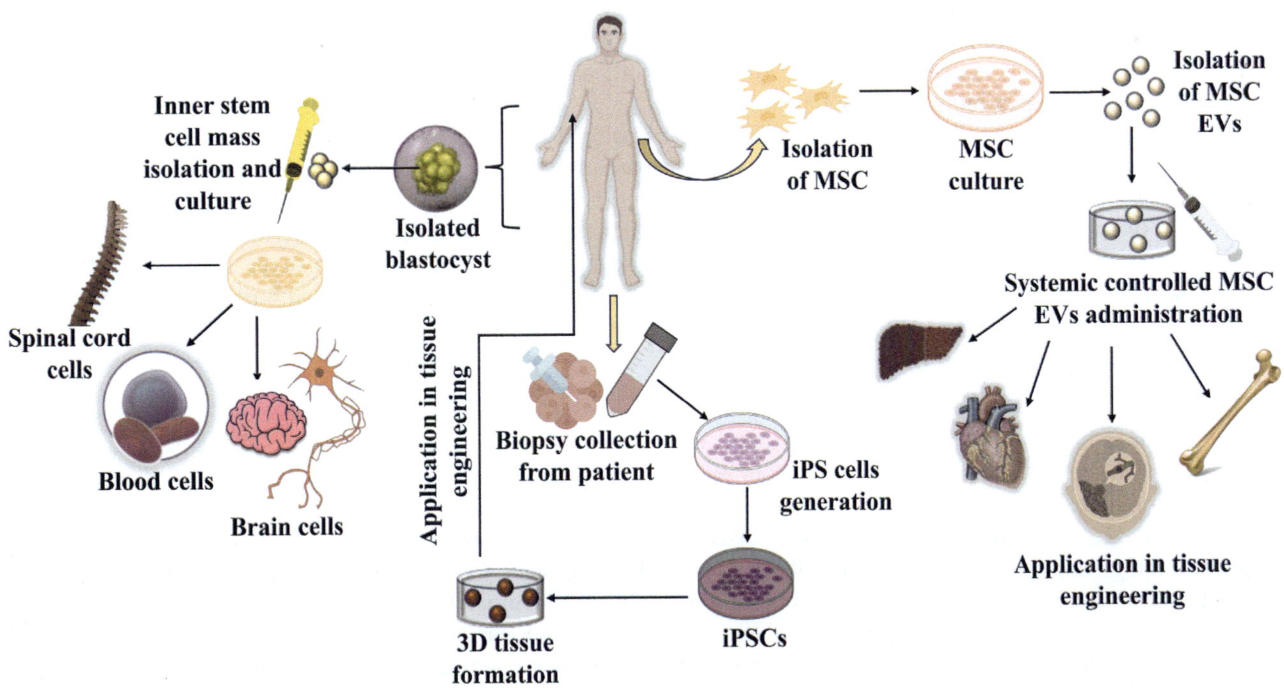

FIGURE 4.2 **Tissue engineering in different applications.** Schematic representation of distinct stem cells in tissue engineering applications.

iPSCs are a potentially useful platform in TE. According to a study, NCSCs made from iPSCs have direct applications in TE and regenerative medicine [94].

4.4.3 Mesenchymal stem cells

MSCs are postnatal stem cells that exhibit a variety of properties, including the capacity to modulate immune responses in addition to multipotent differentiation and self-renewal activity [95]. Human MSCs have been classified into three major categories based on these features: differential potentiality to a triple lineage consisting of chondrogenic, osteogenic, and adipogenic cells; adherence properties on the plastic plate; and presence of multiple surface biomarkers such as CD105, CD73, and CD90 [96,97].

It is interesting to note that MSCs are widely employed for TE and regeneration of bone and cartilage in applications based on MSCs. One technique employed in MSC differentiation is the gene delivery approach, which holds promise for therapeutic applications based on MSCs. To transform MSCs into the triple lineage, for example, a number of transcription factor genes are delivered, such as BMP-2 and RUNX2 for osteogenesis and SOX9 and TGF-β3 for chondrogenic or osteogenic differentiation. However, simultaneous gene delivery may potentially enhance MSC differentiation [1–100] (Fig. 4.2).

4.5 Clinical applications biomanufactured materials for different tissue engineering

4.5.1 Skin tissue engineering

Skin tissue bioengineering is a new area within tissue bioengineering that could unite multidisciplinary teams to advance clinical applications. As an alternative to traditional treatment, skin tissue bioengineering can be used to address a number of medical indications, such as skin loss and significant sections of burned skin. In these situations, skin TE applications employ scaffolds encapsulating biopolymer-based medications to ensure safety. On the other hand, the development of skin TE through bioprinting and bioreactors could potentially speed up the release of therapeutic products [101].

While 3D bioprinting has shown promise, rapid and reliable skin regeneration solutions are needed for skin injuries. Using 3D bioprinting for skin wound recovery could help create bionic skin materials that are reliable and quick to produce by molding cells into the shape of the skin. These products also offer exceptional flexibility, resolution, and high repeatability.

4.5.1.1 Elements for skin bioprinting

A few procedures are involved in skin bioprinting technology, such as prepping the bioink in accordance with the specifications for precise printing of the skin substitutes. The skin substitute is immediately transplanted to the site of the skin wound, where it begins to deteriorate or absorb into the body after being prepared. The body then produces ECM as a result of these broken-down bioactive materials.

To further explore this setting, numerous studies on the materials utilized in skin bioprinting have been published. Collagen, for example, is one of the greatest biopolymers that are used to increase cell attachment, prevent immunological rejection, and act as a safe element for the host [102]. However, depending on its quantities, gelatin, a denatured form of collagen, can likewise be utilized as a crosslinking substance in a gel to enhance the entire fabricating system [103,104]. Hyaluronic acid is also utilized in skin TE to encourage cartilage and tissue regeneration. However, it also has a drawback like poor mechanical properties which may be overcome by modifying it with acrylic acid that further helps to form a 3D scaffold structure [105,106].

4.5.2 Bone and cartilage tissue engineering

Osteoarthritis may be brought on by articular cartilage, one of the various degenerative disorders. TE emerged as a novel therapeutic approach for articular cartilage repair in response to inadequate self-healing capabilities of the cartilage and unsatisfactory clinical interventions. There are two main types of techniques used in TE-based articular cartilage repair: scaffold-based TE and cell-based TE. TE has discovered new cell sources, scaffolds, bioactive compounds, and composites integrating these elements in cell-based articular cartilage. Conversely, scaffold-based TE is a further applicable technique that facilitates the delivery of prochondrogenic bioactive chemicals while also promoting cell differentiation and proliferation. Several types of organic and inorganic carrier systems [107] are also present that have already been used in cancer therapy but have not been explored in bone and cartilage TE. Within this context, Johnson et al. found kartogenin, a tiny molecule that has the potential to function as both a chondrogenic and a chondroprotective agent [108]. Teng et al. demonstrated in a recent study that this tiny chemical is involved in chondrogenesis. The present investigation employed a combination of human bone marrow–derived MSCs, hyaluronic acid, biomimetic peptides (arginine–glycine–aspartate), histidine–alanine–valine, and kartogenin, to stimulate cellular activity and facilitate the differentiation of mesenchymal cells from bone marrow into chondrocytes.

TGF-β was delivered by PLA in a subsequent TE investigation conducted by Shen et al. This study improved the mechanical strength and chondrogenesis process by incorporating graphene oxide nanosheets into PLA and polyethylene glycol hydrogel, loading them with TGF-β3, and producing them from bone marrow mesenchymal cells [109,110].

Because of its low elastic modulus and high friction coefficient, porous tantalum (Ta) has also been identified as a promising biomaterial for orthopedic and dental research. Porous tantalum is utilized in orthopedics and dentistry to maintain primary stability and reduce bone loss because of these beneficial properties. It is utilized in implants during clinical interventions for individuals who specifically require it. It is also used in several in vivo and in vitro investigations to boost antibacterial activity. On the other hand, the Ta implantation has demonstrated remarkable success in a suitable patient [111]. However, Ti_3C_2-MXene has also been explored in drug delivery prospects for photo-thermal-based cancer therapy as it acts as a good biocompatible agent [112]. Therefore, it may also be utilized in bone and cardiac TE to deliver maximum numbers of tissue proliferating agents after coating them on Ti_3C_2-MXene surface.

4.6 Clinical translation advances in tissue engineering

Over the past 25 years, TE has been crucial to the development of cutting-edge treatment options. Numerous products, such as placental tissue allograft products, are on the market and show promise as therapeutic interventions for the treatment of chronic wounds [113,114]. Growth factors, natural tissue ECM, and low immunogenic cellular components make up these products.

Several variations of the TE paradigm are becoming possible thanks to previously approved FDA goods. Products that are currently in various stages of clinical trials, such as Humacyte acellular vascular graft [115], have shown better efficacy and safety than those that are not yet available [116]. To comply with FDA regulations, these goods must, however, be sterile, host-compatible, and functional. Crucially, TE will need the appropriate timeframe to find the intricate systems with difficulties in order to receive FDA approval, just like prospective medications [117]. Nevertheless, this TE therapeutic products should focus on commercial availability, so that these materials can be implanted into the native tissue in a better way and for long-term use [118].

4.7 Conclusion

TE represents a transformative approach in the field of regenerative medicine, employing a combination of bioactive factors, cells, and biomaterial scaffolds to address injuries and defects at the site of injury. Current biomanufacturing methods, including freeze-drying, gas-foaming, 3D bioprinting, extrusion-based bioprinting, and inkjet bioprinting, offer promising solutions for fabricating tissue scaffolds with desired properties. Despite the challenges involved, advances in materials such as collagen, gelatin, chitosan, PLA, and PCL have enhanced the performance of these scaffolds, improving their mechanical strength, biocompatibility, and bioactivity. Stem cells, such as embryonic stem cells, iPSCs, and MSCs, are pivotal in TE, allowing for targeted tissue regeneration. Clinical applications of biomanufactured materials for TE, including skin, bone, and cartilage engineering, have demonstrated significant potential in improving patient outcomes. However, further research and clinical translation efforts are necessary to optimize these technologies for widespread therapeutic use.

AI disclosure

During the preparation of this work, the author(s) used QuillBot to enhance the quality of the language. After using this tool/service, the author(s) reviewed and edited the content as needed and take(s) full responsibility for the content of the publication.

References

[1] K.A. Schwarz, N.M. Daringer, T.B. Dolberg, J.N. Leonard, Rewiring human cellular input-output using modular extracellular sensors, Nature Chemical Biology 13 (2) (2017) 202–209. Available from https://doi.org/10.1038/nchembio.2253, http://www.nature.com/nchembio.

[2] M. Pröschel, R. Detsch, A.R. Boccaccini, U. Sonnewald, Engineering of metabolic pathways by artificial enzyme channels, Frontiers in Bioengineering and Biotechnology 3 (2015). Available from https://doi.org/10.3389/fbioe.2015.00168, http://journal.frontiersin.org/article/10.3389/fbioe.2015.00168/full.

[3] A. Kern, E. Tilley, I.S. Hunter, M. Legiša, A. Glieder, Engineering primary metabolic pathways of industrial micro-organisms, Journal of Biotechnology 129 (1) (2007) 6–29. Available from https://doi.org/10.1016/j.jbiotec.2006.11.021.

[4] C.J. Bashor, A.A. Horwitz, S.G. Peisajovich, W.A. Lim, Rewiring cells: synthetic biology as a tool to interrogate the organizational principles of living systems, Annual Review of Biophysics 39 (1) (2010) 515–537. Available from https://doi.org/10.1146/annurev.biophys.050708.133652.

[5] Y. Liu, H.L. Cai, M. Zelisko, Y. Wang, J. Sun, F. Yan, et al., Ferroelectric switching of elastin, Proceedings of the National Academy of Sciences of the United States of America 111 (27) (2014) E2780. Available from https://doi.org/10.1073/pnas.1402909111, http://www.pnas.org/content/111/27/E2780.full.pdf+html.

[6] S. Roberts, M. Dzuricky, A. Chilkoti, Elastin-like polypeptides as models of intrinsically disordered proteins, FEBS Letters 589 (19) (2015) 2477–2486. Available from https://doi.org/10.1016/j.febslet.2015.08.029, www.elsevier.com/locate/febslet.

[7] K. Ye, D.L. Kaplan, G. Bao, C. Bettinger, G. Forgacs, C. Dong, et al., Advanced cell and tissue biomanufacturing, ACS Biomaterials Science and Engineering 4 (7) (2018) 2292–2307. Available from https://doi.org/10.1021/acsbiomaterials.8b00650, http://pubs.acs.org/journal/abseba.

[8] G.C. Yeo, B. Aghaei-Ghareh-Bolagh, E.P. Brackenreg, M.A. Hiob, P. Lee, A.S. Weiss, Fabricated elastin, Advanced Healthcare Materials 4 (16) (2015) 2530–2556. Available from https://doi.org/10.1002/adhm.201400781, http://onlinelibrary.wiley.com/journal/10.1002/(ISSN)2192-2659.

[9] P.J. Bártolo, C.K. Chua, H.A. Almeida, S.M. Chou, A.S.C. Lim, Biomanufacturing for tissue engineering: present and future trends, Virtual and Physical Prototyping 4 (4) (2009) 203–216. Available from https://doi.org/10.1080/17452750903476288.

[10] M.C. Socci, G. Rodríguez, E. Oliva, S. Fushimi, K. Takabatake, H. Nagatsuka, et al., Polymeric materials, advances and applications in tissue engineering: a review, Bioengineering 10 (2) (2023). Available from https://doi.org/10.3390/bioengineering10020218, www.mdpi.com/journal/bioengineering.

[11] M. Chimerad, A. Barazesh, M. Zandi, I. Zarkesh, A. Moghaddam, P. Borjian, et al., Tissue engineered scaffold fabrication methods for medical applications, International Journal of Polymeric Materials and Polymeric Biomaterials 72 (18) (2023) 1455–1479. Available from https://doi.org/10.1080/00914037.2022.2101112.

[12] B.S. Krishna, K. Vandana, Recent Advances in Scaffold Fabrication Techniques for Tissue Engineering A Holistic and Integrated Approach to Lifestyle Diseases, Apple Academic Press, India, (2022), pp. 251–280. Available from https://www.appleacademicpress.com/a-holistic-and-integrated-approach-to-lifestyle-diseases-/9781774630143.

[13] A. Kumar, A. Jacob, Techniques in scaffold fabrication process for tissue engineering applications: a review, Journal of Applied Biology and Biotechnology 10 (3) (2022) 163–176. Available from https://doi.org/10.7324/JABB.2022.100321, https://www.jabonline.in//admin/php/uploads/711_pdf.pdf.

[14] O. Vesvoranan, A. Anup, K.R. Hixon, Current concepts and methods in tissue interface scaffold fabrication, Biomimetics 7 (4) (2022). Available from https://doi.org/10.3390/biomimetics7040151, www.mdpi.com/journal/biomimetics.

[15] J.Y. Yoon, Tissue Engineering: A Primer with Laboratory Demonstrations Tissue Engineering: A Primer with Laboratory Demonstrations, Springer International Publishing, United States, (2021), pp. 1–267. Available from https://doi.org/10.1007/978-3-030-83696-2, https://link.springer.com/book/10.1007/978-3-030-83696-2.

[16] C. Zhou, K. Yang, K. Wang, X. Pei, Z. Dong, Y. Hong, et al., Combination of fused deposition modeling and gas foaming technique to fabricated hierarchical macro/microporous polymer scaffolds, Materials & Design 109 (2016) 415–424. Available from https://doi.org/10.1016/j.matdes.2016.07.094.

[17] L. Roseti, V. Parisi, M. Petretta, C. Cavallo, G. Desando, I. Bartolotti, et al., Scaffolds for bone tissue engineering: state of the art and new perspectives, Materials Science and Engineering: C 78 (2017) 1246–1262. Available from https://doi.org/10.1016/j.msec.2017.05.017.

[18] S. Soumya, K.M. Sajesh, R. Jayakumar, S.V. Nair, K.P. Chennazhi, Development of a phytochemical scaffold for bone tissue engineering using Cissus quadrangularis extract, Carbohydrate Polymers 87 (2) (2012) 1787–1795. Available from https://doi.org/10.1016/j.carbpol.2011.09.094.

[19] P.J. Bartolo, G. Mitchell, Stereo-thermal-lithography: a new principle for rapid prototyping, Rapid Prototyping Journal 9 (3) (2003) 150–156. Available from https://doi.org/10.1108/13552540310477454.

[20] X. Li, B. Liu, B. Pei, J. Chen, D. Zhou, J. Peng, et al., Inkjet bioprinting of biomaterials, Chemical Reviews 120 (19) (2020) 10793–10833. Available from https://doi.org/10.1021/acs.chemrev.0c00008.

[21] P. Kumar, S. Ebbens, X. Zhao, Inkjet printing of mammalian cells – theory and applications, Bioprinting 23 (2021) e00157. Available from https://doi.org/10.1016/j.bprint.2021.e00157.

[22] H. Gudapati, M. Dey, I. Ozbolat, A comprehensive review on droplet-based bioprinting: past, present and future, Biomaterials 102 (2016) 20–42. Available from https://doi.org/10.1016/j.biomaterials.2016.06.012.

[23] R.E. Saunders, B. Derby, Inkjet printing biomaterials for tissue engineering: bioprinting, International Materials Reviews 59 (8) (2014) 430–448. Available from https://doi.org/10.1179/1743280414Y.0000000040, http://www.maneyonline.com/doi/pdfplus/10.1179/1743280414Y.0000000040.

[24] C.K. Chua, K.F. Leong, K.H. Tan, F.E. Wiria, C.M. Cheah, Development of tissue scaffolds using selective laser sintering of polyvinyl alcohol/hydroxyapatite biocomposite for craniofacial and joint defects, Journal of Materials Science: Materials in Medicine 15 (10) (2004) 1113–1121. Available from https://doi.org/10.1023/B:JMSM.0000046393.81449.a5.

[25] F.E. Wiria, C.K. Chua, K.F. Leong, Z.Y. Quah, M. Chandrasekaran, M.W. Lee, Improved biocomposite development of poly(vinyl alcohol) and hydroxyapatite for tissue engineering scaffold fabrication using selective laser sintering, Journal of Materials Science: Materials in Medicine 19 (3) (2008) 989–996. Available from https://doi.org/10.1007/s10856-007-3176-5.

[26] A. Arslan-Yildiz, R.E. Assal, P. Chen, S. Guven, F. Inci, U. Demirci, Towards artificial tissue models: past, present, and future of 3D bioprinting, Biofabrication 8 (1) (2016) 014103. Available from https://doi.org/10.1088/1758-5090/8/1/014103.

[27] A. Atala, F.K. Kasper, A.G. Mikos, Engineering complex tissues, Science Translational Medicine 4 (160) (2012). Available from https://doi.org/10.1126/scitranslmed.3004890.

[28] A.B. Dababneh, I.T. Ozbolat, Bioprinting technology: a current state-of-the-art review, Journal of Manufacturing Science and Engineering 136 (6) (2014). Available from https://doi.org/10.1115/1.4028512.

[29] L.G. Griffith, M.A. Swartz, Capturing complex 3D tissue physiology in vitro, Nature Reviews. Molecular Cell Biology 7 (3) (2006) 211–224. Available from https://doi.org/10.1038/nrm1858.

[30] F. Guillemot, B. Guillotin, A. Fontaine, M. Ali, S. Catros, V. Kériquel, et al., Laser-assisted bioprinting to deal with tissue complexity in regenerative medicine, MRS Bulletin 36 (12) (2011) 1015–1019. Available from https://doi.org/10.1557/mrs.2011.272.

[31] S. Khalil, W. Sun, Bioprinting endothelial cells with alginate for 3D tissue constructs, Journal of Biomechanical Engineering 131 (11) (2009). Available from https://doi.org/10.1115/1.3128729.

[32] F.P.W. Melchels, W.J.A. Dhert, D.W. Hutmacher, J. Malda, Development and characterisation of a new bioink for additive tissue manufacturing, Journal of Materials Chemistry B 2 (16) (2014) 2282–2289. Available from https://doi.org/10.1039/c3tb21280g, http://pubs.rsc.org/en/journals/journalissues/tb.

[33] I.T. Ozbolat, Y. Yu, Bioprinting toward organ fabrication: challenges and future trends, IEEE Transactions on Biomedical Engineering 60 (3) (2013) 691–699. Available from https://doi.org/10.1109/tbme.2013.2243912.

[34] J. Jang, H.J. Park, S.W. Kim, H. Kim, J.Y. Park, S.J. Na, et al., 3D printed complex tissue construct using stem cell-laden decellularized extracellular matrix bioinks for cardiac repair, Biomaterials 112 (2017) 264–274. Available from https://doi.org/10.1016/j.biomaterials.2016.10.026, http://www.journals.elsevier.com/biomaterials/.

[35] Y.S. Zhang, A. Arneri, S. Bersini, S.R. Shin, K. Zhu, Z. Goli-Malekabadi, et al., Bioprinting 3D microfibrous scaffolds for engineering endothelialized myocardium and heart-on-a-chip, Biomaterials 110 (2016) 45–59. Available from https://doi.org/10.1016/j.biomaterials.2016.09.003, https://www.sciencedirect.com/science/journal/01429612.

[36] V.K. Deb, A. Rabbani, S. Majid, A. Abushouk, J. Khan, B.A. Khawaji, et al., Role of Single Nucleotide Polymorphisms of Inflammatory Molecules in Susceptibility to Cardiovascular Diseases (CVDs), Informa UK Limited, (2022), pp. 421–438. Available from https://doi.org/10.1201/9781003246244-19.

[37] Y.P. Singh, J.C. Moses, B.K. Bhunia, S.K. Nandi, B.B. Mandal, Hierarchically structured seamless silk scaffolds for osteochondral interface tissue engineering, Journal of Materials Chemistry B 6 (36) (2018) 5671–5688. Available from https://doi.org/10.1039/c8tb01344f, http://pubs.rsc.org/en/journals/journal/tb.

[38] D.M.R. Gibbs, M. Vaezi, S. Yang, R.O.C. Oreffo, Hope versus hype: what can additive manufacturing realistically offer trauma and orthopedic surgery? Regenerative Medicine 9 (4) (2014) 535–549. Available from https://doi.org/10.2217/rme.14.20.

[39] X. Lu, Y. Lee, S. Yang, Y. Hao, J.R.G. Evans, C.G. Parini, Fine lattice structures fabricated by extrusion freeforming: process variables, Journal of Materials Processing Technology 209 (10) (2009) 4654–4661. Available from https://doi.org/10.1016/j.jmatprotec.2008.11.039.

[40] C. Mandrycky, Z. Wang, K. Kim, D.H. Kim, 3D bioprinting for engineering complex tissues, Biotechnology Advances 34 (4) (2016) 422–434. Available from https://doi.org/10.1016/j.biotechadv.2015.12.011, www.elsevier.com/inca/publications/store/5/2/5/4/5/5/index.htt.

[41] F. Pati, J. Jang, J.W. Lee, D.W. Cho, Extrusion bioprinting, Essentials of 3D Biofabrication and Translation, Elsevier Inc., South Korea, (2015), pp. 123–152. Available from https://doi.org/10.1016/B978-0-12-800972-7.00007-4, http://www.sciencedirect.com/science/book/9780128009727.

[42] D. Choudhury, S. Anand, M.W. Naing, The arrival of commercial bioprinters–towards 3D bioprinting revolution!, International Journal of Bioprinting 4 (2) (2018) .

[43] S. Boularaoui, G. Al Hussein, K.A. Khan, N. Christoforou, C. Stefanini, An overview of extrusion-based bioprinting with a focus on induced shear stress and its effect on cell viability, Bioprinting 20 (2020) e00093. Available from https://doi.org/10.1016/j.bprint.2020.e00093.

[44] N. Bhardwaj, S.C. Kundu, Silk fibroin protein and chitosan polyelectrolyte complex porous scaffolds for tissue engineering applications, Carbohydrate Polymers 85 (2) (2011) 325–333. Available from https://doi.org/10.1016/j.carbpol.2011.02.027.

[45] Y. Gong, L. He, J. Li, Q. Zhou, Z. Ma, C. Gao, et al., Hydrogel-filled polylactide porous scaffolds for cartilage tissue engineering, Journal of Biomedical Materials Research - Part B Applied Biomaterials 82 (1) (2007) 192–204. Available from https://doi.org/10.1002/jbm.b.30721.

[46] K. Markstedt, A. Mantas, I. Tournier, H. Martínez Ávila, D. Hägg, P. Gatenholm, 3D bioprinting human chondrocytes with nanocellulose-alginate bioink for cartilage tissue engineering applications, Biomacromolecules 16 (5) (2015) 1489–1496. Available from https://doi.org/10.1021/acs.biomac.5b00188, http://pubs.acs.org/journal/bomaf6.

[47] X. Cui, T. Boland, D. D.D'Lima, M. K. Lotz, Thermal inkjet printing in tissue engineering and regenerative medicine, Recent Patents on Drug Delivery & Formulation 6 (2) (2012) 149–155. Available from https://doi.org/10.2174/187221112800672949.

[48] S. Gautam, A.K. Dinda, N.C. Mishra, Fabrication and characterization of PCL/gelatin composite nanofibrous scaffold for tissue engineering applications by electrospinning method, Materials Science and Engineering C 33 (3) (2013) 1228–1235. Available from https://doi.org/10.1016/j.msec.2012.12.015.

[49] R. Raman, B. Bhaduri, M. Mir, A. Shkumatov, M.K. Lee, G. Popescu, et al., High-resolution projection microstereolithography for patterning of neovasculature, Advanced Healthcare Materials 5 (5) (2016) 610–619. Available from https://doi.org/10.1002/adhm.201500721, http://onlinelibrary.wiley.com/journal/10.1002/(ISSN)2192-2659.

[50] T. Miyata, T. Taira, Y. Noishiki, Collagen engineering for biomaterial use, Clinical Materials 9 (3-4) (1992) 139–148. Available from https://doi.org/10.1016/0267-6605(92)90093-9.

[51] K. Gelse, E. Pöschl, T. Aigner, Collagens - structure, function, and biosynthesis, Advanced Drug Delivery Reviews 55 (12) (2003) 1531–1546. Available from https://doi.org/10.1016/j.addr.2003.08.002, www.elsevier.com/locate/drugdeliv.

[52] A.V. Taubenberger, M.A. Woodruff, H. Bai, D.J. Muller, D.W. Hutmacher, The effect of unlocking RGD-motifs in collagen I on pre-osteoblast adhesion and differentiation, Biomaterials 31 (10) (2010) 2827–2835. Available from https://doi.org/10.1016/j.biomaterials.2009.12.051.
[53] G.A. Rico-Llanos, S. Borrego-González, M. Moncayo-Donoso, J. Becerra, R. Visser, Collagen type I biomaterials as scaffolds for bone tissue engineering, Polymers 13 (4) (2021) 599. Available from https://doi.org/10.3390/polym13040599.
[54] Y. Maghdouri-White, N. Sori, S. Petrova, H. Wriggers, N. Kemper, A. Dasgupta, et al., Biomanufacturing organized collagen-based microfibers as a Tissue ENgineered Device (TEND) for tendon regeneration, Biomedical Materials 16 (2) (2021) 025025. Available from https://doi.org/10.1088/1748-605x/abb875.
[55] L.J. Yang, Y.C. Ou, The micro patterning of glutaraldehyde (GA)-crosslinked gelatin and its application to cell-culture, Lab on a Chip 5 (9) (2005) 979–984. Available from https://doi.org/10.1039/b505193b, http://pubs.rsc.org/en/journals/journal/lc.
[56] R. Schrieber, H. Gareis, Gelatine Handbook: Theory and Industrial Practice, Wiley, (2007). Available from https://doi.org/10.1002/9783527610969.
[57] M.I. Avila Rodríguez, L.G. Rodríguez Barroso, M.L. Sánchez, Collagen: a review on its sources and potential cosmetic applications, Journal of Cosmetic Dermatology 17 (1) (2018) 20–26. Available from https://doi.org/10.1111/jocd.12450, http://onlinelibrary.wiley.com/journal/10.1111/(ISSN)1473-2165.
[58] K.B. Djagny, Z. Wang, S. Xu, Gelatin: a valuable protein for food and pharmaceutical industries: Review, Critical Reviews in Food Science and Nutrition 41 (6) (2001) 481–492. Available from https://doi.org/10.1080/20014091091904, www.tandf.co.uk/journals/titles/10408398.asp.
[59] A.B. Bello, D. Kim, D. Kim, H. Park, S.H. Lee, Engineering and functionalization of gelatin biomaterials: from cell culture to medical applications, Tissue Engineering - Part B: Reviews 26 (2) (2020) 164–180. Available from https://doi.org/10.1089/ten.teb.2019.0256, http://www.liebertonline.com/teb.
[60] Y. Wang, Y. Chen, J. Zheng, L. Liu, Q. Zhang, Three-dimensional printing self-healing dynamic/photocrosslinking gelatin-hyaluronic acid double-network hydrogel for tissue engineering, ACS Omega 7 (14) (2022) 12076–12088. Available from https://doi.org/10.1021/acsomega.2c00335.
[61] D. Pathania, D. Gupta, S. Agarwal, M. Asif, V.K. Gupta, Fabrication of chitosan-g-poly(acrylamide)/CuS nanocomposite for controlled drug delivery and antibacterial activity, Materials Science and Engineering C 64 (2016) 428–435. Available from https://doi.org/10.1016/j.msec.2016.03.065.
[62] H.V. Ramos Avilez, D.A. Castilla Casadiego, A.L. Vega Avila, O.J. Perales Perez, J. Almodovar, Production of chitosan coatings on metal and ceramic biomaterials, Chitosan Based Biomaterials 1 (2017) 255–293. Available from https://doi.org/10.1016/B978-0-08-100230-8.00011-X, http://www.sciencedirect.com/science/book/9780081002308.
[63] J.D. Bumgardner, V.P. Murali, H. Su, O.D. Jenkins, D. Velasquez-Pulgarin, J.A. Jennings, et al., Characterization of chitosan matters, Chitosan Based Biomaterials 1 (2017) 81–114. Available from https://doi.org/10.1016/B978-0-08-100230-8.00004-2, http://www.sciencedirect.com/science/book/9780081002308.
[64] A. Panwar, L.P. Tan, Current status of bioinks for micro-extrusion-based 3D bioprinting, Molecules (Basel, Switzerland) 21 (6) (2016). Available from https://doi.org/10.3390/molecules21060685, http://www.mdpi.com/1420-3049/21/6/685/pdf.
[65] Z. Cui, L. Lin, J. Si, Y. Luo, Q. Wang, Y. Lin, et al., Fabrication and characterization of chitosan/OGP coated porous poly(ε-caprolactone) scaffold for bone tissue engineering, Journal of Biomaterials Science, Polymer Edition 28 (9) (2017) 826–845. Available from https://doi.org/10.1080/09205063.2017.1303867.
[66] M. Sahranavard, A. Zamanian, F. Ghorbani, M.H. Shahrezaee, A critical review on three dimensional-printed chitosan hydrogels for development of tissue engineering, Bioprinting 17 (2020) e00063. Available from https://doi.org/10.1016/j.bprint.2019.e00063.
[67] J.O. Bahú, L.R. Melo de Andrade, S. Crivellin, N.G. Khouri, S.O. Sousa, L.M.I. Fernandes, et al., Rotary Jet Spinning (RJS): a key process to produce biopolymeric wound dressings, Pharmaceutics 14 (11) (2022). Available from https://doi.org/10.3390/pharmaceutics14112500, http://www.mdpi.com/journal/pharmaceutics.
[68] S. Castañeda-Rodríguez, M. González-Torres, R.M. Ribas-Aparicio, M.L. Del, Prado-Audelo, G. Leyva-Gómez, et al., Recent advances in modified poly (lactic acid) as tissue engineering materials, Journal of Biological Engineering 17 (1) (2023). Available from https://doi.org/10.1186/s13036-023-00338-8, https://jbioleng.biomedcentral.com/.
[69] E. Lavik, Y.D. Teng, E. Snyder, R. Langer, Seeding neural stem cells on scaffolds of PGA, PLA, and their copolymers, Methods in Molecular Biology (Clifton, N.J.) 198 (2002) 89–97. Available from https://doi.org/10.1385/1-59259-186-8:89.
[70] X. Luo, G. Zhou, W. Liu, W.J. Zhang, L. Cen, L. Cui, et al., In vitro precultivation alleviates post-implantation inflammation and enhances development of tissue-engineered tubular cartilage, Biomedical Materials 4 (2) (2009) 25006. Available from https://doi.org/10.1088/1748-6041/4/2/025006.
[71] J.M. Moran, D. Pazzano, L.J. Bonassar, Characterization of polylactic acid-polyglycolic acid composites for cartilage tissue engineering, Tissue Engineering 9 (1) (2003) 63–70. Available from https://doi.org/10.1089/107632703762687546.
[72] R. Tuli, W.J. Li, R.S. Tuan, Current state of cartilage tissue engineering, Arthritis Research and Therapy 5 (5) (2003) 235–238.
[73] H. Xu, D. Han, J.S. Dong, G.X. Shen, G. Chai, Z.Y. Yu, et al., Rapid prototyped PGA/PLA scaffolds in the reconstruction of mandibular condyle bone defects, International Journal of Medical Robotics and Computer Assisted Surgery 6 (1) (2010) 66–72. Available from https://doi.org/10.1002/rcs.290, http://www3.interscience.wiley.com/cgi-bin/fulltext/123208927/PDFSTART.
[74] S. Mondal, T.P. Nguyen, V.H. Pham, G. Hoang, P. Manivasagan, M.H. Kim, et al., Hydroxyapatite nano bioceramics optimized 3D printed poly lactic acid scaffold for bone tissue engineering application, Ceramics International 46 (3) (2020) 3443–3455. Available from https://doi.org/10.1016/j.ceramint.2019.10.057, https://www.journals.elsevier.com/ceramics-international.
[75] C. Vyas, J. Zhang, Ø. Øvrebø, B. Huang, I. Roberts, M. Setty, et al., 3D printing of silk microparticle reinforced polycaprolactone scaffolds for tissue engineering applications, Materials Science and Engineering: C 118 (2021) 111433. Available from https://doi.org/10.1016/j.msec.2020.111433.

[76] J.C. Silva, M.S. Carvalho, R.N. Udangawa, C.S. Moura, J.M.S. Cabral, C. L. da Silva, et al., Extracellular matrix decorated polycaprolactone scaffolds for improved mesenchymal stem/stromal cell osteogenesis towards a patient-tailored bone tissue engineering approach, Journal of Biomedical Materials Research - Part B Applied Biomaterials 108 (5) (2020) 2153–2166. Available from https://doi.org/10.1002/jbm.b.34554, http://onlinelibrary.wiley.com/journal/10.1002/(ISSN)1552-4965.

[77] L. Koch, A. Deiwick, S. Schlie, S. Michael, M. Gruene, V. Coger, et al., Skin tissue generation by laser cell printing, Biotechnology and Bioengineering 109 (7) (2012) 1855–1863. Available from https://doi.org/10.1002/bit.24455.

[78] L. Penolazzi, E. Tavanti, R. Vecchiatini, E. Lambertini, F. Vesce, R. Gambari, et al., Encapsulation of mesenchymal stem cells from Wharton's Jelly in alginate microbeads, Tissue Engineering Part C: Methods 16 (1) (2010) 141–155. Available from https://doi.org/10.1089/ten.tec.2008.0582.

[79] S. Moon, S.K. Hasan, Y.S. Song, F. Xu, H.O. Keles, F. Manzur, et al., Layer by layer three-dimensional tissue epitaxy by cell-laden hydrogel droplets, Tissue Engineering - Part C: Methods 16 (1) (2010) 157–166. Available from https://doi.org/10.1089/ten.tec.2009.0179, http://www.liebertonline.com/tec.

[80] B. Duan, L.A. Hockaday, K.H. Kang, J.T. Butcher, 3D Bioprinting of heterogeneous aortic valve conduits with alginate/gelatin hydrogels, Journal of Biomedical Materials Research - Part A 101 (5) (2013) 1255–1264. Available from https://doi.org/10.1002/jbm.a.34420.

[81] R. Levato, W.R. Webb, I.A. Otto, A. Mensinga, Y. Zhang, M. van Rijen, et al., The bio in the ink: cartilage regeneration with bioprintable hydrogels and articular cartilage-derived progenitor cells, Acta Biomaterialia 61 (2017) 41–53. Available from https://doi.org/10.1016/j.actbio.2017.08.005, http://www.journals.elsevier.com/acta-biomaterialia.

[82] F. Xu, B.P. Sridharan, N.G. Durmus, S.Q. Wang, A.S. Yavuz, U.A. Gurkan, et al., Living bacterial sacrificial porogens to engineer decellularized porous scaffolds, PLoS One 6 (4) (2011). Available from https://doi.org/10.1371/journal.pone.0019344, http://www.plosone.org/article/fetchObjectAttachment.action?uri=info%3Adoi%2F10.1371%2Fjournal.pone.0019344&representation=PDF.

[83] A. Blaeser, D.F. Duarte Campos, M. Weber, S. Neuss, B. Theek, H. Fischer, et al., Biofabrication under fluorocarbon: a novel freeform fabrication technique to generate high aspect ratio tissue-engineered constructs, BioResearch Open Access 2 (5) (2013) 374–384. Available from https://doi.org/10.1089/biores.2013.0031.

[84] A. Skardal, J. Zhang, L. McCoard, X. Xu, S. Oottamasathien, G.D. Prestwich, Photocrosslinkable hyaluronan-gelatin hydrogels for two-step bioprinting, Tissue Engineering - Part A 16 (8) (2010) 2675–2685. Available from https://doi.org/10.1089/ten.tea.2009.0798, http://www.liebertonline.com/tea.

[85] B. Duan, E. Kapetanovic, L.A. Hockaday, J.T. Butcher, Three-dimensional printed trileaflet valve conduits using biological hydrogels and human valve interstitial cells, Acta Biomaterialia 10 (5) (2014) 1836–1846. Available from https://doi.org/10.1016/j.actbio.2013.12.005.

[86] G. Gao, A.F. Schilling, T. Yonezawa, J. Wang, G. Dai, X. Cui, Bioactive nanoparticles stimulate bone tissue formation in bioprinted three-dimensional scaffold and human mesenchymal stem cells, Biotechnology Journal 9 (10) (2014) 1304–1311. Available from https://doi.org/10.1002/biot.201400305, http://onlinelibrary.wiley.com/journal/10.1002/(ISSN)1860-7314.

[87] A. Skardal, J. Zhang, G.D. Prestwich, Bioprinting vessel-like constructs using hyaluronan hydrogels crosslinked with tetrahedral polyethylene glycol tetracrylates, Biomaterials 31 (24) (2010) 6173–6181. Available from https://doi.org/10.1016/j.biomaterials.2010.04.045.

[88] W. Schuurman, V. Khristov, M.W. Pot, P.R. van Weeren, W.J.A. Dhert, J. Malda, Bioprinting of hybrid tissue constructs with tailorable mechanical properties, Biofabrication 3 (2) (2011) 21001. Available from https://doi.org/10.1088/1758-5082/3/2/021001.

[89] J.Y. Park, J.C. Choi, J.H. Shim, J.S. Lee, H. Park, S.W. Kim, et al., A comparative study on collagen type I and hyaluronic acid dependent cell behavior for osteochondral tissue bioprinting, Biofabrication 6 (3) (2014). Available from https://doi.org/10.1088/1758-5082/6/3/035004, http://iopscience.iop.org/1758-5090/6/3/035004/pdf/1758-5090_6_3_035004.pdf.

[90] W.L. Ng, W.Y. Yeong, M.W. Naing, Polyvinylpyrrolidone-based bio-ink improves cell viability and homogeneity during drop-on-demand printing, Materials 10 (2) (2017). Available from https://doi.org/10.3390/ma10020190, http://www.mdpi.com/1996-1944/10/2/190/pdf.

[91] T.P. Zwaka, J.A. Thomson, Homologous recombination in human embryonic stem cells, Essentials of Stem Cell Biology, Elsevier Inc., United States, (2009), pp. 417–422. Available from https://doi.org/10.1016/B978-0-12-374729-7.00046-9, http://www.sciencedirect.com/science/book/9780123747297.

[92] A. Khademhosseini, N. Ashammakhi, J.M. Karp, S. Gerecht, L. Ferreira, N. Annabi, et al., Embryonic stem cells as a cell source for tissue engineering, Principles of Tissue Engineering, Elsevier, United States, (2020), pp. 467–490. Available from https://doi.org/10.1016/B978-0-12-818422-6.00027-7, https://www.elsevier.com/books/principles-of-tissue-engineering/lanza/978-0-12-398358-9.

[93] K. Takahashi, S. Yamanaka, Induction of pluripotent stem cells from mouse embryonic and adult fibroblast cultures by defined factors, Cell 126 (4) (2006) 663–676. Available from https://doi.org/10.1016/j.cell.2006.07.024.

[94] A. Wang, Z. Tang, I.H. Park, Y. Zhu, S. Patel, G.Q. Daley, et al., Induced pluripotent stem cells for neural tissue engineering, Biomaterials 32 (22) (2011) 5023–5032. Available from https://doi.org/10.1016/j.biomaterials.2011.03.070.

[95] P. Bianco, Mesenchymal" stem cells, Annual Review of Cell and Developmental Biology 30 (1) (2014) 677–704. Available from https://doi.org/10.1146/annurev-cellbio-100913-013132.

[96] M. Dominici, K. Le Blanc, I. Mueller, I. Slaper-Cortenbach, F.C. Marini, D.S. Krause, et al., Minimal criteria for defining multipotent mesenchymal stromal cells. The International Society for Cellular Therapy position statement, Cytotherapy 8 (4) (2006) 315–317. Available from https://doi.org/10.1080/14653240600855905.

[97] D. Kim, A.E. Lee, Q. Xu, Q. Zhang, A.D. Le, Gingiva-derived mesenchymal stem cells: potential application in tissue engineering and regenerative medicine - a comprehensive review, Frontiers in Immunology 12 (2021). Available from https://doi.org/10.3389/fimmu.2021.667221, https://www.frontiersin.org/journals/immunology#.

[98] Y.S. Kim, A.G. Mikos, Emerging strategies in reprogramming and enhancing the fate of mesenchymal stem cells for bone and cartilage tissue engineering, Journal of Controlled Release 330 (2021) 565–574. Available from https://doi.org/10.1016/j.jconrel.2020.12.055, www.elsevier.com/locate/jconrel.

[99] J.S. Park, S.W. Yi, H.J. Kim, S.M. Kim, J.H. Kim, K.H. Park, Construction of PLGA nanoparticles coated with polycistronic SOX5, SOX6, and SOX9 genes for chondrogenesis of human mesenchymal stem cells, ACS Applied Materials and Interfaces 9 (2) (2017) 1361–1372. Available from https://doi.org/10.1021/acsami.6b15354, http://pubs.acs.org/journal/aamick.

[100] R.M. Raftery, A.G. Gonzalez Vazquez, G. Chen, F.J. O'Brien, Activation of the SOX-5, SOX-6, and SOX-9 trio of transcription factors using a gene-activated scaffold stimulates mesenchymal stromal cell chondrogenesis and inhibits endochondral ossification, Advanced Healthcare Materials 9 (10) (2020). Available from https://doi.org/10.1002/adhm.201901827, http://onlinelibrary.wiley.com/journal/10.1002/(ISSN)2192-2659.

[101] B.L. Dearman, S.T. Boyce, J.E. Greenwood, Advances in skin tissue bioengineering and the challenges of clinical translation, Frontiers in Surgery 8 (2021). Available from https://doi.org/10.3389/fsurg.2021.640879, journal.frontiersin.org/journal/surgery.

[102] J. Li, M. Chen, X. Fan, H. Zhou, Recent advances in bioprinting techniques: approaches, applications and future prospects, Journal of Translational Medicine 14 (1) (2016). Available from https://doi.org/10.1186/s12967-016-1028-0, http://www.translational-medicine.com/home/.

[103] S. Huang, B. Yao, J. Xie, X. Fu, 3D bioprinted extracellular matrix mimics facilitate directed differentiation of epithelial progenitors for sweat gland regeneration, Acta Biomaterialia 32 (2016) 170–177. Available from https://doi.org/10.1016/j.actbio.2015.12.039.

[104] W.L. Ng, W.Y. Yeong, M.W. Naing, Development of polyelectrolyte chitosan-gelatin hydrogels for skin bioprinting, Procedia CIRP 49 (2016) 105–112. Available from https://doi.org/10.1016/j.procir.2015.09.002, http://www.sciencedirect.com/science/journal/22128271.

[105] A. Tirella, A. Orsini, G. Vozzi, A. Ahluwalia, A phase diagram for microfabrication of geometrically controlled hydrogel scaffolds, Biofabrication 1 (4) (2009) 045002. Available from https://doi.org/10.1088/1758-5082/1/4/045002.

[106] T. Weng, W. Zhang, Y. Xia, P. Wu, M. Yang, R. Jin, et al., 3D bioprinting for skin tissue engineering: current status and perspectives, Journal of Tissue Engineering 12 (2021). Available from https://doi.org/10.1177/20417314211028574.

[107] V.K. Deb, N. Chauhan, R. Chandra, U. Jain, Recent progression in controlled drug delivery through advanced functional nanomaterials in cancer therapy, BioNanoScience 14 (2) (2024) 2004–2047. Available from https://doi.org/10.1007/s12668-023-01297-6, https://www.springer.com/journal/12668.

[108] K. Johnson, S. Zhu, M.S. Tremblay, J.N. Payette, J. Wang, L.C. Bouchez, et al., A stem cell-based approach to cartilage repair, Science (New York, N.Y.) 336 (6082) (2012) 717–721. Available from https://doi.org/10.1126/science.1215157, http://www.sciencemag.org/content/336/6082/717.full.pdf.

[109] F. Daou, A. Cochis, M. Leigheb, L. Rimondini, Current advances in the regeneration of degenerated articular cartilage: a literature review on tissue engineering and its recent clinical translation, Materials 15 (1) (2022) 31. Available from https://doi.org/10.3390/ma15010031.

[110] H. Shen, H. Lin, A.X. Sun, S. Song, B. Wang, Y. Yang, et al., Acceleration of chondrogenic differentiation of human mesenchymal stem cells by sustained growth factor release in 3D graphene oxide incorporated hydrogels, Acta Biomaterialia 105 (2020) 44–55. Available from https://doi.org/10.1016/j.actbio.2020.01.048, http://www.journals.elsevier.com/acta-biomaterialia.

[111] G. Huang, S.T. Pan, J.X. Qiu, The clinical application of porous tantalum and its new development for bone tissue engineering, Materials 14 (10) (2021). Available from https://doi.org/10.3390/ma14102647, https://www.mdpi.com/1996-1944/14/10/2647/pdf.

[112] V.K. Deb, U. Jain, Ti3C2 (MXene), an advanced carrier system: role in photothermal, photoacoustic, enhanced drugs delivery and biological activity in cancer therapy, Drug Delivery and Translational Research 14 (11) (2024) 3009–3031. Available from https://doi.org/10.1007/s13346-024-01572-3, https://www.springer.com/journal/13346.

[113] Y. Duan-Arnold, A. Gyurdieva, A. Johnson, T.E. Uveges, D.A. Jacobstein, A. Danilkovitch, Retention of endogenous viable cells enhances the anti-inflammatory activity of cryopreserved amnion, Advances in Wound Care 4 (9) (2015) 523–533. Available from https://doi.org/10.1089/wound.2015.0636.

[114] L.A. Lavery, J. Fulmer, K.A. Shebetka, M. Regulski, D. Vayser, D. Fried, et al., The efficacy and safety of Grafix® for the treatment of chronic diabetic foot ulcers: results of a multi-centre, controlled, randomised, blinded, clinical trial, International Wound Journal 11 (5) (2014) 554–560. Available from https://doi.org/10.1111/iwj.12329, http://onlinelibrary.wiley.com/journal/10.1111/(ISSN)1742-481X.

[115] S.L.M. Dahl, A.P. Kypson, J.H. Lawson, J.L. Blum, J.T. Strader, Y. Li, et al., Readily available tissue-engineered vascular grafts, Science Translational Medicine 3 (68) (2011). Available from https://doi.org/10.1126/scitranslmed.3001426, http://stm.sciencemag.org/content/3/68/68ra9.full.pdf.

[116] J.H. Lawson, M.H. Glickman, M. Ilzecki, T. Jakimowicz, A. Jaroszynski, E.K. Peden, et al., Bioengineered human acellular vessels for dialysis access in patients with end-stage renal disease: two phase 2 single-arm trials, The Lancet 387 (10032) (2016) 2026–2034. Available from https://doi.org/10.1016/S0140-6736(16)00557-2, http://www.journals.elsevier.com/the-lancet/.

[117] K. Yano, A.T. Speidel, M. Yamato, Four Food and Drug Administration draft guidance documents and the REGROW Act: a litmus test for future changes in human cell- and tissue-based products regulatory policy in the United States? Journal of Tissue Engineering and Regenerative Medicine 12 (7) (2018) 1579–1593. Available from https://doi.org/10.1002/term.2683, http://onlinelibrary.wiley.com/journal/10.1002/(ISSN)1932-7005.

[118] T. Hoffman, A. Khademhosseini, R. Langer, Chasing the paradigm: clinical translation of 25 years of tissue engineering, Tissue Engineering. Part A 25 (9-10) (2019) 679–687. Available from https://doi.org/10.1089/ten.tea.2019.0032.

Chapter 5

Interactions of tooth-derived stem cells with dental regenerative scaffolds

Manoj Komath and Eva C. Das

Department of Biomaterial Science and Technology, Biomedical Technology Wing, Sree Chitra Tirunal Institute for Medical Sciences and Technology, Thiruvananthapuram, Kerala, India

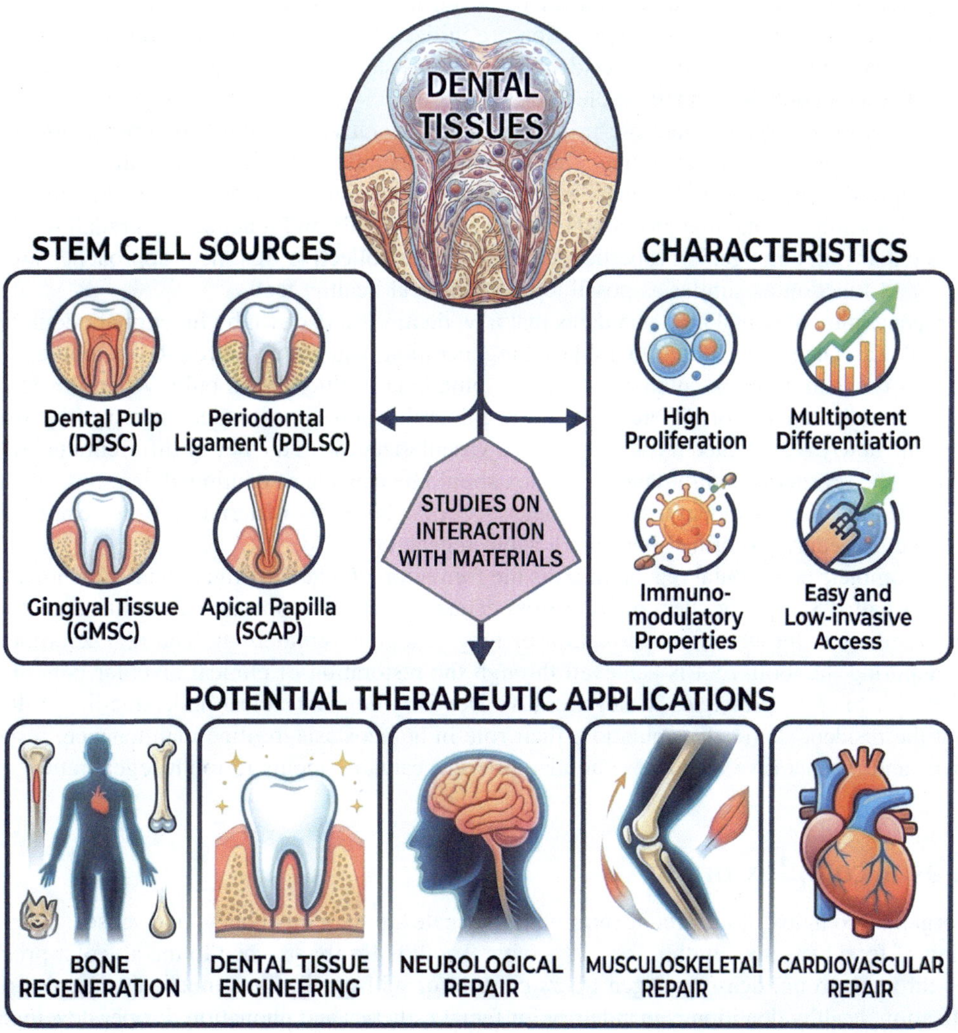

Fundamentals of Translational Tissue Engineering. DOI: https://doi.org/10.1016/B978-0-443-33455-9.00019-5

5.1 Introduction

The application of biomaterial-based regenerative treatment approaches in dentistry has pioneered the emergence and progressive evolution of "Regenerative Dentistry" that strives to achieve predictable regeneration of dental tissues towards reinstating their original structure and function, rather than relying on replacement, repair, or restoration of lost dental tissues. Regenerative dentistry utilizes the advancements in regenerative medicine using suitable biomaterials alone and/or tissue-engineered constructs to adequately translate the most appropriate scientific knowledge on dental diseases and tissue responses—both developmental and regenerative—into therapeutic and regenerative treatment strategies [1]. Advancements in cellular- and molecular-level biological interactions have led to the understanding that regeneration depends on healthy interactions between the changes in the immediate external environments constituting the ECM (extracellular matrix) and the responses of the resident stem cells capable of elucidating a regenerative response. These changes can be brought about by the prudent utilization of carefully designed biomaterials and/or tissue engineering scaffolds.

Regenerative dentistry constitutes the concept of treatment strategies aimed to regenerate dental tissues. Regenerative dentistry offers a novel, revolutionary treatment approach from that of the conventional reparative treatment approaches that ideates mainly structural restoration to a functionally stable regenerative approach, aided by the better understanding of cellular and molecular level biological interactions. The evolution and progression of these regenerative approaches are based on a better perception of developmental biology of dental tissues, and the intricate biology of healing, repair, and regeneration [2]. The biologic principles of regeneration are utilized to engineer biomimetic scaffolds or biomaterials to harness the innate regeneration potential of the resident endogenous stem cells in dental tissues augmented by appropriate growth factors, mediators, and/or biomimetic molecules.

Human tooth is a complete organ. Therefore principles of regenerative medicine attempting organ regeneration can positively influence and augment Regenerative Dentistry. Regenerative Dentistry also makes use of technological advancements in scaffold fabrication techniques, such as bioengineering, electrospinning, directed self-assembly, 3D bioprinting, hydrogel scaffolds containing bioactive molecules, structurally and functionally graded scaffolds consisting of variable porosities enhancing angiogenic properties, etc. [3]. These collective treatment approaches are aimed to restore the tooth structure and function as similar as possible to the natural healthy teeth.

The complete regeneration of dental pulp mandates that new dentin is laid down by functional odontoblasts from within the dental pulp. In the absence of healthy odontoblasts, regeneration is attempted through odontoblast-like cells derived from the endogenous progenitor cell population within the remaining healthy dental pulp. When the dental pulp is absent due to necrosis, stem cells capable of differentiation and mineralization are recruited from nearby tissues such as the periodontium [4]. The newly regenerated tissue needs to be vascularized, with the newly differentiated odontoblasts lining the dentinal tubules. The regenerated pulp-dentin complex should be capable of continued dentin deposition along the old or newly deposited dentin matrix. Prudent use of biomaterials or advanced biomaterial scaffolds can be used concurrently with the regenerative procedures to bring about predictable regeneration [5].

The criterion for complete periodontal regeneration is the formation of a healthy ligamental attachment formed of dense type I collagen fibers called Sharpey's fibers running from cementum to the supporting alveolar bone, devoid of impeding epithelial tissues resulting in the undesired formation of long junctional epithelium. The regeneration of alveolar bone supporting and anchoring the tooth root is achieved through the restoration of clinical alveolar bone height [6].

Applying and establishing the principles of Regenerative Dentistry demand a clear understanding of the tooth structure. Knowledge about the resident stem cell population, their role in homeostasis, routine maintenance, responses to irritants and stimuli, repair, and regenerative responses can also provide valuable inputs to tooth regeneration.

5.2 Tooth as a complex organ

The tooth is an organ of complex tissue architecture with adequate strength, resiliency, and sensory response to aid in the proper mastication of food consumed within the oral cavity. Tooth is firmly attached to the alveolar process of upper and lower jaw bones through strong, dense collagen fibers capable of withstanding the mechanical forces of mastication. In addition to mastication, healthy dentition is mandatory for facial esthetics and phonation. Proper development of teeth and supporting structures are contributing factor for the growth and development of the craniofacial structures, aided by the growth of jaw bones. Diseases affecting teeth and/or supporting structures may cause pain, discomfort, and difficulty in mastication of food. Depending on the severity of disease conditions, reversible or irreversible damage to tooth structure can occur.

5.2.1 Tooth structure

The tooth structure consists of the clinically visible tooth crown and the submerged tooth root that anchors the tooth to the supporting alveolar bone. The outermost part of the tooth crown consists of the hard, inert, acellular outer layer termed enamel, which is supported from within by dentin composed mainly of mineralized collagen fibers. The dentin is more resilient and less mineralized (70%) than enamel and compensates for the brittleness of heavily mineralized (96%) enamel. Unlike enamel, the dentin is a vital hard tissue, deposited continually and maintained by the vital dental pulp. The dental pulp is enclosed within the dentin and constitutes the innermost connective tissue component of the tooth that consists of the vasculature for nutrition, neurons for sensory perception, immune cells, and specialized progenitor cells aimed at maintenance and regeneration of dentin. The pulp and dentin are together known as the pulp-dentin complex, with the ability to lay down dentin matrix, primary, secondary, reactionary, and/or reparative dentin upon appropriate stimuli [7].

Enamel is the hardest mineralized tissue in the human body, with high mineralization (> 96 %) due to minimal organic components, such as proteins, and maximum inorganic components composed of calcium-fluorapatite crystals arranged in a well-organized manner. The cells that lay down enamel are completely lost post tooth eruption, and hence, enamel cannot be regenerated once lost. Currently, enamel loss can only be prevented by remineralization. Once lost after tooth eruption, enamel can only be structurally restored using restorations.

The dentin, in comparison to enamel, consists of a higher percentage of organic component, composed mainly of type I collagen fibers arranged in a specific manner as closely packed dentinal tubules passing throughout the whole thickness of dentin. These dentinal tubules are 70% mineralized with calcium-hydroxyapatite/fluorapatite crystals and provide resilience, thereby compensating for the brittleness of enamel. Odontoblasts comprise the specialized cells responsible for the deposition of dentin matrix and their mineralization. The odontoblast cell bodies are seen lining the pulpal side of the dentin forming the outermost part of the dental pulp, with their extensions extending through the dentinal tubules. The dentinal tubules are also filled with the dentinal fluid secreted by the odontoblasts and aid in sensory perception through fluctuations in fluid dynamics. Since the formative cells of dentin—the odontoblasts are present in the dental pulp, a vital dental pulp enables the dentin to regenerate and repair itself in the presence of healthy odontoblasts or odontoblast progenitor cells. The dental pulp, the vital soft tissue component of the tooth, is enclosed within the dentin, wherein it maintains the homeostasis through formative, nutritive, sensory, and reparative functions [7]. The crown dentin is covered by the dental enamel that is exposed to the oral cavity. The root dentine is covered by the cementum that provides firm attachment of the tooth to the supporting alveolar bone through the Sharpey's fiber running through the periodontal ligament.

The root cementum, also known as cementum, is the mineralized tissue surrounding the root surface of mammalian tooth roots. Cementum is composed of 50%–55% organic component consisting mainly of collagen and glycoproteins, and 45%–50% inorganic component (5% more mineral concentration than bone) made up of calcium-hydroxyapatite crystals. Due to the less mineral content, the cementum is slightly softer than dentin and harder than bone. Cementum is formed and deposited by the cementoblast cells. Cementum is continually deposited throughout life to replace the superficial aging cementum and to maintain the tooth attachment to the supporting alveolar bone. Cementum is classified into acellular cementum, that is, cementum without incorporated cementocytes within, and cellular cementum wherein the cementoblasts are entrapped between layers of deposited cementum as cementocytes. The acellular cementum is usually slowly deposited along the coronal two-thirds of the tooth, and the cellular cementum is found towards the apical part of the tooth. The periodontal ligament attachment fibers are inserted into the cementum of the tooth root and are termed the Sharpey's fibers [8].

The tooth root is attached firmly to the alveolar processes of the maxilla and mandible by the periodontal ligament. The cementum, the periodontal ligament, and the alveolar bone act as a single, dynamic, functional attachment unit [7]. The tooth-root attachment apparatus consists of the periodontium, a specialized supporting structure composed of the periodontal ligament, the cementum covering the root dentin, the alveolar bone, and the gingiva. The tooth is attached to the alveolar bone on each jaw through the periodontal ligament, with strong structural and functional attachment to the alveolar bone and the cementum, on either side, through the Sharpey's fibers. The individual collagen fibers acting as individual units are remodeled, regenerated, and/or repaired individually by the innate periodontal ligament fibroblasts.

Human dentition is succedaneous with a single succession of dentition through the deciduous and permanent dentition that helps to accommodate the proper craniofacial growth and development. The Anatomical division of the tooth into a crown portion in the oral cavity and a root portion anchored within the alveolar bones is demarcated clearly at the cemento-enamel junction (CEJ). Above the CEJ, the dentin is covered by enamel to provide the strength to pulverize the food through mastication by the sharp and strong cusps and ridges. Below the CEJ, dentin is covered by cementum to anchor the tooth firmly and enable proper mastication by withstanding and dissipating the masticatory forces to the supporting alveolar bones.

5.3 The cellular components of tooth

The human tooth is a complex organ with the endogenous cell population varying from stem/progenitor cells, endothelial/vascular cells, neuronal cells, immune cells, and other specialized cells. The cells are harbored within the soft tissue components of the tooth—namely, the dental pulp, periodontium (periodontal ligament, alveolar bone, gingiva), apical papilla, and dental follicle. Among these, the dental follicle regresses after tooth eruption, and the apical papilla regresses about 2 years after tooth eruption. Therefore the periodontal ligament and the dental pulp harbor the cellular component of the teeth. The specialized dental cells are mainly the odontoblasts in dental pulp that lay down dentin, the cementoblasts in periodontal ligament that lay down cementum, and the periodontal ligament fibroblasts that can differentiate to cementogenic and osteogenic lineage, along with remodeling collagen fibers. A detailed description of tooth-derived cells is given below.

5.3.1 Unique resident stem cell population in dental tissues

Stem cells are specialized cells that possess the ability of self-renewal and differentiation. Stem cells are present within all types of the tissues continually replacing damaged or dying cells. Advanced knowledge on stem cells provides a deeper understanding of mechanisms that regulate cellular differentiation, tissue maintenance, and regeneration, and embryonic development [9]. Depending on the source, stem cells can be classified as (1) embryonic stem cells and (2) adult stem cells. Embryonic stem cells are prenatal totipotent stem cells capable of teratoma formations with ethical issues limiting their usage in vitro and in vivo. The adult stem cells are postnatal stem cells that reside in specific tissue niches. They possess self-renewal ability and trans differentiation ability by means of cell division and can differentiate to intermediate progenitor cells or terminally differentiated cells [10]. Adult stem/progenitor cells can remain quiescent until needed, stimulated by injury or regenerative responses to get actively involved in tissue repair and/or regeneration [1].

Gronthos et al. pioneered the stem cell research in dentistry through the successful isolation of the DPSCs (postnatal human dental pulp stem cells) [11]. Isolation of SHED (stem cells of Human Exfoliated Deciduous teeth) [12], hPDLSCs (human periodontal ligament stem cells) [13], and SCAP (Stem Cells of Apical Papilla) [14] followed, reiterating the presence of stem cells within and around human teeth. Since then, tooth-derived stem cells were successfully isolated, characterized, and studied for cellular responses and differentiation potential to aid biomedical, therapeutic, and biological research applications [15].

5.3.2 Origin of dental stem cells

The tooth-derived stem cells hold special importance due to their neural crest cell origin. The dental and periodontal structures other than enamel are derived from the cranial neural crest or the ectomesenchyme also known as the fourth germ layer [16].

The periodontal ligament is the strong fibrous connective tissue that connects the mineralized tissues of periodontium—the cementum on the tooth root and the alveolar bone. Embryologically, the periodontal ligament is derived from the ectomesenchymal cells of the dental follicle, derived from the cranial neural crest cells. The periodontal ligament consists of a heterogeneous cell population of fibroblasts, osteoblasts, and cementoblasts along with type I collagen, vascular, and sensory components [16]. Isolated cell types reveal a predominance of fibroblast cells exhibiting stem cell properties that help to maintain routine homeostasis, collagen and bone formation and remodeling, and cementum formation.

The dental pulp is also embryologically derived from cranial neural crest cells with odontoblasts as specialized cells that secrete dentin matrix. They can differentiate into odontoblasts that secrete dentin, and/or odontoblast-like cells that can lay down dentin-like matrix, mediated by innate noncollagenous proteins such as dentin matrix proteins and/or transcription factors such as Wnt, runx2 etc. [16].

The positioning of stem cells in the immediate niche can also influence the quiescence or activation of dental stem cells. Stem cell division can result in tissue homeostasis, differentiation to terminal specialized cells or intermediate transient amplifying cells capable of migration and site-specific further differentiation [2]. Thus the endogenous periodontal ligament stem cells can differentiate to fibroblasts that deposit and remodel collagen within periodontal ligament, cementoblasts covering root dentin that secrete the cementum matrix, and to osteoblasts towards the alveolar bone end that lay down and remodel the alveolar bone. Similarly, the progenitor cells in dental pulp can differentiate to odontoblasts that secrete true dentin or odontoblast-like cells that lay down reparative dentin.

5.3.3 Isolation of tooth-derived cells

In vitro cell culture systems utilizing tooth-derived cells provide a reliable platform to study the cellular and molecular responses to dental biomaterials. The first step towards cell-material interaction studies is the standardized isolation, culture, and characterization of the relevant tooth-derived cells.

The earlier attempts on isolation of dental stem cells utilized enzymatic tissue digestion techniques of the dental soft tissues. Later on, explant culture techniques were successfully carried out for the isolation of dental stem cells. Upon comparison, the explant culture technique yielded a more heterogeneous population of dental pulp cells and showed that in the enzymatic digestion technique, without much variation in the properties of the cells, including the presence and absence of characterization markers, and differentiation potential did not differ significantly between these techniques [17]. Dental pulp cells were isolated from adult teeth, deciduous teeth, supernumerary teeth, cryopreserved teeth, and even from pulpectomy of traumatized teeth requiring endodontic therapy. The difference in niche and sourcing did not significantly affect the inherent properties of the cells [18]. It was hypothesized that the enzymatic tissue digestion technique might destroy the cellular niche, whereas the tissue explant culture method can preserve the cellular niche environment mimicking intrinsic mechanisms of stem cell activation in conditions such as traumatic dental injuries and provide high purity dental pulp stem cell population repeatedly and in substantial quantities. For these reasons, the explant culture system was recommended to be applied to isolation procedures from other dental tissues such as the dental apical papilla and periodontal ligament [19,20]. Similar findings were reported in isolating the deciduous teeth stem cells also. Similar isolation techniques on periodontal ligament stem cells reported that the enzymatic digestion method yielded early availability of cells in comparison to the time elapsed needed for cell outgrowths from explant culture that necessitate more time for outward cell migration from the explant. Although enzymatic digestion released an adequate number of cells from the tissue samples in lesser time, some degree of cell damage could exist due to the enzymes, which can be negated or minimized in the explant culture system, wherein the intact tissue structure can provide a native ECM—like environment until the cells migrate out [21]. However, the initial heterogeneity was lost upon repeated passaging with the cells exhibiting similar proliferation and differentiation patterns thereafter [22]. Therefore the use of tooth-derived stem cells in early passages enables the maintenance of the unique cellular heterogeneity.

5.3.4 Characterization of tooth-derived cells

The primary cells isolated for in vitro research need to be characterized to validate the properties of proliferation, differentiation potential, and immunomodulation to better understand the responses to biomaterials. The tooth-derived cells constitute a heterogenous population wherein the subpopulation of stem cells and specialized cells is important in cell-material interactions in vitro. Characterization parameters include the presence of specific markers, fibroblast phenotype, and multilineage differentiation [23]

The anatomical site of harvesting the tissue specimens also dictates the characterization of periodontal ligament cells, which is usually done from periodontal ligament collected from the middle third of the tooth root to avoid cross-contamination with gingival fibroblasts coronally, and dental pulp or apical papilla cells from the apical foramen areas. Protein and gene markers used to identify hPDL cells include the ligament-specific marker Scleraxis, Cementum Attachment Protein (CAP), cementum protein-23, periodontal ligament-associated protein (PLAP-1), cementum membrane protein (CEMP), etc. The choice of the markers for periodontal ligament cells can be varied based on the design and purpose of the study. The progenitor/stem cell population within the periodontal ligament cells can be characterized by the presence of the mesenchymal stem cell markers such as Stro1, CD 73, CD 90, and CD 105. Specific markers like Scleraxis, CAP, and CEMP can be used to confirm the specificity of periodontal ligament cells. Stimuli-dependent osteogenic/cementogenic differentiation potential can be studied using variation in the expression of CEMP and Scleraxis [24,25,26]. This helps to differentiate between the inherent cell responses and biomolecule- or biomaterial-based cell responses being tested.

The dental pulp stem cells also consist of a heterogeneous and mixed population with expression of the common MSC markers and dentin-specific markers such as DMP 1 (dentin matrix protein), DSP (dentin sialoprotein), and DSPP (entin sialophosphoprotein) [18]. The deciduous teeth and apical papilla cells also exhibit common mesenchymal stem cell markers such as STRO 1, CD 90, CD 105, CD 44, CD 73, and CD 271 [15,27,28].

5.3.5 Differentiation potential of tooth-derived cells

The tooth-derived stem cells isolated from different sites result in heterogeneous cell populations with an inherent stem cell population and supporting cells that might influence the progenitor cell behavior [29]. Due to the mineralized nature of dental hard tissues, the dental stem cells can exhibit an osteogenic differentiation potential upon induction [30,31]. The presence of type 1 collagen as matrix protein and hydroxyapatite as mineral component, similar to bone, makes difficult to distinguish osteogenic differentiation of dental stem cells from their characteristic cemento/odontogenic differentiation [32,33]. Therefore it is deemed acceptable to use common osteogenic inducing agents, including BMPs, L-ascorbic acid–β-glycerophosphate-dexamethasone cocktail to induce differentiation of tooth-derived stem cells. The use of inherently osteogenic biomaterials such as hydroxyapatite and calcium phosphates is also justified to study cell-material interactions in vitro. The expression of the common osteogenic differentiation marker proteins and genes to evaluate the differentiation potential of tooth-derived cells is also reported [34,35]. The cell density is also found to affect the osteogenic differentiation of dental pulp cells as well as other tooth-derived cells [36].

The periodontal ligament fibroblasts and gingival fibroblasts display osteogenic potential even as a heterogeneous population in vitro [33]. The dental pulp cells exhibit osteogenic differentiation in vitro and in vivo, under suitable conditions, with evidence of mineralization and osteo/odontogenic gene expressions [11,37]. Biomaterial-induced osteogenic differentiation of dental pulp cells is also reported [38]. The differences in isolation techniques did not significantly influence the differentiation potential of dental stem cells. A consistent increase in alkaline phosphatase enzyme activity during the osteogenic differentiation of tooth-derived cells is also reported widely. The presence of collagen matrix deposition, histologically stained mineralized nodules, osteogenic protein and gene expressions, etc., are important parameters to confirm osteo/odonto/cementogenic differentiation of dental stem cells in vitro [17].

5.4 Cellular interactions

The isolated, characterized human tooth-derived cells are ideal primary cells to study the cellular interactions of different aspects of dental biomaterials. The dental tissues are responsive to a wide range of stimuli, including pH changes, mechanical stresses, surface morphology, biologic factors, and calcium and phosphate ions. Different modifications in cell culture strategies are needed to elicit specific cellular responses.

5.4.1 Responses to pH

The changes in optimal pH can bring about degenerative changes in dental tissues, especially the dental pulp. PH is one of the deciding factors in determining whether the dental pulp can shift towards regeneration or degeneration. In fact, the "theory of optimal pH" reiterates the possibility of pulp damage due to pH changes even without microbial infections [39]. Dental caries, inflammation, bacterial toxins, microbial infections, and salivary contamination can reduce the PH of dental pulp, which generally is slightly alkaline in nature. A drop of pH below 7 can cause growth arrest, and further pH reduction below 5.5 can result in apoptosis or necrotic responses in cultured dental pulp cells. Dental pulp cells cultured in an alkaline environment above a pH of 9 result in increased cell proliferation and mineralization in vitro, indicating a regenerative response [40]. This pH dependent regenerative response of resident dental pulp cells within the dental pulp is being utilized for the time-tested success of dental pulp repair and regeneration using alkaline biomaterials such as calcium hydroxide and mineral trioxide aggregate [41,42]. Here, the high alkalinity of the biomaterials (>10) neutralizes the acidic extracellular environment caused by infection and/or inflammation along with eliciting a cell proliferative response resulting in deposition of a new layer of dentin and/or dentin-like hard tissues, preventing further microbial infiltration and possible infection. Studies have shown that an increase in pH resulted in elevated alkaline phosphatase levels, increased BMP2 expression, and promoted mineralization, evidenced through calcified nodule formation. However, a very high pH could initially harm the odontoblasts, resulting in reparative dentin formation rather than the regenerative reactionary dentin formation [43]. The difference may not be evident in in vitro analyses, as histopathologic evaluations are not possible in vitro. The alkaline environment also creates a bactericidal environment, which can be further augmented by the use of antimicrobial agents locally delivered through the biomaterials [44].

pH is also an important parameter in maintaining a healthy periodontium. Alkaline biomaterials used as intracanal regenerative agents in regenerative endodontic treatment approaches, such as revascularization and revitalization promotes the proliferation and differentiation of periodontal ligament cells and alveolar bone cells from the surrounding supporting periradicular tissues to form cementum-like and/or bone-like hard tissue deposition along the inner walls of tooth-root dentin, to allow for continued tooth-root development, further strengthening the tooth-root. Further, the alkaline biomaterials used in endodontic treatment procedures such as root canal treatment also have a proliferative and regenerative effect on the periradicular tissues, thereby regenerating the periradicular alveolar bone and cementum in cases of bone loss [45,46].

5.4.2 Cellular response to calcium and phosphate ions

The dental hard tissues are composed of the biomineral hydroxyapatite in varying concentrations in enamel (97%), dentin (70%), alveolar bone (65%) and cementum (45–50%) [47]. The organic matrix of dental hard tissues, except enamel, predominantly consists of type I collagen, which acts as the template for mineralization with deposition of calcium-hydroxyapatite crystals [48]. Since hydroxyapatite deposition is influenced by calcium and phosphate ions, numerous calcium phosphate and calcium-enriched biomaterials are being continually introduced for dental tissue regeneration applications. Calcium Phosphate biomaterials can undergo dissolution releasing calcium and phosphate ions that can aid in osteogenesis by modulating extracellular and independent metabolic mechanisms. An array of molecular pathways are triggered through different receptor-mediated cellular events that result in elevated osteogenic gene expressions reiterating an osteogenic response rather than mere ossification [49–51]. Inorganic phosphates itself can act as signaling molecules triggering the cascade of cellular and molecular level osteogenic responses [52]. The only occurrence of four phosphate-rich mineralized structures—the enamel, dentin, cementum, and alveolar bone—enables the tooth-derived cells to be ideal candidates to study cellular responses to calcium phosphate-based biomaterials [53]. In this scenario, thorough research on the dental cellular responses to calcium and phosphate ions is mandatory. The calcium and phosphate ions can be substituted in cell culture medium as elutes of dental materials, ions, or as calcium phosphate scaffolds. The cellular responses are analyzed using cellular and molecular methods such as mineralization, gene, and protein expressions.

Commonly used calcium-containing biomaterials, such as calcium hydroxide and mineral trioxide aggregate, along with calcium carbonate, induced the differentiation of in vitro cultured dental pulp stem cells to an odontogenic lineage exhibiting mineralized nodule formation and odontogenic gene expression. Human dental pulp stem cells cultured in medium supplemented with varying concentrations of ionic calcium showed mineralization and gene expression as early as three days [54]. Higher concentrations of extracellular calcium ions showed a reduction in early osteogenic genes such as alkaline phosphatase, RunX2, and collagen 1, but an increase in mineralization or ossification genes such as osteocalcin and osteopontin [55]. Molecular studies on dental pulp stem cell response to calcium hydroxide indicate the early activation of the Mitogen-activated Protein Kinase Pathway (MAPK pathway) as the mechanism for enhanced dental pulp cell migration, proliferation, and osteogenic differentiation in vitro [56]. Calcium hydroxide also increases the secretion of fibronectin, which in turn can induce differentiation of odontoblasts and secretion of dentin matrix. Both elevated pH and the calcium ions can increase BMP2 expression through calcium channel ERK pathway and activation of Notch signaling in dental pulp cells resulting in dentin formation [57]. The cellular responses to calcium ions can be attributed to calcium-sensing receptors in pulp cells, which can respond to accumulated extracellular Ca^{++} that can modulate odontoblastic function. The calcium-dependent cell adhesion molecule N-cadherin can also respond to calcium ions, leading to odontoblast differentiation [43].

Regeneration of alveolar bone with a healthy attachment to the tooth-root cementum through a healthy periodontal ligament is the basis of complete periodontal regeneration. Calcium and phosphate biomaterials in the form of solid granules, injectable pastes, hydrogels, multiphasic scaffolds, etc., constitute regenerative strategies for periodontium [58].

Human periodontal ligament cells cultured with ionic calcium and phosphate ions resulted in increased cellular proliferation and mineralization with RunX2 upregulation, but a reduced alkaline phosphatase enzyme and gene expression was noted [59]. An elevated ionic concentration of calcium and phosphate ions resulted in apoptosis of nondental osteoblast cells, followed by mineralized nodule formation, reiterating the need to differentiate between nonbiologic physicochemical mineralization and true osteogenic differentiation by osteogenic gene and enzyme expression [60]. Intracanal use of calcium hydroxide in in vivo dog models resulted in cementum formation with elevated expression of cementum membrane protein (CEMP) in an ERK-dependent manner [61]. Phosphate ions promoted the proliferation and osteogenic differentiation of cultured periodontal ligament cells through the activation of NF-κB pathway with the expression of osteogenic genes such as RunX2, osterix, osteocalcin, alkaline phosphatase, and odontogenic markers such as dentin sialoprotein and dentin sialophosphoprotein [62]. Human periodontal ligament cells cultured in a solid biphasic calcium phosphate scaffold showed successful cell adhesion and ECM formation. The cells expressed osteogenic genes with elevated alkaline phosphatase enzyme activity with the scaffold alone, without any extraneous osteogenic inducers, reiterating the role of biomaterial-released calcium and phosphate ions in osteoblastic differentiation of periodontal ligament cells, which can clinically correlate to periodontal ligament-mediated alveolar bone formation [63].

5.4.3 Responses to mechanical forces

The tooth within the oral cavity is situated in a dynamic environment, constantly subjected to an array of mechanical stresses due to the occlusal forces. The compressive forces are reported to cause hypoxic conditions that can result in the resorption of the alveolar bone, whereas tensional forces can promote deposition of alveolar bone. This bone resorption–deposition cycle in response to mechanical forces is being utilized in orthodontic tooth movement and dentofacial modifications for functional as well as esthetic purposes.

The occlusal masticatory forces create mechanical signals that can be converted to biochemical signals with gene expression for the maintenance of tissue homeostasis. This is even more important in Periodontal ligament fibroblasts that surround and support the tooth roots anchoring it to the alveolar bone. The conversion of mechanical forces to biochemical signals is elucidated through intracellular calcium signaling resulting in bone resorption and apposition cycles [64].

The cell responses to mechanical forces can be simulated in vitro using modified culture methods that can provide mechanical stimulation through cyclic stresses. Tensional forces can be applied by means of substrate deformation—the periodontal ligament cells are in substrates such as elastic membranes, which can be stretched for uniaxial forces or deformed to shapes such as domes to provide biaxial or multiaxial forces. Compressive forces can be mainly applied by means of weight over cultured cells, hydrostatic pressure in the cell culture system, and/or centrifugation forces [65]. One method of studying resorptive responses under compressive forces in periodontium is by using coculture techniques with osteoclast precursor cells such as isolated peripheral blood monocyte cells (PBMCs) and evaluating osteoclastic enzyme and gene expressions as well as differentiation of precursor cells to multinucleated giant cells. Another method is to directly evaluate enzyme and gene expression from cultured periodontal ligament cells [66].

In vitro cultured human periodontal ligament fibroblasts subjected to a stretching-relaxation pattern of cyclic mechanical forces exhibited a spindle shape morphology with a perpendicular alignment of cytoskeletal actin filaments in relation to the strain force vector. A reduction in alkaline phosphatase enzyme is also reported, which can correlate to the inhibition of bone and cementum forming ability by the periodontal ligament cells [67]. Hydrostatic pressure induced mechanical stresses to cultured periodontal ligament cells resulted in an increased expression of inflammatory cytokines suggesting an inflammatory response in a clinical scenario [68]. Continuous compressive forces can be applied to cultured periodontal ligament cells by placing appropriate cell culture-compatible weights over a monolayer of cultured cells. Such an arrangement with continuous compressive forces resulted in an osteoclastic response correlating to clinical alveolar bone resorption. This was further affirmed with coculture of periodontal ligament cells with peripheral blood mononuclear cells that differentiated into multinucleated giant cells indicative of osteoclastic activity, along with an elevation of TRAP (tissue-resistant acid phosphatase) enzyme activity [69]. Induced mechanical stress through continuous compressive forces resulted in the expression of bone resorptive response with increased osteopontin and RANKL expression by cultured human periodontal ligament cells [70].

Tensional forces applied to cultured periodontal ligament cells result in an osteogenic response, which can be identified based on increased osteogenic enzyme and gene expressions. Mineralization can also be assessed by staining and quantifying the mineralized nodules formed in vitro. Alkaline phosphatase enzyme activity is a potent indicator of osteoblastic differentiation of cultured periodontal ligament cells as it is involved in mineralization and is a valid indicator of osteogenic cells. Other osteogenic genes include Runx2, Osteocalcin, Bone Sialoprotein, and SP7 [66,71].

Human dental pulp stem cells cultured in 3D scaffold with dentin-like pores, when subjected to a compressive force of 19.6 kPa for 9h as a mechanical induction of differentiation, resulted in odontoblastic gene expression (Dentin sialophosphoprotein—DSPP) with upregulation of BMP 7. Mechanical compression also resulted in increased phosphorylation of MAPKs (mitogen-activated protein kinases) and ERKs (extracellular signal-regulated kinases), indicating the role of the MAPK signaling pathway in the mechanical induction of odontoblast differentiation [72].

Compressive forces can cause hypoxic responses in the periodontium, which can be simulated in vitro using human periodontal ligament cells [73]. Since hypoxia can indirectly simulate periodontal inflammation due to mechanical forces, inducing hypoxia in periodontal ligament cell culture systems can be used to study cellular responses in vitro. Subjecting the periodontal ligament cells to brief hypoxia of 2% oxygen saturation followed by normoxia of 20% oxygen saturation resulted in an osteogenic response in vitro [74]. Traumatic injuries induced mechanical stresses can also cut off the blood supply to dental pulp resulting in hypoxia and pulpal necrosis. Simulation of pulpal hypoxia followed by normoxia in vitro using cultured dental pulp cells showed an increased cell growth and alkaline phosphatase activity possibly due to dynamic rebound responses to pulpal hypoxia [75]. Induced hypoxic conditions in cultured human periodontal ligament cells and dental follicle-derived cells resulted in increased alkaline phosphatase activity, mineral deposition, and cementum membrane protein (CEMP1) expression through hypoxia-inducible factor-1 (HIF-1) gene activation [76].

5.4.4 Response to rigidity and stiffness

The dental tissues differ in the rigidity or substrate stiffness with enamel being the hardest (Young's modulus of 83GPa) mineralized tissue in the human body. The elastic moduli of enamel range from 1.338 to 98.3GPa, dentin from 5.5 MPa to

100 GPa, and the dental pulp with 5.5 kPa [77]. The average elastic modulus of human cementum ranges is 2.398 ± 0.455 GPa and that of alveolar bone ranges from 1.5 to 2.5 GPa for trabecular bone and 10 and 20 GPa for cortical bone [78]. The resident cellular component within all tissues respond to the rigidity and stiffness of the extracellular environment through changes in cell morphology, migration ability, proliferation, and differentiation potential brought about by the activation of appropriate signaling pathways [79]. A study of the mechanobiology of resident stem cells within dental tissues can provide a better understanding of the regenerative responses to dental biomaterials ranging from low stiffness injectable hydrogels to high stiffness solid scaffolds [80]. Even in biomaterial substrates with identical chemical compositions, difference flexibility and substrate rigidity influence the cells such that cell spreading and migration is favored on stiff substrates, whereas cell agglomeration and formation of tissue-mimetic structures are favored in soft substrates. The substrate stiffness parameters can be measured by atomic force spectroscopy, and the cellular changes by time-elapse microscopy and immunofluorescence staining of the cytoskeleton [81].

Human dental pulp stem cells (DPSCs) cultured in hydrogel and glass substrates of elastic moduli ranging from ~3 kPa to 50 GPa with and without hydroxyapatite nanoparticles showed elevated ALP activity and increased mineralization on stiffer substrates (>75KPa) even without hydroxyapatite nanoparticles [82]. The DPSCs' migration speed was increased on softer substrates with reduced speed as substrate stiffness increases. An experimental culture set up with a central softer matrix and peripheral stiffer substrate showed the DPSCs maintained a pulp-like tissue architecture in the softer substrate with dentin-like mineralization in the peripheral stiffer matrix. Such constructs can mimic the stiffness variant in human dental pulp (5 kPa) and surrounding dentin matrix (Young's modulus reaches 20 GPa) wherein the odontoblast differentiation and new dentin deposition are made by cells in contact with the stiffer dentin walls [83]. The dental pulp cell senescence is promoted by stiffer substrates (50 kPa), and DPSC quiescence is promoted by softer substrates (2Kpa) with upregulation of quiescence genes, pluripotency, and NF-κB activation correlating to preservation of stemness and reduction of senescence in softer substrates [84]. A thorough study on the effect of stiffness on dental pulp stem cells using a specially formulated hydrogel comprising "Tri-Phase Biomechanical Structure" 3D matrix system resulted in enhanced odontogenic differentiation at a stiffness of 24.12Kpa. A Low stiffness of 0.79 kPa promoted vascular ingrowths and maintained pulp formation in vivo [85].

The human periodontium with the tooth–alveolar bone interface is constantly subjected to mechanical force-induced changes in substrate rigidity, which can impact the maintenance of periodontal width and regulation of remodeling, which are important for the regeneration and maintenance of periodontal health [86]. Human periodontal ligament stem cells (PDLSCs) cultured in substrates of varying stiffness exhibited differences in cell morphology, with reduced cell size in softer substrates. Stiffer substrates promoted the proliferation and differentiation of PDLSCs with upregulation of Notch signaling pathway [87]. An in vitro substrate model with a graded increase in substrate stiffness with cultured PDLCs showed a stiffness-dependent proliferation and apoptosis [88].

5.4.5 Response to surface architecture

The surface architecture greatly influences the cellular responses to biomaterials. For example, the dental pulp cells differentiate to odontoblasts in response to the tubular architecture of dentin. The periodontal ligament fibroblasts can differentiate to osteoblasts and/or cementoblasts depending on tactile stimuli. The nanotopography refers to the surface characterization correlating to intercellular spaces (e.g., Odontoblasts—30–40 nm), collagen fiber diameter (approximately 15 nm at the base of odontoblast and about 50 nm on the calcification area), etc. Nanotopography can effectively modulate cell responses by influencing cell adhesion and spreading, cell distribution and arrangement, morphological changes, differentiation, and gene expression [89]. In addition, a nanofibrous architecture can increase the surface area for better anchorage of filopodial extensions, thereby promoting cell adhesion with better nutrient supply, metabolic waste discharge and enhanced oxygen supply to the cells [90]. Human DPSCs cultured in nanofibrous poly (l-lactic acid) scaffolds with a fiber diameter of 50–500 nm exhibited enhanced cell adhesion, proliferation, and osteo/odontogenic differentiation. Nanofibrous architecture can provide better adsorption of adhesive protein molecules such as collagen, fibronectin, and vitronectin [91]. Human dental pulp cells cultured in porous silicon (~36 nm pore size) exhibited enhanced cell adhesion and mitosis [92]. Human dental pulp stem cells (hDPSCs) cultured in porous 3D calcium phosphate granules successfully induced odontogenic differentiation with increased alkaline phosphatase activity and expression of odontogenic genes such as DSPP and dentin matrix protein 1 (DMP 1) [93].

Macroporous, freeze-dried collagen scaffolds prepared for human periodontal ligament culture showed that a pore size of about 100 μm was necessary for cell recruitment and the even distribution of PDLCs within the scaffold cross-section. The results suggest that the careful control of pore size, pore wall alignment, and percolation diameter needs to be used to fine-tune the efficacy and uniformity of cellular infiltration. Single-cell tracking as well as endpoint bulk cell invasion was

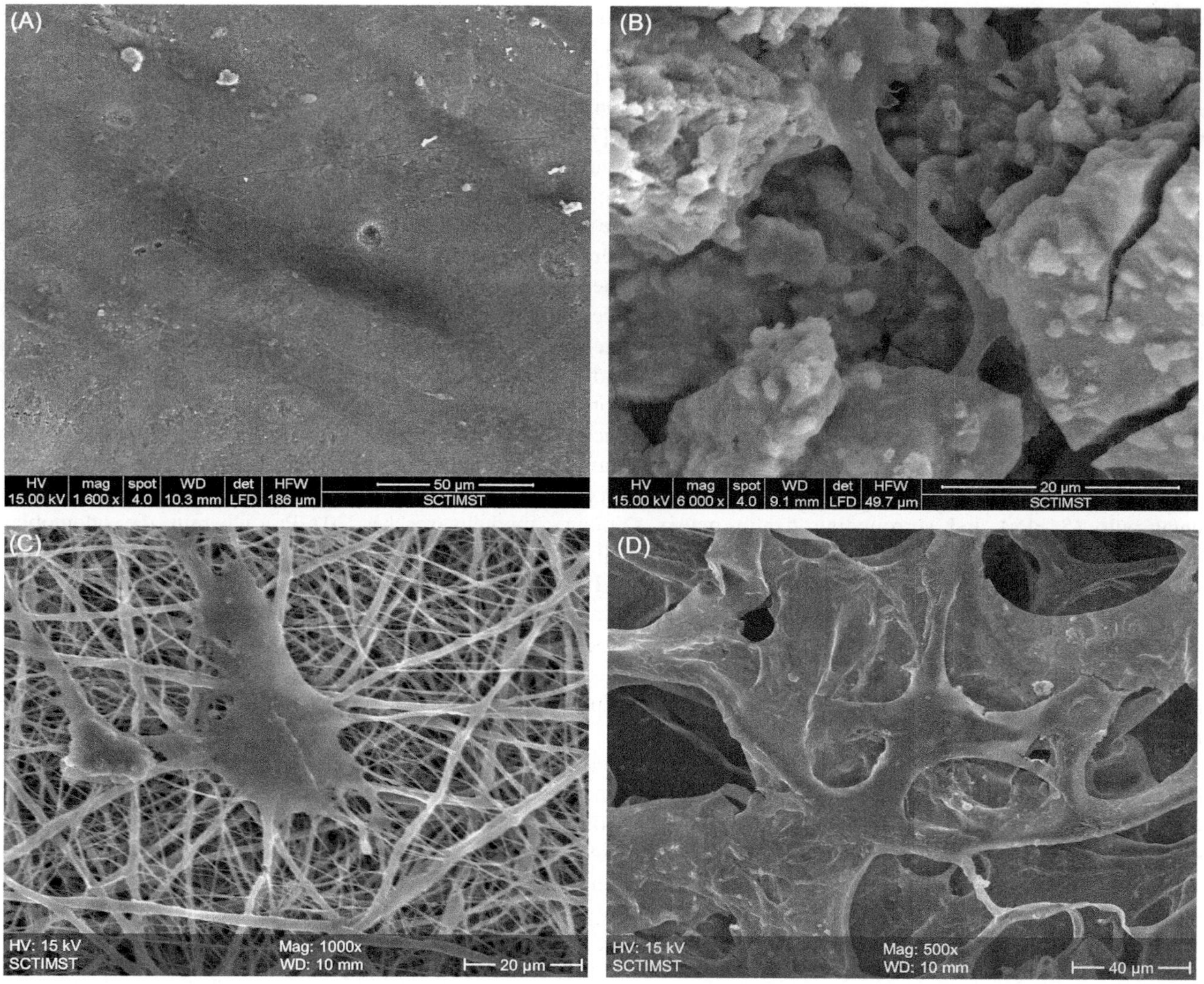

FIGURE 5.1 Human periodontal ligament cells (hPDLCs) on biomaterials. (A) hPDLCs on glass substrate (control). (B) hPDLCs on solid calcium phosphate cement surface. (C) hPDLCs on electrospun PCL scaffold. (D) hPDLCs on freeze dried chitosan scaffold.

carried out to study the cell displacement, direction, and average speed of migration into the porous scaffold [94]. Electrospun nanofibrous scaffolds fabricated using advanced electrospinning techniques can provide membranous scaffolds with controlled surface area, fiber alignment, and porosity capable of protein adsorption, promoting cell adhesion and differentiation. They can additionally function as delivery agents for drugs, antimicrobial agents, adjuvant nanoparticles, and/or growth factors [95,96]. Periodontal ligament cells and macrophage cells cultured in a functionalized (tea polyphenols functional graphene oxide) electrospun scaffold showed cementoblastic differentiation in osteo/cementogenic conditions and suppression of inflammatory responses [97]. Periodontal ligament cell sheets cultured on patterned exhibited improved regenerative response with upregulation of periodontal-specific gene periostin and alkaline phosphatase [98] (Figure 5.1).

5.4.6 Response to released chemicals

The role of calcium and phosphate ions in the regeneration of dental tissues is described in the previous section. With the advancements in biomaterial development and tissue engineering scaffolds, it is able to incorporate various bioactive molecules within the biomaterial preparation [99,100]. These include bioactive molecules and ions, biofunctionalization, drugs, antimicrobial agents, antiinflammatory agents, etc. [101–103]. Current research studies focus on the incorporation of nanoparticles with antimicrobial and antiinflammatory activity, metal nanoparticles such as zinc oxide and gold with antibacterial activity, immunomodulatory nanoparticles (nanoscale liposomes loaded with drugs such as resveratrol), antiinflammatory agent-loaded mesoporous silica nanoparticles, nanoscale exosomes for osteo/odoto/cementogenic responses, and synergistic nanotherapeutic strategies [104].

Calcium silicate-based biomaterials such as Biodentine are reported to induce odontogenic and angiogenic responses in dental pulp due to the presence of calcium and silicates, respectively [105]. Enamel matrix derivative—a biologic derived from developing procine teeth—exhibits angiogenic activity in dental pulp and periodontal regeneration with elevated vascular endothelial growth factor expression [106]. Human dental apical papilla cells cultured in a predominantly collagen-gelatin scaffold showed enhanced cell proliferation, migration, and adhesion with an increase in incorporated fibronectin levels. Human dental apical papilla is the precursor of dental pulp and is regressed after two years of tooth-root formation and eruption. The presence of a healthy apical papilla can be utilized to revitalize an immature permanent tooth with incomplete root formation [107].

In an in vitro periodontal ligament cell culture system, osteogenic induction resulted in downregulation of cementoblastic genes such as cementum attachment protein (CAP) and CEMP1 with elevated osteogenic genes, whereas Vitamin C treatment upregulated the cementogenic CAP and CEMP1 genes. However, a subpopulation of alkaline phosphatase-positive cells showed upregulation of cementogenic genes indicating a differential differentiation of periodontal ligament cells to osteogenic stimuli. This can be correlated to a clinical situation where the quantity of alveolar bone formation is usually more than that of a thin layer of cementum, which is not usually affected by bone resorptive processes. Vitamin C is also a vital element for the maintenance of healthy periodontium [108]. Electrospun scaffold loaded with the silk worm-derived hydrophilic biomolecule sericin showed enhanced proliferation and attachment of gingival fibroblasts and antiinflammatory effect on lipopolysaccharide-treated cultured macrophage cell line (RAW-264.7 cells) [109].

5.4.7 Response to growth factors

Growth factors constitute one of the three core components in tissue engineering, the other two being scaffolds and stem cells. The dental tissues under stimulation secrete certain growth factors and respond to secreted or extraneous growth factors. Most of the growth factors for tooth regeneration are available within the host tissues. However, specific dental cellular responses to these growth factors can be studied by simulating the responses in in vitro cell culture systems. The host blood-derived biologics, such as platelet concentrates, PRP and PRF (Platelet-Rich Plasma / platelet-rich fibrin), growth factor concentrates, and fibrin glue, are rich sources of heterogenous growth factors routinely used in clinical practice to promote periodontal and alveolar bone healing applications [110]. Platelet-derived growth factor (PDGF) enhances angiogenesis and wound healing along with cell proliferation, but the odontogenic differentiation is inhibited by its dimeric form [111].

Several multifunctional cytokines with multiple effects on stem cell differentiation are involved in the odontogenic differentiation of dental pulp stem cells (DPSCs). Transforming growth factor-β1 (TGF-β1) is such a multifunctional cytokine, which, at a concentration range of 0.1–5 ng/mL, inhibited cell growth without apoptotic changes, after three days in a concentration-dependent manner. A dose-dependent increase in alkaline phosphatase activity from day 7 onwards till day 14, corresponding to matrix secretion and subsequently matrix mineralization, succeeded from day 14 to 21 with upregulation of osteogenic and odontogenic genes [112]. A synergistic effect in odontogenic differentiation is noted with fibroblast growth factor [113]. Bone morphogenetic protein 2 (BMP2) induces the odontogenic differentiation of DPSCs with upregulation of DSPPs and elevated ALP activity without affecting cell proliferation [114]. In addition, BMP 4, BMP 7, and BMP 11 also exhibit dentinogenic effect by upregulating the odontogenic gene expressions [115].

Bone morphogenetic protein 7 (BMP-7/osteogenic protein 1) is another multifunctional growth factor that belongs to the TGF-β family. BMP 7 enhances the expression of cementoblast-specific markers such as CEMP1, CAP, and PTPLA (protein tyrosine phosphatase-like, member A) in cultured periodontal ligament cells. The molecular mechanism of cementogenesis was found to be different from odontogenesis as odontogenic markers such as DMP 1and DSPP were not upregulated [116]. Basic fibroblast growth factor (bFGF) also enhances the proliferation of periodontal ligament cells in a dose-dependent manner but reduces the alkaline phosphatase activity, collagen deposition, and mineralization in vitro suggesting an inhibitory effect on the differentiation of periodontal ligament cells [117].

Externally supplemented human recombinant CEMP1 enhanced periodontal ligament cell proliferation and differentiation with bone/cementum—like tissue formation in a 3D cell culture system for the reconstruction of periodontal structures through bioengineering [118]. CEMP1 acts as a marker protein for cementoblastic differentiation as well as the factor that regulates cementum formation [37].

5.5 Implications and conclusions

Tooth-derived stem cells constitute a unique population of adult stem cells that can differentiate into various dental and nondental tissues. Advancements in cell culture techniques, such as organoid culture, have enabled us to simulate even the developmental processes of human tooth in vitro. Such models are capable of mimicking reciprocal interactions between cells and enable detailed study of signaling pathways, mechanotransduction, transcriptome analyses, and lineage commitments [119]. The tooth-derived stem cells are capable of differentiation to nondental cells such as neuronal cells, hepatic cells, pancreatic islet-like cells, myogenic cells, and corneal stromal-like cells [120]. The emulation of cellular responses to dental biomaterials provides a better understanding of the regenerative interactions and enables timely and appropriate material modifications to suit clinical situations. Such research studies have successfully enabled the clinical translation of engineered periodontal ligament cell sheets [121].

The pH of biomaterials is an important factor in eliciting cellular responses within the dental pulp and the periodontium. Once a suitable cell culture system is standardized, inducing pH variations and studying the cellular responses is now possible. Extensive in vitro cell culture studies have revealed that either extremes of pH (whether highly alkaline with pH > 10 or highly acidic with pH < 3.5) can result in apoptotic responses with or without autophagy. However, an alkaline environment above a pH of 9 results in increased cell proliferation and mineralization in vitro, even in cases of initial reduction in cell numbers, indicating a regenerative response ultimately. This is in accordance with clinical healing of dental pulp and/or periradicular tissues following the use of regenerative dental biomaterials that are alkaline in nature. In addition, in vitro cell culture studies have shown bactericidal, neutralizing, and antiinflammatory effects of increased pH in dental biomaterials. Understanding the interactions of tooth-derived cells to variations in pH can aid in fine-tuning regenerative biomaterials for specific regenerative dental applications.

Among the various minerals, calcium and phosphate ions released from biomaterials can aid in osteogenesis by modulating extracellular and through independent metabolic mechanisms. An array of cellular and molecular pathways is triggered through different receptor-mediated cellular events that result in osteogenic, odontogenic, and cementogenic responses. The regenerative dental biomaterials predominantly consist of matrices and scaffolds based on the release of calcium and phosphate ions. The dental pulp cells predominantly show an osteo/odontogenic response, wherein the periodontal ligament cells show an osteo/cementogenic response to the calcium and phosphate ions. Supplementation of calcium and phosphate ions independently in cell culture systems can emulate the different cell signaling pathways, gene and protein expressions leading to the regenerative osteo/odonto/cementogenic responses. It is also possible to simulate resorptive or clastic activity of cells using coculture systems with tooth-derived cells and macrophage cell lines. Coculture with endothelial cells can elicit the angiogenic responses of calcium-containing regenerative dental materials.

The tooth in the oral cavity is subjected to heavy masticatory forces subjecting the tooth to multidirectional mechanical stresses and stimuli. The periodontium consisting of the periodontal ligament, alveolar bone, and cementum is the anchoring unit of the tooth and undergoes constant remodeling under heavy mechanical forces of mastication and/or orthodontic tooth movements. Current advances in cell culture techniques to impart compressive forces, tensile forces, and multidirectional forces in 3D cultures permit the thorough evaluations of cellular responses and interactions simulating clinical situations in the lab. A multitude of cellular and molecular events unveiling the cell–cell and cell-ECM interactions, including cell traction forces, is possible using appropriate cell culture methods. The in vitro simulation of these complex cellular events can help to better fabricate biomaterials that can withstand the complex forces in a way conducive to periodontal regeneration. Alveolar bone regeneration is an important factor in maintaining periodontal health. The periodontal ligament cells upon induction or under the influence of biomaterials predominantly differentiate to an osteogenic lineage and facilitate alveolar bone regeneration.

The dental pulp responds to traumatic injuries through a hypoxic response, which can be simulated in vitro by culturing the dental pulp-derived cells in hypoxic and normoxic conditions simulating the cellular and molecular responses that can help in predicting the changes of dental pulp survival following injuries. In the periodontium, a hypoxic response is triggered by occlusal or orthodontic compressive forces that can result in the resorption of the alveolar bone, whereas tensional forces can promote deposition of alveolar bone. Also, the occlusal masticatory forces create mechanical signals that can be converted to biochemical signals with gene expression for the maintenance of tissue homeostasis.

The rigidity and stiffness of the extracellular environment are important parameters that can directly influence cell viability, morphology, migration ability, proliferation, and differentiation potential brought about by the activation of appropriate signaling pathways. The emergence of tissue engineering has enabled the construction of scaffolds and matrices with varying rigidity, from hydrogels to solid scaffolds, which can be utilized to culture tooth-derived cells to simulate the vast difference in rigidity between the dental hard and soft tissues. Most of the studies showed evidence of preservation of stemness in soft hydrogel-type scaffolds. The increase in rigidity prompted the cultured cells to increase cell adhesion and differentiation. This rigidity induced differentiation can be utilized in biomaterial-based dentin regeneration in dental pulp and alveolar bone regeneration in periodontium. Graded scaffolds with differing stiffness are also used to culture tooth-derived cells to simulate the anatomical complexities of dental pulp and periodontium.

The differences in surface architecture of dental tissues can be simulated in vitro by modifying the micro-to-nanotopography through micro/nano patterning techniques of biomaterial scaffolds or substrates. Electrospun nanofibrous scaffolds also can provide an ECM mimetic extracellular environment with fiber alignment techniques to fine-tune the interactions with cultured tooth-derived cells. This can effectively modulate cell responses by influencing cell adhesion and spreading, cell distribution and arrangement, morphological changes, and differentiation to different lineages with appropriate protein and gene expressions.

Regenerative dental biomaterials and scaffolds can incorporate various bioactive chemical components, including drugs, antimicrobial agents, antiinflammatory agents, nanoparticles, and bioactive moieties and ions. These are mainly intended for therapeutic and regenerative effects through their controlled release into the extracellular environment. These bioactive molecules can promote cell proliferation, adhesion, and differentiation of tooth-derived cells. In addition, silicates and bioactive glass formulations can induce angiogenesis and improve osteogenic differentiation of tooth-derived cells. Similarly, to enhance the bioactivity, growth factors can also be incorporated into the scaffolds. The response of tooth-derived cells to growth factors and autologous biologics, such as platelet concentrates, such as PRP and PRF (Platelet-Rich Plasma/platelet-rich fibrin), can imitate the actual cellular responses to regenerative approaches such as revascularization and guided tissue regeneration techniques providing valuable insights to such treatment approaches.

To conclude, it can be stated that the tooth-derived cells, due to their unique nature and neural crest cell lineage, constitute a valuable cell source to study different cellular responses in vitro, from recreating certain developmental events to multilineage differentiation to various dental and nondental tissues. The tooth-derived cells in culture systems respond well to various mechanical, topographical, physicochemical, and biologic changes in their immediate environment with distinct cellular and molecular events that can be studied by advanced molecular techniques in vitro. The interactions of tooth-derived cells with regenerative dental biomaterials yield predictable and reproducible results that are reliable and extrapolate host clinical responses in vitro. All these advocate well-characterized tooth-derived cells as invaluable tools in the continued development of regenerative dental biomaterials with progressive improvements in their regenerative ability.

AI disclosure

The authors acknowledge the use of generative AI assistance exclusively to improve language readability, streamline phrasing, and check grammatical clarity during the drafting of the chapter. The conceptual framework, scientific analysis, and final manuscript verification were conducted entirely by the authors.

References

[1] M.I. Nicolescu, Regenerative perspective in modern dentistry, Dentistry Journal 4 (2) (2016) 10.

[2] F.M. Watt, R.R. Driskell, The therapeutic potential of stem cells, Philosophical Transactions of the Royal Society B: Biological Sciences 365 (1537) (2010) 155–163.

[3] P.C. Yelick, P.T. Sharpe, Tooth bioengineering and regenerative dentistry, Journal of Dental Research 98 (11) (2019) 1173–1182.

[4] S. Shi, S. Gronthos, Perivascular niche of postnatal mesenchymal stem cells in human bone marrow and dental pulp, Journal of Bone and Mineral Research 18 (4) (2003) 696–704.

[5] A. Angelova Volponi, L.K. Zaugg, V. Neves, Y. Liu, P.T. Sharpe, Tooth repair and regeneration, Current Oral Health Reports 5 (2018) 295–303.

[6] American Academy of Periodontology, Position paper. Periodontal regeneration, The Journal of Periodontology 76 (2005) 1601–1622.

[7] A. Nanci, A.R. Ten Cate, Ten Cate's Oral Histology: Development, Structure, and Function, *9th ed., Elsevier, St. Louis*, (2017).

[8] T. Yamamoto, T. Hasegawa, T. Yamamoto, H. Hongo, N. Amizuka, Histology of human cementum: Its structure, function, and development, Japanese Dental Science Review 52 (3) (2016) 63–74.

[9] M. Ramalho-Santos, H. Willenbring, On the origin of the term "stem cell, Cell Stem Cell 1 (1) (2007) 35–38.

[10] S. Bardelli, M. Moccetti, Remodeling the human adult stem cell niche for regenerative medicine applications, Stem Cells International 2017 (1) (2017) 6406025.

[11] S. Gronthos, M. Mankani, J. Brahim, P.G. Robey, S. Shi, Postnatal human dental pulp stem cells (DPSCs) in vitro and in vivo, Proceedings of the National Academy of Sciences 97 (25) (2000) 13625–13630.

[12] M. Miura, S. Gronthos, M. Zhao, B. Lu, L.W. Fisher, P.G. Robey, et al., SHED: stem cells from human exfoliated deciduous teeth, Proceedings of the National Academy of Sciences 100 (10) (2003) 5807–5812.

[13] B.M. Seo, M. Miura, S. Gronthos, P.M. Bartold, S. Batouli, J. Brahim, et al., Investigation of multipotent postnatal stem cells from human periodontal ligament, The Lancet 364 (9429) (2004) 149–155.
[14] G.T.J. Huang, W. Sonoyama, Y. Liu, H. Liu, S. Wang, S. Shi, The hidden treasure in apical papilla: the potential role in pulp/dentin regeneration and bioroot engineering, Journal of Endodontics 34 (6) (2008) 645–651.
[15] M.T. Saito, K.G. Silvério, M.Z. Casati, E.A. Sallum, F.H. Nociti Jr, Tooth-derived stem cells: update and perspectives, World Journal of Stem Cells 7 (2) (2015) 399.
[16] E. Matalová, V. Lungová, P. Sharpe, Development of tooth and associated structures, Stem Cell Biology and Tissue Engineering in Dental Sciences, Academic Press, (2015), pp. 335–346.
[17] P. Hilkens, P. Gervois, Y. Fanton, J. Vanormelingen, W. Martens, T. Struys, et al., Effect of isolation methodology on stem cell properties and multilineage differentiation potential of human dental pulp stem cells, Cell and Tissue Research 353 (1) (2013) 65–78.
[18] N. Kawashima, Characterisation of dental pulp stem cells: a new horizon for tissue regeneration? Archives of Oral Biology 57 (11) (2012) 1439–1458.
[19] N.F. Lizier, A. Kerkis, C.M. Gomes, J. Hebling, C.F. Oliveira, A.I. Caplan, et al., Scaling-up of dental pulp stem cells isolated from multiple niches, PLoS One 7 (6) (2012) e39885.
[20] B.A. Rodas-Junco, C. Villicana, Dental pulp stem cells: current advances in isolation, expansion and preservation, Tissue Engineering and Regenerative Medicine 14 (4) (2017) 333–347.
[21] I. Kerkis, A.I. Caplan, Stem cells in dental pulp of deciduous teeth, Tissue Engineering Part B: Reviews 18 (2) (2012) 129–138.
[22] H.L.B. Tran, V.N. Doan, H.T.N. Le, L.T.Q. Ngo, Various methods for isolation of multipotent human periodontal ligament cells for regenerative medicine: Tran et al, In Vitro Cellular & Developmental Biology-Animal 50 (7) (2014) 597–602.
[23] T. Lei, X. Zhang, H. Du, Characteristics, classification, and application of stem cells derived from human teeth, Stem Cells International 2021 (1) (2021) 8886854.
[24] T. Iwata, M. Yamato, Z. Zhang, S. Mukobata, K. Washio, T. Ando, et al., Validation of human periodontal ligament-derived cells as a reliable source for cytotherapeutic use, Journal of Clinical Periodontology 37 (12) (2010) 1088–1099.
[25] J.T. Marchesan, C.S. Scanlon, S. Soehren, M. Matsuo, Y.L. Kapila, Implications of cultured periodontal ligament cells for the clinical and experimental setting: a review, Archives of Oral Biology 56 (10) (2011) 933–943.
[26] Z.C. Song, W. Zhou, R. Shu, J. Ni, Hypoxia induces apoptosis and autophagic cell death in human periodontal ligament cells through HIF-1α pathway, Cell Proliferation 45 (2012) 239–248.
[27] K. Avinash, S. Malaippan, J.N. Dooraiswamy, Methods of isolation and characterization of stem cells from different regions of oral cavity using markers: a systematic review, International Journal of Stem Cells 10 (1) (2017) 12–20.
[28] R. Karamzadeh, M.B. Eslaminejad, R. Aflatoonian, Isolation, characterization and comparative differentiation of human dental pulp stem cells derived from permanent teeth by using two different methods, Journal of visualized experiments: JoVE (69) (2012) 4372.
[29] M. Ducret, J.C. Farges, M. Pasdeloup, E. Perrier-Groult, A. Mueller, F. Mallein-Gerin, et al., Phenotypic identification of dental pulp mesenchymal stem/stromal cells subpopulations with multiparametric flow cytometry, Odontogenesis: Methods and Protocols, Springer New York, New York, NY, (2019), pp. 77–90.
[30] Y. Açil, F. Yang, A. Gulses, M. Ayna, J. Wiltfang, M. Gierloff, Isolation, characterization and investigation of differentiation potential of human periodontal ligament cells and dental follicle progenitor cells and their response to BMP-7 in vitro, Odontology / The Society of the Nippon Dental University 104 (2016) 123–135.
[31] R. Monterubbianesi, M. Bencun, P. Pagella, A. Woloszyk, G. Orsini, T.A. Mitsiadis, A comparative in vitro study of the osteogenic and adipogenic potential of human dental pulp stem cells, gingival fibroblasts and foreskin fibroblasts, Scientific Reports 9 (1) (2019) 1761.
[32] L. Winning, I.A. El Karim, F.T. Lundy, A comparative analysis of the osteogenic potential of dental mesenchymal stem cells, Stem Cells and Development 28 (15) (2019) 1050–1058.
[33] S. Proksch, K. Kirsch, K. Vach, E. Hellwig, P. Tomakidi, Comparative differentiation analysis of distinct oral tissue-derived cells in response to osteogenic stimulation, Clinical Oral Investigations 23 (3) (2019) 1077–1089.
[34] J. Wu, G.T.J. Huang, W. He, P. Wang, Z. Tong, Q. Jia, et al., Basic fibroblast growth factor enhances stemness of human stem cells from the apical papilla, Journal of Endodontics 38 (5) (2012) 614–622.
[35] E. Saberi, N. Farhad-Mollashahi, F. Sargolzaei Aval, M. Saberi, Proliferation, odontogenic/osteogenic differentiation, and cytokine production by human stem cells of the apical papilla induced by biomaterials: a comparative study, Clinical, Cosmetic and Investigational Dentistry (2019) 181–193.
[36] S. Noda, N. Kawashima, M. Yamamoto, K. Hashimoto, K. Nara, I. Sekiya, et al., Effect of cell culture density on dental pulp-derived mesenchymal stem cells with reference to osteogenic differentiation, Scientific Reports 9 (1) (2019) 5430.
[37] M. Komaki, K. Iwasaki, H. Arzate, A.S. Narayanan, Y. Izumi, I. Morita, Cementum protein 1 (CEMP1) induces a cementoblastic phenotype and reduces osteoblastic differentiation in periodontal ligament cells, Journal of Cellular Physiology 227 (2) (2012) 649–657.
[38] S.A. Ajlan, N.Y. Ashri, A.M. Aldahmash, M.S. Alnbaheen, Osteogenic differentiation of dental pulp stem cells under the influence of three different materials, BMC Oral Health 15 (1) (2015) 132.
[39] P. Singh, Theory of optimum pH for pulp vitality: a clinical and experimental theory of endodontics, EC Clinical and Medical Case Reports 3 (1) (2020) 01–03.
[40] Y. Hirose, M. Yamaguchi, S. Kawabata, M. Murakami, M. Nakashima, M. Gotoh, et al., Effects of extracellular pH on dental pulp cells in vitro, Journal of Endodontics 42 (5) (2016) 735–741.

[41] A. Covaci, L.T. Ciocan, B. Gălbinaşu, M.V. Bucur, M. Matei, A.C. Didilescu, Dental pulp response to different types of calcium-based materials applied in deep carious lesion treatment—a clinical study, Journal of Functional Biomaterials 13 (2) (2022) 51.

[42] E.A. Saberi, N.F. Mollashahi, F. Ejeian, M. Nematollahi, O. Shahraki, A. Pirhaji, et al., Assessment of cytotoxicity and odontogenic/osteogenic differentiation potential of nano-dentine cement against stem cells from apical papilla, Cell Journal (Yakhteh) 24 (11) (2022) 637.

[43] P. Sangwan, A. Sangwan, J. Duhan, A. Rohilla, Tertiary dentinogenesis with calcium hydroxide: a review of proposed mechanisms, International Endodontic Journal 46 (1) (2013) 3–19.

[44] T.J. Hilton, Keys to clinical success with pulp capping: a review of the literature, Operative Dentistry 34 (5) (2009) 615–625.

[45] F. Banchs, M. Trope, Revascularization of immature permanent teeth with apical periodontitis: new treatment protocol? Journal of Endodontics 30 (4) (2004) 196–200.

[46] M.C. Moreno-Hidalgo, C. Caleza-Jimenez, A. Mendoza-Mendoza, A. Iglesias-Linares, Revascularization of immature permanent teeth with apical periodontitis, International Endodontic Journal 47 (4) (2014) 321–331.

[47] Y. Nakano, P. DenBesten, M. Goldberg, Structure of collagen-derived mineralized tissues (Dentin, Cementum, and Bone) and non-collagenous extra cellular matrix of enamel, Extracellular Matrix Biomineralization of Dental Tissue Structures (2021) 3–34.

[48] W.J. Landis, R. Jacquet, Association of calcium and phosphate ions with collagen in the mineralization of vertebrate tissues, Calcified Tissue International 93 (4) (2013) 329–337.

[49] Y. Li, Q. Zhan, M. Bao, J. Yi, Y. Li, Biomechanical and biological responses of periodontium in orthodontic tooth movement: up-date in a new decade, International Journal of Oral Science 13 (1) (2021) 20.

[50] S. Rui, T. Kubota, Y. Ohata, K. Yamamoto, M. Fujiwara, S. Takeyari, et al., Phosphate promotes osteogenic differentiation through non-canonical Wnt signaling pathway in human mesenchymal stem cells, Bone 164 (2022) 116525.

[51] Y.R.V. Shih, Y. Hwang, A. Phadke, H. Kang, N.S. Hwang, E.J. Caro, et al., Calcium phosphate-bearing matrices induce osteogenic differentiation of stem cells through adenosine signaling, Proceedings of the National Academy of Sciences 111 (3) (2014) 990–995.

[52] G.R. Beck Jr, Inorganic phosphate as a signaling molecule in osteoblast differentiation, Journal of Cellular Biochemistry 90 (2) (2003) 234–243.

[53] B.L. Foster, K.A. Tompkins, R.B. Rutherford, H. Zhang, E.Y. Chu, H. Fong, et al., Phosphate: known and potential roles during development and regeneration of teeth and supporting structures, Birth Defects Research Part C: Embryo Today: Reviews 84 (4) (2008) 281–314.

[54] G. Spagnuolo, I. De Luca, F. Iaculli, E. Barbato, A. Valletta, A. Calarco, et al., Regeneration of dentin-pulp complex: effect of calcium-based materials on hDPSCs differentiation and gene expression, Dental Materials 39 (5) (2023) 485–491.

[55] S. An, J. Ling, Y. Gao, Y. Xiao, Effects of varied ionic calcium and phosphate on the proliferation, osteogenic differentiation and mineralization of human periodontal ligament cells in vitro, Journal of Periodontal Research 47 (3) (2012) 374–382.

[56] L. Chen, L. Zheng, J. Jiang, J. Gui, L. Zhang, Y. Huang, et al., Calcium hydroxide–induced proliferation, migration, osteogenic differentiation, and mineralization via the mitogen-activated protein kinase pathway in human dental pulp stem cells, Journal of Endodontics 42 (9) (2016) 1355–1361.

[57] H. Tada, E. Nemoto, S. Kanaya, N. Hamaji, H. Sato, H. Shimauchi, Elevated extracellular calcium increases expression of bone morphogenetic protein-2 gene via a calcium channel and ERK pathway in human dental pulp cells, Biochemical and Biophysical Research Communications 394 (4) (2010) 1093–1097.

[58] Z. Sheikh, N. Hamdan, Y. Ikeda, M. Grynpas, B. Ganss, M. Glogauer, Natural graft tissues and synthetic biomaterials for periodontal and alveolar bone reconstructive applications: a review, Biomaterials Research 21 (1) (2017) 9.

[59] S. An, Y. Gao, J. Ling, X. Wei, Y. Xiao, Calcium ions promote osteogenic differentiation and mineralization of human dental pulp cells: implications for pulp capping materials, Journal of Materials Science: Materials in Medicine 23 (2012) 789–795.

[60] S. Maeno, Y. Niki, H. Matsumoto, H. Morioka, T. Yatabe, A. Funayama, et al., The effect of calcium ion concentration on osteoblast viability, proliferation and differentiation in monolayer and 3D culture, Biomaterials 26 (23) (2005) 4847–4855.

[61] F.W.G. Paula-Silva, A. Ghosh, H. Arzate, S. Kapila, L.A.B. da Silva, Y.L. Kapila, Calcium hydroxide promotes cementogenesis and induces cementoblastic differentiation of mesenchymal periodontal ligament cells in a CEMP1-and ERK-dependent manner, Calcified Tissue International 87 (2) (2010) 144–157.

[62] Y. Xu, Y. Wang, X. Pang, Z. Li, J. Wu, Z. Zhou, et al., Potassium dihydrogen phosphate promotes the proliferation and differentiation of human periodontal ligament stem cells via nuclear factor kappa B pathway, Experimental Cell Research 384 (1) (2019) 111593.

[63] S. An, Y. Gao, J. Ling, Characterization of human periodontal ligament cells cultured on three-dimensional biphasic calcium phosphate scaffolds in the presence and absence of L-ascorbic acid, dexamethasone and β-glycerophosphate in vitro, Experimental and Therapeutic Medicine 10 (4) (2015) 1387–1393.

[64] E.E.H. Hlaing, Y. Ishihara, Z. Wang, N. Odagaki, H. Kamioka, Role of intracellular Ca2+–based mechanotransduction of human periodontal ligament fibroblasts, The FASEB Journal 33 (9) (2019) 10409.

[65] L. Yang, Y. Yang, S. Wang, Y. Li, Z. Zhao, In vitro mechanical loading models for periodontal ligament cells: from two-dimensional to three-dimensional models, Archives of Oral Biology 60 (3) (2015) 416–424.

[66] M. Li, C. Zhang, Y. Yang, Effects of mechanical forces on osteogenesis and osteoclastogenesis in human periodontal ligament fibroblasts, Bone & Joint Research 8 (1) (2019) 19–31. Available from https://doi.org/10.1302/2046-3758.81.Bjr-2018-0060.R1.

[67] M. Chiba, H. Mitani, Cytoskeletal changes and the system of regulation of alkaline phosphatase activity in human periodontal ligament cells induced by mechanical stress, Cell Biochemistry and Function 22 (2004) 249–256.

[68] T. Yamamoto, M. Kita, I. Kimura, F. Oseko, R. Terauchi, K. Takahashi, et al., Mechanical stress induces expression of cytokines in human periodontal ligament cells, Oral Diseases 12 (2) (2006) 171–175.

[69] H. Kanzaki, M. Chiba, Y. Shimizu, H. Mitani, Periodontal ligament cells under mechanical stress induce osteoclastogenesis by receptor activator of nuclear factor κB ligand up regulation via prostaglandin E2 synthesis, Journal of Bone and Mineral Research 17 (2) (2002) 210–220.

[70] S. Wongkhantee, T. Yongchaitrakul, P. Pavasant, Mechanical stress induces osteopontin expression in human periodontal ligament cells through rho kinase, Journal of Periodontology 78 (2007) 1113–1119.

[71] C. Sun, M. Janjic Rankovic, M. Folwaczny, S. Otto, A. Wichelhaus, U. Baumert, Effect of tension on human periodontal ligament cells: systematic review and network analysis, Frontiers in bioengineering and biotechnology 9 (2021) 695053.

[72] S. Miyashita, N.E.M.B. Ahmed, M. Murakami, K. Iohara, T. Yamamoto, H. Horibe, et al., Mechanical forces induce odontoblastic differentiation of mesenchymal stem cells on three-dimensional biomimetic scaffolds, Journal of Tissue Engineering and Regenerative Medicine 11 (2) (2017) 434–446.

[73] J.S. Song, S.O. Kim, S.H. Kim, H.J. Choi, H.K. Son, H.S. Jung, et al., In vitro and in vivo characteristics of stem cells derived from the periodontal ligament of human deciduous and permanent teeth, Tissue Engineering. Part A 18 (19-20) (2012) 2040–2051.

[74] L. Li, M.X. Han, S. Li, Y. Xu, L. Wang, Hypoxia regulates the proliferation and osteogenic differentiation of human periodontal ligament cells under cyclic tensile stress via mitogen-activated protein kinase pathways, Journal of Periodontology 85 (3) (2014) 498–508.

[75] K. Amemiya, Y. Kaneko, T. Muramatsu, M. Shimono, T. Inoue, Pulp cell responses during hypoxia and reoxygenation in vitro, European Journal of Oral Sciences 111 (4) (2003) 332–338.

[76] H. Choi, H. Jin, J.Y. Kim, K.T. Lim, H.W. Choung, J.Y. Park, et al., Hypoxia promotes CEMP1 expression and induces cementoblastic differentiation of human dental stem cells in an HIF-1-dependent manner, Tissue Engineering. Part A 20 (1-2) (2014) 410–423.

[77] N. Bryniarska-Kubiak, A. Basta-Kaim, A. Kubiak, Mechanobiology of dental pulp cells, Cells 13 (5) (2024) 375.

[78] Y. Delin, N. Haiying, G. Yikang, C. Xing, Elastic modulus of human cementum, Applied Mathematics and Mechanics 20 (10) (1999) 1134–1141.

[79] P.A. Janmey, D.A. Fletcher, C.A. Reinhart-King, Stiffness sensing by cells, Physiological Reviews 100 (2) (2020) 695–724.

[80] P.H. Wu, D.R. Aroush, A. Asnacios, W.C. Chen, M.E. Dokukin, B.L. Doss, et al., A comparison of methods to assess cell mechanical properties, Nature Methods 15 (7) (2018) 491–498. Available from https://doi.org/10.1038/s41592-018-0015-1.

[81] W.H. Guo, M.T. Frey, N.A. Burnham, Y.L. Wang, Substrate rigidity regulates the formation and maintenance of tissues, Biophysical Journal 90 (6) (2006) 2213–2220.

[82] L. Datko Williams, A. Farley, M. Cupelli, S. Alapati, M.S. Kennedy, et al., Effects of substrate stiffness on dental pulp stromal cells in culture, Journal of Biomedical Materials Research. Part A 106 (7) (2018) 1789–1797.

[83] C. Ehlinger, E. Mathieu, M. Rabineau, V. Ball, P. Lavalle, Y. Haikel, et al., Insensitivity of dental pulp stem cells migration to substrate stiffness, Biomaterials 275 (2021) 120969.

[84] M. Ghaffari, A. Shrestha, Optimizing stem cell expansion: the role of substrate stiffness in enhancing dental pulp stem cell quiescence and regeneration, Journal of Endodontics (2025).

[85] X. Li, Y. Xia, Z. Wang, Z. Yin, M. Weng, F. Tian, et al., Three-dimensional matrix stiffness-based stem cell soil: Tri-phase biomechanical structure promoted human dental pulp stem cells to achieve pulpodentin regeneration, Materials Today Bio 31 (2025) 101591.

[86] H. Lin, Y. Zhou, Q. Lei, D. Lin, J. Chen, C. Wu, Effect of inorganic phosphate on migration and osteogenic differentiation of bone marrow mesenchymal stem cells, BMC Developmental Biology 21 (2021) 1–8.

[87] N. Liu, M.I. Zhou, Q.I. Zhang, L. Yong, T. Zhang, T. Tian, et al., Effect of substrate stiffness on proliferation and differentiation of periodontal ligament stem cells, Cell Proliferation 51 (5) (2018) e12478.

[88] Y. Wang, H. Zhang, S. Wu, W. Wan, X. Kang, B. Gao, et al., Substrate stiffness regulates the proliferation and apoptosis of periodontal ligament cells through integrin-linked kinase ILK, ACS Biomaterials Science & Engineering 9 (2) (2023) 662–670.

[89] R. Diana, R. Ardhani, Y. Kristanti, P. Santosa, Dental pulp stem cells response on the nanotopography of scaffold to regenerate dentin-pulp complex tissue, Regenerative Therapy 15 (2020) 243–250.

[90] T.D. Stocco, N.J. Bassous, S. Zhao, A.E. Granato, T.J. Webster, A.O. Lobo, Nanofibrous scaffolds for biomedical applications, Nanoscale 10 (26) (2018) 12228–12255.

[91] J. Wang, H. Ma, X. Jin, J. Hu, X. Liu, L. Ni, et al., The effect of scaffold architecture on odontogenic differentiation of human dental pulp stem cells, Biomaterials 32 (31) (2011) 7822–7830.

[92] P.Y. Collart-Dutilleul, E. Secret, I. Panayotov, D. Deville de Périère, R.J. Martín-Palma, V. Torres-Costa, et al., Adhesion and proliferation of human mesenchymal stem cells from dental pulp on porous silicon scaffolds, ACS Applied Materials & Interfaces 6 (3) (2014) 1719–1728.

[93] S. Nam, J.E. Won, C.H. Kim, H.W. Kim, Odontogenic differentiation of human dental pulp stem cells stimulated by the calcium phosphate porous granules, Journal of Tissue Engineering 2011 (2011) 812547.

[94] J.C. Ashworth, M. Mehr, P.G. Buxton, S.M. Best, R.E. Cameron, Optimising collagen scaffold architecture for enhanced periodontal ligament fibroblast migration, Journal of Materials Science: Materials in Medicine 29 (11) (2018) 166.

[95] P. Zhao, W. Chen, Z. Feng, Y. Liu, P. Liu, Y. Xie, et al., Electrospun nanofibers for periodontal treatment: a recent progress, International Journal of Nanomedicine (2022) 4137–4162.

[96] E.C. Das, J. Babu, S. Dhawan, A.K. PR, T.V. Kumary, V. Haridas, et al., Synthetic osteogenic matrix using polymeric dendritic peptides for treating human periodontal defects–design and in vitro evaluation, Materials Today: Proceedings 15 (2019) 199–216.

[97] X. Han, F. Wang, Y. Ma, X. Lv, K. Zhang, Y. Wang, et al., TPG-functionalized PLGA/PCL nanofiber membrane facilitates periodontal tissue regeneration by modulating macrophages polarization via suppressing PI3K/AKT and NF-κB signaling pathways, Materials Today Bio 26 (2024) 101036.

[98] J.H. Kim, S.Y. Ko, J.H. Lee, D.H. Kim, J.H. Yun, Evaluation of the periodontal regenerative properties of patterned human periodontal ligament stem cell sheets, Journal of Periodontal & Implant Science 47 (6) (2017) 402–415.

[99] E.C. Das, T.V. Kumary, P.A. Kumar, M. Komath, Calcium sulfate-based bioactive cement for periodontal regeneration: an: in vitro: study, Indian Journal of Dental Research 30 (4) (2019) 558–567.

[100] R.K. Adarsh, E.C. Das, G.V. Gopan, S. Selvam, M. Komath, Functionally graded bioactive composites based on poly (vinyl alcohol) made through thiol–ene click reaction, ACS Omega 7 (33) (2022) 29246–29255.

[101] R.K. Adarsh, E.C. Das, G.V. Gopan, R.K. Rajan, M. Komath, Quaternised chitosan composites with in situ precipitated nano calcium phosphate for making bioactive and degradable tissue engineering scaffolds, Journal of Polymer Research 29 (7) (2022) 267.

[102] E.C. Das, S. Dhawan, J. Babu, P.R. Anil Kumar, T.V. Kumary, V. Haridas, et al., Self-assembling polymeric dendritic peptide as functional osteogenic matrix for periodontal regeneration scaffolds—an in vitro study, Journal of Periodontal Research 54 (5) (2019) 468–480.

[103] A. Rajeswari Krishnankutty, S. Najeema Sulaiman, A. Sadasivan, R. Joseph, M. Komath, Porous membranes of quaternized chitosan composited with strontium-based nanobioceramic for periodontal tissue regeneration, Journal of Biomaterials Applications 36 (7) (2022) 1254–1268.

[104] J. Li, Y. Wang, M. Tang, C. Zhang, Y. Fei, M. Li, et al., New insights into nanotherapeutics for periodontitis: a triple concerto of antimicrobial activity, immunomodulation and periodontium regeneration, Journal of Nanobiotechnology 22 (1) (2024) 19.

[105] Z. Luo, M.R. Kohli, Q. Yu, S. Kim, T. Qu, W.X. He, Biodentine induces human dental pulp stem cell differentiation through mitogen-activated protein kinase and calcium-/calmodulin-dependent protein kinase II pathways, Journal of Endodontics 40 (7) (2014) 937–942.

[106] M.A. Saghiri, A. Asatourian, F. Garcia-Godoy, N. Sheibani, Effect of biomaterials on angiogenesis during vital pulp therapy, Dental Materials Journal 35 (5) (2016) 701–709.

[107] M.L. Leite, D.G. Soares, G. Anovazzi, C. Anselmi, J. Hebling, C.A. de Souza Costa, Fibronectin-loaded collagen/gelatin hydrogel is a potent signaling biomaterial for dental pulp regeneration, Journal of Endodontics 47 (7) (2021) 1110–1117.

[108] P. Gauthier, Z. Yu, Q.T. Tran, F.U.R. Bhatti, X. Zhu, G.T.J. Huang, Cementogenic genes in human periodontal ligament stem cells are downregulated in response to osteogenic stimulation while upregulated by vitamin C treatment, Cell and Tissue Research 368 (1) (2017) 79–92.

[109] K. Chachlioutaki, C. Karavasili, E. Adamoudi, N. Bouropoulos, D. Tzetzis, A. Bakopoulou, et al., Silk sericin/PLGA electrospun scaffolds with anti-inflammatory drug-eluting properties for periodontal tissue engineering, Biomaterials Advances 133 (2022) 112723.

[110] Z. Li, L. Liu, L. Wang, D. Song, The effects and potential applications of concentrated growth factor in dentin–pulp complex regeneration, Stem Cell Research & Therapy 12 (1) (2021) 357.

[111] S. Yokose, H. Kadokura, N. Tajima, A. Hasegawa, H. Sakagami, K. Fujieda, et al., Platelet-derived growth factor exerts disparate effects on odontoblast differentiation depending on the dimers in rat dental pulp cells, Cell and Tissue Research 315 (3) (2004) 375–384.

[112] Y. Bai, X. Liu, J. Li, Z. Wang, Q. Guo, M. Xiao, et al., Stage-dependent regulation of dental pulp stem cell odontogenic differentiation by transforming growth factor-β1, Stem Cells International 2022 (1) (2022) 2361376.

[113] H. He, J. Yu, Y. Liu, S. Lu, H. Liu, J. Shi, et al., Effects of FGF2 and TGFβ1 on the differentiation of human dental pulp stem cells in vitro, Cell Biology International 32 (7) (2008) 827–834.

[114] S. Chen, J. Gluhak-Heinrich, M. Martinez, T. Li, Y. Wu, H.H. Chuang, et al., Bone morphogenetic protein 2 mediates dentin sialophosphoprotein expression and odontoblast differentiation via NF-Y signaling, Journal of Biological Chemistry 283 (28) (2008) 19359–19370.

[115] S.G. Kim, J. Zhou, C. Solomon, Y. Zheng, T. Suzuki, M. Chen, et al., Effects of growth factors on dental stem/progenitor cells, Dental Clinics of North America 56 (3) (2012) 563–575.

[116] D. Torii, T.W. Tsutsui, N. Watanabe, K. Konishi, Bone morphogenetic protein 7 induces cementogenic differentiation of human periodontal ligament-derived mesenchymal stem cells, Odontology / The Society of the Nippon Dental University 104 (1) (2016) 1–9.

[117] S. Takayama, S. Murakami, Y. Miki, K. Ikezawa, S. Tasaka, A. Terashima, et al., Effects of basic fibroblast growth factor on human periodontal ligament cells, Journal of Periodontal Research 32 (8) (1997) 667–675.

[118] L. Hoz, E. Romo, M. Zeichner-David, M. Sanz, J. Nunez, L. Gaitan, et al., Cementum protein 1 (CEMP1) induces differentiation by human periodontal ligament cells under three-dimensional culture conditions, Cell Biology International 36 (2) (2012) 129–136.

[119] J. Rosowski, J. Bräunig, A.K. Amler, F.P. Strietzel, R. Lauster, M. Rosowski, Emulating the early phases of human tooth development in vitro, Scientific reports 9 (1) (2019) 7057.

[120] J.Y. Yin, X.H. Luo, W.Q. Feng, S.H. Miao, T.T. Ning, Q. Lei, et al., Multidifferentiation potential of dental-derived stem cells, World Journal of Stem Cells 13 (5) (2021) 342.

[121] T. Iwata, M. Yamato, K. Washio, T. Yoshida, Y. Tsumanuma, A. Yamada, et al., Periodontal regeneration with autologous periodontal ligament-derived cell sheets–a safety and efficacy study in ten patients, Regenerative Therapy 9 (2018) 38–44.

Chapter 6

Growth and differentiation of mammalian embryonic stem cells on three-dimensional nanofibrilar surfaces

Alam Nur-e-kamal[1], Virinder Parmar[2], Ijaz Ahmed[1], and Jabeen Kamal[1]
[1]*Department of Biology, Medgar Evers College, The City University of New York, New York, NY, United States,* [2]*Department of Chemistry and Environmental Science, Medgar Evers College, The City University of New York, Brooklyn, NY, United States*

6.1 Introduction

Scientists have been trying to create a tissue culture condition closer to an in vivo condition to achieve appropriate tissues in need. The composition of molecules in the ECM changes during cell growth, differentiation, and progression of abnormal growth. Multiple proteins, such as ECM structural proteins, matrix-associated enzymes, matrix-bound growth factors, the receptors for ECM molecules, change their composition to direct specific types of tissues. New findings indicating a link between collagen and thrombospondin and tenascin provide additional insight into developing new bioengineered models for developing a therapy for human diseases [11]. As a starting point for tissue engineering, 3D scaffolds have been an attractive approach for mass production of cell products/metabolites, cells, tissues, and organs in the laboratory. In this report, we present studies on the development of an in vivo–like cell culture model with 3D polyamide nanofibers covalently attached with unique polypeptides derived from ECM proteins. Our studies demonstrated that 3D polyamide nanofibers provide in vivo–like mammalian cell culture conditions. We have also demonstrated that bioactive proteins/polypeptides could be covalently attached to 3D polyamide nanofibers, enhancing the growth and differentiation of neurons and mammalian embryonic stem cells.

6.2 Description

The ECM in mammalian cells provides structural support and biochemical cues essential for tissue development, function, and repair in vivo. A specialized form of ECM is called *BM*, which is a thin, dense layer of ECM that separates and supports epithelial, endothelial, and other types of cells. It plays a crucial role in tissue integrity, cell growth, and differentiation. The mammalian ECM and BM consist of a complex mixture of proteins that remain in the fluid component in the intercellular space. Proteins assemble to produce nanofibrillar network providing 3D surfaces for the specific growth and differentiation of specific types of cells [12–14]. The ECM and BM are made of fibrous and adhesive proteins consisting of collagens, elastin, fibronectin, laminin, tenascin, proteoglycans, glycosaminoglycans, and matrix metalloproteinases [12–15]. The ECM is a dynamic structure that undergoes constant remodeling, especially during development, tissue repair, and disease states such as fibrosis, cancer, and arthritis. This remodeling is mediated by enzymes like matrix metalloproteinases (MMPs) and tissue inhibitors of metalloproteinases (TIMPs), which control ECM degradation and synthesis. The composition of the ECM can vary depending on the tissue type and its specific functions (e.g., bone ECM is rich in collagen and hydroxyapatite for rigidity, while cartilage ECM contains more proteoglycans for cushioning). The ECM and BM perform a variety of cell functions, including structural support (provides a scaffold for cells), cellular communication, tissue development and repair, barrier and filtration, and reservoir of growth factors [12,15] (Fig. 6.1).

Fundamentals of Translational Tissue Engineering. DOI: https://doi.org/10.1016/B978-0-443-33455-9.00016-X

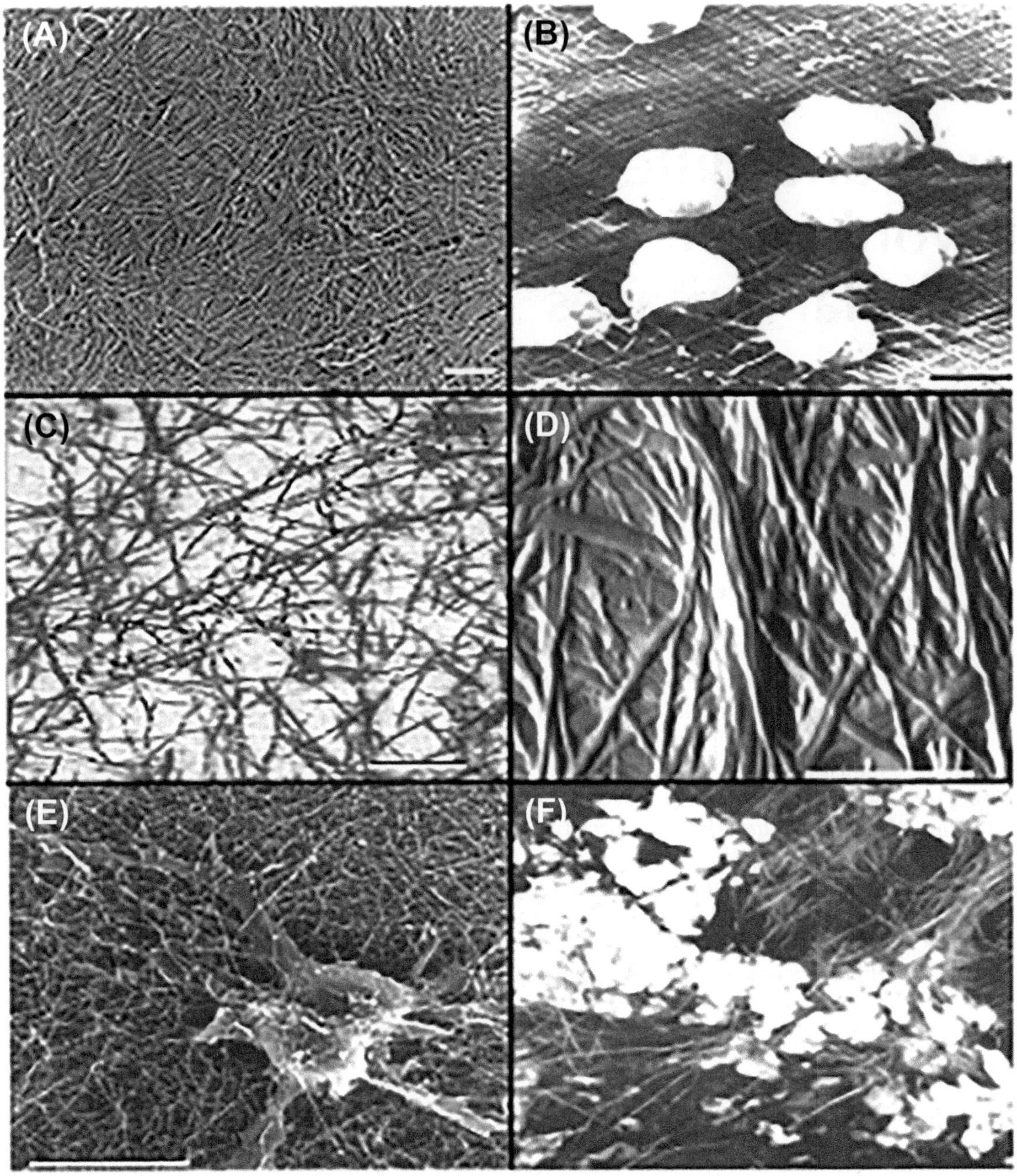

FIGURE 6.1 Architecture of different types of extracellular matrix/basement membranes (20). (A) Scanning electron micrograph of Descemet's membrane side of the substantia propria. Bar, 2 μm. (B) Scanning electron micrograph of epithelial cells that have been isolated from tadpole epidermis and placed on a collagenous substratum. Bar, 5 μm. (C) 3D confocal backscatter microscopy of a migrating T cell in a 3D collagen matrix. Black lines (overlay) indicate the transmigrated path. Bar, 5 μm. (D) AFM image of human sclera. This overview shows scleral collagen fibrils with a variable diameter. Bar, 5 μm. (E) Scanning electron micrograph of fibroblasts adhering to collagen polymers. Bar, 10 μm. (F) Multiphoton intravital image of carcinoma cells moving through extracellular matrix fibers in a mammary tumor. Bar, 5 μm.

6.3 Nanofibrillar surfaces produced by electrospinning

The technique of electrospinning provides an alternative approach to phase separation for the formation of stable, nanostructured materials [19–27]. The electrospinning process uses an electric field to control the formation and deposition of polymers [19,20]. Nanofibrillar matrices produced by electrospinning have an unprecedented porosity (>70%), high spatial interconnectivity, and a high surface-to-volume ratio for cell attachment, all parameters reminiscent of the ECM and BM [14,15] and ideal for the promotion of cellular growth. Biopolymers as well as synthetic polymers may also be electrospun. Work from our laboratory [21,22] has demonstrated that growth of NIH 3T3 fibroblasts on a 3D surface composed of electrospun polyamide nanofibers (Fig. 6.4) promoted structural and proliferative changes that were reminiscent of the changes in cell morphology [28] for cells cultured on natural, cell-free ECMs and for fibroblasts in vivo. The nanofibrillar matrix used in our research had a nanofiber diameter distribution centered at approx. 180 nm and pore diameters ranging from 200 to 1000 nm. The nanofibers were deposited at a thickness of approximately 5 μm [21]. The cellular changes observed for NIH 3T3 fibroblasts cultured on the 3D nanofibrillar surface, in comparison to cells cultured on 2D glass or tissue culture plastic surfaces, included a disruption of actin stress fibers, changes in the patterns of proteins associated with focal adhesions and focal complexes, an enhanced rate of proliferation, and an enhanced rate of fibrillar fibronectin matrix deposition on the apical surface of the cells [21,22]. Of particular importance was the observation that the culture of fibroblasts on nanofibrillar surfaces promoted a significant increase in the activation of the small Rho GTPase Rac [22], an essential regulator of proliferation, differentiation, migration, and morphogenesis in vivo [29–31]. On the other hand, cells cultured on nanofibers that were transfected with dominant-negative Rac exhibited proliferation and fibronectin deposition rates similar to those observed in cells cultured on 2D surfaces [21,22]. This suggests that Rac activation was responsible for at least some of the cellular consequences of culture on nanofibers. The enhanced rate of fibrillar matrix formation for fibroblasts cultured on nanofibers in our work [21,22] was similar to that reported for fibroblasts cultured on cell-free ECMs and for fibroblasts in vivo [27]. Hence, cells synthesizing their own new ECM may be a general consequence of 3D culture and one of the ways in which 3D culture promotes more in vivo–like morphologies. We describe the importance of physical and chemical queues in the laboratory to culture mammalian cells to mimic in vivo–like conditions. This will produce results relevant and close to a real whole-animal system. We provide some convincing evidence below supporting our hypothesis (Figs. 6.2, 6.4 and 6.5).

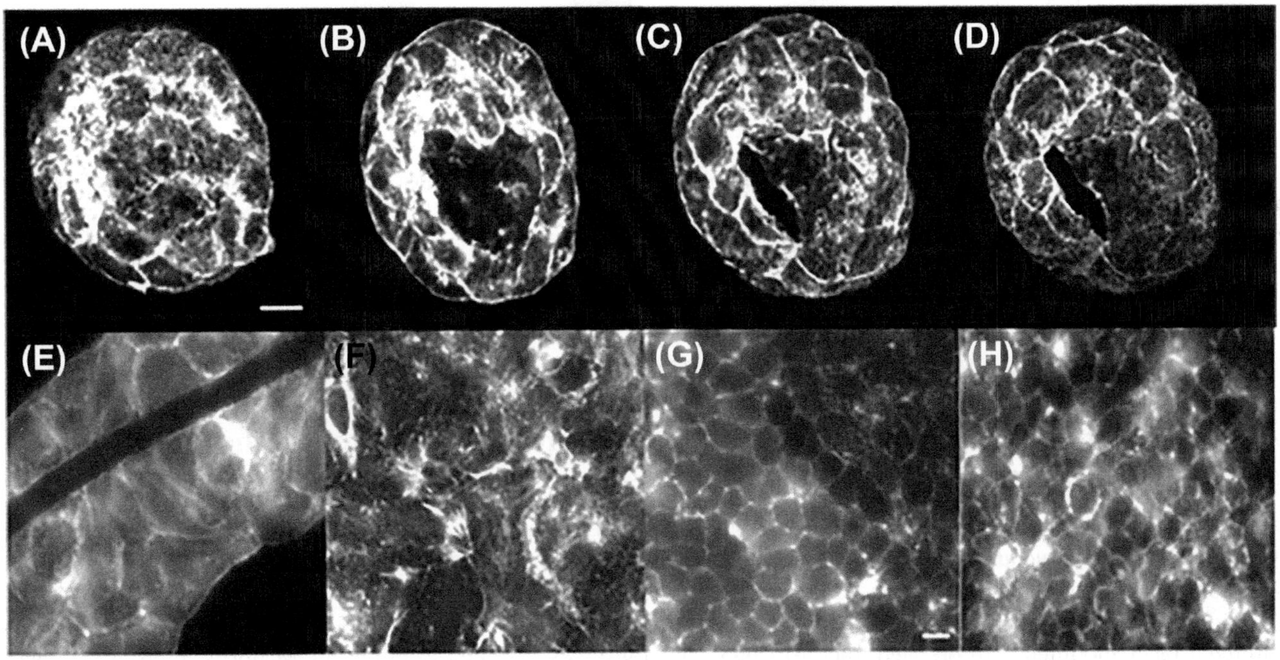

FIGURE 6.2 In vivo–like morphogenesis of T47D breast epithelial cells on 3D polyamide nanofibers. T47D epithelial cells were cultured on nanofibers, stained, and studied. A series of confocal sections of a multicellular spheroid composed of T47D breast epithelial cells grown on nanofibers and stained with phalloidin-FITC (A–D). Images were captured at different micron levels: 0 (A), 20 (B), 34 (C), and 48 (D) from the top. T47D cells after 10 days of culture on a glass 2D surface (F). MCF-7 cells grown on glass and nanofibers, respectively stained with phalloidin-FITC (G and H). Scale bar, 10 μm.

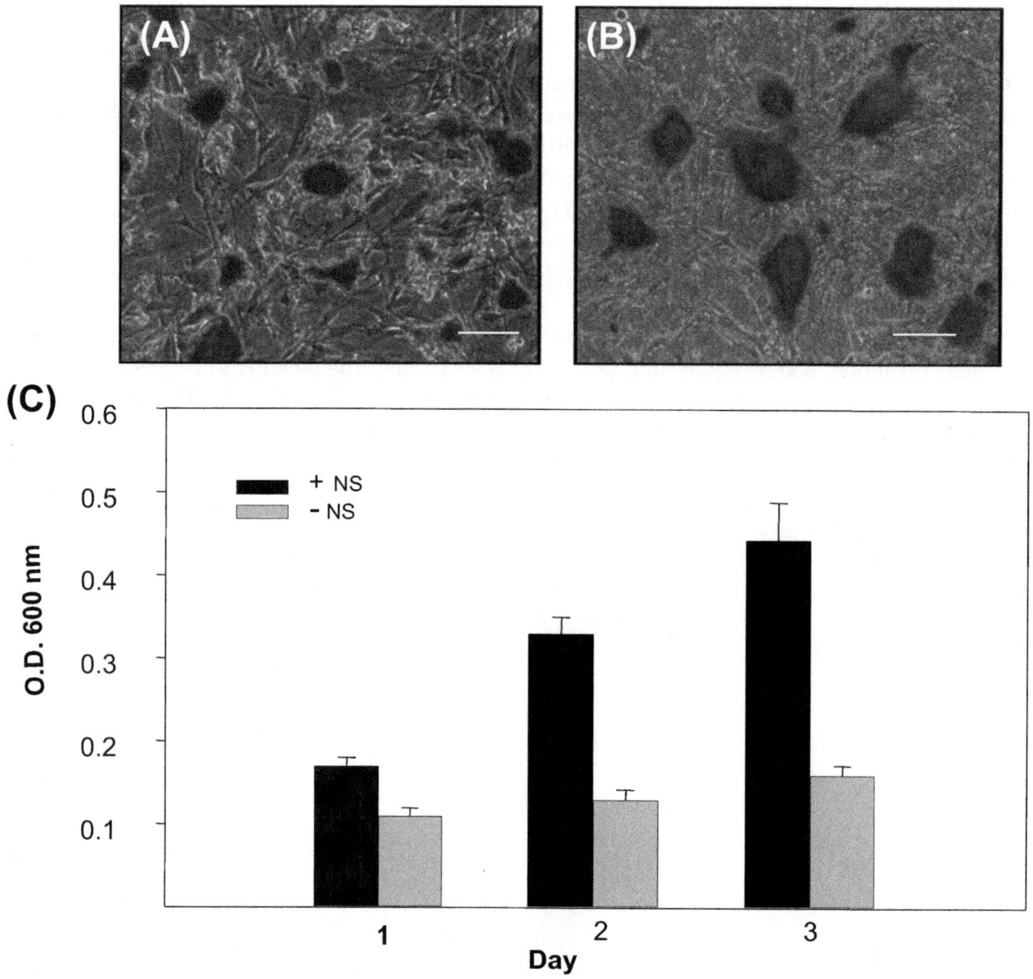

FIGURE 6.3 Growth of mESCs in the presence and absence of 3D nanofibrillar surfaces (Ultra-Web). (A, B): Alkaline phosphatase staining of mESC colonies after 3 days of culture in the absence (A) and presence (B) of the 3D nanofibrillar surface (Ultra-Web). Colonies were significantly larger for mESCs cultured on Ultra-Web than for mESCs cultured on plain coverslips. Scale bar = 10 μm. (C): Proliferation of mESCs was determined at 24-hour intervals for 3 days in the presence and absence of Ultra-Web, as described in Materials and Methods. The rate of proliferation was enhanced for cells cultured on Ultra-Web compared with cells cultured on glass. Data are shown as mean ± standard deviation (n = 3). *mESCs*, Mouse embryonic stem cells.

6.4 Three-dimensional surfaces as a favorable in vivo–like environment for mammalian cell growth and differentiation

Nanofiber 3D scaffolds are widely used in tissue engineering and regenerative medicine due to their ability to mimic the ECM, providing a structure that supports cell attachment, growth, and differentiation. Creating these scaffolds in laboratory conditions involves various fabrication techniques, biomaterial choices, and controlled environmental conditions to promote optimal tissue growth and scaffold functionality. Several studies have demonstrated that cell culture surfaces modified with oriented micro- or nano-sized grooves designed to mimic ECM fibrils. Fibroblasts plated onto such surfaces adjust their shape, orientation, and direction of movement to conform to the alignment of the machined components [16,17]. Topographic or contact-mediated effects were demonstrated on surfaces incorporating neuroactive peptides derived from laminin-1 deposited in patterns guided neurite outgrowth from hippocampal neurons, with neurons preferentially adhering to and extending neurites on the peptide-coated regions [18] as well as morphogenesis of breast epithelial cells (Fig. 6.2).

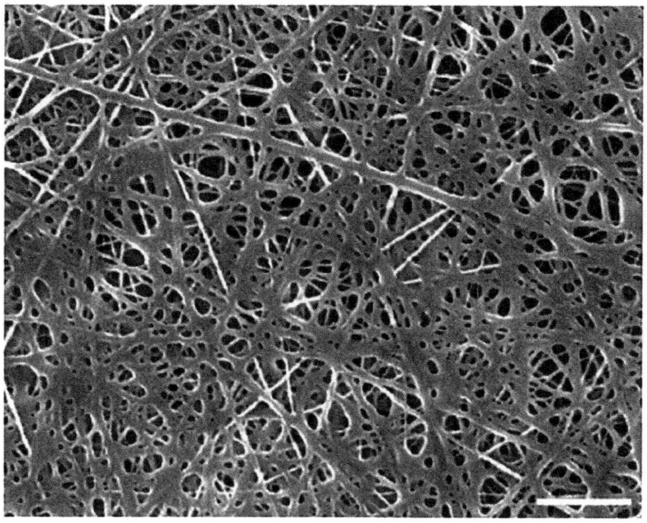

FIGURE 6.4 A Microscopic view of a 3D network of electrospun polyamide nanofibers. Nanofibers were electrospun onto glass coverslips. A SEM view reveals a fibrous network of filaments interspersed with pores. Bar, 5 μm.

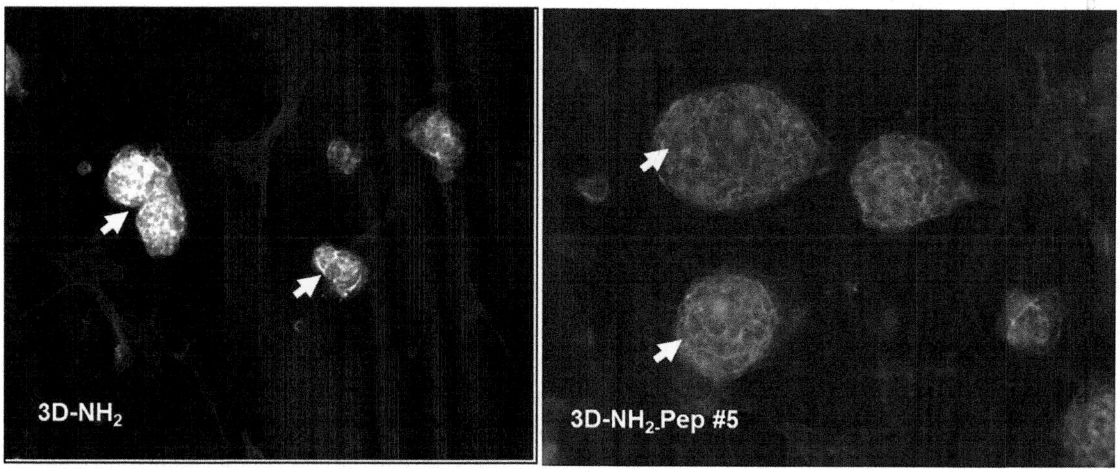

FIGURE 6.5 Stimulation of human embryonic stem cell growth on 3D polyamide nanofiber surfaces. Polypeptides derived from functional domains of ECM protein (Laminin-1) were covalently attached with 3D nanofibers. Human embryonic stem cells were cultured on $-NH_2$ (3D-NH_2) or NH-polypeptide (3D-NH_2-Pep#5) nanofibers on coverslips for 5 days. hESCs were then fixed and stained with phalloidin-FITC. Cells were studied and pictures were captured on a fluorescent microscope. Arrows indicate colonies of hESCs. Scale bar, 10 μm.

6.5 3D nanofiber surfaces induce morphogenesis of breast epithelial cells

We studied whether 3D polyamide nanofibers induce in vivo–like behavior of breast epithelial cells. We cultured T47D breast epithelial cells on glass and nanofibers. These cells were previously demonstrated to form duct-like tubular structures and spheroids under conditions that promote three-dimensional interactions with collagen or Matrigel. We cultured T47D breast epithelial cells and stained the cultured cells with phalloidin-FITC. After 5 days of growth on nanofibers, there was a mixed population of multicellular structures comprised of tubules and spheroids. Following 10 days in culture, the multicellular spheroids were dominant (A–D), although some tubules persisted (E). The lowest portion of the spheroid, in contact with the nanofibers (D), appears to have an opening. A multicellular tubule with an elongated lumen is shown in E. In contrast, the growth of T-47D cells on glass demonstrated a monolayer with groups of F-actin fibers (F). Interestingly, while MCF-7 cells (a human breast tumor line) proliferated as a monolayer on glass (G), these cells grew as more complex multilayers on nanofibers (H). We have demonstrated that synthetic three-dimensional polyamide nanofibers can provide a chemically and physically stable environment for NRK cells to structurally form more in vivo morphologies and functionally, for epithelial cells, to undergo morphogenesis and form multicellular spheroids containing a central lumen (Fig. 6.2) [21]. Our study provided evidence that electrospun polyamide nanofibers provide a permissive surface that promotes signaling programs normally associated with cell attachment to BM/ECM. The general availability of uniform and well-characterized three-dimensional surfaces should have major benefits for applications in tissue engineering, high-throughput drug screening, and studies of tissue function and pathology in basic research, biotechnology, and the biomedical and pharmaceutical sciences (Fig. 6.4).

6.6 3D nanofiber surfaces enhance growth of mouse enbryonic stem cells

Recent experiments using mesenchymal stem cells have demonstrated an important role for mechanical cues in regulating stem cell fate. Investigations demonstrating that mechanical signals are transduced to the cell cytoskeleton through the activation of Rho, a small GTPase, and Rho kinase (ROCK) have provided important evidence that stem cell fate may not only depend on soluble factors, but may require attachment to 3D ECM/BM surfaces whose physical, mechanical, and chemical properties can directly or indirectly regulate the pathways controlling stem cell fate. Culture on or within 3D as opposed to 2D arrays of matrix molecules promotes cellular phenotypes that display more in vivo–like structure and function [28]. We studied the pluripotency of mESCs cultured for 3 days on Ultra-Web 3D nanofibrillar surfaces by determining the expression of alkaline phosphatase, a marker for undifferentiated stem cells. As shown in A, mESCs cultured on glass coverslips had small colonies of undifferentiated cells. In contrast, mESCs cultured on 3D nanofibrillar surfaces had significantly larger colonies of undifferentiated cells (B). Significantly, cells did not adhere to two-dimensional films composed of polyamide (data not shown), demonstrating the importance of the nanofibrillar geometry for stem cell proliferation. Measurements of colony area for 50 colonies after 3 days of culture showed a mean projected colony area of 4590 + 730 μm2 on glass coverslips and 9930 + 1035 μm2 on 3D nanofibrillar surfaces. We determined cell proliferation every 24 hours for 3 days for mESCs cultured on glass coverslips with and without Ultra-Web coatings. It was found that proliferation of mESCs was stimulated when cultured on the 3D nanofibrillar surfaces (C). There was little stimulation in the growth of mESCs that were cultured on coverslips in the absence of Ultra-Web (C) (Fig. 6.3).

6.7 Attachment of bioactive polypeptides/proteins to create novel in vivo–like cell culture surfaces

We introduced a novel nanofibrillar surface for cell culture composed of electrospun nanofibers. These surfaces were designed to mimic the scale and porosity of the nanofibrillar networks of polymers comprising the BM/ECM [32]. Work with a variety of cell types demonstrated that these surfaces provide an environment that promotes more in vivo–like phenotypes for cells than conventional two-dimensional tissue culture surfaces. This work was followed by the demonstration that synthetic peptides derived from sequences within the ECM proteins had significantly greater neurite outgrowth capacity for different neuronal types when these peptides were linked to the nanofibrillar surfaces [33]. We have attempted to further create a chemical environment on 3D nanofibrillar surfaces. To insert protein/peptide on nanofibers, we have attached amine (–NH_2) on nanofibers. Then we covalently attached FGF-2 to the nanofibers through the amine group and studied whether FGF-2 retains its biological activity [34]. Two novel observations from this work are that crosslinking FGF-2 to nanofibrillar surfaces significantly enhances the potency of FGF-2 and also increases the stability of the molecule. Of particular note is that covalently bound FGF-2 retains its biological activity under conditions of dry storage. We have demonstrated that FGF-2 was at least 100-fold more active in covalently bound form than when it remains in solution [34]. Increased activity may be due to the increase in local concentration of FGF-2, as well as an increase in the stability of FGF-2.

6.8 Covalently attached bioactive peptides on 3D nanofiber surfaces enhance growth of human embryonic stem cells

To further develop cell culture surfaces chemically, we attempted to covalently attach functional polypeptide domains of ECM proteins on 3D polyamide nanofibers. Polyamide nanofibers were electrospun onto glass coverslips and covalently attached with a proprietary polyamine. Nanofiber surfaces were then washed to remove excess crosslinker. ECM-derived polypeptide solution was incubated to covalently link with polyamide nanofibers [35].

Synthetic polyamide scaffolds that mimic the 3D nanotopography and porosity of the ECM are believed to play a crucial and important role in applications of regenerative medicine as they are close to the physiologically relevant in vivo systems. Adding functional peptides to the 3D polyamide peptides demonstrated an even better physiological environment in terms of neuron growth in laboratory conditions [35]. It was found that hippocampal neurons attached to polyamide nanofibers and extended neurites, with significantly enhanced neurite generation in comparison to neurite generation on PLL-coated coverslips [35], suggesting that the nanofibers might provide a good grafting surface for neuronal replacement therapies (e.g., with hippocampal progenitor cells) even in the absence of neuroactive peptides. It was found that a synthetic peptide derived from the second fibronectin type III module of the Neural Cell Adhesion Molecule, called the FGL peptide, promoted neurite outgrowth from hippocampal neurons through a direct interaction with fibroblast growth factor receptor. Thus nanofibers modified with the neural cell adhesion molecule–derived FGL peptide might be an even more efficacious growth surface for the replacement of hippocampal neurons than unmodified nanofibers. The 3D polyamide nanofibrillar surface supported neuronal attachment and neurite outgrowth to a similar extent as 2D PLL-coated glass. A covalent modification of the nanofibers with neuroactive ECM peptides further enhanced their ability to support neuronal attachment, neurite generation, and neurite outgrowth. We demonstrated that covalently attached polypeptides (derived from ECM) enhanced the growth of human embryonic stem cells (hESCs) compared to their growth on amine-linked nanofibers (Fig. 6.5). These findings suggest that 3D nanofibrillar matrix mimics the geometry and topography of the ECM to provide a more in vivo–like environment for neurons and hESCs, particularly in conjunction with ECM-derived chemical cues. Further studies in this field might provide an appropriate physical and chemical environment closer to in vivo conditions to manipulate the fate of cells critical for tissue engineering.

6.9 Conclusion

Increasing evidence now supports the idea that cells in each tissue type are located in a unique niche that controls their specific growth and differentiation. It is beyond doubt that growing cells in an in vivo–like environment in a laboratory is a great challenge to the scientific community. To develop an in vivo–like condition in the laboratory, attempts have been made by considering two general conditions of high importance: (1) 3D cell culture conditions to grow cells and tissues [21,29–33,36] and (2) the chemical composition of the environment surrounding cells in culture conditions [13,32,34,36,37]. In our studies, we developed 3D polyamide nanofibers that provide in vivo–like environment [35,37] and also developed methods to covalently attach polypeptides/proteins on 3D polyamide nanofibers. We demonstrated that covalently attached proteins retain and enhance, at least in our studies, biological activity [34,35]. These conditions allow cells to behave in vivo–like conditions such as the development and growth of breast tissues, neurons, embryonic stem cells. It has become increasingly clear that the development of more precise physical and chemical environments similar to the specific type of ECM and BM might provide more physiologically relevant conditions for cells in the laboratory. Further studies are required to develop specific chemical environments that might direct differentiation of stem cells to specific types of tissue cells. We have reviewed current progress in developing synthetic conditions similar to in vivo–like ECM and BM environment, hoping to achieve conditions to generate cells/tissues required for regenerative medicine.

References

[1] P.J. Hayden, J.W. Harbell, Special review series on 3D organotypic culture models: introduction and historical perspective, In Vitro Cellular and Developmental Biology - Animal 57 (2) (2021) 95–103. Available from https://doi.org/10.1007/s11626-020-00500-2, http://www.springerlink.com/content/1543-706X.

[2] J.P. Stegemann, R.M. Nerem, Altered response of vascular smooth muscle cells to exogenous biochemical stimulation in two- and three-dimensional culture, Experimental Cell Research 283 (2) (2003) 146–155. Available from https://doi.org/10.1016/S0014-4827(02)00041-1, http://www.journals.elsevier.com/experimental-cell-research/.

[3] G. Jose, K.T. Shalumon, J.P. Chen, Natural polymers based hydrogels for cell culture applications, Current Medicinal Chemistry 27 (16) (2020) 2734–2776. Available from https://doi.org/10.2174/0929867326666190903113004, http://www.eurekaselect.com/182480/article.

[4] C. Liu, M. Pei, Q. Li, Y. Zhang, Decellularized extracellular matrix mediates tissue construction and regeneration, Frontiers of Medicine 16 (1) (2022) 56–82. Available from https://doi.org/10.1007/s11684-021-0900-3, http://www.springer.com/medicine/journal/11684.

[5] O. Chaudhuri, J. Cooper-White, P.A. Janmey, D.J. Mooney, V.B. Shenoy, Effects of extracellular matrix viscoelasticity on cellular behaviour, Nature 584 (7822) (2020) 535–546. Available from https://doi.org/10.1038/s41586-020-2612-2, http://www.nature.com/nature/index.html.

[6] R. Edmondson, J.J. Broglie, A.F. Adcock, L. Yang, Three-dimensional cell culture systems and their applications in drug discovery and cell-based biosensors, Assay and Drug Development Technologies 12 (4) (2014) 207–218. Available from https://doi.org/10.1089/adt.2014.573, www.liebertonline.com/adt.

[7] F. Berthiaume, P.V. Moghe, M. Toner, M.L. Yarmush, Effect of extracellular matrix topology on cell structure, function, and physiological responsiveness: hepatocytes cultured in a sandwich configuration, FASEB Journal 10 (13) (1996) 1471–1484. Available from https://doi.org/10.1096/fasebj.10.13.8940293, https://onlinelibrary.wiley.com/journal/15306860.

[8] G. Knedlitschek, F. Schneider, E. Gottwald, T. Schaller, E. Eschbach, K.F. Weibezahn, A tissue-like culture system using microstructures: influence of extracellular matrix material on cell adhesion and aggregation, Journal of Biomechanical Engineering 121 (1) (1999) 35–39. Available from https://doi.org/10.1115/1.2798040.

[9] S.I. Ertel, A. Chilkoti, T.A. Horbett, B.D. Ratner, Endothelial cell growth on oxygen-containing films deposited by radio-frequency plasmas: the role of surface carbonyl groups, Journal of Biomaterials Science, Polymer Edition 3 (2) (1992) 163–183. Available from https://doi.org/10.1163/156856291X00269.

[10] M. Hojo, S. Inokuchi, M. Kidokoro, N. Fukuyama, E. Tanaka, C. Tsuji, et al., Induction of vascular endothelial growth factor by fibrin as a dermal substrate for cultured skin substitute, Plastic and Reconstructive Surgery 111 (5) (2003) 1638–1645. Available from https://doi.org/10.1097/01.PRS.0000053842.90564.26.

[11] D. Osuna de la Peña, S.M.D. Trabulo, E. Collin, Y. Liu, S. Sharma, M. Tatari, et al., Bioengineered 3D models of human pancreatic cancer recapitulate in vivo tumour biology, Nature Communications 12 (1) (2021). Available from https://doi.org/10.1038/s41467-021-25921-9, http://www.nature.com/ncomms/index.html.

[12] A. Abbott, Biology's new dimension, Nature 424 (6951) (2003) 870–872. Available from https://doi.org/10.1038/424870a.

[13] R. Kalluri, Basement membranes: structure, assembly and role in tumour angiogenesis, Nature Reviews. Cancer 3 (6) (2003) 422–433. Available from https://doi.org/10.1038/nrc1094.

[14] J. Ashkenas, J. Muschler, M.J. Bissell, The extracellular matrix in epithelial biology: shared molecules and common themes in distant phyla, Developmental Biology 180 (2) (1996) 433–444. Available from https://doi.org/10.1006/dbio.1996.0317, http://www.elsevier.com/inca/publications/store/6/2/2/8/1/6/index.htt.

[15] E.D. Hay, The mesenchymal cell, its role in the embryo, and the remarkable signaling mechanisms that create it, Developmental Dynamics 233 (3) (2005) 706–720. Available from https://doi.org/10.1002/dvdy.20345.

[16] C. Oakley, N.A.F. Jaeger, D.M. Brunette, Sensitivity of fibroblasts and their cytoskeletons to substratum topographies: Topographic guidance and topographic compensation by micromachined grooves of different dimensions, Experimental Cell Research 234 (2) (1997) 413–424. Available from https://doi.org/10.1006/excr.1997.3625, http://www.journals.elsevier.com/experimental-cell-research/.

[17] A.I. Teixeira, G.A. Abrams, P.J. Bertics, C.J. Murphy, P.F. Nealey, Epithelial contact guidance on well-defined micro- and nanostructured substrates, Journal of Cell Science 116 (10) (2003) 1881–1892. Available from https://doi.org/10.1242/jcs.00383.

[18] S. Saneinejad, M.S. Shoichet, Patterned poly(chlorotrifluoroethylene) guides primary nerve cell adhesion and neurite outgrowth, Journal of Biomedical Materials Research 50 (4) (2000) 465–474. Available from https://doi.org/10.1002/(SICI)1097-4636(20000615)50:4##465::AID-JBM1$$3.0.CO;2-K.

[19] H.Y. Chung, J.R.B. Hal, M.A. Gogins, D.G. Crofoot, T.M. Weik, Polymer, polymer microfiber, polymer nanofiber and applications including filter structures. (2004).

[20] J. Doshi, D.H. Reneker, Electrospinning process and applications of electrospun fibers, Journal of Electrostatics 35 (2-3) (1995) 151–160. Available from https://doi.org/10.1016/0304-3886(95)00041-8.

[21] M. Schindler, I. Ahmed, J. Kamal, A. Nur-E-Kamal, T.H. Grafe, H. Young Chung, et al., A synthetic nanofibrillar matrix promotes in vivo-like organization and morphogenesis for cells in culture, Biomaterials 26 (28) (2005) 5624–5631. Available from https://doi.org/10.1016/j.biomaterials.2005.02.014.

[22] A. Nur-E-Kamal, I. Ahmed, J. Kamal, M. Schindler, S. Meiners, Three dimensional nanofibrillar surfaces induce activation of Rac, Biochemical and Biophysical Research Communications 331 (2) (2005) 428–434. Available from https://doi.org/10.1016/j.bbrc.2005.03.195.

[23] W.J. Li, K.G. Danielson, P.G. Alexander, R.S. Tuan, Biological response of chondrocytes cultured in three-dimensional nanofibrous poly(ε-caprolactone) scaffolds, Journal of Biomedical Materials Research - Part A 67 (4) (2003) 1105–1114. Available from https://doi.org/10.1002/jbm.a.10101, http://onlinelibrary.wiley.com/journal/10.1002/(ISSN)1552-4965.

[24] H. Yoshimoto, Y.M. Shin, H. Terai, J.P. Vacanti, A biodegradable nanofiber scaffold by electrospinning and its potential for bone tissue engineering, Biomaterials 24 (12) (2003) 2077–2082. Available from https://doi.org/10.1016/S0142-9612(02)00635-X.

[25] E.D. Boland, J.A. Matthews, K.J. Pawlowski, D.G. Simpson, G.E. Wnek, G.L. Bowlin, Electrospinning collagen and elastin: preliminary vascular tissue engineering, Frontiers in Bioscience 9 (2004) 1422–1432. Available from https://doi.org/10.2741/1313, https://www.fbscience.com/archives.

[26] M. Li, M.J. Mondrinos, M.R. Gandhi, F.K. Ko, A.S. Weiss, P.I. Lelkes, Electrospun protein fibers as matrices for tissue engineering, Biomaterials 26 (30) (2005) 5999–6008. Available from https://doi.org/10.1016/j.biomaterials.2005.03.030.

[27] J.J. Stankus, J. Guan, W.R. Wagner, Fabrication of biodegradable elastomeric scaffolds with sub-micron morphologies, Journal of Biomedical Materials Research - Part A 70 (4) (2004) 603–614. Available from https://doi.org/10.1002/jbm.a.30122, http://onlinelibrary.wiley.com/journal/10.1002/(ISSN)1552-4965.

[28] E. Cukierman, R. Pankov, D.R. Stevens, K.M. Yamada, Taking cell-matrix adhesions to the third dimension, Science (New York, N.Y.) 294 (5547) (2001) 1708–1712. Available from https://doi.org/10.1126/science.1064829.

[29] K. Burridge, K. Wennerberg, Rho and Rac take center stage, Cell 116 (2) (2004) 167–179. Available from https://doi.org/10.1016/S0092-8674(04)00003-0, www.cell.com.

[30] C.D. Nobes, A. Hall, Rho, Rac, and Cdc42 GTPases regulate the assembly of multimolecular focal complexes associated with actin stress fibers, lamellipodia, and filopodia, Cell 81 (1) (1995) 53–62. Available from https://doi.org/10.1016/0092-8674(95)90370-4.

[31] S. Etienne-Manneville, A. Hall, Rho GTPases in cell biology, Nature 420 (6916) (2002) 629–635. Available from https://doi.org/10.1038/nature01148.

[32] J.H. Miner, P.D. Yurchenco, Laminin functions in tissue morphogenesis, Annual Review of Cell and Developmental Biology 20 (2004) 255–284. Available from https://doi.org/10.1146/annurev.cellbio.20.010403.094555.

[33] H.K. Kleinman, D. Philp, M.P. Hoffman, Role of the extracellular matrix in morphogenesis, Current Opinion in Biotechnology 14 (5) (2003) 526–532. Available from https://doi.org/10.1016/j.copbio.2003.08.002, http://www.elsevier.com/locate/copbio.

[34] A. Nur-E-Kamal, I. Ahmed, J. Kamal, A.N. Babu, M. Schindler, S. Meiners, Covalently attached FGF-2 to three-dimensional polyamide nanofibrillar surfaces demonstrates enhanced biological stability and activity, Molecular and Cellular Biochemistry 309 (1-2) (2008) 157–166. Available from https://doi.org/10.1007/s11010-007-9654-8.

[35] I. Ahmed, H.Y. Liu, P.C. Mamiya, A.S. Ponery, A.N. Babu, T. Weik, et al., Three-dimensional nanofibrillar surfaces covalently modified with tenascin-C-derived peptides enhance neuronal growth in vitro, Journal of Biomedical Materials Research - Part A 76 (4) (2006) 851–860. Available from https://doi.org/10.1002/jbm.a.30587.

[36] N.J. Boudreau, Organized living: from cell surfaces to basement membranes, Science's STKE signal transduction knowledge environment 2003 (196) (2003) pe34. Available from https://doi.org/10.1126/scisignal.1962003pe34.

[37] A. Nur-E-Kamal, I. Ahmed, J. Kamal, M. Schindler, S. Meiners, Three-dimensional nanofibrillar surfaces promote self-renewal in mouse embryonic stem cells, Stem Cells 24 (2) (2006) 426–433. Available from https://doi.org/10.1634/stemcells.2005-0170, http://stemcells.alphamedpress.org/cgi/reprint/24/2/426.pdf.

Chapter 7

Silk routes to healing: engineering biomaterials for mesenchymal stem cell guidance in bone tissue remodeling and repair

Saumya Dash, and Sujata Mohanty

Stem Cell Facility, DBT-Centre of Excellence for Stem Cell Research, All India Institute of Medical Sciences, New Delhi, India

7.1 Introduction

Silk is a natural fibrous protein biopolymer, produced by arthropods such as spiders, mites, fleas, and silkworms. Silkworms are silk-spinning insects widely researched and applied in tissue engineering. The production of silk, known as sericulture, has a long history associated with China. The silk was first discovered in 2640 BCE by His-Ling-Chi, who stated that silk fibers can be loosened and unwound in hot water. He also developed various methods of raising the silkworms and reeling the fibers. Further, silk became a very precious commodity, and thus sericulture spread within China and to other countries through the trading route known as the Silk Road.

The structural composition and properties of silk vary depending on its source. In tissue engineering, silk from silkworms is mostly used due to their better bioactivity and greater availability than other sources [1]. The silk from the silkworm is classified into two types, that is, mulberry and nonmulberry types. The mulberry silk is produced from the silkworm *Bombyx mori*. The nonmulberry silk can be obtained from three sources, that is, Tasar silk, Muga silk, and Eri silk from the silkworm *Antheraea myllita, Antheraea assamensis*, and *Philosamia cynthia ricini*, respectively. The silk cocoon is made up of 95% protein and 5% impurities, that is, minerals, salts, and waxes. The protein consists of filamentous core fibroin and the glue-like coat sericin, which have varied amino acid types and compositions. Fibroin content is higher in mulberry silk than in nonmulberry silk. Mulberry silk has a fibroin-to-sericin ratio of roughly 70%–30% [2]. The protein fibroin has drawn a lot of interest in tissue engineering due to its exceptional biocompatibility, permeability, nonirritation, and superior mechanical properties. Conversely, sericin elicits immunogenic responses [3]. Thus the high yield of fibroin from mulberry silk has attracted researchers to use it as a potential silk protein source for designing biomaterials. The primary microstructure of silk protein contains 18 different amino acids with the sequence Gly–Ala–Gly–Ala–Gly–X, with X representing the Ser residue or containing Ser residue of another amino acid, such as GAGAGS, GAGAGA, GAGAGY [4]. The results of a thorough investigation of silk protein show that it consists of two polypeptide chains, one heavy (H chain, approximately 340 kDa) and one light (L chain, about 26 kDa), joined by a disulfide link. To complete the silk protein molecule, the glycoprotein P25 links to the chains via a noncovalent hydrophobic contact. The H chain's GAGAGS/RGD sequence often forms a stable β-sheet in the silk fiber's crystalline region. On the contrary, an amorphous region with low mechanical stability is formed by the L chain's irregular and nonrepetitive polypeptide fragments. Two typical confirmations of the silk protein are found: the water-soluble, amorphous, random coil, α-helix, type I confirmation; and the stable, regular, crystalline, β-sheet, type II confirmation. Depending on the level of chemical and physical exposure, the two confirmations can be taken interchangeably. The crystalline domain of the silk contains the bioactive RGD (Arg–Gly–Asp) sequence or RGD-like sequences such as (GAGAGS)n, TGRGDSPA, which aids in better cellular response [5]. It is also worth noting that nonmulberry silk fibroin contains an RGD sequence that serves as a recognition site for integrin-like cell surface receptors and shows better

Fundamentals of Translational Tissue Engineering. DOI: https://doi.org/10.1016/B978-0-443-33455-9.00001-8

elasticity, making it promising in tissue engineering [6]. However, the easy availability due to extensive farming, effortless processing, high fibroin content, better hydrophilicity, and presence of GAGAGS and TGRGDSPA sequences of fibronectin containing RGD-like triplet of the mulberry silk made it a widely used silk source for biomaterial designing to date [7]. Recently, spider silk protein spidroin from the species *Nephila clavipes* and *Nephalia edulis* are also being studied in tissue engineering applications. It contains amino acids such as Glycine and Alanine, but lacks sericin protein, thus eliminating the tedious sericin removal step. It has a similar arrangement of fibroin as the *B. mori*, with high strength and large elongation at break. The three major spidroin proteins studied so far are major ampullate spidroin 1, major ampullate spidroin 2, and major ampullate spidroin 3. The major ampullate spidroin 1 and 2 have repeated sequences of polyalanine sandwiched between glycine-rich sequences [8], whereas the major ampullate spidroin 3 lacks the polyalanine and glycine–proline–glycine motifs. The major ampullate spidroins assemble to form hierarchically structured fibers [9]. However, the lack of large-scale production of spider silk-based materials due to low production yield and difficulties in cultivation [10] limited the use of the protein for extensive studies. Furthermore, the excellent mechanical strength, optical properties, biocompatibility, sustained biodegradability, tuneable molecular structure, moldable morphology, implant ability, and presence of specific cell surface recognition receptor sites of silk fibroin have been found to increase application in biomedical material designing.

7.2 Silk fibroin in tissue engineering

Tissue engineering is a multidisciplinary field that aims for tissue or organ repair by integrating the science and technology of materials, cells, and biochemical factors. The persistent shortage of donor tissue and organ transplantation, in addition to allograft rejection, has prompted the development of tissue engineering as a gold-standard alternative therapy. Scaffolds in tissue engineering serve as a 3-dimensional (3D) template for cellular growth, proliferation, and attaining specific tissue/organ function. The 3D scaffold helps as a starter matrix and degrades eventually, and further gets resorbed or metabolized. The scaffold provides structural support to the cells for subsequent tissue formation for guided tissue regeneration. Thus, the physicochemical and biological property of the scaffold plays a crucial role in guiding the cell fate. Given that the silk fibroin polymer attracts attention owing to its extracellular matrix (ECM)-like nature, along with high mechanical strength and biodegradability [11]. At first, the silkworm silk entered into biomedical use as silk suture in 150 A.D by Aelius Galenus—a Greek physician. Further, *B. mori* silk fibroin-based porous scaffolds came into human clinical trials in the 1980s. In the 2000s, the recombinant spider silk hydrogels and particles entered the global market. And nowadays silk-coated implants and silk-based implants with enhanced biomechanical properties are in use [12].

Until now, silk fibroin-based films, fibers, nets, meshes, membranes, yarns, sponges, hydrogels, etc. have been explored in a variety of tissue regeneration. However, most research focuses on silk fibroin-based constructs for hard tissue regeneration. Furthermore, the most used type of silk is *B. mori* silk fibroin, as evidenced from preclinical study reports (Table 7.1).

In tissue engineering silk fibroin is used in various forms such as silk solution, silk particle, and silk microfiber and it should be noted that the form of silk fibroin imparts a major role in physicochemical properties of the biomaterial. The amphiphilic silk solution is the fragmented form of silk fibers dispersed in the solvent and can be converted into various morphologies such as sponges, films, gels, fibers, and particles with tuneable degradation properties (Fig. 7.1). Silk particles possess a high impact in tissue engineering due to their ability to bind with various ionic species for their controlled release [61]. Whereas, silk microfibers are microstructural protein fibers with higher mechanical strength than the other silk forms. The silk microfiber-reinforced biomaterials have a high impact on hard tissue regeneration owing to their additive advantage of strong protein–protein interfacial bonding [62] and high surface area which helps in cell–matrix interaction. Additionally, the microfibers provide surface roughness and low immune response mimicking the native bone structure. Here we are highlighting the use of silk in various tissue engineering applications.

***Skin*:** Studies reported that silk supports the growth, proliferation, and migration of fibroblast and keratinocyte-like skin cells [63]. Additionally, silk shows antimicrobial properties and the effectiveness of silk as an antimicrobial agent can be comparable to the topical ointment (corticosteroid) [64]. Studies reported the use of silk fibroin sheet alone or in combination with chitosan, recombinant spider silk, polycaprolactone, and polyglycerol sebacate as a skin patch for skin wound healing with better moisture and vapor permeability, fluid handling capacity, waterproofness, biocompatibility, antimicrobial activity, and wound healing [15,65,66].

***Ocular*:** The silk fibroin showed remarkable efficiency in ocular therapy by keratinocyte and corneal epithelial cell attachment, growth, sprouting, and migration. The silk fibroin-based films have thin transparency, neovascularization, insignificant inflammatory response, unaltered tear formation, and properties like those of the native cornea. It was also demonstrated that silk fibroin-based corneal implants are better in transparency as compared to widely used amniotic membrane or fibrin sealers [28,67].

TABLE 7.1 Table showing the details of silk-based biomaterials in tissue engineering.

S. No.	Type of silk	Type of biomaterial	Fabrication technique	Cell model	Application	References
1.	Antheraea *assamensis* silk fibroin	Mat	Electrospinning	Fibroblast	Skin regeneration	[13]
2.	*Antheraea mylitta* silk fibroin	Mat	Electrospinning	Fibroblast	Skin regeneration	[14]
3.	*Bombyx mori* silk fibroin	Mat	Electrospinning	Fibroblast	Skin regeneration	[15]
4.	*B. mori* silk fibroin	3D porous scaffold	Freeze drying	Fibroblast	Skin regeneration	[16]
5.	*B. mori* silk fibroin	Mat	Electrospinning	Fibroblast, melanocytes	Skin regeneration	[17]
6.	*B. mori* silk fibroin	Mat	Electrospinning	MSC-derived keratinocytes	Skin regeneration	[18]
7.	*B. mori* silk fibroin	Nanofiber scaffold	Electrospinning	Human umbilical vein endothelial cell	Skin regeneration	[19]
8.	*B. mori* silk fibroin	Injectable Hydrogel	Self-assembly	Endothelial cell	Skin regeneration	[20]
9.	*B. mori* silk fibroin	Hydrogel	3D printing	NA	Scarless wound healing	[21]
10.	*B. mori* silk fibroin	3D porous scaffold	Freeze drying	Fibroblast cell	Skin regeneration	[22]
11.	*A. mylitta* silk fibroin	Film	Electrospinning	NA	Wound healing	[23]
12.	*B. mori* silk fibroin	Film/Mat	Electrospinning	Fibroblast	Skin regeneration	[24]
13.	*B. mori* silk fibroin and silk sericin	Film	Self-assembly	Fibroblast	Skin regeneration	[25]
14.	*B. mori* silk fibroin	Scaffold	Freeze drying	NA	Skin regeneration	[26]
15.	*B. mori* silk fibroin	Bi-layer scaffold	Electrospinning	Human umbilical vein endothelial cell	Skin regeneration	[27]
16.	*A. mylitta* silk fibroin	Film	Casting	Corneal epithelial cell and keratocytes	Cornea regeneration	[28]
17.	*B. mori* silk fibroin	Solubilized protein	NA	Human corneal limbal-epithelial cells	Cornea regeneration	[29]

(*Continued*)

TABLE 7.1 (Continued)

S. No.	Type of silk	Type of biomaterial	Fabrication technique	Cell model	Application	References
18.	*B. mori* silk fibroin	Silk fibroin solution	NA	Corneal epithelial cell	Dry eye treatment	[30]
19.	*B. mori* silk fibroin	hydrogel	Casting	Human corneal epithelial cells	Corneal wound healing	[31]
20.	D9L transgenic silkworm strain	Transgenic silk protein	Genetic engineering	Retinal ganglion cell	Retina regeneration	[32]
21.	*B. mori* silk fibroin	Microparticle	Coprecipitation	NA	Corneal neovascularization	[33]
22.	*B. mori* silk fibroin	Injectable hydrogel	Sonication	NA	Maxillary bone regeneration	[34]
23.	Silk fibroin	Scaffold	Freeze drying	Dental pulp stem cell	Regenerative endodontics	[35]
24.	Silk fibroin	Scaffold	3D printing	Human dental pulp cell	Dentine–pulp complex regeneration	[36]
25.	*B. mori* silk fibroin	Scaffold	Porogen leaching	Human dental pulp stem cell	Osteo/odontogenic differentiation	[37]
26.	Spider silk	Composite	Compression molding	NA	Dental regeneration	[38]
27.	Silk fibroin	Scaffold	Freeze drying	Mesenchymal stem cell	Bone regeneration	[39]
28.	*B. mori* silk fibroin	Hydrogel	Sonication	Mesenchymal stem cell	Bone regeneration	[40]
29.	*B. mori* silk fibroin	3D scaffold	Freeze drying	Osteosarcoma cells	Bone regeneration	[41]
30.	*B. mori* silk fibroin	3D scaffold	3D printing	Mesenchyme stem cell	Osteoregeneration	[42]
31.	*B. mori* silk fibroin	3D scaffold	Freeze drying	Mesenchymal stem cell	Bone regeneration	[43]
32.	*B. mori* silk fibroin	3D scaffold	Freeze drying	Mesenchymal stem cell	Bone regeneration	[44]
33.	*B. mori* silk fibroin	Spheroid	Culture	Mesenchymal stem cell	Osteogenesis	[45]
34.	*B. mori* silk fibroin	3D scaffold	3D printing	Osteoblast cell	Maxillofacial regeneration	[46]
35.	Silk fibroin	Nanofiber scaffold	Electrospinning	Mesenchymal stem cell	Bone regeneration	[47]
36.	Silk fibroin	Scaffold	Vacuum freeze drying and chemical crosslinking	Mesenchymal stem cell	Bone regeneration	[48]
37.	*B. mori* silk fibroin	Scaffold	Cryogelation	Mesenchymal stem cell	Bone regeneration	[49]
38.	*B. mori* silk fibroin	Injectable hydrogel	Sonication	Mesenchymal stem cell	Bone regeneration	[50]
39.	*B. mori* silk fibroin	Composite membrane	Vacuum filtering	Osteoblast	Bone regeneration	[51]

40.	Methacrylated silk fibroin of *B. mori*	Sponge scaffold	Solvent casting & salt leaching	Osteosarcoma cells	Bone regeneration	[52]
41.	*B. mori* silk fibroin	3D scaffold	Freeze drying	Mesenchymal stem cell	Cardiac regeneration	[53]
42.	*B. mori* silk fibroin	Nanofibrous scaffold	Electrospinning	Cardiofibroblast	Improved cardiac function and reduced cardiac fibrosis after Acute myocardial infarction	[54]
45.	*B. mori* silk fibroin	Sponge scaffold	Solvent casting & salt leaching	Induced pluripotent stem cell	Cardiomyogenic differentiation	[55]
46.	*N. edulis* spider silk	Silk fibers	Coating	Schwann cell, fibroblast, rat dorsal root ganglion neuron	Neuronal regeneration	[56]
47.	*B. mori* silk fibroin	hydrogel	Click reaction	Schwann cell, fibroblast	Nerve regeneration	[57]
48.	*B. mori* silk fibroin	Matrices	Electrospinning	Mesenchymal stem cell	Liver regeneration	[58]
49.	*B. mori* & *A assamensis* silk fibroin	Porous scaffold	Blending	Human hepatocarcinoma cells	Liver regeneration	[59]
50.	*B. mori* silk fibroin	Scaffolds	Freeze drying	Human hepatoma cell	Liver regeneration	[60]

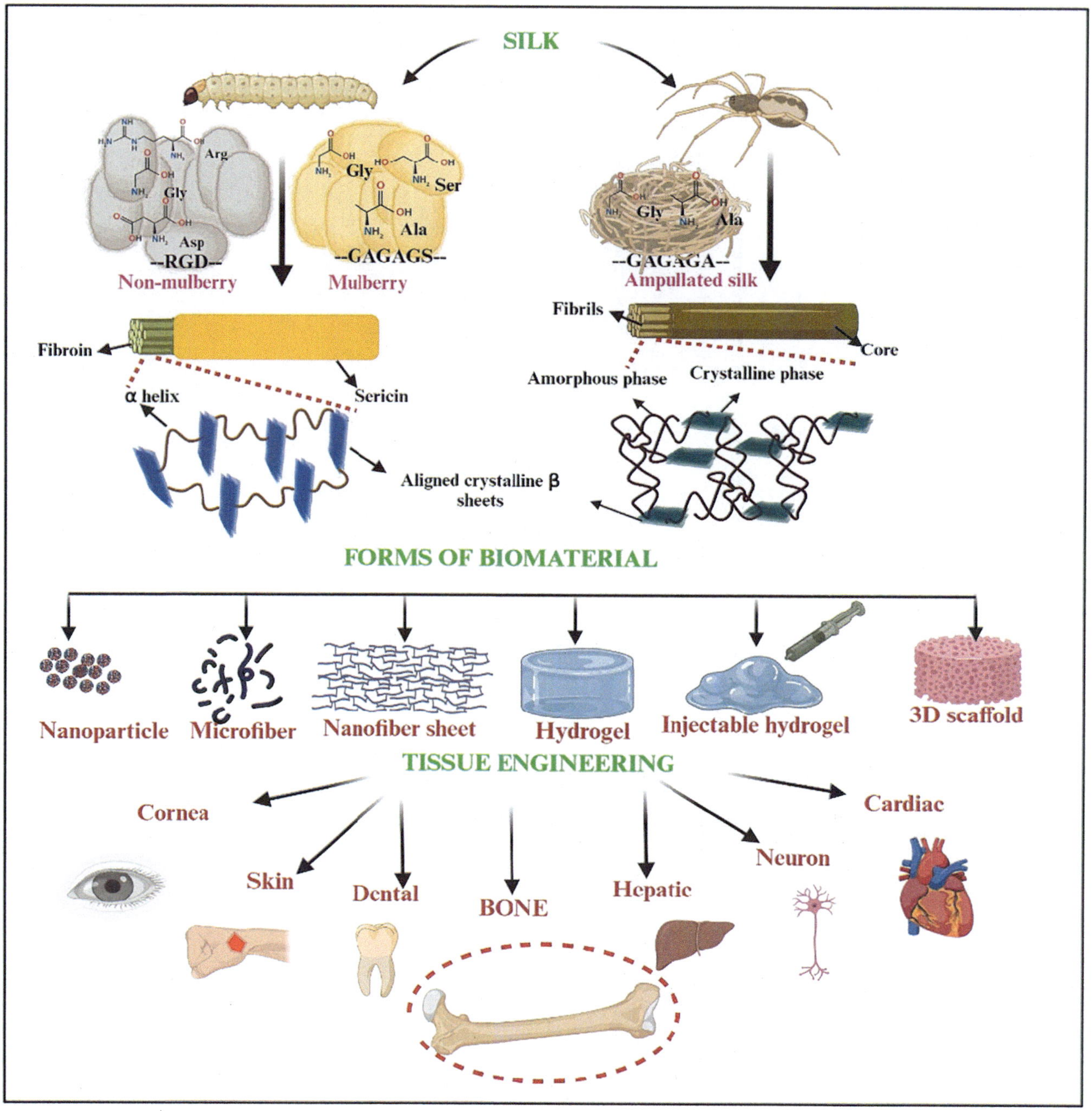

FIGURE 7.1 Detail outline of silk source, form, and application in tissue engineering. Created in BioRender.com.

Dental: The silk scaffolds are also observed to guide tooth bud cells to form robust bioengineered mineralized osteodentin. The implantation of silk scaffold into root fragments of the tooth helped the dental pulp mesenchymal stem cells (MSCs) to form dentin-like tissue and new matrix deposition along with vascularization, proving the silk-based biomaterial a promising candidate for regenerative endodontics [35,68].

Bone: Tissue engineering of bone deals with the repair and regeneration of natural bone using artificial methods. Herein, silk-based biomaterial implants have achieved great attention owing to the robust mechanical strength of silkworm silk fibroin, spider silk, and silk microfibers. Various studies evaluated the osteogenesis potential of bone implant scaffolds prepared using silk and found that the scaffold provides biomimetic cancellous bone-like compatibility, compressive strength, toughness, and porosity. Additionally, the silk fibroin supports stem cell proliferation and bone cell formation. Moreover, the silk provides osteoconduction, osteoinduction, and osteointegration [69,70].

***Cardiac*:** In cardiac tissue regeneration, the major challenge is to design a bioactive scaffold capable of providing adequate microenvironment for the cell to function. Given that silk protein grabs attention, researchers have studied the silk fibroin as well as sericin scaffolds to improve cardiomyocyte functionality. Additionally, the silk proteins also helped in the reduction of inflammation, improved microvessel density, and formation of cardiac tissue with enhanced features [71–73].

***Neuron*:** In terms of neuronal growth silk fibroin fibers and fluid showed good biocompatibility towards the dorsal ganglia, hippocampal neurons, and Schwann cells. Moreover, the tough template of silk fibroin enhances neurite guidance and extension for better neuronal differentiation [74–76].

***Hepatic*:** The liver is another major organ in our body and its metabolic, immune, and regulatory functions are regulated by the hepatocytes. The hepatocytes engage in release, storage, synthesis, and metabolism. Abnormalities in any of these functions lead to liver failure. Hence, liver tissue engineering is the gold standard to save lives. Studies reported that silk protein supports human hepatocellular carcinoma cells to become multicellular spheroids, promotes albumin secretion, HNF-4, and transferrin-like liver-specific gene expression. Moreover, the silk-based sponge helps to develop a bioartificial liver [77].

7.3 Methods of silk extraction and their challenges

Silk fibroin has the utmost advantages over other natural biopolymers, including huge commercial availability (Table 7.2), facile processing and purification, low production cost, and most importantly, US FDA approval of fibroin for drug delivery, surgical suture, and tissue engineering applications. Hence, the extraction of silk fibroin easily and cost-effectively needs special attention, wherein the cocoons pass through the standard degumming process by using Na_2CO_3. Briefly, 10 g of cut cocoon pieces were added to boiling water containing 0.02 M Na_2CO_3 for 30 min to remove sericin. After that, the fibers undergo washing with distilled water followed by overnight oven drying at 60°C. Sericin removal is a major step in silk fibroin extraction. Any remnant of sericin could reduce/alter the fibroin yield [78].

TABLE 7.2 Commercially available silk-based products in biomedical applications.

S. No.	Company Name	Place	Products
1.	Sofregen Inc.	Medford, MA, United States	SERI Surgical Scaffold, Silk fillers
2.	Silk Therapeutics Inc.	Medford, MA, United States	Antiaging skin care
3.	Cocoon Biotech Inc.	Cambridge, MA, United States	Silk-based drug delivery, hydrogel, microspheres for ophthalmic and osteoarthritis
4.	Silk Technologies Ltd.	Plymouth, MN, United States	Topical ocular therapy
5.	Oxford Biomaterials Ltd.	Oxford, United Kingdom	Vascular grafts
6.	Orthox Ltd.	Oxford, United Kingdom	FibroFix Meniscus, Spidrex tissue scaffold for hard tissue regeneration
7.	Neurotex Ltd.	Oxford, United Kingdom	Spidrex nerve conduit
8.	Zhejiang Xingyue Biotechnology Co., Ltd.	Hangzhou, China	Medical wound dressing, Soft tissue mesh
9.	SilkTech Biopharmaceuticals	Plymouth, Minnesota, United States	Eye drops
10.	Evolved by nature	Needham, MA, United States	Silk for skin care
11.	AMSilk	Neuried, Munich, Germany	Medical device coatings, gels for wound care
12.	FIBROHEAL Wound care pvt.ltd.	Bengaluru, Karnataka, India	Wound care materials

7.3.1 Extraction of silk fibroin particles

Silk particles have been used in the biomedical field owing to their structural stability. Ball milling is the silk nanoparticle fabrication technique that preserves the primary structure of silk fibers. In the ball milling process, silk fibers are charge-stabilized by keeping the pH at 10 to prevent particle aggregation due to electrostatic repulsion. In ball milling at pH 10, hydrolysis of silk fibers occurs under alkaline conditions; further, the increased temperature also enhances the breakdown of silk fibers. Briefly, the degummed silk fibers are chopped with a cutter mill until the snippets can pass through a 1 mm grid. Further, the wet milling of the snippets was carried out in an attritor. There, the mixture ratio should be 200 g of snippet and 1500 mL of distilled water. Then the slurry should be mixed with balls of size 5 mm in diameter for 7 hours. Furthermore, the slurry collected should be processed with a spray dryer to get silk micro powder. To prepare silk nanoparticle powder, 2 g of the collected micro powder should be milled in 100 mL of deionized water by bead milling. The pH 10 buffer or Tween 80 at 30% concentration should be added to avoid particle aggregation. Here, the cerium-doped zirconium oxide balls with a diameter of 0.5–0.6 mm would be used. The milling would be done at 2000 rpm for 10 or 7 hours for pH 10 or Tween 80, respectively. After milling, the alkali would be removed by dialysis, followed by washing and centrifugation with deionized water. Further, the washed particles would be freeze-dried to get nanoparticle powders [79]. However, ball milling is an expensive and time-consuming conventional method to get silk fibroin particles.

Another strategy of silk fibroin nanoparticle extraction is through lysis. The silk fibroin solution is prepared after breaking down the strong intermolecular bonds of silk fibers into smaller fragments by dissolving in a chaotropic agent.

Herein, the degummed silk fibers would be added to a 9.6 M LiBr solution at 60°C for 3 hours. Furthermore, this solution would dialyze against deionized water using a dialysis cassette with six water changes over 3 days. Then the silk solution would be collected and centrifuged for 20 minutes twice at 5°C and 9500×g. Further, the solution would be used to prepare films or scaffolds [80].

7.3.2 Extraction of silk microfiber

In the era of nanoparticles, microfibers have established a distinct role in tissue engineering due to their high specific surface area, which facilitates better interaction with cells. Silk fibroin microfibers are introduced as an artificial ECM to enhance cell–cell/ cell–matrix interactions in biomaterials. In addition to this, the easy manufacturing technique and tunable fiber size of silk fibroin microfibers make it a promising component to fabricate biomaterials. To manufacture the silk fibroin microfibers, 1 g dried and degummed silk fibers need to be submerged in 15 mL of 17.5 M NaOH solution for 20 minutes (12 minutes vigorous stirring and 8 minutes rest) for *B. mori* and 5 h for Tasar silk. Followed by the addition of 150 mL of deionized water to stop the alkaline hydrolysis reaction. Further, the solutions require centrifugation at 48,384×*g* for 20 minutes. the centrifugation step needs to be repeated 5 times, followed by adjustment of pH 7 by adding HCL. Then the neutralized silk microfiber solution was centrifuged again for 20 minutes to collect the microfiber pellets. This pellet was further lyophilized to get the silk microfiber powder [81]. Silk microfibers are highly important in bone tissue engineering owing to their enhanced cellular interactions and mechanical strength; however, the major challenge in silk microfiber synthesis is the time-dependent lysis of particles. Much care should be given to maintain a particular time to initiate and terminate the alkaline hydrolysis of the fibers to avoid randomness in microfiber size.

Among other biopolymers, such as collagen and chitosan, the extraction in acidic conditions is a major challenge when dealing with the delivery of biomolecules or cells. In contrast, the two-step extraction of silk fibroin (1) degumming using Sodium carbonate, and (2) dissolution/breakdown by using chaotropic/ alkaline agents, does not compromise its bioactivity to exploit as the best tissue engineering agent to carry biomolecules.

7.4 Silk fibroin and silk fibroin reinforced scaffold in bone tissue engineering

Guided bone regeneration (GBR) is the process of new bone formation in a bone defect site using the barrier membrane [82]. Wherein, the barrier membrane prevents infiltration of epithelial cells and inhibits the growth of connective tissue in the defect site, and secures the place for osteoblast and osteogenic cell infiltration. Thus the barrier membrane must be biocompatible, biodegradable, nonimmunogenic, cell-occlusive to prevent epithelial cell migration, mechanically strong for space-making, tissue-integrative, and easy to handle clinically. Recently, silk has emerged as an ideal barrier membrane for hard tissue regeneration owing to its outstanding mechanical properties, biocompatibility, immunomodulatory properties, and mineralization efficiency [83]. Silk fibroin nanofiber membrane showed increased osteoblast-like cell attachment and enhanced osteocalcin expression and ALP activity. The nanofiber mat showed satisfactory mechanical stability and new bone formation in rat calvarial defects [84]. Though the silk nanofiber structure is a very good candidate as a GBR membrane, its weak mechanical strength and brittleness have shifted the interest of researchers to fabricate other biomaterial forms, such as sponges/scaffolds. Whereas the amphiphilic nature of silk fibroin gives the liberty to design the material into various forms. The various advantages of 3D porous scaffolds, owing to their 3D structure, pore size, and mechanical strength, make them ideal for bone regeneration. The silk fibroin scaffolds show good mechanical strength; however, a study showed that the incorporation of silk gel into silk sponge enhanced stem cell survival in in vivo conditions and helped in more targeted and sustained delivery of stem cells at defect sites for calvarial bone defect treatment. The hydrogel-containing stem cell exhibited the best osteogenic effect in vivo with new bone formation within 8 weeks after implantation [40]. In another study, silk microfibers were incorporated into the silk fibroin sponge scaffolds. Wherein, the microfibers enhance the mechanical strength due to strong protein–protein interfacial bonding and also aid in MSCs-biomaterial interaction for guided bone regeneration [62]. Furthermore, the silk microfibers show better osteogenic differentiation of MSCs when cultured in a microfiber-reinforced spheroid by providing the cells with a void space to prevent hypoxia and better nutrient permeability [45]. Many studies are also performed by adding other composite materials with silk to enhance the osteogenic potential of the silk to make it a seedbed for bone regeneration. Hydroxyapatite is a bone mineral added to silk protein to prepare composite scaffolds where the seeded bone marrow stem cells promoted bone regeneration mainly through signaling pathways associated with cell–biomaterial interaction and showed augmented bone regeneration in calvarial defects in rats [39]. In another study, magnesium oxide was added along with hydroxyapatite to the silk scaffold to use the osteoconductive property of hydroxyapatite and the osteoregeneration property of magnesium ions. The composite scaffold showed better mechanical strength, biocompatibility, osteoconductivity, and differentiation potential of MSCs to bone cells. The *in vivo* study in rat distal lateral femoral bone defect showed enhanced bone regeneration [43]. In another study, magnesium ions were mixed with silk fibroin to design a scaffold with a shape memory function. Here, the scaffold also promoted cell proliferation, adhesion, and differentiation of MSCs into osteoblasts. After in vivo implantation, the scaffolds showed very good mineralization and osteogenesis with low foreign body reaction. The important outcome of the study is that they were able to design a smart water-shape memory-based scaffold by using the silk polymer to treat irregular bone defects with minimal surgical trauma [49]. The addition of platelet-rich plasma into silk hydrogel to prepare a bioactive injectable hydrogel has been performed recently to enhance vascularization and bone regeneration. The bioactive gel functions via the PI3K-Akt signaling pathway and promotes mitochondrial calcium uptake and converting exogenous amorphous calcium phosphate to endogenous amorphous calcium phosphate. The gel acted as an in situ biomineralization seedbed to provide new ideas for critical-size bone defect treatment [50]. Further to improve the functioning of silk, the methacrylation of silk is also performed, wherein the methacrylated silk (SilMA) showed significant bone regeneration [52]. The combination of magnetic nanoparticles in silk polymer to get magnetic-responsive 4D scaffolds is another approach for better bone regeneration. The presence of an external static magnetic field imparts a positive effect on osteoblast cells. Hence, the silk-based bone scaffolds incorporated with magnetic nanoparticles were studied by researchers and showed positive results in bone regeneration *in vitro* as well as in *in vivo* conditions [41]. However, more extensive studies should be done to make bioengineered silk scaffolds with enhanced osteoregeneration ability to avoid complex composite biomaterial design.

7.5 The interaction between silk fibroin to extracellular matrix and mesenchymal stem cells in tissue engineering

The ECM is a complex network of proteins and polysaccharides surrounded by cells. The ECM provides the extracellular microenvironment for cell adhesion. It further maintains the tensile strength of the microarchitecture to facilitate cell migration, guides tissue morphogenesis and repair, regulates protein activity, and, most importantly, is involved in cell–cell communication. A biomaterial that can restore several natural ECM characteristics and is further absorbed and replaced by de novo ECM proteins is a suitable candidate for tissue engineering. The important properties of a biomaterial, such as the presence of cell adhesion motifs, wettability, charge, and elasticity, make it an ECM-like biomaterial. In the case of silk protein, the RGD sequence of nonmulberry silk acts as the adhesion motif for better cellular binding and cell–matrix anchorage. However, the mulberry silk protein and spider silk protein have RGD-like sequences instead of RGD, hence showing comparatively lower cell attachment than the nonmulberry type. However, this issue could be eliminated by blending natural ECM proteins with silk or by functionalizing silk proteins with tripeptide RGD or other adhesion domains. Further, it was observed that the positive surface charge of a biomaterial plays an important role in cell attachment. Although a direct correlation of surface charge and cell attachment with silk matrices has not been studied. It is interesting to note that low cell adhesion was observed in *B. mori,* which has a high density of negative charges. Additionally, the wettability or the hydrophilicity of a material enhances cell adhesion. A biomaterial with a water contact angle of about 60° supports cell adhesion due to increased ECM protein adsorption. Otherwise, extremely hydrophilic or extremely hydrophobic matrices are mostly avoided for cell adhesion. In terms of silk, it shows the water contact angle between 20° and 100°, whereas the water contact angle of silk can be modulated by functionalizing with polar/nonpolar groups that is, plasma treatment. The surface topography, which is the substrate alignment and patterning of silk, can be modulated by its fabrication technique. The electrospun silk fibers have rough surface topography and spaces between the fibers that matches with the cytoskeleton for better cell adhesion. Furthermore, the cellular response to silk surface roughness is controversial due to the inconsistent methodology of studies. Studies reported that a moderate surface roughness (10–45 nm) of a biomaterial with a nearly Brownian fractal dimension (~ 2.5) promotes maximum cell proliferation. The mulberry silk showed positive surface roughness; however, this aspect requires thorough studies by considering the nanostructure profile, irregular features, grooves, and broadness peaks toward cellular response. Another intriguing property of silk is the low immune response upon engraftment owing to low macrophage adhesion and spreading. Cells can recognize the mechanical properties of a material's surface, which affects their metabolic activity, proliferation, migration, and differentiation. In tissue engineering, it is therefore necessary to design matrices with similar mechanical properties to the particular tissue/organ. Another important aspect of an ideal biomaterial is degradation. The study reported that the degradation rate of silk is very low owing to its crystal structure. Moreover, silk interacts with the ECM by mimicking the ECM properties. Silk can mimic the 3D network of the fibrous structure of the ECM to provide a natural environment for cell growth and guided regeneration. Furthermore, silk surface topography and physicochemical properties can regulate cell adhesion and can dictate cell fate for guided tissue regeneration.

In recent years, MSCs have constituted the most frequently used cell population in regenerative medicine due to their self-renewal and differentiation into one or more specialized cell types. Furthermore, the immunosuppressive behavior of MSCs due to unique cellular cross-talk makes them ideal candidates for autologous and allogenic transplantation. In tissue engineering, silk has been used in a variety of ways to enhance the use of MSCs for guided tissue formation, better immunomodulation as a form of scaffold, a delivery tool, and as a hydrogel substrate. Silk Fibroin scaffolds mimic the natural ECM structure and promote MSC function by their guided regeneration to a particular tissue or by their improved delivery and attachment at the target site. The electrospun silk fibroin matrix is reported to be the appropriate biomaterial for stem cell homing due to its natural ECM-like morphology and architecture. Furthermore, the silk fibroin scaffold structure changes dynamically, allowing the MSCs to proliferate and produce their own ECM. Silk biomaterials can trigger specific stem cell differentiation based on their intrinsic toughness, mechanical strength, biocompatibility and binding motif, molecular tunability, topography, geometry, chemical functionality, degradability, and degradation by-products (Fig. 7.2). Silk scaffolds can withstand high compressive loads to support guided bone regeneration. Its high tensile strength can be helpful for ligament tissue repair. The study also reported that silk fibroin enhances the immunomodulatory function of MSCs by increasing IFN-γ by providing a suitable natural ECM-like niche to MSCs for

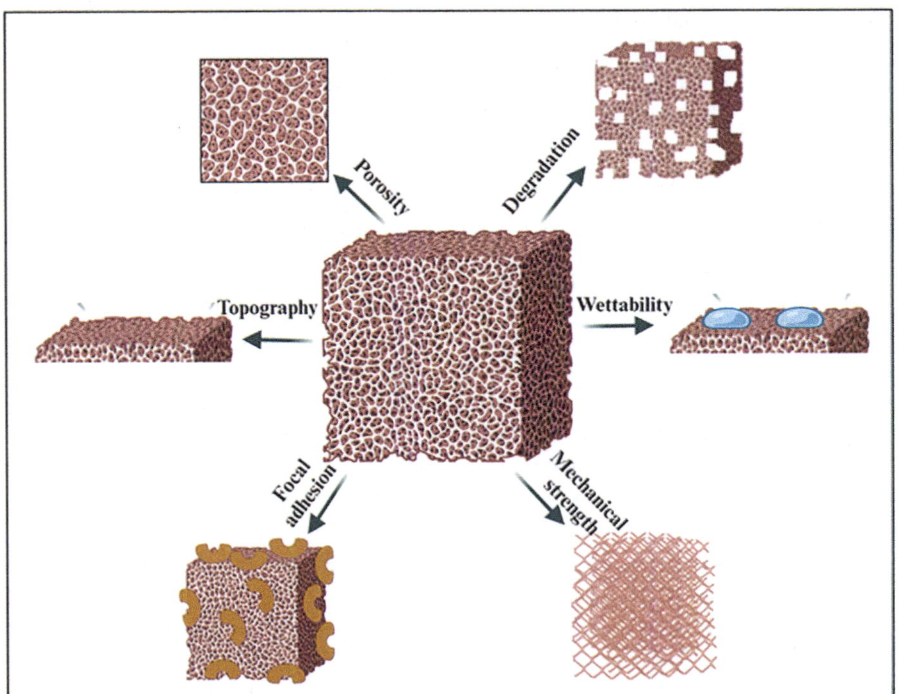

FIGURE 7.2 Silk scaffold and its biophysical properties on guided MSC-based tissue engineering. *MSC*, Mesenchymal stem cell. Created in BioRender.com.

adhesion, proliferation, air and water permeability, and low inflammatory response [85]. Such mechanotransduction by silk promotes the outside-in and inside-out signaling of MSCs for gene/protein expression. Moreover, the biomimetic structure of silk improves MSC differentiation and immunomodulation ability, which is essential for tissue engineering.

7.6 Conclusion and future perspective

Silk proteins are highly malleable and have been widely used in tissue engineering due to their superior physicochemical and biological properties. The biophysical properties of silk are useful to guide cellular fate, including cellular adhesion, proliferation, migration, and differentiation. Beyond its inherent characteristics, silk can be combined with other functional materials or undergo surface modification to enhance its biophysical influence on cells and tissue regeneration. The excellent biophysical properties of silk make it a suitable biopolymer for tissue engineering, skin care, drug delivery, and biomolecule delivery. The myriads of desirable properties make the silk-based biomaterials a natural ECM-like 3D matrix. Silk biomaterials have proven to be good candidates for regenerating tissues, including skin, bone, tooth, cornea, cardiac, hepatic, and nerve.

Nevertheless, the silk-based biomaterial faces a few challenges as a biomedical implant. For example, the binding site, or focal adhesion point, of silk protein varies with the type of silk. Hence, it is necessary to pay attention to activating the biomaterial as a natural ECM. Despite several studies on the interaction of silk with MSC-guided tissue engineering, in-depth research is required to elucidate the mechanisms by which different silk forms modulate MSCs. Genetic engineering of silk-producing organisms should be promoted to synthesize silk with customized amino acid sequences and desired functionalities. Studies have reported that the rates of biodegradation and immunogenicity of silk-based biomaterials depend on the scaffold type, a finding that requires further research. Several smarter silk-based materials, such as 4D smart scaffold/hydrogel, are expected in biomedical engineering. More insight into the 3D printing of silk biomaterial and methacrylate silk fibroin should be performed. So far, the silkworm silk is likely the dominant silk source in tissue engineering. However, various other sources, such as spider silk, may be taken into consideration. More attention should be given to the synthetic manufacturing of artificial silk to overcome the challenges of the lack of availability of spider silk and other silk varieties. Furthermore, the use of silk microfiber for hard tissue construction may be encouraged due to its better cellular interaction, facile synthesis, and enhanced mechanical strength. Despite several in vitro and in vivo studies, the translational benefit of silk-based implants is mostly restricted to ectopic or superficial applications. More clinical trials of silk-based biomaterials as bone/dental implants should be conducted. Translational research for a silk-based biomaterial is of immediate importance for consideration and collaborative action to overcome the gap in the future.

References

[1] A.C. MacIntosh, V.R. Kearns, A. Crawford, P.V. Hatton, Skeletal tissue engineering using silk biomaterials, Journal of Tissue Engineering and Regenerative Medicine 2 (2008) 71–80. Available from https://doi.org/10.1002/term.68.

[2] D. Jao, X. Mou, X. Hu, Tissue regeneration: a silk road, Journal of Functional Biomaterials 7 (2016) 22. Available from https://doi.org/10.3390/jfb7030022.

[3] B. Panilaitis, G.H. Altman, J. Chen, H.-J. Jin, V. Karageorgiou, D.L. Kaplan, Macrophage responses to silk, Biomaterials 24 (2003) 3079–3085. Available from https://doi.org/10.1016/s0142-9612(03)00158-3.

[4] J. Liu, X. Ge, L. Liu, W. Xu, R. Shao, Challenges and opportunities of silk protein hydrogels in biomedical applications, Materials Advances 3 (2022) 2291–2308. Available from https://doi.org/10.1039/d1ma00960e.

[5] A. Datta, A.K. Ghosh, S.C. Kundu, Differential expression of the fibroin gene in developmental stages of silkworm, *Antheraea mylitta* (Saturniidae), Comparative Biochemistry and Physiology Part B Biochemistry and Molecular Biology 129 (2001) 197–204. Available from https://doi.org/10.1016/s1096-4959(01)00377-3.

[6] Y. Maghdouri-White, G.L. Bowlin, C.A. Lemmon, D. Dréau, Bioengineered silk scaffolds in 3D tissue modeling with focus on mammary tissues, Materials Science and Engineering C 59 (2015) 1168–1180. Available from https://doi.org/10.1016/j.msec.2015.10.007.

[7] S.C. Kundu, Silk Biomaterials for Tissue Engineering and Regenerative Medicine, Elsevier, (2014).

[8] P. Bhattacharjee, B. Kundu, D. Naskar, H.-W. Kim, T.K. Maiti, D. Bhattacharya, et al., Silk scaffolds in bone tissue engineering: an overview, Acta Biomaterialia 63 (2017) 1–17. Available from https://doi.org/10.1016/j.actbio.2017.09.027.

[9] S. Salehi, K. Koeck, T. Scheibel, Spider silk for tissue engineering applications, Molecules 25 (2020) 737. Available from https://doi.org/10.3390/molecules25030737.

[10] M. Branković, F. Zivic, N. Grujovic, I. Stojadinovic, S. Milenkovic, N. Kotorcevic, Review of spider silk applications in biomedical and tissue engineering, Biomimetics 9 (2024) 169. Available from https://doi.org/10.3390/biomimetics9030169.

[11] Y. Wang, H.-J. Kim, G. Vunjak-Novakovic, D.L. Kaplan, Stem cell-based tissue engineering with silk biomaterials, Biomaterials 27 (2006) 6064–6082. Available from https://doi.org/10.1016/j.biomaterials.2006.07.008.

[12] C. Holland, K. Numata, J. Rnjak-Kovacina, F.P. Seib, The biomedical use of silk: past, present, future, Advanced Healthcare Materials 8 (2018). Available from https://doi.org/10.1002/adhm.201800465.

[13] C.M. Srivastava, R. Purwar, Fabrication of robust *Antheraea assama* fibroin nanofibrous mat using ionic liquid for skin tissue engineering, Materials Science and Engineering C 68 (2016) 276–290. Available from https://doi.org/10.1016/j.msec.2016.05.020.

[14] C.M. Srivastava, R. Purwar, A.P. Gupta, Enhanced potential of biomimetic, silver nanoparticles functionalized *Antheraea mylitta* (tasar) silk fibroin nanofibrous mats for skin tissue engineering, International Journal of Biological Macromolecules 130 (2019) 437–453. Available from https://doi.org/10.1016/j.ijbiomac.2018.12.255.

[15] A. Keirouz, M. Zakharova, J. Kwon, C. Robert, V. Koutsos, A. Callanan, et al., High-throughput production of silk fibroin-based electrospun fibers as biomaterial for skin tissue engineering applications, Materials Science and Engineering C 112 (2020) 110939. Available from https://doi.org/10.1016/j.msec.2020.110939.

[16] H. Khalid, H. Iqbal, R. Zeeshan, M. Nasir, F. Sharif, M. Akram, et al., Silk fibroin/collagen 3D scaffolds loaded with TiO2 nanoparticles for skin tissue regeneration, Polymer Bulletin 78 (2020) 7199–7218. Available from https://doi.org/10.1007/s00289-020-03475-y.

[17] N. Varshney, A.K. Sahi, S. Poddar, S.K. Mahto, Soy protein isolate supplemented silk fibroin nanofibers for skin tissue regeneration: fabrication and characterization, International Journal of Biological Macromolecules 160 (2020) 112–127. Available from https://doi.org/10.1016/j.ijbiomac.2020.05.090.

[18] A. Fathi, M. Khanmohammadi, A. Goodarzi, L. Foroutani, Z.T. Mobarakeh, J. Saremi, et al., Fabrication of chitosan-polyvinyl alcohol and silk electrospun fiber seeded with differentiated keratinocyte for skin tissue regeneration in animal wound model, Journal of Biological Engineering 14 (2020). Available from https://doi.org/10.1186/s13036-020-00249-y.

[19] J. Yin, Y. Fang, L. Xu, A. Ahmed, High-throughput fabrication of silk fibroin/hydroxypropyl methylcellulose (SF/HPMC) nanofibrous scaffolds for skin tissue engineering, International Journal of Biological Macromolecules 183 (2021) 1210–1221. Available from https://doi.org/10.1016/j.ijbiomac.2021.05.026.

[20] L. Wang, Z. Chen, Y. Yan, C. He, X. Li, Fabrication of injectable hydrogels from silk fibroin and angiogenic peptides for vascular growth and tissue regeneration, Chemical Engineering Journal 418 (2021) 129308. Available from https://doi.org/10.1016/j.cej.2021.129308.

[21] R. Liang, X. Shen, C. Xie, Y. Gu, J. Li, H. Wu, et al., Silk gel recruits specific cell populations for scarless skin regeneration, Applied Materials Today 23 (2021) 101004. Available from https://doi.org/10.1016/j.apmt.2021.101004.

[22] S.S. Shera, R.M. Banik, Development of tunable silk fibroin/xanthan biopolymeric scaffold for skin tissue engineering using L929 fibroblast cells, Journal of Bionic Engineering 18 (2021) 103–117. Available from https://doi.org/10.1007/s42235-021-0004-4.

[23] S. Gupta, P. Dutta, V. Acharya, P. Prasad, A. Roy, A. Bit, Accelerating skin barrier repair using novel bioactive magnesium-doped nanofibers of non-mulberry silk fibroin during wound healing, Journal of Bioactive and Compatible Polymers 37 (2021) 38–52. Available from https://doi.org/10.1177/08839115211061737.

[24] P. Farshi, R. Salarian, M. Rabiee, S. Alizadeh, M. Gholipourmalekabadi, S. Ahmadi, et al., Design, preparation, and characterization of silk fibroin/carboxymethyl cellulose wound dressing for skin tissue regeneration applications, Polymer Engineering and Science 62 (2022) 2741–2749. Available from https://doi.org/10.1002/pen.26057.

[25] P.K. Byram, M. Mukherjee, M. Rahaman, H. Bora, M. Kaushal, S. Dhara, et al., Bioactive self-assembling silk fibroin–sericin films for skin tissue engineering, Biomedical Materials 19 (2024) 025009. Available from https://doi.org/10.1088/1748-605x/ad1c9d.

[26] R. Fan, J. Zhao, L. Yi, J. Yuan, A. McCarthy, B. Li, et al., Anti-inflammatory peptide-conjugated silk fibroin/cryogel hybrid dual fiber scaffold with hierarchical structure promotes healing of chronic wounds, Advanced Materials 36 (2024). Available from https://doi.org/10.1002/adma.202307328.

[27] S. Zhou, Q. Wang, W. Yang, L. Wang, J. Wang, R. You, et al., Development of a bioactive silk fibroin bilayer scaffold for wound healing and scar inhibition, International Journal of Biological Macromolecules 255 (2023) 128350. Available from https://doi.org/10.1016/j.ijbiomac.2023.128350.

[28] S. Hazra, S. Nandi, D. Naskar, R. Guha, S. Chowdhury, N. Pradhan, et al., Non-mulberry silk fibroin biomaterial for corneal regeneration, Scientific Reports 6 (2016). Available from https://doi.org/10.1038/srep21840.

[29] W. Abdel-Naby, B. Cole, A. Liu, J. Liu, P. Wan, V.H. Guaiquil, et al., Silk-derived protein enhances corneal epithelial migration, adhesion, and proliferation, Investigative Ophthalmology & Visual Science 58 (2017) 1425. Available from https://doi.org/10.1167/iovs.16-19957.

[30] C.E. Kim, J.H. Lee, Y.K. Yeon, C.H. Park, J. Yang, Effects of silk fibroin in murine dry eye, Scientific Reports 7 (2017). Available from https://doi.org/10.1038/srep44364.

[31] B. Pourjabbar, E. Biazar, S.H. Keshel, A. Baradaran-Rafii, Improving the properties of fish skin collagen/silk fibroin dressing by chemical treatment for corneal wound healing, International Wound Journal 20 (2022) 484–498. Available from https://doi.org/10.1111/iwj.13896.

[32] W. Lan, W. Geng, X. Jiang, F. Chen, M. Zhou, G. Shen, et al., Transgenic silkworm expressing bioactive human ciliary neurotrophic factor for biomedical application, Insect Science (2024). Available from https://doi.org/10.1111/1744-7917.13442.

[33] H.K. Polat, E. Aytekin, N.F. Karakuyu, Y.A. Çaylı, S. Çalamak, N. Demirci, et al., Harnessing silk fibroin microparticles for metformin delivery: a novel approach to treating corneal neovascularization, Journal of Drug Delivery Science and Technology 96 (2024) 105625. Available from https://doi.org/10.1016/j.jddst.2024.105625.

[34] W. Zhang, X. Wang, S. Wang, J. Zhao, L. Xu, C. Zhu, et al., The use of injectable sonication-induced silk hydrogel for VEGF165 and BMP-2 delivery for elevation of the maxillary sinus floor, Biomaterials 32 (2011) 9415–9424. Available from https://doi.org/10.1016/j.biomaterials.2011.08.047.

[35] J.-W. Yang, Y.-F. Zhang, Z.-Y. Sun, G.-T. Song, Z. Chen, Dental pulp tissue engineering with bFGF-incorporated silk fibroin scaffolds, Journal of Biomaterials Applications 30 (2015) 221–229. Available from https://doi.org/10.1177/0885328215577296.

[36] S. Jiang, Z. Yu, L. Zhang, G. Wang, X. Dai, X. Lian, et al., Effects of different aperture-sized type I collagen/silk fibroin scaffolds on the proliferation and differentiation of human dental pulp cells, Regenerative Biomaterials 8 (2021). Available from https://doi.org/10.1093/rb/rbab028.

[37] S. López-García, S.D. Aznar-Cervantes, A. Pagán, C. Llena, L. Forner, J.L. Sanz, et al., 3D Graphene/silk fibroin scaffolds enhance dental pulp stem cell osteo/odontogenic differentiation, Dental Materials 40 (2023) 431–440. Available from https://doi.org/10.1016/j.dental.2023.12.009.

[38] J. Kukkila, P.K. Ramesh, K.H.K. Shetty, N.B. Thopegowda, R. Malayathodi, S. Rao, et al., Fabrication of hair and spider silk fiber-reinforced polymethyl methacrylate composites and evaluation of their mechanical properties for dental applications, World Journal of Dentistry 15 (2024) 372–376. Available from https://doi.org/10.5005/jp-journals-10015-2439.

[39] H. Liu, G.W. Xu, Y.F. Wang, H.S. Zhao, S. Xiong, Y. Wu, et al., Composite scaffolds of nano-hydroxyapatite and silk fibroin enhance mesenchymal stem cell-based bone regeneration via the interleukin 1 alpha autocrine/paracrine signaling loop, Biomaterials 49 (2015) 103–112. Available from https://doi.org/10.1016/j.biomaterials.2015.01.017.

[40] X. Ding, G. Yang, W. Zhang, G. Li, S. Lin, D.L. Kaplan, et al., Increased stem cells delivered using a silk gel/scaffold complex for enhanced bone regeneration, Scientific Reports 7 (2017). Available from https://doi.org/10.1038/s41598-017-02053-z.

[41] S. Aliramaji, A. Zamanian, M. Mozafari, Super-paramagnetic responsive silk fibroin/chitosan/magnetite scaffolds with tunable pore structures for bone tissue engineering applications, Materials Science and Engineering C 70 (2016) 736–744. Available from https://doi.org/10.1016/j.msec.2016.09.039.

[42] L. Wei, S. Wu, M. Kuss, X. Jiang, R. Sun, P. Reid, et al., 3D printing of silk fibroin-based hybrid scaffold treated with platelet rich plasma for bone tissue engineering, Bioactive Materials 4 (2019) 256–260. Available from https://doi.org/10.1016/j.bioactmat.2019.09.001.

[43] Z. Wu, Z. Meng, Q. Wu, D. Zeng, Z. Guo, J. Yao, et al., Biomimetic and osteogenic 3D silk fibroin composite scaffolds with nano MgO and mineralized hydroxyapatite for bone regeneration, Journal of Tissue Engineering 11 (2020). Available from https://doi.org/10.1177/2041731420967791.

[44] L.-M. Yu, T. Liu, Y.-L. Ma, F. Zhang, Y.-C. Huang, Z.-H. Fan, Fabrication of silk-hyaluronan composite as a potential scaffold for tissue repair, Frontiers in Bioengineering and Biotechnology 8 (2020). Available from https://doi.org/10.3389/fbioe.2020.578988.

[45] W. Lee, J.H. Choi, S. Lee, J.E. Song, G. Khang, Fabrication and characterization of silk fibroin microfiber-incorporated bone marrow stem cell spheroids to promote cell–cell interaction and osteogenesis, ACS Omega 5 (2020) 18021–18027. Available from https://doi.org/10.1021/acsomega.0c01415.

[46] S. Sangkert, S. Kamolmatyakul, M. Gelinsky, J. Meesane, 3D printed scaffolds of alginate/polyvinylalcohol with silk fibroin based on mimicked extracellular matrix for bone tissue engineering in maxillofacial surgery, Materials Today Communications 26 (2021) 102140. Available from https://doi.org/10.1016/j.mtcomm.2021.102140.

[47] L. Xiao, M. Wu, F. Yan, Y. Xie, Z. Liu, H. Huang, et al., A radial 3D polycaprolactone nanofiber scaffold modified by biomineralization and silk fibroin coating promote bone regeneration in vivo, International Journal of Biological Macromolecules 172 (2021) 19–29. Available from https://doi.org/10.1016/j.ijbiomac.2021.01.036.

[48] Y. Zhou, X. Liu, H. She, R. Wang, F. Bai, B. Xiang, A silk fibroin/chitosan/nanohydroxyapatite biomimetic bone scaffold combined with autologous concentrated growth factor promotes the proliferation and osteogenic differentiation of BMSCs and repair of critical bone defects, Regenerative Therapy 21 (2022) 307–321. Available from https://doi.org/10.1016/j.reth.2022.08.006.

[49] Z. Mao, X. Bi, C. Yu, L. Chen, J. Shen, Y. Huang, et al., Mechanically robust and personalized silk fibroin-magnesium composite scaffolds with water-responsive shape-memory for irregular bone regeneration, Nature Communications 15 (2024). Available from https://doi.org/10.1038/s41467-024-48417-8.

[50] Y. Zhu, H. Gu, J. Yang, A. Li, L. Hou, M. Zhou, et al., An injectable silk-based hydrogel as a novel biomineralization seedbed for critical-sized bone defect regeneration, Bioactive Materials 35 (2024) 274–290. Available from https://doi.org/10.1016/j.bioactmat.2024.01.024.

[51] X. Li, A.H. Hirad, A.A. Alarfaj, H. Li, R. Santhanam, A convergent fabrication of graphene oxide/silk fibroin/hydroxyapatite nanocomposites delivery improved early osteoblast cell adhesion and bone regeneration, Arabian Journal of Chemistry 17 (2023) 105468. Available from https://doi.org/10.1016/j.arabjc.2023.105468.

[52] E. Spessot, S. Passuello, L.V. Shah, D. Maniglio, A. Motta, Nanocomposite methacrylated silk fibroin-based scaffolds for bone tissue engineering, Biomimetics 9 (2024) 218. Available from https://doi.org/10.3390/biomimetics9040218.

[53] Y. Cetin, M.G. Sahin, F.N. Kok, Application potential of three-dimensional silk fibroin scaffold using mesenchymal stem cells for cardiac regeneration, Journal of Biomaterials Applications 36 (2021) 740–753. Available from https://doi.org/10.1177/08853282211018529.

[54] Y. Feng, G. Zhao, M. Xu, X. Xing, L. Yang, Y. Ma, et al., RGO/silk fibroin-modified nanofibrous patches prevent ventricular remodeling via YAP/TAZ-TGFB1/SMADS signaling after myocardial infarction in rats, Frontiers in Cardiovascular Medicine 8 (2021). Available from https://doi.org/10.3389/fcvm.2021.718055.

[55] Y. Tufan, H. Öztatlı, D. Doganay, A. Buyuksungur, M.O. Cicek, İ.T. Döş, et al., Multifunctional silk fibroin/carbon nanofiber scaffolds for in vitro cardiomyogenic differentiation of induced pluripotent stem cells and energy harvesting from simulated cardiac motion, ACS Applied Materials & Interfaces 15 (2023) 42271–42283. Available from https://doi.org/10.1021/acsami.3c08601.

[56] F. Millesi, T. Weiss, A. Mann, M. Haertinger, L. Semmler, P. Supper, et al., Defining the regenerative effects of native spider silk fibers on primary Schwann cells, sensory neurons, and nerve-associated fibroblasts, The FASEB Journal 35 (2020). Available from https://doi.org/10.1096/fj.202001447r.

[57] Y. Gao, Y. Wang, J. Zhang, M. Zhang, C. Dai, Y. Zhang, et al., Advancing neural regeneration via adaptable hydrogels: enriched with Mg2+ and silk fibroin to facilitate endogenous cell infiltration and macrophage polarization, Bioactive Materials 33 (2023) 100–113. Available from https://doi.org/10.1016/j.bioactmat.2023.10.026.

[58] L. Xu, S. Wang, X. Sui, Y. Wang, Y. Su, L. Huang, et al., Mesenchymal stem cell-seeded regenerated silk fibroin complex matrices for liver regeneration in an animal model of acute liver failure, ACS Applied Materials & Interfaces 9 (2017) 14716–14723. Available from https://doi.org/10.1021/acsami.7b02805.

[59] G. Janani, S.K. Nandi, B.B. Mandal, Functional hepatocyte clusters on bioactive blend silk matrices towards generating bioartificial liver constructs, Acta Biomaterialia 67 (2017) 167–182. Available from https://doi.org/10.1016/j.actbio.2017.11.053.

[60] L. Guo, Z. Zhu, C. Gao, K. Chen, S. Lu, H. Yan, et al., Development of biomimetic Hepatic Lobule-Like constructs on silk-collagen composite scaffolds for liver tissue engineering, Frontiers in Bioengineering and Biotechnology 10 (2022). Available from https://doi.org/10.3389/fbioe.2022.940634.

[61] C. Acharya, S.K. Ghosh, S.C. Kundu, Silk fibroin protein from mulberry and non-mulberry silkworms: cytotoxicity, biocompatibility and kinetics of L929 murine fibroblast adhesion, Journal of Materials Science: Materials in Medicine 19 (2008) 2827–2836. Available from https://doi.org/10.1007/s10856-008-3408-3.

[62] B.B. Mandal, A. Grinberg, E.S. Gil, B. Panilaitis, D.L. Kaplan, High-strength silk protein scaffolds for bone repair, Proceedings of the National Academy of Sciences 109 (2012) 7699–7704. Available from https://doi.org/10.1073/pnas.1119474109.

[63] Q. Zhang, S. Yan, M. Li, Silk fibroin based porous materials, Materials 2 (2009) 2276–2295. Available from https://doi.org/10.3390/ma2042276.

[64] G. Senti, L.S. Steinmann, B. Fischer, R. Kurmann, T. Storni, P. Johansen, et al., Antimicrobial silk clothing in the treatment of atopic dermatitis proves comparable to topical corticosteroid treatment, Dermatology 213 (2006) 228–233. Available from https://doi.org/10.1159/000095041.

[65] W. Luangbudnark, J. Viyoch, W. Laupattarakasem, P. Surakunprapha, P. Laupattarakasem, Properties and biocompatibility of chitosan and silk fibroin blend films for application in skin tissue engineering, The Scientific World Journal 2012 (2012) 1–10. Available from https://doi.org/10.1100/2012/697201.

[66] D. Chouhan, N. Thatikonda, L. Nilebäck, M. Widhe, M. Hedhammar, B.B. Mandal, Recombinant spider silk functionalized silkworm silk matrices as potential bioactive wound dressings and skin grafts, ACS Applied Materials & Interfaces 10 (2018) 23560–23572. Available from https://doi.org/10.1021/acsami.8b05853.

[67] K. Higa, J. Shimazaki, Recent advances in cultivated epithelial transplantation, Cornea 27 (2008) S41–S47. Available from https://doi.org/10.1097/ico.0b013e31817f358e.

[68] W.-P. Xu, W. Zhang, R. Asrican, H.-J. Kim, D.L. Kaplan, P.C. Yelick, Accurately shaped tooth bud cell–derived mineralized tissue formation on silk scaffolds, Tissue Engineering: Part A 14 (2008) 549–557. Available from https://doi.org/10.1089/tea.2007.0227.

[69] P. He, S. Sahoo, K.S. Ng, K. Chen, S.L. Toh, J.C.H. Goh, Enhanced osteoinductivity and osteoconductivity through hydroxyapatite coating of silk-based tissue-engineered ligament scaffold, Journal of Biomedical Materials Research: Part A 101A (2012) 555–566. Available from https://doi.org/10.1002/jbm.a.34333.

[70] J.-X. He, W.-L. Tan, Q.-M. Han, S.-Z. Cui, W. Shao, F. Sang, Fabrication of silk fibroin/cellulose whiskers–chitosan composite porous scaffolds by layer-by-layer assembly for application in bone tissue engineering, Journal of Materials Science 51 (2016) 4399–4410. Available from https://doi.org/10.1007/s10853-016-9752-7.

[71] S. Tandon, B. Kandasubramanian, S.M. Ibrahim, Silk-based composite scaffolds for tissue engineering applications, Industrial & Engineering Chemistry Research 59 (2020) 17593–17611. Available from https://doi.org/10.1021/acs.iecr.0c02195.

[72] Y. Song, C. Zhang, J. Zhang, N. Sun, K. Huang, H. Li, et al., An injectable silk sericin hydrogel promotes cardiac functional recovery after ischemic myocardial infarction, Acta Biomaterialia 41 (2016) 210–223. Available from https://doi.org/10.1016/j.actbio.2016.05.039.

[73] G. Zhao, Zhang, L., B. Xu, G. Huang, F. Xu, X. Zhang, Solvent-free fabrication of carbon nanotube/silk fibroin electrospun matrices for enhancing cardiomyocyte functionalities, ACS Biomaterials Science & Engineering 6 (2020) 1630–1640. Available from https://doi.org/10.1021/acsbiomaterials.9b01682.

[74] Y. Yang, X. Chen, F. Ding, P. Zhang, J. Liu, X. Gu, Biocompatibility evaluation of silk fibroin with peripheral nerve tissues and cells in vitro, Biomaterials 28 (2007) 1643–1652. Available from https://doi.org/10.1016/j.biomaterials.2006.12.004.

[75] X. Tang, F. Ding, Y. Yang, N. Hu, H. Wu, X. Gu, Evaluation on in vitro biocompatibility of silk fibroin-based biomaterials with primarily cultured hippocampal neurons, Journal of Biomedical Materials Research: Part A 91A (2008) 166–174. Available from https://doi.org/10.1002/jbm.a.32212.

[76] J.D. White, S. Wang, A.S. Weiss, D.L. Kaplan, Silk–tropoelastin protein films for nerve guidance, Acta Biomaterialia 14 (2014) 1–10. Available from https://doi.org/10.1016/j.actbio.2014.11.045.

[77] Y. Gotoh, Y. Ishizuka, T. Matsuura, S. Niimi, Spheroid formation and expression of liver-specific functions of human hepatocellular carcinoma-derived FLC-4 cells cultured in lactose–silk fibroin conjugate sponges, Biomacromolecules 12 (2011) 1532–1539. Available from https://doi.org/10.1021/bm101495c.

[78] I.F. Cengiz, H. Pereira, J. Espregueira-Mendes, I.K. Kwon, R.L. Reis, J.M. Oliveira, Suturable regenerated silk fibroin scaffold reinforced with 3D-printed polycaprolactone mesh: biomechanical performance and subcutaneous implantation, Journal of Materials Science: Materials in Medicine 30 (2019). Available from https://doi.org/10.1007/s10856-019-6265-3.

[79] M. Kazemimostaghim, R. Rajkhowa, K. Patil, T. Tsuzuki, X. Wang, Structure and characteristics of milled silk particles, Powder Technology 254 (2014) 488–493. Available from https://doi.org/10.1016/j.powtec.2014.01.060.

[80] B. Kundu, N.E. Kurland, V.K. Yadavalli, S.C. Kundu, Isolation and processing of silk proteins for biomedical applications, International Journal of Biological Macromolecules 70 (2014) 70–77. Available from https://doi.org/10.1016/j.ijbiomac.2014.06.022.

[81] J. Kaewchuchuen, N. Roamcharern, S. Phuagkhaopong, L.M. Bimbo, F.P. Seib, Microfibre-functionalised silk hydrogels, Cells 13 (2023) 10. Available from https://doi.org/10.3390/cells13010010.

[82] M.C. Bottino, V. Thomas, G. Schmidt, Y.K. Vohra, T.-M.G. Chu, M.J. Kowolik, et al., Recent advances in the development of GTR/GBR membranes for periodontal regeneration—a materials perspective, Dental Materials 28 (2012) 703–721. Available from https://doi.org/10.1016/j.dental.2012.04.022.

[83] G. Griffanti, M.D. McKee, S.N. Nazhat, Mineralization of bone extracellular matrix-like scaffolds fabricated as silk sericin-functionalized dense collagen–fibrin hybrid hydrogels, Pharmaceutics 15 (2023) 1087. Available from https://doi.org/10.3390/pharmaceutics15041087.

[84] S. Lu, P. Wang, F. Zhang, X. Zhou, B. Zuo, X. You, et al., A novel silk fibroin nanofibrous membrane for guided bone regeneration: a study in rat calvarial defects, PubMed 7 (2015) 2244–2253.

[85] O.-H. Kim, O.J. Yoon, H.J. Lee, Silk fibroin scaffolds potentiate immunomodulatory function of human mesenchymal stromal cells, Biochemical and Biophysical Research Communications 519 (2019) 323–329. Available from https://doi.org/10.1016/j.bbrc.2019.09.006.

Chapter 8

Bioinspired scaffolds as ideal platform in engineering artificial organs and tumor microenvironment: importance in regenerative and reproductive medicine

Somoshree Sengupta[1], Souvik Das[2], Tanushri Ghosh[1], Asima Mukhopadhay[1], and Biswarup Basu[2]
[1]*Translational and Clinical Trials Unit, Kolkata Gynecological Oncology Trials and Translational Research Group (KolGOTrg), Kolkata, West Bengal, India,*
[2]*Department of Neuroendocrinology and Experimental Hematology, Chittaranjan National Cancer Institute, Kolkata, West Bengal, India*

8.1 Introduction

A growing tumor requires the surrounding tissue microenvironment that directly plays a complex and dynamic role in shaping the heterogeneous tumor mass. This evolving ecosystem, known as the tumor microenvironment (TME), is crucial in cancer progression, metastasis, and response to therapy. It comprises various cellular and noncellular components that interact with cancer cells, influencing their behavior and survival. As oncological research progresses, addressing this complex molecular behavior of continuously evolving heterogeneous tumor mass requires accurate models to replicate the TME, which becomes essential for enabling a deeper understanding of cancer progression and the development of effective therapeutic strategies. To replicate key characteristics of the TME—such as porosity, surface topology, cell migration, nutrient exchange, and oxygen diffusion—there is a growing demand for artificial architectures that closely mimic the complexity of native tissues. Tissue engineering (TE) is a multidisciplinary field integrating biology, engineering, and medicine, providing prominent innovative approaches to address these challenges by designing artificial architectures that actively mimic native tissues. TE primarily relies on scaffolds—a three-dimensional (3D) structures that support cell growth, attachment, and differentiation by replicating the extracellular matrix (ECM) and its structural, biochemical, and mechanical properties. Scaffolds can be composed of natural materials (e.g., collagen, fibrin), synthetic polymers (e.g., polylactic acid, polyethylene glycol [PEG]), or hybrid materials. The design of scaffolds is crucial to ensure mechanical stability, biocompatibility, and appropriate degradation rates. Bioinspired scaffolds mainly contain materials that exactly replicate natural tissue characteristics, such as cell-adhesion ligands, biodegradable components, and 3D structures, making them ideal for applications in TE, regenerative medicine, as well as in cancer research by enabling advanced multidimensional *in vitro* models that mimic the TME and help to understand cancer progression Fig. 8.1. For example, in regenerative medicine, 3D scaffolds accelerate the natural healing processes by providing better cellular adhesion, proliferation, and differentiation, aiding tissue anchorage and integration, reducing risks like rejection and disease transmission compared to traditional grafts. In oncology, advanced scaffolds integrated with cell-adhesion ligands and biodegradable cross-linkers enhance the simulation of cancer niches, support cellular growth and angiogenesis, improve drug-delivery research, and help in understanding gene expression, metastasis, and resistance mechanisms, advancing targeted therapies for ovarian and breast cancer. Biomaterials such as collagen, hyaluronic acid, and fibrin are often employed in designing 3D scaffolds to mimic the ECM, while synthetic polymers like poly(lactic-co-glycolic acid) (PLGA) and PEG offer tunable mechanical and degradation properties.

The most common form of ovarian carcinoma is high-grade serous (HGS), which is typically diagnosed at an advanced stage (stage III or IV, 70% of cases) and shows inherently aggressive malignancy, causing the majority of ovarian cancer deaths due to resistance development towards chemotherapy. The TME of a growing ovarian tumor mass exerts several

Fundamentals of Translational Tissue Engineering. DOI: https://doi.org/10.1016/B978-0-443-33455-9.00009-2

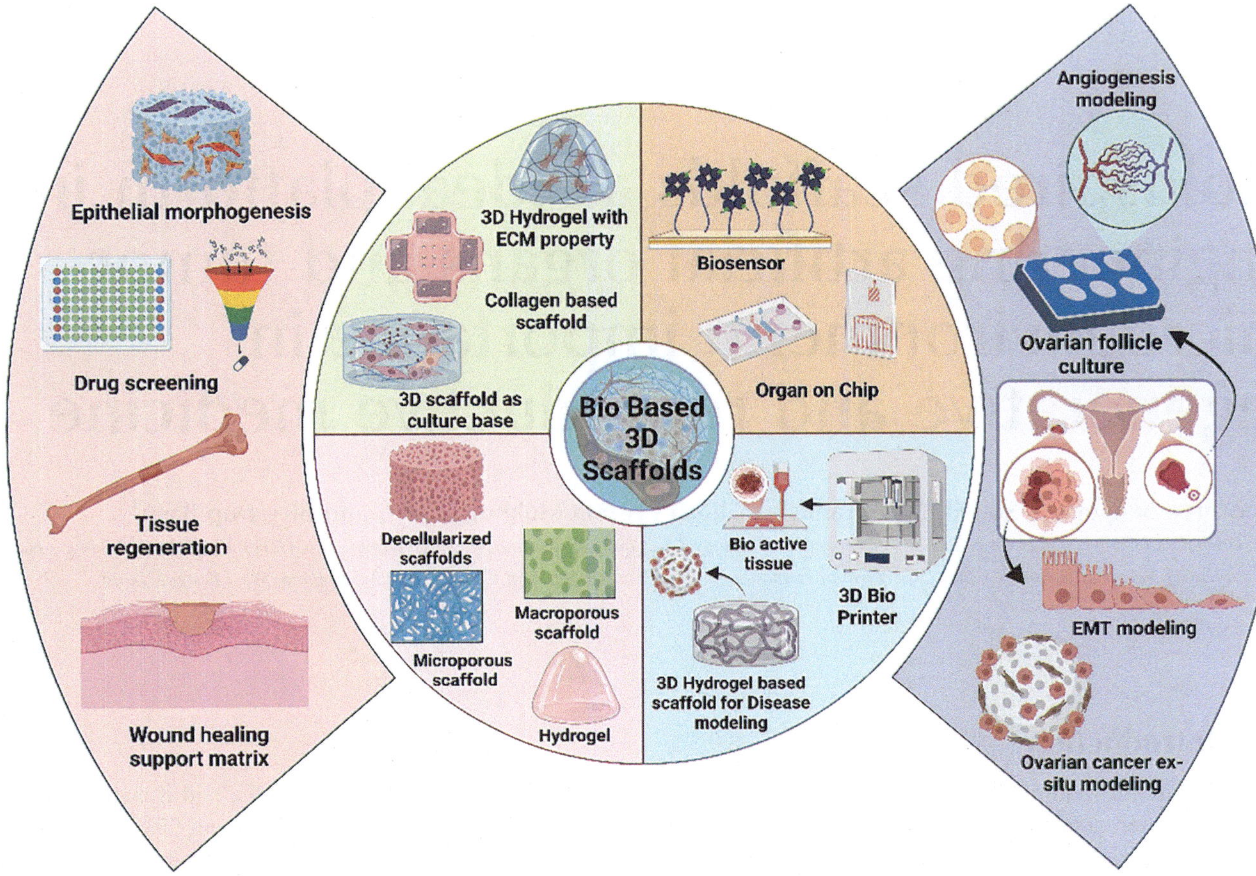

FIGURE 8.1 **Applications and classification of bio-based 3D scaffolds in biomedical research.** Various types of bio-based 3D scaffolds, including hydrogels, decellularized matrices, collagen-based structures, macroporous and microporous designs with their diverse biomedical applications in disease modeling, drug screening, and reproductive medicine, including ovarian tissue remodeling and cancer research, showcasing advancements of tissue engineering in personalized therapeutics.

physical forces like compression, tension, and shear, ultimately altering the ECM architecture. These changes further impact the molecular signaling, influencing cancer cell proliferation, migration, gene expression, and chemoresistance, ultimately affecting tumor cells' responses towards treatment [1]. Cancer cells often secrete vascular endothelial growth factor (VEGF) to promote blood vessel formation around them [2]. The process of tumorigenesis and its mechanical constraints and interaction between cells, and ECM in epithelial ovarian cancer are being well studied and have become impactful in designing more efficient bio-based scaffolds that can appropriately mimic the natural tumor tissue environment. Enhanced scaffolds with cell-adhesion ligands and cross-linkers can better mimic the ovarian cancer niche [3]. For example, integrin adhesion, tissue architecture, stiffness-regulated growth, and apoptosis play a vital role in controlling cell–ECM interactions. Progress in modifying the physical and chemical properties of 3D scaffolds has significantly improved their ability to integrate with tissues, achieving remarkable success in regenerative medicine. Biodegradable materials like PEG, PLA, PGA, PLGA, and PCL are commonly used for scaffold generation but struggle to replicate the TME. Local tissue growth relies on the dynamic interaction between cells and the ECM, where the remodeled ECM responds to the scaffold's properties, like elasticity and structure. By developing clinically relevant 3D scaffold systems, TE not only supports cancer research but also develops new drug delivery methods [4]. Since 1970, 3D scaffold-based cancer research (3DCR) has revealed the advantages like self-aggregation and molecular interaction with ECM [5]. Therefore biomaterial-based scaffold application has an important aspect in mimicking the cell-ECM interaction which helps in enlighten the following parameters i.e., hydration, stiffness and bifunctionality [6]. The biomaterial-based scaffold will provide structural support for cell attachment. Lastly, the chemical and biophysical signals help in crosstalk and recreating the tissues [7]. The 3D scaffold-based research in ovarian cancer shows a fivefold enhancement of drug resistance (paclitaxel, curcumin, 5-fluorouracil) compared to 2D monolayers. Therefore 3D scaffold-based trials for evaluating the possible therapeutics for ovarian cancer can be a good representation to simulate the in vivo characteristics of ovarian cancer cells.

8.2 Properties of tumor microenvironment to consider for applied tissue engineering research

TME has a complex structure of abundant blood vessels, ECM, and supporting cells dependent upon a few things like the stiffness of ECM, the pressure of the growing tumor mass, secreted cytokines, growth factors, O_2 level inside the mass, pH condition (acidic or basic), nutrient concentration, etc. So TME exhibits several key properties that are critical for TE applications in cancer research. These include biochemical cues, such as growth factors (VEGF, TGF-β) and cytokines that regulate cell behavior, as well as biophysical characteristics like ECM stiffness, porosity, and surface topology, which influence cell adhesion, migration, and differentiation. Additionally, hypoxia and nutrient gradients within the TME drive metabolic reprogramming, affecting tumor progression and drug response. The presence of immune cells, fibroblasts, and vascular networks further modulates tumor growth and resistance mechanisms. Altogether, these characteristics will determine whether cancer cells grow or migrate and further contribute to cancer development and metastasis. To create effective tumor models, engineered scaffolds must replicate these dynamic cellular interactions, biomechanical forces, and molecular signaling pathways, ensuring a physiologically relevant microenvironment for studying cancer progression and therapeutic responses in a 3D environment, which makes the experimental platform completely different from 2D culture.

8.2.1 Extracellular matrix properties

Proteins like fibronectin, collagen, laminin, glycoprotein, hyaluronic acid (HA), chondroitin sulfate, heparan sulfate have been the predominant building blocks of ECM [8]. The 30% of the total protein content of ECM, that is, collagen, mainly deals with the tensile strength and cellular adhesion of ECM. Whereas the fibronectin (FN), mainly helps in wound healing, fibrosis and tumor progression as well. It connects to the cells through integrins and other receptors, helping in cell adhesion, migration, and differentiation [9]. Laminin is a glycoprotein that promotes cell adhesion and differentiation, particularly in the basal lamina of epithelial cells. Similarly, chondroitin sulfate is a different class of glycosaminoglycan that contributes to the elasticity and hydration of cartilage and connective tissues. Heparan sulfate, a glycosaminoglycan, plays a role in cell signaling, angiogenesis, and the regulation of growth factors. These components collectively form a dynamic microenvironment that influences tissue development, repair, and cellular responses to stimuli.

8.2.2 Cancer-associated fibroblast and immune cells

Cancer-associated fibroblasts (CAFs) play a key role in tumor progression by transforming the ECM into a fibrillar structure and producing collagens (types I–IV), fibronectin, and laminin. They secrete proteases like matrix metalloproteinases (MMPs) and ADAMTSs (a disintegrin and metalloproteinase with thrombospondin motifs), which modify the ECM to promote cancer growth. CAFs are activated by signals like TGF-β, PDGF, and proteases, remodeling the TME and supplying growth factors that support tumor development [10]. In cancer models, tumor-associated macrophages (TAMs) and regulatory T cells (Tregs) contribute to immune suppression and tumor progression, making their incorporation essential for mimicking the TME. Engineered scaffolds must be designed to modulate immune responses, ensuring biocompatibility while replicating immune-cancer cell interactions, cytokine signaling, and inflammation dynamics. By integrating immune components, advanced TE models can better simulate in vivo conditions, improving cancer research, drug testing, and regenerative therapies.

8.2.3 O_2 and pH

Cellular growth is significantly influenced by two key factors, that is, pH and oxygen (O_2) concentration. These factors play a crucial role in the development and migration of cancer cells. Within the TME, as the distance from the blood vessel increases, the O_2 concentration drops below 2%, a condition known as hypoxia. This triggers angiogenesis and enhances the metastatic potential of cancer cells. In severe hypoxic conditions (1.5% O_2), the transcription factor hypoxia-inducible factor (HIF)-1α rapidly accumulates in the cells. Furthermore, hypoxia affects various factors such as proangiogenesis and ECM modification. Under hypoxic conditions, cancer cells activate glycolysis to maintain high levels of adenosine triphosphate (ATP), which leads to a decrease in pH, causing an acidic environment supporting further tumor aggressiveness (pH 5.6–6.8) [11]. 3D scaffolds designed to mimic these gradients, enabling realistic cancer models to study hypoxia-induced gene expression, metabolic reprogramming, and therapeutic resistance.

8.2.4 Tissue stiffness

The ECM, mainly composed of collagen type I, influences ECM production and interactions with CAFs and cancer cells. It has been found that increased LOX and PLOD2 expression promote tumor progression [12]. In 3D matrices with 400 Pa stiffness, cancer cells show increased size, α5β1-integrin clustering, FAK phosphorylation, RhoA activation, and cytoskeleton contractility. ECM stiffness enhances cancer cell migration, particularly on polyacrylamide pillars, and supports cell differentiation. In denser collagen gels, stronger cell–cell junctions lead to higher migration. Stiff ECM also aids epithelial cancer cell differentiation and nuclear translocation of TWIST1 via the mechano-transduction pathway [8].

8.3 Popular biomaterial and fabrication techniques for designing bio-mimicking scaffold

Achieving consistency between in vitro and in vivo models is a key challenge in cancer translational research. Advanced 3D and 4D culture systems are crucial for replicating the TME and understanding tissue complexity. Several fabrication techniques have been developed to address these needs.

Fabrication techniques for scaffolding in TE include electrospinning, 3D bioprinting, freeze-drying, solvent casting, and hydrogel formation, enabling precise scaffold architecture, advancing studies on cancer behavior and drug resistance. Electrospinning refers to a technique to create nanofibrous scaffolds using an electric field to draw polymer solutions into fine fibers that create nanofiber scaffolds with high surface area. 3D bioprinting demonstrates a layer-by-layer deposition process to create complex tissue structures using bioinks containing cells and biomaterials applicable in customizable designs for complex tissue structures. Freeze-drying involves a drying method that removes water via sublimation, preserving the structure and porosity of biomaterials. Solvent casting refers to A process where a polymer solution is cast into a mold, followed by solvent evaporation to form a thin film or porous scaffold. Important for creating porous scaffolds with good structural integrity, often for bone or cartilage applications. Among them, hydrogel is the most widely used technique, where synthesis of hydrophilic polymer networks that can retain large amounts of water, is often used for drug delivery and TE. Hydrogel formation is ideal for soft tissues, providing scaffolds with high water content that mimic natural ECM. These methods offer various advantages, depending on the material and application, and are crucial for advancing regenerative medicine and cancer research. The most commonly used biomaterials that are used to generate bio-based scaffolds are discussed below:

8.3.1 Hydrogel

It is a water-based material that is made up of a 3D network of crosslinked polymeric chains that can absorb and retain water molecules. The properties of the polymers used, as well as the density and nature of the network joints, determine how much water a hydrogel can contain. The 3D gels of hydrophilic polymer-based hydrogel used in cancer research, TE, drug delivery, and wound healing have their ability to mimic the ECM and to recreate TME [13]. Bio or synthetic polymers like proteins and natural molecules are the building blocks of hydrogels. Natural hydrogels like collagen, gelatin, hyaluronic acid, and alginate offer excellent bioactivity, promoting cell adhesion and proliferation, while synthetic hydrogels like PEG and polyvinyl alcohol (PVA) provide mechanical strength and tenable properties. Collagen commonly used to achieve the best body compatibility and to identify and analyze cancer cell proliferation [14]. Smart hydrogels, responsive to stimuli like pH and temperature, and nanocomposite hydrogels, enhanced with nanoparticles, further expand their applications by improving the mimicry of native tissues and supporting advanced drug delivery systems. The hydrophilic nature of the hydrogel aids in drug testing, resulting in a wide range of biomedical applications [15]. Hydrogel can be found in various forms, including membranes, microparticles, solid molded shapes, and even liquid forms [13]. The hydrophilic nature of polymers in hydrogel helps in the formation of different forms, that is, membranes, microparticles, solid molded shapes, and liquid forms. Natural or synthetic source–derived hydrogel shows mechanical and porous characteristic variation by using differences in temperature, pH, and ionic strength at the time of synthesis [16,17]. The various classified hydrogels and their applications are listed in Table 8.1:

TABLE 8.1 Different sources of hydrogel materials, their sources, and applications.

Natural hydrogel (animal based)	Row source	Characteristics	Applications
1. Alginate [18]	Brown seaweed	Downregulate toxicity, biodegradable, and cross-linking properties.	Cell-laden bioinks in 3D bioprinting
2. Gelatin [19]	Fish, cattle, bone, skin, pig	Biocompatible, biodegradable, low-toxic, low mechanical stability	Bone tissue engineering
3. Hyaluronic acid [20]	Human and bacterial fermentation	Immunoneutral, polysaccharide, nonadhesive	Dynamic cell patterning
4. Silk fibroin [21]	Silkworm (Bombyx mori)	High strength and stiffness, biocompatible, high cell affinity, and rapid biodegradability. weak mechanical strength	Fibrin-hybrid constructs are useful for cartilage, cardiac, smooth muscle, cultures, and drug delivery systems
5. Fibrin [22]	Human plasma	Biocompatible, high cell affinity, rapid durability, weak mechanical strength	Fibrin-hybrid construct useful for cartilage, cardiac, smooth muscle cultures, and drug delivery systems
Natural (plant based)			
1. Plant–lignocellulose materials based			
a. Cellulose [23]	Plant cell and natural fibers	Highly crystalline, low water solubility, and slow biodegradability in vivo	Application in biomedical
b. Hemicellulose [24]	Plant cell wall	Biocompatible Nontoxic Biodegradable	Application in biomedical
c. Lignin [25]	Plant cell wall	Biodegradable Hydrophobic Antioxidant Antimicrobial High global availability	Drug delivery and tissue engineering
2. Plant–polysaccharide based			
a. Starch [26]	Chloroplast and amyloplast of the plant	Hydrophilic, biocompatible	Tissue engineering, drug delivery, agricultural usage, water treatment, food industry
b. Pectin [26]	Fruit peels and pulp, sugar beet, sunflower heads	High gelling capacity hydrophilic	Drug delivery and tissue engineering
3. Plant—gum based			
a. Natural gums [27]	Plant exudation, seed endosperm, tree exudation	Excellent biocompatibility and biodegradability Mimic ECM Renewable Edible	Tissue engineering, drug delivery, can be used for oral delivery platform
4. Plant protein based			
a. Soy [28]	Soyabean	Biocompatible Bioreactive Biodegradable High water holding capacity	Tissue engineering, drug delivery, regenerative medicine

(Continued)

TABLE 8.1 (Continued)

Natural hydrogel (animal based)	Row source	Characteristics	Applications
b. Zein [29]	Corn kernels	Hydrophobic	Tissue engineering, drug delivery, regenerative medicine
5. Synthetic			
a. Polyethylene glycol [30]	By-products of petroleum	Compositionally consistent Nontoxic Hydrophilic	In combination with a natural polymer suitable as a scaffold
b. Polyvinyl alcohol [31]	Hydrolysis of polyvinyl acetate	Lack biologically active sites	In combination with a natural polymer suitable as a scaffold

ECM, Extracellular matrix.

8.3.2 Bioink

Bioink is a blend of living cells and biomaterials that imitates the ECM environment. It can contain natural polymers like gelatin, collagen, and hyaluronic acid, or synthetic polymers like polycaprolactone (PCL), PEG, and polyvinylpyrrolidone (PVP). The main purpose of blending cells with natural/synthetic polymer is to give a surface/platform for cells to attach, proliferate, and differentiate, to cross-link for developing the desired product, and ultimately to prevent cell damage during printing. Table 8.2 shows below various bioink applications in cancer research.

8.4 Physicochemical properties of 3D scaffold to enhance cellular adhesion, interaction, and overcoming systemic toxicity

The interaction of cells with the surrounding 3D TME is the primary arena for the pathogenesis of diseases like fibrosis and cancer. The term "matrisome" refers to the interaction of molecules that build the ECM during disease progression. These concepts of 3D scaffolding extend the comprehensiveness beyond traditional studies based on collagen, elastin, proteoglycan, and fibronectin, along with matrix-associated enzymes, inhibitors, growth factors, and, in some cases, receptors for ECM molecules. Multifunctional biomaterial-based Drug Delivery Systems (DDS) improve the effectiveness and precision of cancer immunotherapy. The instant drug delivery improves the efficiency of immunotherapeutic cargo, controls their spatiotemporal release, and improves targeting and accumulation of the drug at the tumor site, while reducing off-target effects and systemic toxicity. Overcoming challenges in immunotherapy, like systemic toxicity and less accumulation of delivery agents like nanoparticles at the tumor site with traditional bolus administration, hinders drug delivery biomaterials from achieving satisfactory results. 3D scaffold-based biomaterials show excellent biocompatibility and environmental responsiveness; they feature macroscale volume, 3D porous structures, swelling properties, flexibility, and elasticity over nano and micro biomaterials. These scaffold-based regional therapies actually minimize the systemic toxicity and are suitable for regional therapies as well. Their adaptive shapes conform better to irregular lesions, allowing localized drug delivery that enhances immunotherapy efficacy at lower dosages. Furthermore, these 3D biomaterials often mimic physiological matrices, serving as artificial immune tissues to recruit, house, and program host immune cells or as incubation depots for engineered immune cells. Currently, 3D scaffold biomaterials are used in various fields, including tumor immunotherapy, TE, bone-related diseases, and infectious diseases.

8.4.1 Surface properties

Surface chemistry and properties of scaffolds enhance cell attachment, proliferation, and differentiation. Research indicates that within textured surfaces, variations in cellular behavior depend solely on texture size. Furthermore, the interplay between surface characteristics and cellular dynamics extends beyond mere texture; the chemical composition and functionalization of scaffolds play a pivotal role in modulating cellular responses. Specific surface coatings can be engineered to promote better adsorption of bioactive molecules, thereby fostering a conducive microenvironment in the case of osteogenic differentiation. For instance, the incorporation of hydrophilic or hydrophobic elements can influence protein adsorption patterns, subsequently affecting cell adhesion and morphology. Studies demonstrate that a harmonious

TABLE 8.2 Different sources of BioInk materials and their application in cancer research.

Name of the bioink	Bioprinting technology	3D model	Types of action	References
Protein	Extrusion-based printing	3D model of ovarian cancer	Cell response to different chemotherapies	[32]
Polysaccharides	Laser-assisted bioprinting	Triple-negative breast cancer and lung carcinoma	High-throughput drug screening	[33]
ECM	Robotic dispensing	Human lung	Generating human functional tissue	[34]
Human skin-derived natural bioink	Robotic dispensing	Human skin	Bone formation	[35]
Thermo-responsive hydrogel (NOVOGEL)	Custom-made inkjet bioprinter Inkjet printing	AORTA	Tissue regeneration	[36]
Antibody drug conjugate-based Bioink (Kite, a Gilead company and Oxford BioTherapeutics Ltd.)	Bioprinter with a temperature control unit and a UV light source	Solid tumor	Cell therapy for the treatment of blood cancer and solid tumors	[37]
Biomolecule-loaded bioink	Extrusion-based bioprinting	Complex cell-laden construct	In vitro research for clinical applications	[38]
Alginate-based natural polysaccharide (brown algae)	Drop-based filament-based	Vascular, cartilage, bone, neural tissue, fibroblast, and many more	Drug delivery, cancer stem cell research, breast cancer, melanoma, and many more cancers, tumor spheroids	[39]
Gelatin-based natural protein	Drop-based Filament-based Plane-based	Vascular, cartilage, bone, muscle, fibroblast, and many more	Cholangiocarcinoma, bladder cancer, and many more cancers, tumor spheroids	[40]
Natural polysaccharides obtained from the biosynthesis of plants or bacteria (cellulose and hemicellulose)	Filament-based	Cartilage and bone	Drug delivery gastric, cervical, pancreatic, and many more cancers	[41]

ECM, Extracellular matrix.

balance between mechanical strength and surface topography (stiffness and elasticity) is essential, as it significantly impacts cellular behavior and tissue integration. Overly rigid scaffolds may hinder cellular infiltration, while softer matrices could fail to sustain adequate mechanical support for newly formed tissues. Creating vascular networks within scaffolds by incorporating angiogenic factors, or designing porous structures that promote blood vessel infiltration, are being actively researched. Biological response to scaffold is further influenced by the incorporation of growth factors and signaling molecules and their controlled release to enhance the regenerative process. For example, scaffolds containing osteogenic growth factors (BMPs and FGFs) have been shown to stimulate the early phases of bone healing, thereby accelerating the overall repair process.

8.4.2 Physical properties

3D scaffold must maintain appropriate porosity and pore structure to support cell growth, and also be essential for enabling progenitor and adult cells to populate the entire matrix. Therefore, considering the scaffold's structural integrity, pore size design is crucial. The mechanical properties of the scaffold provide sufficient stiffness to withstand physiological loads but also possess the flexibility to accommodate cellular activities. This balance ensures that the mechanical signals can effectively influence cell behavior, guiding differentiation and maturation. The biochemical cues present within the scaffold play a pivotal role in TE. Ultimately, the integration of these design principles—adequate porosity, appropriate pore size, mechanical stability, and surface modifications—constitutes a holistic strategy for developing effective scaffolds.

8.4.3 Biological properties

The ideal scaffold provides a framework that induces cell attachment, proliferation, and differentiation to form the ECM. The cell will interact with the scaffold through focal adhesion, where protein absorption will occur from the scaffold to the cellular environment through the adhesion receptor integrin. Cell adhesion to the surface of the scaffold is facilitated by the topography and architecture of the scaffold. The hydrophobicity or hydrophilicity of the scaffold surface also influences the cellular attachment and proliferation which was analyzed by the contact angle measurement of the scaffold. An angle more than 90 degrees refer hydrophobic and less than 90 degrees consider hydrophilic. The hydrophilic surface facilitates better adherence of the cell, followed by its proliferation. Different cells behave differently to the surface roughness of scaffold which can be differentiate into three categories, that is, macroroughness (more than 100 μm), microroughness (100 nm–100μm), and nano roughness (100 nm). For example, primary rat osteoblasts usually facilitate their proliferation with high alkaline phosphatase activity along the microscale rough surface of 0.81 μm of metallic scaffold [42]. Scaffold pores are crucial to enhance the biological activities of cell migration, infiltration, nutrient supply, oxygen, and waste product exchange. The details of pore size along with tissue migration are described in Table 8.3.

8.4.4 Electrical conductivity

Cellular signaling was enabled and fictionalized by an electrical conduction mechanism, in which the piezoelectric properties of the apatite crystal are hypothesized to generate electric fields involved in bone remodeling during the bone regeneration process. The scaffold manufacturing process is highly dependent upon the ability to interface with the bioelectrical field, which can mimic the human body's electrophysiology. The conductive polymers such as poly(pyrrole) (PPy), poly(aniline), and polythiophene derivatives are the most studied conductive polymers. They were utilized in scaffold shows lesser biocompatibility, whereas conductive materials, such as PPy, and piezoelectric materials, including poly(vinylidene fluoride) and poly(tetrafluoroethylene), generate transient charge in response to mechanical deformation [52]. Matrigel application for the promotion of cellular response and integration shows electrical stimulation, which helps in promoting the maturation of stem cells. Conductive polymer results in alteration of the shape and function of bovine endothelial cells, including DNA synthesis and extension [53].

The mechanistic basis of electrical interactions with tissue remains incomplete and will continue to be a key area for future research. Existing literature presents several hypotheses. Enhanced functionality of engineered cardiac tissues may result from improved ultrastructural organization in response to electric fields. Increased neurite outgrowth from electrical stimulation is likely due to better ECM protein adsorption, although direct effects on the cells also exist. The electrophoretic redistribution of cell surface receptors likely mediates the galvanotropic responses of neurons to horizontally applied two-dimensional fields. However, these mechanisms alone do not fully account for changes in cell function when stimulation is applied to substrates versus media or distant grounds. Depolarization signaling through voltage-gated calcium channels can activate second messengers like cyclic adenosine monophosphate (cAMP),

TABLE 8.3 Summary of representative studies on skin tissue generation within porous scaffolds.

Cell type	Preparing method	Materials	Pore size (μm)	Results	References
NIH-3T3 fibroblast	Electrospinning	Silk fibroin, silk sericin, PVA, PCL	0.3	The microporous layer prevents bacterial invasion and the nanofibers inhibit fibroblast adhesion and proliferation, wound closure, and skin tissue formation	[43]
NIH-3T3 fibroblast	Freeze-drying	Silk fibroin, hyaluronic acid, sodium alginate	≈93	The scaffold facilitated cell adhesion and proliferation, supported increased collagen deposition, and demonstrated the fastest wound healing rate, with enhanced angiogenesis, and the formation of new epidermis closely aligned with the neonatal dermis layer	[44]
HUVEC, human splenic fibroblast, HaCaT	SLg peptide, GelMA	3D printing, freeze-drying	200–400	Inflammatory cells filled the pores, where later collagen fibers and blood vessels were formed, and thus, led to the growth of skin tissue	[45]
Fibroblast, Keratinocytes	PCl, collagen	3D bioprinting	195.8±9.1	The scaffold supported the growth of fibroblasts and keratinocytes and the further formation of dermal and epidermal layers, with tight junctions between them.	[46]
No cell	Electrospinning, freeze-drying	Gelation, GelMA/ Laaponite	≈10	The bilayer scaffold, made of hydrogel and nanofibers, showed superior performance in wound closure, tissue granulation, and skin appendage formation compared to the individual layers	[47]
HFBSCs, keratinocytes and fibroblast	Electrospinning	Gelatin, PEGDMA	1.6–2.5	Small pore sizes regulate keratinocyte growth and promote epidermis formation	[48]
Human keratinocytes and fibroblast	Freeze-drying	NaCMC, collagen	75–123	Keratinocytes in small pores and fibroblasts in large pores achieved paracrine signaling, but the high-water content in the large pores deteriorated fibroblast attachment	[49]
L929 fibroblast	Electrospinning, freeze-drying	PAAm, aloe vera (powder)	≈230	The combination of macropores and nanopores supported fibroblast proliferation, collagen secretion, angiogenesis, and the formation of epidermal and keratin layers.	[50]
HFBSCs, keratinocytes, fibroblasts	Electrospinning casting	Gelatin-PEG methacrylate, PVA	1.5–2.5	Gradient in pore size facilitates the targeted delivery of distinct cell types specific to each skin appendage	[51]
Human dermal fibroblast human epidermal keratinocytes	Electrospinning, 3D printing	PCL, keratin	4.5–6.4	Fibroblasts demonstrated enhanced growth on microfibers, whereas keratinocytes proliferated better on nanofibers with a smaller pore size	[51]

influencing gene transcription that affects learning, memory, survival, and growth. Secreted gene products may enhance the survival of host cells near an implanted scaffold. Recent studies show that electrical preconditioning of human neural progenitor cells on PPy scaffolds can improve stroke recovery outcomes by regulating angiogenesis-related pathways, such as VEGF-A and hypoxia-inducible factor 1α (HIF-1α). As electric fields significantly influence cell function, conductive 3D scaffolds will be essential for creating functional tissues.

8.5 Functionalization of 3D scaffold

The high-quality, functional, and biocompatible product design and preformulation for drug delivery utilize the physicochemical properties and thermal response of the scaffold. The interaction of scaffold highly depends on the tissue type, for bone TE, parameters such as weight-bearing ability during the repair period, cell attachment, lack of inflammatory reaction and toxicity, biodegradability and biocompatibility need to be considered [54], whereas a malleable polymeric scaffold requires soft tissues like nerves, and a flexible polymeric scaffold requires flexible tissue like skin (or blood vessels) [55]. The mesenchymal stem cells always require harder material, but on the other hand, neural stem cells require low modulus materials [56]. The clever strategy involves the application of growth factors to functionalize bioscaffolds. Printing growth factors or delivery vehicles containing them at specific locations within a scaffold can help minimize dosage. Depending upon the geometry of the scaffold, the growth factors needed to be added, which further modify or influence the tissue migration toward the scaffolds. The growth factors release the active ions, which help in the chronological development of tissues. The involvement of growth factors for scaffold formation was mentioned as follows (Table 8.4).

The main goals and strategies for three-dimensional printing with growth factors include:

- Bioactivity preservation involves providing a nondenaturing condition for the scaffold, which helps protect cells and tissues by growth factors.
- Spatial Localization involves material printing in distinct regions.
- Control release involves the degradation of construct materials.

8.6 Application of bio-based scaffold in precision medicine

Scaffolds may be used in several techniques to improve cell adhesion ECM deposition, and mobile–scaffold interactions, frequently used in conjunction with stem cells (SCs) and increase growth factors (GFs) [80]. Cell–scaffold interactions are decisive in developing and delivering precision medicine, notably in areas like regenerative medicine and targeted drug delivery [81]. In precision medicine, TE is used to produce, based on the patient's biological characteristics, the necessary tissue or organ. Scaffolds provide a three-dimensional framework that imitates the ECM and supports cell adhesion, proliferation, and differentiation within the cells [82]. Cells and scaffolds collaborate to ensure proper tissue formation, enhancing the suitability of engineered tissues for individual patients. In other cases, the aim of precision medicine is the development of a particular treatment, and cell–scaffold interrelations can be designed for the individual’s cells for biochemical and mechanical use for effective TE [83,84].

Controlled drug release and site-specific delivery can also be attained by using scaffolds at the required concentrations and amounts over the expected time frame [85]. Concentrating active substances in the correct place at the prescribed time with concise pharmaceutical formulations improves treatment effectiveness. Scaffolds equipped with cellular tissues can also work as drug delivery systems responding to biological stimuli via appropriate drug timing and contours. Customized scaffolds and PEG-based hydrogels are designed to release drugs according to the cellular response to the stimulation with signals integrated into target cells [86,87]. Scaffolds serve a supportive role to the cells, which aids in upholding their structure and function. Optimal cell–scaffold interactions enable cells to remain viable and active, which is fundamental to achieving precise pharmacological effects in precision medicine [88,89]. The precision of treatments like cancer treatment or stem cell transplantation can be increased by designing scaffolds that maximize interactions with particular biological kinds, such as cancer cells or stem cells. Biocompatibility is vital for scaffolds used in precision medicine. The interaction between the scaffold cloth and the body’s cells must avoid triggering immune responses or toxic effects. Proper cellular-scaffold interactions ensure that scaffolds are well tolerated by the body and do not cause infection or rejection, which is vital for customized medical treatments, particularly in organ regeneration and long-term implants. Biodegradability is another important feature, which enables degradation at a rate that matches tissue regeneration, allowing the body to naturally take over the scaffold's function without the need for removal surgery. Precision medicine can benefit from scaffolds to degrade according to a patient’s healing rate [90,91].

TABLE 8.4 Different growth factors that are used and involved in 3D scaffold formation for biomedical applications.

Primary application	Growth factors	Construct material	Co-printed cells	Evaluation method
Cartilage formation	Transforming growth factor-β	Alginate, poly(ε-caprolactone)	Chondrocytes	In vitro, in vivo [57]
		Poly(D, L-lactic-*co*-glycolic acid) microparticles in poly(ε-caprolactone)	None	In vitro, in vivo [58]
		Hyaluronic acid, polyurethane	None	In vitro [58]
		Poly(D, L-lactic-*co*-glycolic acid) nanoparticles in polyethylene glycol diacrylate with nanohydroxyapatite	None	In vitro [59]
		Poly(D,L-lactic-*co*-glycolic acid) nanoparticles in gelatin methacrylamide, poly(ethylene glycol) diacrylate	Mesenchymal stem cells	In vitro [60]
		Poly(D,L-lactic-*co*-glycolic acid) microparticles in poly (ε-caprolactone)	None	In vitro and in vivo [61]
Dental tissue formation	Bone morphogenetic protein-7	Poly(D,L-lactic-*co*-glycolic acid) microparticles in poly (ε-caprolactone)	None	In vitro [62]
Neural tissue formation	Nerve growth factor	Poly(D,L-lactic-*co*-glycolic acid) nanoparticles in poly (ethylene glycol), poly(ethylene glycol) diacrylate	None	In vitro [63]
	Vascular endothelial growth factor	Collagen, fibrin	Neural stem cells	In vitro [64]
Sweat gland regeneration	Epidermal growth factor	Alginate, dermal homogenate, gelatin	Epithelial progenitors	In vitro, in vivo [65]
Vascularization	Vascular endothelial growth factor	Poly(D,L-lactic-co-glycolic acid) microparticles in alginate, gelatin with β-tricalcium phosphate	None	In vitro [66]
		Alginate, alginate gellan gum, calcium phosphate cement	None	In vitro [67]
		Gelatin microparticles in alginate, Matrigel®	Endothelial progenitors	In vitro and in vivo [68]
		Chitosan/dextran sulfate microparticles in calcium phosphate cement	None	In vitro [69]
		Alginate, Laponite, methylcellulose	None	In vitro [70]
Demonstration of growth factor release	Bone morphogenetic protein-2	Poly(D,L-lactic acid-cotrimethylene carbonate)	None	None [71]
	Vascular endothelial growth factor	Alginate	None	None [72]

(*Continued*)

TABLE 8.4 (Continued)

Primary application	Growth factors	Construct material	Co-printed cells	Evaluation method
Bone Formation	Bone morphogenic protein-2	Collagen, methacrylamide gelatin	Bone mesenchymal stem cells	In vitro [73]
		Alginate, collagen, gelatin, poly(€ caprolactone)	Dental pulp stem cells	In vitro [74]
		Collagen, gelatin, poly(€ caprolactone), poly(D,L-lactic-co-glycolic acid) with β-tricalcium phosphate	None	In vivo [74]
		Bone extracellular matrix with poly(D,L-lactic-co-glycolic acid) with β-tricalcium phosphate	None	In vitro and In vivo [75]
		Hyaluronic acid, poly(ethylene glycol), poly(D,L-lactic-*co*-glycolic acid)	None	In vitro and In vivo [76]
		Poly(D,L-lactic-*co*-glycolic acid) microparticles in poly (propylenefumarate), diethyl fumarate	None	In vitro and In vivo [77]
		Calcium silicate, poly(ε-caprolactone)	None	In vitro [78]
		Calcium phosphate cement, poly(L-lactic acid)	None	In vitro [79]

Thus cell-scaffold interactions are advancing precision medication through the improvement of customized remedies, helping tissue regeneration, and optimizing drug delivery. These interactions permit the introduction of more patient-specific therapeutic systems that align with the desire for precision remedies to tailor treatments for the development of biological and genetic profiles.

8.6.1 Implication of cell–scaffold dynamics in applied tissue engineering

To maximize the results of TE, it is critical to comprehend the dynamics of cellular attachment and differentiation and to direct the development of novel tissue relationships [92]. Understanding the dynamics of mobile-scaffold interactions includes reading each biophysical and biochemical aspect, in addition to the temporal (time-related) and spatial (area-related) behavior of cells in response to their scaffold environment. These factors play a critical role in TE, regenerative medicine, and drug delivery systems, as they dictate how efficaciously cells adhere, grow, and differentiate on scaffolds [93,94].

Biophysical interactions include scaffold stiffness, surface topography, and mechanical forces that influence manual cell responses. Cells are sensitive to the stiffness or elasticity of the scaffold material. Different tissues inside the body showcase specific stiffness stages, and scaffolds want to mimic these to elicit the right cell behavior [95]. For instance, bone cells (osteoblasts) select stiffer surfaces, while neurons thrive on softer substrates. Mechano-transduction is a process by which cells convert mechanical signals into biochemical responses, influencing their growth and differentiation. Nano-patterned surfaces can guide mobile orientation, migration, and proliferation by way of mimicking the natural ECM. Scaffold porosity plays a role in mobile infiltration and tissue integration. High porosity allows for higher nutrient and oxygen diffusion, while also presenting pathways for cellular migration. However, porosity must be balanced with mechanical power to aid cells without collapsing [96,97]. Biochemical interactions involve the molecular signals exchanged between cells and the scaffold. These interactions guide cell adhesion, migration, proliferation, and differentiation, which are essential for successful tissue regeneration and repair [98]. The ECM is a complex network of proteins (including collagen, fibronectin, and laminin) that cells interact with in their natural surroundings. Scaffolds frequently mimic the ECM by incorporating or displaying adhesive molecules, such as integrins and ligands, that bind to receptors on the cell surface. This cellular–ECM interplay is essential for maintaining the right cellular attachment and initiating downstream signaling pathways that affect cell behavior [99,100].

Functionalized scaffold which includes bioactive molecules (e.g., growth factors, peptides) that bind to precise cellular receptors to promote desired outcomes like proliferation or differentiation [101]. For example, including RGD peptides (arginylglycylaspartic acid: which sell integrin binding) enhances mobile adhesion. Growth elements are proteins that modify cell capabilities, which include proliferation and differentiation. Scaffolds can be designed to release increased factors in a controlled way, presenting biochemical cues that guide manual cells. The release can be spatiotemporally controlled, delivering increased amounts at specific locations and times, which is important in complex tissues such as bone, cartilage, or vascular structures. In a few instances, scaffolds are engineered to respond to cell signals or environmental adjustments (e.g., pH, temperature) to release growth elements only when needed, enhancing the precision of tissue regeneration [102]. Cells interact with scaffolds through various signaling pathways that influence their behavior. For example, Wnt signaling is vital for cell proliferation and differentiation in many tissues. Notch signaling influences mobile fate choices and is mainly vital in stem cell biology. These pathways are activated while cells sense changes in their environment, such as changes in scaffold stiffness or biochemical ligand density, which in turn modulate cellular responses like differentiation or apoptosis [103].

8.6.2 Cellular interaction with scaffolds

The procedures that are crucial for successful tissue formation include adhesion, migration, proliferation, and differentiation [104]. It has been seen that cells shape preliminary bonds with the scaffold through adhesion molecules like integrins, pivotal to cell death (anoikis) or inefficient tissue formation [105]. After adhesion, cells begin to proliferate (increase in variety) and migrate across the scaffold. The scaffold's surface residences, such as topography and porosity, influence the speed and course of migration. This section is time-based, and the scaffold needs to offer the proper cues for cells to proliferate and cover the floor effectively. In later degrees, cells begin to differentiate into unique mobile kinds primarily based on the biochemical and mechanical cues provided by the scaffold. For instance, mesenchymal stem cells (MSCs) may additionally differentiate into osteoblasts on stiff scaffolds or into chondrocytes on softer, more elastic scaffolds [106,107]. As cells proliferate and convey their personal ECM, the scaffold starts to degrade (if biodegradable), allowing the herbal tissue to replace the scaffold material over the years. This temporal alignment of scaffold degradation with tissue formation is crucial for a hit tissue integration [106].

The spatial organization of cells on a scaffold is equally important as the temporal aspects. Cells must organize in specific ways to form functional tissues. Key spatial elements that affect cellular behavior consist of the following sequences [108,109]. Proper seeding techniques ensure that cells are lightly allotted during the scaffold. High densities may additionally promote quicker growth, however, can cause overcrowding and nutrient boundaries, at the same time as low densities might result in slower tissue formation [110]. Scaffolds with interconnected pores allow cells to transport more freely, which promotes uniform tissue formation and higher integration of the scaffold with local tissue. Aligned scaffolds can guide mobile growth and orientation. For example, aligned fibers in a scaffold can induce a directional increase of cells, which is critical for tissues like muscles or nerves that have an inherent directional shape [111].

Scaffold areas may be functionalized with special bioactive molecules to create wonderful zones for unique cell behaviors, which include better proliferation in one place and differentiation in some other. Spatial capabilities of the scaffold, consisting of aligned fibers or directional pores, have an impact on the orientation of cells. Cells like fibroblasts and myoblasts generally tend to align with scaffold fibers, which is essential for tissues with anisotropic mechanical properties like tendons and muscular tissues. Scaffold geometry can induce mobile polarity, wherein cells have an awesome front and back. This is critical for processes such as migration and the development of epithelial tissues, where cells should form organized layers [112].

8.6.3 Technologies effecting the fabrication of scaffolds

In TE and regenerative medicine, cell-scaffold technology may be personalized to a patient's particular tissue needs. For example, scaffolds can be engineered to launch growth factors in a spatiotemporally managed manner that supports tissue regeneration in a patient-precise context, including in wound healing or organ restoration [113]. The customization of scaffolds in TE and regenerative medicine is central to advancing precision medicine. By making scaffolds to patient-specific factors such as genetics, tissue architecture, and disease state, more personalized, effective therapies could be delivered. Emerging technologies like organ-on-a-chip systems, 3D bioprinting, and AI further enhance the ability to design patient-specific scaffolds and integrate them into complex biological systems. This customization involves altering scaffold materials, architecture, and bioactive components to match the patient's tissue-specific needs. Selecting biomaterials that are compatible with the patient's specific immune profile reduces the risk of rejection and inflammation [114]. Biodegradable polymers like PLGA or natural materials like collagen and hyaluronic acid are often altered to suit the patient's tissue type [115]. Decellularized ECM (dECM) from a patient's tissues is used as scaffolds, minimizing immune reactions. Customization of scaffold stiffness, elasticity, and porosity is essential to match the mechanical properties of the tissue, for e.g., bone tissue scaffolds require high mechanical strength, whereas neural scaffolds need greater flexibility. 3D printing allows for precision in designing mechanical properties based on a patient's bone density or muscle elasticity. Autologous cells (cells derived from the patient) can be seeded into customized scaffolds to promote tissue regeneration and reduce the risk of rejection. For example, chondrocytes from a patient's cartilage can be cultured and integrated into a scaffold designed for cartilage repair in joints. Stem cells such as mesenchymal stem cells (MSCs) can be isolated from the patient's bone marrow or adipose tissue and incorporated into scaffolds to regenerate damaged tissues like skin, bone, or heart muscle. Using the genetic information, scaffolds can be designed to interact with the patient's unique molecular signaling pathways [116]. If a patient has a genetic predisposition contributing to a slow healing process, scaffolds can be embedded with growth factors (e.g., VEGF, BMP) or gene therapy vectors to upregulate the healing responses. CRISPR and other gene-editing technologies can also be integrated into scaffolds for targeted therapy in genetic diseases [117].

8.7 Bioscaffolds in reproductive medicine and fertility prevention

Cell–scaffold interactions have emerged as key tools in the fields of reproductive medicine, particularly in ovarian TE and fertility preservation. These advanced biomaterials support the growth, differentiation, and functionalization of cells, which is crucial for developing personalized therapies and regenerative approaches. One of the primary packages of scaffolds in reproductive medicinal drugs includes the development of bioengineered constructs for ovarian TE, toward fertility preservation for patients' present remedies like postchemotherapy or radiotherapy, which could compromise ovarian function. At the same time, an artificial ovary should be capable of accepting the transplant of a previously isolated follicle, supporting its survival and proper development, ensuring the secretion of sex hormones and generation of a mature oocyte. Engineered ovaries, built from biocompatible scaffolds, can house remoted follicles or ovarian tissue fragments. These scaffolds provide the structural and biochemical surroundings important for follicle maturation and hormone production. Choosing the right matrix is essential for supporting ovarian follicle survival, growth, and hormone production. Both synthetic and natural polymers have been tested for this purpose, each with unique advantages, along with this, a scaffold requires careful consideration and fine balance of mechanical and biochemical cues to support follicle survival, growth, and hormone production as well as differentiation by replicating the ECM of the ovary, providing an optimized microenvironment for cell–scaffold interactions, allowing better *in vitro* models for reproductive biology. Natural polymeric materials over synthetic materials better resemble human ECM and its bioactive components. It has strong biocompatibility and bioactivity and is crucial for cell adhesion, migration, proliferation, and differentiation. Collagen, fibrin, or alginate have been used to recreate the ECM of the ovary. These constructs can help in vitro follicular-genesis or may be transplanted to decrease the lower back pain in the affected person to restore fertility. For instance, collagen was one of the first natural matrices used to encapsulate follicles, while fibrin has been modified to enhance follicle survival and development [118]. Fibrin, combined with growth factors like VEGF, has yielded successful outcomes in follicle transplantation, leading to ovulation and hormone production in animal models. These natural matrices better mimic the ovarian microenvironment but have limitations, such as weaker mechanical strength and less vascularization. Advancements in 3D matrices have improved the potential of artificial ovaries, making them promising for fertility preservation and restoring ovarian function in clinical settings [119]. Similar to the situation of reproductive medicine, the isolated preantral follicles of distinct sizes (30–150 μm) require an exact match for them in a 3D matrix. A thermo-responsive hydrogel provides perfect encapsulation of isolated follicles and ovarian cells while preserving the configuration of the decellularized. Several other natural polymers have become popular for the generation of artificial ovary models, viz., fibrin protein, gelatine, agarose, hyaluronic acid, chitosan, to name a few [120]. Live births have been achieved using encapsulated follicles or ovarian tissue within 3D hydrogels. 3D printing has emerged as a powerful tool, enabling precise scaffold designs to promote vascularization and follicle survival. Laronda et al. developed a bioprosthetic ovary using 3D-printed microporous scaffolds of gelatin, which helped mice regain ovarian function and foster follicle maturation [112]. Jiwon Kim et al. highlighted the use of PEG hydrogels to enhance ovarian function by mimicking the rigidity of stromal ovarian tissue while supporting follicle architecture. Other efforts by E. Kniazeva et al. included creating biomaterial grafts to minimize follicle loss, enhance engraftment, and restore endocrine function using fibrin-based combinations with alginate or with collagen. Peter D. Rios and colleagues demonstrated "in vivo follicle maturation" through alginate hydrogel-encapsulated ovarian follicles transplanted at heterotopic sites. This approach promoted follicle survival and antral follicle growth, while preventing tumor development after implant removal [121]. Chitin-based scaffolds have shown advantages over alginate hydrogels in sustaining follicle growth *in vitro* and improving reproductive and bone health, *in vivo*. Despite these advancements, significant challenges remain in translating these approaches to human application, particularly due to diffusion limitations, insufficient vascularization in larger implants, and the complex biomechanical properties required to mimic native ovarian tissue. Decellularized ovarian matrices have been used to create herbal scaffolds that retain the ovarian ECM's complexity, improving follicle survival and functionality after transplantation. Prepubescent cancer sufferers and women present with untimely ovarian failure (POF) gain from cryopreservation of ovarian tissue and subsequent transplantation onto scaffolds [112]. Bioengineered ovaries can support the long-term survival and improvement of the ovarian follicles upon transplantation, offering a potential solution for fertility restoration. In addition to fertility maintenance, cell scaffold generation is being used to expand bioartificial ovaries for hormone substitute treatments (HRT) in women who are afflicted by situations like early menopause or POF. These ovaries could produce hormones (e.g., estrogen and progesterone) in a regulated manner, restoring endocrine characteristics without the need for synthetic hormones. Granulosa cells or other hormone-generating cells are seeded onto scaffolds designed to imitate ovarian tissue. These bioartificial ovaries may be transplanted to regulate hormone manufacturing in women, imparting natural alternatives to artificial HRT. Studies have shown that those bioengineered ovaries can efficiently launch hormones and stimulate the development of secondary sexual characteristics in preclinical models [122,123].

Ovarian cancer treatments frequently result in reduced ovarian function, leading to complications such as early menopause and infertility. Current reproductive technologies like IVF and ovarian transplants have limited efficacy, especially for pediatric cancer patients with metastatic diseases [108,109]. The intersection of reproductive medicine and oncology, known as onco-fertility, benefits greatly from scaffold-based technologies seeking to address these issues by creating an artificial ovary that restores long-term hormone function and provides fertility benefits [110]. Ovarian follicles require a 3D environment to maintain essential cell–cell interactions, proper shape, survival, and maturation, which are supported by female hormones [111]. Follicle culture systems using scaffolds have turned out to be a promising alternative for preserving and growing ovarian follicles *ex vivo*. This approach is important for sufferers prone to infertility due to most cancer treatments. 3D Scaffolds for in vitro follicular-genesis allow the lifestyle of remoted primordial follicles in 3D surroundings that mimic the ovarian area of interest, helping their growth and maturation into fertilizable oocytes. Scaffolds made of hydrogels are used to encapsulate and support follicles, supplying a managed microenvironment with key signals for the boom [124]. Growth elements, like VEGF and FSH (follicle-stimulating hormone), are frequently integrated into those scaffolds to promote follicle survival and differentiation. Bioengineered scaffolds can also facilitate ovarian regeneration for patients with broken ovarian tissues due to aging or disease. By mimicking the ovarian ECM, those scaffolds aid stem cell recruitment and tissue restoration, restoring ovarian features and potentially reversing infertility. The use of mesenchymal stem cells (MSCs) or pluripotent stem cells (iPSCs) in combination with scaffolds can aid the regeneration of ovarian tissue. The scaffold offers essential structural and signaling support for the differentiation of those stem cells into granulosa cells or other ovarian-unique cells [125]. In cases where ovarian tissue is transplanted after cryopreservation, scaffolds can be engineered to take away most cancer cells that could have survived in the ovarian tissue, reducing the risk of reintroducing cancer during fertility. Scaffolds can create supportive environments for follicle maturation, focused cancer treatments, and regenerative strategies, imparting new avenues for personalized remedies [126].

8.8 Application in cancer research

Mimicking the artificial ovary using bioinspired 3D scaffolds has become crucial in oncology and cancer research due to its ability to mimic the ovarian tissue niche, ability in hormone production, and ovarian cancer modeling. Engineered scaffolds that can mimic the ovarian ECM are useful to create 3D in vitro models of ovarian cancers to look into tumor progression, metastasis, and cellular reaction to chemotherapy. These trends can help in elucidating the interactions among growing aggressive cancer cells and the TME, which includes surrounding tumor stroma, immune cells, and blood vessels. Scaffolds are also being explored as transport systems for most cancer therapeutics [127]. Nanoparticles loaded with chemotherapeutic agents (e.g., cisplatin or paclitaxel) can be integrated into biodegradable scaffolds, ensuring that healing concentrations are maintained within the TME without affecting healthy tissues, thus minimizing systemic toxicity. This approach is helpful in treating aggressive chemo-resistant ovarian tumors because the scaffold can provide local awareness of the drug at the same time as sparing ordinary tissues [128,129]. Scaffolds can also serve as platforms for gene delivery systems, such as siRNA or CRISPR/Cas9, targeting specific oncogenes or pathways involved in ovarian cancer progression [130].

On the other front of the battle against cancer, immunotherapy has become a promising frontier in ovarian cancer treatment. Scaffolds designed to release immune-activating agents or recruit immune cells to the tumor site can boost the natural defenses against cancer. These scaffolds can release cytokines or immune checkpoint inhibitors, enhancing the infiltration and activation of T cells within the tumor. By growing a scaffold that mimics the immune-stimulating microenvironment, those techniques intend to conquer the immunosuppressive nature of ovarian tumors and enhance the admission of immune cells to most cancer cells [131]. In cases of advanced ovarian cancers, scaffolds are used to enhance the efficacy of tumor ablation techniques, such as radiofrequency ablation (RFA) or photothermal therapy. By incorporating heat-sensitive or radiation-enhancing materials, scaffolds can increase the precision and effectiveness of tumor destruction. For photothermal therapy, scaffolds infused with materials like gold nanoparticles can absorb infrared light and generate heat, which selectively destroys cancer cells while sparing surrounding healthy tissues [132]. After ovarian cancer surgery, there is a high risk of tumor recurrence. Scaffolds loaded with antiinflammatory agents, growth factor inhibitors, or chemotherapeutics can be used postsurgery to prevent ovarian cancer cells from migrating and reestablishing themselves at the tumor site [133].

Organs-on-chips (OOC) are microfluidic gadgets that mimic the physiological environment of human organs, allowing for dynamic interactions among cells and tissues in a controlled way [99]. Conventional studies often rely on large sample volumes and lack physiological relevance. In contrast, OOC systems utilize microliter-scale samples, improving efficiency and reducing costs. OOC technology offers a superior simulation of biological phenomena, allowing for improved research practices, reduced costs, and better efficiency. OOC devices can mimic different generations' organ systems, such as the lung, heart, liver, ovary, or kidney. OOC sensor components allow pathways to examine and monitor biological events in real time. They include microsensors and sensing elements fully embedded into a specific device system or imaging systems that can capture stereospecific motional parameters such as fluid compressibility, ionologic signals, and biochemical transformations, capture the complexity of human organ microenvironments, supersede conventional preclinical testing models, and thus enhance drug development processes [123].

Ovarian cancer-on-a-chip can comprise scaffolds mimicking the ECM of a TME. This setup enables them to have a look at patient-specific cancer progression and drug responses [134]. 3D bioprinting enables the fabrication of scaffolds with precise control over architecture, material composition, and the inclusion of bioactive molecules. 3D bioprinting allows the creation of scaffolds that precisely replicate the anatomy of a patient's damaged organ or tissue. For example, complex tissues like bone, the liver, the kidney, or the heart, bio-printed scaffolds can utilize MRI or CT-generated information to include multiple cell types and blood vessel networks to replicate the functional architecture of the target tissue. This level of customization is essential for organ regeneration and for creating functional tissue constructs for transplant [135]. Bioprinting can also create smart scaffolds that change properties in response to environmental signals, such as pH, temperature, or pressure. These scaffolds can adjust drug release rates or mechanical properties based on the patient's healing progress. Artificial Intelligence (AI) is playing an increasingly important role in designing custom scaffolds by analyzing patient data and predicting optimal scaffold configurations for tissue regeneration or drug delivery. Machine learning models can predict how different scaffold designs will interact with patient-specific cells and tissues, optimizing the scaffold structure and composition for better integration and regeneration. AI can predict how a patient's tissue will heal based on their genetic makeup, age, and health status, guiding the selection of scaffold materials, growth factors, and drugs. AI could recommend slower-degrading scaffolds for elderly patients with slower regenerative capabilities [136]. A cardiac patch scaffold integrated with a heart-on-a-chip can replicate the patient's cardiac cycle, medication, or other conditions before implantation [137]. A liver-on-a-chip combined with a patient-specific scaffold can model liver fibrosis progression in patients with chronic liver diseases, enabling targeted drug testing. 3D-printed scaffolds made from a patient's cartilage cells for a joint replacement can be integrated with OOC systems to monitor tissue regeneration and interactions with surrounding tissues. A nerve-on-a-chip integrated with a personalized nerve regeneration scaffold could apply controlled electrical impulses to guide nerve growth and function [138]. A kidney scaffold designed to elicit a minimal immune response in a transplant patient can be integrated with a kidney-on-a-chip, which monitors time-specific immune responses. A 3D-printed scaffold with vascular networks tailored to match a patient's circulatory structure, integrated with a vascular-on-a-chip system, could simulate blood flow and monitor tissue perfusion. This can enhance preclinical studies for transplant viability. For a diabetic patient with poor vascularization, a scaffold can be designed to include proangiogenic factors and tested in vitro using a vascular OOC to ensure effective blood vessel formation before implantation [139].

Advanced OCC includes "multi-organs-on-a-chip" technology, which integrates several organ models connected by bionic blood channels to mimic whole-body interactions. Systems like liver-intestine chips for bile acid synthesis and pumpless multiorgan chips offer significant advances. These platforms improve drug safety assessments by simulating how hepatic metabolism affects off-target organs such as the heart, surpassing traditional 2D cultures and animal models in accuracy. OOC technology enables researchers to model diseases with greater physiological relevance, offering promising insights into disease mechanisms and drug discovery, revolutionizing preclinical testing and personalized medicine.

8.9 Bio-based 3D scaffolds in clinical trials

The use of bio-based 3D scaffolds in clinical trials has served as platforms for drug delivery, immune activation, tissue regeneration, and disease modeling. Ongoing trials explore their applications in conditions such as melanoma, breast reconstruction, bone defects, and ovarian TE. Table 8.5 provides an overview of key clinical trials involving bio-based 3D scaffolds, highlighting their objectives, study types, and phases. The information collected from https://clinicaltrials.gov/, search keyword includes bio-based 3D scaffolds and cancer.

8.10 Discussion and conclusion

Bioinspired 3D scaffolds represent a transformative opportunity in both regenerative medicine and cancer research, particularly by addressing the limitations of traditional 2D models. They offer a multidimensional environment that closely mimics the native ECM, facilitating crucial cell–ECM interactions. Such advancements are pivotal in enhancing cell adhesion, migration, and differentiation, which are critical for TE and disease modeling. Encapsulation of ovarian follicles within a biocompatible scaffold fosters replicating the complex microenvironment of the ovarian niche, promoting follicle maturation and hormone secretion. Notably, the use of natural polymers such as collagen and fibrin improves bioactivity and compatibility, ensuring better outcomes in follicular-genesis. However, issues such as mechanical strength and vascularization remain challenges that need to be addressed to optimize long-term function in human applications. On the oncology front, ovarian cancer remains a focus, with 3D scaffolds enabling the creation of *in vitro* tumor models that closely mimic the ovarian cancer microenvironment. These models have advanced the study of tumor progression, metastasis, and drug resistance. While the application of 3D scaffolds has shown promise in drug delivery and gene therapy, further refinement is required to translate these results into clinically relevant therapies. Furthermore, bioinspired 3D scaffolds represent a significant leap forward in precision medicine too, offering solutions for complex challenges in TE and oncology. Despite the advancements, several challenges remain, particularly in mimicking

TABLE 8.5 Clinical trials carried out involving bio-based 3D scaffold.

NCT number	Study title	Objective	Study type	Phase
NCT03698721	Urothelium Tissue Engineering Using Biopsies	Porous sodium alginate hydrogels will be evaluated for structural, impermeability, and stitchability properties. Cellularized scaffolds will be implanted subcutaneously in nude mice for 28 days.	Observational	NA
NCT01057745	Ovarian Freezing Before Cancer Treatment	Investigates the impact of 3D bio gel scaffolds, hormones, and culture conditions on ovarian follicle maturation prior to cancer treatment.	Observational	NA
NCT05623748	Establishment and Application of 3D Culture Models for Circulating Tumor Cells	Evaluates the efficiency of 3D culture scaffolds in isolating, purifying, and proliferating clinical CTCs.	Observational	NA
NCT01753089	Dendritic Cell Activating Scaffold in Melanoma	A tumor-lysate-based scaffold incorporating GM-CSF and CpG for immune activation will be tested in melanoma patients.	Interventional	Phase 1
NCT03348293	3D Printing Personalized Biodegradable Implants for Breast Reconstruction	Personalized porous scaffolds fabricated using 3D printing will be implanted following tumor resection to optimize cosmetic breast reconstruction outcomes.	Interventional	Phase 1
NCT02059252	Smart-Matrix Dermal Replacement Scaffold	A porous dermal scaffold (Smart-Matrix) will be evaluated for its role in promoting cellular infiltration, tissue regeneration, and wound healing after excision of basal and squamous cell carcinoma.	Interventional	Phase 1/2
NCT02734979	Surgical Solution for Breast Cancer-Associated Lymphedema	The biobridge scaffold will be evaluated as an adjunct to vascularized lymph node transfer to improve outcomes in lymphedema treatment.	Interventional	NA
NCT03066245	Stem Cells on Scaffold for Aneurysmal Bone Cysts	PLGA scaffolds seeded with bone marrow-derived mesenchymal stem cells will be tested for bone lesion healing in aneurysmal bone cyst patients.	Interventional	Phase 1/2
NCT01970306	Esophageal Reinforcement Using a Cell Matri-Stem Scaffold	Evaluates Matri-Stem biological scaffolds to reduce postoperative complications after stomach and esophageal surgeries.	Interventional	Phase 2
NCT02293031	Gene-Activated Matrix for Maxillofacial Bone Tissue Repair	Investigates the safety and efficacy of a gene-activated matrix (Nucleostim) for regenerating bone tissue in maxillofacial defects such as injuries, surgeries, or alveolar atrophy.	Interventional	NA
NCT01389661	Maxillary Bone Cyst Treatment Using MSC-Seeded Scaffolds	Mesenchymal stem cells cultivated in autologous plasma scaffolds will be implanted into cyst cavities following surgery to evaluate bone tissue regeneration.	Interventional	Phase 1/2
NCT03742726	Safety and Performance of Smart-Matrix Scaffold	Evaluates Smart-Matrix scaffolds for promoting wound healing and tissue regeneration postsurgery.	Interventional	NA
NCT04904185	MASE-T Cells Expanded via Artificial Antigen Scaffolds in Metastatic Melanoma	Artificial antigen-presenting scaffolds are used for antigen-driven T cell expansion to generate MASE-T cells targeting melanoma antigens, mimicking dendritic-T cell interactions for optimal immune stimulation.	Interventional	Phase 1

CTCs, Circulating tumor cells.

the intricate architecture and biological functionality of native tissues. The need for enhanced scaffold materials that can balance biocompatibility, mechanical strength, and bioactivity is critical. Achieving the correct microarchitecture to mimic native tissues, particularly the ECM, is a key technical challenge. This is crucial for directing cell behavior, adhesion, and nutrient exchange. Biocompatibility, appropriate cellular interactions, and controlled degradation rates are essential to support cell proliferation and tissue regeneration without adverse immune reactions. Replicating the dynamic and complex nature of tissue microenvironments, such as the ovarian niche, adds another layer of difficulty. Moreover, future research should focus on overcoming issues related to immune response, scaffold degradation, and vascularization to translate these technologies effectively into clinical applications. Advances in materials science and bioprinting offer potential solutions, enabling the creation of customizable scaffolds with precise microarchitectures and bioactive functionalities. In summary, while the emerging field is still evolving, bioinspired 3D scaffolds hold considerable promise in advancing both reproductive health and cancer research, paving the way for more effective and personalized therapeutic strategies.

Declaration

The authors declare that there is no conflict of interest related to this study. All authors have read and approved the final version of the manuscript and agree to its submission for publication.

Acknowledgment

BB likes to thank the Director, CNCI, for intramural funding. BB also like to acknowledge funding agencies, that is, Department of Biotechnology, Govt of India (Sanction No. BT/PR38493/TRM/120/465/2020), Indian Council of Medical Research (File No. 17X(3)/Adhoc/87/2022-ITR), and West Bengal Department of Science Technology and Biotechnology (sanction no. 2402 (Sanc.)/STBT-11012(12)/2/2022-ST SEC) pertaining to the work.

References

[1] F. Paradiso, S. Lenna, S.A. Gazze, J. Garcia Parra, K. Murphy, L. Margarit, et al., Mechanomimetic 3D scaffolds as a humanized in vitro model for ovarian cancer, Cells 11 (5) (2022) 824. Available from https://doi.org/10.3390/cells11050824.

[2] Z. Azari, S. Nazarnezhad, T.J. Webster, S.J. Hoseini, P. Brouki Milan, F. Baino, et al., Stem cell-mediated angiogenesis in skin tissue engineering and wound healing, Wound Repair and Regeneration 30 (4) (2022) 421–435. Available from https://doi.org/10.1111/wrr.13033.

[3] A.K. Mendoza-martinez, D. Loessner, A. Mata, H.S. Azevedo, Modeling the tumor microenvironment of ovarian cancer: the application of self-assembling biomaterials, Cancers 13 (22) (2021). Available from https://doi.org/10.3390/cancers13225745, https://www.mdpi.com/2072-6694/13/22/5745/pdf.

[4] T. Meng, X. Wang, S. Jiang, S.-rui Chen, S. Zhou, Y. Zhu, et al., Delivery of small-molecule drugs and protein drugs by injectable acid-responsive self-assembled COF hydrogels for combinatorial lung cancer treatment, ACS Applied Materials & Interfaces 15 (36) (2023) 42354–42368. Available from https://doi.org/10.1021/acsami.3c10074.

[5] M. Zanoni, M. Cortesi, A. Zamagni, C. Arienti, S. Pignatta, A. Tesei, Modeling neoplastic disease with spheroids and organoids, Journal of Hematology & Oncology 13 (1) (2020). Available from https://doi.org/10.1186/s13045-020-00931-0.

[6] S.R. Caliari, J.A. Burdick, A practical guide to hydrogels for cell culture, Nature methods 13 (5) (2016) 405–414. Available from https://doi.org/10.1038/nmeth.3839, http://www.nature.com/nmeth/.

[7] M. Krishani, W.Y. Shin, H. Suhaimi, N.S. Sambudi, Development of scaffolds from bio-based natural materials for tissue regeneration applications: a review, Gels 9 (2) (2023) Available from https://doi.org/10.3390/gels9020100, http://www.mdpi.com/journal/gels.

[8] K.M. Park, D. Lewis, S. Gerecht, Bioinspired hydrogels to engineer cancer microenvironments, Annual Review of Biomedical Engineering 19 (2017) 109–133. Available from https://doi.org/10.1146/annurev-bioeng-071516-044619, http://www.annualreviews.org/journal/bioeng.

[9] A.D. Bradshaw, The extracellular matrix, Encyclopedia of Cell Biology: Volume 1-6, Second Edition 3 (2022) 211–221. Available from https://doi.org/10.1016/B978-0-12-821618-7.00088-2, https://www.sciencedirect.com/book/9780128216248.

[10] D. Yang, J. Liu, H. Qian, Q. Zhuang, Cancer-associated fibroblasts: from basic science to anticancer therapy, Experimental & Molecular Medicine 55 (7) (2023) 1322–1332. Available from https://doi.org/10.1038/s12276-023-01013-0.

[11] H. Zhang, Q. Yang, X. Lian, P. Jiang, J. Cui, Hypoxia-inducible factor-1α (HIF-1α) promotes hypoxia-induced invasion and metastasis in ovarian cancer by targeting matrix metallopeptidase 13 (MMP13), Medical Science Monitor 25 (2019) 7202–7208. Available from https://doi.org/10.12659/msm.916886.

[12] K. Sato, K. Parag-Sharma, M. Terajima, A.M. Musicant, R.M. Murphy, M.R. Ramsey, et al., Lysyl hydroxylase 2-induced collagen cross-link switching promotes metastasis in head and neck squamous cell carcinomas, Neoplasia 23 (6) (2021) 594–606. Available from https://doi.org/10.1016/j.neo.2021.05.014.

[13] A.S. Hoffman, Hydrogels for biomedical applications, Advanced Drug Delivery Reviews 54 (1) (2002) 3–12. Available from https://doi.org/10.1016/S0169-409X(01)00239-3, http://www.elsevier.com/locate/drugdeliv.

[14] L. Rezakhani, M. Alizadeh, A. Alizadeh, A three dimensional in vivo model of breast cancer using a thermosensitive chitosan-based hydrogel and 4 T1 cell line in Balb/c, Journal of biomedical materials research. Part A 109 (7) (2021) 1275–1285. Available from https://doi.org/10.1002/jbm.a.37121.

[15] M. Mahmoodi, S. Ebrahimi-Barough, S. Kamian, M. Azami, M. Mehri, M. Abdi, et al., Fabrication and characterization of a three-dimensional fibrin gel model to evaluate ant-proliferative effects of astragalus hamosus plant extract on breast cancer cells, Asian Pacific Journal of Cancer Prevention 23 (2) (2022) 731–741. Available from https://doi.org/10.31557/apjcp.2022.23.2.731.

[16] O. Wichterle, D. Lím, Hydrophilic gels for biological use, Nature 185 (4706) (1960) 117–118. Available from https://doi.org/10.1038/185117a0.

[17] J. Redmond, H. McCarthy, P. Buchanan, T.J. Levingstone, N.J. Dunne, Advances in biofabrication techniques for collagen-based 3D in vitro culture models for breast cancer research, Materials Science and Engineering C 122 (2021). Available from https://doi.org/10.1016/j.msec.2021.111944, https://www.journals.elsevier.com/materials-science-and-engineering-c.

[18] S. Swetha, K. Lavanya, R. Sruthi, N. Selvamurugan, An insight into cell-laden 3D-printed constructs for bone tissue engineering, Journal of Materials Chemistry B 8 (43) (2020) 9836–9862. Available from https://doi.org/10.1039/d0tb02019b.

[19] X. Jiang, Z. Du, X. Zhang, F. Zaman, Z. Song, Y. Guan, et al., Gelatin-based anticancer drug delivery nanosystems: a mini review, Frontiers in Bioengineering and Biotechnology 11 (2023). Available from https://doi.org/10.3389/fbioe.2023.1158749.

[20] K. Goodarzi, S.S. Rao, Hyaluronic acid-based hydrogels to study cancer cell behaviors, Journal of Materials Chemistry B 9 (31) (2021) 6103–6115. Available from https://doi.org/10.1039/d1tb00963j, http://pubs.rsc.org/en/journals/journal/tb.

[21] B. Yu, Y. Li, Y. Lin, Y. Zhu, T. Hao, Y. Wu, et al., Research progress of natural silk fibroin and the application for drug delivery in chemotherapies, Frontiers in Pharmacology 13 (2023). Available from https://doi.org/10.3389/fphar.2022.1071868.

[22] Q.Q. Wang, C. Tan, G. Qin, S.K. Yao, Promising clinical applications of hydrogels associated with precise cancer treatment: a review, Technology in Cancer Research and Treatment 22 (2023). Available from https://doi.org/10.1177/15330338221150322, https://journals.sagepub.com/home/TCT.

[23] R. Mohammadinejad, H. Maleki, E. Larrañeta, A.R. Fajardo, A.B. Nik, A. Shavandi, et al., Status and future scope of plant-based green hydrogels in biomedical engineering, Applied Materials Today 16 (2019) 213–246. Available from https://doi.org/10.1016/j.apmt.2019.04.010.

[24] R. Kumar, A. Butreddy, N. Kommineni, P.G. Reddy, N. Bunekar, C. Sarkar, et al., Lignin: drug/gene delivery and tissue engineering applications, International journal of nanomedicine 16 (2021) 2419–2441. Available from https://doi.org/10.2147/IJN.S303462, https://www.dovepress.com/getfile.php?fileID=68043.

[25] R. Chavooshi, M.R. Ranjkesh, B. Hashemi, L. Roshangar, Cellulose and lignin-derived scaffold and their biological application in tissue engineering, drug delivery, and wound healing: a review, Cell Journal 25 (3) (2023) 158–164. Available from https://doi.org/10.22074/CELLJ.2023.559011.1095, https://www.celljournal.org/article_703191_5ef8677f7f6a341920532006e21a479c.pdf.

[26] C.S. Lee, H.S. Hwang, Starch-based hydrogels as a drug delivery system in biomedical applications, Gels 9 (12) (2023) Available from https://doi.org/10.3390/gels9120951, http://www.mdpi.com/journal/gels.

[27] S. Gajbhiye, S. Dhoble, D. Tobin, Natural Gum (Xanthan, Gellan, Arabic, Guar, Ghatti Gum, Etc.)-based bio-scaffold and their application in tissue engineering, Natural product inspired scaffolds: applications in tissue engineering, Springer Nature, India, (2024), pp. 61–91. Available from https://link.springer.com/book/9789819731114, https://doi.org/10.1007/978-981-97-3111-4_4.

[28] S. Feng, S. Dai, Z. Wei, J. Wang, N. Xiang, P. Shao, Soy conglycinin amyloid fibril and chitosan complex scaffold for cultivated meat application, Food Hydrocolloids 153 (2024) 110017. Available from https://doi.org/10.1016/j.foodhyd.2024.110017.

[29] S. Preetam, D. Duhita Mondal, N. Mukerjee, S.S. Naser, T.A. Tabish, N. Thorat, Revolutionizing cancer treatment: the promising horizon of zein nanosystems, ACS Biomaterials Science & Engineering 10 (4) (2024) 1946–1965. Available from https://doi.org/10.1021/acsbiomaterials.3c01540.

[30] K. Xu, Y. Fu, W.J. Chung, X. Zheng, Y. Cui, I.C. Hsu, et al., Thiol–ene-based biological/synthetic hybrid biomatrix for 3-D living cell culture, Acta biomaterialia 8 (7) (2012) 2504–2516. Available from https://doi.org/10.1016/j.actbio.2012.03.049.

[31] F. Gomi, N. Sasaki, Y. Shichi, F. Minami, S. Shinji, M. Toyoda, et al., Polyvinyl alcohol increased growth, migration, invasion, and sphere size in the PK-8 pancreatic ductal adenocarcinoma cell line, Heliyon 7 (2) (2021) e06182. Available from https://doi.org/10.1016/j.heliyon.2021.e06182.

[32] C.L. Hedegaard, Peptide-protein coassembling matrices as a biomimetic 3D model of ovarian cancer, Science Advances 6 (2024) eabb3298.

[33] A. Gebeyehu, S.K. Surapaneni, J. Huang, A. Mondal, V.Z. Wang, N.F. Haruna, et al., Polysaccharide hydrogel based 3D printed tumor models for chemotherapeutic drug screening, Scientific Reports 11 (1) (2021). Available from https://doi.org/10.1038/s41598-020-79325-8.

[34] L.P. Ferreira, V.M. Gaspar, J.F. Mano, Decellularized extracellular matrix for bioengineering physiomimetic 3D in vitro tumor models, Trends in Biotechnology 38 (12) (2020) 1397–1414. Available from https://doi.org/10.1016/j.tibtech.2020.04.006, http://www.elsevier.com/locate/tibtech.

[35] A.N. Leberfinger, D.J. Ravnic, A. Dhawan, I.T. Ozbolat, Concise review: bioprinting of stem cells for transplantable tissue fabrication, Stem Cells Translational Medicine 6 (10) (2017) 1940–1948. Available from https://doi.org/10.1002/sctm.17-0148, http://stemcellsjournals.onlinelibrary.wiley.com/hub/journal/10.1002/(ISSN)2157-6580.

[36] H. Zhang, 3D bioprinting of multi-biomaterial/crosslinked bioink for skin tissue engineering, Front. Bioeng. Biotechnol. Conference Abstract: 10th World Biomaterials Congress (2016) .

[37] I.T. Ozbolat, Scaffold-based or scaffold-free bioprinting: competing or complementing approaches? Journal of Nanotechnology in Engineering and Medicine 6 (2) (2015). Available from https://doi.org/10.1115/1.4030414.

[38] B.B. Mendes, M. Gómez-Florit, A.G. Hamilton, M.S. Detamore, R.M.A. Domingues, R.L. Reis, et al., Human platelet lysate-based nanocomposite bioink for bioprinting hierarchical fibrillar structures, Biofabrication 12 (1) (2020) 015012. Available from https://doi.org/10.1088/1758-5090/ab33e8.

[39] F. Abasalizadeh, S.V. Moghaddam, E. Alizadeh, E. Akbari, E. Kashani, S.M.B. Fazljou, et al., Alginate-based hydrogels as drug delivery vehicles in cancer treatment and their applications in wound dressing and 3D bioprinting, Journal of biological engineering 14 (1) (2020). Available from https://doi.org/10.1186/s13036-020-0227-7.

[40] M.A. Miranda, P.D. Marcato, A. Mondal, N. Chowdhury, A. Gebeyehu, S.K. Surapaneni, et al., Cytotoxic and chemosensitizing effects of glycoalkaloidic extract on 2D and 3D models using RT4 and patient derived xenografts bladder cancer cells, Materials Science and Engineering: C 119 (2021) 111460. Available from https://doi.org/10.1016/j.msec.2020.111460.

[41] R. Olmos-Juste, B. Alonso-Lerma, R. Pérez-Jiménez, N. Gabilondo, A. Eceiza, 3D printed alginate-cellulose nanofibers based patches for local curcumin administration, Carbohydrate Polymers 264 (2021) 118026. Available from https://doi.org/10.1016/j.carbpol.2021.118026.

[42] Y. Zhang, S.E. Chen, J. Shao, J.J.J.P. Van Den Beucken, Combinatorial surface roughness effects on osteoclastogenesis and osteogenesis, ACS Applied Materials and Interfaces 10 (43) (2018) 36652–36663. Available from https://doi.org/10.1021/acsami.8b10992, http://pubs.acs.org/journal/aamick.

[43] V. Kumar, A. Kumar, N.S. Chauhan, G. Yadav, M. Goswami, G. Packirisamy, Design and fabrication of a dual protein-based trilayered nanofibrous scaffold for efficient wound healing, ACS Applied Bio Materials 5 (6) (2022) 2726–2740. Available from https://doi.org/10.1021/acsabm.2c00200.

[44] W. Yang, H. Xu, Y. Lan, Q. Zhu, Y. Liu, S. Huang, et al., Preparation and characterisation of a novel silk fibroin/hyaluronic acid/sodium alginate scaffold for skin repair, International Journal of Biological Macromolecules 130 (2019) 58–67. Available from https://doi.org/10.1016/j.ijbiomac.2019.02.120.

[45] B. Chu, J.-mei He, Z. Wang, L.-lan Liu, X.-li Li, C.-X. Wu, et al., Proangiogenic peptide nanofiber hydrogel/3D printed scaffold for dermal regeneration, Chemical Engineering Journal 424 (3) (2021) 128146. Available from https://doi.org/10.1016/j.cej.2020.128146.

[46] S. Ramasamy, P. Davoodi, S. Vijayavenkataraman, J.H. Teoh, A.M. Thamizhchelvan, K.S. Robinson, et al., Optimized construction of a full thickness human skin equivalent using 3D bioprinting and a PCL/collagen dermal scaffold, Bioprinting 21 (2021) e00123. Available from https://doi.org/10.1016/j.bprint.2020.e00123.

[47] N. Zandi, B. Dolatyar, R. Lotfi, Y. Shallageh, M.A. Shokrgozar, E. Tamjid, et al., Biomimetic nanoengineered scaffold for enhanced full-thickness cutaneous wound healing, Acta biomaterialia 124 (2021) 191–204. Available from https://doi.org/10.1016/j.actbio.2021.01.029.

[48] Y.-J. Sun, C.-H. Hsu, T.-Y. Ling, L. Liu, T.-C. Lin, S. Jakfar, et al., The preparation of cell-containing microbubble scaffolds to mimic alveoli structure as a 3D drug-screening system for lung cancer, Biofabrication 12 (2) (2020) 025031. Available from https://doi.org/10.1088/1758-5090/ab78ee.

[49] M. Tavakoli, M. Mirhaj, J. Varshosaz, S. Salehi, S.M. Mohanna, S. Salehi, et al., Asymmetric tri-layer sponge-nanofiber wound dressing containing insulin-like growth factor-1 and multi-walled carbon nanotubes for acceleration of full-thickness wound healing, Biomaterials Advances 151 (2023) 213468. Available from https://doi.org/10.1016/j.bioadv.2023.213468.

[50] C. Kilic Bektas, I. Kimiz, A. Sendemir, V. Hasirci, N. Hasirci, A bilayer scaffold prepared from collagen and carboxymethyl cellulose for skin tissue engineering applications, Journal of Biomaterials Science, Polymer Edition 29 (14) (2018) 1764–1784. Available from https://doi.org/10.1080/09205063.2018.1498718.

[51] B. Sumathy, S. Velayudhan, Fabrication and evaluation of a bi-layered gelatin based scaffold with arrayed micro-pits for full-thickness skin construct, International Journal of Biological Macromolecules 251 (2023) 126360. Available from https://doi.org/10.1016/j.ijbiomac.2023.126360.

[52] M.T. Chorsi, E.J. Curry, H.T. Chorsi, R. Das, J. Baroody, P.K. Purohit, et al., Piezoelectric biomaterials for sensors and actuators, Advanced Materials 31 (1) (2019). Available from https://doi.org/10.1002/adma.201802084, http://onlinelibrary.wiley.com/journal/10.1002/(ISSN)1521-4095.

[53] J.Y. Wong, R. Langer, D.E. Ingber, Electrically conducting polymers can noninvasively control the shape and growth of mammalian cells, Proceedings of the National Academy of Sciences 91 (8) (1994) 3201–3204. Available from https://doi.org/10.1073/pnas.91.8.3201.

[54] L. Roseti, V. Parisi, M. Petretta, C. Cavallo, G. Desando, I. Bartolotti, et al., Scaffolds for bone tissue engineering: state of the art and new perspectives, Materials Science and Engineering: C 78 (2017) 1246–1262. Available from https://doi.org/10.1016/j.msec.2017.05.017.

[55] L. Suamte, A. Tirkey, J. Barman, P. Jayasekhar Babu, Various manufacturing methods and ideal properties of scaffolds for tissue engineering applications, Smart Materials in Manufacturing 1 (2023) 100011. Available from https://doi.org/10.1016/j.smmf.2022.100011.

[56] M.S. Shoichet, Polymer scaffolds for biomaterials applications, Macromolecules 43 (2) (2010) 581–591. Available from https://doi.org/10.1021/ma901530r.

[57] H.H. Abdelkader, M. Mabrouk, A.A. Salama, E.M.A. Hamzawy, Alginate-based composite scaffolds loaded with streptomycin sulfate for bone regeneration: in vitro studies, Journal of Drug Delivery Science and Technology 94 (2024). Available from https://doi.org/10.1016/j.jddst.2024.105486, https://www.sciencedirect.com/science/journal/17732247.

[58] S. Tarafder, A. Koch, Y. Jun, C. Chou, M.R. Awadallah, C.H. Lee, Micro-precise spatiotemporal delivery system embedded in 3D printing for complex tissue regeneration, Biofabrication 8 (2) (2016) 025003. Available from https://doi.org/10.1088/1758-5090/8/2/025003.

[59] N.J. Castro, J. O'Brien, L.G. Zhang, Integrating biologically inspired nanomaterials and table-top stereolithography for 3D printed biomimetic osteochondral scaffolds, Nanoscale 7 (33) (2015) 14010–14022. Available from https://doi.org/10.1039/c5nr03425f, http://pubs.rsc.org/en/journals/journal/nr.

[60] W. Zhu, H. Cui, B. Boualam, F. Masood, E. Flynn, R.D. Rao, et al., 3D bioprinting mesenchymal stem cell-laden construct with core–shell nanospheres for cartilage tissue engineering, Nanotechnology 29 (18) (2018) 185101. Available from https://doi.org/10.1088/1361-6528/aaafa1.

[61] K. Legemate, S. Tarafder, Y. Jun, C.H. Lee, Engineering human TMJ discs with protein-releasing 3D-printed scaffolds, Journal of Dental Research 95 (7) (2016) 800–807. Available from https://doi.org/10.1177/0022034516642404.

[62] H. Cho, S. Tarafder, M. Fogge, K. Kao, C.H. Lee, Periodontal ligament stem/progenitor cells with protein-releasing scaffolds for cementum formation and integration on dentin surface, Connective Tissue Research 57 (6) (2016) 488–495. Available from https://doi.org/10.1080/03008207.2016.1191478.

[63] S.J. Lee, W. Zhu, L. Heyburn, M. Nowicki, B. Harris, L.G. Zhang, Development of novel 3-D printed scaffolds with core-shell nanoparticles for nerve regeneration, IEEE Transactions on Biomedical Engineering 64 (2) (2017) 408–418. Available from https://doi.org/10.1109/TBME.2016.2558493, http://ieeexplore.ieee.org/xpl/RecentIssue.jsp?reload=true&punumber=10.

[64] Y.B. Lee, S. Polio, W. Lee, G. Dai, L. Menon, R.S. Carroll, et al., Bio-printing of collagen and VEGF-releasing fibrin gel scaffolds for neural stem cell culture, Experimental Neurology 223 (2) (2010) 645–652. Available from https://doi.org/10.1016/j.expneurol.2010.02.014.

[65] S. Huang, B. Yao, J. Xie, X. Fu, 3D bioprinted extracellular matrix mimics facilitate directed differentiation of epithelial progenitors for sweat gland regeneration, Acta Biomaterialia 32 (2016) 170–177. Available from https://doi.org/10.1016/j.actbio.2015.12.039.

[66] F. Fahimipour, M. Rasoulianboroujeni, E. Dashtimoghadam, K. Khoshroo, M. Tahriri, F. Bastami, et al., 3D printed TCP-based scaffold incorporating VEGF-loaded PLGA microspheres for craniofacial tissue engineering, Dental Materials 33 (11) (2017) 1205–1216. Available from https://doi.org/10.1016/j.dental.2017.06.016.

[67] T. Ahlfeld, A.R. Akkineni, Y. Förster, T. Köhler, S. Knaack, M. Gelinsky, et al., Design and fabrication of complex scaffolds for bone defect healing: combined 3D plotting of a calcium phosphate cement and a growth factor-loaded hydrogel, Annals of Biomedical Engineering 45 (1) (2017) 224–236. Available from https://doi.org/10.1007/s10439-016-1685-4.

[68] M.T. Poldervaart, H. Gremmels, K. Van Deventer, J.O. Fledderus, F.C. Öner, M.C. Verhaar, et al., Prolonged presence of VEGF promotes vascularization in 3D bioprinted scaffolds with defined architecture, Journal of Controlled Release 184 (1) (2014) 58–66. Available from https://doi.org/10.1016/j.jconrel.2014.04.007, http://www.elsevier.com/locate/jconrel.

[69] A.R. Akkineni, Y. Luo, M. Schumacher, B. Nies, A. Lode, M. Gelinsky, 3D plotting of growth factor loaded calcium phosphate cement scaffolds, Acta biomaterialia 27 (2015) 264–274. Available from https://doi.org/10.1016/j.actbio.2015.08.036, http://www.journals.elsevier.com/acta-biomaterialia.

[70] T. Ahlfeld, G. Cidonio, D. Kilian, S. Duin, A.R. Akkineni, J.I. Dawson, et al., Development of a clay based bioink for 3D cell printing for skeletal application, Biofabrication 9 (3) (2017) 034103. Available from https://doi.org/10.1088/1758-5090/aa7e96.

[71] C. Wang, Y. Zhou, M. Wang, In situ delivery of rhBMP-2 in surface porous shape memory scaffolds developed through cryogenic 3D plotting, Materials Letters 189 (2017) 140–143. Available from https://doi.org/10.1016/j.matlet.2016.11.039.

[72] F.E. Freeman, D.J. Kelly, Tuning alginate bioink stiffness and composition for controlled growth factor delivery and to spatially direct MSC Fate within bioprinted tissues, Scientific Reports 7 (1) (2017) Available from https://doi.org/10.1038/s41598-017-17286-1, http://www.nature.com/srep/index.html.

[73] M. Du, B. Chen, Q. Meng, S. Liu, X. Zheng, C. Zhang, et al., 3D bioprinting of BMSC-laden methacrylamide gelatin scaffolds with CBD-BMP2-collagen microfibers, Biofabrication 7 (4) (2015) 044104. Available from https://doi.org/10.1088/1758-5090/7/4/044104.

[74] J.Y. Park, J.H. Shim, S.A. Choi, J. Jang, M. Kim, S.H. Lee, et al., 3D printing technology to control BMP-2 and VEGF delivery spatially and temporally to promote large-volume bone regeneration, Journal of Materials Chemistry B 3 (27) (2015) 5415–5425. Available from https://doi.org/10.1039/c5tb00637f, http://pubs.rsc.org/en/journals/journalissues/tb.

[75] E.B. Bae, K.H. Park, J.H. Shim, H.Y. Chung, J.W. Choi, J.J. Lee, et al., Efficacy of rhBMP-2 loaded PCL/ β -TCP/bdECM scaffold fabricated by 3D printing technology on bone regeneration, BioMed Research International 2018 (2018). Available from https://doi.org/10.1155/2018/2876135, http://www.hindawi.com/journals/biomed/.

[76] J.K. Park, J.H. Shim, K.S. Kang, J. Yeom, H.S. Jung, J.Y. Kim, et al., Solid free-form fabrication of tissue-engineering scaffolds with a poly (lactic-co-glycolic acid) grafted hyaluronic acid conjugate encapsulating and intact bone morphogenetic protein-2/poly(ethylene glycol) complex, Advanced Functional Materials 21 (15) (2011) 2906–2912. Available from https://doi.org/10.1002/adfm.201100612.

[77] J.W. Lee, K.S. Kang, S.H. Lee, J.Y. Kim, B.K. Lee, D.W. Cho, Bone regeneration using a microstereolithography-produced customized poly (propylene fumarate)/diethyl fumarate photopolymer 3D scaffold incorporating BMP-2 loaded PLGA microspheres, Biomaterials 32 (3) (2011) 744–752. Available from https://doi.org/10.1016/j.biomaterials.2010.09.035.

[78] C. Wang, Q. Zhao, M. Wang, Cryogenic 3D printing for producing hierarchical porous and rhBMP-2-loaded Ca-P/PLLA nanocomposite scaffolds for bone tissue engineering, Biofabrication 9 (2) (2017) 025031. Available from https://doi.org/10.1088/1758-5090/aa71c9.

[79] K.H. Huang, Y.H. Lin, M.Y. Shie, C.P. Lin, Effects of bone morphogenic protein-2 loaded on the 3D-printed MesoCS scaffolds, Journal of the Formosan Medical Association 117 (10) (2018) 879–887. Available from https://doi.org/10.1016/j.jfma.2018.07.010, http://www.journals.elsevier.com/journal-of-the-formosan-medical-association/.

[80] F. Wang, X. Cai, Y. Shen, L. Meng, Cell–scaffold interactions in tissue engineering for oral and craniofacial reconstruction, Bioactive Materials 23 (2023) 16–44. Available from https://doi.org/10.1016/j.bioactmat.2022.10.029.

[81] A. Zielińska, J. Karczewski, P. Eder, T. Kolanowski, M. Szalata, K. Wielgus, et al., Scaffolds for drug delivery and tissue engineering: the role of genetics, Journal of Controlled Release 359 (2023) 207–223. Available from https://doi.org/10.1016/j.jconrel.2023.05.042.

[82] A. Mansour, M. Romani, A.B. Acharya, B. Rahman, E. Verron, Z. Badran, Drug delivery systems in regenerative medicine: an updated review, Pharmaceutics 15 (2) (2023) 695. Available from https://doi.org/10.3390/pharmaceutics15020695.

[83] P. Martinez-Redondo, J.C. Izpisua Belmonte, Tailored chromatin modulation to promote tissue regeneration, Seminars in Cell and Developmental Biology 97 (2020) 3–15. Available from https://doi.org/10.1016/j.semcdb.2019.04.015, http://www.elsevier.com/inca/publications/store/6/2/2/9/4/4/index.htt.
[84] I.M. Zurina, V.S. Presniakova, D.V. Butnaru, A.A. Svistunov, P.S. Timashev, Y.A. Rochev, Tissue engineering using a combined cell sheet technology and scaffolding approach, Acta biomaterialia 113 (2020) 63–83. Available from https://doi.org/10.1016/j.actbio.2020.06.016.
[85] C. Li, J. Wang, Y. Wang, H. Gao, G. Wei, Y. Huang, et al., Recent progress in drug delivery, Acta Pharmaceutica Sinica B 9 (6) (2019) 1145–1162. Available from https://doi.org/10.1016/j.apsb.2019.08.003.
[86] A. Gaikwad, M. Waje, V. Tile, P. Patharkar, Targeted drug delivery -from magic bullet to nanomedicine introduction, Journal of Xidian University 17 (2023) 1540–1551.
[87] J. Yoo, C. Park, G. Yi, D. Lee, H. Koo, Active targeting strategies using biological ligands for nanoparticle drug delivery systems, Cancers 11 (5) (2019) 640. Available from https://doi.org/10.3390/cancers11050640.
[88] H.A.E. Benson, J.E. Grice, Y. Mohammed, S. Namjoshi, M.S. Roberts, Topical and transdermal drug delivery: from simple potions to smart technologies, Current drug delivery 16 (5) (2019) 444–460. Available from https://doi.org/10.2174/1567201816666190201143457.
[89] A. Mazumder, C. Cerella, M. Diederich, Natural scaffolds in anticancer therapy and precision medicine, Biotechnology advances 36 (6) (2018) 1563–1585. Available from https://doi.org/10.1016/j.biotechadv.2018.04.009.
[90] K.N. Bitar, E. Zakhem, Design strategies of biodegradable scaffolds for tissue regeneration, Biomedical Engineering and Computational Biology 6 (2014). Available from https://doi.org/10.4137/becb.s10961, BECB.S10961.
[91] R. Zeinali, L.J. del Valle, J. Torras, J. Puiggalí, Recent progress on biodegradable tissue engineering scaffolds prepared by thermally-induced phase separation (TIPS), International Journal of Molecular Sciences 22 (7) (2021) 3504. Available from https://doi.org/10.3390/ijms22073504.
[92] H.S. Kim, S.G. Kumbar, S.P. Nukavarapu, Biomaterial-directed cell behavior for tissue engineering, Current Opinion in Biomedical Engineering 17 (2021). Available from https://doi.org/10.1016/j.cobme.2020.100260, https://www.journals.elsevier.com/current-opinion-in-biomedical-engineering.
[93] Z. Su, K. Dhusia, Y. Wu, Understand the functions of scaffold proteins in cell signaling by a mesoscopic simulation method, Biophysical Journal 119 (10) (2020) 2116–2126. Available from https://doi.org/10.1016/j.bpj.2020.10.002, http://www.sciencedirect.com/science/journal/00063495.
[94] R. Podlipec, J. Štrancar, Cell-scaffold adhesion dynamics measured in first seconds predicts cell growth on days scale – optical tweezers study, ACS Applied Materials & Interfaces 7 (12) (2015) 6782–6791. Available from https://doi.org/10.1021/acsami.5b00235.
[95] N. Almouemen, H.M. Kelly, C. O'Leary, Tissue engineering: understanding the role of biomaterials and biophysical forces on cell functionality through computational and structural biotechnology analytical methods, Computational and Structural Biotechnology Journal 17 (2019) 591–598. Available from https://doi.org/10.1016/j.csbj.2019.04.008, http://www.csbj.org.
[96] J.C. Valdoz, B.C. Johnson, D.J. Jacobs, N.A. Franks, E.L. Dodson, C. Sanders, et al., The ECM: to scaffold, or not to scaffold, that is the question, International Journal of Molecular Sciences 22 (23) (2021) 12690. Available from https://doi.org/10.3390/ijms222312690.
[97] G. Lutzweiler, A.N. Halili, N.E. Vrana, The overview of porous, bioactive scaffolds as instructive biomaterials for tissue regeneration and their clinical translation, Pharmaceutics 12 (7) (2020) 1–29. Available from https://doi.org/10.3390/pharmaceutics12070602, https://www.mdpi.com/1999-4923/12/7/602/pdf.
[98] D. Yari, M.H. Ebrahimzadeh, J. Movaffagh, A. Shahroodi, M. Shirzad, D. Qujeq, et al., Biochemical aspects of scaffolds for cartilage tissue engineering; from basic science to regenerative medicine, Archives of Bone and Joint Surgery 10 (3) (2022) 229–244. Available from https://doi.org/10.22038/ABJS.2022.55549.2766, https://abjs.mums.ac.ir/issue_2136_2159.html.
[99] B. Yue, Biology of the extracellular matrix: an overview, Journal of Glaucoma 23 (8) (2014) S20. Available from https://doi.org/10.1097/IJG.0000000000000108, http://journals.lww.com/glaucomajournal.
[100] J.M. Muncie, V.M. Weaver, The physical and biochemical properties of the extracellular matrix regulate cell fate, Current Topics in Developmental Biology 130 (2018) 1–37. Available from https://doi.org/10.1016/bs.ctdb.2018.02.002, http://www.elsevier.com/wps/find/bookdescription.cws_home/678093/description#description.
[101] M. Yang, Z.-C. Zhang, Y. Liu, Y.-R. Chen, R.-H. Deng, Z.-N. Zhang, et al., Function and mechanism of RGD in bone and cartilage tissue engineering, Frontiers in Bioengineering and Biotechnology 9 (2021). Available from https://doi.org/10.3389/fbioe.2021.773636.
[102] B. Zhao, Q. Peng, R. Zhou, H. Liu, S. Qi, R. Wang, Precision medicine in tissue engineering on bone, Methods in Molecular Biology 2204 (2020) 207–215. Available from https://doi.org/10.1007/978-1-0716-0904-0_18, http://www.springer.com/series/7651.
[103] A. Acar, A. Hidalgo-Sastre, M.K. Leverentz, C.G. Mills, S. Woodcock, M. Baron, et al., Inhibition of Wnt signalling by Notch via two distinct mechanisms, Scientific Reports 11 (1) (2021). Available from https://doi.org/10.1038/s41598-021-88618-5.
[104] R.G.M. Breuls, T.U. Jiya, T.H. Smit, Scaffold stiffness influences cell behavior: opportunities for skeletal tissue engineering, The Open Orthopaedics Journal 2 (1) (2008) 103–109. Available from https://doi.org/10.2174/1874325000802010103.
[105] B.E. Housden, N. Perrimon, Spatial and temporal organization of signaling pathways, Trends in Biochemical Sciences 39 (10) (2014) 457–464. Available from https://doi.org/10.1016/j.tibs.2014.07.008, http://www.elsevier.com/locate/tibs.
[106] H. Vilaça-Faria, J. Noro, R.L. Reis, R.P. Pirraco, Extracellular matrix-derived materials for tissue engineering and regenerative medicine: a journey from isolation to characterization and application, Bioactive Materials 34 (2024) 494–519. Available from https://doi.org/10.1016/j.bioactmat.2024.01.004.
[107] M. Ahmed, C. Ffrench-Constant, Extracellular matrix regulation of stem cell behavior, Current Stem Cell Reports 2 (3) (2016) 197–206. Available from https://doi.org/10.1007/s40778-016-0056-2.

[108] M. Akhmanova, E. Osidak, S. Domogatsky, S. Rodin, A. Domogatskaya, Physical, spatial, and molecular aspects of extracellular matrix of in vivo niches and artificial scaffolds relevant to stem cells research, Stem Cells International 2015 (2015) 1–35. Available from https://doi.org/10.1155/2015/167025.

[109] M. Cámara-Torres, R. Sinha, P. Scopece, T. Neubert, K. Lachmann, A. Patelli, et al., Tuning cell behavior on 3D scaffolds fabricated by atmospheric plasma-assisted additive manufacturing, ACS Applied Materials & Interfaces 13 (3) (2021) 3631–3644. Available from https://doi.org/10.1021/acsami.0c19687.

[110] E.M. Bueno, G. Laevsky, G.A. Barabino, Enhancing cell seeding of scaffolds in tissue engineering through manipulation of hydrodynamic parameters, Journal of Biotechnology 129 (3) (2007) 516–531. Available from https://doi.org/10.1016/j.jbiotec.2007.01.005.

[111] D. Ali, Effect of scaffold architecture on cell seeding efficiency: a discrete phase model CFD analysis, Computers in Biology and Medicine 109 (2019) 62–69. Available from https://doi.org/10.1016/j.compbiomed.2019.04.025.

[112] J. Chi, M. Wang, J. Chen, L. Hu, Z. Chen, L.J. Backman, et al., Topographic orientation of scaffolds for tissue regeneration: recent advances in biomaterial design and applications, Biomimetics 7 (3) (2022) 131. Available from https://doi.org/10.3390/biomimetics7030131.

[113] Y. Esmaeili, E. Bidram, A. Bigham, M. Atari, R. Nasr Azadani, M. Tavakoli, et al., Exploring the evolution of tissue engineering strategies over the past decade: from cell-based strategies to gene-activated matrix, Alexandria Engineering Journal 81 (2023) 137–169. Available from https://doi.org/10.1016/j.aej.2023.08.080.

[114] P.S. Poh, T. Lingner, S. Kalkhof, S. Märdian, J. Baumbach, P. Dondl, et al., Enabling technologies towards personalization of scaffolds for large bone defect regeneration, Current Opinion in Biotechnology 74 (2022) 263–270. Available from https://doi.org/10.1016/j.copbio.2021.12.002, http://www.elsevier.com/locate/copbio.

[115] C.V. Rocha, V. Gonçalves, M.C. da Silva, M. Bañobre-López, J. Gallo, PLGA-based composites for various biomedical applications, International Journal of Molecular Sciences 23 (4) (2022). Available from https://doi.org/10.3390/ijms23042034, https://www.mdpi.com/1422-0067/23/4/2034/pdf.

[116] F. Shang, Y. Yu, S. Liu, L. Ming, Y. Zhang, Z. Zhou, et al., Advancing application of mesenchymal stem cell-based bone tissue regeneration, Bioactive Materials 6 (3) (2021) 666–683. Available from https://doi.org/10.1016/j.bioactmat.2020.08.014.

[117] É.R. Oliveira, L. N, D. P, A. A, J. R, D.L. B, et al., Advances in growth factor delivery for bone tissue engineering, International Journal of Molecular Sciences 22 (2021) 903.

[118] B. Fisch, R. Abir, Female fertility preservation: past, present and future, Reproduction (Cambridge, England) 156 (1) (2018) F11. Available from https://doi.org/10.1530/rep-17-0483.

[119] R. Nair, M. Kasturi, V. Mathur, R.N. Seetharam, K.S. Vasanthan, Strategies for developing 3D printed ovarian model for restoring fertility, Clinical and Translational Science 17 (7) (2024). Available from https://doi.org/10.1111/cts.13863, http://onlinelibrary.wiley.com/journal/10.1111/(ISSN)1752-8062.

[120] S. Canosa, A. Revelli, G. Gennarelli, G. Cormio, V. Loizzi, F. Arezzo, et al., Innovative strategies for fertility preservation in female cancer survivors: new hope from artificial ovary construction and stem cell-derived neo-folliculogenesis, Healthcare 11 (20) (2023) 2748. Available from https://doi.org/10.3390/healthcare11202748.

[121] X. Liu, K. Wu, L. Gao, L. Wang, X. Shi, Biomaterial strategies for the application of reproductive tissue engineering, Bioactive Materials 14 (2022) 86–96. Available from https://doi.org/10.1016/j.bioactmat.2021.11.023, http://www.keaipublishing.com/en/journals/bioactive-materials/.

[122] N. Zaninovic, O. Elemento, Z. Rosenwaks, Artificial intelligence: its applications in reproductive medicine and the assisted reproductive technologies, Fertility and Sterility 112 (1) (2019) 28–30. Available from https://doi.org/10.1016/j.fertnstert.2019.05.019.

[123] J.S. Del Valle, S.M. Chuva de Sousa Lopes, Bioengineered 3D ovarian models as paramount technology for female health management and reproduction, Bioengineering 10 (7) (2023) Available from https://doi.org/10.3390/bioengineering10070832, http://www.mdpi.com/journal/bioengineering.

[124] S. Han, J. Wu, Three-dimensional (3D) scaffolds as powerful weapons for tumor immunotherapy, Bioactive Materials 17 (2022) 300–319. Available from https://doi.org/10.1016/j.bioactmat.2022.01.020.

[125] Y. Jahanbani, S. Davaran, M. Ghahremani-Nasab, L. Aghebati-Maleki, M. Yousefi, Scaffold-based tissue engineering approaches in treating infertility, Life Sciences 240 (2020) 117066. Available from https://doi.org/10.1016/j.lfs.2019.117066.

[126] T.K. Woodruff, Oncofertility: a grand collaboration between reproductive medicine and oncology, Reproduction (Cambridge, England) 150 (3) (2015) S1. Available from https://doi.org/10.1530/rep-15-0163.

[127] S. Braccini, C. Tacchini, F. Chiellini, D. Puppi, Polymeric hydrogels for in vitro 3D ovarian cancer modeling, International Journal of Molecular Sciences 23 (6) (2022) 3265. Available from https://doi.org/10.3390/ijms23063265.

[128] H. Ghazal, A. Waqar, F. Yaseen, M. Shahid, M. Sultana, M. Tariq, et al., Role of nanoparticles in enhancing chemotherapy efficacy for cancer treatment, Next Materials 2 (2024) 100128. Available from https://doi.org/10.1016/j.nxmate.2024.100128.

[129] S. Gavas, S. Quazi, T.M. Karpiński, Nanoparticles for cancer therapy: current progress and challenges, Nanoscale Research Letters 16 (1) (2021). Available from https://doi.org/10.1186/s11671-021-03628-6, http://www.springer.com/materials/nanotechnology/journal/11671.

[130] S. Aghamiri, S. Talaei, A.A. Ghavidel, F. Zandsalimi, S. Masoumi, N.H. Hafshejani, et al., Nanoparticles-mediated CRISPR/Cas9 delivery: recent advances in cancer treatment, Journal of Drug Delivery Science and Technology 56 (2020) 101533. Available from https://doi.org/10.1016/j.jddst.2020.101533.

[131] Westwood IJNEEL. The road after cancer: biomaterials and tissue engineering approaches to mediate the tumor microenvironment post-cancer treatment. Frontiers in Biomaterials Science. 2023;3:1347324.

[132] Y. Qu, K. Lu, Y. Zheng, C. Huang, G. Wang, Y. Zhang, et al., Photothermal scaffolds/surfaces for regulation of cell behaviors, Bioactive Materials 8 (2022) 449–477. Available from https://doi.org/10.1016/j.bioactmat.2021.05.052.

[133] V. Mansouri, N. Beheshtizadeh, M. Gharibshahian, L. Sabouri, M. Varzandeh, N. Rezaei, Recent advances in regenerative medicine strategies for cancer treatment, Biomedicine & Pharmacotherapy 141 (2021) 111875. Available from https://doi.org/10.1016/j.biopha.2021.111875.

[134] Y. Zhao, E.Y. Wang, F.B.L. Lai, K. Cheung, M. Radisic, Organs-on-a-chip: a union of tissue engineering and microfabrication, Trends in Biotechnology 41 (3) (2023) 410–424. Available from https://doi.org/10.1016/j.tibtech.2022.12.018.

[135] H. Mao, L. Yang, H. Zhu, L. Wu, P. Ji, J. Yang, et al., Recent advances and challenges in materials for 3D bioprinting, Progress in Natural Science: Materials International 30 (5) (2020) 618–634. Available from https://doi.org/10.1016/j.pnsc.2020.09.015.

[136] M. Gharibshahian, M. Torkashvand, M. Bavisi, N. Aldaghi, A. Alizadeh, Recent advances in artificial intelligent strategies for tissue engineering and regenerative medicine, Skin Research and Technology 30 (9) (2024). Available from https://doi.org/10.1111/srt.70016.

[137] K.W. Cho, W.H. Lee, B.S. Kim, D.H. Kim, Sensors in heart-on-a-chip: a review on recent progress, Elsevier B.V., South Korea Talanta 219 (2020). Available from https://doi.org/10.1016/j.talanta.2020.121269, https://www.journals.elsevier.com/talanta.

[138] M. Kasturi, V. Mathur, M. Gadre, V. Srinivasan, K.S. Vasanthan, Three dimensional bioprinting for hepatic tissue engineering: from in vitro models to clinical applications, Tissue Engineering and Regenerative Medicine 21 (1) (2024) 21–52. Available from https://doi.org/10.1007/s13770-023-00576-3, https://www.springer.com/journal/13770.

[139] J. Zhu, Application of organ-on-chip in drug discovery, Journal of Biosciences and Medicines 08 (03) (2020) 119–134. Available from https://doi.org/10.4236/jbm.2020.83011.

Chapter 9

Nonmedical applications of biomaterials

Khushboo Razdan[1] and Nishant Tyagi[2]
[1]*Amity Institute of Biotechnology, Amity University, Noida, Uttar Pradesh, India,* [2]*Voxvess Biotech Lab Private Limited, New Delhi, India*

9.1 Introduction

Biomaterial, as a term, was first defined in 1986 as "a nonviable material used in a medical device, intended to interact with biological systems" [1]. Initially, biomaterials were primarily developed and explored as a means for medical applications in regenerative medicine due to their enhanced biocompatibility. However, as time has passed, this definition cannot properly encapsulate the evolution in the field. Biomaterials traditionally associated with implanted devices, drug delivery systems, imaging agents, and tissue engineering constructs have expanded their use into various industries [2].

Biomaterials may be derived naturally from plants, animals, microorganisms, or fungi. They are mostly biological polymers, such as polysaccharides. However, synthetic biomaterials are available as well. In the context of nonmedical applications, biomaterials are materials of biological origin [3].

Modern biomaterials science seeks to address challenges that arise due to environmental pollution and sustainable development, and to create novel inventions by creating alternatives that combine functionality while being eco-friendly and biodegradable. This shift is driven by growing awareness of the ecological damage caused by petroleum-based synthetics and enabled by advances in synthetic biology, microbial engineering, and nanotechnology.

Consumer preferences and regulatory frameworks now favor renewable, ethically sourced materials that are biodegradable. This shift reshapes product design and process chains across industries. Recent surveys indicate that the majority of global consumers have been moving towards purchasing sustainable products over the past five years, as seen in Fig. 9.1 [4].

This chapter explores the emerging nonmedical uses of biomaterials, including in agriculture, food packaging, pollution control, radiation mitigation, energy, aerospace, cosmetics, and forensics. Fig. 9.2 summarizes these diverse areas. Understanding these applications is essential to aid the armed forces and provides future directions. Examining the current technologies, the need for innovation, and their future potential gives insight into creating circular economies and aligning with the UN Sustainable Development Goals.

9.2 Agriculture and sustainable packaging

The concept of green technology has been gaining relevance in recent years to combat climate change. One of the most important frontiers in this change is the agriculture and packaging sector. The use of technology and innovations that help preserve soil health and reduce damage to the ecosystem while also maintaining crop viability are being prioritized. A major problem threatening the future of food is plastic consumption in these industries. Plastic has deeply integrated itself into food production from seedling trays, product packaging, and irrigation, which contribute to the biomagnification of microplastics, waste accumulation, and reduction in soil quality. The United Nations Food and Agriculture Organization released a report that calculated the use of plastics in agricultural value chains and packaging in 2019, which was 12.5 million tonnes and 37.3 million tonnes, respectively [5]. This number is said to increase by 50% by 2030, according to the same report. With an increased need for sustainable agriculture, biomaterials originating from biological sources such as cellulose, chitosan, starch, mycelium, and even microbial polymers like polyhydroxyalkanoates (PHAs) have emerged as potential solutions, as these materials integrate into the natural cycles without leaving behind harmful residues, as shown in Table 9.1.

Fundamentals of Translational Tissue Engineering. DOI: https://doi.org/10.1016/B978-0-443-33455-9.00004-3

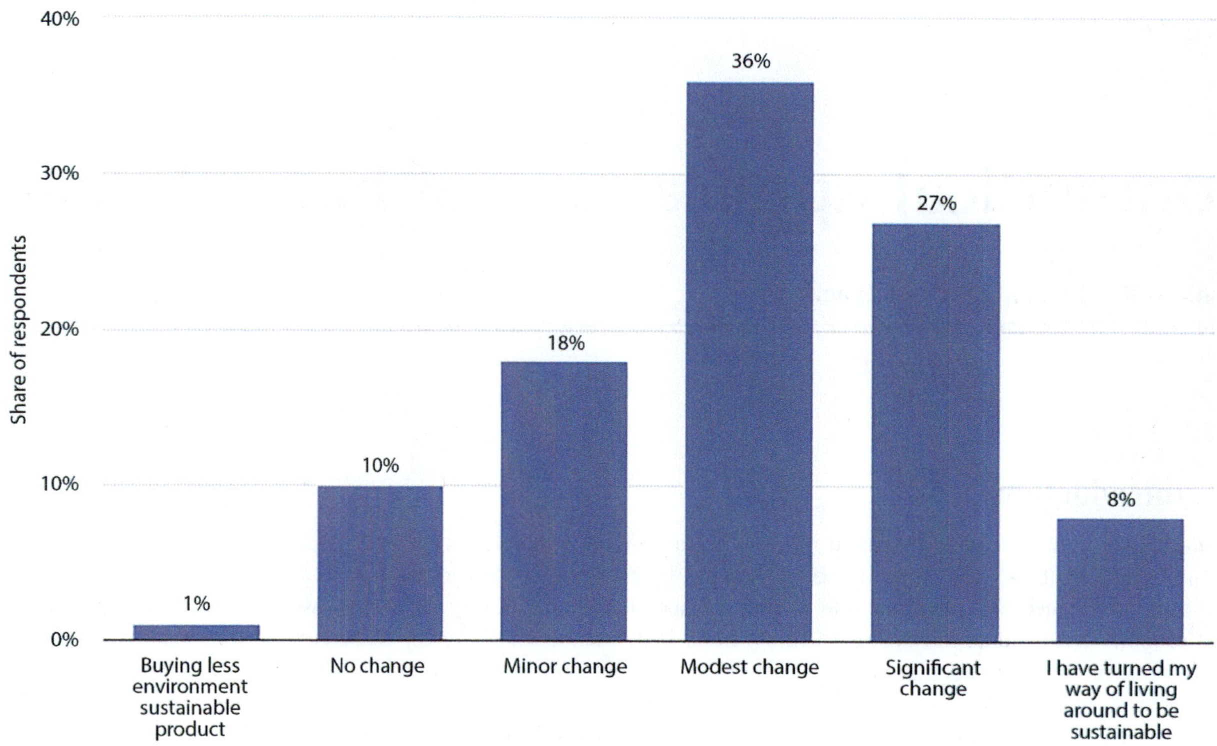

FIGURE 9.1 Degree to which consumers' purchasing behavior and choices shifted towards buying more sustainable products over the past five years worldwide in 2022 [4].

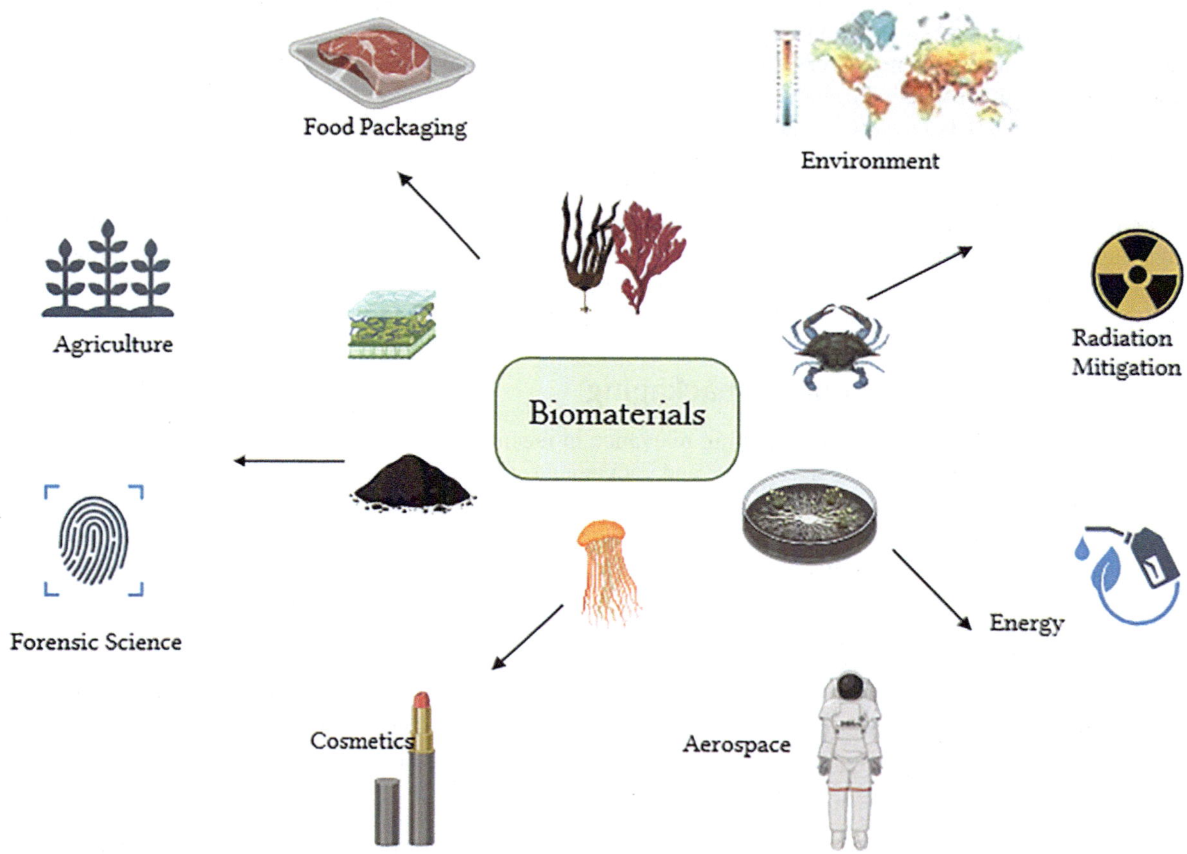

FIGURE 9.2 Overview of nonmedical applications of biomaterials. Map lines delineate study areas and do not necessarily depict accepted national boundaries.

TABLE 9.1 Specific applications of biomaterials in agriculture.

Area	Application	Description
Agriculture	Biodegradable mulch films and seed coatings	Polymers derived from cellulose, chitosan, starch, and PHAs are utilized to make mulch films and seed coatings. Films derived from starch have shown efficiency in weed suppression, gradual decomposition, moisture retention, and soil enrichment [6]. Chitosan coatings provide a dual advantage due to their antimicrobial properties and physical protection, which helps support seed germination [7].
	Controlled-release fertilizers and soil conditioners	Composites of starch and PHAs are being used to encapsulate fertilizers, ensuring nutrient availability, which is in line with the plant growth stage [8]
	Soil health enhancers and hydrogels	Hydrogels derived from cellulose and alginate are used to enhance water retention in arid soils. These hydrogels swell and release water over time, which supports crops that have limited water available [6].
	Integration of biosensors in agricultural biopolymers	Biosensors help in real-time monitoring of soil conditions, empowering farmers to make the best choices for their crops [8]

PHAs, Polyhydroxyalkanoates.

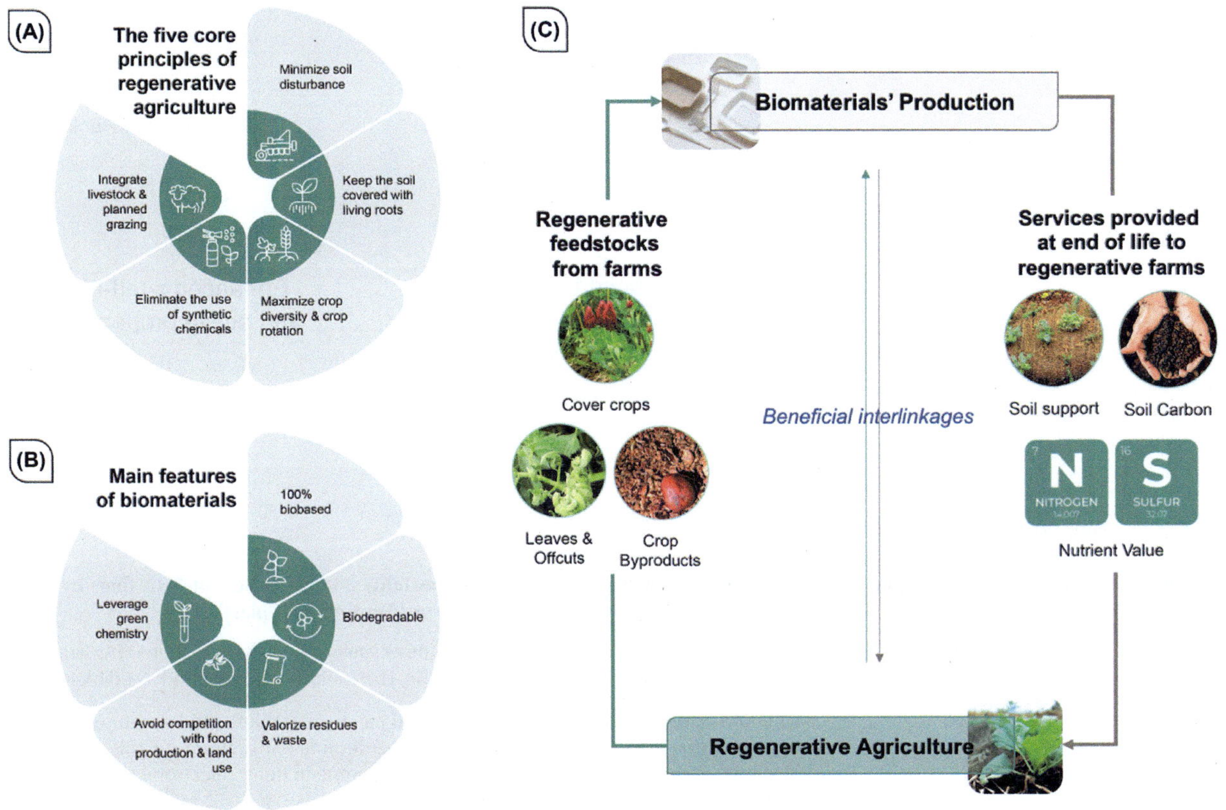

FIGURE 9.3 Concepts of regenerative agriculture and the role of Biomaterials within it. (A) Core principles of regenerative agriculture. (B) Main features of biomaterials. (C) Cycle of biomaterials in regenerative agriculture [7].

9.2.1 Applications in agriculture and sustainable packaging

Conventional methods in agriculture have become reliant on plastics for seed coatings, seedling trays, controlled release formulations, and mulching films. These synthetic plastics, while cost-effective, lead to microplastic contamination, ecosystem imbalance, and soil degradation. Fig. 9.3 displays the use of biomaterials for regenerative agriculture. Food packaging accounts for nearly 40% of total plastic use worldwide, making it one of the largest contributors to global plastic wastage. Biomaterials can help reduce or phase out the use of petroleum-based plastics. These materials are not only eco-friendly but also match the functional properties required, opening possibilities for both active packaging as seen in Fig. 9.4 and intelligent packaging [8]. Fig. 9.1 describes the various methods that can be utilized to enhance materials for food packaging. Biomaterials are suitable alternatives given their biodegradability, as shown in Table 9.2 and Fig. 9.3).

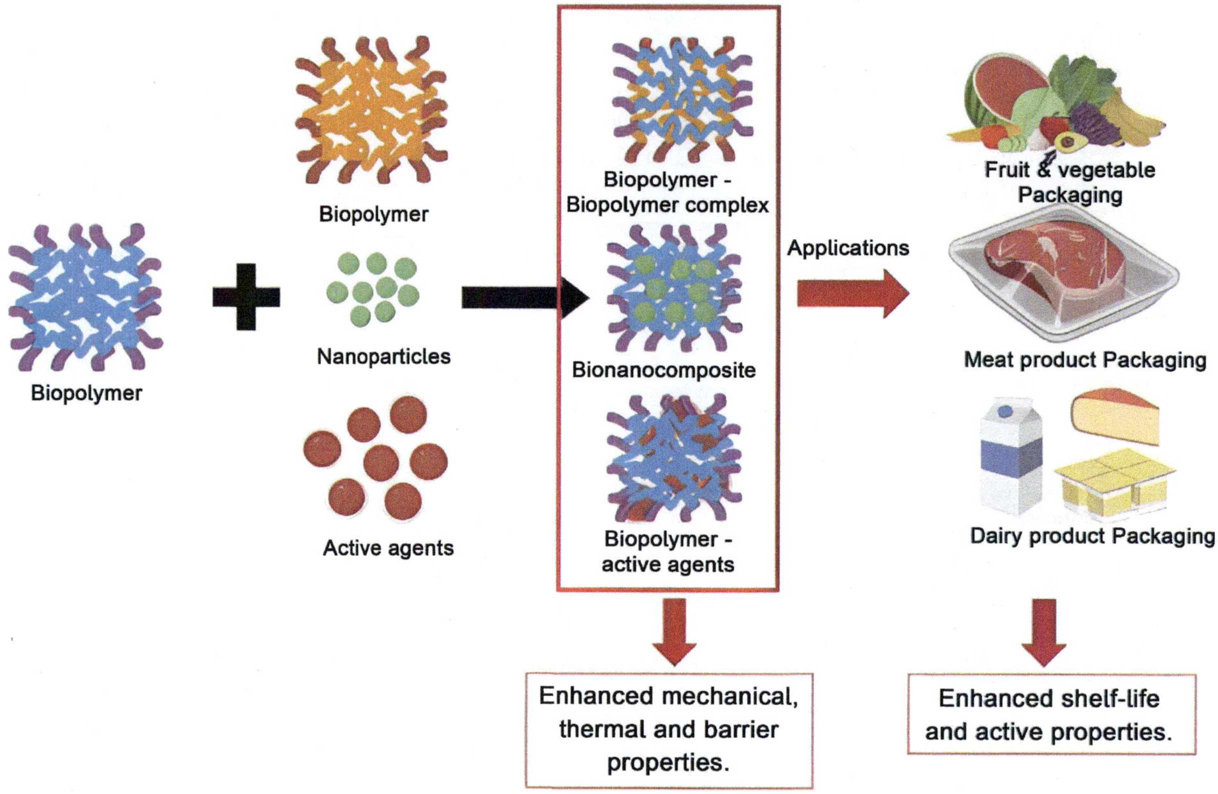

FIGURE 9.4 Overview of biomaterials in food packaging [10].

TABLE 9.2 Usage of biomaterials for packaging.

Packaging	PHAs in food packaging	Advancements focus on improving permeability and brittleness of PHA films by reinforcing with lignin or nanocellulose or blending with plasticizers
	Starch-based materials	Starch-based films can be combined with other materials such as PLA or PHAs to improve mechanical strength and water resistance [6]. These films are often used as edible coatings.
	Mycelium-based materials	Mycelium-based materials are lightweight and can be made into rigid structures which are compostable and shock absorbent. These properties make mycelium suitable as a packing alternative for luxury packaging, replacing polystyrene in items, electronics, or wine bottle cushioning [8].
	Biopolymer films	Bacterial cellulose films derived from the Acetobacter species offer mechanical flexibility and enhanced oxygen impermeability making them useful for packaging high-barrier foods. Chitosan-based films are often used in active packaging due to their antimicrobial properties to reduce food spoilage. Carrageenan when combined with natural dyes like curcumin or anthocyanin can provide visual indicators of food freshness by showing spoilage through color change. This makes them useful for intelligent packaging purposes [9]. Smart packaging by integrating biosensors and biopolymer blends helps not only protect food quality but also monitor it.

PHAs, Polyhydroxyalkanoates; *PLA*, polylactic acid.

9.3 Environment sustainability

Environmental sustainability has become central to tackling modern-day challenges of climate change, pollution, and ecosystem damage. Accumulation of plastics in the ecosystems and environments due to pollution disrupts wildlife and causes biomagnification of microplastics. To address synthetic plastic pollution, renewable biobased materials such as chitosan, mycelium, PHAs, and alginate are gaining traction as described in Table 9.3.

9.3.1 Applications in pollution control

Bioremediation is prioritized for environmental sustainability and pollution control. Genetically engineered strains of microbes that break down pollutants such as polyaromatic hydrocarbons (PAHs), pesticides, and plastics, as well as detoxifying or immobilizing heavy metals through bioaccumulation and biosorption mechanisms [11]. To complement this, phytoremediation uses plants to capture, metabolize, or sequester soil-based toxins, including volatile organic compounds (VOCs) and metals [12].

Solutions to decrease increasing air pollution have been done through biofiltration technologies utilizing microbial biofilms. These films react with harmful gases and degrade, reducing emissions of NOx, SOx, and VOCs.

Biosensors enhanced with nanomaterials provide real-time monitoring of pollutants in air, water, and soil, enabling rapid response and smarter pollution management strategies [13].

Bioplastics derived from microbial and plant sources offer a promising replacement for petroleum-based plastics. Examples include PHAs and polylactic acid (PLA), which help reduce persistent plastic pollution.

Researchers extracted lye from eucalyptus ash, and they activated coffee-ground biochar to create fluoride adsorbents for water treatment, as they showed favorable fluoride removal kinetics and porosity in water purification tests.

9.4 Radiation protection

As the world makes advances in nuclear science and energy, especially with the development of nuclear weapons and reactors, there is a pressing need for radiation mitigation. In recent years, radiation protection has evolved from bulky lead aprons into the realm of biomaterials. Biomaterials can offer dynamic, biodegradable, and site-specific developing interfaces with biological systems in real-time. Current innovations have led to various patented technologies and upcoming clinical applications.

TABLE 9.3 Biomaterials in environmental remediation.

Application	Description
Bioremediation (soil/water)	Engineered microbes (e.g., *Pseudomonas* and *Rhodococcus*) degrade organic pollutants (hydrocarbons, PAHs, pesticides) and immobilize heavy metals through biosorption and bioaccumulation [11]
Phytoremediation (air/soil)	Plants absorb, sequester, or transform pollutants (e.g., heavy metals and VOCs) through root uptake and enzymatic degradation [12]
Biosensors	Microbial, enzymatic, and DNA-based biosensors detect heavy metals, pesticides, and air pollutants in situ; nanomaterials enhance sensitivity [13]
Biofilters (air pollution)	Biofilms of microbes (e.g., *Thiobacillus*) degrade airborne VOCs, NOx, and sulfur compounds in industrial emissions [11]
Biodegradable polymers	Bioplastics (PHA, PLA, bacterial cellulose) replace conventional plastics and degrade safely after use, reducing persistent pollution [12]
Bioleaching and biosorption	Microbes solubilize or capture toxic metals from mining or industrial waste for recovery or safe disposal [11]
Biofuel production from waste	Organic wastes are converted by microbes into biofuels, reducing landfill load and air pollution from open burning [11]

9.4.1 Applications for radiation protection and mitigation

Orally deliverable nanoparticle capsules have been made that can be injected into hydrogels and 3D-printable shielding materials. For example, BIO300, a nanoparticle formulated from genistein, which is a soy-derived antioxidant, has shown promising results for mitigating radiation-induced pulmonary fibrosis and pneumonitis in cancer patients. It is currently in phase 1B/2A clinical trials. The nanocarrier system makes oral delivery possible, improving bioavailability in lung tissues. This offers noninvasive prevention against radiation-induced injury [14].

Oral radioprotective capsules have also been developed, as described in recent patents (e.g., US 12,076,307). These pills have encapsulated microspheres and slow-release hydrogels to deliver a cocktail of radioprotective agents such as antioxidants, DNA-repair promoters, and immune modulators. This offers protection with minimal intervention. These oral capsules can be helpful for astronauts or in case of radiation emergencies.

FDA has approved SpaceOAR PEG, a hydrogel that temporarily acts as a bioresorbable spacer. It is inserted between the rectum and prostate during radiotherapy. By doing so, a physical distance is created between healthy tissues and the radiation beam, decreasing off-target damage.

Researchers have developed bismuth-nanoparticle reinforced cellulose films and 3D-printed hydrogels that act as radiation shields against X-rays and gamma radiation. They offer improvements from conventional lead materials by being lighter and offering flexibility for integration into clothing, space suits, and other mobile structures. These technologies are under exploration by space agencies for use in long-duration missions to reduce the threat caused by cosmic rays and solar energetic particles [15].

9.5 Energy and fuels

Another metric to move forward towards sustainable development is novel fuel and energy technologies. Combining living systems and advanced materials to create cutting-edge innovations. Beyond standard bioethanol or biodiesel, applications showcasing advancements in synthetic biology, nanotech, and biohybrid devices are ushering in the next generation of biomaterials for energy.

9.5.1 Applications in energy

The "Bionic Leaf," developed at Harvard and MIT, combines an artificial-leaf water-splitting catalyst (silicon with cobalt-phosphide alloy and cobalt-phosphate anode) with engineered bacteria (*Ralstonia eutropha).* It splits water into hydrogen and oxygen photolytically, feeding hydrogen and CO_2 to bacteria. This generates isopropanol (and other alcohol fuels) with ~10% solar-to-chemical efficiency, which is on par with the best bioenergy systems and far surpasses natural photosynthesis [16].

This bionic leaf can be self-healing (the catalyst regenerates in situ), is scalable, and can be adapted to produce diverse organic compounds—including biofuels, plastics precursors, and fertilizers. "Cyberplasm," a concept introduced in Wired, is a micro-biohybrid robot built from tunable hydrogels that house microfluidic channels and genetically optimized microbial fuel cells. These robots derive electricity directly from bacterial metabolism. There are no traditional membranes. Making tiny autonomous systems capable of environmental sensing or self-powered tasks [17].

Peptide-based supercapacitors, waste-derived carbon electrodes, and biochar-integrated microbial fuel cells (MFCs) show immense promise for the fuel and energy sector. These systems offer a sustainable route to decentralized energy generation and transient electronics while addressing e-waste concerns. Table 9.4 highlights energy capacity, biodegradation profiles, and potential application areas.

9.6 Aerospace

Biomaterial-inspired designs are propelling aerospace design into the future. They combine lightweight structure, self-healing capabilities, and incorporate living components once thought exclusive to biology into aircraft.

9.6.1 Applications in aerospace

A 3D-printed cabin bulkhead created by Airbus, Autodesk, and The Living was designed for the A320, a low-cost and full-service mainline carrier. It utilized the human bone to make a stress-adaptive lattice on the microscale and optimized with a slime mold network on a macroscale. They found the cabin bulkhead to be 45% lighter (66 lb vs 143 lb) using 95%

TABLE 9.4 Applications of biomaterial-based technologies and their energy capacity.

S. No.	Technology	Key findings	Energy metric
1.	Dipeptide supercapacitor (Fmoc-EF-NH_2)	Self-assembling peptide forms stable films; robust cycling	125 F/g; 95% retention (5,000 cycles) [18]
2.	Fully biodegradable peptide/ionic liquid device	Biocompatible peptide electrodes + amino-acid ionic liquids	~20× energy density of planar systems (simulated) [19]
3.	Polypyrrole–peptide supercapacitor	Biodegradable prototype; stable for 5,000 cycles	94.7% retention; substantial degradation in 90 days [20]
4.	Lignin-activated carbon electrodes	Upcycled from anaerobic digester waste; long life	114 F/g; 84.5% retention (15,000 cycles) [21]
5.	Forestry residue biochar in MFCs	Biochar electrodes match commercial activated carbon	457–532 mW/m^2 power density reported under MFC conditions. [22]
6.	Carbon nanomaterials for MFCs	CNTs, graphene, doped carbons improve conductivity, EASA (Electrochemically Active Surface Area)	High efficiency, sustainable sourcing [23]

less raw material. This could save fuel and reduce carbon emissions. This 3D-printed airoplane component, inspired by and mimicking nature, could open new avenues for biomaterials in airoplane manufacturing [24].

The EU-funded ECO-COMPASS is an initiative that explores bio-based composites. Flax and ramie fiber reinforcements and resins derived from rosin are being used for secondary aircraft structures such as interior panels and winglets. These biomaterials offer advantages over traditional glass and carbon composites in vibrational dampening and are more eco-friendly [25].

NASA's Glenn Research Centre is developing biomimetic polymers with reversible ionic cross-links and self-healing dielectric properties to autonomously repair damage in flight electronics [26]. Ice-resistant surface coatings emulate the antifreeze glycoproteins found in cold-adapted fish and insects to mitigate airframe icing. It does this without needing chemical sprays.

Encouraged by bone and coral microstructures, researchers are designing porous carbon ablators that proficiently scatter thermal radiation during spacecraft re-entry. In laser heating experiments this reduces char depth by up to 42% in insulators [27]. Adaptive architected lattices offer dynamic energy absorption for crash protection and micrometeoroid impact resistance [28].

Biomimetic designs have also been explored for space exploration. Developments have been made for telescope optics, which are like the eyes of a lobster, gecko-inspired adhesive grippers for on-orbit servicing, and wood wasp-inspired ovipositor drills. They use slender, biomimetic geometry for low-impact penetration [29]. These innovations improve functionality in microgravity and can be an asset in extraterrestrial environments.

9.7 Cosmetics

In 2023 the global beauty market retail sales was estimated at $446 billion, approximately a 10 percent increase from 2022. With further growth projected, the cosmetics industry establishes itself as an important sector for growth. Fueled by massive investment and consumer demand, companies are leveraging biotechnology to develop smart formulations in which biomaterials can play a variety of roles.

9.7.1 Application in cosmetics

Peptides (serilesince, decorinyl, and rubixyl) that mimic signalling motifs associated with natural proteins like SNAP 25 and decorin activate collagen and tighten skin as well as strengthen the dermal-epidermal junction [30].

Research harnessing the skin's ability to achieve antiaging effects with the development of biomolecules by firms, including Lipotec, Sederma, and Induchem. Oils emulating sebum (jojoba, argan, squalane) are patented as having skin reconstructing, inflammation-reducing, and regenerative potential [31].

Plant stem cells (such as those of Vitis vinifera, argan, tomato, and ginger) are mined. Callus cultures are made to create growth factors and powerful antioxidants that help in repairing wrinkles and enhancing skin elasticity. Similarly, microalgae such as Spirulina, Chlorella, and Dunaliella salina produce pigments (examples: choincyto, 8 carotene) and polysaccharides with antioxidant, photoprotective, and skin firming capacity [32].

Givaudan's biomimetic vegan silk is inspired by silk fibroin, which supports barrier repair, reduction of skin irratation, and improvement of visible skin qualities. It does this by mimicking skin's extracellular matrix, improving skin healing, reducing irritation, and diminishing visible skin imperfections [33]. Silk fibroin is often used in skin regeneration.

9.8 Forensic science

Forensic science is a field that is heavily reliant on analyzing evidence which is biological in nature. These techniques used for analysis can be time-consuming and need to be highly sensitive, specific, and accurate. Biomaterials can help optimize the process and techniques for better and faster results.

9.8.1 Applications in forensic Science

An innovation from the University of Bath and Shanghai Normal University developed a spray using green fluorescent protein (GFP) derived from *Aequorea Victoria, which is a bioluminescent jellyfish*. The spray, available in two fluorescent variants (LFP-Yellow and LFP-Red), binds to sweat/oil residues, revealing fingerprints on a variety of surfaces within 10 seconds without damaging any of the DNA evidence. This outdoes traditional powders and solvents, which are also often toxic [34]. This spray promises to be safer, faster, and more eco-friendly, with commercialization underway.

A diverse range of nanomaterial-powered optical and electrochemical biosensors, which have been tailored for forensic use, are being created. These biosensors detect bodily fluids, secretions, toxins, drugs, and explosives with high sensitivity. They are additionally portable, making them useful for analyzing at a crime scene [35].

For example, Aptamer-based nanoflares, which are gold nanoparticle cores with a fluorophore aptamer that precisely detects human blood. This is done through FRET mechanisms, enabling nondestructive on-site testing with real-time fluorescence [35]. Carbon-based immunosensors, which combine carbon nanotubes, fullerene (C_{60}), gold nanoparticles, and polymers to form an ultrasensitive detector for highly specific detection of substances like mycotoxins, illicit drugs, and botulinum toxin [36].

The Florida International University (Abuzar Kabir, Kenneth Furton) holds multiple patents on polymer/sol–gel sorbents, microcrystalline cellulose, and fabric phase sorptive extractors. These are intended for trace detection of body fluids, explosives, and narcotics by optimizing sample collection, preservation, and transport for downstream analysis [37].

9.9 Challenges and future prospects

Biomaterials, while holding significant promise in agriculture and food packaging regarding sustainability, face challenges that hinder their worldwide adoption. While being eco-friendly, the production costs for biopolymers like PHAs, bacterial cellulose, or chitosan are expensive, especially when compared to synthetic plastics. This is due to needing a more extensive and complex extraction, fermentation, and production processes [38].

Scalability remains a pressing issue, as consistency in quality and a need to establish supply chains for natural polymers are difficult to establish, particularly when relying on agricultural waste and microbial fermentation [8].

Additionally, the mechanical strength of these materials remains to be further optimized. Brittleness and poor water resistance limit utility under humid or variable environmental conditions. As an example, starch-based packaging can absorb moisture from the environment, losing integrity during storage. Films could also fragment before crop cycles are complete. Balancing biodegradability with functional longevity remains a key technical challenge for utilizing biopolymers [6].

Integration can also be difficult, as existing packaging and agricultural equipment are designed for synthetic plastics. Processing of biomaterials often requires specialization in manufacturing systems, which adds to cost and complexity. Bioplastics may also display incompatibility with established recycling streams and need complicated waste management systems [38].

In the environmental domain, practical challenges faced by biomaterials in the environment and pollution control include scalability, cost competitiveness with synthetic materials, and durability in harsh conditions. Biopolymer-based adsorbents can lose capacity or even degrade under extreme pH or temperature.

For radiation protection technologies, the long-term safety of nanoparticle-based materials remains a question, especially those using or containing heavy metals. They require further toxicological studies and profiling. Oral and injectable systems to become commercially viable must overcome hurdles related to pharmacokinetics, tissue-specific accumulation, and clearance. Scalability, cost, and integration into available frameworks and regulatory compliance will require extensive interdisciplinary collaboration.

In sustainable energy, despite laboratory success, devices like sugar batteries and peptide superconductors require improvements in longevity, voltage, and mass manufacturing. Cyberplasm and MFCs need to function reliably in real-world environments and be safe to use. These materials often require complicated multistep synthesis, making production costly. Deploying microbial robots, synthetic biology, and enzymatic implants involves overcoming biosafety, ecological, and regulatory barriers.

Modern bio-inspired aerospace materials face both intense regulatory barriers and demanding aerospace conditions. Moreover, they are not sufficiently standardized in terms of the complexity of their production and the reliability of their operations.

With cosmetics, issues concerning safety need to be sorted through toxicological and regulatory assessment, especially on biomimetic peptide extracts or marine-based extracts. The same case applies to the processes of stem cell cultures and microbial cellulose production that require specialized process operations, thus adding complexity and cost to production.

Barriers such as court admissibility of forensic biomaterials are also extremely high, requiring formal accreditation in the legal context, and at the same time retaining the stability needed in varying conditions of the environment. Decreased specificity may stimulate false positives, and how it is applied in practice during field debugging depends on the development of procedures that are user-friendly because, irrespective of the specificity achieved, field operatives cannot afford on-the-job error.

Despite the challenges, advances in biopolymers, synthetic biology, material engineering, and nanotechnology will pave the way for next generation biomaterials. Innovations focused on biopolymer blends (e.g., PHA-starch nanocomposites) will enhance mechanical properties and help maintain longevity while being biodegradable.

Integrating smart functions such as biosensors to indicate the freshness of food or environmental monitoring are features that add value, which cannot be offered by traditional packaging.

Conductive or responsive plasmonic nanoparticles that change color reversibly with pH over many cycles can be utilized to make reusable smart packaging.

Self-healing packages can be made by incorporating dynamic covalent bonds (e.g., imine and disulfide bonds) or supramolecular structures that can heal micro-tears or pinholes, prolonging packaging life and barrier integrity.

Environment-responsive packaging can also be created using biosensors to change the structure or functionality of the package in response to environmental stressors.

Soil-integrated biofilms that are sensitive to a pollutant or a group of pollutants can change color to serve as real-time indicators of soil. Nano-sponge materials capable of absorbing only oil or chemical spills can be developed. RO filters utilizing biomaterials for their membrane can be developed to improve the concentration of needed minerals while still filtering out hazardous substances.

Work focusing on hybrid biomaterials and nanoparticle blends can improve resilience and regeneration capacity. As synthetic biology advances with time, waste valorization could reduce production costs while aligning with circular economy goals [39].

Ongoing research into smart biomaterials that dynamically respond to radiation exposure can be proposed. For example, hydrogels with self-healing properties after damage or microspheres that release antioxidants only when triggered by reactive oxygen species (ROS).

Proposed gene-circuit bioshields involve bioengineering the human microbiome, creating gene circuits which act as sensors and produce protective enzymes or immune responses to radiation, potentially creating an internal biological defense system that can adapt in real-time.

Space medicine research has been trying to utilize melanin-producing fungi, which can be integrated into habitat walls or spacecraft, absorbing and converting ionizing radiation. These living materials could be self-healing and self-replicating, reducing the need for resupply and generation.

"Fuel farms" composed of Bionic Leaves that are distributed across panels can be utilized to create a solar farm. This can lead to the production of liquid fuels on-site with minimal infrastructure.

Multiple Cyberplasm derivatives can be used collectively as sensors for pollutants, to deliver therapeutic payloads in situ, or to generate electricity. Enzyme batteries or microbial cells can be embedded into wearables. They can be used as self-powered clothes, harvesting sweat or body heat to power small electronics. Sugar batteries using physiological glucose can be implanted within the body to power sensors or neural interfaces.

Self-repairing aircraft surfaces have been suggested in which bacteria or fungi can be added to the surface for self-healing. Morphing Wings that can change shape during a flight would streamline the process of lift generation and reduce the frequency of noise, and bio batteries that use photosynthesis as a system would allow stability in fuel generation. Plant fibers would be used more widely in the cabin interiors, where they are more recyclable at the end of the service. Companies that are using fungal neural networks in the form of biosensors can assist in the maintenance of the aircraft, as sensors can identify a fault, balance load, and monitor pressure.

Meanwhile, living skin prints that take full functionality and sensory potential are on the way to individual testing and restoration intents and personified skin-care formulations.

In the future, forensic biomaterials can be based on intelligent responsive platforms, including living microbe patches as well as surface-enhanced Raman spectroscopy (SERS) films, hence a quick, noninvasive sensing and safe sampling. New technologies, such as dissolvable nanofiber mats and bioencrypted DNA chips, are on the horizon that will enhance sensitivity and protect the privacy and integrity of evidentiary samples.

9.10 Conclusion

Major roles are being projected by biomaterials in nonmedical fields, and their transformative potential is observed in sectors of agriculture, environmental sustainability, radiation protection, energy, aerospace, cosmetics, and forensics. However, the current limitations in terms of scalability, costs, and complexity still hinder large-scale incorporation. The future of biomaterials looks bright since developments in synthetic biology, nanotechnology, and material science will help curb some of these barriers, hence making biomaterials viable alternatives to synthetic versions of the same product, as well as nudging the trajectory towards smart, adaptable, and environmentally friendly solutions. The future of the sector thus depends on the intersection of bioengineering and real-world implementation, and the aim of this is to work towards creating resilient, regenerative technologies that will support the idea of sustainable development of a world whose resources are finite.

References

[1] L. Ghasemi-Mobarakeh, D. Kolahreez, S. Ramakrishna, D. Williams, Key terminology in biomaterials and biocompatibility, Current Opinion in Biomedical Engineering 10 (2019) 45–50. Available from https://doi.org/10.1016/j.cobme.2019.02.004.

[2] N. Huebsch, D.J. Mooney, Inspiration and application in the evolution of biomaterials, Nature 462 (7272) (2009) 426–432. Available from https://doi.org/10.1038/nature08601.

[3] What is a biomaterial? (Center for Professional Personnel Development), n.d. Department of agricultural economics, sociology, and education. https://aese.psu.edu/teachag/curriculum/modules/biomaterials/what-is-a-biomaterial.

[4] Evolution of sustainable shopping worldwide 2022| Statista, 2025. Statista. https://www.statista.com/statistics/1377869/global-shift-to-buying-sustainable-products/#:~:text=Consumer%20shift%20towards%20buying%20sustainable%20products%20over%20the%20past%20five%20years%202022&text=In%202022%2C%20the%20vast%20majority,order%20to%20be%20environmentally%20sustainable.

[5] Plastics in Agriculture | Climate change | Food and Agriculture Organization of the United Nations, n.d. ClimateChange. https://www.fao.org/climate-change/areas-of-work/plastics-in-agriculture/en#:~:text=In%20late%202021%2C%20FAO%20released,million%20tonnes%20in%20food%20packaging.)

[6] E.M. Gonçalves, M. Silva, L. Andrade, J. Pinheiro, From fields to films: exploring starch from agriculture raw materials for biopolymers in sustainable food packaging, Agriculture (2024). Available from https://doi.org/10.3390/agriculture14030453.

[7] P.M. Stathatou, L. Corbin, J.C. Meredith, A. Garmulewicz, Biomaterials and regenerative agriculture: a methodological framework to enable circular transitions, Sustainability 15 (19) (2023) 14306. Available from https://doi.org/10.3390/su151914306.

[8] M.Z. Mahmud, M.H. Mobarak, N. Hossain, Emerging trends in biomaterials for sustainable food packaging: a comprehensive review, Heliyon 10 (2024). Available from https://doi.org/10.1016/j.heliyon.2024.e24122.

[9] S.S. Mathew, A.K. Jaiswal, S. Jaiswal, Carrageenan-based sustainable biomaterials for intelligent food packaging: a review, Carbohydrate Polymers (2024). Available from https://doi.org/10.1016/j.carbpol.2024.122267.

[10] K.Y. Perera, A.K. Jaiswal, S. Jaiswal, Biopolymer-based sustainable food packaging materials: challenges, solutions, and applications, Foods 12 (12) (2023) 2422. Available from https://doi.org/10.3390/foods12122422.

[11] R.C. Sobti, A. Sharma, S.K. Soni, Applications of Biotechnological Techniques in Mitigating Environmental Concerns, Informa, (2022), pp. 249–312. Available from https://doi.org/10.1201/9781003220831-17.

[12] S. Ghosh, M. Bera, Use of environmental biotechnology to reduce environmental pollutions, 1 (2) (2020) 1–6. Available from https://doi.org/10.15864/AJABTR.125.

[13] L. Fahmideh, E. Khodadadi. A review of applications of biotechnology in the environment, 2014.

[14] P.G.S. Prasanna, D. Narayanan, K. Hallett, E.J. Bernhard, M.M. Ahmed, G. Evans, et al., Radioprotectors and radiomitigators for improving radiation therapy: the small business innovation research (SBIR) gateway for accelerating clinical translation, Radiation Research 184 (3) (2015) 235–248. Available from https://doi.org/10.1667/rr14186.1.

[15] P. Tiamduangtawan, C. Kamkaew, S. Kuntonwatchara, E. Wimolmala, K. Saenboonruang, Comparative mechanical, self-healing, and gamma attenuation properties of PVA hydrogels containing either nano- or micro-sized Bi2O3 for use as gamma-shielding materials, Radiation Physics and Chemistry 177 (2020) 109164. Available from https://doi.org/10.1016/j.radphyschem.2020.109164.

[16] J.P. Torella, C.J. Gagliardi, J.S. Chen, D.K. Bediako, B. Colón, J.C. Way, et al., Efficient solar-to-fuels production from a hybrid microbial–water-splitting catalyst system, Proceedings of the National Academy of Sciences 112 (8) (2015) 2337–2342. Available from https://doi.org/10.1073/pnas.1424872112.

[17] J. Ayers, Biohybrid Robots are Synthetic Biology Systems, Oxford University Press eBooks, (2018). Available from https://doi.org/10.1093/oso/9780199674923.003.0051.

[18] K. Hu, C. Zheng, M. An, X. Ma, L. Wang, A peptide-based supercapacitor and its performance improvement via TiO2 coating, Journal of Materials Chemistry A 6 (17) (2018) 8047–8052. Available from https://doi.org/10.1039/c8ta01102h.

[19] G. Colherinhas, T. Malaspina, E.E. Fileti, Storing energy in biodegradable electrochemical supercapacitors, ACS Omega 3 (10) (2018) 13869–13875. Available from https://doi.org/10.1021/acsomega.8b01980.

[20] Z. Li, K. Hu, Z. Li, C. Li, Y. Deng, Polypyrrole-Stabilized polypeptide for eco-friendly supercapacitors, International Journal of Molecular Sciences 24 (3) (2023) 2497. Available from https://doi.org/10.3390/ijms24032497.

[21] B.K. Mutuma, N.F. Sylla, A. Bubu, N.M. Ndiaye, C. Santoro, A. Brilloni, et al., Valorization of biodigestor plant waste in electrodes for supercapacitors and microbial fuel cells, Electrochimica Acta 391 (2021) 138960. Available from https://doi.org/10.1016/j.electacta.2021.138960.

[22] T. Petrova, I. Naydenova, J. Ribau, A.F. Ferreira, Biochar from agro-forest residue: application perspective based on decision support analysis, Applied Sciences 13 (5) (2023) 3240. Available from https://doi.org/10.3390/app13053240.

[23] M. Abedi, V. Alimardani, Y. Ghasemi, R. Kazemi, M. Shafiee, Waste-derived carbon nanomaterials for microbial fuel cells, Nanostructure Science and Technology (2025) 341–358. Available from https://doi.org/10.1007/978-3-031-75247-6_14.

[24] K. Losey. Your next flight may be designed by slime mold and human bones. The Biomimicry Institute, 2024. https://biomimicry.org/your-next-flight-may-be-designed-by-slime-mold-and-human-bones/.

[25] Bio-based materials for aircraf. EC R&I Success Stories, 2018. https://projects.research-and-innovation.ec.europa.eu/en/projects/success-stories/all/bio-based-materials-aircraft.

[26] T.S. Williams. Bio-inspired materials and processes for aerospace environments. NASA Technical Reports Server (NTRS), 2023. https://ntrs.nasa.gov/citations/20230014747.

[27] E. Poloni, F. Bouville, A.L. Schmid, P.I. Pelissari, V.C. Pandolfelli, M.L. Sousa, et al., Carbon ablators with porosity tailored for aerospace thermal protection during atmospheric re-entry, Carbon 195 (2022) 80–91. Available from https://doi.org/10.1016/j.carbon.2022.03.062.

[28] Y. Wang, B. Ramirez, K. Carpenter, C. Naify, D.C. Hofmann, C. Daraio. Architected lattices with adaptive energy absorption. arXiv (Cornell University), 2019. https://doi.org/10.48550/arxiv.1909.05231.

[29] E. Banken, J. Oeffner, Biomimetics for innovative and future-oriented space applications - a review, Frontiers in Space Technologies 3 (2023). Available from https://doi.org/10.3389/frspt.2022.1000788.

[30] A. Pintea, A. Manea, C. Pintea, R. Vlad, M. Bîrsan, P. Antonoaea, et al., Peptides: emerging candidates for the prevention and treatment of skin senescence: a review, Biomolecules 15 (1) (2025) 88. Available from https://doi.org/10.3390/biom15010088.

[31] S. Abdalla, M.K. Aroua, L.T. Gew, A comprehensive review of plant-based cosmetic oils (Virgin coconut oil, olive oil, argan oil, and jojoba oil): chemical and biological properties and their cosmeceutical applications, ACS Omega 9 (44) (2024) 44019–44032. Available from https://doi.org/10.1021/acsomega.4c04277.

[32] C. Gomes, A.C. Silva, A.C. Marques, J.S. Lobo, M.H. Amaral, Biotechnology applied to cosmetics and aesthetic medicines, Cosmetics 7 (2) (2020) 33. Available from https://doi.org/10.3390/cosmetics7020033.

[33] D.N. Breslauer, Current progress on scale-up and commercialization of microbially-produced silk, Advanced Functional Materials (2024). Available from https://doi.org/10.1002/adfm.202408386.

[34] N. Ruan, Q. Qiu, X. Wei, J. Liu, L. Wu, N. Jia, et al., De novo green fluorescent protein chromophore-based probes for capturing latent fingerprints using a portable system, Journal of the American Chemical Society 146 (3) (2024) 2072–2079. Available from https://doi.org/10.1021/jacs.3c11277.

[35] H. Costanzo, J. Gooch, N. Frascione, Nanomaterials for optical biosensors in forensic analysis, Talanta 253 (2022) 123945. Available from https://doi.org/10.1016/j.talanta.2022.123945.

[36] A.M. Díez-Pascual, D.L. Cruz, A.L. Redondo, Advanced carbon-based polymeric nanocomposites for forensic analysis, Polymers 14 (17) (2022) 3598. Available from https://doi.org/10.3390/polym14173598.

[37] A. Kabir, K.G. Furton. US20140274660A1 - Fabric phase sorptive extractors (FPSE) - Google Patents, 2013. https://patents.google.com/patent/US20140274660A1/en.

[38] J.C.C. Yeo, J.K. Muiruri, X. Fei, T. Wang, X. Zhang, Y. Xiao, et al., Innovative biomaterials for food packaging: unlocking the potential of polyhydroxyalkanoate (PHA) biopolymers, Biomaterials Advances 163 (2024) 213929. Available from https://doi.org/10.1016/j.bioadv.2024.213929.

[39] K.G. Şensoy, M. Muti, The Novel Nanomaterials Based Biosensors and Their Applications, IntechOpen eBooks, (2020). Available from https://doi.org/10.5772/intechopen.94930.

Section II

Biological aspects of tissue engineering

Chapter 10

Quantitative in vitro bioassays for scaffold evaluation

Sthuthi Jose M., and S. Sumathi
Department of Chemistry, School of Advanced Sciences, Vellore Institute of Technology, Vellore, Tamil Nadu, India

10.1 Introduction

Human organs and tissues have an inbuilt capacity to heal. However, a serious injury cannot be dealt with. Thus organ transplantation has always been the solution to the problem. However, the shortage of organ donors, as well as the danger of rejection and infection, has led to the conclusion that this technique has limited promise. Tissue engineering has emerged as a potential advance in this field of biomaterial research. It is a multidisciplinary field of study that helps to reconstruct and regenerate damaged tissues and organs [1,2]. Engineering of tissues has extended its impact on skin, nerve, bone, etc., as an effective and efficient approach [3,4]. In the science of tissue engineering, translational tissue engineering focuses on the clinical application of biomaterials for patient care and therapy. In vitro assays are crucial for testing a material's biological efficiency and providing more practical options for translational tissue engineering.

Owing to its high incidence of orthopedic disorders and fractures, bone is the second most transplanted tissue worldwide. This marks the necessity of the development of scaffolds to repair bone defects as a key focus in tissue engineering strategies [5]. BTE promotes the growth of new functional bone by the combined use of cells, biomaterials, and factor treatment. Its primary goal is to create biological replacements that closely resemble the composition and structure of native bone by supporting bone regeneration in response to illness, damage, or other defects. The major challenge in BTE is the development of materials with appropriate biocompatibility, biodegradability, and mechanical properties [6]. By functioning as a protective layer, skin tissue engineering improves the healing of wounds and returns the skin's mechanical integrity and flexibility [7]. A biomaterial is suited for skin applications if it has suitable porosity, a large surface area, mechanical strength comparable to the skin tissue, effective cell adhesion properties, good proliferation of cells, and the capacity to resemble the extracellular matrix [8]. Nerve tissue engineering treats damaged and diseased nerves that have a limited ability to regenerate spontaneously [9] and uses tissue-specific biomaterials with controlled pore structures that promote cell growth by enhancing nerve healing [10]. Treatments using nerve tissue engineering have been recently recognized as viable approaches for treating central and peripheral nervous system injuries [11,12]. Damage to the nerves generates an environment that is not conducive to tissue regeneration. Thus the difficult challenge in nerve tissue engineering is to create a microenvironment that is conducive to nerve regeneration [9].

To generate functional organs and tissues for applications in in vitro systems or transplants is the main objective of tissue engineering [13]. Tissue engineering has the promise of future clinical applications, although precise assessment of the therapeutic efficacy of the biomaterials is still required. But it is a fact that, so far, only a few tissue-like systems, such as in vitro engineered skin and cartilage, are used in clinical translation [14–18]. The success of a biomaterial in clinical translation is fundamentally based on the in vitro analysis.

This chapter provides an overview of different in vitro methods used to assess the compatibility of scaffolds, films, and fibers in the cellular environment for tissue engineering applications. A detailed examination is conducted on the cell viability and proliferation assays, hemocompatibility assessment, antimicrobial investigations, and cell differentiation studies.

10.2 In vitro assays

Prior to preclinical and clinical research, in vitro methods offer important information on cell and tissue compatibility. Because of this, they are a crucial component of the medical device biological safety evaluation process [19]. In vitro

Fundamentals of Translational Tissue Engineering. DOI: https://doi.org/10.1016/B978-0-443-33455-9.00017-1

experiments play a significant role in preclinical research by upholding the principles of animal ethics—reduce, replace, and refine (3 R). In vitro assays lessen the harm in clinical translation and boost the success of in vivo analysis by evaluating the safety, effectiveness, and functionality of the engineered constructs. In vitro assays are performed as fundamental screening tests before the in vivo assessment. Through the use of in vitro procedures, it is easy to determine whether a substance is suitable for clinical translation. It serves as a foundation for determining how a biomaterial promotes the growth of tissues, the division and proliferation of cells, and the induction of toxicity. Depending on the function and the use of each in vitro method, they can be categorized as follows.

10.2.1 Cell viability and proliferation assays

These tests are necessary in biomedical applications to evaluate a biomaterial's capacity to maintain cell growth and viability. They measure the amount of toxicity inflicted upon living cells as well as their ability to proliferate and flourish on a scaffold that has been built [20]. These are commonly referred to as cytotoxicity assays. They are primarily used to assess the toxicity of materials by subjecting them to in vitro tissue cells and evaluating resultant effects [21]. Typically, the tests are performed in vitro using grown mammalian cells, like those seen in mice.

10.2.1.1 MTT assay

The primary factor determining a material's efficacy in clinical translation is its toxicity. MTT assay evaluates toxicity and suitability for application in the cellular environment. It is one of the most widely used and adaptable viability tests in biological research. This test relies on the process of conversion of a substrate to a colored product by the live cells. This is a simple, reliable, and reproducible test that uses the dehydrogenase of mitochondria to measure how living cells appear. 3-(4,5-dimethylthiazol-2-yl)-2,5-diphenyltetrazolium bromide, commonly known as MTT, is the essential component associated with this test. This salt of tetrazolium salt is water soluble and becomes yellow in the absence of phenol red. The chemistry involved is that a living cell's mitochondrial dehydrogenase enzymes break down the tetrazolium ring, which transforms water-soluble MTT into an insoluble purple formazan (Fig. 10.1).

Formazan is then solubilized using DMSO or other solvents like acidified isopropanol, Pure propanol, or ethanol, and the resultant purple-colored solution is used for determining the concentration using the optical density at 570 nm. The amount of formazan produced varies as the number of cells varies, revealing information on the cytotoxicity of the test substance [22–24]. The percentage of growth inhibition is calculated using the following formula:

$$\%\text{Inhibition} = 100 - (\boldsymbol{OD}_{\boldsymbol{Sample}} \div \boldsymbol{OD}_{\boldsymbol{Control}}) * 100 \quad (10.1)$$

Using the dose-response curves for each cell line, the concentration of test material required to inhibit cell growth by 50% (IC_{50}) can be determined.

MTT is found to be a practical, affordable, and appropriate method among the various cytotoxic tests for determining cell viability. Yet, several factors influence cellular metabolism, which in turn affects MTT activity and can lead to erroneous positive or negative counts of cells. Therefore it is crucial to consider assay parameters carefully by selecting appropriate controls for each cell line. Parameters that are to be taken into consideration include optimal cell concentration, growth medium, MTT quantity, and duration of exposure for MTT. It is possible to estimate the number

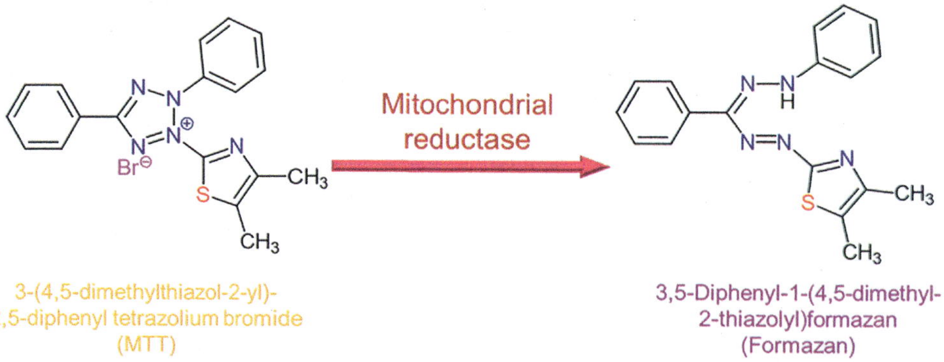

FIGURE 10.1 Mechanism involved in the conversion of MTT to formazan. Schematic illustration of the conversion of MTT to formazan in the presence of mitochondrial dehydrogenase.

of living cells with accuracy and dependability by managing these crucial factors [25]. Table 10.1 shows different cell lines and polymer-ceramic composites used for MTT cell viability evaluation in bone, skin, and nerve tissue applications.

10.2.1.2 WST-1 assay

Similar to MTT, the WST-1 assay is used to assess the vitality and proliferation of the cells. This is a colorimetric assay that uses the colored formazan product as a basis for evaluation. Succinate tetrazolium reductase present in mitochondria of the living cells converts 2-(4-iodophenyl)-3-(4-nitrophenyl)-5-(2,4-disulfophenyl)-2*H*-tetrazolium, monosodium salt, which is commonly called WST-1, to a formazan product that is soluble in water [36]. The presence of negatively charged sulfonate groups in WTS-1 makes it unable to cross the cell membrane. Therefore the process occurs outside the cell, whereas MTT occurs intracellularly. However, it is advantageous over MTT because it eliminates the need for the extra step of formazan solubilization [37]. The quantity of formazan produced is exactly proportionate to the number of viable cells. OD value can be measured using a spectrophotometer, and conclusions regarding cell viability can be drawn similarly to that of the MTT assay [23].

10.2.1.3 XTT assay

This colorimetric assay measures the vitality and proliferation of cells using sodium 3′-[1-(phenylamino carbonyl)-3,4-tetrazolium]-bis (4-methoxy6-nitro) benzene sulfonic acid hydrate. The basis of XTT, like MTT and WST, is the reduction of tetrazolium using mitochondrial reductase found in biologically active cells, which results in a colored, water-soluble product [38]. Formed formazan is quantified to measure the viability using a spectrophotometer at 450 and 620 nm, where 620 nm is used as the reference [39]. All cell types cannot be successfully tested with the XTT assay. For a thorough comprehension of the cells' viability, it is therefore always utilized in conjunction with other cytotoxic tests [38].

10.2.1.4 Trypan blue exclusion assay

Dyes are employed in studies to distinguish between cells that are alive and deceased. Live cells have healthy plasma membranes that block certain dyes, whereas dead cells do not [40]. Trypan blue exclusion assay is a cytotoxicity assay that uses trypan blue dye to gauge the cell vitality [41]. This dye enters living cells inactively, but it permeates dead cells due to the ruptured plasma membrane. Dye reacts with intracellular proteins as it enters the plasma membrane to give a blue color [42]. The shape and dimensions of living cells can be used to distinguish them from dead ones. Live cells are tiny and spherical, whereas dead ones are huge and blue in color [43]. The assay method is very simple; cells are suspended in trypan blue-containing PBS and subsequently analyzed to identify the proportion of cells with viable (clear cytoplasm) versus nonviable (blue cytoplasm) cytoplasm. In circumstances where a spectrophotometer is not available, this can be an

TABLE 10.1 Cell lines and material combination for MTT assay in bone, skin, and nerve tissue applications

MTT assay				
S. No.	**Cell line**	**Material**	**Application**	**Reference**
1.	L929	PCL-gelatin	Tissue engineering	[26]
2.	MG63	Borate substituted bioactive glass	BTE	[27]
3.	SH-SY5Y	PCL-gelatin-nanoclay	Nerve Tissue	[28]
4.	G-292	PCL-octa Calcium phosphate	BTE	[29]
5.	Skin Fibroblast	PCL-PU-graphite oxide	Skin Tissue	[30]
6.	L929 & MG63	Gellan-guar gum-HAP	BTE	[31]
7.	MG63	PCL-chitosan and PCL-carboxy methyl chitosan	BTE	[32]
8.	MSCs	Nano HAP-PLLA	BTE	[33]
9.	L929	Chitosan-SiO_2	Skin Tissue	[34]
10.	B65	PHB-chitosan	Nerve	[35]

advantageous method [41]. However, there are some disadvantages as well, such as labor-intensive and time-consuming sample handling, computational errors, and difficulties differentiating dead cells from stained debris [42]. Long-term incubation with the dye can sometimes cause the live cells' slow dye uptake, which might result in a faded appearance. Cell vitality can be evaluated by counting living cells and dead cells under an optical microscope [20,43].

$$\text{Cell Viability}(\%) = \frac{\textit{Number of live cells}}{\textit{Total number of cells}} * 100 \tag{10.2}$$

Or

$$\text{Dead cell}(\%) = \frac{\textit{Number of dead cells}}{\textit{Total number of cells}} * 100 \tag{10.3}$$

10.2.1.5 Lactate dehydrogenase assay

An approach used to assess the vitality of cells includes the evaluation of cytoplasmic enzymes generated by injured cells. Lactate dehydrogenase (LDH) is a cytoplasmic oxidoreductase [20] found in live cells and is released to the cytoplasm during cell death. LDH assay quantifies the level of LDH to evaluate the viability or cytotoxicity. Compared to dead cells, normal cells have a lower level of LDH. LDH is generated when apoptosis or necrosis causes damage to the plasma membrane. This is a colorimetric test that assesses the integrity of the membrane [44]. Fig. 10.2 illustrates the process of LDH enzyme release and cell viability assessment.

The conversion of lactate to pyruvate is catalyzed by LDH [45]. Under normal cell conditions, LDH remains within the cell. But when cell damage occurs, LDH is released from the cytoplasm into the culture medium. Following the release of LDH, pyruvate is produced from lactate and nicotinamide adenine dinucleotide from NAD+. The NADH thus produced is utilized to transform an INT (tetrazolium salt) into a colorful formazan product. Absorbance corresponding to the intensity of the colored product is calculated by a spectrophotometer at a range between 490 and 520 nm [46]. The quantity of LDH in the culture is related to the number of dead cells, and the amount of formazan formed [44]. While reduced LDH values indicate good cell viability, greater LDH values correlate with enhanced cytotoxicity and cell death.

10.2.1.6 Alamar Blue assay/resazurin assay

Cytotoxicity investigations from last 50 years are carried out using Alamar Blue (AB) assay [47]. The oxidized form of blue-colored, nonfluorescent redox indicator resazurin (7-hydroxy-3*H*-phenoxazin-3-one 10-oxide) is used to quantify cellular proliferation [33]. The AB assay is useful in determining a cell's cytotoxicity, rate of cell division, and metabolic activity when it comes into contact with a synthetic foreign substance. Fig. 10.3 shows a schematic representation of the process of the alamar blue assay. Resazurin is a nontoxic dye that is used to quantify cellular invasion and migration without sacrificing any cells during the analysis procedure. Fluorometric and colorimetric changes in this dye are correlated with cellular metabolic activity [49,50]. Reduction of the dye resazurin to resorufin and dihydro-resorufin by living, functioning cells is the basis for determining cell viability [51]. This process takes place within the cell, and

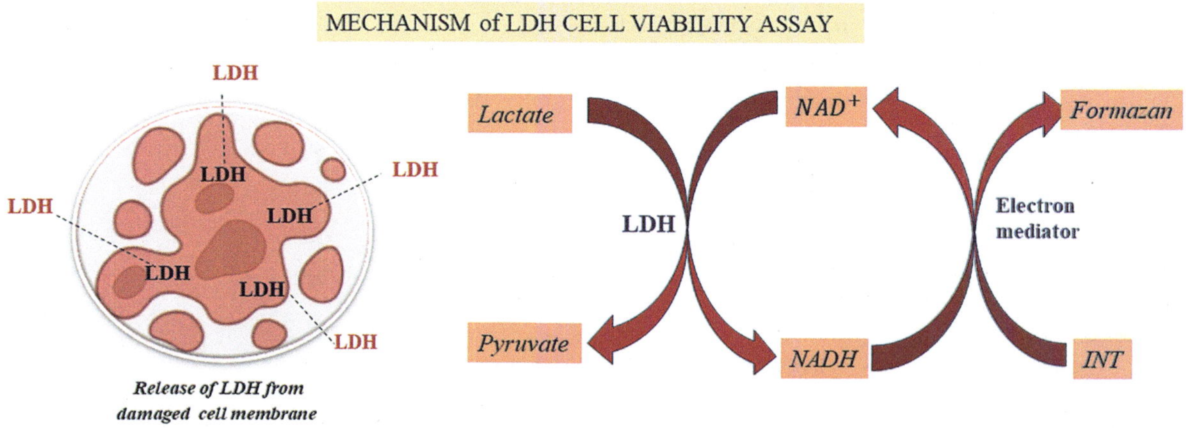

FIGURE 10.2 **Mechanism of LDH cell viability assay.** Representation of the mechanism underlying the LDH assay. *LDH*, Lactate dehydrogenase.

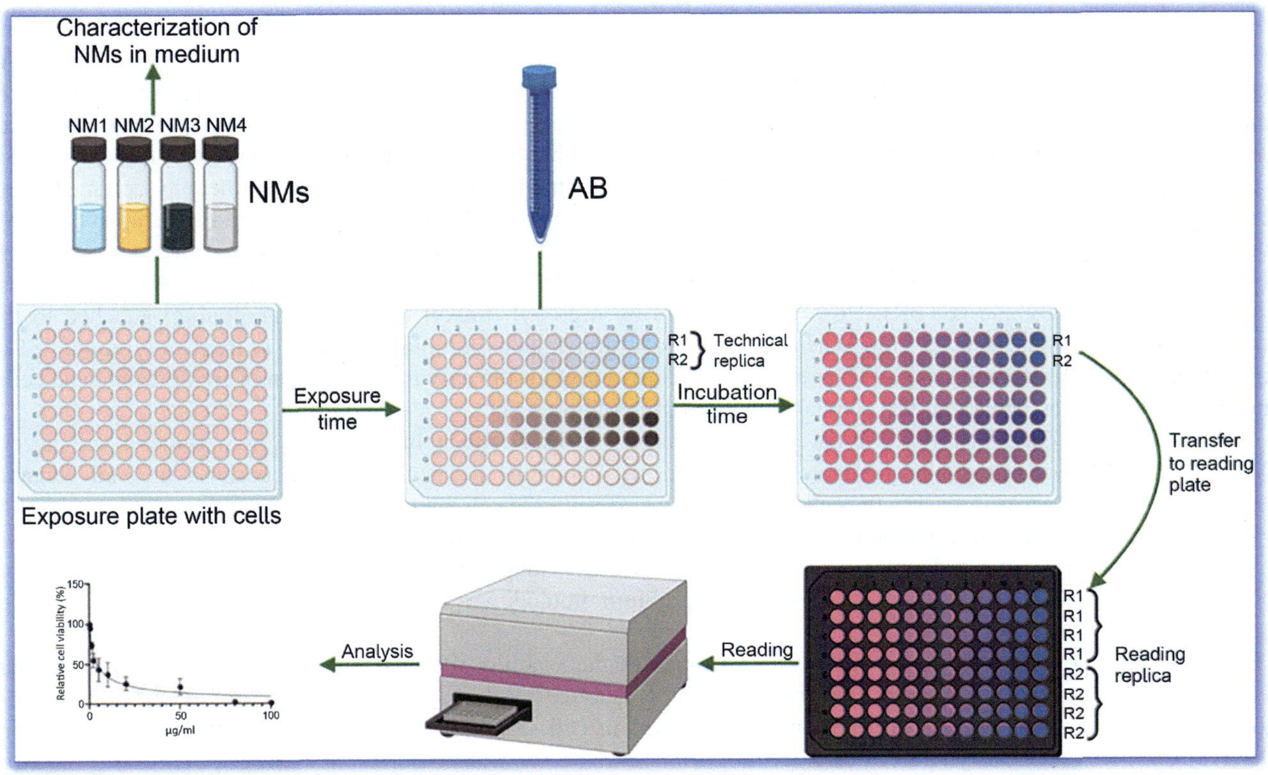

FIGURE 10.3 A graphical illustration of the ALAMAR BLUE test. A graphical illustration of the ALAMAR BLUE test, which measures the vitality of cells using nanomaterials. Reproduced by permission under copyright, © 2022 Longhin, El Yamani, Rundén-Pran, and Dusinska [48].
E.M. Longhin, N. El Yamani, E. Rundén-Pran, M. Dusinska. The alamar blue assay in the context of safety testing of nanomaterials, Frontiers in Toxicology 4 (2022). https://doi.org/10.3389/ftox.2022.981701.

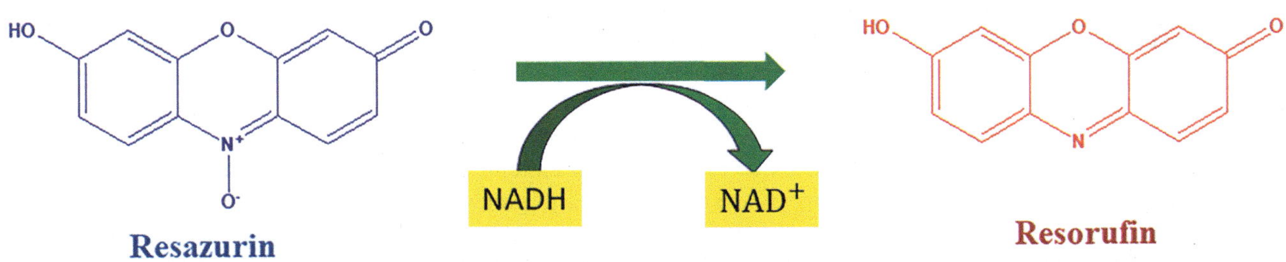

FIGURE 10.4 Mechanism of the conversion of resazurin to resorufin. Structural illustration showing the conversion of resazurin to resorufin.

resazurin reaches the cytoplasm and undergoes conversion to resorufin by mitochondrial enzymes (Fig. 10.4) that accept electrons from NADPH or NADH [52]. Resazurin in its reduced form appears red and is fluorescent [53].

These red fluorescent signals can be evaluated between 530 and 570 nm, which is the excitation range, and 580–610 nm as their emission range [48]. Absorbance is measured at 540–630 nm [52]. By detecting changes related to the fluorescence appearance in the intracellular membrane, the quantity of active cells is identified [54].

Resazurin is nontoxic, water-soluble, permeable through cell membranes, and stable in culture conditions in both its oxidized and reduced forms. Compared to other assays for cell viability and proliferation of cells, the AB assay is reliable, advantageous, and easy to perform [47]. Similar knowledge about cell growth and viability can be obtained with tetrazolium salts (MTT). However, in these situations, formazan salts insolubility in water requires a solubilization phase that uses toxic solvents like DMSO [47] and increases the possibility of damage to the cells during the solubilization stage. As a result, the ALAMAR BLUE assay is recommended over MTT [54,55].

10.2.1.7 Live/dead staining assay

Live/dead staining assay is used for determining the cell viability and cytotoxicity of the cells. It distinguishes live and dead cells to evaluate the viability of cells. Cellular membrane integrity is the basis for evaluating a material's capacity to stimulate cell growth and proliferation [56]. Stains of nucleic acid are extensively utilized for membrane integrity assessments due to their significant fluorescence amplification upon binding and the high concentration of these molecules in the cells [57]. Fluorescent dyes are used as indicators to predict membrane integrity because a healthy cell membrane inhibits fluorescence dye absorption. Stains like propidium iodide and SYTO 9 are commonly used in this approach [58]. Nucleic acid stain SYTO 9 is cell-permeable, green fluorescent, with maximum RNA emission at 501 nm, maximum DNA excitation at 498 nm, and maximum blue excitation at 485 nm [59]. Propidium iodide is an intercalating nucleic acid similar to ethidium bromide (ETBR). However, it differs by n-propyl trialkyl quaternary ammonium attached to the nitrogen atom of the heterocyclic ring. When used with additional stains, propidium iodide is frequently employed to eliminate dead cells from analyses [60]. These combined dyes allow for the easy differentiation of living from dead cells, with the viable cells as green and the nonviable cells as green fluorescent.

Additional fluorescent dyes utilized in the measurement of live/dead cells include ethidium homodimer-1 and calcein AM. Calcein AM mostly exhibits nonfluorescence and cell membrane permeability. And because the cell contains esterases, it breaks down as it reaches the cell membrane. As a result, the cell emits green fluorescence, which is a sign of living cells. However, ethidium homodimer-1 cannot pass through the membranes of living cells. As a result, dead cells absorb the dye and release red fluorescence [61]. The microscopic image representing the evaluation of the toxicity of IONPs to CCD-18Co cells using calcein AM and ethidium homodimer-1, reported by Raval et al., is given in Fig. 10.5 [61].

Ethidium monoazide (EMA) is another stain that assesses the viability of cells by analyzing the integrity of their membranes. Only cells with damaged membranes can be penetrated by this dye [62]. Propidium monoazide (PMA) is another dye that is membrane impermeant and can be used in cell viability testing. Results show that PMA is more selective than EMA at identifying dead cells [63].

10.2.2 Hemocompatibility assay

Hemocompatibility is a key requirement in translational tissue engineering. It is a crucial prerequisite for designed materials to avoid blood coagulation after implantation in its medical use. This assay helps to identify how a material interacts with blood. It evaluates the interactions between foreign substances and the different components of the blood [64,65]. Hemocompatibility assesses the thrombotic reaction of a material in its interaction with the blood, initiating the blood coagulation cascade, which includes platelet response, the stimulation of complement, and the coagulation pathway [66]. When implanted in the body, a scaffold should be hemocompatible by preventing complicated effects like inflammation, immune response, and infection [67]. When a scaffold is introduced into the body, it should be hemocompatible by preventing complex effects such as inflammatory processes, immune-mediated response, and infection [26]. If it is not, it can cause negative side effects that, in extreme cases, can be fatal to the patient.

When a biomaterial interacts with the biological system, initially, protein adsorption occurs in the blood-material interaction. This layer of adsorbed protein has a different composition and molecular interactions than plasma proteins, which in turn trigger immunological response, thrombolytic, and platelet binding pathways and result in varying degrees of blood clot formation [68–70]. The transformation of fibrinogen into fibrin is catalyzed by thrombin, in which fibrin develops into a blood clot. Severe immunological response could result from poor hemolysis, which is due to the breakdown of RBC, and low adhesion inclination for substances promoting coagulation and molecules linked to the complement activation pathway [71]. According to the ISO standard (ISO 10993-4), hemocompatibility comprehensively assesses thrombosis, coagulation, immunology, hematology, and platelets [72].

Hemocompatibility of a substance is assessed using a diluted blood solution in acid citrate dextrose (ACD). To conduct the assay, the material to be studied is exposed to this ACD in a static state for a predetermined time. Following this, the solution is taken apart, and the undamaged blood cells are separated by centrifugation. The supernatant thus formed is taken for measuring OD values using a spectrophotometer at around 540 nm. During this process, broken blood cells release free hemoglobin, which accounts for the variation in the spectral data. Using both positive and negative controls, the percentage of hemolysis is calculated from the acquired results [73–77].

$$Hemolysis\,(\%) = \frac{OD_{Sample} - OD_{-veControl}}{OD_{+VeControl} - OD_{-VeControl}} * 100 \tag{10.4}$$

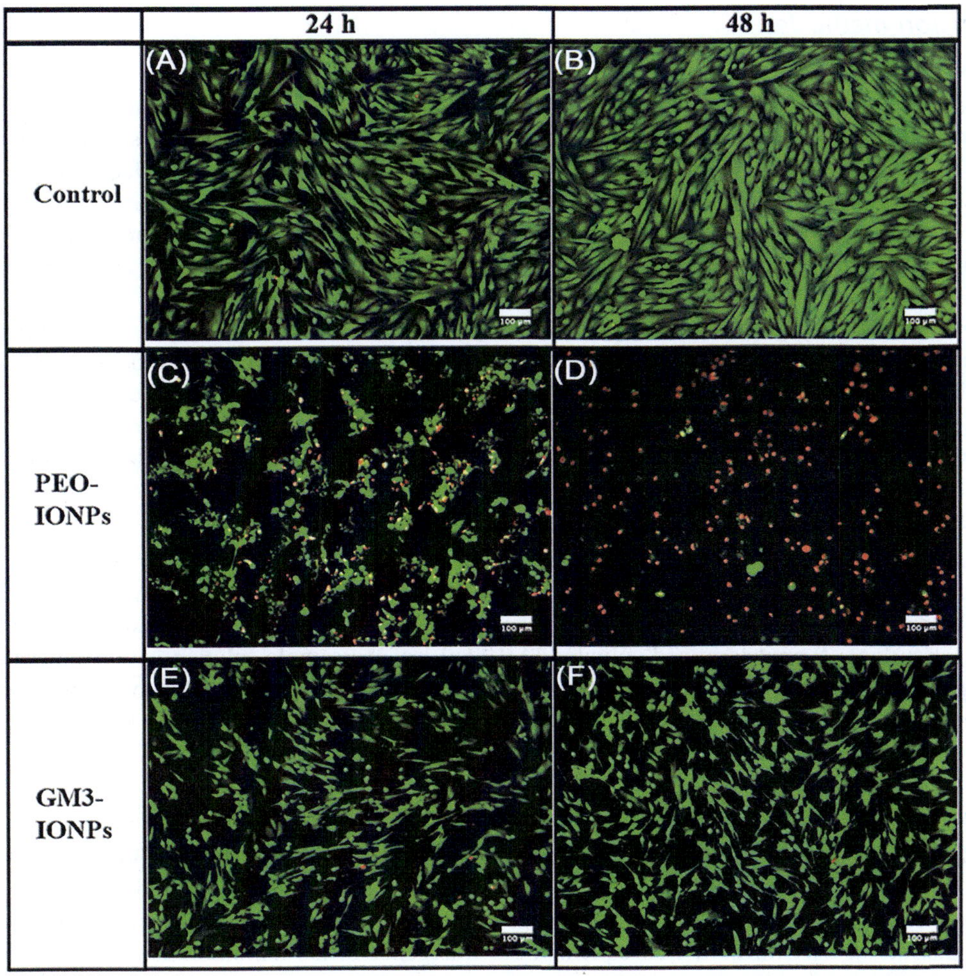

FIGURE 10.5 Image of live/dead test with ethidium bromide-1 and calcein-AM stains. Microscopic images showing the live/dead test using ethidium bromide-1 and calcein-AM stains. PEO-IONPs and GM3-IONPs were added to CCD-18Co cells and incubated for 24 and 48 hours (500 μg/mL). Dead cells look red, and live cells appear green. Reproduced with permission Copyright: © 2021 by the authors. Licensee MDPI, Basel, Switzerland [61].
From, Y.S. Raval, A. Samstag, C. Taylor, G. Huang, O.T. Mefford, T.R.J. Tzeng, Assessing the biocompatibility of multi-anchored glycoconjugate functionalized iron oxide nanoparticles in a normal human colon cell line ccd-18co. Nanomaterials, 11 (10) (2021). https://doi.org/10.3390/nano11102465.

According to the ASTM standard, hemolysis percentage (%) of less than 5 is regarded as extremely hemocompatible, and greater than 10 is hemocompatible, and above 20 is nonhemocompatible [77–79]. Material within this permissible limit of 5% hemolysis is allowable for the in vivo examination [80]. Hemocompatibility is a reliable, reproducible, and accurate technique that closely resembles the in vivo conditions [81].

10.2.3 Antimicrobial studies

Antimicrobial analysis is very crucial in the clinical translation of biomaterials. The risk of microbial infections is too high when an implant or an external material is inserted into the body. These infections adversely affect cell survival, raise the possibility of implant failure, and worsen patient outcomes [82]. Therefore the use of antimicrobial screening and evaluation is essential. These tests help to identify the microbiological resistance and the required effective treatment approach of an engineered material [83]. Even now, methods based on the 1929 AST (Antimicrobial Susceptibility Test) protocols continue to be the most successful for evaluating microbial resistance. It is still possible to determine the effectiveness of antibiotics or other agents against different microbes using these techniques [84]. There are several approaches for determining microbial resistance. However, diffusion and dilution methods are the basic techniques among them. Here, Methods for antimicrobial evaluation will be discussed.

10.2.3.1 Diffusion methods

10.2.3.1.1 Agar disk method

One of the most common methods for finding microbial inhibition is the agar disk method. This approach entails introducing a standard inoculum of the test microorganism onto agar plates. Then, test material with a specific concentration is distributed across a 6 mm-diameter paper disk. Then it is put on the agar surface and allowed to incubate under the necessary circumstances. The component with antimicrobial properties diffuses and inhibits microorganisms from growing if it shows signs of microbial resistance. Throughout the incubation time, this creates inhibitory zones in the plates, which can be assessed to determine a material's antimicrobial activity. This approach is used to evaluate bacterial pathogens such as *streptococci, Haemophilus parainfluenzae, Haemophilus influenzae, Neisseria meningitidis*, and *Neisseria gonorrhoeae* according to Clinical and Laboratory Standards Institute [85]. This is a simple technique that doesn't require any laborious equipment setup. There is also ease in changing the antibiotic type by just switching the disc. But the major drawback is that it is inappropriate for anaerobic and slow-growing microbes [86]. Additionally, a key consideration in this procedure is the molecular weight of the material to be analyzed. False results can also arise from uneven agar media distribution. This method cannot yield quantitative values; it can only yield qualitative outcomes [87].

10.2.3.1.2 Well diffusion

Agar well diffusion is also a commonly employed method for antimicrobial studies, which is similar to the agar disk method. The fundamental idea behind this is the diffusion of antimicrobial substances to prevent microbial growth. Agar plates are initially inoculated with the necessary test microorganism inoculum. Then, using a sterile borer, a well is cut on the inoculated agar surface. A predetermined volume (20–100 μL) of the sample solution is then introduced into the cut well. Depending on the microorganism being employed, different conditions are used for plate incubation. During the incubation period, by the diffusion of antimicrobial agents, a zone of inhibitory growth of the microorganism is produced. This is measured and statistically analyzed to assess the microbial resistance of the material/substance used [88,89].

10.2.3.1.3 Plug diffusion

This method emphasizes the competitive relationship between microbial species. In this method, initially, agar plates are inoculated with appropriate microbial strains and are incubated. Followed by incubation, plugs are cut on the cultured medium using a sterile borer and placed on another fresh inoculated agar medium [90]. During this process, diffusion occurs from the plug to the agar medium. Subsequently, the emergence of the inhibition zone surrounding the agar plug indicates the antimicrobial action [85,91].

10.2.3.1.4 E Test

This test is an expanded variant of the agar diffusion, which is carried out in the same way as the disc diffusion approach and uses diffusion and dilution principles to obtain corresponding MIC (minimum inhibitory concentration) values [85]. Rather than using paper disks, an antimicrobial agent is introduced exponentially to a plastic strip at the bottom. This strip is then kept over the medium of agar to diffuse the antimicrobial agent. Following incubation, the MIC value associated with the microbial resistance can be identified on the strip [83]. The elliptical inhibitory growth zone and the strip intersect to determine the MIC value.

10.2.3.2 Dilution methods

The dilution method, which yields the MIC value, is the most efficient quantitative analysis for the assessment of antimicrobial efficiency. These techniques precisely determine the quantity of antimicrobial agent present in a culture. Broth and agar dilutions are the two most used dilution techniques for in vitro antimicrobial research. The minimal amount of substance required to prevent the growth of the microorganism is recorded as MIC. Usually, the MIC value is reported in micrograms per milliliter or milligrams per liter [85].

10.2.3.2.1 Broth dilution

One of the simplest techniques to assess antimicrobial sensitivity is the broth dilution method. The process entails utilizing two liquid medium dilutions by either microdilution (96-well plate) or macrodilution (2 ml- min.vol) method. These diluted tubes are inoculated with the required microbial suspension and are incubated under suitable conditions. Following incubation, the tubes are carefully inspected for turbidity, which is a sign of apparent microbial growth [85,92,93]. Using this information, the MIC value is computed, and the level of microbial resistance is quantitatively assessed [94].

10.2.3.2.2 Agar dilution

The agar dilution method is effectively employed to test both antifungal and antibacterial resistance. This method of dilution is typically preferred over the broth dilution to determine MIC when multiple strains are evaluated against one substance or when the substance being examined obscures the identification of growth of the microbe in the medium of fluid with its coloration [85]. This is frequently suggested as a standard approach for organisms that are picky eaters, like anaerobes and species of Helicobacter. Additionally, it is used against *Fusarium, Candida sp., Aspergillus*, and *dermatophytes* [95,96].

10.2.4 Cell differentiation assay

Cells undergo a number of modifications as they come into contact with the ECM, or extracellular matrix, during development. Differentiation is the process through which cells start to specialize in their functions. The intricate process of differentiation results in variations in the characteristics of cells within the same group. The ECM influences cell movement, division, and differentiation of cells. Cell shape can be changed by the structural integrity of the ECM, and matrix-dependent morphological alterations can promote differentiation and also impede growth. Differentiation assays are therefore employed to provide insights into cell differentiation and reveal how biomaterials interact with the extracellular matrix [97].

10.2.4.1 Alkaline phosphatase assay

An early indicator for differentiation of osteoblasts is alkaline phosphatase. The alkaline phosphatase assay is a crucial tool in evaluating osteogenesis. It detects how quickly stem cells become bone cells. To ensure the effectiveness of designed materials, ALP is utilized in bone tissue engineering applications. For bone tissue engineering, mesenchymal stem cells' osteogenic differentiation on a scaffold is essential. A rise in ALP levels denotes active bone production since ALP is a byproduct of osteoblast activity [98,99]. Alkaline phosphatase converts para-nitrophenyl phosphate into para-nitrophenol in a 2-amino, 2-methyl, 1-propanol buffer at a pH of 10.5 [100]. This technique uses a simple reaction in which activators of magnesium and zinc are added to PNPP, and alkaline phosphatase acts upon it to produce p-nitrophenol, which is evaluated at 450 nm. The amount of alkaline phosphatase present in the samples has a direct correlation with the rate of p-nitrophenol produced.

10.2.4.2 Western blotting analysis

Western blotting is a method to study protein–protein interactions, the availability, abundance, mass, and existence of posttranslational modifications (PTM). It depends on how precisely antibodies engage with target antigens that are mixed into the sample. The original goal of the Western blot is to identify whether the target protein is present or not present in a sample. Using an eye assessment, this qualitative method confirms the presence of target bands [101].

10.3 Conclusion

In vitro assays for tissue engineering are the core techniques used to assess the safety, functionality, and suitability of a biomaterial for use in a clinical setting. Cell viability, the capacity for proliferation and differentiation, hemocompatibility, and antimicrobial qualities are the essential qualities that a biomaterial must possess to be suitable for tissue engineering applications. Providing similar biological circumstances, in vitro tests can help to determine the nature and potential of an engineered tissue sample before it is examined in vivo. As tissue engineering advances, in vitro assays for scaffold evaluation remain a crucial component in translational engineering by creating compatible functional materials. This chapter thus successfully aids in the researcher's selection of the best in vitro assessment technique depending on the application of the engineered tissue.

References

[1] C. Li, L. Ouyang, J.P.K. Armstrong, M.M. Stevens, Advances in the fabrication of biomaterials for gradient tissue engineering, Trends in Biotechnology 39 (2) (2021) 150–164. Available from https://doi.org/10.1016/j.tibtech.2020.06.005, www.elsevier.com/locate/tibtech.

[2] Y. Gao, Y. Wang, J. Zhang, M. Zhang, C. Dai, Y. Zhang, et al., Advancing neural regeneration via adaptable hydrogels: enriched with Mg2+ and silk fibroin to facilitate endogenous cell infiltration and macrophage polarization, Bioactive Materials 33 (2024) 100–113. Available from https://doi.org/10.1016/j.bioactmat.2023.10.026, https://www.sciencedirect.com/science/journal/2452199X.

[3] Y. Zhao, E.Y. Wang, F.B.L. Lai, K. Cheung, M. Radisic, Organs-on-a-chip: a union of tissue engineering and microfabrication, Trends in Biotechnology 41 (3) (2023) 410–424. Available from https://doi.org/10.1016/j.tibtech.2022.12.018, www.elsevier.com/locate/tibtech.

[4] A.R. Amini, C.T. Laurencin, S.P. Nukavarapu, Bone tissue engineering: recent advances and challenges, Critical Reviews in Biomedical Engineering 40 (5) (2012) 363–408. Available from https://doi.org/10.1615/CritRevBiomedEng.v40.i5.10UnitedStates, http://www.dl.begellhouse.com/download/article/20e3cde6200d53aa/CRB_5064.pdf.

[5] S.S. Lee, X. Du, I. Kim, S.J. Ferguson, Scaffolds for bone-tissue engineering, Matter 5 (9) (2022) 2722–2759. Available from https://doi.org/10.1016/j.matt.2022.06.003, www.cell.com/matter.

[6] D. Subramaniam, S. Sekaran, In vitro biocompatibility assessment of a novel membrane containing magnesium-chitosan/carboxymethyl cellulose and alginate intended for bone tissue regeneration, Cureus 16 (2) (2024). Available from https://doi.org/10.7759/cureus.54597.

[7] M. Ayran, A.Y. Dirican, E. Saatcioglu, S. Ulag, A. Sahin, B. Aksu, et al., 3D-printed PCL scaffolds combined with juglone for skin tissue engineering, Bioengineering 9 (9) (2022). Available from https://doi.org/10.3390/bioengineering9090427, www.mdpi.com/journal/bioengineering.

[8] N. Kaur, B. Singh, R.K. Sharma, S. Sharma, S. Singh, Applications of natural product-loaded scaffold for wound healing, Natural Product Inspired Scaffolds: Applications in Tissue Engineering, Springer Nature, India, (2024), pp. 213–245. Available from https://link.springer.com/book/9789819731114, https://doi.org/10.1007/978-981-97-3111-4_9.

[9] S. Li, T. Qian, X. Wang, J. Liu, X. Gu, Noncoding RNAs and their potential therapeutic applications in tissue engineering, Engineering 3 (1) (2017) 3–15. Available from https://doi.org/10.1016/J.ENG.2017.01.005, http://www.journals.elsevier.com/engineering/.

[10] M. Kun, C. Chan, S. Ramakrishna, A. Kulkarni, K. Vadodaria, Textile-Based Scaffolds for Tissue Engineering, Elsevier BV, (2019), pp. 329–362. Available from 10.1016/b978-0-08-102192-7.00012-6.

[11] J.P. Harris, L.A. Struzyna, P.L. Murphy, D.O. Adewole, E. Kuo, D.K. Cullen, Advanced biomaterial strategies to transplant preformed micro-tissue engineered neural networks into the brain, Journal of Neural Engineering 13 (1) (2016). Available from http://iopscience.iop.org/article/10.1088/1741-2560/13/1/016019/pdf, https://doi.org/10.1088/1741-2560/13/1/016019, http://iopscience.iop.org/article/10.1088/1741-2560/13/1/016019/pdf.

[12] B.S. Spearman, V.H. Desai, S. Mobini, M.D. McDermott, J.B. Graham, K.J. Otto, et al., Tissue-engineered peripheral nerve interfaces, Advanced Functional Materials 28 (12) (2018). Available from https://doi.org/10.1002/adfm.201701713, http://onlinelibrary.wiley.com/journal/10.1002/(ISSN)1616-3028.

[13] E.C. Novosel, C. Kleinhans, P.J. Kluger, Vascularization is the key challenge in tissue engineering, Advanced Drug Delivery Reviews 63 (4) (2011) 300–311. Available from https://doi.org/10.1016/j.addr.2011.03.004.

[14] S. MacNeil, Progress and opportunities for tissue-engineered skin, Nature 445 (7130) (2007) 874–880. Available from https://doi.org/10.1038/nature05664, http://www.nature.com/nature/index.html.

[15] S. Macneil, Biomaterials for tissue engineering of skin, Tissue-engineered skin includes cells delivered on their own, cells delivered within two-or three-dimensional biomaterials, biomaterials for replacement of the skin's dermal layer (both with and without cells), and biomaterials to support the replacement Open access under CC BY-NC-ND license, 2008.

[16] C. Clar, E. Cummins, L. McIntyre, S. Thomas, J. Lamb, L. Bain, et al., Clinical and cost-effectiveness of autologous chondrocyte implantation for cartilage defects in knee joints: systematic review and economic evaluation, Health Technology Assessment 9 (47) (2005). Available from https://doi.org/10.3310/hta9470, www.journalslibrary.nihr.ac.uk/hta.

[17] A. Ruano-Ravina, M.J. Díaz, Autologous chondrocyte implantation: a systematic review, Osteoarthritis and Cartilage 14 (1) (2006) 47–51. Available from https://doi.org/10.1016/j.joca.2005.07.017, http://www.elsevier.com/inca/publications/store/6/2/3/0/5/5/index.htt.

[18] M.R. Safran, K. Seiber, The evidence for surgical repair of articular cartilage in the knee, American Academy of Orthopaedic Surgeon 18 (5) (2010) 259–266. Available from https://doi.org/10.5435/00124635-201005000-00002.

[19] M.F. Wolf, K.P. Coleman, E.A. Rankin, G.M. Lewerenz, In vitro assessment of cell and tissue compatibility, Biomaterials Science: An Introduction to Materials in Medicine, Elsevier, United States, (2020), pp. 851–868. Available from https://www.sciencedirect.com/book/9780128161371, https://doi.org/10.1016/B978-0-12-816137-1.00057-X.

[20] P.S.T. Shanmugam, T. Sampath, I. Jagadeeswaran, S. Thamizharasan, S. Fathima, V. Krithaksha, Cytotoxicity Biocompatibility Protocols for Medical Devices and Materials, Elsevier, United States, (2023), pp. 1–18. Available from https://www.sciencedirect.com/book/9780323919524, https://doi.org/10.1016/B978-0-323-91952-4.00020-9.

[21] L. Khalef, R. Lydia, K. Filicia, B. Moussa, Cell viability and cytotoxicity assays: biochemical elements and cellular compartments, Cell Biochemistry and Function 42 (3) (2024). Available from https://doi.org/10.1002/cbf.4007, http://onlinelibrary.wiley.com/journal/10.1002/(ISSN)1099-0844.

[22] F. Denizot, R. Lang, Rapid colorimetric assay for cell growth and survival, Journal of Immunological Methods 89 (2) (1986) 271–277. Available from https://doi.org/10.1016/0022-1759(86)90368-6.

[23] C. Sarı, A comparative study of MTT and WST-1 assays in cytotoxicity analysis, Haydarpasa Numune Training and Research Hospital Medical Journal (2019). Available from https://doi.org/10.14744/hnhj.2019.16443.

[24] P.R. Twentyman, M. Luscombe, A study of some variables in a tetrazolium dye (MTT) based assay for cell growth and chemosensitivity, British Journal of Cancer 56 (3) (1987) 279–285. Available from https://doi.org/10.1038/bjc.1987.190.

[25] P.W. Sylvester, Optimization of the tetrazolium dye (MTT) colorimetric assay for cellular growth and viability, Methods in Molecular Biology 716 (2011) 157–168. Available from https://doi.org/10.1007/978-1-61779-012-6_9, http://www.springer.com/series/7651.

[26] S. Gautam, A.K. Dinda, N.C. Mishra, Fabrication and characterization of PCL/gelatin composite nanofibrous scaffold for tissue engineering applications by electrospinning method, Materials Science and Engineering C 33 (3) (2013) 1228–1235. Available from https://doi.org/10.1016/j.msec.2012.12.015.

[27] S. Murimadugula, A. Prasad, A.M. Lakshmi, S. Kumari, P.V. Rao, G.N. Raju, et al., Biomimetic Bi3+ substituted borate bioactive glasses for bone tissue engineering via enhanced biological and antimicrobial activity, Surfaces and Interfaces 51 (2024). Available from https://doi.org/10.1016/j.surfin.2024.104697, https://www.sciencedirect.com/science/journal/24680230.

[28] S. Ahmadi, S.S. Shafiei, F. Sabouni, Electrospun nanofibrous scaffolds of polycaprolactone/gelatin reinforced with layered double hydroxide nanoclay for nerve tissue engineering applications, ACS Omega 7 (32) (2022) 28351–28360. Available from https://doi.org/10.1021/acsomega.2c02863, pubs.acs.org/journal/acsodf.

[29] Z. Heydari, D. Mohebbi-Kalhori, M.S. Afarani, Engineered electrospun polycaprolactone (PCL)/octacalcium phosphate (OCP) scaffold for bone tissue engineering, Materials Science and Engineering C 81 (2017) 127–132. Available from https://doi.org/10.1016/j.msec.2017.07.041.

[30] A. Sadeghianmaryan, Y. Karimi, S. Naghieh, H. Alizadeh Sardroud, M. Gorji, X. Chen, Electrospinning of scaffolds from the polycaprolactone/polyurethane composite with graphene oxide for skin tissue engineering, Applied Biochemistry and Biotechnology 191 (2) (2020) 567–578. Available from https://doi.org/10.1007/s12010-019-03192-x, http://www.springer.com/humana+press/journal/12010.

[31] D. Anandan, G. Madhumathi, N.A. Nambiraj, A.K. Jaiswal, Gum based 3D composite scaffolds for bone tissue engineering applications, Carbohydrate Polymers 214 (2019) 62–70. Available from https://doi.org/10.1016/j.carbpol.2019.03.020, http://www.elsevier.com/wps/find/journaldescription.cws_home/405871/description#description.

[32] F. Sharifi, S.M. Atyabi, D. Norouzian, M. Zandi, S. Irani, H. Bakhshi, Polycaprolactone/carboxymethyl chitosan nanofibrous scaffolds for bone tissue engineering application, International Journal of Biological Macromolecules 115 (2018) 243–248. Available from https://doi.org/10.1016/j.ijbiomac.2018.04.045, www.elsevier.com/locate/ijbiomac.

[33] E. Nejati, V. Firouzdor, M.B. Eslaminejad, F. Bagheri, Needle-like nano hydroxyapatite/poly(l-lactide acid) composite scaffold for bone tissue engineering application, Materials Science and Engineering: C 29 (3) (2009) 942–949. Available from https://doi.org/10.1016/j.msec.2008.07.038.

[34] L. Mei, D. Hu, J. Ma, X. Wang, Y. Yang, J. Liu, Preparation, characterization and evaluation of chitosan macroporous for potential application in skin tissue engineering, International Journal of Biological Macromolecules 51 (5) (2012) 992–997. Available from https://doi.org/10.1016/j.ijbiomac.2012.08.004, www.elsevier.com/locate/ijbiomac.

[35] A. Karimi Tar, S. Karbasi, E. Naghashzargar, H. Salehi, Biodegradation and cellular evaluation of aligned and random poly (3-hydroxybutyrate)/chitosan electrospun scaffold for nerve tissue engineering applications, Materials Technology 35 (2) (2020) 92–101. Available from https://doi.org/10.1080/10667857.2019.1658170, https://www.tandfonline.com/loi/ymte20.

[36] K. Präbst, H. Engelhardt, S. Ringgeler, H. Hübner, Basic colorimetric proliferation assays: MTT, WST, and resazurin, Germany Methods in Molecular Biology 1601 (2017) 1–17. Available from https://doi.org/10.1007/978-1-4939-6960-9_1, http://www.springer.com/series/7651.

[37] K.M. Kim, S.B. Lee, S.H. Lee, Y.K. Lee, K.N. Kim, Comparison of validity between WST-1 and MTT test in bioceramic materials, Key Engineering Materials 284-286 (2005) 585–588. Available from https://doi.org/10.4028/0-87849-961-x.585, https://www.scientific.net/KEM.

[38] D.M. Kuhn, M. Balkis, J. Chandra, P.K. Mukherjee, M.A. Ghannoum, Uses and limitations of the XTT assay in studies of candida growth and metabolism, Journal of Clinical Microbiology 41 (1) (2003) 506–508. Available from https://doi.org/10.1128/jcm.41.1.506-508.2003.

[39] S. Mani, G. Swargiary, In Vitro Cytotoxicity Analysis: MTT/XTT, Trypan Blue Exclusion, Springer Science and Business Media LLC, (2023), pp. 267–284. Available from 10.1007/978-3-031-19485-6_18.

[40] B.A. Avelar-Freitas, V.G. Almeida, M.C.X. Pinto, F.A.G. Mourão, A.R. Massensini, O.A. Martins-Filho, et al., Trypan blue exclusion assay by flow cytometry, Brazilian Journal of Medical and Biological Research 47 (4) (2014) 307–315. Available from https://doi.org/10.1590/1414-431X20143437.

[41] W. Strober, Trypan blue exclusion test of cell viability, Current Protocols in Immunology 111 (1) (2015). Available from https://doi.org/10.1002/0471142735.ima03bs111A3.B.1, http://onlinelibrary.wiley.com/book/10.1002/0471142735.

[42] R. Gupta, K. Rajpoot, M. Tekade, M.C. Sharma, R.K. Tekade, Methods and models for in vitro toxicity, Pharmacokinetics and Toxicokinetic Considerations, Elsevier, India, (2022), pp. 145–174. Available from https://www.sciencedirect.com/book/9780323983679, https://doi.org/10.1016/B978-0-323-98367-9.00006-8, vol. II.

[43] M. Shibata, H. Kanetaka, M. Furuya, K. Yokota, T. Ogawa, M. Kawashita, Cytotoxicity evaluation of iron nitride nanoparticles for biomedical applications, Journal of Biomedical Materials Research - Part A 109 (10) (2021) 1784–1791. Available from https://doi.org/10.1002/jbm.a.37171, http://onlinelibrary.wiley.com/journal/10.1002/(ISSN)1552-4965.

[44] P. Kumar, A. Nagarajan, P.D. Uchil, Analysis of cell viability by the lactate dehydrogenase assay, Cold Spring Harbor Protocols 2018 (6) (2018). Available from https://doi.org/10.1101/pdb.prot095497.

[45] S. Kaja, A.J. Payne, Y. Naumchuk, P. Koulen, Quantification of lactate dehydrogenase for cell viability testing using cell lines and primary cultured astrocytes, Current Protocols in Toxicology 2017 (2017) 1–10. Available from https://doi.org/10.1002/cptx.21, http://onlinelibrary.wiley.com/store/10.1002/cptx.21/asset/tx0226.pdf?v=1&t=j5p9053z&s=2c3540952ebd1c7ec5ea823b685c93497ea0e216.

[46] B. Gu, D.J. Burgess, Polymeric materials in drug delivery, Natural and Synthetic Biomedical Polymers, Elsevier Inc., United States, (2014), pp. 333–349. Available from http://www.sciencedirect.com/science/book/9780123969835, https://doi.org/10.1016/B978-0-12-396983-5.00021-1.

[47] S.N. Rampersad, Multiple applications of alamar blue as an indicator of metabolic function and cellular health in cell viability bioassays, Sensors (Basel, Switzerland) 12 (9) (2012) 12347–12360. Available from https://doi.org/10.3390/s120912347.

[48] E.M. Longhin, N. El Yamani, E. Rundén-Pran, M. Dusinska, The alamar blue assay in the context of safety testing of nanomaterials, Frontiers in Toxicology 4 (2022). Available from https://doi.org/10.3389/ftox.2022.981701, https://www.frontiersin.org/journals/toxicology/about.

[49] R.J. Gonzalez, J.B. Tarloff, Evaluation of hepatic subcellular fractions for Alamar blue and MTT reductase activity, Toxicology In Vitro 15 (3) (2001) 257–259. Available from https://doi.org/10.1016/s0887-2333(01)00014-5.

[50] S. Ansar Ahmed, R.M. Gogal, J.E. Walsh, A new rapid and simple non-radioactive assay to monitor and determine the proliferation of lymphocytes: an alternative to [3H]thymidine incorporation assay, Journal of Immunological Methods 170 (2) (1994) 211–224. Available from https://doi.org/10.1016/0022-1759(94)90396-4.

[51] K. Yadav, N. Singhal, V. Rishi, H. Yadav, Cell proliferation assays, Encyclopedia of Life Sciences, Wiley, (2014). Available from 10.1002/9780470015902.a0002566.

[52] S. Al-Nasiry, N. Geusens, M. Hanssens, C. Luyten, R. Pijnenborg, The use of Alamar Blue assay for quantitative analysis of viability, migration and invasion of choriocarcinoma cells, Human Reproduction 22 (5) (2007) 1304–1309. Available from https://doi.org/10.1093/humrep/dem011.

[53] E.M. Czekanska, Assessment of cell proliferation with resazurin-based fluorescent dye, Methods in Molecular Biology 740 (2011) 27–32. Available from https://doi.org/10.1007/978-1-61779-108-6_5, http://www.springer.com/series/7651.

[54] Z. Ghasemi, S. Labbaf, M.H. Enayati, M. Mohammadzadeh, Feasibility of magnetite-poly(n-isopropylacrilamide) smart nanocarriers containing doxorubicin for cancer therapeutic applications, Materials Chemistry and Physics 309 (2023) 128377. Available from https://doi.org/10.1016/j.matchemphys.2023.128377.

[55] M. Ghasemi, T. Turnbull, S. Sebastian, I. Kempson, The MTT assay: utility, limitations, pitfalls, and interpretation in bulk and single-cell analysis, International Journal of Molecular Sciences 22 (23) (2021) 12827. Available from https://doi.org/10.3390/ijms222312827.

[56] A. Adan, Y. Kiraz, Y. Baran, Cell proliferation and cytotoxicity assays, Current Pharmaceutical Biotechnology 17 (14) (2016) 1213–1221. Available from https://doi.org/10.2174/1389201017666160808160513, http://www.benthamdirect.org/pages/all_b_bypublication.php.

[57] J.P. Acker, L.E. McGann, Cell-cell contact affects membrane integrity after intracellular freezing, Cryobiology 40 (1) (2000) 54–63. Available from https://doi.org/10.1006/cryo.1999.2221, http://www.elsevier.com/inca/publications/store/6/2/2/8/1/4/index.htt.

[58] R. Lencastre Fernandes, M. Nierychlo, L. Lundin, A.E. Pedersen, P.E. Puentes Tellez, A. Dutta, et al., Experimental methods and modeling techniques for description of cell population heterogeneity, Biotechnology Advances 29 (6) (2011) 575–599. Available from https://doi.org/10.1016/j.biotechadv.2011.03.007.

[59] C. McGoverin, J. Robertson, Y. Jonmohamadi, S. Swift, F. Vanholsbeeck, Species dependence of SYTO 9 staining of bacteria, Frontiers in Microbiology 11 (2020). Available from https://doi.org/10.3389/fmicb.2020.545419, https://www.frontiersin.org/journals/microbiology#.

[60] B.T. Grimberg, Methodology and application of flow cytometry for investigation of human malaria parasites, Journal of Immunological Methods 367 (1-2) (2011) 1–16. Available from https://doi.org/10.1016/j.jim.2011.01.015.

[61] Y.S. Raval, A. Samstag, C. Taylor, G. Huang, O.T. Mefford, T.R.J. Tzeng, Assessing the biocompatibility of multi-anchored glycoconjugate functionalized iron oxide nanoparticles in a normal human colon cell line CCD-18Co, Nanomaterials 11 (10) (2021). Available from https://doi.org/10.3390/nano11102465, https://www.mdpi.com/2079-4991/11/10/2465/pdf.

[62] M.C. Riedy, K.A. Muirhead, C.P. Jensen, C.C. Stewart, Use of a photolabeling technique to identify nonviable cells in fixed homologous or heterologous cell populations, Cytometry 12 (2) (1991) 133–139. Available from https://doi.org/10.1002/cyto.990120206.

[63] A. Nocker, C.Y. Cheung, A.K. Camper, Comparison of propidium monoazide with ethidium monoazide for differentiation of live vs. dead bacteria by selective removal of DNA from dead cells, Journal of Microbiological Methods 67 (2) (2006) 310–320. Available from https://doi.org/10.1016/j.mimet.2006.04.015.

[64] N. Beheshtizadeh, M. Mohammadzadeh, M. Mostafavi, A.A. Seraji, F. Esmaeili Ranjbar, S.Z. Tabatabaei, et al., Improving hemocompatibility in tissue-engineered products employing heparin-loaded nanoplatforms, Pharmacological Research 206 (2024). Available from https://doi.org/10.1016/j.phrs.2024.107260, https://www.sciencedirect.com/science/journal/10436618.

[65] M. Ul-Islam, K.F. Alabbosh, S. Manan, S. Khan, F. Ahmad, M.W. Ullah, Chitosan-based nanostructured biomaterials: synthesis, properties, and biomedical applications, Advanced Industrial and Engineering Polymer Research 7 (1) (2024) 79–99. Available from https://doi.org/10.1016/j.aiepr.2023.07.002, https://www.sciencedirect.com/science/journal/25425048.

[66] P. Mulinti, J.E. Brooks, B. Lervick, J.E. Pullan, A.E. Brooks, Strategies to Improve the Hemocompatibility of Biodegradable Biomaterials, Elsevier BV, (2018), pp. 253–278. Available from 10.1016/b978-0-08-100497-5.00017-3.

[67] M.W. Ullah, K.F. Alabbosh, A. Fatima, S.U. Islam, S. Manan, M. Ul-Islam, et al., Advanced biotechnological applications of bacterial nanocellulose-based biopolymer nanohybrids: a review, Advanced Industrial and Engineering Polymer Research 7 (1) (2024) 100–121. Available from https://doi.org/10.1016/j.aiepr.2023.07.004, https://www.sciencedirect.com/science/journal/25425048.

[68] Z.M. Ruggeri, G.L. Mendolicchio, Adhesion mechanisms in platelet function, Circulation Research 100 (12) (2007) 1673–1685. Available from https://doi.org/10.1161/01.RES.0000267878.97021.ab.

[69] M. Mezger, H. Nording, R. Sauter, T. Graf, C. Heim, N. von Bubnoff, et al., Platelets and immune responses during thromboinflammation, Frontiers in immunology 10 (2019) 1731. Available from https://doi.org/10.3389/fimmu.2019.01731.

[70] N. Huang, P. Yang, Y.X. Leng, J.Y. Chen, H. Sun, J. Wang, et al., Hemocompatibility of titanium oxide films, Biomaterials 24 (13) (2003) 2177–2187. Available from https://doi.org/10.1016/s0142-9612(03)00046-2.

[71] M.-C. Bélanger, Y. Marois, Hemocompatibility, biocompatibility, inflammatory and in vivo studies of primary reference materials low-density polyethylene and polydimethylsiloxane: a review, Journal of Biomedical Materials Research 58 (5) (2001) 467–477. Available from https://doi.org/10.1002/jbm.1043.

[72] M. Nalezinková, In vitro hemocompatibility testing of medical devices, Thrombosis Research 195 (2020) 146–150. Available from https://doi.org/10.1016/j.thromres.2020.07.027, www.elsevier.com/locate/thromres.

[73] N.S. Murthy, V.B. Damodaran, S.H. Lee, A.S. Hwang, H.J. Sung, Characterization of thin films for biomedical applications, Thin Film Coatings for Biomaterials and Biomedical Applications, Elsevier, United States, (2016), pp. 81–115. Available from https://www.sciencedirect.com/book/9781782424536, https://doi.org/10.1016/B978-1-78242-453-6.00005-5.

[74] J.P. Singhal, A.R. Ray, Synthesis of blood compatible polyamide block copolymers, Biomaterials 23 (4) (2002) 1139–1145. Available from https://doi.org/10.1016/S0142-9612(01)00228-9.

[75] J. Li, W. Zheng, Y.F. Zheng, X. Lou, Cell responses and hemocompatibility of g-HA/PLA composites, Science China Life Sciences 54 (4) (2011) 366–371. Available from https://doi.org/10.1007/s11427-011-4155-0.

[76] H. Nan, Yr Chen, J.M. Luo, Y. Jin, L. Rong, X. Jing, et al., In vitro investigation of blood compatibility of Ti with oxide layers of rutile structure, Journal of Biomaterials Applications 8 (4) (1994) 404–412. Available from https://doi.org/10.1177/088532829400800406.

[77] N. Valarmathi, S. Sumathi, Biomimetic hydroxyapatite/silkfibre/methylcellulose composites for bone tissue engineering applications, New Journal of Chemistry 44 (11) (2020) 4647–4663. Available from https://doi.org/10.1039/c9nj05592d, http://pubs.rsc.org/en/journals/journal/nj.

[78] V. Sarath Chandra, G. Baskar, R.V. Suganthi, K. Elayaraja, M.I. Ahymah Joshy, W. Sofi Beaula, et al., Blood compatibility of iron-doped nanosize hydroxyapatite and its drug release, ACS Applied Materials & Interfaces 4 (3) (2012) 1200–1210. Available from https://doi.org/10.1021/am300140q.

[79] S. Kalidas, S. Sumathi, Mechanical, biocompatibility and antibacterial studies of gelatin/polyvinyl alcohol/silkfibre polymeric scaffold for bone tissue engineering, Heliyon 9 (6) (2023) e16886. Available from https://doi.org/10.1016/j.heliyon.2023.e16886.

[80] W.H. De Jong, J.W. Carraway, R.E. Geertsma, In vivo and in vitro testing for the biological safety evaluation of biomaterials and medical devices, Biocompatibility and Performance of Medical Devices, Elsevier, Netherlands, (2019), pp. 123–166. Available from http://www.sciencedirect.com/science/book/9780081026434, https://doi.org/10.1016/B978-0-08-102643-4.00007-0.

[81] M. Weber, H. Steinle, S. Golombek, L. Hann, C. Schlensak, H.P. Wendel, et al., Blood-contacting biomaterials: in vitro evaluation of the hemocompatibility, Frontiers in Bioengineering and Biotechnology 6 (2018). Available from https://doi.org/10.3389/fbioe.2018.00099, http://journal.frontiersin.org/journal/bioengineering-and-biotechnology#archive.

[82] M. Godoy-Gallardo, U. Eckhard, L.M. Delgado, Y.J.D. de Roo Puente, M. Hoyos-Nogués, F.J. Gil, et al., Antibacterial approaches in tissue engineering using metal ions and nanoparticles: from mechanisms to applications, Bioactive Materials 6 (12) (2021) 4470–4490. Available from https://doi.org/10.1016/j.bioactmat.2021.04.033, http://www.keaipublishing.com/en/journals/bioactive-materials/.

[83] A. Schumacher, T. Vranken, A. Malhotra, J.J.C. Arts, P. Habibovic, In vitro antimicrobial susceptibility testing methods: agar dilution to 3D tissue-engineered models, European Journal of Clinical Microbiology & Infectious Diseases 37 (2) (2018) 187–208. Available from https://doi.org/10.1007/s10096-017-3089-2.

[84] J.A. Poupard, S.F. Rittenhouse, L.R. Walsh, The evolution of antimicrobial susceptibility testing methods, Advances in Experimental Medicine and Biology 349 (1994) 3–14. Available from https://doi.org/10.1007/978-1-4757-9206-5_2, http://www.springer.com/series/5584.

[85] M. Balouiri, M. Sadiki, S.K. Ibnsouda, Methods for in vitro evaluating antimicrobial activity: a review, Journal of Pharmaceutical Analysis 6 (2) (2016) 71–79. Available from https://doi.org/10.1016/j.jpha.2015.11.005, http://www.journals.elsevier.com/journal-of-pharmaceutical-analysis.

[86] R. Schwalbe, L. Steele-Moore, A.C. Goodwin, Antimicrobial Susceptibility Testing Protocols Antimicrobial Susceptibility Testing Protocols, CRC Press, United States, (2007), pp. 1–432. Available from https://www.routledge.com/Antimicrobial-Susceptibility-Testing-Protocols/Schwalbe-Steele-Moore-Goodwin/p/book/9780824741006.

[87] J.H. Jorgensen, M.J. Ferraro, Antimicrobial susceptibility testing: a review of general principles and contemporary practices, Clinical Infectious Diseases 49 (11) (2009) 1749–1755. Available from https://doi.org/10.1086/647952.

[88] C. Erhonyota, G.I. Edo, F.O. Onoharigho, Comparison of poison plate and agar well diffusion method determining the antifungal activity of protein fractions, Acta Ecologica Sinica 43 (4) (2023) 684–689. Available from https://doi.org/10.1016/j.chnaes.2022.08.006, https://www.journals.elsevier.com/acta-ecologica-sinica/.

[89] S. Magaldi, S. Mata-Essayag, C. Hartung De Capriles, C. Perez, M.T. Colella, C. Olaizola, et al., Well diffusion for antifungal susceptibility testing, International Journal of Infectious Diseases 8 (1) (2004) 39–45. Available from https://doi.org/10.1016/j.ijid.2003.03.002.

[90] A. Rütten, T. Kirchner, E.M. Musiol-Kroll, Overview on strategies and assays for antibiotic discovery, Pharmaceuticals 15 (10) (2022). Available from https://doi.org/10.3390/ph15101302, https://www.mdpi.com/journal/pharmaceuticals.

[91] H.G.M. Saleem, U. Aftab, I. Sajid, Z. Abbas, A.N. Sabri, Effect of crude extracts of selected actinomycetes on biofilm formation of A. schindleri, M. aci, and B. cereus, Journal of Basic Microbiology 55 (5) (2015) 645–651. Available from https://doi.org/10.1002/jobm.201400358, http://onlinelibrary.wiley.com/journal/10.1002/(ISSN)1521-4028.

[92] I. Wiegand, K. Hilpert, R.E.W. Hancock, Agar and broth dilution methods to determine the minimal inhibitory concentration (MIC) of antimicrobial substances, Nature Protocols 3 (2) (2008) 163–175. Available from https://doi.org/10.1038/nprot.2007.521.

[93] C. Dwivedi, I. Pandey, H. Pandey, P.W. Ramteke, A.C. Pandey, S.B. Mishra, et al., Electrospun nanofibrous scaffold as a potential carrier of antimicrobial therapeutics for diabetic wound healing and tissue regeneration, Nano- and Microscale Drug Delivery Systems: Design and Fabrication, Elsevier, India, (2017), pp. 147–164 , http://www.sciencedirect.com/science/book/9780323527279 10.1016/B978-0-323-52727-9.00009-1.

[94] I.O. Chikezie, Determination of minimum inhibitory concentration (MIC) and minimum bactericidal concentration (MBC) using a novel dilution tube method, African Journal of Microbiology Research 11 (23) (2017) 977–980. Available from https://doi.org/10.5897/AJMR2017.8545.

[95] A. Imhof, S.A. Balajee, K.A. Marr, New methods to assess susceptibilities of Aspergillus isolates to caspofungin, Journal of Clinical Microbiology 41 (12) (2003) 5683–5688. Available from https://doi.org/10.1128/JCM.41.12.5683-5688.2003.

[96] T. Menon, K. Umamaheswari, N. Kumarasamy, S. Solomon, S.P. Thyagarajan, Efficacy of fluconazole and itraconazole in the treatment of oral candidiasis in HIV patients, Acta Tropica 80 (2) (2001) 151–154. Available from https://doi.org/10.1016/S0001-706X(01)00170-X.

[97] J.A. Cooper, B.R. Mintz, S.L. Palumbo, W.J. Li, Assays for determining cell differentiation in biomaterials, Characterization of Biomaterials, Elsevier Ltd., United States, (2012), pp. 101–137. Available from http://www.sciencedirect.com/science/book/9781845698102, 10.1533/9780857093684.101.

[98] S. Trivedi, K. Srivastava, A. Gupta, T.S. Saluja, S. Kumar, D. Mehrotra, et al., A quantitative method to determine osteogenic differentiation aptness of scaffold, Journal of Oral Biology and Craniofacial Research 10 (2) (2020) 158–160. Available from https://doi.org/10.1016/j.jobcr.2020.04.006, http://www.journals.elsevier.com/journal-of-oral-biology-and-craniofacial-research/.

[99] H.J. Prins, A.K. Braat, D. Gawlitta, W.J.A. Dhert, D.A. Egan, E. Tijssen-Slump, et al., In vitro induction of alkaline phosphatase levels predicts in vivo bone forming capacity of human bone marrow stromal cells, Stem Cell Research 12 (2) (2014) 428–440. Available from https://doi.org/10.1016/j.scr.2013.12.001, http://www.elsevier.com/wps/find/journaldescription.cws_home/711630/description#description.

[100] P. Kanta, T. Ghosh, A. Kaur, T. Muthukumarappa, An innovative and cost-effective way to estimate alkaline phosphatase activity in in vitro cellular model systems, International Journal of Biochemistry and Molecular Biology 12 (1) (2021) 1–7. Available from http://www.ijbmb.org/files/ijbmb0116766.pdf.

[101] L. Pillai-Kastoori, A.R. Schutz-Geschwender, J.A. Harford, A systematic approach to quantitative Western blot analysis, Analytical Biochemistry 593 (2020). Available from https://doi.org/10.1016/j.ab.2020.113608, http://www.elsevier.com/inca/publications/store/6/2/2/7/8/1/index.htt.

Chapter 11

In vitro platforms for advancing tissue engineering research

Manasa Nune, Maya Lakshmanan*, and Ashitha Mariam Leslie*
Manipal Institute of Regenerative Medicine, Manipal Academy of Higher Education, Manipal, Karnataka, India

11.1 Introduction

Tissue engineering is a multidisciplinary field encompassing a broad range of experimental approaches dedicated to improving, maintaining, and restoring tissue function. Biomaterials play a crucial role in tissue engineering, providing structural and functional support depending on the intended use. In vitro assays are fundamental for the success of tissue engineering, as they focus on evaluating cellular behavior, biomaterial interactions, and functional outcomes within controlled laboratory settings before their eventual implantation. It enables carrying out experiments to optimize the developed construct and test hypotheses and thereby predict material performance prior to the in vivo applications. Beyond being nontoxic, a biomaterial must promote beneficial interactions with surrounding cells. In this chapter, various in vitro assays used to assess biomaterials and their potential for clinical applications are explored, highlighting their role in advancing safe and effective tissue engineering solutions. Key techniques covered in the chapter include cell viability and proliferation assays to identify metabolically active cells, biochemical assays to assess antioxidant scavenging properties in vitro, and ELISA to identify proteins expressed by cells on a scaffold. The cell migration assay involves the ability of cells to move within a construct using several assays, such as pore bridging, physical barrier, transwell, and microfluidics based assessments. Cell differentiation assays include differentiation factors that could be assessed and techniques to detect these genes that pertain to immunofluorescence, alkaline phosphatase, osteocalcin, and osteopontin assays for bone differentiation and periodic acid Schiff (PAS) staining for glycogen synthesis. Angiogenesis and vascularization involve the detection of sprouting of the blood vessels using assays like the aortic ring assay, tube formation assay, and several of the staining techniques that include blood vessel staining, anti CD31 staining, and many more. These in vitro techniques assist in minimizing the need for in vivo testing and animal experiments and solely depend on the synergistic conditions of cells, biomaterials, and precisely controlled incubation conditions.

11.2 Cell proliferation and viability assays

Assessing the proliferation and viability of cells in the developed constructs plays a crucial role in monitoring cell response to the scaffold over time. It is a cornerstone for basic translational clinical research for observing cell division, DNA synthesis, their metabolic activity, and the cytotoxic effect of tissue engineered constructs in vitro. It is a robust tool for identifying new materials with seamless integration of cytotoxic assays to monitor the cellular response to various materials used in tissue engineering. Some of the most feasible assays in this category can be summarized as

- Metabolic cell proliferation assays
- DNA synthesis based proliferation assays
- Fluorescent dye based proliferation assays

*. Both the authors contributed to the work equally.

Fundamentals of Translational Tissue Engineering. DOI: https://doi.org/10.1016/B978-0-443-33455-9.00014-6

11.2.1 Metabolic cell proliferation assays

Some of the most common metabolic proliferation assays include MTT, MTS, XTT, WST-1, and Alamar blue, lactate dehydrogenase (LDH) and crsytal violet assays, which are widely used where there is bioreduction of tetrazolium salts like MTT, XTT, WST-1 that gives colored formazan compounds only in the metabolically active cells. There is decreased activity in the cells exposed to toxins.

11.2.1.1 MTT (3-[4,5-dimethylthiazol-2-2yl]-2,5-diphenyltetrazolium bromide) assay

MTT is a water soluble tetrazolium salt that gives a yellow color when prepared in media or salt solutions. MTT is converted to insoluble purple formazan when the succinate dehydrogenase enzyme in the mitochondria of metabolically active cells cleaves the tetrazolium ring. The formazan accumulates in the healthy cells as they are not permeable to the cell membrane. The formazan crystals are solubilized using organic solvents like isopropanol/Dimethyl sulfoxide (DMSO), and the absorbance is spectrophotometrically measured. The absorbance correlates directly with the quantity of formazan crystals, which reflect the metabolic activity of the cells [1,2] (Fig. 11.1.)

MTT assay can be carried out in 96 well plates by incubating the cell seeded scaffold in a replicated manner with the desired test factors. Replace the media with 100 μL of 0.5 mg mL^{-1} MTT solution at desired time points and incubate for 3-4 hours at 37°C in a CO_2 incubator. Remove MTT reagent and replace it with 100-200 μL organic solvent, followed by gentle shaking and incubation for 15-20 minutes in the dark at room temperature to ensure complete dissolution of formazan crystals. Measure the absorbance at 570 nm using a spectrophotometer and record the viable cell number. Cytocompatibility and cytotoxicity are among the key characterization techniques routinely performed on tissue engineered constructs for various applications.

In a study conducted by Ijaola et al. [3] to identify the in vitro cytotoxicity of a trilayered scaffold for wound healing, an MTT assay was done by seeding a 24 well plate with 2 x 10^4 3T3 and L929 cells, and after 24 hours of culture, UV irradiated trilayered scaffold was immersed and cocultured for 4 days. The goal was to assess the cytocompatibility of the scaffold consisting of heat treated electrospun superhydrophobic layer at the top, an intermediate polyethylene glycol cross linked with chitosan, and a bottom electrospun hydrophilic layer useful for skin wound healing. Cell proliferation and biocompatibility are major requirements for the developed scaffold, and the in vitro cytotoxicity was assessed with the colorimetric reduction of MTT to formazan crystals by metabolically active cells in the scaffold. The assay identified the cytocompatibility of the scaffold and demonstrated that the scaffold supported cell proliferation and had no cytotoxicity, making it suitable for wound healing applications.

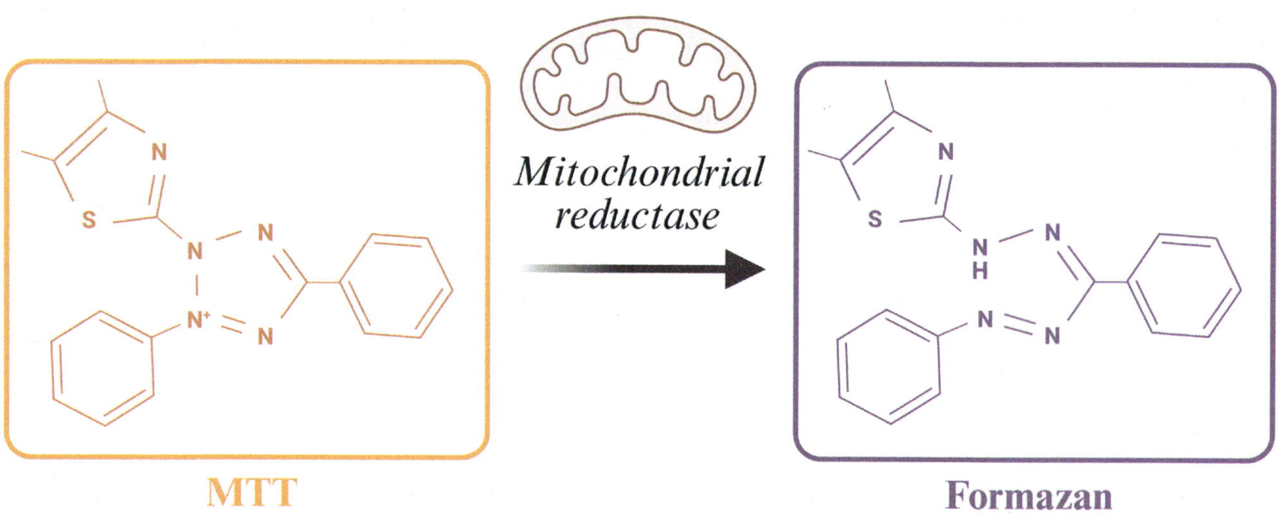

FIGURE 11.1 MTT reaction and formation of formazan crystals. The figure illustrates the mechanism of reduction of water soluble yellow tetrazolium salt (MTT) to insoluble purple formazan crystals by mitochondrial reductase of metabolically active cells.
Created using BioRender.com.

11.2.1.2 MTS (3-(4,5-dimethylthiazol-2-yl)-5-(3-carboxymethoxyphemyl)-2-(4-sulfophenyl)-2H-tetrazolium) assay

MTS, being a negatively charged compound, is directly soluble in water, facilitating the reduction of the tetrazolium into formazan by the NADPH dependent dehydrogenase enzyme. It is known as a one step simplified MTT as there are no intermittent steps involved [4]. MTS was performed to assess the ability of the biomaterials to integrate cells and living tissue for bone tissue engineering. The biocompatibility of citrate modified bacterial cellulose/hydroxyapatite scaffold to support the proliferation and bioactivity of osteoblast cells was analyzed by immersing the scaffold in media and seeding it with human fetal osteoblast and culturing them for 7 days. After incubation, MTS reagent was added and incubated for 3–4 hrs at 37°C, and absorbance was measured at 490 nm. The assay reported the survival and proliferation of osteoblasts in the scaffold, suggesting its potential for bone tissue regeneration [5].

11.2.1.3 XTT (2,3-bis-(2-methoxy-4-nitro-5-sulfophenyl)-2H-tetrazolium-5-carboxanilide) and water-soluble tetrazolium 1 (WST-1) assay

In contrast to MTT, XTT and WST-1 are colorimetric assays for cell proliferation where the tetrazolium salt is cleaved to form water soluble formazan by cellular dehydrogenase and directly correlate to metabolically active cells [6]. Wang et al. [7] quantitatively evaluated the cytotoxicity during in vitro degradation of 3D printed polypropylene fumarate scaffold using MTS for bone tissue engineering. Wacker et al. [8] carried out WST-1 assay to determine the in vitro metabolic activity and functionality of human saphenous vein endothelial cells and endothelial progenitor cells on bacterial nanocellulose, small diameter vascular grafts with albumin, fibronectin, and heparin/chitosan. The assay was done under dynamic and static conditions to identify a promising material to improve the endothelialization.

11.2.1.4 Lactate dehydrogenase assay

Lactate dehydrogenase (LDH) assay is one of the common techniques to determine the cytotoxicity of cells on a scaffold by measuring the cytoplasmic enzyme in the damaged cells. LDH is a stable cytoplasmic enzyme that is present in all cells. Necrotic and apoptotic cells with compromised plasma membranes release LDH into the extracellular environment. LDH assay is based on the enzymatic coupling reaction where the LDH oxidizes lactate and generates NADH, which then reacts with WST and forms colored formazan dye. The intensity of color measured using a microplate reader directly correlates with the damaged cells [9]. The cytotoxicity of a scaffold developed using 3D printed polycaprolactone and electrospun poly (lactic coglycolic acid) was evaluated using the LDH assay. The culture supernatant was used to detect the presence of the intracellular enzyme lactate dehydrogenase, which indicates cell death. Cells treated with Triton X were used as a positive control due to high LDH release. The supernatant of the cell seeded scaffold showed lower LDH release, indicating cell membrane integrity, suggesting good biocompatibility and proliferation of cells. The results demonstrated the potential of scaffolds for hard tissue regeneration and cell therapy applications [10].

11.2.1.5 Crystal violet staining

It is a nonenzymatic assay for quantitative estimation of cell adhesion on scaffolds. The cationic dye is attached to the negatively charged DNA, RNA, and proteins of viable adherent cells. The nonviable and nonadherent cells are washed away. The stained cells can be imaged using light microscopy, revealing the cell attachment areas. The dye bound to cells can be solubilized using an appropriate solvent like ethanol/acetic acid and quantified spectrophotometrically at absorbance 570–590 nm [11]. The cell adhesion capacity of thermosensitive hydrogel made of polycaprolactone and poly (lactic acid) in combination with pluronic, chitosan, and gelatin was performed with the crystal violet assay [12]. The 3D printed scaffolds were placed onto a 24 well plate and seeded with cells and incubated for 20 minutes at 37°C, followed by fixing cells in paraformaldehyde for 10 minutes. The cells attached to the scaffold were stained with crystal violet at room temperature for 30 minutes, followed by the addition of Triton X-100. The optical density measured showed better cell adhesion on the scaffold than on the control cell culture plate plastic surface. The in vitro study demonstrated that the thermoresponsive hydrogel is nontoxic and exhibits enhanced cell adhesion capacity, making it a promising material for tissue engineering.

11.2.2 DNA synthesis based proliferation assays

The proliferation of cells in the scaffold material is assessed by detecting DNA replication and cell division using monoclonal antibodies that targets antigens associated with cell proliferation. The exogenous or endogenous substances that could be taken up during specific stages of the cell cycle are detected by the antibodies. Some of the widely used DNA synthesis based proliferation assays include 5'-Bromo-2'-deoxyuridine (BrdU) and (5-ethynyl-2'-deoxyuridine) EdU assays, which are discussed below.

11.2.2.1 BrdU cell proliferation assay

The assay incorporate thymidine analog BrdU as a supplement in the culture media of cell seeded scaffolds. The replicating cells incorporate BrdU in place of thymidine into their newly synthesized DNA. This BrdU is detected with an anti-BrdU antibody along with a fluorochrome conjugated secondary antibody. Fluorescence microscopy validates the cell proliferation in the developed scaffolds. To enhance staining and improve antibody accessibility, a DNA denaturation step is also performed. The BrdU incorporated cells can be directly correlated to the cell proliferation rate [13] (Fig. 11.2.)

To perform the BrdU cell proliferation assay on scaffolds, prepare the cell seeded scaffolds in culture. Dilute the BrdU reagent to cell culture media according to the manufacturer's instructions to get a final concentration of 10 μM. Replace the media with BrdU containing medium and incubate for a specific duration depending on the cell proliferation rate at 37°C. Remove and rinse with phosphate buffered saline (PBS) and fix the scaffold using PFA/ethanol at room temperature. Treat the scaffold with DNA denaturation solution for 30 minutes at room temperature to expose BrdU incorporated DNA. Block nonspecific binding sites using a blocking solution and incubate the scaffolds with anti-BrdU primary antibody overnight at 4°C. Wash the scaffold with PBS to remove unbound primary antibodies. Incubate in a fluorescently labeled secondary antibody for 2 hours and image cells with a fluorescent microscope.

In a study conducted by Mogha et al. [14] to mimic the elasticity of human skin by designing a polyacrylamide based hydrogel, BrdU incorporation assay was done to identify the human primary keratinocytes proliferation on the hydrogel of varying stiffness. The quantitative measurement of DNA was carried out by incorporating the labeling agent BrdU onto the human primary keratinocytes (HPKs) seeded on the scaffold and tissue culture plate for 24 hours, followed by fixing, permeabilizing, and blocking. Subsequently, after the addition of anti-BrdU antibody and incubation at 4°C overnight, the quantification of cells was done. The results demonstarted increased DNA synthesis of HPKs on the softer hydrogel than on the tissue culture plate, indicating a favorable environment for keratinocyte proliferation.

11.2.2.2 EdU cell proliferation assay

EdU is another thymidine nucleoside analog dye that does not require antibodies. The detection of the analog is based on copper induced click chemistry, where the fluorescent probe is attached to EdU and can be quantified. The assay can be carried out for microscopic imaging, flow cytometry, and high throughput screening techniques [15]. Zhou et al. [16] developed a composite scaffold made with acellular vascular matrix with 0.25% human like collagen 1, which was seeded with human gingival fibroblasts, gingival epithelial cells, and vascular endothelial like cells to create a vascularized oral

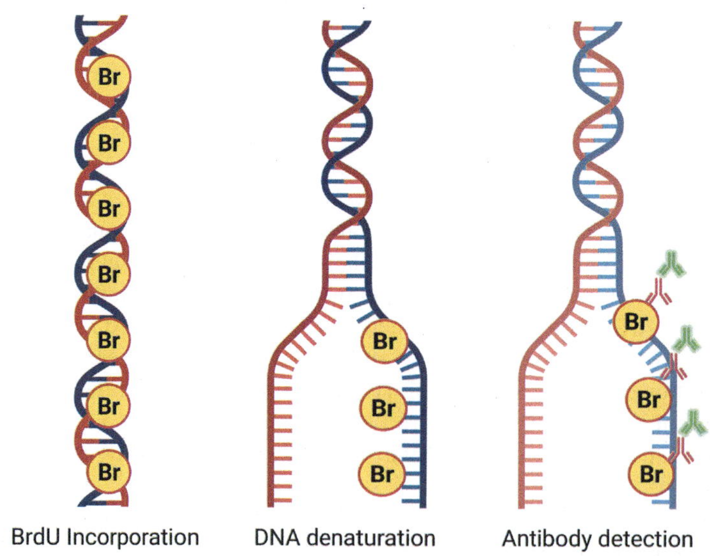

FIGURE 11.2 Schematic representation of BrdU cell proliferation assay. The figure illustrates the integration of BrdU into the replicating DNA of cells and subsequent antibody based detection.
Created using BioRender.com.

mucosa like structure both in vitro and in vivo. 10 mM of EdU was incorporated into the cells for 24 hours to detect the proliferation and attachment of cells in the scaffold. The compatibility of the cells makes it a stable scaffold for oral tissue regeneration.

11.2.3 Fluorescent dye based proliferation assays

This method includes fluorescent dye based assays such as live/dead staining, resazurin (Alamar blue) and 5(6)-Carboxyfluorescein diacetate succinimidyl ester (CFDA-SE) assays, which uses fluorescence to track the cellular components, metabolic activity, and DNA synthesis, thereby assessing cell proliferation and viability.

11.2.3.1 Live/dead cell double staining

It is a widely used method to assess cell viability based on the cell membrane integrity and metabolic activity, allowing differentiation between live and dead cells. The assay uses 2 fluorescent dyes, Calcein acetomethoxy (calcein-AM) for live cells and propidium iodide (PI) or Ethidium homodimer-1 for dead cells [17] (Fig. 11.3).

Calcein-AM: It is a nonfluorescent, hydrophobic derivative of calcein, with a highly lipophilic nature that facilitates cell permeability. Once inside the cell, cytoplasmic esterase hydrolyzes the acetomethoxy group, converting it into hydrophilic calcein, which emits a strong green fluorescence in viable cells. In contrast, dead or nonviable cells lack active cytoplasmic esterase, so the acetomethoxy group remains intact, and no fluorescence is generated that can be detected between 490 and 515 nm. This property allows the detection of signals from live cells in scaffolds, enabling cell tracing, labeling, and sorting.

PI/ethidium homodimer: These are membrane impermeable dyes that bind to the nucleus of cells with a compromised cell membrane. The dye enters the necrotic cells that do not have cell membrane integrity and emits a bright red fluorescence when it intercalates with the DNA. It can be detected between 535 and 617 nm. The simultaneous staining with both dyes enables visualization of viable and nonviable cells.

The protocol can be carried out by preparing scaffolds in culture. Prepare the staining solution as per the manufacturer's instructions, typically add 1–2 μM calcein-AM and 1–4 μM ethidium homodimer/propidium iodide in PBS, and gently wash the scaffold. Add enough stain to ensure the stain penetrates the scaffold. Wrap the plate and protect it from light. Incubate at 37°C for 30–45 minutes, incubate for 60 minutes for hydrogels, followed by washing to remove excess dye, if required. Image using fluorescence or confocal microscopy and quantify the proportion of live and dead cells, if required, using imaging software.

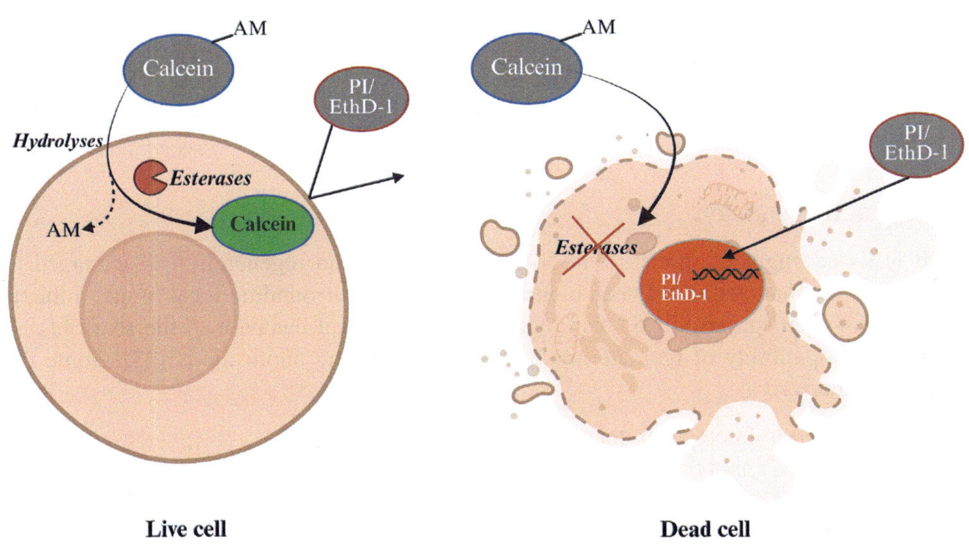

FIGURE 11.3 Mechanism of action of calcein-AM *(green fluorescence)* and PI/ethidium homodimer-1 *(red fluorescence)* in live/dead assay. The figure describes the mechanism behind the detection of live and dead cells using two fluorescent dyes. The live cells are stained with calcein-AM, which is converted to green fluorescent calcein by intracellular esterase. Dead cells are stained with PI/EthD-1, which penetrates damaged cell membranes and binds to DNA emitting red fluorescence.
Created using BioRender.com.

Zheng et al. [18] developed bioink using decellularized extracellular matrix, gelatin, and sodium alginate, and printed a porous 3D scaffold by adding primary ovarian cells. The viability of the printed cells was assessed using a live/dead assay on days 1, 7, and 14 and observed under a confocal microscope. The visual and quantitative measures of cell viability revealed 95% viability on day 1 and 90% viability on the 14th day. Despite the shear force during printing and residual chemicals from decellularization, the cells remained viable throughout the construct as indicated by the proportion of live versus dead cells from the assay.

11.2.3.2 Resazurin/Alamar blue assay

It is a colorimetric indicator for detecting cell proliferation. Alamar blue is a blue colored nonfluorescent resazurin (7-hydroxy-3H-phenoxazin-3-one10-oxide) that gets reduced to red fluorescent resorufin (7-hydroxy-3H-phenoxazin-3-one), which is water soluble and nontoxic after cellular uptake. The fluorescence emitted directly correlates with the cell viability [19]. A study conducted by Yadav et al. aimed to develop a bioengineered liver by heparin modifying the decellularized rat liver scaffolds to enhance hemocompatibility, re endothelialization, and angiogenesis. The resazurin perfusion reduction assay was carried out to evaluate the proliferation and viability of endothelial cells seeded on a heparinized scaffold. 1:10 dilution of resazurin was perfused at a 2 ml/min flow rate for 1 hour at 37°C on days 1, 3, 5, and 7 of recellularization, and absorbance was measured at 570 nm and 600 nm. The metabolically active cells in the constructs convert blue to red in the presence of resazurin, indicating active proliferation of cells [20].

11.2.3.3 CFDA-SE assay

The CFDA-SE assay utilizes a nonfluorescent dye that, upon activation by cellular esterase, becomes highly fluorescent as the acetate groups are removed. This dye is highly cell permeable and provides long lasting fluorescence, making it an excellent tool for tracking cell proliferation over time. It is beneficial for evaluating cellular behavior on the scaffolds [21]. When human adipose derived hydrogel was cultured with adipose tissue derived stromal/stem cells (ASCs) to identify their adhesion and migration for potential application in 3D culture and bioprinting. The immortalized human ASCs were incubated in the hydrogel for 7 days, and a live/dead assay was carried out using CFDA, propidium iodide, and DAPI, followed by fluorescent microscopy. The fluorescence emitted by CFDA staining confirmed the viability and migration of cells within the hydrogel, additionally demonstrating biocompatibility and the nontoxic nature of the extracellular matrix (ECM) based hydrogel [22].

11.3 Biochemical assays

Biochemical assays play critical role in evaluating the functional response of biomaterials to cells. They provide insights into the antioxidant activity, oxidative stress, protein expression and overall biochemical behaviour of the cells and scaffolds. Some of the commonly employed biochemical assays discussed below includes antioxidant assays and enzyme linked immunosorbent assay (ELISA).

11.3.1 Antioxidant activity assays

Biomaterials utilized in tissue engineering contribute to the restoration and regeneration of damaged tissue. Oxidative stress plays a crucial role in the interaction between biomaterials and the surrounding local tissue, influencing the success of implantation. Reactive oxygen species (ROS) also regulate the biological behavior of the scaffold, as excessive ROS levels can impact the biocompatibility of biomaterials in tissue engineering applications [23]. Some of the antioxidant scavenging assays are summarized as

- DCFH-DA (2',7'-dichlorodihydrofluorescein diacetate) assay
- Total antioxidant capacity (TAC) assay
- Malondialdehyde (MDA) assay

11.3.1.1 DCFH-DA assay

DCFH-DA is a nonfluorescent, cell permeable dye. When it enters the cell, cellular esterase cleaves the acetyl group and forms 2'-7, dichlorodihydrofluorescein (DCFH), which interacts with thiols and is entrapped in the cell. DCFH then oxidized to highly fluorescent 2'-7, dichlorofluorescein (DCF), interacting with free radicals and ROS like superoxide anion, hydroxyl radical, nitric oxide, superoxide anion, hydrogen peroxide, and hydroxyl radical. The green fluorescence emitted is directly proportional to the ROS produced, which can be detected using fluorescent microscopy and flow cytometry [24,25] (Fig. 11.4).

DCFH-DA assay on a cell seeded scaffold can be carried out by preparing the cell seeded scaffold in culture. Prepare the DCFH-DA 10 mM stock solution in DMSO and store at –20°C. Make DCFH-DA 10 μM working solution in culture media. Add the working solution to the scaffold and incubate at 37°C for 30–60 minutes protected from light. Wash with PBS to remove unincorporated dye. ROS can be added to induce oxidative stress before measuring fluorescence. Measure the fluorescence and image using a fluorescent microscope. The cell lysate can be analyzed for fluorescence using a spectrophotometer with an excitation wavelength of 485 nm and an emission wavelength of 530 nm. A study was conducted to identify the ROS scavenging property of nanoceria (Ce) infused electrospun polycaprolactone (PCL) nanofibers on bone cells subjected to ionizing radiation induced senescence. The scaffold's ability to modulate high ROS due to the presence of senescent cells was studied. DCFH-DA was carried out to measure the ROS in senescent cells cultured on the developed PCL and PCL-Ce scaffolds. As the positive control, nonsenescent cells were cultured on both scaffolds and treated with hydrogen peroxide for 30 minutes at 37°C. The results demonstrated that senescent cells cultured on nanoceria infused PCL nanofibers exhibited significantly lower fluorescence intensity, indicating a reduction of ROS within the cells. The study suggested that ceria infused PCL scaffolds had a protective role against oxidative stress as well as improved osteogenesis, promising a therapeutic strategy in orthopedic complications [26].

11.3.1.2 TAC assay

It is used to evaluate the antioxidant capacity of biomaterials. Antioxidants released from the digested biomaterial are reacted with stable radicals like ABTS (2,2'-azino-bis-3 ethylbenzothiazoline-6-sulfonic acid), DPPH (2,2'-diphenyl-1-picrylhydrazyl), or through ferric reducing antioxidant power (FRAP) assay. These reactions emits colorimetric changes, upon interaction with antioxidants, which can be quantified spectrophotometrically to determine the total antioxidant capacity of the material [27]. In a study conducted by Luong et al. [28], a bilayered scaffold for wound dressing made of poly (vinyl alcohol), Sodium alginate infused with *Calophyllum inophyllum* oil and *Platostoma palustre* extract was developed by electrospinning and polymerization techniques. The combination had a synergistic effect to enhance wound healing. The total antioxidant properties of the scaffold were assessed using the DPPH and ABTS assays. To evaluate the DPPH scavenging activity, the scaffolds were immersed in distilled water for 30 minutes at 37°C. This extract, when mixed with DPPH for 60 minutes at 25°C, showed a color change, which was measured at 517 nm. Similarly, for ABTS analysis, the extract, when added to the ABTS solution, was incubated, and the absorbance was measured at 734 nm,

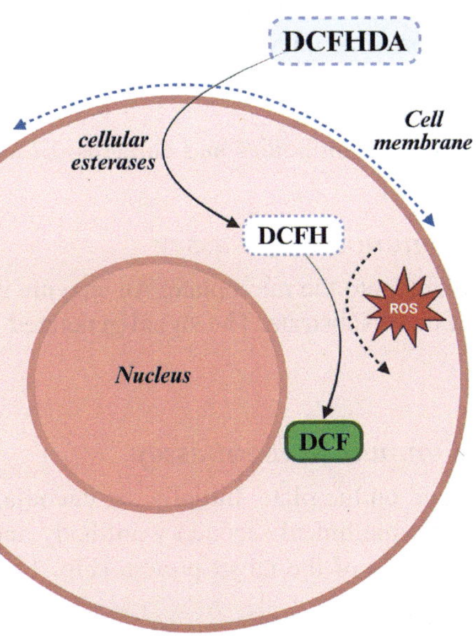

FIGURE 11.4 Schematic representation of DCFDA assay for ROS detection. The figure illustrates the mechanism by which nonfluorescent DCFDA diffuses into cells and is deacetylated to nonfluorescent DCFH by intracellular esterase, which is rapidly oxidized to fluorescent DCF by ROS, emitting green fluorescence.
Created using BioRender.com.

followed by calculation of the antioxidant capacity. A strong antioxidant activity was observed in the scaffold, confirming its protective role in wound healing. The significant scavenging activity of the developed scaffold suggests its broad application in treating various types of wounds.

11.3.1.3 MDA assay

MDA assay is used to assess lipid peroxidation, a critical marker of oxidative stress. The assay quantifies malondialdehyde, a byproduct generated during peroxidation of polyunsaturated fatty acids in the cellular membrane. During oxidative stress conditions, MDA reacts with thiobarbituric acid and forms a pink colored MDA-TBA complex with a strong absorbance that can be quantified, thereby assessing the oxidative stress of biomaterials in various experimental conditions [29]. MDA assay was used for detecting the oxidative stress by measuring the lipid peroxidation rate of cells cultured on an electrospun polycaprolactone/zinc acetate scaffold postgamma irradiation. The concentration of MDA is a reliable marker for detecting oxidative damage to cells. Ismail et al. [30] exposed vero cells cultured on the scaffold to various doses of gamma irradiation, like 0.5, 1, 2.5, and 5 Gy. To detect the MDA levels, a mixture of sample, 20% trichloroacetic acid, 0.67% thiobarbituric acid was heated for 30 minutes at 95°C, cooled, and mixed with butanol to extract the pink colored chromogen, followed by measurement of absorbance at 532 nm. The study observed decreased MDA levels for cells exposed to 0.5 Gy gamma radiation, with a dose dependent increase in MDA levels and a decrease in cell viability at higher doses, indicating elevated lipid peroxidation and oxidative stress. The MDA assay highlighted the detrimental effect of gamma radiation on scaffolds seeded with cells, encouraging further investigation of the scaffold for various applications.

11.3.2 Enzyme linked immunosorbent assay

Enzyme linked immunosorbent assay (ELISA) is a biochemical technique used to detect and quantify specific proteins, including cytokines, antibodies, peptides, growth factors, ECM proteins, and hormones expressed by cells on a scaffold. This assay is based on the highly specific interaction between antigens and antibodies, enabling the binding of the target protein, with quantification facilitated by an enzyme linked detection system. In ELISA, specific antibodies bind to target proteins, followed by the introduction of a secondary antibody conjugated with an enzyme such as horseradish peroxidase or alkaline phosphatase. This enzyme catalyzes a colorimetric reaction, and the resulting enzymatic activity is measured to determine signal intensity, which correlates with the protein concentration. The assay involves immobilizing a specific antigen or antibody onto a solid surface, such as a microplate. The sample containing the target molecules is then added to allow binding, followed by the introduction of an enzyme-linked antibody specific to the target. Upon adding a substrate, the enzyme catalyzes a reaction that generates a signal proportional to the concentration of the target molecule in the sample.

Depending on the method used for binding antibodies and antigens, ELISA is categorized into four main types.

11.3.2.1 Direct enzyme linked immunosorbent assay

The antigen/sample is directly immobilized onto the microplate. An enzyme linked antibody is then added, which binds to the target protein, followed by the addition of substrate. The signal generated is used to directly measure the concentration of the target protein (Fig. 11.5.)

11.3.2.2 Indirect enzyme linked immunosorbent assay

In this method, the antigen is immobilized on the plate. Initially, an unconjugated primary antibody is added to bind the target, followed by the addition of an enzyme linked secondary antibody and substrate. The resulting signal intensity is then measured to determine the concentration of the target protein (Fig. 11.6).

11.3.2.3 Sandwich enzyme linked immunosorbent assay

This method involves a "sandwich" technique where two antibodies are used to bind the antigen. A capture antibody is first immobilized on a microtiter plate, allowing the sample protein to bind to it. Following this, an enzyme linked secondary antibody is added, which binds to the antigen. The signal generated after adding the substrate is measured, providing the concentration of the target protein (Fig. 11.7).

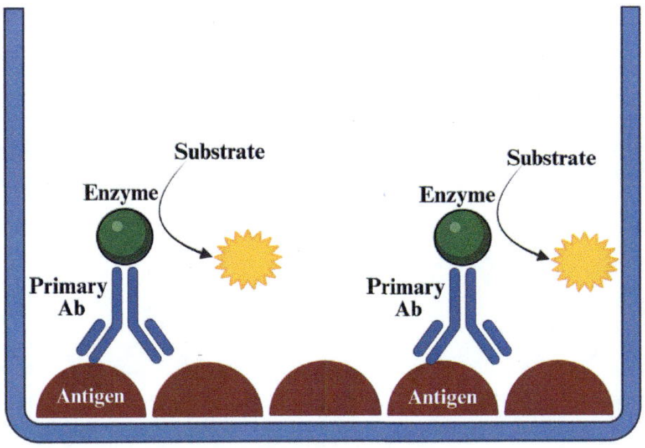

FIGURE 11.5 Schematic representation of direct ELISA. Illustration of direct ELISA depicting antigen immobilization on the plate surface and detection using an enzyme conjugated primary antibody. *ELISA*, Enzyme linked immunosorbent assay.
Created using BioRender.com.

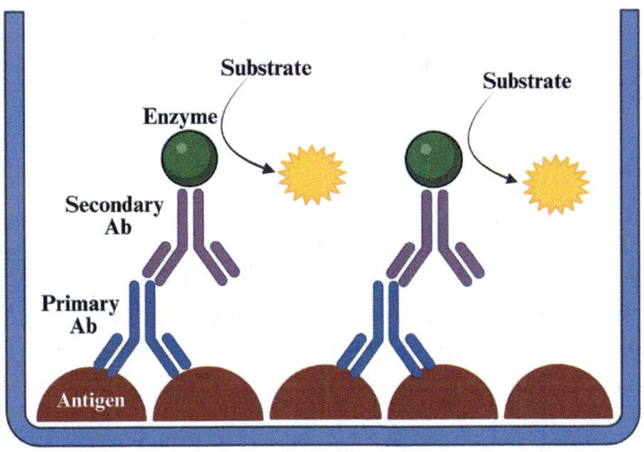

FIGURE 11.6 Schematic representation of Indirect ELISA. Illustration of Indirect ELISA depicting antigen binding, followed by detection using primary antibody and enzyme linked secondary antibody. *ELISA*, Enzyme linked immunosorbent assay.
Created using BioRender.com.

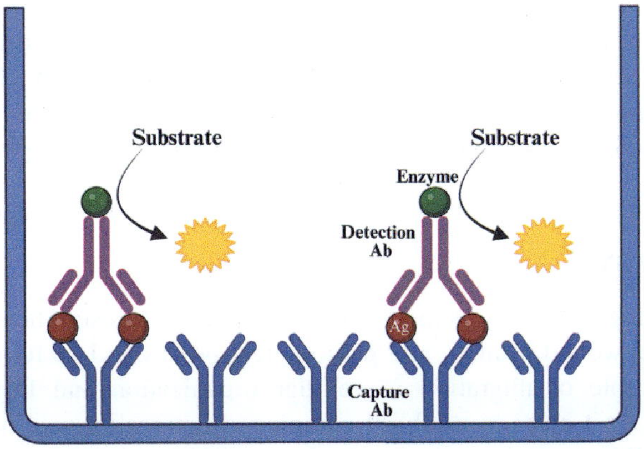

FIGURE 11.7 Schematic representation of sandwich ELISA. Illustration of sandwich ELISA depicting the highly specific capture of target antigen between a capture antibody and an enzyme linked detection antibody. *ELISA*, Enzyme linked immunosorbent assay.
Created using BioRender.com.

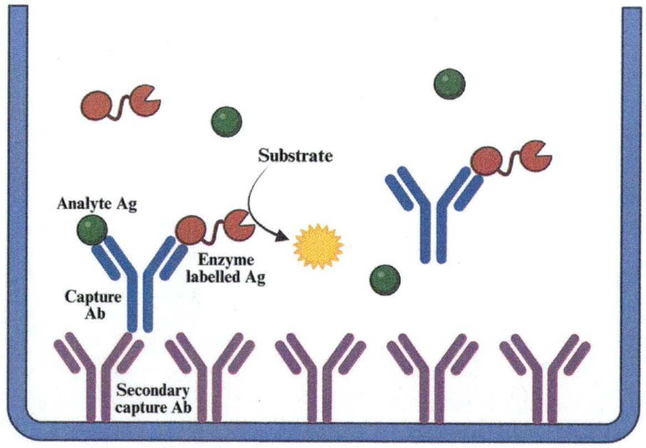

FIGURE 11.8 **Schematic representation of competitive ELISA.** Illustration of sandwich ELISA depicts the sample analyte antigen competing with enzyme labeled antigen for the limited antibody binding site offering inverse signals. *ELISA*, Enzyme linked immunosorbent assay.
Created using BioRender.com.

11.3.2.4 Competitive enzyme linked immunosorbent assay

This method is used when the protein of interest is too small. In this approach, the sample antigen and enzyme linked antigen compete for binding to a specific antibody. The secondary capture antibody is immobilized on the microtiter plate, followed by the addition of enzyme labeled antigen, sample containing the target antigen, and primary capture antibody. Both antigens compete to bind to the antibody. A substrate specific to the labeled antigen is then added, catalyzing a reaction that produces a detectable signal. The intensity of the signal is inversely proportional to the concentration of the target antigen in the sample [31] (Fig. 11.8).

ELISA is a commonly used in vitro evaluation technique for various tissue engineered constructs. Various ELISA kits are available, specific to the target protein, hormone, and molecules. In a study conducted by Perez et al. [32], they developed a protocol for 3D bioprinting by incorporating mesenchymal stem cells, followed by differentiation. The goal was to create functional neural tissue models. Competitive ELISA was carried out to quantify the dopamine levels secreted by the differentiating neurons. The culture media from 3D and 2D control groups were collected on day 12 in triplicate and analyzed in the ELISA kit following the manufacturer's instructions. The quantitative assessment of dopamine levels released by differentiated neurons using ELISA also validates the functional maturation of the bioprinted constructs into dopaminergic neurons. Similarly, the sandwich ELISA technique was used by Elshabrawy et al. [33] to quantitatively measure the levels of pro inflammatory cytokines like interleukin-6 and tumor necrosis factor alpha in a double layered transdermal patch developed to treat rheumatoid arthritis (RA). The reduction in the levels of these cytokines, determined by ELISA, highlighted the effectiveness and advantages of the patch and its role in reducing the symptoms of RA. The osteocalcin levels of a 3D scaffold developed using ceria-doped mesoporous bioactive glass and *Spongia agaricina*, which is functionalized with vitamin D_3 were detected using an ELISA kit. Grama et al. [34] prepared scaffold extracts in culture media and quantified osteocalcin, which is an osteogenic marker expressed by mature osteoblasts, regarded as an indicator of osteogenic differentiation. The assay was done for 14 and 21 days by treating dental pulp stem cells with the extract. The increased osteocalcin expression by the functionalized scaffold was detected by ELISA, which confirms the synergistic effect of vitamin D_3 and ceria and their promising role in bone tissue engineering applications.

11.4 Cell migration assays

Understanding the crucial role of migration of cells in 2D and 3D constructs has significantly contributed to the knowledge of tissue repair, immune response, wound healing, and pathological processes like tumor metastasis, inflammation, and vascular diseases. The apparent role of migration in cellular organization and formulating precise architecture is customary in organ development, as it helps to establish complex structures by guiding cells to specific locations. The migration of the cells along the ECM occurs through peptidoglycan and integrin-containing receptors on the surface of the cell. The decade-old practice of understanding the role of cell migration through 2D models might prove insufficient in understanding certain cellular migration behaviors like multiaxial movement that explores the realistic migration patterns, cell–cell interaction, collective migration, cell polarity, and tumor modeling [35]. There are several assays that help understand and elucidate the aforementioned cellular migration patterns in 3D.

A few of the cell migration assays are listed below.

11.4.1 Fluorescence/immunofluorescence-based assays

11.4.1.1 Pore bridging experiments

This experiment enables the use of porous structures to evaluate the ability of the cells to migrate through them, thereby evaluating the haptotropic, chemotactic, and mechanotactic properties of the cell. A study conducted by Vandenheuvel et al. with PCL electrospun fibers [36] and osteoblast cell line (MC3T3-E1) extensively explores the effect of cell growth in different geometries. Electrospun fibers are used for this study, fabricated with the irregularly shaped pore of variable measurement using a melt electrowriting printer [37]. Appropriate cell lines (osteoblast/cancer cell lines/fibroblast cell lines) are seeded onto the scaffold and are evaluated for cell viability and cell migration at different time points and are assessed through fluorescence microscopy using DAPI and phalloidin.

Microscopic analysis of cell migration is assessed by

$A(t)/ A(0)=1 - (t/Tb)^{v}$

A(t)=Area at experimental time course

A(0)=Area at initial pore size

Tb=Time taken for bridging

v=Migration rate exponent

11.4.1.2 Physical barrier mediated experiment

The physical barrier method is used to assess the ability of cells to migrate by creating an obstruction through barriers like silicon inserts, stencils, or removable devices on a cell culture substrate with a scaffold, and the removal of the barrier allows the cells to migrate across the previously obstructed area. It is carried out by placing a barrier of 1 mm x 13 mm size perpendicular barrier and fixed onto the scaffold using a rubbery-O-ring. After the set confluency has reached, the barrier is removed. The cell viability and migration are assessed at set day points (1, 2, 12, 24, 36, and 48 hrs) using fluorescent staining. In addition, chemoattractants (VEGF) can be used to further evaluate directional migration [38].

11.4.2 Transwell migration assays

This assay involves the use of chemotactic stimuli to study migration/motility. It comprises a transwell insert (microporous membrane) that separates the upper chamber containing/low serum media containing cells from the lower chamber, which comprises a chemoattractant that guides cell migration. Recent studies showed an integration of hydrogel into the transwell apparatus in the upper and lower chambers to mimic an ECM-like environment to study 3D migrations of cells. CXCL12-based chemotaxis entrapment of glioblastoma cells using a thiocure-based hydrogel was proven to be efficient in luring glioblastoma cells into it [39]. This technique could be explored to develop superior assays to evaluate cell migration within 3D structures (Fig. 11.9).

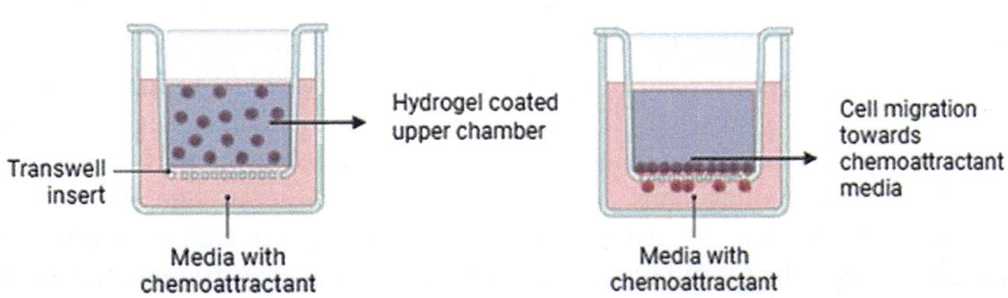

FIGURE 11.9 **Trans well migration apparatus with integrated hydrogels.** The figure illustrates an integrative hydrogel model incorporated into a transwell migration apparatus with a scaffold placed in the upper chamber to understand the three-dimensional migration of cells.
Created using BioRender.com.

11.4.3 Microfluidics-based experiments

The emerging field of microfluidics has been proven to be efficient in many realms of tissue engineering and drug development. It could also be explored to evaluate cell migration with significant precision. It also mimics the physiological conditions making it superior in comparison to other cell migration assays. Specifically engineered microchannels and microbarriers effectively assess cell migration. A partial digestion of a cell monolayer was employed in a microfluidics system using trypsin and Ethylenediaminetetraacetic Acid (EDTA), by Nie et al, demonstrating the efficiency of the cell migration assay evaluated in microchannels within a microfluidics system [40].

11.4.4 Preclinical studies

Gene expression is often used as the state-of–the-art technique to evaluate cell migration in vivo. CD44 expression is often associated with survival and migration of glioma cells in a biphasic manner. CD44 is often called a "molecular clutch" in assisting cell migration and is simultaneously associated with substrate adhesiveness [41]. Techniques like intravital microscopy that illuminate the gene of interest through paired fluorescent proteins assist with evaluating cell migration for days. Furthermore, other techniques like magnetic resonance imaging (MRI), positron emission tomography (PET), and histological studies are also conducted to establish cell migration [42].

11.5 Cell differentiation assays

Cell differentiation is an intricate process through which a change in certain biomolecules results in an alteration of the structure or behavior of a cell. Primarily by specialization through gene expression, daughter cells can synthesize these macromolecules that the mother cells are incapable of synthesizing [43]. Once differentiated, the daughter cells will have limited capability for further division. The benefits of studying cell differentiation in a 3D construct are vast, including valuable insights into the role of guided differentiation using different biomolecules through a 3D construct that could potentially map out the metabolic and cellular processes associated with differentiation, a more accurate perception of cell polarity and morphology could be obtained in a 3D scaffold, and it can provide an accurate replica of tumor microenvironment thereby providing insights into the differentiation of cancer stem cells, metastasis, and more targeted therapies. Furthermore, the impact of mechanical and biophysical cues, including mechanical forces and shear stress, on cell differentiation can only be studied in 3D constructs. There are several assays that are specific to certain cells and some that could be performed in all types of cells to evaluate cell differentiation within a scaffold.

There are several differentiation markers that are used to study different types of differentiation.

11.5.1 Immunofluorescence analysis of cell differentiation factors

The primary antibody is incubated (overnight at 4°C) on previously fixed cells (4% paraformaldehyde for 20 minutes) that are further washed to remove excess of antibody, followed by incubation with fluorescent-tagged secondary antibody for 5 hrs. Stain the cells with DAPI to visualize the cell nuclei. The total number of cells can be counted using NIS element D program and analysis through GSA Image Analyzer program [44].

11.5.2 Alkaline Phosphatase activity, Osteocalcin, and Osteopontin assays

These experiments are specifically performed to confirm osteogenic differentiation, as alkaline phosphatase enzymes, osteopontin, and osteocalcin are usually expressed in the early stages of osteoblast formation. The substrate nitrophenyl phosphate (pNPP) gets activated in the presence of alkaline phosphatase and releases p-nitrophenol (yellow), which could be detected using quantifiable colorimetric change. The status of osteoblast activity could be analyzed through the osteocalcin assay. Osteocalcin, a noncollagenous protein, tends to increase in rate when there is bone formation and mineralization. Likewise, osteopontin, a glycoprotein, enables the identification of differentiated bone cells by assessing osteoclast attachment. The cells seeded on the scaffold with osteogenic differentiative media are used for further analysis.

Osteocalcin analysis: At set time points of 7, 14, 21, and 28 days, the media is replaced with DMEM and antibiotics only. Incubate it for 24 hrs. Quantify the amount of osteocalcin secreted using GLA-type Osteocalcin EIA (Enzyme Immunoassay) Kit and carry out the experiment based on the manufacturer's manual. The secretion was normalized to the total secreted proteins.

Osteopontin analysis: The set time points and the step of serum starvation remain the same as mentioned above in the osteocalcin analysis. The quantification of osteopontin is done using the Osteopontin EIA Kit as per the manufacturer's manual. The secretion was normalized to the total secreted proteins.

Alkaline phosphatase enzyme assay: At different time points (preferred 3, 7, 14, 21, and 28 days) after the cells reach the desired confluency, the cells are lysed and assessed for alkaline phosphatase using the Alkaline Phosphatase kit and further evaluation of absorbance measurement at 405 nm. Total protein estimation can also be carried out using the Bradford assay to normalize alkaline phosphatase activity [45].

11.5.3 Glycogen estimation using PAS staining

This assay is done specifically to evaluate the presence of glycogen expression (expression indicates the differentiation of hepatocytes) in hepatocytes. It is an immunohistochemical assay that stains polysaccharides like glycogen, glycolipids, and glycoproteins. The cell-seeded/cell-laden scaffolds containing anticipated hepatocytes are embedded in paraffin to obtain horizontal and vertical sections. After deparaffinization, the sections are immersed in periodic acid solution (0.5%) at room temperature (5 min) followed by Schiff's reagent for 15 min at room temperature. The cytoplasmic pink color at different time points will give a quantifiable difference in intensity. Simultaneously, hematoxylin and eosin staining can also be done to morphologically view the hepatocytes [46].

11.5.4 Protein analysis through western blotting

This technique could be used for qualitative and quantitative evaluation of protein expression seen in differentiation. Especially the protein associated with transcription (MyoD, PPARγ, Neurogenin2, Oct4), signaling (Wnt, hedgehog proteins, notch signaling pathway, epidermal growth factor, fibroblast growth factor), epigenetic (histone modifiers, DNA methyltransferases), cell cycle regulators (cyclins and cyclin-dependent kinases), and structural regulators. The procedure involves the lysis of cells on the scaffold followed by the isolation of proteins and the separation of proteins using electrophoresis. The separated proteins are transferred to a nitrocellulose paper, and immunoblotting is carried out. Other techniques could also be adopted to assess the protein expression, using flow cytometry.

11.5.5 RNA in situ hybridization technique

This technique uses complementary base pairing between nucleic acids with a labeled probe that targets RNA of interest, followed by hybridization and signal detection (fluorescent dye allows the detection and visualization of the hybridized probe).

11.5.6 Preclinical studies

Like in vitro studies, gene expression studies are employed in in vivo as well where the detection of markers indicated in Table 11.1 is used to assess different types of cell differentiation using techniques like qPCR. Furthermore, histopathological analysis is also employed to study cell differentiation.

11.6 Angiogenesis and vascularization assays

Crucial for cellular metabolism and survival angiogenesis, a process of development of new blood vessels from the preexisting capillary structures is thoroughly explored especially the role of ECM in angiogenesis. When the newly formed blood vessels penetrate through tissues and organs to supply oxygen and nutrients, then the process is known as vascularization. Angiogenesis and vascularization play a fundamental role in wound healing, tissue repair, embryonic development, and maintaining reproductive functions. It also aids in tumor growth and metastasis [50]. In 2D models, there is no accurate replication of the microenvironment, a lack of tissue-like architecture, and altered cell behavior as well as the morphology can hinder the accurate representation of angiogenesis. 3D assays like tube formation assays, sprouting assays, and many more have been proven to be proficient in providing a better understanding of angiogenesis and vascularization. Some of the assays are discussed below.

11.6.1 Blood vessel staining

This staining technique is used to stain von Willebrand Factor (Vwf) in the elongated rod-shaped organelle (Weibel-Palade bodies) in the cytoplasm, and since some of them are excreted out, they can be seen on the cell surface as well.

TABLE 11.1 Types of differentiation markers.

S. No.	Types of cells	Differentiation markers	Method of study
1.	Human fetal neural progenitor cells (ReNcell VM) [44]	Beta-III-tubulin Tyrosine hydroxylase	Confocal microscopy
2.	Human Adipose Stem cells (fibroblast-like adipose tissue-derived stromal cells—ADSC-FL) to osteogenic cells [45]	Alkaline phosphatase Osteocalcin Osteopontin	ELISA
3.	Rat bone marrow mesenchymal stem cells to hepatocytes [47]	ALB AFP rt-PCR CK18 CK19 G6PD HNF-4 Western blot CytochromeP450 (P450) Cytokeratin 18 (CK18) Cytokeratin (CK19) L-glutamate dehydrogenase assay: Biochemical test Glycogen assay: PAS staining	RT-PCR, western blot, confocal laser scanning microscopy, PAS staining, biochemical assay
4.	Differentiation of human embryonic stem cells towards renal lineages [48]	WT1 PAX2 AQP1 AQP2 SIX2 qPCR SYN SMA UPD EMA CK Vimentin	qPCR fluorescence microscopy
5.	Differentiation of human induced pluripotent stem cells to cardiomyocytes [49]	TNNT2 MYH6 MYH7 rt-PCR SCN5A KCNJ4 GJA1 ACTA2 FLIPR Calcium 5 Assay	Rt-PCR FLIPR Tetra system

Cell-seeded/cell-laden scaffold can be stained using a blood vessel staining kit and can be followed up with a similar protocol as that of immunofluorescence [51].

11.6.2 CD31 and lectin staining

CD31 is expressed in endothelial and hematopoietic cells when there is angiogenic and vasculogenic activation. Functional blood vessels can be stained with lectin. After the initial procedure for immunofluorescence assay, the sample is incubated with rabbit anti-CD31 overnight at 4°C. Then the rinsed sample is incubated with a secondary antibody (1:200 dilution of goat antirabbit IgG conjugated with Alexa 488) and 10 μg mL^{-1} of BSL (lectin) in PBS for 1 hour at room temperature. The fluorescence can be detected using fluorescence/confocal microscopy [52].

11.6.3 Tube formation assay

Tube formation assay evaluates the early stage of angiogenesis by endothelial cells in arranging themselves in a tubelike structure. This self-assembling property can be elucidated effectively in 3D tissue-engineered constructs. This assay is similar to CD31 and lectin-based immunofluorescence assay. The analysis must include parameters like overall length of the tube, number of branching points, and number of tubes [53].

11.6.4 Aortic ring assay

Aortic ring assay involves the use of a 3D scaffold, especially hydrogel, to embed aortic rings, which are obtained by dissecting the aorta to obtain concentric rings. Exposing the aortic rings to angiogenic factors and antiangiogenic factors can potentially evaluate the formation of capillary-like networks in a scaffold. The scaffold will provide structural and biochemical support [54].

11.6.5 Preclinical studies

CAM assay, where the hydrogels and other tissue-engineered scaffolds can be placed inside the chick membrane through a window cut. After a specific period of time, angiogenesis is evaluated by removing the xenograft, and the number of blood vessels, branch points, and vascularity scores is observed under a dissection microscope. In xenograft-based studies, injecting the animal with FITC (fluorescein isothiocyanate)-dextran and quantifying its concentration inside the xenograft was also proven to be efficient in evaluating angiogenesis [55].

11.7 Regulatory testing standards for material-based scaffolds

The advancements in development and application of biomaterials demand continuous innovation, making biomaterial testing an evolving and complex field. To ensure the safety, quality, and efficacy of biomaterials, established guidelines and standards are rigorously and consistently followed. These standards enhance the reliability of the materials and ensure confidence in the results obtained by adhering to the established procedures. Effective testing requires understanding the biological interactions, specific properties, and functions of the biomaterial. Some of the important standards for biomaterial testing are discussed below.

11.7.1 International Organization for Standardization standards

The ISO is a worldwide federation of national standards to develop and publish voluntary standards across a wide range of fields. The technical committees under ISO cover various biomaterial features like sterilization, biological evaluation, clinical investigations, and nanotechnology. Some of the major sets of standards that cover the preclinical evaluation of biomaterials and medical devices include the ISO 10993 series, which focuses on the biological evaluation of medical devices, conducting various tests for in vitro and in vivo cytotoxicity after implantation. ISO 14971 identifies and addresses the risks associated with medical devices. ISO 22442 series apply to risk management and biological safety of medical devices that use animal tissue and their derivatives. The biological safety and risk management for biomaterials are covered by ISO [56,57].

11.7.2 American Society for Testing and Materials standards

American Society for Testing and Materials (ASTM) is a globally recognized international organization that develops standards for various biomaterials and products used in tissue engineering. These standards offer guidance to select and conduct tests and evaluate the appropriate biomaterials for their intended application, ensuring the quality and safety of materials. The guidelines provide standards for mechanical testing of materials, characterization techniques, biocompatibility, and sterility. The ASTM committee F04 covers guidelines on medical and surgical devices used for tissue-engineered medical products. Some of the relevant standards, like ASTM F2150, cover the standards for characterization and testing of biomaterials used in tissue engineering. ASTM F2883-11 is a guide for surgical implants like ceramic and mineral-based scaffolds, and the F2900-11 is a standard guide for hydrogel characterization. These guidelines are necessary to assess the scaffold performance in tissue-engineered biomaterials [58,59].

11.7.3 US Food and Drug Administration guidance

The Food and Drug Administration (FDA) provides guidelines for assessing the safety of metals and biomaterials, especially used in medical devices and implants, to ensure that they are effective and safe to be used in the body. The materials used in these implants come in contact with the body for an extended period, and the FDA identifies the properties of the material used in the device and only gives approval for the final medical device, not the material used in its production. The goal is to identify their intended use and function, assessing the safety of these devices in the human body. The FDA oversees decisions regarding the approval of medical devices and implants for marketing and continues monitoring their manufacturing and performance postmarketing to ensure the safety of the product. They regularly inspect whether the manufacturers adhere to good manufacturing practice regulations. They also analyze the reported failures through multiple monitoring systems to evaluate whether the manufacturer has addressed the issues and taken suitable measures to minimize the risk to users [60].

11.8 Future of in vitro assays: Integration of artificial intelligence and 3D-dimensional tissue-engineering-based constructs

Artificial Intelligence (AI)-based assays have been in existence for more than a decade, and the extensive research associated with AI has paved the way into the biological realms thereby increasing the precision of results. Many studies have been reported on the use of AI, where the cells are evaluated to be live or dead based on their morphological characteristics, including the area and perimeter. Machine learning based on ANN and SHAP model was used to assess the cell viability of Schwann cells in *pc*-MXene/GelMA (gelatin methacrylate) hydrogels in a study conducted by Chung et al, and this model was validated using fluorescence microscopy to be a suitable alternative to preclinical studies [61]. Cutting-edge techniques of evaluating the deformation index and taper index to assess cell migration using artificial intelligence in a transwell model have been demonstrated to be efficient in evaluating cell migration by negating the difficulty by providing steadfast and quantitative results. This algorithm raised the accuracy of conventional evaluation techniques from 54.7% to 84% [62]. In a 3D model, cell migration can be evaluated without the need for a transwell apparatus, and a varying degree of gel porosity can create a suitable barrier to evaluate cell migration. An article written by Solbu et al. depicts the use of hydrogels as a medium to evaluate cell migration by creating a chemotactic gradient. Collagen or matrigel was said to be useful as it is biomimetic with suitable structural and biochemical properties. The cell persistence time and cell persistence length can be evaluated using algorithms, and moreover, the paper also conveys the idea of integrating biosensors to enhance the accuracy of the movement of the cell within the hydrogel and towards the media [63]. A novel study conducted by Salvante et al. evaluated the integration of AI tool *IKOSA* in Chick Chorioallantoic Membrane (CAM) assay to assess the effect of collagen-based scaffolds on neovascularization. This approach was found to be effective in reducing the reliance on animal experiments in evaluating the effect of different types of hydrogels [64].

11.9 Conclusion

Even though there are typical limitations associated with the use of in vitro assays in evaluating certain physiological parameters, the use of scaffold-based study inducted a significant leap towards mimicking the normal physiological response within a 3D construct. The crucial adaptation of two-dimensional in vitro assays to assist the detection of cytochemical changes, metabolic activities, and physiological changes has been reverberating its importance across the realms of tissue engineering and theranostics. Regeneration of different tissues, drug screening, and establishing disease models are the hallmarks of these adaptations. The set protocols could guide the ongoing and future experiments to yield optimum results with precision. The integration of AI-based technologies to assist these assays could further enhance the effectiveness of these experiments and could negate the role of reliance on preclinical models.

List of abbreviations

PAS	Periodic acid-Schiff
MTT	(3-[4,5-dimethylthiazol-2-2yl]-2,5-diphenyltetrazolium bromide)
DMSO	Dimethyl sulfoxide
MTS	(3-(4,5-dimethylthiazol-2-yl)-5-(3-carboxymethoxyphemyl)-2-(4-sulfophenyl)-2H-tetrazolium)
XTT	(2,3-bis-(2-methoxy-4-nitro-5-sulfophenyl)-2H-tetrazolium-5-carboxanilide)
WST-1	water-soluble tetrazolium 1
LDH	Lactate dehydrogenase

PBS	Phosphate buffer saline
BrdU	5'-bromo-2'-deoxyuridine
HPKs	human primary keratinocytes
EdU	5-ethynyl-2'-deoxyuridine
PI	propidium iodide
Calcein-AM	Calcein acetoxymethyl ester
ASCs	adipose tissue-derived stromal/stem cells
ECM	extracellular matrix
ROS	Reactive oxygen species
DCFH-DA	(2',7'-Dichlorodihydrofluorescein diacetate)
DCF	2'-7, Dichlorofluorescein
DCFH	2'-7, dichlorodihydrofluorescein
TAC	Total antioxidant capacity
MDA	Malondialdehyde
PCL	Polycaprolactone
DPPH	(2,2'-diphenyl-1-picrylhydrazyl)
ABTS	(2,2'-azino-bis-3 ethylbenzothiazoline-6-sulfonic acid)
ELISA	Enzyme-linked immunosorbent assay
RA	Rheumatoid arthritis
ISO	International Organization for Standardization
ASTM	American Society for Testing and Materials
FDA	US Food and Drug Administration
DAPI	4′,6-diamidino-2-phenylindole
AI	Artificial Intelligence

References

[1] J. Niu, M. Li, Y. Wang, Cell proliferation and cytotoxicity assays, the fundamentals for drug discovery, International Journal of Drug Discovery and Pharmacology 3 (3) (2024) 100013. Available from https://doi.org/10.53941/ijddp.2024.100013.

[2] S. Supriya Bhatt, G. Thakur, M. Nune, Preparation and characterization of PVA/Chitosan cross-linked 3D scaffolds for liver tissue engineering, Materials Today: Proceedings (2023). Available from https://doi.org/10.1016/j.matpr.2023.02.251.

[3] A.O. Ijaola, B. Subeshan, A. Pham, M.N. Uddin, S.-Y. Yang, E. Asmatulu, Fabrication, characterization, and in vitro cytotoxicity assessment of tri-layered multifunctional scaffold for effective chronic wound healing, Bioengineering 10 (10) (2023) 1148. Available from https://doi.org/10.3390/bioengineering10101148.

[4] S. Teixeira, M.P. Ferraz, F.J. Monteiro, Biocompatibility of highly macroporous ceramic scaffolds: cell adhesion and morphology studies, Journal of Materials Science: Materials in Medicine 19 (2) (2008) 855–859. Available from https://doi.org/10.1007/s10856-007-3005-x.

[5] R. Salihu, S. Izwan, A. Razak, M. Helmi Sani, M. Ahmad Wsoo, N. Ahmad Zawawi, et al., Citrate-modified bacterial cellulose as a potential scaffolding material for bone tissue regeneration, PLoS ONE 19 (12) (2024) e0312396. Available from https://doi.org/10.1371/journal.pone.0312396.

[6] M.V. Berridge, P.M. Herst, A.S. Tan, Tetrazolium dyes as tools in cell biology: new insights into their cellular reduction, Biotechnology Annual Review 11 (2005) 127–152. Available from https://doi.org/10.1016/S1387-2656(05)11004-7.

[7] M.O. Wang, C.M. Piard, A. Melchiorri, M.L. Dreher, J.P. Fisher, Evaluating changes in structure and cytotoxicity during in vitro degradation of three-dimensional printed scaffolds, Tissue Engineering - Part A 21 (9-10) (2015) 1642–1653. Available from https://doi.org/10.1089/ten.tea.2014.0495, http://www.liebertonline.com/tea.

[8] M. Wacker, J. Riedel, H. Walles, M. Scherner, G. Awad, S. Varghese, et al., Comparative evaluation on impacts of fibronectin, heparin–chitosan, and albumin coating of bacterial nanocellulose small-diameter vascular grafts on endothelialization in vitro, Nanomaterials 11 (8) (2021) 1952. Available from https://doi.org/10.3390/nano11081952.

[9] P. Kumar, A. Nagarajan, P.D. Uchil, Analysis of cell viability by the lactate dehydrogenase assay, Cold Spring Harbor Protocols 2018 (6) (2018). Available from https://doi.org/10.1101/pdb.prot095497.

[10] N. Maurmann, D.P. Pereira, D. Burguez, F.D.A. de S Pereira, P. Inforçatti Neto, R.A. Rezende, et al., Mesenchymal stem cells cultivated on scaffolds formed by 3D printed PCL matrices, coated with PLGA electrospun nanofibers for use in tissue engineering, Biomedical Physics & Engineering Express 3 (4) (2017) 045005. Available from https://doi.org/10.1088/2057-1976/aa6308.

[11] S. Wu, Y. Shuai, G. Qian, S. Peng, Z. Liu, C. Shuai, et al., A spatiotemporal drug release scaffold with antibiosis and bone regeneration for osteomyelitis, Journal of Advanced Research 54 (2023) 239–249. Available from https://doi.org/10.1016/j.jare.2023.01.019.

[12] I. Koumentakou, A. Michopoulou, M.J. Noordam, Z. Terzopoulou, D.N. Bikiaris, 3D-printed scaffolds for tissue engineering applications using thermosensitive hydrogels based on biopolymer blends, Journal of Materials Science 59 (20) (2024) 9021–9041. Available from https://doi.org/10.1007/s10853-024-09707-0, https://www.springer.com/journal/10853.

[13] B.L. Cavanagh, T. Walker, A. Norazit, A.C.B. Meedeniya, Thymidine analogues for tracking DNA synthesis, Molecules 16 (9) (2011) 7980–7993. Available from https://doi.org/10.3390/molecules16097980Australia, http://www.mdpi.com/1420-3049/16/9/7980/pdf.

[14] P. Mogha, A. Srivastava, S. Kumar, S. Das, S. Kureel, A. Dwivedi, et al., Hydrogel scaffold with substrate elasticity mimicking physiological-niche promotes proliferation of functional keratinocytes, RSC Advances 9 (18) (2019) 10174–10183. Available from https://doi.org/10.1039/c9ra00781d.

[15] T.J. Mead, V. Lefebvre, Proliferation assays (BrdU and EdU) on skeletal tissue sections, Methods in Molecular Biology 1130 (2014) 233–243. Available from https://doi.org/10.1007/978-1-62703-989-5_17, http://www.springer.com/series/7651.

[16] M. Zhou, X. Chen, Y. Qiu, H. Chen, Y. Liu, Y. Hou, et al., Study of tissue engineered vascularised oral mucosa-like structures based on ACVM-0.25% HLC-I scaffold in vitro and in vivo, Artificial Cells, Nanomedicine, and Biotechnology 48 (1) (2020) 1167–1177. Available from https://doi.org/10.1080/21691401.2020.1817055.

[17] M.J. Stoddart, Cell viability assays: introduction, Methods in Molecular Biology 740 (2011) 1–6. Available from https://doi.org/10.1007/978-1-61779-108-6_1, http://www.springer.com/series/7651.

[18] J. Zheng, Y. Liu, C. Hou, Z. Li, S. Yang, X. Liang, et al., Ovary-derived decellularized extracellular matrix-based bioink for fabricating 3D primary ovarian cells-laden structures for mouse ovarian failure correction, International Journal of Bioprinting 8 (3) (2022) 597. Available from https://doi.org/10.18063/ijb.v8i3.597.

[19] E.M. Longhin, N. El Yamani, E. Rundén-Pran, M. Dusinska, The alamar blue assay in the context of safety testing of nanomaterials, Frontiers in Toxicology 4 (2022). Available from https://doi.org/10.3389/ftox.2022.981701.

[20] C.J. Yadav, U. Yadav, S. Afrin, J.Y. Lee, J. Kamel, K.M. Park, Heparin immobilization enhances hemocompatibility, re-endothelization, and angiogenesis of decellularized liver scaffolds, International Journal of Molecular Sciences 25 (22) (2024) Available from https://doi.org/10.3390/ijms252212132, http://www.mdpi.com/journal/ijms.

[21] X.Q. Wang, X.M. Duan, L.H. Liu, Y.Q. Fang, Y. Tan, Carboxyfluorescein diacetate succinimidyl ester fluorescent dye for cell labeling, Acta biochimica et biophysica Sinica 37 (6) (2005) 379–385. Available from https://doi.org/10.1111/j.1745-7270.2005.00051.x.

[22] V.E. Getova, J.A. van Dongen, L.A. Brouwer, M.C. Harmsen, Adipose tissue-derived ECM hydrogels and their use as 3D culture scaffold, Artificial Cells, Nanomedicine and Biotechnology 47 (1) (2019) 1693–1701. Available from https://doi.org/10.1080/21691401.2019.1608215, http://www.tandfonline.com/loi/ianb20#.VmugQbfovcs.

[23] M. Lakshmanan, M. Saini, M. Nune, Exploring the innovative application of cerium oxide nanoparticles for addressing oxidative stress in ovarian tissue regeneration, Journal of Ovarian Research 17 (1) (2024). Available from https://doi.org/10.1186/s13048-024-01566-2.

[24] N. Rizwana, N. Maslekar, K. Chatterjee, Y. Yao, V. Agarwal, M. Nune, Dual crosslinked antioxidant mixture of poly(vinyl alcohol) and cerium oxide nanoparticles as a bioink for 3D bioprinting, ACS Applied Nano Materials 7 (16) (2024) 18177–18188. Available from https://doi.org/10.1021/acsanm.3c02962.

[25] H. Kim, X. Xue, Detection of total reactive oxygen species in adherent cells by 2',7'-dichlorodihydrofluorescein diacetate staining, Journal of Visualized Experiments 2020 (160) (2020). Available from https://doi.org/10.3791/60682.

[26] S. Nilawar, P. Yadav, N. Jain, D.K. Saini, K. Chatterjee, Protective role of nanoceria-infused nanofibrous scaffold toward bone tissue regeneration with senescent cells, Biomacromolecules 25 (7) (2024) 4074–4086. Available from https://doi.org/10.1021/acs.biomac.4c00184.

[27] İ. Gulcin, S.H. Alwasel, DPPH radical scavenging assay, Processes 11 (8) (2023). Available from https://doi.org/10.3390/pr11082248, http://www.mdpi.com/journal/processes.

[28] A.H. Luong, W.-C. Lin, Enhancement of wound healing by a bilayer hydrogel and nanofiber scaffold infused with Calophyllum inophyllum oil and Platostoma palustre aqueous extract, Biomaterials Advances 172 (2025) 214247. Available from https://doi.org/10.1016/j.bioadv.2025.214247.

[29] D. Tsikas, Assessment of lipid peroxidation by measuring malondialdehyde (MDA) and relatives in biological samples: analytical and biological challenges, Analytical Biochemistry 524 (2017) 13–30. Available from https://doi.org/10.1016/j.ab.2016.10.021.

[30] N.M. Ismail, S. Korraa, M.B.S. Osman, E.E. Sheikh, Effect of low doses of gamma radiation on cell growth over an electrospun polycaprolactone/zinc acetate scaffold for biomedical applications, Biophysical Reviews 17 (1) (2025) 139–149. Available from https://doi.org/10.1007/s12551-024-01256-9, https://www.springer.com/journal/12551.

[31] S. Aydin, A short history, principles, and types of ELISA, and our laboratory experience with peptide/protein analyses using ELISA, Peptides 72 (2015) 4–15. Available from https://doi.org/10.1016/j.peptides.2015.04.012.

[32] M. Restan Perez, N. Masri, J. Walters-Shumka, S. Kahale, S. Willerth, Protocol for 3D bioprinting mesenchymal stem cell–derived neural tissues using a fibrin-based bioink, Bio-Protocol 13 (9) (2023). Available from https://doi.org/10.21769/bioprotoc.4663.

[33] H.A. Elshabrawy, A.S. Abo Dena, M.H. Sobhy, I.M. El-Sherbiny, Integrating 3D-printed hydrogels and polymeric nanofibers in a double-layered transdermal patch for the treatment of rheumatoid arthritis, RSC Advances 14 (54) (2024) 40187–40197. Available from https://doi.org/10.1039/d4ra07306a, http://pubs.rsc.org/en/journals/journal/ra.

[34] A.M. Seciu-Grama, S.E. Lazăr, S. Petrescu, O.C. Mocioiu, O. Crăciunescu, I. Atkinson, Impact of vitamin D3 functionalization on the osteogenic capacity of bioinspired 3D scaffolds based on Ce-doped bioactive glass and *Spongia agaricina*, Journal of Functional Biomaterials 16 (4) (2025) Available from https://doi.org/10.3390/jfb16040141, http://www.mdpi.com/journal/jfb.

[35] P.J. Keely, Cell Migration within Three-Dimensional Matrices, Elsevier BV, (2013), pp. 436–445. Available from 10.1016/b978-0-12-378630-2.00495-3.

[36] D.J. VandenHeuvel, B.L. Devlin, P.R. Buenzli, M.A. Woodruff, M.J. Simpson, New computational tools and experiments reveal how geometry affects tissue growth in 3D printed scaffolds, Chemical Engineering Journal 475 (2023) Available from https://doi.org/10.1016/j.cej.2023.145776, www.elsevier.com/inca/publications/store/6/0/1/2/7/3/index.htt.

[37] M. Lanaro, M.P. Mclaughlin, M.J. Simpson, P.R. Buenzli, C.S. Wong, M.C. Allenby, et al., A quantitative analysis of cell bridging kinetics on a scaffold using computer vision algorithms, Acta Biomaterialia 136 (2021) 429–440. Available from https://doi.org/10.1016/j.actbio.2021.09.042, http://www.journals.elsevier.com/acta-biomaterialia.

[38] M. Ahmed, T. Ramos, P. Wieringa, C.V. Blitterswijk, J.D. Boer, L. Moroni, Geometric constraints of endothelial cell migration on electrospun fibres, Scientific Reports 8 (1) (2018) Available from https://doi.org/10.1038/s41598-018-24667-7, www.nature.com/srep/index.html.

[39] Z.M. Khan, J.M. Munson, T.E. Long, E. Vlaisavljevich, S.S. Verbridge, Development of a synthetic, injectable hydrogel to capture residual glioblastoma and glioblastoma stem-like cells with CXCL12-mediated chemotaxis, Advanced Healthcare Materials 12 (14) (2023). Available from https://doi.org/10.1002/adhm.202300671, http://onlinelibrary.wiley.com/journal/10.1002/(ISSN)2192-2659.

[40] F. Nie, M. Yamada, J. Kobayashi, M. Yamato, A. Kikuchi, T. Okano, On-chip cell migration assay using microfluidic channels, Biomaterials 28 (27) (2007) 4017–4022. Available from https://doi.org/10.1016/j.biomaterials.2007.05.037.

[41] R.L. Klank, S.A. Decker Grunke, B.L. Bangasser, C.L. Forster, M.A. Price, T.J. Odde, et al., Biphasic dependence of glioma survival and cell migration on CD44 expression level, Cell Reports 18 (1) (2017) 23–31. Available from https://doi.org/10.1016/j.celrep.2016.12.024, http://www.sciencedirect.com/science/journal/22111247.

[42] K. Govindaraj, P. Padmanaban, In Vivo Models of Cell Migration, Springer Science and Business Media LLC, (2024), pp. 299–314. Available from https://doi.org/10.1007/978-3-031-64532-7_14.

[43] H. Holtzer, H. Weintraub, R. Mayne, B. Mochan, Chapter 6 The Cell Cycle, Cell Lineages, and Cell Differentiation, Elsevier BV, (1972), pp. 229–256. Available from https://doi.org/10.1016/s0070-2153(08)60073-3.

[44] S. Ortinau, J. Schmich, S. Block, A. Liedmann, L. Jonas, D.G. Weiss, et al., Effect of 3D-scaffold formation on differentiation and survival in human neural progenitor cells, Biomedical Engineering Online 9 (1) (2010). Available from https://doi.org/10.1186/1475-925x-9-70.

[45] G. Marino, F. Rosso, G. Cafiero, C. Tortora, M. Moraci, M. Barbarisi, et al., β-Tricalcium phosphate 3D scaffold promote alone osteogenic differentiation of human adipose stem cells: in vitro study, Journal of Materials Science: Materials in Medicine 21 (1) (2010) 353–363. Available from https://doi.org/10.1007/s10856-009-3840-z.

[46] F. Aleahmad, S. Ebrahimi, M. Salmannezhad, M. Azarnia, M. Jaberipour, M. Hoseini, et al., Heparin/collagen 3D scaffold accelerates hepatocyte differentiation of Wharton's jelly-derived mesenchymal stem cells, Tissue Engineering and Regenerative Medicine 14 (4) (2017) 443–452. Available from https://doi.org/10.1007/s13770-017-0048-z.

[47] J. Li, R. Tao, W. Wu, H. Cao, J. Xin, J. Li, et al., 3D PLGA scaffolds improve differentiation and function of bone marrow mesenchymal stem cell–derived hepatocytes, Stem Cells and Development 19 (9) (2010) 1427–1436. Available from https://doi.org/10.1089/scd.2009.0415.

[48] C.A. Batchelder, M.L. Martinez, A.F. Tarantal, D. Long, Natural scaffolds for renal differentiation of human embryonic stem cells for kidney tissue engineering, PLoS ONE 10 (12) (2015) e0143849. Available from https://doi.org/10.1371/journal.pone.0143849.

[49] M. Ding, H. Andersson, S. Martinsson, A. Sabirsh, A. Jonebring, Q.D. Wang, et al., Aligned nanofiber scaffolds improve functionality of cardiomyocytes differentiated from human induced pluripotent stem cell-derived cardiac progenitor cells, Scientific Reports 10 (1) (2020) Available from: https://doi.org/10.1038/s41598-020-70547-4, www.nature.com/srep/index.html.

[50] K. Moeinabadi-Bidgoli, A. Babajani, G. Yazdanpanah, B. Farhadihosseinabadi, E. Jamshidi, S. Bahrami, et al., Translational insights into stem cell preconditioning: from molecular mechanisms to preclinical applications, Biomedicine & Pharmacotherapy 142 (2021) 112026. Available from https://doi.org/10.1016/j.biopha.2021.112026.

[51] S. Bose, S. Tarafder, A. Bandyopadhyay, Effect of chemistry on osteogenesis and angiogenesis towards bone tissue engineering using 3D printed scaffolds, Annals of Biomedical Engineering 45 (1) (2017) 261–272. Available from https://doi.org/10.1007/s10439-016-1646-y.

[52] C. Galindo-Pumariño, A. Herrera, A. Muñoz, A. Carrato, M. Herrera, C. Peña, Fibroblast-derived 3D matrix system applicable to endothelial tube formation assay, Journal of Visualized Experiments 2019 (154) (2019). Available from https://doi.org/10.3791/60304.

[53] J.W. Lee, Y.J. Choi, W.J. Yong, F. Pati, J.H. Shim, K.S. Kang, et al., Development of a 3D cell printed construct considering angiogenesis for liver tissue engineering, Biofabrication 8 (1) (2016). Available from https://doi.org/10.1088/1758-5090/8/1/015007, http://iopscience.iop.org/article/10.1088/1758-5090/8/1/015007/pdf.

[54] A. Kapoor, C.G. Chen, R.V. Iozzo, A simplified aortic ring assay: a useful ex vivo method to assess biochemical and functional parameters of angiogenesis, Matrix Biology Plus 6-7 (2020) 100025. Available from https://doi.org/10.1016/j.mbplus.2020.100025.

[55] C.A. Staton, M.W.R. Reed, N.J. Brown, A critical analysis of current in vitro and in vivo angiogenesis assays, International Journal of Experimental Pathology 90 (3) (2009) 195–221. Available from https://doi.org/10.1111/j.1365-2613.2008.00633.x.

[56] A. Bruinink, R. Luginbuehl, Evaluation of Biocompatibility Using In Vitro Methods: Interpretation and Limitations, Springer Science and Business Media LLC, (2011), pp. 117–152 vol. 126.

[57] B. Huzum, B. Puha, R. Necoara, S. Gheorghevici, G. Puha, A. Filip, et al., Biocompatibility assessment of biomaterials used in orthopedic devices: an overview (Review), Experimental and Therapeutic Medicine 22 (5) (2021). Available from https://doi.org/10.3892/etm.2021.10750.

[58] C.G. Simon, M.J. Yaszemski, A. Ratcliffe, P. Tomlins, R. Luginbuehl, J.A. Tesk, ASTM international workshop on standards and measurements for tissue engineering scaffolds, Journal of Biomedical Materials Research - Part B Applied Biomaterials 103 (5) (2015) 949–959. Available from https://doi.org/10.1002/jbm.b.33286, http://onlinelibrary.wiley.com/journal/10.1002/(ISSN)1552-4981.

[59] M.N. Helmus, D.F. Gibbons, D. Cebon, Biocompatibility: meeting a key functional requirement of next-generation medical devices, Toxicologic Pathology 36 (1) (2008) 70–80. Available from https://doi.org/10.1177/0192623307310949.

[60] E. Mueller, R. Kammula, D. Marlowe, Regulation of "biomaterials" and medical devices, MRS Bulletin 16 (9) (1991) 39–41. Available from https://doi.org/10.1557/s0883769400056037.

[61] T.C. Chung, Y.H. Hsu, T. Chen, Y. Li, H. Yang, J.X. Yu, et al., Machine learning integrated workflow for predicting Schwann cell viability on conductive MXene biointerfaces, ACS Applied Materials and Interfaces 15 (39) (2023) 46460–46469. Available from https://doi.org/10.1021/acsami.3c08070, http://pubs.acs.org/journal/aamick.

[62] F. Zhang, R. Zhang, M. Wei, G. Li, A machine learning based approach for quantitative evaluation of cell migration in Transwell assays based on deformation characteristics, The Analyst 148 (6) (2023) 1371–1382. Available from https://doi.org/10.1039/d2an01882a.

[63] A.A. Solbu, D. Caballero, S. Damigos, S.C. Kundu, R.L. Reis, Ø. Halaas, et al., Assessing cell migration in hydrogels: an overview of relevant materials and methods, Materials Today Bio 18 (2023) Available from https://doi.org/10.1016/j.mtbio.2022.100537, www.journals.elsevier.com/materials-today-bio.

[64] E.R.G. Salvante, A.V. Popoiu, A.C. Barb, A.A. Cosma, M.P. Fenesan, A.K. Saxena, et al., Artificial intelligence (AI) based analysis of in vivo polymers and collagen scaffolds inducing vascularization, In Vivo (Athens, Greece) 38 (2) (2024) 620–629. Available from https://doi.org/10.21873/invivo.13481, https://iv.iiarjournals.org/content/38/2/620.

Chapter 12

Harnessing in vitro models and assays for tissue engineering

E. Pranshu Rao[1], and Misba Majood[2]
[1]*Stem Cell Facility, All India Institute of Medical Sciences, New Delhi, India,* [2]*Department of Physiology, Johns Hopkins School of Medicine, Baltimore, MD, United States*

12.1 Introduction to tissue engineering

A variety of diseases and injuries impact tissues and organs, potentially leading to a reduction in functional capacity. These disorders are primarily linked to acute injuries or persistent degenerative alterations. In the absence of medical intervention, the body's response is mostly constrained and primarily focused on repair mechanisms. Repair may result in the reestablishment of continuity in the afflicted area by the formation of scar tissue, which is primarily collagenous and does not resemble the original injured tissue. This may serve as an effective first response to injury; nevertheless, it does not facilitate the restoration of normal structure and function and may, if left unchecked, result in adverse implications for the patient [1,2]. At present, five distinct techniques are utilized for the replacement of missing tissues: Replacement of missing tissue with tissue from another bodily part, such as substituting a lost thumb with a big toe. This method is not universally applicable; for instance, myocardium cannot be substituted with another tissue. Replacement with synthetic materials, such as substituting the aorta artery with Dacron. The issue is that the synthetic material may not adequately conform to all physiological conditions. Furthermore, in a young patient of developmental age, the substance may not be capable of expanding in proportion to the body's growth. Xenografts and allografts represent alternative methodologies. The limitations of these approaches include their restricted availability and immunological incompatibility. Utilization of an artificial apparatus, such as a hemodialyzer, to reinstate the impaired physiological function. However, the constraints of prolonged device usage and the restricted capacity to accommodate the body's diverse physiological requirements under varying settings are significant problems. The utilization of living cells to substitute missing tissue, becoming the foundation of tissue engineering (TE) [3–5].

Tissue regeneration seeks to restore normal structure and function by producing new tissue that precisely replicates the lost tissue. Yannas explains in his monograph on tissue and organ regeneration that adult mammals do not spontaneously regenerate many damaged organs and possess only a limited capacity to regenerate specific tissues [6]. To persuade adult humans to regenerate whole organs or tissues that do not naturally regenerate, it is necessary to provide cues or signals and implement a mechanism that differs from the natural response to those conditions. Induced regeneration is fundamental to TE, distinct from tissue repair or replacement. Repair processes in most tissues typically do not result in fully functional tissues, leading to clinical outcomes that are significantly inferior to those achieved through functional regeneration. TE involves the creation of new tissue, where engineering refers to the act of creation [7]. The creative process is facilitated by cells, which enter an unnatural regenerative mode due to various factors. These factors can be categorized as biomolecules, providing molecular signals, and supporting structures, offering mechanical signals. While various broad definitions of TE exist, I favor the perspective that it involves encouraging the body to heal itself by delivering cells, biomolecules, and/or supporting structures to the appropriate site [8,9].

12.1.1 Historical evolution and milestones

The evolution of TE spans centuries and integrates insights from mythology, biology, and technology. Whole timeline of TE could be summarized as follows:

- Ancient Roots: TE has philosophical roots in myths like Prometheus's liver regeneration (800 BCE) and Aristotle's observations of salamander tail regrowth (300 BCE). Early practices like cranial trepanation showcased human curiosity about bodily repair [10].

Fundamentals of Translational Tissue Engineering. DOI: https://doi.org/10.1016/B978-0-443-33455-9.00006-7

- 18th–19th Century: Foundations of Regeneration and Transplantation: Abraham Trembley's 1740 s studies on hydra regeneration and Spallanzani's 1768 experiments on amphibians advanced understanding of epigenesis [11,12]. John Hunter's transplantation experiments paved the way for organ transplantation. In 1838–1839 the cell theory established cells as life's fundamental units, and the term "stem cell" emerged in 1868 [13,14].
- 20th Century: Birth of Modern Tissue Engineering: The 1900s witnessed critical developments, including the first *in vitro* tissue cultures and Alexis Carrel's Nobel-winning cell culture techniques. The 1950s introduced bone marrow transplantation and the concept of cellular plasticity through nuclear reprogramming by John Gurdon [15,16]. The 1970s-80s marked the establishment of TE, with pioneers like Joseph Vacanti and Robert Langer creating hybrid tissues using cells, scaffolds, and growth factors. Innovations like the first synthetic skin products and genetically modified stem cells revolutionized regenerative possibilities [17].
- 21st Century: Integration and Expansion: The 2000s heralded stem cell breakthroughs like induced pluripotent stem cells (iPSCs) by Shinya Yamanaka in 2006, which enabled patient-specific regenerative therapies [18]. Advances in 3D bioprinting, biomaterials, and organoid technology expanded the horizons of TE. Clinical applications included artificial organs, drug screening, and personalized medicine.

TE, a pillar of regenerative medicine, continues evolving with interdisciplinary research and technological integration [19].

12.1.2 Importance in regenerative medicine and healthcare

TE is essential in advancing regenerative medicine and transforming healthcare by tackling significant challenges in disease treatment, organ transplantation, and drug development. TE is a multidisciplinary field that combines biology, material science, and engineering to repair, replace, or regenerate damaged tissues and organs. This has become fundamental in regenerative medicine, emphasizing curative methods over symptomatic treatments, and has the potential to transform healthcare [17,20].

TE is crucial for offering sustainable, patient-specific solutions in tissue repair. Conventional methods, like organ transplants, encounter obstacles such as a restricted donor pool, elevated expenses, and the potential for immune rejection. TE addresses these challenges by developing functional tissues through the use of biocompatible scaffolds, living cells, and bioactive molecules. Recent advancements in stem cell research and 3D bioprinting have enhanced the field, facilitating the creation of tissues customized for individual patient requirements. Engineered skin has revolutionized burn treatment by providing effective solutions that markedly decrease recovery time and enhance quality of life. Bioengineered cartilage and bone are similarly targeting age-related degenerative conditions, improving mobility, and reducing the necessity for invasive surgeries [21,22].

12.1.3 Applications in disease modeling, drug testing, and transplantation

The application of TE in disease modeling is highly significant. Engineered tissues and organoids replicate human physiological conditions, offering a controlled setting for the investigation of disease progression and mechanisms. These models are essential for comprehending complex diseases like cancer, neurodegenerative disorders, and cardiovascular conditions, which are challenging to replicate using traditional animal models. Brain organoids are enhancing research on Alzheimer's and Parkinson's diseases, and heart-on-a-chip systems are yielding insights into cardiac dysfunction. These advancements enable researchers to more accurately predict disease outcomes and develop targeted therapies, thus narrowing the translational gap between preclinical studies and clinical applications [23].

TE has transformed drug testing by providing more reliable and human-relevant platforms, alongside its role in disease modeling. Conventional approaches, which depend on two-dimensional (2D) cell cultures or animal models, frequently do not adequately represent the intricacies of human physiology. Engineered three-dimensional (3D) tissues closely mimic human organ function, allowing for accurate assessment of drug efficacy and toxicity. Liver organoids are utilized to investigate drug metabolism and hepatotoxicity, thereby minimizing the risks of adverse effects in clinical trials. TE enhances the predictability of drug responses, thereby expediting the drug development process and lowering the time and costs involved in introducing new pharmaceuticals to the market [24].

The use of TE in organ transplantation tackles a significant healthcare issue: the increasing gap between the need for donor organs and their supply. Tissue-engineered constructs, including skin grafts, cartilage, and complex organs such as kidneys and livers, are becoming viable alternatives to donor organs. These constructs can be created from a patient's own cells, reducing the risk of immune rejection and eliminating the necessity for lifelong immunosuppressive therapy. Recent developments in 3D bioprinting show significant potential, allowing for the layer-by-layer construction of functional tissues and organs that replicate the complex architecture of natural structures. This technology has the potential to address organ shortages and enhance transplant outcomes through customized, biocompatible solutions [1,2].

The impact of TE on healthcare and regenerative medicine is significant, demonstrated by its various applications and capacity to meet critical medical demands. TE has positioned itself as a crucial area for the future of medicine through innovative solutions in disease modeling, drug testing, and transplantation. Advancements in biomaterials, stem cell technologies, and bioprinting are set to enable TE to achieve breakthroughs that will transform healthcare, providing hope to millions of patients globally and initiating a new era of curative treatments.

12.2 Role of in vitro models and assays

In vitro models and assays are essential instruments in research, offering controlled settings to investigate biological processes, disease mechanisms, and therapeutic strategies. These models simulate physiological or pathological settings externally, allowing researchers to examine cellular and molecular interactions comprehensively. Their main advantage is their capacity to offer replicable and ethically sound alternatives to animal research, consistent with the 3Rs principles (replacement, reduction, and refinement) [25].

In research, in vitro models, including organoids, 2D cell cultures, and lab-on-a-chip technologies, provide meticulous examination of intricate biological processes. Organoids replicate the structural and functional characteristics of human tissues, rendering them essential for investigating disease development, pharmacological responses, and regenerative therapies. Likewise, assays such as biochemical, viability, and gene expression tests are employed to evaluate the impacts of therapies, toxicity, and cellular activity in controlled environments [26,27]. In vitro models and assays expedite drug discovery by facilitating high-throughput screening of compounds for efficacy and safety, hence diminishing development time and costs. Furthermore, these technologies enable personalized treatment by permitting researchers to evaluate medications on cells originating from patients. Their adaptability and accuracy render them fundamental to contemporary biomedical research, connecting laboratory investigations with clinical applications.

In research, both in vitro and in vivo systems are essential for comprehending biological processes, evaluating therapeutic efficacy, and investigating disease etiology. In vitro systems, which entail the examination of cells or tissues outside their native context, have become increasingly significant due to their accuracy and ethical benefits. In contrast, in vivo approaches, which depend on whole-organism research, provide insights into intricate systemic relationships but have their own limits. By evaluating the advantages of in vitro systems against the limitations of *in vivo* approaches, researchers can make educated choices regarding the most suitable strategy for a certain study. The table below delineates the comparative advantages of *in vitro* systems and the obstacles inherent in *in vivo* methodologies, underscoring the increasing preference for in vitro models in numerous domains of contemporary research.

12.2.1 Advantages of in vitro systems over traditional in vivo methods

This comparison highlights why in vitro systems are increasingly favored for early-stage research while acknowledging that in vivo models are still crucial for studying systemic effects (Table 12.1; Fig. 12.1).

In TE, diverse in vitro models are utilized to replicate the physiological conditions of tissues and organs, facilitating the examination of cellular activity, tissue growth, and disease causes. These models consist of 2D monolayer cultures, wherein cells are cultivated on flat surfaces, offering a straightforward and economical technique for preliminary investigations; however, they lack the intricacy characteristic of tissue. Conversely, 3D culture techniques, like spheroids and organoids, replicate the 3D structure of tissues, enhancing cell-to-cell interactions and more accurately simulating in vivo circumstances [33–36]. Hydrogel-based scaffolds are frequently employed to offer a biomimetic matrix that facilitates cell proliferation, differentiation, and tissue development. Moreover, bioreactors are sophisticated systems that utilize mechanical, chemical, or electrical stimulation to cells, replicating in vivo circumstances for tissue development [37]. Microfluidic devices, known as organ-on-chip systems, incorporate miniature tissues with fluid dynamics to simulate organ-level activities such as nutrition exchange and medication processing. Finally, coculture models facilitate the simultaneous growth of many cell types, simulating tissue heterogeneity and intercellular interactions. These models jointly connect uncomplicated 2D studies with intricate in vivo systems, enhancing TE research [38,39].

The ethical ramifications of animal experimentation have prompted an increasing focus on minimizing animal utilization in research by adopting alternative models, such as in vitro systems. Although animal models have significantly contributed to biological and medical research, they pose ethical dilemmas related to animal pain, mortality, and intricate moral issues. The principles of the 3Rs are fundamental to resolving these challenges. In vitro models adhere to the principle of Replacement by utilizing cell cultures, tissues, or organoids in lieu of animal experimentation. This substantially reduces

TABLE 12.1 A comparative table listing the advantages of in vitro systems alongside the disadvantages of in vivo methods [28–32].

Aspect	Advantages of in vitro systems	Disadvantages of in vivo methods
Control	Precise control of experimental variables like temperature and pH	Limited control over physiological variables in live organisms
Reproducibility	Highly reproducible conditions, reducing variability	Biological variability between animals leads to inconsistent results
Cost	Cost-effective; does not require animal housing or care	Expensive due to animal care, housing, and ethical approvals
Ethics	Avoids ethical concerns associated with animal testing	Raises ethical concerns and regulatory hurdles due to animal use
Human relevance	Uses human cells, better mimicking human physiology.	Animal models may not always accurately replicate human biology
Accessibility	Enables high-throughput screening and faster experimentation	Limited throughput due to time-intensive processes with animals
Customization	Easily adaptable to specific research needs (e.g., genetic editing)	Difficult to customize biological systems in live organisms
Imaging and analysis	Direct observation of cellular and molecular processes in real-time	Limited real-time access to internal processes without invasive methods
Simplified complexity	Isolates specific pathways without interference from other systems	Complexity of whole-organism responses can obscure results
Safety testing	Early identification of toxicity or side effects	Late identification of toxicity, often after significant time and cost
Time	Experiments are faster due to simpler setups	Time-consuming procedures, such as breeding or waiting for results

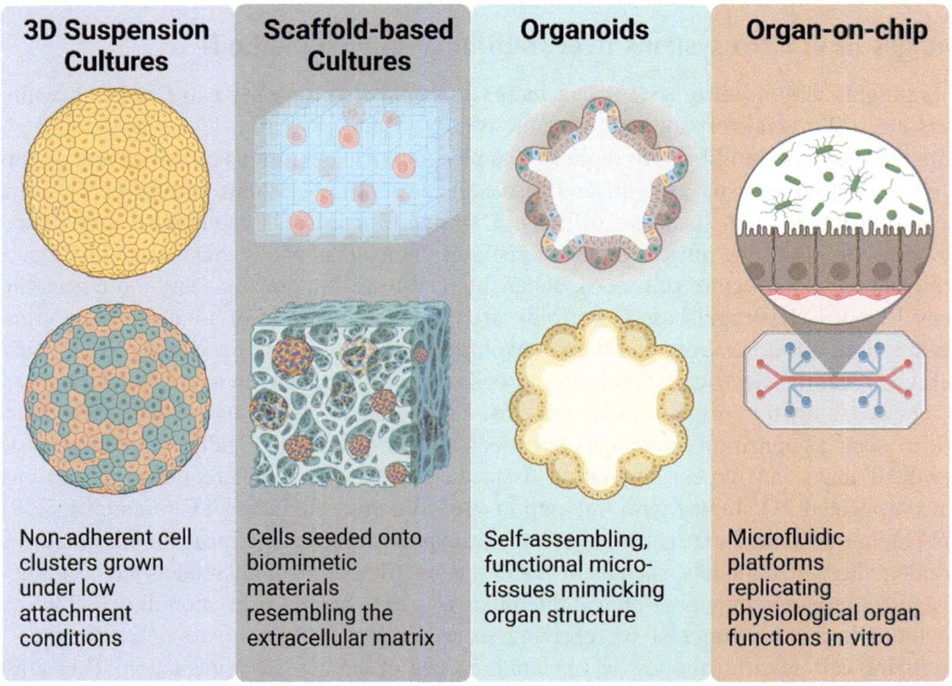

FIGURE 12.1 Key 3D culture systems used in tissue engineering, including 3D suspension cultures, scaffold-based cultures, organoids, and organ-on-chip platforms. These systems represent advanced methodologies for replicating physiological environments, supporting cell growth, mimicking organ structures, and enabling precise modeling of organ-specific functions in vitro, thereby advancing research in tissue engineering, regenerative medicine, and drug discovery.

animal testing while generating more pertinent human data [40]. The incorporation of in vitro systems facilitates Reduction by necessitating fewer animals when validated alternatives are utilized. Furthermore, progress in computer models and microfluidic devices enables researchers to assess pharmaceuticals, investigate diseases, and analyze toxicity without inflicting harm on real organisms. Notwithstanding these gains, difficulties endure, including restricted complexity in existing in vitro models and inadequate modeling of systemic interconnections seen in vivo. Nevertheless, ongoing advancements in in vitro research uphold ethical standards, diminish reliance on animal experimentation, and respond to increasing societal and regulatory calls for humane scientific methodologies [41,42].

In vitro models are essential in contemporary research, especially in TE, pharmacological testing, and disease modeling. Nonetheless, their existing constraints encompass an incapacity to entirely emulate the intricacies of the human body, including complex tissue architecture, systemic reactions, and interorgan communication evident in vivo. Moreover, numerous 3D in vitro systems remain in the developmental phase and have technological obstacles, including scalability, reproducibility, and the requirement for sophisticated biomaterials to more accurately replicate native tissues. Notwithstanding these constraints, forthcoming progress in bioengineering, stem cell technologies, and microfluidic organ-on-chip systems presents considerable potential. The incorporation of artificial intelligence and machine learning can enhance data analysis and experimental design, while novel technologies such as 3D bioprinting and organoids seek to connect in vitro and in vivo systems. By surmounting these obstacles, in vitro models are positioned to diminish dependence on animal research, enhance human-relevant results, and revolutionize personalized treatment.

12.3 Fundamentals of in vitro models

12.3.1 2D culture systems

2D cell cultures are a conventional and extensively utilized model in TE and biological research. Cells in 2D cultures proliferate as monolayers, adhering to the planar surfaces of culture dishes or flasks. This method was initially investigated in the early 20th century and has since developed into an essential instrument in cell biology, pharmacological research, and disease investigation [43]. The creation of 2D cultures is quite straightforward. Cells are inoculated onto planar plastic or glass substrates, where they attach, multiply, and establish a monolayer. The system is economical, user-friendly, and reproducible, rendering it optimal for high-throughput drug screening and molecular investigations. It enables the fast multiplication of cells in regulated environments and promotes the examination of cell shape, signaling pathways, and gene expression. The culture medium provides constant availability of nutrients, oxygen, and signaling chemicals, facilitating uniform growth [44,45].

The principal advantages of 2D cultures encompass simplicity, cost-effectiveness, and the convenience of performing functional experiments, including drug testing and genetic manipulation. These models have greatly enhanced the comprehension of essential cell biology, cancer processes, and therapeutic responses. Moreover, 2D cultures are compatible with several analytical techniques, including microscopy, immunostaining, and molecular tests [46]. Nonetheless, despite their advantages, 2D cultures lack the intricacy of original tissues, prompting the advancement of 3D models that more accurately replicate in vivo situations. Nevertheless, 2D systems continue to be vital for initial explorations and act as essential instruments in research [47,48].

12.4 Transition to 3D culture systems

The shift from 2D to 3D cell culture methods has become increasingly essential in enhancing TE and medicinal research. Although 2D cultures have been the conventional method for numerous years owing to their simplicity and cost-efficiency, they inadequately replicate the intricacies of in vivo tissue settings. Cells cultivated in 2D cultures develop on planar surfaces and encounter consistent environmental circumstances, markedly differing from the varied habitats seen in tissues. This leads to a diminished accuracy in depicting cellular processes, including migration, differentiation, and drug response [45,49]. Conversely, 3D culture techniques more accurately mimic the natural extracellular matrix (ECM) and facilitate cellular interactions in a manner that is more physiologically relevant. These techniques facilitate cellular growth in aggregates, spheroids, or within scaffolds, offering more authentic models for investigating tissue functions, disease progression, and pharmacological responses. 3D cultures provide more precise modeling of intricate circumstances, such as tumor microenvironments, where gradients of nutrients, oxygen, and waste may be monitored. Furthermore, 3D models can replicate the mechanical pressures encountered by cells in vivo, thereby affecting cellular behavior in manners unattainable with 2D cultures [50].

Nevertheless, the transition to 3D cell culture has its own array of obstacles. A significant obstacle is the technical intricacy associated with establishing 3D cultures. These systems frequently necessitate sophisticated technologies, including scaffold-based methods (hydrogels, polymers, etc.) or scaffold-free techniques (such as spheroids and organoids), each demanding distinct handling procedures and protocols. Scaffold-based cultures are particularly challenging to produce consistently, because scaffolds must adequately replicate the ECM to facilitate cell growth and differentiation without compromising the system [51,52]. A notable concern is the escalated expenses and time demands associated with 3D culture systems. These models frequently incur more costs for establishment and maintenance compared to conventional 2D cultures, attributable to the intricacy of materials, the necessity for specialized apparatus, and the increased duration required for cellular proliferation and testing. Moreover, imaging and analysis in 3D systems present difficulties owing to the requirement for sophisticated microscopy techniques. In contrast to 2D cultures, which only necessitate single-plane imaging, 3D models demand approaches like z-stack imaging, hence augmenting the time and computer resources needed for data analysis [53].

Moreover, repeatability is an additional barrier in 3D culture systems. Although 2D cultures exhibit great reproducibility owing to their simplicity, 3D models may differ in structure, cell density, and organization, hence complicating results and interpretations. This diversity is particularly evident in scaffold-free models, where controlling the aggregation of cells into uniform structures poses challenges. Ultimately, sustaining long-term cultures in 3D systems can be challenging, since cells located in the interior of spheroids or substantial tissue constructs may not obtain sufficient nutrients and oxygen, resulting in necrosis in the central areas. This replicates the obstacles encountered in vivo, although also hampers the evaluation of therapeutic efficacy and the investigation of cellular behavior over prolonged durations [54–56]. Notwithstanding these hurdles, the shift to 3D cultures presents significant potential for enhancing TE, disease modeling, and drug development. As technologies advance and automation and imaging techniques enhance, the constraints of 3D culture systems are expected to be surmounted, rendering them essential in biological research and therapeutic applications.

12.5 3D models and scaffolds

3D models have transformed TE and biomedical research by offering platforms that more accurately replicate in vivo tissue architecture, microenvironments, and cellular activities than traditional 2D cultures. The primary categories of 3D models encompass spheroids, scaffold-based systems, organoids, and microfluidic systems (organs-on-chips). Each possesses distinct traits, applications, and limitations, rendering them appropriate for various experimental objectives.

12.5.1 Spheroids

Spheroids are basic 3D cell aggregates created in nonadhesive or low-adhesion environments. They autonomously aggregate owing to inherent cell-to-cell adhesion, emulating solid tissue microenvironments. Spheroids generate gradients of nutrients, oxygen, and waste as they increase in size, mirroring the hypoxic center found in solid tumors. Spheroids are extensively utilized in cancer biology, pharmacological testing, and stem cell research due to their cost-effectiveness and ease of generation [52,57].

12.5.2 Scaffold-based models

Scaffold-based systems utilize natural or synthetic biomaterials, like hydrogels, collagen, or synthetic polymers, to create a 3D framework. This replicates the ECM, facilitating cell attachment, migration, proliferation, and differentiation. These systems are optimal for building tissues, including skin, cartilage, and bone, because of their structural support and biocompatibility. Nonetheless, attaining homogenous cell seeding and nutrient dispersion continues to be a hurdle [51].

12.5.3 Organoids

Organoids are self-organized, miniature 3D entities originating from stem cells or primary tissues. They emulate the essential structural and functional characteristics of organs, like the liver, kidney, brain, or intestines. Organoids are essential for disease modeling, medication screening, and personalized treatment because they can replicate patient-specific responses. They are particularly beneficial in oncological research and regenerative medicine [58,59].

12.6 Microfluidic systems (organ-on-a-chip)

Microfluidic systems, commonly referred to as organs-on-chips, amalgamate 3D cultures with microfluidics to simulate dynamic physiological circumstances such as blood circulation and mechanical forces. These models provide accurate regulation of nutrition perfusion, waste elimination, and oxygen supply. They are optimal for examining intricate organ functions, medication pharmacokinetics, and interorgan interactions. Despite being very advanced, their intricacy and expense restrict widespread implementation [38,39] (Table 12.2).

Every 3D model presents unique benefits and constraints based on its application. Spheroids are straightforward and economical, rendering them optimal for drug screening. Scaffold-based models offer structural reinforcement for the creation of tissues such as bone and cartilage. Organoids provide unparalleled physiological relevance, especially for disease models and tailored therapy. Ultimately, microfluidic systems incorporate dynamic flow, facilitating advanced organ-on-chip research. Integrating these models with future technologies such as bioengineering and sophisticated biomaterials has the potential to address existing difficulties and enhance TE and biomedical research.

12.7 Fabrication methods for scaffolds

The production of scaffolds is essential in TE, as they provide a 3D structure that facilitates cell adhesion, proliferation, and tissue regeneration. Diverse approaches are employed for scaffold creation, each offering unique advantages based on the material, tissue type, and application. The principal techniques encompass electrospinning, freeze-drying, solvent casting, particle leaching, 3D printing, and gas foaming [61–63].

12.7.1 Electrospinning

Electrospinning is extensively employed to fabricate nanofiber scaffolds that replicate the ECM. The procedure entails the application of a high-voltage electric field to a polymer solution, resulting in the ejection of a charged jet that solidifies into fibers. This technique generates scaffolds characterized by elevated surface-to-volume ratios, superior porosity, and mechanical pliability. Electrospun scaffolds are very beneficial for the regeneration of brain, skin, and bone tissues. Nonetheless, the consistency of fiber deposition and scalability continues to pose issues [64,65].

12.7.2 Lyophilization (freeze-drying)

Freeze-drying entails the freezing of a polymer solution, succeeded by the sublimation of the solvent under vacuum conditions, resulting in extremely porous scaffolds. This method facilitates regulated pore dimensions and interlinked porosity, essential for nutrition diffusion and cellular penetration. It is extensively utilized for the fabrication of scaffolds for cartilage and bone regeneration. The primary restriction is its sluggish process and insufficient mechanical strength [66,67].

12.7.3 Solvent casting and particle leaching

This technique integrates the casting of a polymer solution into a mold with the incorporation of a porogen, such as salt or sugar particles. The porogen is subsequently leached away, resulting in a porous scaffold. This approach is straightforward and economical for achieving porosity; nevertheless, it is constrained to thin, 2D scaffolds because of solvent evaporation challenges [68–70].

12.7.4 3D printing (additive manufacturing)

3D printing is an advanced method that facilitates the accurate production of intricate, individualized scaffolds through computer-aided design (CAD). Methods such as fused deposition modeling (FDM), stereolithography (SLA), and selective laser sintering (SLS) provide the customization of pore size, shape, and structure. This approach is especially beneficial for scaffolds of bone and organs. Challenges encompass elevated expenses and restricted material compatibility [71,72].

12.7.5 Gas foaming

Gas foaming uses high-pressure gas, usually carbon dioxide, to generate porous scaffolds. It obviates the necessity for hazardous solvents and is optimal for biodegradable polymers. The resultant scaffolds demonstrate significant porosity; nevertheless, the absence of pore interconnectivity may restrict cellular penetration [73].

TABLE 12.2 Key differences, advantages, and limitations of these 3D models [47,50,55,60].

Aspect	Spheroids	Scaffold-based models	Organoids	Microfluidic systems (organs-on-chips)
Description	Cell aggregates formed in suspension or low-adhesion environments	Cells grown on 3D biomaterial scaffolds mimicking ECM	Self-organized mini-tissues derived from stem cells	3D cultures integrated with microfluidic systems to mimic organ dynamics
Structure	Simple, spherical cell aggregates	Complex, scaffold-based architecture	Organ-like structure with functional features	Simulates organ functions under flow and mechanical stimuli
Cell-ECM Interaction	Limited	High	High	High
Physiological relevance	Moderate	High	Very high	Extremely high
Cost	Low	Moderate	High	Very high
Ease of use	Easy to generate and maintain	Moderate; requires biomaterial preparation	Moderate; requires stem cell cultures	Complex; requires microfluidic expertise
Applications	Drug testing, cancer biology, stem cell studies	Tissue engineering, regeneration, and drug testing	Disease modeling, drug screening, and personalized medicine	Organ simulation, drug pharmacokinetics, and toxicology studies
Challenges	Limited nutrient diffusion in large spheroids	Nonuniform cell seeding, diffusion issues	Expensive, difficult to maintain long-term	Complex setup, high cost, and limited scalability

TABLE 12.3 Advantages and limitations of different fabrication methods.

Method	Advantages	Limitations	Applications
Electrospinning	High porosity, ECM mimicry, flexibility	Fiber uniformity, scalability issues	Neural, skin, and bone tissue
Freeze-drying	Controlled pore size, interconnected pores	Weak mechanical strength, slow process	Cartilage and bone regeneration
Solvent casting and leaching	Cost-effective, simple process	Limited to thin scaffolds	Bone and skin tissue
3D printing	Customizable, complex structures	High cost, limited materials	Bone, organ scaffolds, patient-specific
Gas foaming	Solvent-free, high porosity	Poor pore interconnectivity	Biodegradable polymer scaffolds
Hydrogels	Biocompatibility, ECM mimicry	Limited mechanical strength	Soft tissue (e.g., cartilage and wounds)

12.7.6 Hydrogel-infused frameworks

Hydrogels, composed of natural or synthetic polymers, have superior biocompatibility and replicate the ECM. Techniques such as cross-linking and self-assembly facilitate the production of hydrogel scaffolds. These scaffolds are appropriate for soft TE applications, including cartilage regeneration and wound healing [22,74] (Table 12.3).

Each scaffold fabrication method offers unique advantages and limitations depending on the target application. The combination of techniques and continuous advancements in materials and technologies holds the potential to overcome existing challenges in scaffold fabrication for TE [75–78].

12.8 Impact of fabrication method on cell behavior and tissue development

Scaffolds are essential in TE as they offer a 3D structure that facilitates cellular adhesion, proliferation, migration, and differentiation. These synthetic platforms seek to replicate the native ECM, thereby affecting cellular activity and promoting tissue growth. The mechanical strength, surface chemistry, topography, porosity, and biodegradability of scaffolds are essential factors influencing their biological efficacy and success in tissue regeneration [79,80].

The mechanical properties of scaffolds, including stiffness, elasticity, and tensile strength, are critical determinants of cellular responses. Cells detect mechanical signals from their substrate by mechanotransduction, a process that transforms external mechanical inputs into biochemical signals. Substrate stiffness regulates stem cell differentiation, with softer substrates favoring neuronal differentiation and stiffer materials promoting osteogenic or chondrogenic lineages. Scaffolds intended for neural tissue regeneration necessitate flexibility akin to brain tissue to promote the development of functional neural networks, while scaffolds for bone tissue must demonstrate superior mechanical strength to endure physiological pressures [81,82].

The topography and architecture of scaffolds directly influence cell adhesion, migration, and alignment. Micro- and nanoscale topographical features, including aligned fibers, grooves, and microchannels, are essential in directing tissue formation. Aligned scaffolds facilitate the directional proliferation of cells and tissues, crucial for applications such as nerve regeneration, where axonal guidance is vital for restoring connections. Electrospun fibrous scaffolds emulate the fibrous characteristics of the original ECM and have demonstrated the ability to enhance neurite outgrowth in neural stem cells. Likewise, scaffolds with structured porosity promote cell infiltration and vascularization, essential for tissue integration and healing [83–85]. Porosity and interconnectivity are critical scaffold attributes that facilitate nutrition and oxygen diffusion, waste elimination, and cellular penetration. Highly porous scaffolds offer sufficient area for cellular migration and proliferation, facilitating tissue ingrowth and vascularization. Achieving the optimal pore size is essential—excessively big pores may impede cell adhesion, whereas excessively small pores can restrict cell movement and nutrient exchange. Methods such as freeze-drying and gas foaming have been employed to create scaffolds with interconnected porosity that promote cell viability and tissue growth [86,87].

The surface chemistry of scaffolds is crucial in modulating cell adherence and functionality. The functionalization of scaffold surfaces with bioactive molecules, like collagen, laminin, or growth factors, improves cellular adhesion and stimulates targeted cellular responses. Laminin-coated scaffolds enhance neural cell adhesion and facilitate neurite extension, essential for neural tissue regeneration. Surface changes, including plasma treatment or peptide grafting, can enhance scaffold wettability, hence increasing protein adsorption and improving biological connections [88]. Biodegradability is a crucial characteristic of scaffolds, as they must decompose at a rate commensurate with the development of new tissue. Scaffolds constructed from biodegradable polymers, including poly(L-lactic acid) (PLLA) and poly(ε-caprolactone) (PCL), are extensively utilized due to their progressive degradation into nontoxic byproducts, facilitating the replacement of the scaffold by new tissue. This characteristic guarantees that the scaffold offers temporary support during tissue regeneration without inducing long-term problems. Nonetheless, attaining equilibrium between deterioration rate and mechanical integrity continues to pose a problem, especially for load-bearing tissues such as bone [89].

Ultimately, cell-scaffold interactions are regulated by the interplay of these features, establishing a dynamic microenvironment that affects cellular behavior. Mechanical cues, biochemical signals, and topographical characteristics of scaffolds collaboratively influence cell adhesion, proliferation, migration, and differentiation. Anisotropic scaffolds featuring aligned microstructures are recognized for facilitating directed axonal growth in neural TE, whereas scaffolds with regulated porosity improve vascularization in bone and cartilage regeneration. In summary, scaffold characteristics are crucial in TE as they influence cellular behavior and promote tissue formation. By meticulously adjusting mechanical qualities, surface chemistry, topography, and degradation rates, scaffolds can be engineered to replicate the natural ECM and create an ideal environment for cellular proliferation and tissue regeneration. Ongoing improvements in scaffold fabrication methods, including electrospinning, 3D printing, and surface functionalization, are facilitating the creation of next-generation scaffolds with enhanced biological efficacy and clinical applicability [90–94].

12.9 Cutting-edge technologies in in vitro modeling

The landscape of in vitro modeling has undergone a striking transformation, evolving from simple two-dimensional (2D) monolayer cultures to highly sophisticated three-dimensional (3D) systems that more closely recapitulate the complexity of native human tissues. Traditional 2D cultures, while foundational for exploring basic cell biology, fail to capture the intricate architecture, mechanical cues, and biochemical gradients that characterize living organs. In contrast, cutting-edge in vitro models—such as organoids, organ-on-a-chip platforms, and 3D bioprinted constructs—offer unprecedented insight into tissue development, disease progression, and drug responses under conditions that emulate the human body more closely than ever before [24,95–98].

- *Three-dimensional (3D) cell cultures and organoids*: Central to this revolution is the emergence of organoids, stem cell-derived 3D structures that self-organize into tissue-like entities [23,95,98,99]. Organoids derived from human induced pluripotent stem cells (hiPSCs) have been established for a variety of tissues: brain, liver, kidney, intestine, lung, and heart, among others [24,95,98,100,101]. These miniature organ analogs capture key functional and structural features of their in vivo counterparts, making them exceptionally valuable for studying human development, modeling genetic diseases, and testing therapeutics [24,95,99,102]. Kidney organoids have shed light on nephrogenesis and drug-induced nephrotoxicity [103], while brain organoids have provided novel insights into cortical layering, neuronal network formation, and neurological disorders [95,104,105].

By deriving organoids from patient-specific stem cells, researchers create "avatars" of human tissues that retain the genetic and phenotypic characteristics of the donor [24,106]. Such patient-derived organoids enable personalized testing of drugs, study of rare mutations, and design of individualized treatment strategies [24,106,107]. This approach is transforming research on cystic fibrosis, cancer, and other complex diseases [24,98].

- *Organ-on-a-chip platforms and microfluidics*: While organoids have advanced tissue complexity, they often lack vasculature and dynamic fluid flow. Organ-on-a-chip devices address this gap by integrating living cells into microfluidic chips that simulate tissue-tissue interfaces, fluid flow, mechanical forces, and chemical gradients [96,108–110]. For example, lung-on-a-chip systems recreate (air liquid interface) alveolar-capillary interfaces and breathing motions, while heart-on-a-chip platforms replicate cardiac contractility, hemodynamic stresses, and sometimes electrical signals [96,108,111]. These microfluidic devices allow precise environmental control and facilitate high-throughput experimentation, making it easier to run parallel tests on multiple conditions [108,110,112].
- *3D bioprinting and scaffold engineering*: 3D bioprinting introduces a new level of control, enabling the fabrication of tissues with defined architectures, spatial gradients, and multiple cell types [100,101]. Complex scaffolds that mimic the ECM can guide cell alignment and tissue maturation [100,101,113]. For example, gradients of growth factors or specific

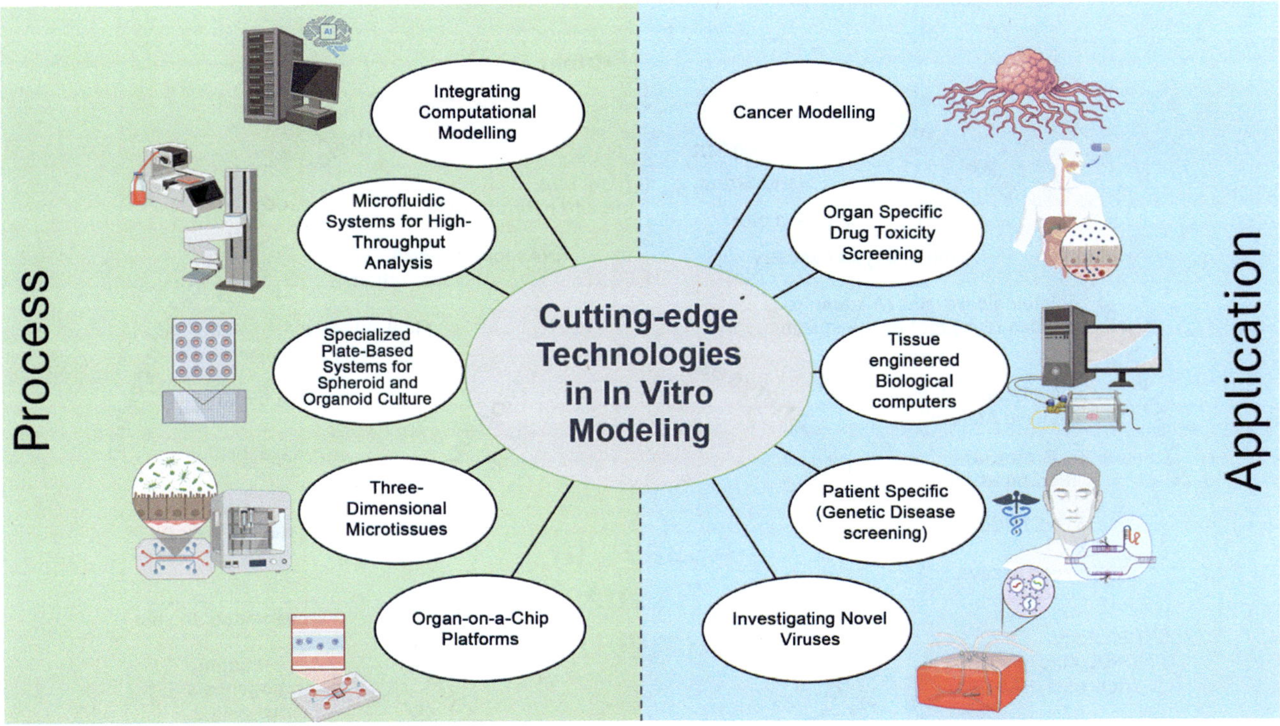

FIGURE 12.2 Cutting-edge technologies in in vitro modeling, highlighting innovative processes such as computational modeling, microfluidic systems, 3D microtissues, and organ-on-a-chip platforms, alongside their applications in areas like cancer modeling, organ-specific drug toxicity screening, genetic disease research, and investigating novel viruses. These advancements bridge the gap between laboratory research and clinical applications, offering precise, scalable, and ethically sound approaches for biomedical innovation.

ECM proteins can direct stem cell differentiation, while microstructured scaffolds can promote vascularization [100,101]. Coupled with organoids or organ-on-a-chip systems, 3D bioprinting creates hybrid models that better approximate native tissues [100,101].

- *Advanced imaging and computational modelling*: High-resolution imaging techniques, such as confocal and multiphoton microscopy, combined with fluorescence-based reporters, allow deep visualization of 3D constructs and dynamic cell behaviors [98,102,114]. These massive datasets are increasingly analyzed by machine learning (ML) and artificial intelligence (AI) algorithms that detect subtle phenotypes, track cell lineages, and integrate morphological, molecular, and mechanical data [98,114,115]. Computational modeling can predict how changes in fluid flow, nutrient delivery, or substrate stiffness influence cell fate, guiding more rational experimental design [98,114,115].
- *Applications in disease modelling and drug discovery*: Cutting-edge models are revolutionizing biomedical research. Patient-derived tumor organoids (tumoroids) allow screening of chemotherapeutics and targeted therapies [24,106]. Brain organoids enable studies of Alzheimer's or Parkinson's disease from early developmental stages [95,104,105]. Organ-on-a-chip platforms model organ-specific toxicity, reducing reliance on animal models and improving predictive accuracy [96,108,110,111]. These models accelerate drug discovery and foster precision medicine (Fig. 12.2).

In sum, these technologies represent a paradigm shift in our ability to model human biology in vitro. By closely replicating human organ structure and function, they bridge the gap between lab-based research and clinical applications.

12.10 Assays for tissue engineering

As TE aims to produce functional, biologically relevant tissues and organs, the importance of robust assays is paramount. Evaluating engineered tissues requires methods to assess viability, differentiation, mechanical properties, biochemical composition, metabolic activity, and functionality [98,102,106,113,116]. Modern assays integrate classical histology with molecular diagnostics, imaging, and high-throughput screening to provide a comprehensive understanding of tissue health and maturity, thus dividing all assays into four subsection, such as (Fig. 12.3).

1. Cell level analyses
2. Biochemical and molecular

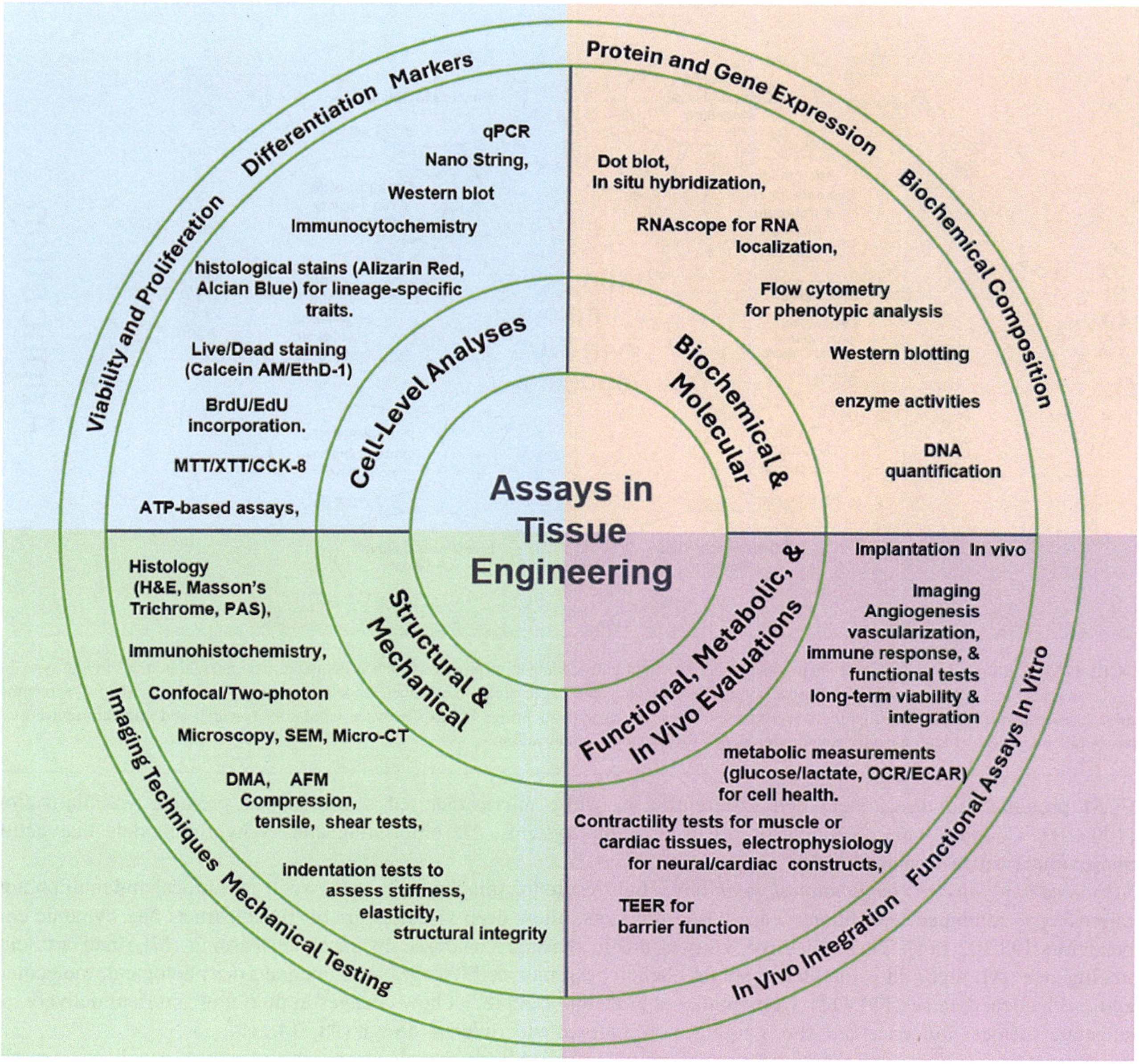

FIGURE 12.3 A comprehensive overview of assays utilized in tissue engineering, categorized into different analytical domains. It encompasses cell-level analyses such as viability, proliferation, differentiation, and protein/gene expression markers; structural and mechanical evaluations, including imaging techniques, stiffness, and elasticity testing; biochemical and molecular assessments like enzyme activity and DNA quantification; and functional/metabolic evaluations, including in vivo assays for vascularization, integration, and functional tests. This integrated framework highlights the diverse methodologies employed to assess the functionality and biocompatibility of engineered tissues, offering insights into their structural, biochemical, and physiological properties.

3. Functional metabolic and in vivo evaluation
4. Structural and mechanical
   - *Cell viability and proliferation assays*: Foundational methods like MTT, XTT, WST-1, and CCK-8 assays measure metabolic activity as a proxy for viability and proliferation [115,116]. Fluorescence-based live/dead assays (e.g., Calcein AM and EthD-1) differentiate viable from nonviable cells within 3D constructs [98,115]. ATP-based assays quantify energy reserves, while BrdU/EdU incorporation tracks DNA synthesis, indicating active proliferation [115,116].
   - *Differentiation markers and lineage commitment*: Quantitative PCR (qPCR), NanoString, and single-cell RNA sequencing measure lineage-specific gene expression to confirm that cells follow the correct developmental program [23,24,98,102]. Western blotting, immunocytochemistry, and immunohistochemistry detect lineage-specific proteins.

Histological stains (Alizarin red, Alcian blue) highlight tissue-specific ECM components, verifying osteogenic or chondrogenic differentiation [95,102].

- *Biochemical and molecular analyses*: Enzyme-linked immunosorbent assays (ELISAs) quantify secreted cytokines, growth factors, or hormones [102,116]. DNA quantification estimates cell density, while flow cytometry characterizes cell populations and heterogeneity [95,115]. Western blotting, dot blot, in situ hybridization, and RNA localization techniques (e.g., RNAscope) provide deeper molecular insights [24,98,102].
- *Structural and mechanical characterization*: Mechanical properties are critical, particularly for load-bearing tissues like bone or cartilage [100,101]. Compression, tensile, and shear tests evaluate stiffness, strength, and elasticity [101,113]. Atomic force microscopy (AFM) measures surface stiffness at the nanoscale, while dynamic mechanical analysis (DMA) reveals viscoelastic properties [9]. Imaging methods, including microcomputed tomography (micro-CT), scanning electron microscopy (SEM), confocal, and two-photon microscopy, visualize internal architecture and ECM organization [98,101,102].
- *Functional, metabolic, and in vivo evaluations*: Functional assays tailored to the tissue's intended role. For cardiac constructs, contractility and electrophysiological measurements confirm cardiomyocyte functionality [111]. For epithelial barriers, transepithelial/endothelial electrical resistance (TEER) tests ensure barrier integrity [98,102]. Metabolic assays monitoring glucose or lactate consumption gauge tissue health. In vivo implantation studies assess vascularization, immune responses, and long-term viability [106,117]. Angiogenesis assays (e.g., choriallantoic membrane models) quantify vascular integration [106,117].

Thus this holistic toolkit ensures that engineered tissues are viable, phenotypically correct, mechanically sound, and functionally competent—a crucial step toward bench to bedside clinical translation [24,98,101,115].

12.11 Integration of high-throughput and automated assays

As advanced in vitro models proliferate, so does the need for efficient, reproducible, and scalable assays leading to mass-produced experiments. High-throughput and automated systems integrated with AI-driven analysis address this challenge, enabling rapid data acquisition, standardized protocols, and dynamic experimental optimization [108,110,112,114,115].

- *Automated liquid handling and robotics*: Robotic liquid handling systems (e.g., Tecan Freedom EVO and Hamilton Microlab STAR) dispense reagents, seed cells, and manage tiny fluid volumes with precision, reducing human error and variability [114,115]. Automated incubators, integrated plate readers, and robotic arms streamline the experimental pipeline, allowing large-scale screens of drugs, biomaterials, or growth conditions [114,115].
- *High-content imaging and analysis*: High-content imaging platforms (e.g., PerkinElmer Opera Phenix, Molecular Devices ImageXpress, and Sartorius Incucyte) capture morphological, viability, and differentiation markers across thousands of samples rapidly [114]. Combining centaur-like interactions between human, ML, and AI algorithms helps to better analyze these images, detecting subtle phenotypic shifts, classifying cell states, and identifying rare events [114,115]. For instance, deep learning can segment organoids, measure their size or complexity, and distinguish drug-resistant from drug-sensitive populations [114].
- *Combining molecular and functional assays with automation:* Gene expression profiling (e.g., NanoString), protein assays, and metabolic measurements can all be automated, providing large-scale molecular and biochemical data [24,106,114]. Integrating these datasets with high-content imaging creates a multidimensional view of tissue state. ML models then correlate morphological changes with molecular signatures, revealing biomarkers that predict tissue performance [114,115].
- *AI-driven predictive modelling and adaptive experimentation*: One of the most transformative applications of automation is closed-loop experimentation. Early results guide predictive models to propose new conditions to test, and robotic systems implement these changes in real-time [110,114,115]. This iterative process converges on optimal parameters—such as ideal nutrient concentrations or drug dosages—faster than traditional trial-and-error approaches.
- *Droplet microfluidics and single-cell analysis*: Droplet microfluidics processes thousands of individuals "microreactors" in parallel, perfect for single-cell genomics, proteomics, or drug response assays [114]. AI-driven image analysis and ML algorithms rapidly classify droplets, linking cellular phenotypes to genomic or transcriptomic profiles [114,115]. Single-cell resolution is invaluable for tissues with diverse cell populations, like brain organoids, where minority cell subtypes can influence overall function [95,104,105,114].
- *Enhancing reproducibility and speed*: Standardized, automated workflows minimize user-to-user variability and enhance reproducibility. Large datasets are processed rapidly, enabling rigorous statistical analysis and uncovering subtle patterns. By improving predictability, these systems guide smarter experimental design and more ethical research, reducing the need for animal testing and accelerating drug development pipelines [96,106,108,110].

The integration of robotics, high-throughput assays, and AI analysis allows researchers to pose more complex questions and tackle larger parameter spaces. Instead of weeks of manual labor, entire experimental campaigns can be executed in days, with adaptive feedback loops homing in on the most promising variables. This synergy transforms the research landscape, bringing TE and in vitro modeling closer to clinical relevance.

12.12 Conclusion

TE has developed as a transformative discipline that combines biology, materials science, and engineering to tackle the significant issues of tissue repair, organ regeneration, and disease modeling. The field has significantly advanced in mimicking native tissue architecture, improving comprehension of cellular function, and offering sustainable solutions to healthcare challenges. This work demonstrates that TE has gone beyond conventional limits by creating unique methodologies, including in vitro models, sophisticated tests, and state-of-the-art scaffold manufacturing procedures.

The significance of TE resides in its capacity to offer patient-specific, regenerative therapies that markedly surpass traditional treatments. TE addresses the constraints of donor organ scarcity, immune rejection, and the difficulties associated with synthetic implants, providing a means to create functional tissues that integrate harmoniously with the human body. Its utilization in regenerative medicine illustrates the transition from symptomatic treatment to curative interventions, facilitating substantial enhancements in patient outcomes and quality of life. Engineered tissues, including skin grafts, cartilage constructs, and hybrid systems that integrate biological and synthetic components, have proven the viability of mending and substituting damaged tissues with biocompatible alternatives.

TE significantly contributes to the advancement of in vitro models and tests, hence transforming the field of biomedical research. From basic 2D cell cultures to advanced 3D organoids and organ-on-chip systems, these models have connected laboratory discoveries with clinical applications. Organoids, originating from patient-specific stem cells, emulate the structural and functional intricacies of organs, providing exceptional prospects for disease modeling, medication development, and personalized therapy. Microfluidic devices replicate dynamic physiological settings, facilitating high-throughput research and real-time study of biological responses. These advancements diminish dependence on animal models while offering more precise and ethically responsible alternatives for investigating human diseases.

The progress in scaffold construction methods underscores the revolutionary potential of TE. Scaffolds serve as essential structures in TE, offering the requisite framework for cell adhesion, proliferation, and differentiation. Methods that include electrospinning, freeze-drying, 3D printing, and hydrogel engineering have facilitated the development of biomimetic scaffolds that closely replicate the ECM. These scaffolds are engineered with specific mechanical, chemical, and biological qualities to direct cellular behavior and promote tissue growth. Electrospun nanofibers emulate the fibrous characteristics of native ECM, facilitating brain regeneration, while 3D-printed structures enable the customization of bone and organ scaffolds to meet individual patient requirements.

Notwithstanding these significant achievements, TE persists in encountering obstacles that require continual research and innovation. The scalability and reproducibility of sophisticated models continue to be significant challenges, especially in converting laboratory discoveries into clinical applications. The expense and intricacy of developing organ-on-chip systems and 3D bioprinted tissues restrict their extensive utilization, underscoring the necessity for more accessible and economical alternatives. The creation of scaffolds exhibiting ideal degradation rates, mechanical characteristics, and functional integration with host tissues continues to be a focus of ongoing research. Anticipating the future, the incorporation of emerging technologies like artificial intelligence (AI), robotics, and high-throughput systems presents significant potential for the advancement of TE. AI-driven predictive modeling, along with automated experimental procedures, can enhance parameters for scaffold manufacturing, cell culture, and medicinal applications. These technologies facilitate the swift analysis of extensive information, revealing intricate patterns and informing more intelligent experimental designs. The integration of droplet microfluidics with single-cell analysis improves the resolution and accuracy of TE methodologies, facilitating the examination of diverse cell populations and intricate tissue connections. As TE advances, its applications are set to extend beyond regenerative medicine into fields such as organ transplantation, cancer research, and biotechnology. The capacity to manufacture functional tissues and organs introduces novel possibilities in fulfilling unmet medical requirements, ranging from the production of bioengineered kidneys and livers for transplantation to the creation of tumoroids for customized cancer treatment. Moreover, the ethical ramifications of TE, especially its capacity to diminish animal testing and enhance the translational relevance of preclinical studies, highlight its significance in progressing healthcare while upholding ethical standards. TE is at the vanguard of contemporary biomedical research, providing innovative answers to critical healthcare concerns. Innovative in vitro models, enhanced scaffold fabrication techniques, and interdisciplinary approaches have positioned TE to establish individualized, regenerative, and ethically sound therapies as the standard of care. Despite ongoing hurdles, the persistent integration of technology, innovation, and collaboration is poised to realize the full potential of TE, heralding a new epoch in medical science that redefines the limits of possibility.

AI disclosure

During the preparation of this work, the author(s) used ChatGpt to improve the readability and language of the work. After using this tool/service, the author(s) reviewed and edited the content as needed and take(s) full responsibility for the content of the publication.

References

[1] B.L. Seal, T.C. Otero, A. Panitch, Polymeric biomaterials for tissue and organ regeneration, Materials Science and Engineering: R: Reports 34 (4-5) (2001) 147–230. Available from https://doi.org/10.1016/S0927-796X(01)00035-3.

[2] S. Agarwal, S. Saha, V.K. Balla, A. Pal, A. Barui, S. Bodhak, Current developments in 3D bioprinting for tissue and organ regeneration–a review, Frontiers in Mechanical Engineering 6 (2020). Available from https://doi.org/10.3389/fmech.2020.589171.

[3] A. Samadikuchaksaraei, Scientific and industrial status of tissue engineering, African Journal of Biotechnology 6 (25) (2007) 2897–2909. Available from https://doi.org/10.5897/ajb2007.000-2459, http://www.academicjournals.org/AJB/PDF/pdf2007/28Dec/Samadikuchaksaraei.pdf.

[4] C. Laurent, X. Liu, N. De Isla, X. Wang, R. Rahouadj, Defining a scaffold for ligament tissue engineering: what has been done, and what still needs to be done, Journal of Cellular Immunotherapy 4 (1) (2018) 4–9. Available from https://doi.org/10.1016/j.jocit.2018.09.002.

[5] F. Akter, What is Tissue Engineering? Tissue Engineering Made Easy, Elsevier Inc., United Kingdom, (2016), pp. 1–2. Available from http://www.sciencedirect.com/science/book/9780128053614, 10.1016/B978-0-12-805361-4.00001-1.

[6] M. Spector, An interview with i v Yannas. Tracing one of the deepest roots of biomaterials in tissue engineering/regenerative medicine, Biomedical Materials (Bristol) 8 (4) (2013). Available from https://doi.org/10.1088/1748-6041/8/4/040401, http://iopscience.iop.org/1748-605X/8/4/040401/pdf/1748-605X_8_4_040401.pdf.

[7] D.F. Williams, To engineer is to create: the link between engineering and regeneration, Trends in Biotechnology 24 (1) (2006) 4–8. Available from https://doi.org/10.1016/j.tibtech.2005.10.006, www.elsevier.com/locate/tibtech.

[8] T.M. Freyman, I.V. Yannas, L.J. Gibson, Cellular materials as porous scaffolds for tissue engineering, Progress in Materials Science 46 (3-4) (2001) 273–282. Available from https://doi.org/10.1016/s0079-6425(00)00018-9.

[9] D.E. Ingber, V.C. Mow, D. Butler, L. Niklason, J. Huard, J. Mao, et al., Tissue engineering and developmental biology: going biomimetic, Tissue Engineering 12 (12) (2006) 3265–3283. Available from https://doi.org/10.1089/ten.2006.12.3265.

[10] K.K. Shyang See, Liver regeneration: the myth of Prometheus in modern medicine a case report, Journal of Regenerative Biology and Medicine (2023). Available from https://doi.org/10.37191/mapsci-2582-385x-5(5)-140.

[11] E. Heber-Katz, J.M. Leferovich, K. Bedelbaeva, D. Gourevitch, Spallanzani's mouse: a model of restoration and regeneration, Current topics in Microbiology and Immunology 280 (2004) 165–189. Available from https://doi.org/10.1007/978-3-642-18846-6_5.

[12] H.M. Lenhoff, S.G. Lenhoff, Tissue grafting in animals: its discovery in 1742 by Abraham Trembley as he experimented with Hydra, The Biological Bulletin 166 (1) (1984) 1–10. Available from https://doi.org/10.2307/1541425.

[13] L. Armstrong, M. Lako, N. Buckley, T.R.J. Lappin, M.J. Murphy, J.A. Nolta, et al., Editorial: our top 10 developments in stem cell biology over the last 30 years, Stem Cells 30 (1) (2012) 2–9. Available from https://doi.org/10.1002/stem.1007, http://onlinelibrary.wiley.com/doi/10.1002/stem.1007/pdf.

[14] J.L. Turk, Inflammation: John Hunter's "a treatise on the blood, inflammation and gun-shot wounds, International Journal of Experimental Pathology 75 (1994).

[15] R. Williams, Sir John Gurdon: godfather of cloning, Journal of Cell Biology 181 (2) (2008) 178–179. Available from https://doi.org/10.1083/jcb.1812piundefined, http://www.jcb.org/cgi/reprint/181/2/178.

[16] H.W. Park, Constructing failure: leonard hayflick, biomedicine, and the problems with tissue culture, Annals of Science 73 (3) (2016) 303–327. Available from https://doi.org/10.1080/00033790.2015.1057764, www.tandf.co.uk/journals/titles/00033790.asp.

[17] J.P. Vacanti, R. Langer, Tissue engineering: the design and fabrication of living replacement devices for surgical reconstruction and transplantation, Lancet 354 (1) (1999) 32–34. Available from https://doi.org/10.1016/s0140-6736(99)90247-7, http://www.journals.elsevier.com/the-lancet/.

[18] K. Takahashi, S. Yamanaka, A developmental framework for induced pluripotency, Development (Cambridge) 142 (19) (2015) 3274–3285. Available from https://doi.org/10.1242/dev.114249, http://dev.biologists.org/content/142/19/3274.full.pdf.

[19] M.H.A. Santana, S.C. Huber, History and Evolution of Regenerative Medicine Nanotechnology and Regenerative Medicine: History, Techniques, Frontiers, and Applications, Elsevier, Brazil, (2022), pp. 23–44. Available from https://www.sciencedirect.com/book/9780323904711, 10.1016/B978-0-323-90471-1.00011-6.

[20] A. Curtis, M. Riehle, Tissue engineering: the biophysical background, Physics in Medicine and Biology 46 (4) (2001) R47. Available from https://doi.org/10.1088/0031-9155/46/4/201.

[21] S. Kim, M. Uroz, J.L. Bays, C.S. Chen, Harnessing mechanobiology for tissue engineering, Developmental Cell 56 (2) (2021) 180–191. Available from https://doi.org/10.1016/j.devcel.2020.12.017, https://www.cell.com/developmental-cell/home.

[22] A. Khademhosseini, R. Langer, Microengineered hydrogels for tissue engineering, Biomaterials 28 (34) (2007) 5087–5092. Available from https://doi.org/10.1016/j.biomaterials.2007.07.021.

[23] M.A. Lancaster, J.A. Knoblich, Organogenesis in a dish: modeling development and disease using organoid technologies, Science (New York, N.Y.) 345 (6194) (2014). Available from https://doi.org/10.1126/science.1247125.

[24] H. Clevers, Modeling development and disease with organoids, Cell 165 (7) (2016) 1586–1597. Available from https://doi.org/10.1016/j.cell.2016.05.082.

[25] I. Kmetič, M. Roller, M. Miletić, T. Murati, Scientific research in the 21st century: ethical aspects for the use of laboratory animals and alternative models. Croatian Veterinary Institute, Croatia, Veterinarska Stanica 53 (5) (2022). Available from https://doi.org/10.46419/VS.53.5.4, https://hrcak.srce.hr/264509.

[26] W. Agnes, L.L. Sonia, Intestinal organoids of farm animals: new in vitro models for a better understanding of intestinal functions, Inra Productions Animales 36 (2) (2023). Available from https://doi.org/10.20870/productions-animales.2023.36.2.7562, https://productions-animales.org/article/view/7562.

[27] A. Hinterberger, S. Bea, How do scientists model humanness? A qualitative study of human organoids in biomedical research, Social Science & Medicine 320 (2023) 115676. Available from https://doi.org/10.1016/j.socscimed.2023.115676.

[28] K. Bowers, L. Amelse, A. Bow, S. Newby, A. MacDonald, X. Sun, et al., Mesenchymal stem cell use in acute tendon injury: in vitro tenogenic potential vs. in vivo dose response, Bioengineering 9 (8) (2022) 407. Available from https://doi.org/10.3390/bioengineering9080407.

[29] P.I. Petkov, H. Ivanova, M. Honma, T. Yamada, T. Morita, A. Furuhama, et al., Differences between in vitro and in vivo genotoxicity due to metabolism: the role of kinetics, Computational Toxicology 22 (2022) 100222. Available from https://doi.org/10.1016/j.comtox.2022.100222.

[30] B.P. Enright, P. Lonergan, A. Dinnyes, T. Fair, F.A. Ward, X. Yang, et al., Culture of in vitro produced bovine zygotes in vitro vs in vivo: implications for early embryo development and quality, Theriogenology 54 (5) (2000) 659–673. Available from https://doi.org/10.1016/s0093-691x(00)00381-2.

[31] S. Pereira, M. Tettamanti, Testing times in toxicology – in vitro vs in vivo testing, ALTEX Proceedings 2 (2013).

[32] V.G. Kokich, In-vitro vs in-vivo materials research, American Journal of Orthodontics and Dentofacial Orthopedics 143 (4) (2013) S11. Available from https://doi.org/10.1016/j.ajodo.2013.02.002, http://www.sciencedirect.com/science/journal/08895406.

[33] R. Cai, N. Gimenez-Camino, M. Xiao, S. Bi, K.A. Divito, Technological advances in three-dimensional skin tissue engineering, Reviews on Advanced Materials Science 62 (1) (2023). Available from https://doi.org/10.1515/rams-2022-0289, https://www.degruyter.com/view/j/rams.

[34] O. Habanjar, M. Diab-Assaf, F. Caldefie-Chezet, L. Delort, 3D cell culture systems: tumor application, advantages, and disadvantages, International Journal of Molecular Sciences 22 (22) (2021) 12200. Available from https://doi.org/10.3390/ijms222212200.

[35] X. Ma, X. Qu, W. Zhu, Y.S. Li, S. Yuan, H. Zhang, et al., Deterministically patterned biomimetic human iPSC-derived hepatic model via rapid 3D bioprinting, Proceedings of the National Academy of Sciences of the United States of America 113 (8) (2016) 2206–2211. Available from https://doi.org/10.1073/pnas.1524510113, http://www.pnas.org/content/113/8/2206.full.pdf.

[36] J. Barrila, A.L. Radtke, A. Crabbé, S.F. Sarker, M.M. Herbst-Kralovetz, C.M. Ott, et al., Organotypic 3D cell culture models: using the rotating wall vessel to study hostĝ€"pathogen interactions, Nature Reviews. Microbiology 8 (11) (2010) 791–801. Available from https://doi.org/10.1038/nrmicro2423.

[37] D.S.A. Al Maruf, Y.A. Ghosh, H. Xin, K. Cheng, P. Mukherjee, J.M. Crook, et al., Hydrogel: a potential material for bone tissue engineering repairing the segmental mandibular defect, Polymers 14 (19) (2022). Available from https://doi.org/10.3390/polym14194186, http://www.mdpi.com/journal/polymers.

[38] A.G. Monteduro, S. Rizzato, G. Caragnano, A. Trapani, G. Giannelli, G. Maruccio, Organs-on-chips technologies – a guide from disease models to opportunities for drug development, Biosensors and Bioelectronics 231 (2023). Available from https://doi.org/10.1016/j.bios.2023.115271, www.elsevier.com/locate/bios.

[39] Jy Shoji, R.P. Davis, C.L. Mummery, S. Krauss, Global literature analysis of organoid and organ-on-chip research, Advanced Healthcare Materials 13 (21) (2024). Available from https://doi.org/10.1002/adhm.202301067, http://onlinelibrary.wiley.com/journal/10.1002/(ISSN)2192-2659.

[40] N. Sass, Humane endpoints and acute toxicity testing, ILAR Journal 41 (2) (2000) 114–123. Available from https://doi.org/10.1093/ilar.41.2.114.

[41] M. Brinkmann, T.G. Preuss, H. Hollert, Advancing in vitro–in vivo extrapolations of mechanism-specific toxicity data through toxicokinetic modeling, Advances in Biochemical Engineering/Biotechnology 157 (2017) 293–317. Available from https://doi.org/10.1007/10_2015_5015, http://www.springer.com/series/10.

[42] S. Wahyuwardani, S.M. Noor, B. Bakrie, Animal welfare ethics in research and testing: implementation and its barrier, Indonesian Bulletin of Animal and Veterinary Sciences 30 (4) (2020) 211. Available from https://doi.org/10.14334/wartazoa.v30i4.2529.

[43] C.H. Hsieh, Y.D. Chen, S.F. Huang, H.M. Wang, M.H. Wu, The effect of primary cancer cell culture models on the results of drug chemosensitivity assays: the application of perfusion microbioreactor system as cell culture vessel, BioMed Research International 2015 (2015). Available from https://doi.org/10.1155/2015/470283, http://www.hindawi.com/journals/biomed/.

[44] J. Paul, Methods in enzymology, vol. LVIII, cell culture, Trends in Biochemical Sciences 5 (4) (1980) XX. Available from https://doi.org/10.1016/0968-0004(80)90267-4.

[45] M. Kapałczyńska, T. Kolenda, W. Przybyła, M. Zajączkowska, A. Teresiak, V. Filas, et al., 2D and 3D cell cultures – a comparison of different types of cancer cell cultures, Archives of Medical Science 14 (4) (2018) 910–919. Available from https://doi.org/10.5114/aoms.2016.63743, https://www.termedia.pl/2D-and-3D-cell-cultures-a-comparison-of-different-types-of-cancer-cell-cultures,19,28752,1,1.html.

[46] C. Li, M. Kato, L. Shiue, J.E. Shively, M. Ares, R.J. Lin, Cell type and culture condition-dependent alternative splicing in human breast cancer cells revealed by splicing-sensitive microarrays, Cancer Research 66 (4) (2006) 1990–1999. Available from https://doi.org/10.1158/0008-5472.CAN-05-2593.

[47] F. Pampaloni, E.G. Reynaud, E.H.K. Stelzer, The third dimension bridges the gap between cell culture and live tissue, Nature Reviews. Molecular Cell Biology 8 (10) (2007) 839–845. Available from https://doi.org/10.1038/nrm2236.

[48] B.B. Aggarwal, D. Danda, S. Gupta, P. Gehlot, Models for prevention and treatment of cancer: problems vs promises, Biochemical Pharmacology 78 (9) (2009) 1083–1094. Available from https://doi.org/10.1016/j.bcp.2009.05.027, www.elsevier.com/locate/biochempharm.

[49] C. Jensen, Y. Teng, Is it time to start transitioning from 2D to 3D cell culture? Frontiers in Molecular Biosciences 7 (2020). Available from https://doi.org/10.3389/fmolb.2020.00033.

[50] L.P. Ferreira, V.M. Gaspar, J.F. Mano, Design of spherically structured 3D in vitro tumor models -advances and prospects, Acta Biomaterialia 75 (2018) 11–34. Available from https://doi.org/10.1016/j.actbio.2018.05.034.

[51] A. Eltom, G. Zhong, A. Muhammad, Scaffold techniques and designs in tissue engineering functions and purposes: a review, Advances in Materials Science and Engineering 2019 (2019) 1–13. Available from https://doi.org/10.1155/2019/3429527.

[52] Z. Cesarz, K. Tamama, R.-K. Li, Spheroid culture of mesenchymal stem cells, Stem Cells International 2016 (1) (2016). Available from https://doi.org/10.1155/2016/9176357.

[53] T.H. Booij, L.S. Price, E.H.J. Danen, 3D cell-based assays for drug screens: challenges in imaging, image analysis, and high-content analysis, SLAS Discovery 24 (6) (2019) 615–627. Available from https://doi.org/10.1177/2472555219830087, http://journals.sagepub.com/loi/jbx.

[54] A.G. Souza, I.C. C Ferreira, Advances in cell culture: more than a century after cultivating cells, Journal of Biotechnology & Biomaterials 6 (2) (2016). Available from https://doi.org/10.4172/2155-952X.1000221.

[55] E. Richiardone, V. Van den Bossche, C. Corbet, Metabolic studies in organoids: current applications, opportunities and challenges, Organoids 1 (1) (2022) 85–105. Available from https://doi.org/10.3390/organoids1010008.

[56] D. Anton, H. Burckel, E. Josset, G. Noel, Three-dimensional cell culture: a breakthrough in vivo, International Journal of Molecular Sciences 16 (3) (2015) 5517–5527. Available from https://doi.org/10.3390/ijms16035517, http://www.mdpi.com/1422-0067/16/3/5517/pdf.

[57] D. Sztankovics, D. Moldvai, G. Petővári, R. Gelencsér, I. Krencz, R. Raffay, et al., 3D bioprinting and the revolution in experimental cancer model systems—a review of developing new models and experiences with in vitro 3D bioprinted breast cancer tissue-mimetic structures, Pathology and Oncology Research 29 (2023). Available from https://doi.org/10.3389/pore.2023.1610996.

[58] S. Hockney, J. Parker, J.E. Turner, X. Todd, S. Todryk, R.G. Gieling, et al., Next generation organoid engineering to replace animals in cancer drug testing, Biochemical Pharmacology 213 (2023). Available from https://doi.org/10.1016/j.bcp.2023.115586, www.elsevier.com/locate/biochempharm.

[59] G.T. de Dios-Figueroa, J.D.R. Aguilera-Marquez, T.A. Camacho-Villegas, P.H. Lugo-Fabres, 3d cell culture models in covid-19 times: a review of 3d technologies to understand and accelerate therapeutic drug discovery, Biomedicines 9 (6) (2021). Available from https://doi.org/10.3390/biomedicines9060602, https://www.mdpi.com/2227-9059/9/6/602/pdf.

[60] N. Shahrubudin, T.C. Lee, R. Ramlan, An overview on 3D printing technology: technological, materials, and applications, Procedia Manufacturing 35 (2019) 1286–1296. Available from https://doi.org/10.1016/j.promfg.2019.06.089.

[61] M. Chimerad, A. Barazesh, M. Zandi, I. Zarkesh, A. Moghaddam, P. Borjian, et al., Tissue engineered scaffold fabrication methods for medical applications, International Journal of Polymeric Materials and Polymeric Biomaterials 72 (18) (2023) 1455–1479. Available from https://doi.org/10.1080/00914037.2022.2101112.

[62] J.W. Jang, K.E. Min, C. Kim, J. Shin, J. Lee, S. Yi, Review: scaffold characteristics, fabrication methods, and biomaterials for the bone tissue engineering, International Journal of Precision Engineering and Manufacturing 24 (3) (2023) 511–529. Available from https://doi.org/10.1007/s12541-022-00755-7, https://www.springer.com/journal/12541.

[63] O. Vesvoranan, A. Anup, K.R. Hixon, Current concepts and methods in tissue interface scaffold fabrication, Biomimetics 7 (4) (2022). Available from https://doi.org/10.3390/biomimetics7040151, www.mdpi.com/journal/biomimetics.

[64] Application of biosystems engineering to tissue engineering - electrospining technology to fabricate scaffolds for bioartificial organs, Journal of Biosystems Engineering 29 (2004). Available from https://doi.org/10.5307/jbe.2004.29.4.367.

[65] Y.P. Afsharian, M. Rahimnejad, Bioactive electrospun scaffolds for wound healing applications: a comprehensive review, Polymer Testing 93 (2021). Available from https://doi.org/10.1016/j.polymertesting.2020.106952, https://www.journals.elsevier.com/polymer-testing.

[66] Z. Fereshteh, Freeze-Drying Technologies for 3D Scaffold Engineering Functional 3D Tissue Engineering Scaffolds: Materials, Technologies, and Applications, Elsevier, United States, (2018), pp. 151–174. Available from http://www.sciencedirect.com/science/book/9780081009796, 10.1016/B978-0-08-100979-6.00007-0.

[67] A. Mirtaghavi, J. Luo, R. Muthuraj, Recent advances in porous 3D cellulose aerogels for tissue engineering applications: a review, Journal of Composites Science 4 (4) (2020) 152. Available from https://doi.org/10.3390/jcs4040152.

[68] I. Allijn, N. du Preez, M. Tasior, R. Bansal, D. Stamatialis, One-step fabrication of porous membrane-based scaffolds by air-water interfacial phase separation: opportunities for engineered tissues, Membranes 12 (5) (2022) 453. Available from https://doi.org/10.3390/membranes12050453.

[69] A. Prasad, M.R. Sankar, V. Katiyar, State of art on solvent casting particulate leaching method for orthopedic scaffolds fabrication, Materials Today: Proceedings 4 (2) (2017) 898–907. Available from https://doi.org/10.1016/j.matpr.2017.01.101, https://www.sciencedirect.com/journal/materials-today-proceedings.

[70] L. Suamte, A. Tirkey, J. Barman, P. Jayasekhar Babu, Various manufacturing methods and ideal properties of scaffolds for tissue engineering applications, Smart Materials in Manufacturing 1 (2023) 100011. Available from https://doi.org/10.1016/j.smmf.2022.100011.

[71] M.P. Mani, M. Sadia, S.K. Jaganathan, A.Z. Khudzari, E. Supriyanto, S. Saidin, et al., A review on 3D printing in tissue engineering applications, Journal of Polymer Engineering 42 (3) (2022) 243–265. Available from https://doi.org/10.1515/polyeng-2021-0059, http://www.reference-global.com/loi/polyeng.

[72] A. Zaszczyńska, M. Moczulska-Heljak, A. Gradys, P. Sajkiewicz, Advances in 3D printing for tissue engineering, Materials 14 (12) (2021) 3149. Available from https://doi.org/10.3390/ma14123149.

[73] P. Song, C. Zhou, H. Fan, B. Zhang, X. Pei, Y. Fan, et al., Novel 3D porous biocomposite scaffolds fabricated by fused deposition modeling and gas foaming combined technology, Composites Part B: Engineering 152 (2018) 151–159. Available from https://doi.org/10.1016/j.compositesb.2018.06.029.

[74] F. Xu, C. Dawson, M. Lamb, E. Mueller, E. Stefanek, M. Akbari, et al., Hydrogels for tissue engineering: addressing key design needs toward clinical translation, Frontiers in Bioengineering and Biotechnology 10 (2022). Available from https://doi.org/10.3389/fbioe.2022.849831.

[75] A. Babbar, V. Jain, D. Gupta, S. Singh, C. Prakash, C. Pruncu, Biomaterials and Fabrication Methods of Scaffolds for Tissue Engineering Applications Materials Horizons: From Nature to Nanomaterials, Springer Nature, India, (2020), pp. 167–186. Available from www.springer.com/series/16122, 10.1007/978-981-15-5424-7_8.

[76] V. Liu Tsang, S.N. Bhatia, Three-dimensional tissue fabrication, Advanced Drug Delivery Reviews 56 (11) (2004) 1635–1647. Available from https://doi.org/10.1016/j.addr.2004.05.001.

[77] A. Wubneh, E.K. Tsekoura, C. Ayranci, H. Uludağ, Current state of fabrication technologies and materials for bone tissue engineering, Acta Biomaterialia 80 (2018) 1–30. Available from https://doi.org/10.1016/j.actbio.2018.09.031, http://www.journals.elsevier.com/acta-biomaterialia.

[78] P. Zhao, H. Gu, H. Mi, C. Rao, J. Fu, Ls Turng, Fabrication of scaffolds in tissue engineering: a review, Frontiers of Mechanical Engineering 13 (1) (2018) 107–119. Available from https://doi.org/10.1007/s11465-018-0496-8, http://www.springerlink.com/content/2095-0233/.

[79] A. Gandaglia, A. Bagno, F. Naso, M. Spina, G. Gerosa, Cells, scaffolds and bioreactors for tissue-engineered heart valves: a journey from basic concepts to contemporary developmental innovations, European Journal of Cardio-thoracic Surgery 39 (4) (2011) 523–531. Available from https://doi.org/10.1016/j.ejcts.2010.07.030.

[80] X. Lu, Z. Ding, F. Xu, Q. Lu, D.L. Kaplan, Subtle regulation of scaffold stiffness for the optimized control of cell behavior, ACS Applied Bio Materials 2 (7) (2019) 3108–3119. Available from https://doi.org/10.1021/acsabm.9b00445.

[81] Z. Miri, H.J. Haugen, D. Loca, F. Rossi, G. Perale, A. Moghanian, et al., Review on the strategies to improve the mechanical strength of highly porous bone bioceramic scaffolds, Journal of the European Ceramic Society 44 (1) (2024) 23–42. Available from https://doi.org/10.1016/j.jeurceramsoc.2023.09.003, http://www.journals.elsevier.com/journal-of-the-european-ceramic-society/.

[82] Z. Abpeikar, P.B. Milan, L. Moradi, M. Anjomshoa, S. Asadpour, Influence of pore sizes in 3D-scaffolds on mechanical properties of scaffolds and survival, distribution, and proliferation of human chondrocytes, Mechanics of Advanced Materials and Structures 29 (26) (2022) 4911–4922. Available from https://doi.org/10.1080/15376494.2021.1943077, http://www.tandf.co.uk/journals/titles/15376494.html.

[83] A. Prasopthum, K.M. Shakesheff, J. Yang, Direct three-dimensional printing of polymeric scaffolds with nanofibrous topography, Biofabrication 10 (2) (2018). Available from https://doi.org/10.1088/1758-5090/aaa15b, http://iopscience.iop.org/article/10.1088/1758-5090/aaa15b/pdf.

[84] A. Lopez Marquez, I.E. Gareis, F.J. Dias, C. Gerhard, M.F. Lezcano, How fiber surface topography affects interactions between cells and electrospun scaffolds: a systematic review, Polymers 14 (1) (2022). Available from https://doi.org/10.3390/polym14010209, https://www.mdpi.com/2073-4360/14/1/209/pdf.

[85] N. Vanawati, A. Barlian, H. Judawisastra, I. Wibowo, The combinatory effect of scaffold topography and culture condition: an approach to nucleus pulposus tissue engineering, Future Science OA 8 (7) (2022) 243–256. Available from https://doi.org/10.2144/fsoa-2021-0157, http://www.future-science.com/loi/fso?open=2015.

[86] M. Majood, A. Shakeel, A. Agarwal, S. Jeevanandham, R. Bhattacharya, D. Kochhar, et al., Hydrogel nanosheets confined 2D rhombic ice: a new platform enhancing chondrogenesis, Biomedical Materials 17 (6) (2022) 065004. Available from https://doi.org/10.1088/1748-605x/ac8e43.

[87] M. Majood, O. Agrawal, P. Garg, A. Selvam, S.K. Yadav, S. Singh, et al., Carbon quantum dot-nanocomposite hydrogel as denovo nexus in rapid chondrogenesis, Biomaterials Advances 157 (2024) 213730. Available from https://doi.org/10.1016/j.bioadv.2023.213730.

[88] C. Liu, Z. Xia, J.T. Czernuszka, Design and development of three-dimensional scaffolds for tissue engineering, Chemical Engineering Research and Design 85 (7) (2007) 1051–1064. Available from https://doi.org/10.1205/cherd06196.

[89] A. Pryadko, M.A. Surmeneva, R.A. Surmenev, Review of hybrid materials based on polyhydroxyalkanoates for tissue engineering applications, Polymers 13 (11) (2021). Available from https://doi.org/10.3390/polym13111738, https://www.mdpi.com/2073-4360/13/11/1738/pdf.

[90] A. Randhawa, S.D. Dutta, K. Ganguly, T. V. Patil, R. Luthfikasari, K.T. Lim, Understanding cell-extracellular matrix interactions for topology-guided tissue regeneration, Biocell: Official Journal of the Sociedades Latinoamericanas de Microscopia Electronica. et. al 47 (4) (2023) 789–808. Available from https://doi.org/10.32604/biocell.2023.026217, https://www.techscience.com/biocell/v47n4/51787.

[91] S.H. Shin, O. Purevdorj, O. Castano, J.A. Planell, H.W. Kim, A short review: recent advances in electrospinning for bone tissue regeneration, Journal of Tissue Engineering 3 (1) (2012) 1–11. Available from https://doi.org/10.1177/2041731412443530, http://tej.sagepub.com/.

[92] V.A. da Silva, B.C. Bobotis, F.F. Correia, T.H. Lima-Vasconcellos, G.M.D. Chiarantin, L. De La Vega, et al., The impact of biomaterial surface properties on engineering neural tissue for spinal cord regeneration, International Journal of Molecular Sciences 24 (17) (2023). Available from https://doi.org/10.3390/ijms241713642, http://www.mdpi.com/journal/ijms.

[93] N. Khuu, S. Kheiri, E. Kumacheva, Structurally anisotropic hydrogels for tissue engineering, Trends in Chemistry 3 (12) (2021) 1002–1026. Available from https://doi.org/10.1016/j.trechm.2021.09.009.

[94] N. Amiryaghoubi, M. Fathi, A. Barzegari, J. Barar, H. Omidian, Y. Omidi, Recent advances in polymeric scaffolds containing carbon nanotube and graphene oxide for cartilage and bone regeneration, Materials Today Communications 26 (2021) 102097. Available from https://doi.org/10.1016/j.mtcomm.2021.102097.

[95] M.A. Lancaster, J.A. Knoblich, Generation of cerebral organoids from human pluripotent stem cells, Nature Protocols 9 (10) (2014) 2329–2340. Available from https://doi.org/10.1038/nprot.2014.158, http://www.natureprotocols.com/.

[96] D. Huh, B.D. Matthews, A. Mammoto, M. Montoya-Zavala, H.Y. Hsin, D.E. Ingber, Reconstituting organ-level lung functions on a chip, Science (New York, N.Y.) 328 (5986) (2010) 1662–1668. Available from https://doi.org/10.1126/science.1188302.

[97] A. Bein, W. Shin, S. Jalili-Firoozinezhad, M.H. Park, A. Sontheimer-Phelps, A. Tovaglieri, et al., Microfluidic organ-on-a-chip models of human intestine, Cellular and Molecular Gastroenterology and Hepatology 5 (4) (2018) 659–668. Available from https://doi.org/10.1016/j.jcmgh.2017.12.010.

[98] A. Fatehullah, S.H. Tan, N. Barker, Organoids as an in vitro model of human development and disease, Nature Cell Biology 18 (3) (2016) 246–254. Available from https://doi.org/10.1038/ncb3312, http://www.nature.com/ncb/index.html.

[99] T. Takebe, J.M. Wells, Organoids by design, Science (New York, N.Y.) 364 (6444) (2019) 956–959. Available from https://doi.org/10.1126/science.aaw7567, https://science.sciencemag.org/content/364/6444/956/tab-pdf.

[100] S.V. Murphy, A. Atala, 3D bioprinting of tissues and organs, Nature Biotechnology 32 (8) (2014) 773–785. Available from https://doi.org/10.1038/nbt.2958, http://www.nature.com/nbt/index.html.

[101] A.S. Mao, D.J. Mooney, Regenerative medicine: current therapies and future directions, Proceedings of the National Academy of Sciences of the United States of America 112 (47) (2015) 14452–14459. Available from https://doi.org/10.1073/pnas.1508520112, http://www.pnas.org/content/112/47/14452.full.pdf.

[102] E.R. Shamir, A.J. Ewald, Three-dimensional organotypic culture: experimental models of mammalian biology and disease, Nature Reviews. Molecular Cell Biology 15 (10) (2014) 647–664. Available from https://doi.org/10.1038/nrm3873, http://www.nature.com/molcellbio.

[103] B.S. Freedman, C.R. Brooks, A.Q. Lam, H. Fu, R. Morizane, V. Agrawal, et al., Modelling kidney disease with CRISPR-mutant kidney organoids derived from human pluripotent epiblast spheroids, Nature Communications 6 (1) (2015). Available from https://doi.org/10.1038/ncomms9715.

[104] I. Kelava, M.A. Lancaster, Stem cell models of human brain development, Cell Stem Cell 18 (6) (2016) 736–748. Available from https://doi.org/10.1016/j.stem.2016.05.022, http://www.journals.elsevier.com/cell-stem-cell/.

[105] M.A. Lancaster, N.S. Corsini, S. Wolfinger, E.H. Gustafson, A.W. Phillips, T.R. Burkard, et al., Guided self-organization and cortical plate formation in human brain organoids, Nature Biotechnology 35 (7) (2017) 659–666. Available from https://doi.org/10.1038/nbt.3906, http://www.nature.com/nbt/index.html.

[106] T. Sato, R.G. Vries, H.J. Snippert, M. Van De Wetering, N. Barker, D.E. Stange, et al., Single Lgr5 stem cells build crypt-villus structures in vitro without a mesenchymal niche, Nature 459 (7244) (2009) 262–265. Available from https://doi.org/10.1038/nature07935.

[107] X. Yin, H.F. Farin, J.H. Van Es, H. Clevers, R. Langer, J.M. Karp, Niche-independent high-purity cultures of Lgr5 + intestinal stem cells and their progeny, Nature Methods 11 (1) (2014) 106–112. Available from https://doi.org/10.1038/nmeth.2737.

[108] S.N. Bhatia, D.E. Ingber, Microfluidic organs-on-chips, Nature Biotechnology 32 (8) (2014) 760–772. Available from https://doi.org/10.1038/nbt.2989, http://www.nature.com/nbt/index.html.

[109] S. Bersini, J.S. Jeon, G. Dubini, C. Arrigoni, S. Chung, J.L. Charest, et al., A microfluidic 3D invitro model for specificity of breast cancer metastasis to bone, Biomaterials 35 (8) (2014) 2454–2461. Available from https://doi.org/10.1016/j.biomaterials.2013.11.050.

[110] P.J. Lee, P.J. Hung, L.P. Lee, An artificial liver sinusoid with a microfluidic endothelial-like barrier for primary hepatocyte culture, Biotechnology and Bioengineering 97 (5) (2007) 1340–1346. Available from https://doi.org/10.1002/bit.21360.

[111] K. Ronaldson-Bouchard, G. Vunjak-novakovic, organs-on-a-, chip: a fast track for engineered human tissues in drug development, Cell Stem Cell 22 (3) (2018) 310–324. Available from https://doi.org/10.1016/j.stem.2018.02.011.

[112] H.N. Joensson, H. Andersson Svahn, Droplet microfluidics—a tool for single-cell analysis, Angewandte Chemie International Edition 51 (49) (2012) 12176–12192. Available from https://doi.org/10.1002/anie.201200460.

[113] M. Ravi, V. Paramesh, S.R. Kaviya, E. Anuradha, F.D.P. Solomon, 3D cell culture systems: advantages and applications, Journal of Cellular Physiology 230 (1) (2015) 16–26. Available from https://doi.org/10.1002/jcp.24683, http://onlinelibrary.wiley.com/journal/10.1002/(ISSN)1097-4652.

[114] P.H. Wu, A. Giri, D. Wirtz, Statistical analysis of cell migration in 3D using the anisotropic persistent random walk model, Nature Protocols 10 (3) (2015) 517–527. Available from https://doi.org/10.1038/nprot.2015.030, http://www.natureprotocols.com/.

[115] P. Held, An Introduction to Reactive Oxygen Species Measurement of ROS in Cells, BioTek Instruments, (2012).

[116] M. Baker, Reproducibility crisis: blame it on the antibodies, Nature 521 (7552) (2015) 274–276. Available from https://doi.org/10.1038/521274a.

[117] B. Palikuqi, D.H.T. Nguyen, G. Li, R. Schreiner, A.F. Pellegata, Y. Liu, et al., Adaptable haemodynamic endothelial cells for organogenesis and tumorigenesis, Nature 585 (7825) (2020) 426–432. Available from https://doi.org/10.1038/s41586-020-2712-z, http://www.nature.com/nature/index.html.

Chapter 13

Precision in vitro tools for tissue engineering applications

Mukesh Saran[1], Shefali Pareek[2], and Ashima Bagaria[2]

[1]*Directorate of Research, Manipal University Jaipur, Jaipur, Rajasthan, India,* [2]*Department of Physics, Manipal University Jaipur, Jaipur, Rajasthan, India*

13.1 Introduction

Tissue engineering stands at the forefront of regenerative medicine, aiming to restore, maintain, and enhance tissue functions by combining scaffolds, cells, and bioactive molecules. Here comes the in vitro models that provide foundational insights; the complexity of living organisms necessitates the use of in vivo models to fully understand the interaction and efficacy of engineered tissues.

In vitro assays are a requisite area in the field of tissue engineering that serves as a structural and foundational tool for assessing the biocompatibility, functionality, and reliability of engineered tissues before their biomedical applications. These assays provide a controlled environment to accelerate the extensive study of cellular behaviors, interactions, and responses to different stimuli, accelerating the development of effective tissue-engineered constructs.

In vitro assays are providing us with cost-effective, rapid, and ethical alternatives for in vivo models (i.e., animal testing). For detailed evaluation of the interactions between cells and biomaterials, these assays provide a controlled environment and gauge the activities, including cell death and cell growth. This monitors the biomaterial behavior inside the cell without inserting the constructs into in vivo models. It facilitates scaffold property assessments and the efficacy of growth factors working inside the cell for further processes.

Through in vitro methods, scientists can thoroughly characterize the mechanical properties of any scaffold, which is of utmost importance for projects like transplanting grafts and orthopedic treatments [1].

In vitro models play a crucial role in enabling the precise observation of cell proliferation, differentiation, and phenotype stability in response to various biomaterials. These models provide insights into bioactivity and optimizing biomaterials for tissue engineering applications. By manipulating factors such as surface topography, chemistry, and mechanical properties, in vitro assays help better understand cell-material interactions, further advancing tissue engineering strategies [2].

Additionally, scaffolds with tunable stiffness significantly influence cell adhesion, migration, and differentiation, offering critical insights into the optimal conditions required for various tissue types [3]. For instance, in vitro testing can reveal scaffold immunogenicity and cytocompatibility, significantly influencing scaffold selection for clinical applications. It enhances the relevance of studies by refining the scaffold properties to understand how tissue might behave in real-world settings, which allows us to strengthen the modeling of physiological conditions.

For drug testing, in vitro systems have become indispensable for testing drug toxicity and pharmacological effects, providing insights that may often prove more relevant than those gained through traditional animal models [4]. However, in vitro assays offer invaluable insights, but they may not fully replicate the complex in vivo environment, potentially leading to discrepancies in outcomes when moving from laboratory research to clinical applications. Controlled environments are crucial for scaffold testing and cellular analysis as they provide optimum conditions that impact cell behavior and interactions, eventually aiding tissue engineering research.

The role of in vitro assays extends beyond tissue engineering research to align with global health goals, particularly the United Nations Sustainable Development Goal 3 (i.e., Good Health and Well-being), by enhancing disease prevention, improving health outcomes, and promoting universal health coverage. These assays support broader health objectives by providing well-founded data on drug efficacy and safety [5]. In vitro assays support the development of essential medicines, as emphasized in target 3. B of SDG 3, and help tailor treatments to specific populations, promoting equitable access to healthcare.

Fundamentals of Translational Tissue Engineering. DOI: https://doi.org/10.1016/B978-0-443-33455-9.00007-9

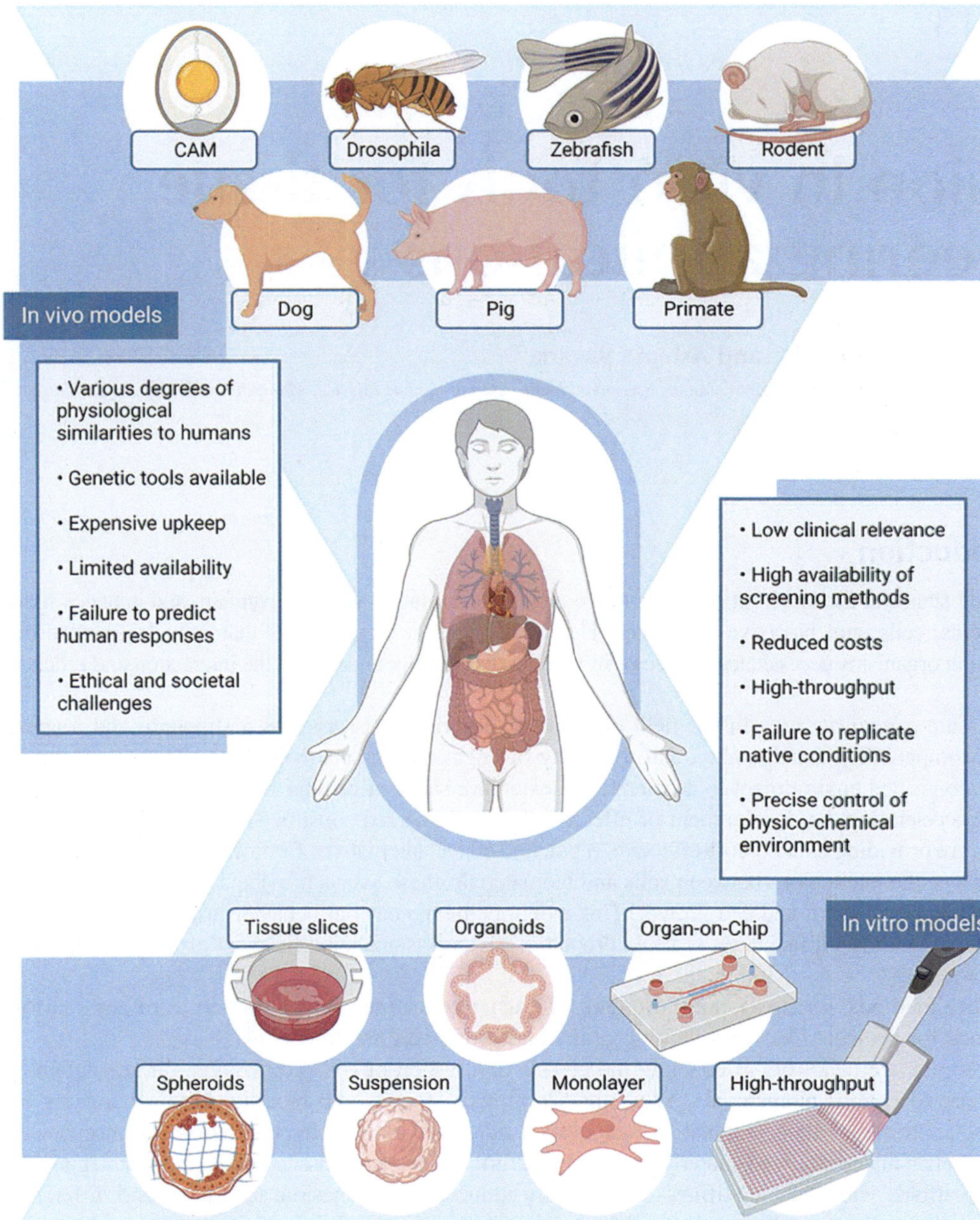

FIGURE 13.1 Represents the basic difference between in vitro and in vivo models.
Retrieved from: Ref. [6].

The use of in vitro assays presents opportunities to hammer out healthcare goals and biomedical applications despite new challenges such as accessing technology that may hinder implementation and funding disproportion among countries, particularly in low and middle-income countries. Addressing such barriers, collaboration across sectors is crucial to ensure that the benefits of in vitro research translate into tangible health improvements worldwide.

Just like that, cell behavior and interaction assays are equally vital for understanding cellular responses during wound healing and regeneration. Some assays, like cell migration assays (CMA), scratch assay, and trans well assays, give an edge to observe and quantify cell movements, and adhesion is crucial for understanding wound healing and regeneration mechanisms.

Collectively, in vitro assays offer valuable insights into cell-material interactions, scaffold efficacy, and biological responses in a controlled environment and serve as a cornerstone in tissue engineering research. This chapter will explore key types of in vitro assays, highlighting their applications and recent advancements that continue to shape the future of regenerative medicine (Fig. 13.1).

13.2 Types of in vitro assays

Various types of in vitro assays serve different purposes and contribute significantly to the development and assessment of tissue constructs, enabling scientists to assess important factors such as cell viability, proliferation, and behavior. The following paragraphs detail key types of in vitro assays utilized in tissue engineering, covering essential aspects and recent advances in the field.

Assay	Purpose	Strengths	Limitations
Live/Dead Assay	Cell membrane integrity	Easy to use	May not detect early apoptosis
MTT/AlamarBlue	Metabolic activity	Quantitative	Depends on mitochondrial function
Trypan Blue	Cell viability (live vs dead cells)	Rapid and cost-effective	Low specificity
Confocal Microscopy	3D cell interaction and visualization	High resolution, 3D imaging	Expensive, requires training

1. **Cell viability assay**: Cell viability and cell proliferation assays are fundamental in tissue engineering for assessing scaffold efficacy and bio-printed constructs. These assays help determine cell survival and proliferation rates crucial for successful tissue regeneration [7]. Cell viability is assessed by measuring moles of cellular adenosine triphosphate (ATP) per gram of cellular deoxyribonucleic acid (DNA). Cell viability assays are important as they allow assessing cell health under diverse conditions, such as stress or injury, providing insights into cell proliferation potential. These insights are particularly valuable for regenerative medicine and tissue engineering applications [8].

Researchers use live/dead staining and metabolic assays to measure cell viability in 3D cultures, including hydrogels and bio-printed constructs. Automated techniques, like confocal microscopy, enhance the accuracy of assessing cell viability and cell interactions in complex 3D environments, promoting and advancing tissue engineering research.

Additional tools used in viability and proliferation assays include MTT and AlamarBlue Tests that measure responses to biomaterials, such as poly(L-lactic acid) (PLLA) fibrous membranes, indicating biocompatibility [9]. The Live/Dead assay evaluates cell membrane integrity, offering insights into the cytotoxicity of materials. The Trypan Blue Exclusion Assay is a widely used, straightforward method for assessing cell viability in tissue engineering. This assay provides rapid quantification of live versus dead cells, which is essential for evaluating cell health under various treatments or conditions. However, potential inaccuracies in distinguishing live from dead cells under certain conditions may necessitate complementary assays for comprehensive viability analysis.

Computational models that predict cell viability in bio-printed constructs further support the design of optimized tissue engineering strategies [10]. Nevertheless, accurately measuring cell viability in complex 3D environments remains a challenge, with variability across hydrogel formulations underscoring the importance of carefully selecting and validating assay methods [11,12].

General biological response	ISO or OECD standards	Biological subcategory response	Assays
Cytotoxicity	ISO 10993-5	Viability	Dye assays, MTT, AlamarBlue, Celltiter Glo, (3H)-thymidine assay
		Cell death	Annexin-V staining, TUNEL assay
Hemocompatibility	ISO 10993-4	Thrombogenicity	ELISA
		Coagulation	Microscopy
		Platelets	Flow cytometry
			Prothrombin time assay, partial thrombo-plastin time assay
		Leukocytes	Blood cell counters
		Hematology	Hemoglobinometer, Cyanmethemoglobin method
Inflammation and foreign body reaction for implantable biomaterials	ISO 10993-6, ISO 10993-20	Blood plasma proteins adsorption	ELISA
		Recruitment, activation, and fusion of macrophages	qPCR, Immunohistochemistry, Microscopy, Flow cytometry
			Phagocytosis assay (Cytoselect 96-well), Chemiluminescence assay
		Fibrotic tissue formation	Fibrosis assay
Genotoxicity	ISO 10993-3, OECD 471	Gene mutational assays	Ames test, Mouse lymphoma gene mutation assay
		Chromosomal aberration assays	In vitro micronucleus assay
		DNA damage-based assays	Single-cell gel electrophoresis assay

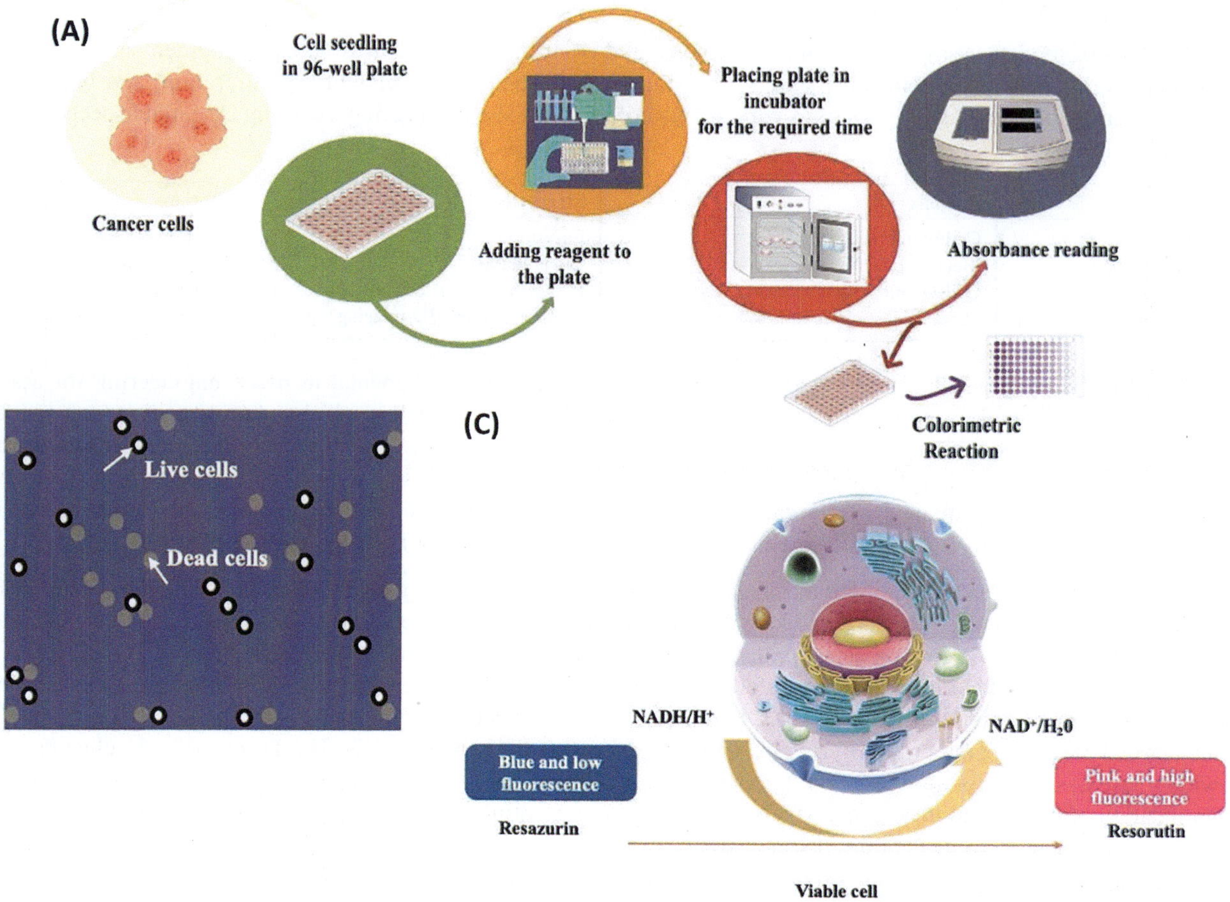

FIGURE 13.2 Depicts the methodology of performing an in vitro assay and its validation by observing the absorption, checking live and dead cells. Retrieved from G. Chaudhry, Zeenia, N. Safdar, S. Begum, A.M. Akim, Y.Y. Sung et al., Cytotoxicity assays for cancer drug screening: methodological insights and considerations for reliable assessment in drug discovery, Brazilian Journal of Biology 84 (2024) e284409.

13.3 Biocompatibility assays

1. **Cytotoxicity assays:** Cytotoxicity assays are employed to assess the potential toxic effects of substances, drugs, or materials and combinations, making them crucial in drug development, safety assessment of chemicals, and evaluation of biomedical applications. These cytotoxic assays enable the identification of toxicity at the cellular level before proceeding with higher animal studies or in vivo studies.

Providing the toxicity reports at the early stage reduces the cost and risks of later-stage failure. These assays are the best to adopt for allowing rapid screening and high-throughput formats that eventually help in increasing the drug discovery process [13]. For example, LDH assay is used to detect the lactate dehydrogenase enzyme, which is released from the damaged cells and serves as an indicator of cell toxicity by dead cells [14]. These cytotoxicity testing assays provide an edge by testing the biocompatibility of materials used in medical devices to eradicate the risks of adverse cellular reactions, ultimately leading to cell death [13].

Cell viability and proliferation assays are fundamental in tissue engineering for assessing scaffold efficacy and bio-printed constructs [15]. Cell survival and proliferation rates are specifically crucial for successful tissue regeneration. Similarly, cell viability assays are important to check cell health under diverse conditions, providing insights into cell proliferation potential.

These insights are particularly valuable for regenerative medicine and tissue engineering applications [8] (Fig. 13.2).

13.4 Genotoxicity assay

Even before the DNA structure was illustrated, it was established that chemical, physical, and biological agents may interact with the genetic material, causing mutations [16]. The terms "genotoxic" and "genotoxicity" refer to "agents or processes that alter the structure, information content, or segregation of DNA, including those that cause DNA damage by interfering with normal replication processes, or that alter its replication in a nonphysiological manner temporarily." Genotoxic compounds alter a cell's genetic makeup through interactions with the DNA sequence and structure, resulting in mutations and malignancies [17]. Common examples are the RSMN assay, CBMN assay, Comet assay, and Micronucleus assay. Genotoxic chemicals affect the structure, information content, and segregation of DNA in various ways. "DNA-reactive" chemicals predominantly damage DNA through the formation of covalent DNA adducts and cross-linking.

13.5 Cell adhesion assays

Cell adhesion assay is an essential in the growth and maintenance of tissues. It is important in cell regulation and communication. Cell functions and behavior are regulated by their interaction with the extracellular matrix [18].

Cell adhesion assays measure how cells attach to surfaces, other cells, and biomaterials. Maintaining the tissue architecture within the cell can be assessed in detail to find out the potential therapy targets to enhance the treatment quality. To perform dynamic adhesion processes, flow-based adhesion assays are considered. Static adhesion assays are suitable for studying cell-matrix interactions under controlled cellular conditions.

1. **Migration and invasion assays:** Wound and scratch healing monitoring can be done using migration assays, as it measures the ability of a cell to move across the surface through chemotaxis or could be random migration. These CMA are also named chemotaxis assays. Trans well migration assay and microfluidic device assay are used to understand the immune cell movement in inflammation or wound healing, tissue repair, evaluating drug motility, and studying cancer cells metastasis by using the Boyden chamber assay [19].
2. **Genetic expression assays:** Gene expression assays help to determine how much a gene is transcribed into mRNA. These assays are used to indicate the genetic activity, commonly used in genetic regulation in diseases such as neurodegenerative disorders and cancer. Some techniques like RT-PCR, PCR, RNA sequencing, and microarray analysis are used to determine the changes in genetic material [20]. These techniques can further check the genetic changes in certain diseases like cancer and provide detailed insights into alternative splicing.
3. **Protein expression assays:** Protein expression assays are used for the visualization and quantification of the protein inside the cell. It helps in analyzing posttranslational effects and protein-protein interactions. Studying the cellular communication with the help of using this assay chain also called signal transduction of the cell is how they help in identifying the key biomarkers for specific diseases like cancer and metabolic disorders [21]. Some commonly used protein expression assays are ELISA, mass spectroscopy, flow cytometry, and Immunohistochemistry. These protein assays are specially used to diagnose diseases using biomarkers. Through these protein assays, monitoring of immune responses generated by cells in regenerative medicine can be observed and studied. Similarly, the detection of viral and bacterial infections through RNA profiling can be performed, as they are highly proficient in giving us an insight into functional protein states, understanding cell-specific gene expression, and also some inflammatory reactions in response to a particular drug or treatment, giving us a real picture of the inside cell events more clearly.

1. **Biochemical assays:** Researchers have unlocked the ability to peer into the heart of the cell up to the molecular level, studying its shifts, revealing its secrets, understanding the signaling, and comparing the internal changes with precision [22]. These molecular levels of cells were revealed by some assessing techniques named biochemical assays, which help us understand the cell behavior in different conditions, like optimum and unfavorable conditions of a cell, hence providing a clear outlook of cell viability, cell proliferation, and behavior change.

13.6 Applications of in vitro assays

1. **Immune research and disease modeling:** Immune research and disease modeling can be used to assess and check the immune cell functions for cytokine production and immune signaling pathways. For newly generated immune responses by cells when introduced to biomaterials or pathogens, these assays play a vital role in research by offering detailed insights that contribute to reliable results and ultimately drive high-quality scientific outcomes connected to better inventions like personalized medicine, therapy, theranostics, and other trending biomedical applications. Recreating disease-like environments for cells or the creation of organoids has been its major advantage, as this helps in understanding disease progression at cellular and molecular levels [23].
2. **Drug development and pharmacology:** Assessing the interactions between the drug and the targets is considered the pivotal step in drug targeting and, ultimately, new drug development and drug discovery, ultimately laying the

foundation of therapeutic safety [24]. Since pharmacology deals with the study of interactions between drugs and biological systems, it is a dynamic process where pharmacokinetics and pharmacodynamics are both used. These techniques are used to analyze how the drug is affecting the body by analyzing receptor binding and cellular responses.

3. **Cell biology studies and cancer research:** These assays are used to provide high precision and a controlled environment to study cell behavior and signaling pathways at the molecular level. Studying the tumor microenvironment and replicating cancer-stromal interactions can significantly enhance cancer research, leading to more precise and reliable outcomes [25]. To evaluate drug resistance and immune invasion, assays like Transwell Invasion assay and wound healing assays are used. A few more examples that help determine long-term survival and proliferation assays are trypan blue, colony formation assay, 3D tumor spheroid assay, and caspase activity assays assessing apoptosis.
4. **Regenerative medicine:** Detailed assessment of tissue engineering models and differentiation studies. To determine the cell adhesion ability of the material, personalized medicine is all about drugs and their wise usage; hence, toxicity testing and drug screening are the biggest parts of personalized therapy and regenerative medicine. CRISPR-based regeneration is a new technology that acts as a powerful gene editing tool, which is revolutionizing regenerative medicine by enabling the precise genetic modification for enhanced healing, tissue repair, and disease correction [26].
5. **Cardiovascular disease research:** This encompasses a broad range of disorders that affect blood vessels and the heart, like coronary heart disease, stroke, hypertension, and heart failure. These in vitro assays offer reproducible, controlled, and ethical platforms for studying the cellular and molecular mechanisms of CVDs. These assays are essential in the early stage of toxicity analysis, biomaterial screening, drug discovery, and regenerative strategies before moving to clinical trials [27]. Out of all the assays listed above, the assays that are used in CVD research are AlamarBlue Assays, Trypan Blue exclusion, MTT and WST-1, and similarly endothelial cell behavior could be studied by Tran-well migration assays.

13.7 Conclusion

In vitro assays have an important role in tissue engineering. It offers a controlled, ethical, and cost-effective platform to study cell-material interactions and assess biocompatibility, cytotoxicity, and tissue regeneration potential. These models play an important role in optimizing scaffold performance before clinical application. It also reduces reliance on animal models thereby aligning with the 3Rs (reduce, refine, replace).

Physiological relevance of in vitro studies can be enhanced by using advanced techniques like bioreactors, 3D cultures, and organ-on-chip systems. It also simulates mechanical forces and biochemical signaling. However, they still fall short in replicating complex in vivo conditions.

Future progress depends on integrating in vitro models with high-throughput screening, in silico modeling, bio-sensing, and AI-driven analysis. This convergence will improve predictive accuracy and enable patient-specific solutions and also accelerate clinical translation and advance personalized regenerative medicine.

AI disclosure

During the preparation of this work, the author(s) used Grammarly and ChatGPT to check grammar and spelling. After using this tool/service, the author(s) reviewed and edited the content as needed and take(s) full responsibility for the content of the publication.

List of abbreviations

ELISA Enzyme-linked immunosorbent assay
MTT 3-(4,5-dimethylthiazol-2-yl)-2,5-diphenyltetrazolium bromide assay
DAPI 4',6-diamidino-2-phenylindole
CMA GenoType Mycobacterium CM/AS assay
OECD The organization for economic cooperation and development
LDH Lactate Dehydrogenase
RT-PCR Reverse transcription polymerase chain reaction
PCR Polymerase chain reaction
RNA Ribonucleic acid
CRISPR Clustered Regularly Interspaced Short Palindromic Repeats
WST- I Water-Soluble Tetrazolium-1

References

[1] A. Johnston, A. Callanan, Recent methods for modifying mechanical properties of tissue-engineered scaffolds for clinical applications, Biomimetics 8 (2) (2023) 205.

[2] E. Azuaje-Hualde, M. García-Hernando, J. Etxebarria-Elezgarai, M.M. De Pancorbo, F. Benito-Lopez, L. Basabe-Desmonts, Microtechnologies for cell microenvironment control and monitoring, Micromachines 8 (6) (2017) 166.

[3] M. d'Angelo, E. Benedetti, M.G. Tupone, M. Catanesi, V. Castelli, A. Antonosante, et al., The role of stiffness in cell reprogramming: a potential role for biomaterials in inducing tissue regeneration, Cells 8 (9) (2019) 1036.

[4] A. Vashishat, P. Patel, G. Das Gupta, B. Das Kurmi, Alternatives of animal models for biomedical research: a comprehensive review of modern approaches, Stem Cell Reviews and Reports 20 (4) (2024) 881–899.

[5] C.H. Emmerich, L.M. Gamboa, M.C. Hofmann, M. Bonin-Andresen, O. Arbach, P. Schendel, et al., Improving target assessment in biomedical research: the GOT-IT recommendations, Nature Reviews Drug Discovery 20 (1) (2021) 64–81.

[6] D. Soto Veliz, K.L. Lin, C. Sahlgren, Organ-on-a-chip technologies for biomedical research and drug development: A focus on the vasculature, Smart Medicine 2 (1) (2023) e20220030.

[7] D. Arora, G. Babakhanova, C.G. Simon Jr, , Tissue engineering measurands, ACS Biomaterials Science & Engineering 6 (10) (2020) 5368–5376.

[8] E.A. Pierce, T.S. Aleman, K.T. Jayasundera, B.S. Ashimatey, K. Kim, A. Rashid, et al., Gene editing for CEP290-associated retinal degeneration, New England Journal of Medicine 390 (21) (2024) 1972–1984.

[9] M. Dhayer, V. Barral, D. Cleret, A. Jordao, A.S. Drucbert, N. Germain, et al., Material and biological characterization of 3D knitted bioresorbable poly (D, L-lactide)(PLA) and polycaprolactone (PCL) scaffolds for soft tissue regeneration: from fabrication to in vivo performance, Journal of Biological Engineering 19 (1) (2025) 53.

[10] Y.F. Wu, Y.T. Wen, E. Salamanca, L.M. Aung, Y.Q. Chao, C.Y. Chen, et al., 3D-bioprinted alginate-based bioink scaffolds with β-tricalcium phosphate for bone regeneration applications, Journal of Dental Sciences 19 (2) (2024) 1116–1125.

[11] S. Avnet, G. Di Pompo, G. Borciani, T. Fischetti, G. Graziani, N. Baldini, Advantages and limitations of using cell viability assays for 3D bioprinted constructs, Biomedical Materials 19 (2) (2024) 025033.

[12] A.J. Dominijanni, M. Devarasetty, S.D. Forsythe, K.I. Votanopoulos, S. Soker, Cell viability assays in three-dimensional hydrogels: a comparative study of accuracy, Tissue Engineering Part C: Methods 27 (7) (2021) 401–410.

[13] M.A.M. Shafiee, M.A.M. Asri, S.S.S. Alwi, Review on the in vitro cytotoxicity assessment in accordance to the international organization for standardization (ISO), Malaysian Journal of Medicine & Health Sciences 17 (2) (2021).

[14] A. Gambacurta, V. Tullio, I. Savini, M.V. Catani, V. Gasperi, In vitro cell-based MTT and crystal violet assays for drug toxicity screening, Computational Toxicology: Methods and Protocols, Springer US, New York, NY, (2024), pp. 293–301.

[15] G. Chaudhry, Zeenia, N. Safdar, S. Begum, A.M. Akim, Y.Y. Sung, et al., Cytotoxicity assays for cancer drug screening: methodological insights and considerations for reliable assessment in drug discovery, Brazilian Journal of Biology 84 (2024) e284409.

[16] R.P. Araldi, T.C. de Melo, T.B. Mendes, P.L. de Sá Júnior, B.H.N. Nozima, E.T. Ito, et al., Using the comet and micronucleus assays for genotoxicity studies: a review, Biomedicine & Pharmacotherapy 72 (2015) 74–82.

[17] A.J. Seukep, J.A. Noumedem, D.E. Djeussi, V. Kuete, Genotoxicity and teratogenicity of African medicinal plants, Toxicological survey of African medicinal plants, Elsevier, (2014), pp. 235–275.

[18] A. Ahmad Khalili, M.R. Ahmad, A review of cell adhesion studies for biomedical and biological applications, International Journal of Molecular Sciences 16 (8) (2015) 18149–18184.

[19] T. Tadokoro, M. Kato, T. Kobayashi, H. Taniguchi, Optimizing cell migration assays: critical roles of fluorescent labeling and chemoattractant gradients, Biochemical and Biophysical Research Communications 739 (2024) 150998.

[20] O. Martínez-Iglesias, V. Naidoo, J.C. Carril, S. Seoane, N. Cacabelos, R. Cacabelos, Gene expression profiling as a novel diagnostic tool for neurodegenerative disorders, International Journal of Molecular Sciences 24 (6) (2023) 5746.

[21] A. Ahmad, M. Imran, H. Ahsan, Biomarkers as biomedical bioindicators: approaches and techniques for the detection, analysis, and validation of novel biomarkers of diseases, Pharmaceutics 15 (6) (2023) 1630.

[22] L. Khalef, R. Lydia, K. Filicia, B. Moussa, Cell viability and cytotoxicity assays: biochemical elements and cellular compartments, Cell Biochemistry and Function 42 (3) (2024) e4007.

[23] E. Morrocchi, S. van Haren, P. Palma, O. Levy, Modeling human immune responses to vaccination in vitro, Trends in Immunology 45 (1) (2024) 32–47.

[24] N. Jain, S. Kaul, K.A. Triveni, A. Sinha, T. Mathur, M. Pandey, et al., 8 QbD and artificial intelligence in nanoparticulate drug delivery systems: recent advances, Computational Drug Delivery (2024) 163–182.

[25] A. Albini, A. Bruno, C. Gallo, G. Pajardi, D.M. Noonan, K. Dallaglio, Cancer stem cells and the tumor microenvironment: interplay in tumor heterogeneity, Connective Tissue Research 56 (5) (2015) 414–425.

[26] B.M. Hussen, M. Taheri, R.K. Yashooa, G.H. Abdullah, S.R. Abdullah, R.K. Kheder, et al., Revolutionizing medicine: recent developments and future prospects in stem-cell therapy, International Journal of Surgery 110 (12) (2024) 8002–8024.

[27] G. Palano, A. Foinquinos, E. Müllers, In vitro assays and imaging methods for drug discovery for cardiac fibrosis, Frontiers in Physiology 12 (2021) 697270.

Chapter 14

In vivo models for tissue engineering

Asheesh Gupta, and Divya Yadav
Department of Pharmacology, Defence Institute of Physiology and Allied Sciences (DIPAS), DRDO, Delhi, India

14.1 Introduction

Tissue engineering (TE) based biomaterial matrix acts as a biomimetic scaffold to accelerate the repair of different types of injuries/defects. Animal models play a pivotal role in TE by bridging basic and applied research work. They are essential for understanding the multifaceted interactions between engineered scaffolds and the biological microenvironment in a living organism [1]. Animal models allow us to investigate the safety, efficacy, and functionality of tissue-engineered constructs in a regulated manner that meticulously mimics human physiology. Furthermore, the preclinical trials using large in vivo models (dog, pigs, horse, monkey) provide more realistic approaches to evaluate the safety and efficacy of novel TE constructs, as these models closely resemble heterogeneity with humans. These animal models enable the evaluation of how cells, scaffolds, and biomaterials behave in a dynamic, living system, offering insights into processes like vascularization, immunological response, and tissue integration [2]. Different animal models, varying from small rodents to larger mammals (pigs, nonhuman primates), are used according to the specific translational applications. Rodents, for instance, are commonly used in early-stage studies due to their genetic tractability and cost-effectiveness. In contrast, larger animals like pigs or goats are used for preclinical efficacy and safety studies that require anatomical and physiological relevance to humans [3]. Other animal models, like zebra fish and ferrets, are also used in TE, but their role is restrained due to various limitations.

Furthermore, animal models help in identifying potential issues such as immune rejection, improper integration, or unexpected side effects, which are crucial to address before moving to clinical trials [4]. Despite their limitations, animal models are indispensable for advancing TE technologies and ensuring their safety and effectiveness in clinical settings. This chapter describes the various animal models used for TE, their advantages, and drawbacks. Additionally, the utilization of in vivo models in diverse disciplines of TE will be elucidated.

14.2 Considerations before animal model selection

Selecting an appropriate animal model for research requires a vigilant balance of scientific, ethical, and practical considerations. The chosen model must be scientifically relevant, closely mimicking human biology and disease pathology, to ensure that study outcomes might be easily applicable to human conditions [1]. This relevance is a key factor for studies focusing on disease mechanisms, therapeutic testing, or TE. Ethical concerns are paramount, guided by the 3Rs principle—replacement, reduction, and refinement, confirming that alternatives to animal models are explored, the number of animals used is minimized, and procedures are refined to reduce suffering [5]. Compliance with major national and international regulations, including obtaining approval from ethical review boards like the Institutional Animal Care and Use Committee (IACUC) guidelines, is mandatory to protect animal welfare [6,7]. Fig. 14.1 depicts the prime factors and criteria for selecting the appropriate in vivo models.

Practical considerations also play a crucial role; these include the availability of the animal species, the cost of the animal, and the feasibility of housing, handling, and caring for the animals. The lifespan, reproductive cycle, and genetic consistency of animal species are important to ensure that the study's duration and outcomes are manageable and reliable [1]. Additionally, the model's prior use in similar studies provides context and enhances the reliability of the research. Public perception is another significant factor, especially when using animals that are culturally significant or widely regarded as pets, such as dogs [8]. Transparency in communicating the scientific necessity and ethical considerations of

Fundamentals of Translational Tissue Engineering. DOI: https://doi.org/10.1016/B978-0-443-33455-9.00024-9

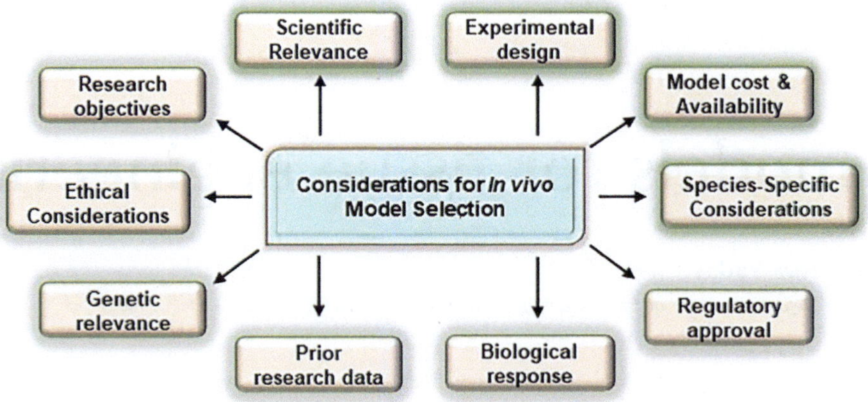

FIGURE 14.1 **Criteria for selecting in vivo model.** Depicts the prime factors for selecting an appropriate in vivo model.

the research is vital to maintaining public trust and support. Researchers must ensure that the animal model chosen aligns with the study's goals, complies with ethical standards, and is practical for the research environment, ultimately contributing to meaningful and responsible scientific advancements [7]. Fig. 14.1 depicts the prime factors for selecting an appropriate in vivo model.

14.3 Different in vivo models for tissue engineering

This section describes the most frequently used animal models in the TE field for the evaluation of efficacy, safety, molecular, and cellular action mechanisms.

14.3.1 Murine model

Preclinical TE research extensively uses rats and mice. Rodents are preferred due to their small size, ease of handling, short lifespan, and low maintenance requirements [9]. Commercially available rodent strains provide a homogenous population to study anatomical and physiological changes during TE interventions. In addition, they can be adapted to a particular environment through genetic modification and inbreeding. Rodents allow the study of the biological behavior of engineered cells and tissues efficiently without compromising the outcome [10].

However, because of their small size, rodents are challenging to operate on during intricate surgical procedures and restrict the scalability of implants. Rodents also have physiological and anatomical differences from humans, such as distinct immune responses and metabolic rates, which can affect the translatability of results [11]. This could limit the experiment's anticipated outcome, necessitating the use of larger animals with a higher similarity rate. Nevertheless, in the early stages of research, rodents are very helpful in understanding basic biological processes, evaluating the safety and effectiveness of biomaterials, and improving TE methods [12]. Thereby, TE studies using rodent models serve as a preliminary proof-of-concept to move forward to studies with larger animals. Various animal models used in TE have been shown in Fig. 14.2.

14.3.2 Leporine model

Rabbits are comparatively larger than rodents, allowing for easy surgical alteration and implantation of tissue-engineered constructs. Owing to convenient scaffold implantation, rabbits can be used to investigate TE scaffolds for growth plate injuries [13]. In addition, rabbits have more consistent and reproducible results in experiments that aid in obtaining reliable information.

Leporine models also have short reproductive phases and life cycles, which allows for the study of multiple generations and long-term responses in a shorter period. Rabbits are more cost-effective compared to larger animals like dogs, sheep, or pigs. Transgenic alterations (knock-out and knock-in studies) in rabbits are feasible to unravel the role of particular genetic modification and their implication on TE [14].

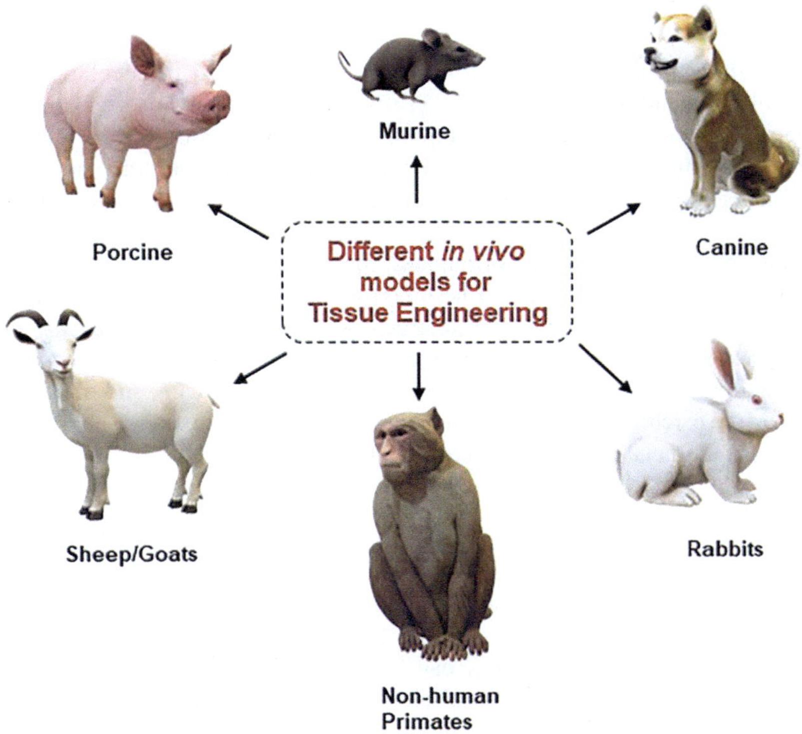

FIGURE 14.2 **Various in vivo models.** Diagrammatic representation of various TE in vivo models (rodents, pigs, rabbits, etc.). *TE*, Tissue engineering.

14.3.3 Porcine model

The porcine model plays a significant role in TE studies due to its anatomical and physiological similarities to humans. Pigs are predominantly valuable for research involving cardiovascular, dermal, and organ regeneration because their organ size, skin structure, and cardiovascular system closely resemble those of humans [15]. This makes them an excellent choice for preclinical testing of large-scale implants, such as heart valves, vascular grafts, and skin grafts, as well as for developing bioengineered organs like kidneys and lungs. The size and architecture of pig organs allow for more accurate testing of human-sized implants, providing critical data on the functionality, durability, and integration of tissue-engineered constructs [16]. Additionally, pigs have a relatively long lifespan, which is beneficial for long-term studies on tissue development, regeneration, and chronic conditions.

However, the porcine model also has a few drawbacks. High costs and the need for specialized facilities and care make large-scale and long-term studies challenging [17]. Ethical concerns arise due to the use of large animals, particularly when considering the potential for pain and suffering during surgical procedures. Furthermore, while pigs are physiologically similar to humans, there are still species-specific variances, particularly in immunological responses and metabolism, that can affect the interpretation of results compared to humans [3]. Despite these limitations, the porcine model remains a critical tool in TE, particularly for bridging the gap between small animal studies and clinical trials.

14.3.4 Ovine model

The ovine model shares similarities with humans in terms of bone structure, joint mechanics, and cardiovascular systems. Sheep are preferred for orthopedic and musculoskeletal research, as their bone and joint sizes closely resemble those of humans, making them suitable for studies on bone grafts, spinal disc regeneration, and joint replacement [18]. Similar to pigs, sheep are also used in cardiovascular research to develop and test vascular grafts, heart valves, and other cardiovascular devices [19]. Sheep have a relatively long lifespan, making them suitable for long-term studies that monitor the durability and integration of TE constructs over time [20]. Moreover, their docile nature and well-understood physiology make them manageable subjects for experimental procedures. The high cost of maintaining and housing sheep, as well as the need for specialized care and facilities, can limit the scale of research [20]. Ethical concerns also arise due to the use of large animals, particularly in invasive procedures. Despite these challenges, the ovine model remains a critical tool in TE, especially for preclinical testing in orthopedic and cardiovascular research [19].

14.3.5 Caprine model

Goats are large animals with bone and joint structures similar in size and biomechanical properties to those of humans, making them highly relevant for studying bone grafts, joint replacements, and spinal disc regeneration [21]. Their size allows for the assessment of human-sized implants and devices, providing a more accurate assessment of constructs inside the body. Additionally, goats have a relatively fast growth rate and well-understood skeletal physiology, which facilitates the study of bone healing, remodeling, and the integration of tissue-engineered constructs over time [22].

Their immune system also closely parallels that of humans, making them suitable for evaluating the biocompatibility and immunogenicity of biomaterials used in TE. Furthermore, their larger body size compared to rodents permits the harvesting of sufficient tissue samples for detailed histological and biomechanical analysis, providing comprehensive insights into the tissue integration and function of engineered grafts [23]. However, the use of caprine models comes with certain challenges, including higher costs, the need for specialized housing and care, and ethical considerations related to the use of larger animals in research. Despite these challenges, the caprine model's ability to closely mimic human orthopedic conditions makes it a crucial tool in advancing TE [23].

14.3.6 Canine model

Dogs tend to have a longer lifespan compared to rodents and other small animal models, which supports long-term studies for TE. Dogs closely resemble humans in several physiological and anatomical features. Therefore dogs are an advantageous model for preclinical studies in TE, especially for applications involving bone, joint, and cardiovascular tissues [24,25]. Dogs are quite easy to handle as they can be trained for research, cooperate during experiments, and follow-up studies, leading to more coherent and rational results. The canine immune system is similar to our immune system; it helps to evaluate biocompatibility and immunological responses of tissue-engineered constructs [26]. In addition, dogs exhibit greater genetic diversity, which allows them to study a wide range of responses of tissue-engineered constructs and biomaterials [27].

14.3.7 Primate model

Primates are the most suitable models for TE due to their close genetic, anatomical, and physiological similarities to humans, making them highly pertinent for studies focusing on developing therapies that closely mimic the human tissue microenvironment [28]. Their genetic makeup resembles about 95–98% of the DNA of humans, providing a reliable platform for testing TE constructs, particularly in complex tissues like skin, bone, cartilage, and organ systems, including the cardiovascular and nervous systems [29]. The primary advantage of using the common marmoset (*Callithrix jacchus*) and the squirrel monkey (*Saimiri* sp.) in TE lies in their ability to model human-like responses to biomaterials and engineered tissues, allowing for more accurate assessments of biocompatibility, immunological reactions, and the long-term durability of implants [30].

On the other hand, the use of primates in research is heavily regulated and ethically sensitive due to their great cognitive abilities, sophisticated social behavior, and ability to feel pain and distress [31]. The ethical justification for using primates must be robust, with a clear demonstration of the potential scientific benefits outweighing the ethical costs. Logistically, the use of primates is challenging due to their high maintenance requirements, including the need for specialized housing, care, and social enrichment, which are resource-intensive and expensive [31]. Moreover, their longer lifespans and slower reproductive cycles compared to other animal models can limit their use in studies requiring quick results or multigenerational data [32]. Regardless of these challenges, primates remain a critical model when the research demands high fidelity to human biology. The use of primates in TE research provides critical data on the performance and safety of novel TE therapeutic approaches in a model that accurately reflects human anatomy and biomechanics [33]. Differences between various in vivo models frequently used in TE are described in Table 14.1.

14.3.8 Other animal models for tissue engineering

14.3.8.1 Equine model

Horses (*Equus caballus*) are widely used for musculoskeletal, tendinopathy, focal cartilage injuries, and osteoarthritis TE studies because of their large size; multiple or critical-size defects can be created easily [3]. Horses are advantageous in nonterminal studies in which thorough examination is carried out via advanced diagnostic techniques such as radiographs,

TABLE 14.1 Comparison between different in vivo models for tissue engineering (TE).

Type of in vivo models	Advantages	Limitations	Major application
Rodents	• Small size • Short lifespan • Cost-effective • Well-characterized genetics and immune systems	• Small body size limits surgical complexity • Limited in size for grafts and implants • Physiological differences with humans • Murine wound model heals differently to human's due to the process of contraction	• Widely used for pioneer and transgenic studies • Used for cardiovascular and neural TE
Leporine	• Larger size compared to rodents, enabling more complex grafts • Cost-effective	• Variations in immune response compared to humans • Intermediate size limits direct scaling to human models	• Good model for musculoskeletal and dental TE
Porcine	• Large size, similar physiology, and organ size to humans • Pigs have large arteries that can be accessed surgically • Long-term survival studies possible	• Expensive and require specialized care • Pigs have a higher rate of plasma coagulation	• Used in cardiovascular, skin, and organ TE, including heart valves, vascular grafts, and liver
Ovine	• Larger size allows complex implants and surgical procedures • Thickness of knee cartilage is similar to humans	• Huge maintenance costs • Differences in wound healing and immune responses compared to humans	• Excellent model for orthopedic and joint TE studies
Caprine	• Similar to sheep but more agile • Relatively easy surgical access	• Huge maintenance costs • Long gestation time • Polygastric stomach, which varies from human	• Useful for bone and joint TE, including long-term orthopedic implant studies
Canine	• Similar anatomical structures to humans, predominantly for cardiovascular and skeletal research • Long-term survival studies feasible • Similar size to human arteries and bones	• High ethical concerns and costs • Variability in immune response and physiology compared to humans	• Cardiovascular TE (heart valves, blood vessels), bone, and joint repair studies, especially in long-term implant studies
Primates	• Closest anatomical, physiological, and immunological resemblances to humans • Mimic human disease and tissue responses more accurately	• Very high ethical concerns • Extremely costly and require specialized facilities • Limited availability and regulatory restrictions.	• Advanced research in neural, cardiovascular, orthopedic, and complex TE for translational studies

ultrasounds, and CT scans. However, the limited availability of specific antibodies and molecular markers constrained the usage of horses as TE models.

14.3.8.2 Feline model

Cats have been used in corneal TE. Researchers have used a live feline model to transplant the TE-corneal stromas, which, within two months, have shown effective sensory reinnervation.

14.4 Application of animal model in tissue engineering

14.4.1 Dermal tissue engineering construct

Recent developments in dermal TE have introduced a variety of innovative constructs tested in in vivo models, showcasing significant advancements in wound healing and skin regeneration. Some well-known examples are hydrogel-based constructs, such as hydrogels loaded with platelet-rich plasma (PRP), which have demonstrated accelerated wound closure and enhanced vascularization in rodent models [34]. Electrospun scaffolds, made from poly-l-lactide-co-caprolactone (PLCL) combined with fibrinogen, have shown effective re-epithelialization, dermal reconstruction, and maturation of mature blood vessels in diabetic porcine models [35]. Additionally, 3D bioprinted dermal scaffolds containing human fibroblasts and keratinocytes have successfully created stratified skin layers in mice, closely mimicking natural skin architecture [36]. Decellularized dermal matrices, derived from porcine or other sources, support cell infiltration, neovascularization, and tissue remodeling, making them suitable for reconstructive surgeries [37]. Composite scaffolds, like silk fibroin constructs loaded with stem cells, have promoted rapid wound closure and collagen deposition and reduced the formation of fibrotic scar tissue in a murine excisional wound splinting model [38]. Furthermore, gene-activated scaffolds delivering platelet-derived growth factor and mesenchymal stem cells have enhanced early angiogenesis, neovascularization, and burn wound healing in a full-thickness skin burn wound model [39]. These pioneering constructs highlight the diverse strategies in TE, with each offering unique benefits for specific clinical applications, pushing the boundaries of translational research in dermal repair.

14.4.2 Cardiovascular tissue engineering construct

Current advancements in cardiovascular TE have led to the generation of several constructs tested in animal models, demonstrating significant potential for clinical application. Decellularized scaffolds, such as porcine pulmonary heart valves, have been successfully implanted in sheep, showing reduced immunogenicity and improved mechanical function [40]. Electrospun nanofiber scaffolds made from PLCL, silk fibroin, and heparin have been used to create small-diameter vascular grafts in rabbit models, enhancing cell alignment and vascular function [41]. 3D bioprinted cardiac patches containing human-induced pluripotent stem cell-derived cardiomyocytes, fibroblasts, and endothelial cells have been applied to rodent models of myocardial infarction, leading to improved cardiac function, elevated angiogenesis, and reduced scarring [42]. Stem cell-seeded constructs, including fibrin-based cardiac patches with embryonic stem cell-derived cardiomyocytes, have also shown promising therapeutic potential in mice and porcine models by improving heart function and integrating well with host tissue [43]. Injectable hydrogels, such as those made from alginate and loaded with vascular endothelial growth factor (VEGF) and insulin-like growth factor-1, have been tested in mice and rabbit models of peripheral artery disease, promoting angiogenesis and restoring blood flow [44]. Gelatin-based hydrogel microsphere sheets targeting microRNA-92a have shown enhanced cardiac repair in rat models by promoting angiogenesis and cardiomyogenesis, reducing infarct size, and improving left ventricular function [45]. Additionally, biomimetic vascular grafts made from silk fibroin, gelatin, and PLCL have been implanted in rat models to create a compliance-matched TE vascular graft that can closely resemble the native vessel [46]. These diverse approaches highlight the rapid progress in cardiovascular TE, each offering unique advantages for specific clinical applications and paving the way for future human trials.

14.4.3 Bone tissue engineering construct

Modern bone TE constructs have shown substantial advancements through various innovative methods tested in animal models to enhance bone repair and regeneration. One method involves 3D-printed scaffolds made from biocompatible materials like calcium phosphate, hydroxyapatite, or bioactive glass, often combined with growth factors such as VEGF and BMP-2 (Bone Morphogenetic Protein 2). These scaffolds have been evaluated in large animal models like sheep and rabbits, demonstrating improved bone formation, vascularization, and mechanical properties, closely mimicking natural bone [47]. Nanocomposite scaffolds, incorporating nanoparticles such as graphene oxide or carbon nanotubes, have been tested in rodent models, showing enhanced osteoinductivity and faster bone healing in critical-sized defects [48]. Injectable hydrogels that are loaded with stem cells or osteoinductive factors provide a minimally invasive option, conforming to complex bone defects and promoting new bone formation, as shown in studies with rabbit and rat models [49]. Likewise, cell-laden scaffolds using mesenchymal stem cells or induced pluripotent stem cells have been employed in bone defect or fracture models, substantially improving bone regeneration and integration with host tissue [49]. Furthermore, gene-activated and RNAi-activated scaffolds, delivering plasmids encoding osteogenic factors or RNA that downregulate expression of antiosteogenic proteins, have been tested in rats and dogs and have successfully enhanced bone regeneration by promoting stem cell differentiation and matrix mineralization [50]. Lastly, biomimetic scaffolds that replicate the native bone extracellular matrix using materials like collagen or silk fibroin have shown promising results in large animal models by supporting bone regeneration and remodeling [51]. These varied TE constructs are at the forefront of bone repair and regeneration research, offering encouraging approaches for future clinical applications in orthopedics and craniofacial reconstruction.

14.4.4 Intestinal tissue engineering construct

Recent advancements in intestinal TE have led to the development of various constructs tested in animal models, aiming to address challenges in regenerating functional intestinal tissue. Notably, decellularized scaffolds implanted in rat models have shown promising results, including the development of crypt-villus structures, proper epithelialization, and maintenance of biomechanical, adhesion, and angiogenic properties [52]. Another approach involves the use of synthetic biodegradable scaffolds made from materials like polyglycolic acid (PGA) and polylactic acid (PLA), which are often seeded with intestinal epithelial cells and tested in large animal models, such as pigs [53]. These constructs have demonstrated the ability to support cell growth, neovascularization, and the formation of functional intestinal segments capable of nutrient absorption. Studies have also demonstrated the formation of engineered intestines and stomachs with intact stem cell niches in large animal models [54]. Additionally, organoid-based TE approaches using stem cells and polyglycolic/poly-L-lactic acid have been explored in small animal models like mice [55]. These scaffolds have successfully replicated the complex architecture of the intestine, including the formation of functional crypts and villi.

Furthermore, tissue-adhesive molecular coatings, for example, o-nitrobenzaldehyde-modified gelatin, have also been tested in models of intestinal injury, showing potential for minimally invasive repair by promoting the regeneration of the intestinal mucosa, reducing intestinal inflammation, and restoring barrier function [56]. These innovative approaches emphasize the ongoing progress in intestinal TE, with each construct offering distinct advantages for treating conditions such as short bowel syndrome, inflammatory bowel disease, and other intestinal disorders, paving the way for future clinical applications.

14.4.5 Dental tissue engineering construct

Current progress in dental TE has led to the expansion of innovative constructs tested in animal models, focusing on the regeneration of dental pulp, periodontal tissue, and alveolar bone. For instance, biomimetic scaffolds composed of materials like nanohydroxyapatite and collagen, loaded with dental follicle cells, have been used in rodent models to promote the regeneration of alveolar bone defects, showing enhanced bone formation and integration [57]. In dental pulp regeneration, injectable hydrogels incorporating stem cells derived from dental pulp or mesenchymal stem cells have demonstrated significant potential in mouse models, leading to the formation of vascularized pulp-like tissue and the deposition of dentin [58]. Moreover, 3D bioprinted constructs that precisely arrange dental stem cells and ECM components have been explored in rodent and minipig models, successfully regenerating complex dental structures, including dentin and pulp tissues [59].

Periodontal regeneration has also seen progress with the use of bioactive membranes made from materials like chitosan or polylactic acid, often loaded with stem cells or bioactive molecules [60]. These membranes have been tested in canine models with periodontal defects, showing improved attachment of periodontal ligament fibers and enhanced bone regeneration [61]. Additionally, gene-activated scaffolds delivering plasmids encoding growth factors have been employed in rodent and canine models to promote periodontal tissue regeneration, resulting in significant improvements in dental tissue repair and function [62]. The aforementioned TE constructs illustrate the ongoing efforts to address various dental tissue regeneration challenges, each providing unique benefits and pushing the field closer to clinical application in restorative dentistry and oral surgery.

14.5 In vivo bioreactor for tissue engineering

An in vivo tissue bioreactor is a cutting-edge method in TE that involves using the body's natural microenvironment to cultivate and grow engineered tissues or organs directly within the body [63]. This technique leverages the physiological conditions present in the body, such as blood supply, immune responses, and mechanical forces, to support the growth, maturation, and integration of engineered tissues more naturally and efficiently [64]. One of the significant advantages of in vivo tissue bioreactors is the ability to achieve better vascularization of the engineered tissues, as the implanted constructs can integrate with the host's vascular system, ensuring sufficient nutrient supply and waste removal, which is crucial for the survival of larger, more complex tissues like bone, cartilage, and vascularized structures [65,66]. Additionally, the dynamic mechanical environment of the body, such as movement and blood flow, can enhance the development of tissues that require mechanical stimulation, like bone and muscle, leading to stronger and more functional tissues [67]. This approach also allows for more seamless integration of the engineered tissue with the host's native tissues, potentially lessening the possibility of rejection and enhancing the overall success of tissue regeneration [64].

However, despite having huge advantages, there are a few notable limitations to the in vivo tissue bioreactor approach. Governing the tissue development process within the body is inherently more complex than in a controlled in vitro environment, making it challenging to precisely regulate factors such as cell differentiation, scaffold degradation, and tissue architecture [68]. There is also the risk of immune rejection, particularly if nonautologous cells or materials are used, which can compromise the success of the engineered tissue. Furthermore, ethical and regulatory hurdles must be navigated, especially when transitioning this technology from animal models to human clinical applications. The need for specialized surgical techniques for scaffold implantation and the potential for complications during tissue growth also add to the complexity of this approach [68]. Despite these obstacles, in vivo tissue bioreactors offer a promising avenue for advancing TE, particularly in areas where traditional methods strive to achieve the necessary vascularization and integration required for functional tissue regeneration [69].

14.6 Limitations of in vivo models

Using in vivo models for TE comes with several limitations that can influence the translatability of research findings. One significant challenge is the species-specific differences between animal models and humans, including variations in immune responses, tissue regeneration rates, and metabolic processes. These differences can lead to discrepancies in how engineered tissues behave in animals compared to humans, making it difficult to predict clinical outcomes precisely [70]. Another limitation is the complexity of monitoring environmental variables within a living organism [71]. Factors such as immune system interactions, mechanical forces, and the body's response to biomaterials can vary widely and are difficult to standardize, introducing variability into the results. This inconsistency can complicate the interpretation of data and the reproducibility of experiments.

Additionally, ethical and logistical constraints limit the use of certain in vivo models, particularly larger animals or those with high cognitive abilities, such as primates [33]. These constraints can restrict the scope of research and necessitate the use of less clinically relevant models. Lastly, in vivo studies are expensive and time-consuming, making it hard to conduct large-scale or long-term research. These limitations highlight the need for careful model selection and the development of alternative approaches to complement in vivo research in TE [72].

14.7 Conclusion

Animal models are vital in TE because they bridge the gap between laboratory research and human clinical applications. They provide a complex, living environment where researchers can assess the safety, biocompatibility, and efficacy of engineered tissues and biomaterials in ways that cannot be replicated in vitro. By mimicking human physiological and pathological conditions, in vivo models allow for the evaluation of how well TE constructs integrate with host tissues, support cell growth, and restore function. These models are essential for testing the mechanical properties and long-term stability of engineered tissues under real-life conditions, such as in load-bearing bones or contractile heart tissues. Additionally, animal studies help identify potential immunological responses, toxicity, and other adverse effects, which are critical for ensuring patient safety. The insights gained from animal models guide the design of human clinical trials and are often required by regulatory agencies to demonstrate that TE products are safe, efficacious, and effective. Thus animal models are indispensable in the field of TE, providing essential insights into the interaction between engineered constructs and living systems. The future improved animal models for efficacy and safety studies for a diverse range of ailments are highly warranted. However, ongoing research strives to refine these models and develop complementary alternative methods to advance the field responsibly and ethically. Ultimately, the combination of genetic tractability, biological relevance, and practical advantages makes rodents imperative models in TE, providing a critical bridge between basic research and the development of human TE therapies.

Acknowledgments

The authors are grateful to the Director of DIPAS for his constant support and guidance. Divya Yadav is grateful to the University Grant Commission (UGC), New Delhi, India, for providing the Junior Research Fellowship.

References

[1] P. Mukherjee, S. Roy, D. Ghosh, S.K. Nandi, Role of animal models in biomedical research: a review, Laboratory Animal Research 38 (1) (2022). Available from https://doi.org/10.1186/s42826-022-00128-1.

[2] S.E. Koch, B.J. de Kort, N. Holshuijsen, H.F.M. Brouwer, D.C. van der Valk, P.Y.W. Dankers, et al., Animal studies for the evaluation of in situ tissue-engineered vascular grafts — a systematic review, evidence map, and meta-analysis, npj Regenerative Medicine 7 (1) (2022). Available from https://doi.org/10.1038/s41536-022-00211-0, https://www.nature.com/npjregenmed/.

[3] I. Ribitsch, P.M. Baptista, A. Lange-Consiglio, L. Melotti, M. Patruno, F. Jenner, et al., Large animal models in regenerative medicine and tissue engineering: to do or not to do, Frontiers in Bioengineering and Biotechnology 8 (2020). Available from https://doi.org/10.3389/fbioe.2020.00972, http://journal.frontiersin.org/journal/bioengineering-and-biotechnology#archive.

[4] L. West-Livingston, J.W. Lim, S.J. Lee, Translational tissue-engineered vascular grafts: from bench to bedside, Biomaterials 302 (2023). Available from https://doi.org/10.1016/j.biomaterials.2023.122322, http://www.journals.elsevier.com/biomaterials/.

[5] J. Shim, J. Kim, Considerations for experimental animal ethics in the research planning and evaluation process, Kosin Medical Journal 37 (4) (2022) 271–277. Available from https://doi.org/10.7180/kmj.22.139.

[6] M.J. Brown, C. Symonowicz, L.V. Medina, N.A. Bratcher, C.A. Buckmaster, H. Klein, et al., Culture of care: organizational responsibilities, Management of Animal Care and Use Programs in Research, Education, and Testing, second ed. CRC Press, United States, (2017), pp. 11–26. Available from https://www.taylorfrancis.com/books/9781498748452.

[7] L. Díaz, E. Zambrano, M.E. Flores, M. Contreras, J.C. Crispín, G. Alemán, et al., Ethical considerations in animal research: the principle of 3R, Revista de Investigacion Clinica: Organo del Hospital de Enfermedades de la Nutricion 73 (4) (2020) 199–209. Available from https://doi.org/10.24875/RIC.20000380.

[8] F. Petetta, R. Ciccocioppo, Public perception of laboratory animal testing: historical, philosophical, and ethical view, Addiction Biology 26 (6) (2021). Available from https://doi.org/10.1111/adb.12991.

[9] J.A. Scholz, C.C. Ramos, C.J. VanLith, R.E. De La Vega, S.L. Boschen, C.H. Evans, et al., Use of the rat as a model in regenerative medicine, The Laboratory Rat, Elsevier, United States, (2019), pp. 1077–1105. Available from http://www.sciencedirect.com/science/book/9780128143384.

[10] E.C. Bryda, The mighty mouse: the impact of rodents on advances in biomedical research, Missouri Medicine 110 (3) (2013) 207–211.

[11] C. Riehle, J. Bauersachs, Small animal models of heart failure, Cardiovascular Research 115 (13) (2019) 1838–1849. Available from https://doi.org/10.1093/cvr/cvz161.

[12] A. da Silva Morais, J.M. Oliveira, R.L. Reis, Small animal models, Advances in Experimental Medicine and Biology 1059 (2018) 423–439. Available from https://doi.org/10.1007/978-3-319-76735-2_19, http://www.springer.com/series/5584.

[13] M. Fan, Y. Wang, Y. Liu, L. Qiang, R. Guo, H. Zhuang, et al., A new method for modeling rabbit growth plate injury for the study of tissue engineering scaffolds, Tissue Engineering Part C: Methods 28 (9) (2022) 489–497. Available from https://doi.org/10.1089/ten.tec.2022.0105.

[14] Y. Mu, Y. Zhang, L. Wei, L. Chen, F. Hao, A. Shao, et al., GTKO rabbit: a novel animal model for preclinical assessment of decellularized xenogeneic grafts via in situ implantation, Materials Today Bio 18 (2023) 100505. Available from https://doi.org/10.1016/j.mtbio.2022.100505.

[15] S.G. Cone, P.B. Warren, M.B. Fisher, Rise of the pigs: utilization of the porcine model to study musculoskeletal biomechanics and tissue engineering during skeletal growth, Tissue Engineering Part C: Methods 23 (11) (2017) 763–780. Available from https://doi.org/10.1089/ten.tec.2017.0227.

[16] C. Tapking, D. Popp, L.K. Branski, Pig model to test tissue-engineered skin, Methods in Molecular Biology 1993 (2019) 239–249. Available from https://doi.org/10.1007/978-1-4939-9473-1_19, http://www.springer.com/series/7651.

[17] J.K. Lunney, A. Van Goor, K.E. Walker, T. Hailstock, J. Franklin, C. Dai, Importance of the pig as a human biomedical model, Science Translational Medicine 13 (621) (2021). Available from https://doi.org/10.1126/scitranslmed.abd5758, https://www.science.org/doi/10.1126/scitranslmed.abd5758.

[18] Y.A. Ghosh, H. Xin, D.S.A. Al Maruf, K. Cheng, I. Wise, C. Burrows, et al., Novel sheep model to assess critical-sized bone regeneration with periosteum for in vivo bioreactors, Tissue Engineering - Part C: Methods 30 (4) (2024) 159–169. Available from https://doi.org/10.1089/ten.tec.2023.0345, http://www.liebertonline.com/tec.

[19] J. Boltze, Rschler, Nitzsche, D. Waldmin, A. Hoffmann, C.M. Boltze, et al., The sheep as a large animal model for the investigation and treatment of human disorders, Biology 2022 (9) (2022). Available from https://doi.org/10.3390/BIOLOGY11091251.

[20] S.C. Sartoretto, M.J. Uzeda, F.B. Miguel, J.R. Nascimento, F. Ascoli, M.D. Calasans-Maia, Sheep as an experimental model for biomaterial implant evaluation, Acta Ortopedica Brasileira 24 (5) (2016) 262–266. Available from https://doi.org/10.1590/1413-785220162405161949, http://www.scielo.br/pdf/aob/v24n5/1413-7852-aob-24-05-00262.pdf.

[21] I.E. Dias, C.A. Viegas, J.F. Requicha, M.J. Saavedra, J.M. Azevedo, P.P. Carvalho, et al., Mesenchymal stem cell studies in the goat model for biomedical research—a review of the scientific literature, Biology 11 (9) (2022). Available from https://doi.org/10.3390/biology11091276, www.mdpi.com/journal/biology.

[22] M. Abraham, I. Kori, U. Vishwakarma, S. Goel, Comprehensive assessment of goat adipose tissue-derived mesenchymal stem cells cultured in different media, Scientific Reports 14 (1) (2024). Available from https://doi.org/10.1038/s41598-024-58465-1, https://www.nature.com/srep/.

[23] R.D. Alvites, M.V. Branquinho, A.C. Sousa, B. Lopes, P. Sousa, C. Mendonça, et al., Small ruminants and its use in regenerative medicine: recent works and future perspectives, Biology 10 (3) (2021). Available from https://doi.org/10.3390/biology10030249, https://www.mdpi.com/2079-7737/10/3/249/pdf.

[24] M. Atari, A. Saroukhani, M. Manshaei, P. Bateni, A. Zargar kharazi, E. Vatankhah, et al., Preclinical in vivo assessment of a cell-free multi-layered scaffold prepared by 3D printing and electrospinning for small-diameter blood vessel tissue engineering in a canine model, Biomaterials Science 11 (20) (2023) 6871–6880. Available from https://doi.org/10.1039/d3bm00642e.

[25] E. Perrier-Groult, A. Pagnon, C. Boulocher, E. Viguier, T. Roger, D. Hartmann, et al., The canine model to test functional tissue-engineered cartilage, Osteoarthritis and Cartilage 22 (2014) S489. Available from https://doi.org/10.1016/j.joca.2014.02.936.

[26] J. Hernandez, E. Rouillé, F. Chocteau, M. Allard, K. Haurogné, F. Lezin, et al., Nonhypoalbuminemic inflammatory bowel disease in dogs as disease model, Inflammatory Bowel Diseases 27 (12) (2021) 1975–1985. Available from https://doi.org/10.1093/ibd/izab064, https://academic.oup.com/ibdjournal/issue.

[27] R.W. Bianco, K.R. Wasiluk, J.M. Voight, M.T. Lahti, A.L. Rivard, R.P. Gallegos, Large animal models in cardiac and vascular biomaterials research and assessment, Biomaterials Science: An Introduction to Materials, third ed. Elsevier Inc., United States, (2013), pp. 653–676. Available from http://www.sciencedirect.com/science/book/9780123746269.

[28] J. Wu, Y. Shi, S. Yang, Z. Tang, Z. Li, Z. Li, et al., Current state of stem cell research in non-human primates: an overview, Medical Review 3 (4) (2023) 277–304. Available from https://doi.org/10.1515/mr-2023-0035, https://www.degruyter.com/journal/key/mr/html.

[29] T. Nakamura, K. Fujiwara, M. Saitou, T. Tsukiyama, Non-human primates as a model for human development, Stem Cell Reports 16 (5) (2021) 1093–1103. Available from https://doi.org/10.1016/j.stemcr.2021.03.021.

[30] J.M. Ward, E.J. Vallender, The resurgence and genetic implications of new world primates in biomedical research, Trends in Genetics 28 (12) (2012) 586–591. Available from https://doi.org/10.1016/j.tig.2012.09.003.

[31] J.D. Harding, Nonhuman primates and translational research: progress, opportunities, and challenges, ILAR Journal 58 (2) (2017) 141–150. Available from https://doi.org/10.1093/ilar/ilx033, http://www.oxfordjournals.org/our_journals/ilarjournal/.

[32] K.A. Phillips, K.L. Bales, J.P. Capitanio, A. Conley, P.W. Czoty, B.A. 't Hart, et al., Why primate models matter, American Journal of Primatology 76 (9) (2014) 801–827. Available from https://doi.org/10.1002/ajp.22281, http://www.interscience.wiley.com/jpages/0275-2565/.

[33] A.J. Cauvin, C. Peters, F. Brennan, Advantages and limitations of commonly used nonhuman primate species in research and development of biopharmaceuticals, The Nonhuman Primate in Nonclinical Drug Development and Safety Assessment, Elsevier Inc., Belgium, (2015), pp. 379–395. Available from http://www.sciencedirect.com/science/book/9780124171442.

[34] Y.G. Park, I.H. Lee, E.S. Park, J.Y. Kim, Hydrogel and platelet-rich plasma combined treatment to accelerate wound healing in a nude mouse model, Archives of Plastic Surgery 44 (3) (2017) 194–201. Available from https://doi.org/10.5999/aps.2017.44.3.194, https://e-aps.org/Synapse/Data/PDFData/2023APS/aps-44-194.pdf.

[35] G. Wang, S. Ju, X. Li, Y. Cai, Y. Li, W. Li, et al., Preclinical animal study of electrospun poly (L-lactide-co-caprolactone) and formulated porcine fibrinogen for full-thickness diabetic wound regeneration, Biomedicine and Pharmacotherapy 162 (2023). Available from https://doi.org/10.1016/j.biopha.2023.114734, https://www.journals.elsevier.com/biomedicine-and-pharmacotherapy.

[36] N. Cubo, M. Garcia, J.F. Del Cañizo, D. Velasco, J.L. Jorcano, 3D bioprinting of functional human skin: production and in vivo analysis, Biofabrication 9 (1) (2017). Available from https://doi.org/10.1088/1758-5090/9/1/015006, http://iopscience.iop.org/article/10.1088/1758-5090/9/1/015006/pdf.

[37] M. Rosadas, I.V. Silva, J.B. Costa, V.P. Ribeiro, A.L. Oliveira, Decellularized dermal matrices: unleashing the potential in tissue engineering and regenerative medicine, Frontiers in Materials 10 (2023). Available from https://doi.org/10.3389/fmats.2023.1285948, journal.frontiersin.org/journal/materials.

[38] J.E. Millán-Rivero, C.M. Martínez, P.A. Romecín, S.D. Aznar-Cervantes, M. Carpes-Ruiz, J.L. Cenis, et al., Silk fibroin scaffolds seeded with Wharton's jelly mesenchymal stem cells enhance re-epithelialization and reduce formation of scar tissue after cutaneous wound healing, Stem Cell Research and Therapy 10 (1) (2019). Available from https://doi.org/10.1186/s13287-019-1229-6, http://stemcellres.com.

[39] T. Paramasivam, S.K. Maiti, S. Palakkara, R.D. Mohan, H.V. Manjunthaachar, K. Karthik, et al., Effect of PDGF-B gene-activated acellular matrix and mesenchymal stem cell transplantation on full thickness skin burn wound in rat model, Tissue Engineering and Regenerative Medicine. 18 (2) (2021) 235–251. Available from https://doi.org/10.1007/s13770-020-00302-3, http://www.springer.com/biomed/journal/13770.

[40] R. Ramm, T. Goecke, K. Theodoridis, K. Hoeffler, S. Sarikouch, K. Findeisen, et al., Decellularization combined with enzymatic removal of N-linked glycans and residual DNA reduces inflammatory response and improves performance of porcine xenogeneic pulmonary heart valves in an ovine in vivo model, Xenotransplantation 27 (2) (2020). Available from https://doi.org/10.1111/xen.12571.

[41] D. Jin, J. Hu, D. Xia, A. Liu, H. Kuang, J. Du, et al., Evaluation of a simple off-the-shelf bi-layered vascular scaffold based on poly(L-lactide-co-ε-caprolactone)/silk fibroin in vitro and in vivo, International Journal of Nanomedicine 14 (2019) 4261–4276. Available from https://doi.org/10.2147/IJN.S205569, https://www.dovepress.com/getfile.php?fileID=50425.

[42] E. Yeung, T. Fukunishi, Y. Bai, D. Bedja, I. Pitaktong, G. Mattson, et al., Cardiac regeneration using human-induced pluripotent stem cell-derived biomaterial-free 3D-bioprinted cardiac patch in vivo, Journal of Tissue Engineering and Regenerative Medicine 13 (11) (2019) 2031–2039. Available from https://doi.org/10.1002/term.2954, http://onlinelibrary.wiley.com/journal/10.1002/(ISSN)1932-7005.

[43] Q. Xiong, K.L. Hill, Q. Li, P. Suntharalingam, A. Mansoor, X. Wang, et al., A fibrin patch-based enhanced delivery of human embryonic stem cell-derived vascular cell transplantation in a porcine model of postinfarction left ventricular remodeling, Stem Cells 29 (2) (2011) 367–375. Available from https://doi.org/10.1002/stem.580, http://onlinelibrary.wiley.com/doi/10.1002/stem.580/pdf.

[44] E.M. Anderson, E.A. Silva, Y. Hao, K.D. Martinick, S.A. Vermillion, A.G. Stafford, et al., VEGF and IGF delivered from alginate hydrogels promote stable perfusion recovery in ischemic hind limbs of aged mice and young rabbits, Journal of Vascular Research 54 (5) (2017) 288–298. Available from https://doi.org/10.1159/000479869, http://www.karger.com/JVR.

[45] M. Fujita, H. Otani, M. Iwasaki, K. Yoshioka, T. Shimazu, I. Shiojima, et al., Antagomir-92a impregnated gelatin hydrogel microsphere sheet enhances cardiac regeneration after myocardial infarction in rats, Regenerative Therapy 5 (2016) 9–16. Available from https://doi.org/10.1016/j.reth.2016.04.002.

[46] K.J. Furdella, S. Higuchi, A. Behrangzade, K. Kim, W.R. Wagner, J.P. Vande Geest, In-Vivo Assessment of a Tissue Engineered Vascular Graft Computationally Optimized for Target Vessel Compliance, SSRN, United States SSRN, (2020). Available from https://doi.org/10.2139/ssrn.3640710, https://www.ssrn.com/index.cfm/en/.

[47] G. Turnbull, J. Clarke, F. Picard, P. Riches, L. Jia, F. Han, et al., 3D bioactive composite scaffolds for bone tissue engineering, Bioactive Materials 3 (3) (2018) 278–314. Available from https://doi.org/10.1016/j.bioactmat.2017.10.001.

[48] S. Farjaminejad, R. Farjaminejad, F. Garcia-Godoy, Nanoparticles in bone regeneration: a narrative review of current advances and future directions in tissue engineering, Journal of Functional Biomaterials 15 (9) (2024) 241. Available from https://doi.org/10.3390/jfb15090241.

[49] C. Wang, S. Min, Y. Tian, Injectable and cell-laden hydrogel in the contained bone defect animal model: a systematic review, Tissue Engineering and Regenerative Medicine 20 (6) (2023) 829–837. Available from https://doi.org/10.1007/s13770-023-00569-2.

[50] N.Z. Laird, T.M. Acri, K. Tingle, A.K. Salem, Gene- and RNAi-activated scaffolds for bone tissue engineering: current progress and future directions, Advanced Drug Delivery Reviews 174 (2021) 613–627. Available from https://doi.org/10.1016/j.addr.2021.05.009, www.elsevier.com/locate/drugdeliv.

[51] R. Liang, R. Li, W. Mo, X. Zhang, J. Ye, C. Xie, et al., Engineering biomimetic silk fibroin hydrogel scaffolds with "organic-inorganic assembly" strategy for rapid bone regeneration, Bioactive Materials 40 (2024) 541–556. Available from https://doi.org/10.1016/j.bioactmat.2024.06.024.

[52] G. Totonelli, P. Maghsoudlou, M. Garriboli, J. Riegler, G. Orlando, A.J. Burns, et al., A rat decellularized small bowel scaffold that preserves villus-crypt architecture for intestinal regeneration, Biomaterials 33 (12) (2012) 3401–3410. Available from https://doi.org/10.1016/j.biomaterials.2012.01.012.

[53] E. Elia, D. Brownell, S. Chabaud, S. Bolduc, Tissue engineering for gastrointestinal and genitourinary tracts, International Journal of Molecular Sciences 24 (1) (2023) 9. Available from https://doi.org/10.3390/ijms24010009.

[54] F.G. Sala, S.M. Kunisaki, E.R. Ochoa, J. Vacanti, T.C. Grikscheit, Tissue-engineered small intestine and stomach form from autologous tissue in a preclinical large animal model, Journal of Surgical Research 156 (2) (2009) 205–212. Available from https://doi.org/10.1016/j.jss.2009.03.062.

[55] S.R. Finkbeiner, J.J. Freeman, M.M. Wieck, W. El-Nachef, C.H. Altheim, Y.H. Tsai, et al., Generation of tissue-engineered small intestine using embryonic stem cell-derived human intestinal organoids, Biology Open 4 (11) (2015) 1462–1472. Available from https://doi.org/10.1242/bio.013235, http://bio.biologists.org/content/biolopen/4/11/1462.full.pdf.

[56] Q. Mao, H. Pan, Y. Zhang, Y. Zhang, Q. Zhu, Y. Hong, et al., GelNB molecular coating as a biophysical barrier to isolate intestinal irritating metabolites and regulate intestinal microbial homeostasis in the treatment of inflammatory bowel disease, Bioactive Materials 19 (2023) 251–267. Available from https://doi.org/10.1016/j.bioactmat.2022.04.001, http://www.keaipublishing.com/en/journals/bioactive-materials/.

[57] A.C. Costa, P.M. Alves, F.J. Monteiro, C. Salgado, Interactions between dental MSCs and biomimetic composite scaffold during bone remodeling followed by in vivo real-time bioimaging, International Journal of Molecular Sciences 24 (3) (2023). Available from https://doi.org/10.3390/ijms24031827, http://www.mdpi.com/journal/ijms.

[58] Y. Han, J. Xu, H. Chopra, Z. Zhang, N. Dubey, W.L. Dissanayaka, et al., Injectable tissue-specific hydrogel system for pulp–dentin regeneration, Journal of Dental Research 103 (4) (2024) 398–408. Available from https://doi.org/10.1177/00220345241226649.

[59] Y. Qian, J. Gong, K. Lu, Y. Hong, Z. Zhu, J. Zhang, et al., DLP printed hDPSC-loaded GelMA microsphere regenerates dental pulp and repairs spinal cord, Biomaterials 299 (2023). Available from https://doi.org/10.1016/j.biomaterials.2023.122137, http://www.journals.elsevier.com/biomaterials/.

[60] M.S. Santos, A.B. dos Santos, M.S. Carvalho, New insights in hydrogels for periodontal regeneration, Journal of Functional Biomaterials 14 (11) (2023). Available from https://doi.org/10.3390/jfb14110545, http://www.mdpi.com/journal/jfb.

[61] S. Zang, G. Dong, B. Peng, J. Xu, Z. Ma, X. Wang, et al., A comparison of physicochemical properties of sterilized chitosan hydrogel and its applicability in a canine model of periodontal regeneration, Carbohydrate Polymers 113 (2014) 240–248. Available from https://doi.org/10.1016/j.carbpol.2014.07.018.

[62] O. Krasilnikova, A. Yakimova, S. Ivanov, D. Atiakshin, A.A. Kostin, D. Sosin, et al., Gene-activated materials in regenerative dentistry: narrative review of technology and study results, Federation International Journal of Molecular Sciences 24 (22) (2023). Available from https://doi.org/10.3390/ijms242216250, http://www.mdpi.com/journal/ijms.

[63] C.I. Chamorro, S. Zeiai, N. Juul, O. Willacy, J. Huo, J. Hilborn, et al., Exploring the concept of in vivo guided tissue engineering by a single-stage surgical procedure in a rodent model, International Journal of Molecular Sciences 23 (20) (2022). Available from https://doi.org/10.3390/ijms232012703, http://www.mdpi.com/journal/ijms.

[64] X.G. Li, I.S. Park, B.H. Choi, U.J. Kim, B.H. Min, In vivo bioreactor using cellulose membrane benefit engineering cartilage by improving the chondrogenesis and modulating the immune response, Tissue Engineering and Regenerative Medicine. 17 (2) (2020) 165–181. Available from https://doi.org/10.1007/s13770-019-00236-5, http://www.springer.com/biomed/journal/13770.

[65] B. Akar, A.M. Tatara, A. Sutradhar, H.Y. Hsiao, M. Miller, M.H. Cheng, et al., Large animal models of an in vivo bioreactor for engineering vascularized bone, Tissue Engineering - Part B: Reviews 24 (4) (2018) 317–325. Available from https://doi.org/10.1089/ten.teb.2018.0005, http://www.liebertonline.com/teb.

[66] V. Borisov, L. Gili Sole, G. Reid, G. Milan, G. Hutter, M. Grapow, et al., Upscaled skeletal muscle engineered tissue with in vivo vascularization and innervation potential, Bioengineering 10 (7) (2023). Available from https://doi.org/10.3390/bioengineering10070800, www.mdpi.com/journal/bioengineering.

[67] A.H. Huang, Y.U. Lee, E.A. Calle, M. Boyle, B.C. Starcher, J.D. Humphrey, et al., Design and use of a novel bioreactor for regeneration of biaxially stretched tissue-engineered vessels, Tissue Engineering - Part C: Methods 21 (8) (2015) 841–851. Available from https://doi.org/10.1089/ten.tec.2014.0287, http://www.liebertonline.com/tec.

[68] E. Watson, A.G. Mikos, Advances in in vitro and in vivo bioreactor-based bone generation for craniofacial tissue engineering, BME Frontiers 4 (2023). Available from https://doi.org/10.34133/bmef.0004, https://spj.science.org/doi/full/10.34133/bmef.0004.

[69] E. Watson, A.M. Tatara, J.J.J.P. Van Den Beucken, J.A. Jansen, M.E. Wong, A.G. Mikos, An ovine model of in vivo bioreactor-based bone generation, Tissue Engineering - Part C: Methods 26 (7) (2020) 384–396. Available from https://doi.org/10.1089/ten.tec.2020.0125, http://www.liebertonline.com/tec.

[70] S.C. Wistner, L. Rashad, G. Slaughter, Advances in tissue engineering and biofabrication for in vitro skin modeling, Bioprinting 35 (2023) e00306. Available from https://doi.org/10.1016/j.bprint.2023.e00306.

[71] C.M. Moysidou, C. Barberio, R.M. Owens, Advances in engineering human tissue models, Frontiers in Bioengineering and Biotechnology 8 (2021). Available from https://doi.org/10.3389/fbioe.2020.620962, http://journal.frontiersin.org/journal/bioengineering-and-biotechnology#archive.

[72] P. Bédard, S. Gauvin, K. Ferland, C. Caneparo, È. Pellerin, S. Chabaud, et al., Innovative human three-dimensional tissue-engineered models as an alternative to animal testing, Bioengineering 7 (3) (2020) 115. Available from https://doi.org/10.3390/bioengineering7030115.

Chapter 15

Comprehending the role of in vivo models in tissue engineering

Shefali Pareek[1], Mukesh Saran[2], and Ashima Bagaria[1]
[1]*Department of Physics, Manipal University Jaipur, Jaipur, Rajasthan, India,* [2]*Directorate of Research, Manipal University Jaipur, Jaipur, Rajasthan, India*

15.1 Introduction

Tissue engineering (TE) applies engineering, medical, and physiological methods to create new tissues that function similarly to natural ones, serving as biological substitutes for invasive techniques such as dental implants, orthopedic implants, cardiovascular implants, soft tissue implants and nerve implants, and tedious surgeries. Its goal is to restore, repair, or enhance the function of organs within the body . New tissue can be created from human and animal sources, hybrid materials, or entirely synthetic components, for example, PLA (polylactic acid), PGA (polyglycolic acid), and PCA (polycaprolactone), falling under the category of synthetic material. Chitosan-PCL and collagen-PLGA are hybrid materials; these material plays an important role in maintaining mechanical properties while supporting cellular functions, also leading to an ideal condition within cell-scaffold interaction. The TE process begins with the extraction of cells from a cell source, which are then cultured and expanded. These cells are then embedded in a scaffold structure to prepare for the final implantation stage, where the engineered tissue is integrated into the patient's body [1]. Tissue engineering is an important aspect when understanding cellular physiology, cell division, cell growth, and cell death. For that, some models have been used to perform tests and visualize results, and these are called in vivo models.

These are living organisms used for experimental setups to study the biological processes and to get positive tissue interaction, disease progression, and therapeutic interventions in a controlled laboratory environment [2]. In TE, in vivo models are essential for evaluating how engineered tissues or biomaterials interact within a living organism, assessing factors such as immune response, vascularization, and functional integration over time [3].

Animal models like mice, rats, rodents, rabbits, pigs, dogs, hamsters, and nonhuman primates help replicate human-like physiological conditions, enabling scientists to study healing, immune reaction, cell–cell interaction, cell–drug interaction, and tissue regeneration within an intact biological system, as shown in Fig. 15.1.

These insights are pivotal for understanding the feasibility and safety of biomaterials and engineered tissues, thus bridging laboratory research and clinical application very strongly (Fig. 15.2 and Fig. 15.3). This figure was created in Biorender.com.

15.2 Historical approaches to tissue engineering

The early onset of making up tissues is an interesting phenomenon where, in the onset of 1970s, a pediatric orthopedic surgeon, Dr. W.T. Green, at the Children's Hospital, conducted experiments to generate a completely new cartilage. He embedded a flexible connective tissue, chondrocytes upon bone spicules, and inserted them in nude mice, stating pioneering efforts in TE for cartilage regeneration. This came out to be unsuccessful, but then he finally concluded it with the development of biocompatible materials [4]. It would be possible to form a whole new tissue by embedding cells on structured scaffolds.

Years later, Drs. Burke and Yannas from Massachusetts General Hospital and MIT collaborated with each other to create a tissue-engineered skin substitute. They used a collagen matrix to support the growth of dermal fibroblasts, which was tested both in the laboratory and in humans [5]. The rise of this emerging field happened in the mid-1980s when Dr.

Fundamentals of Translational Tissue Engineering. DOI: https://doi.org/10.1016/B978-0-443-33455-9.00005-5

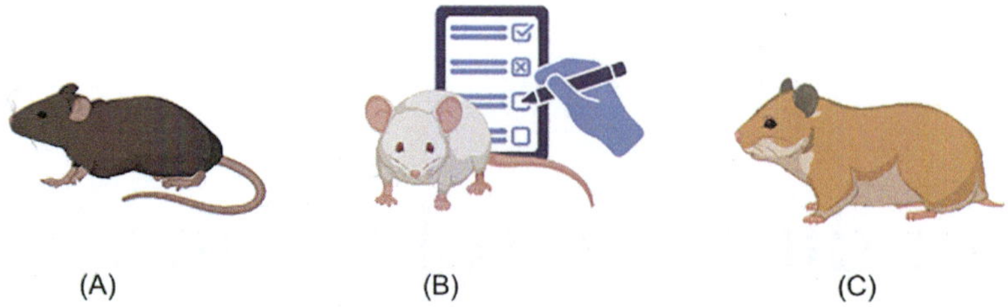

FIGURE 15.1 The figure represents the in vivo animal models for laboratory research.

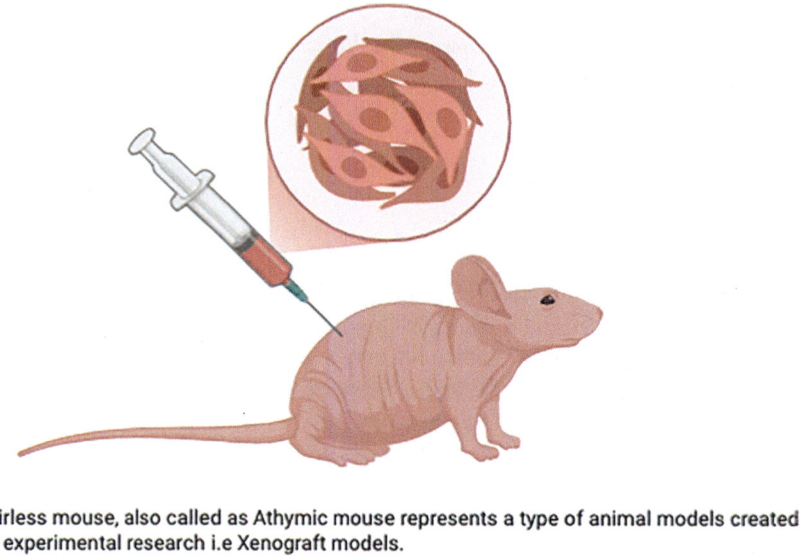

FIGURE 15.2 This figure represents one of the types of animal models created for experimental research, that is, xenograft models commonly known as nude mice or Athymic mice. These are a type of immunodeficient mouse that lack thymus that prevents the formation of T cells in the body. This image was created by biorender.com.

Joseph Vacanti approached Dr. Robert Langer with an idea to design custom scaffolds for cell delivery, instead of using naturally occurring scaffolds that were unpredictable.

Dr. Vacanti then conducted extensive studies to create functional tissue equivalents. He developed scaffolds made from synthetic, biocompatible, and biodegradable polymers, which were seeded with viable cells to generate tissue. These pioneering efforts laid the foundation for the emerging field of TE [1].

The development of functional tissues relies not only on the biological components but also on the structural and supportive material, that is, biomaterials. These materials have the ability to enhance cell growth, cell attachment, and tissue regeneration to form cartilage or a desired microenvironment that closely resembles the natural ones for medical purposes.

They can be derived from natural or synthetic sources and are also classified based on their origin. These are polymer-based materials, ceramics, metal-based materials, and composites [3]. Naturally derived biomaterials include collagen-based materials and chitosan, which mimic biological functions, while synthetic biomaterials, like polylactic acid (PLA) and hydroxyapatite, are designed for enhanced control and compatibility with TE needs. Natural biomaterials provide a friendly environment for cells to grow and provide the space for cell adhesion due to their bioactive nature[4]. They ease and support cellular responses like cell proliferation and cell differentiation.

They contain signals that cells recognize and respond to, helping the overall procedure of TE [6]. These materials excel in biocompatibility; hence, they are less likely to get triggered by Immune responses, thereby supporting in vivo testing. Biomaterials are defined broadly by their ability to integrate with living tissues and support cell growth, adhesion, and functional performance in medical applications [7]. Many natural biomaterials, like collagen, chitosan, gelatin, and alginate, are found to mimic the extracellular matrix.

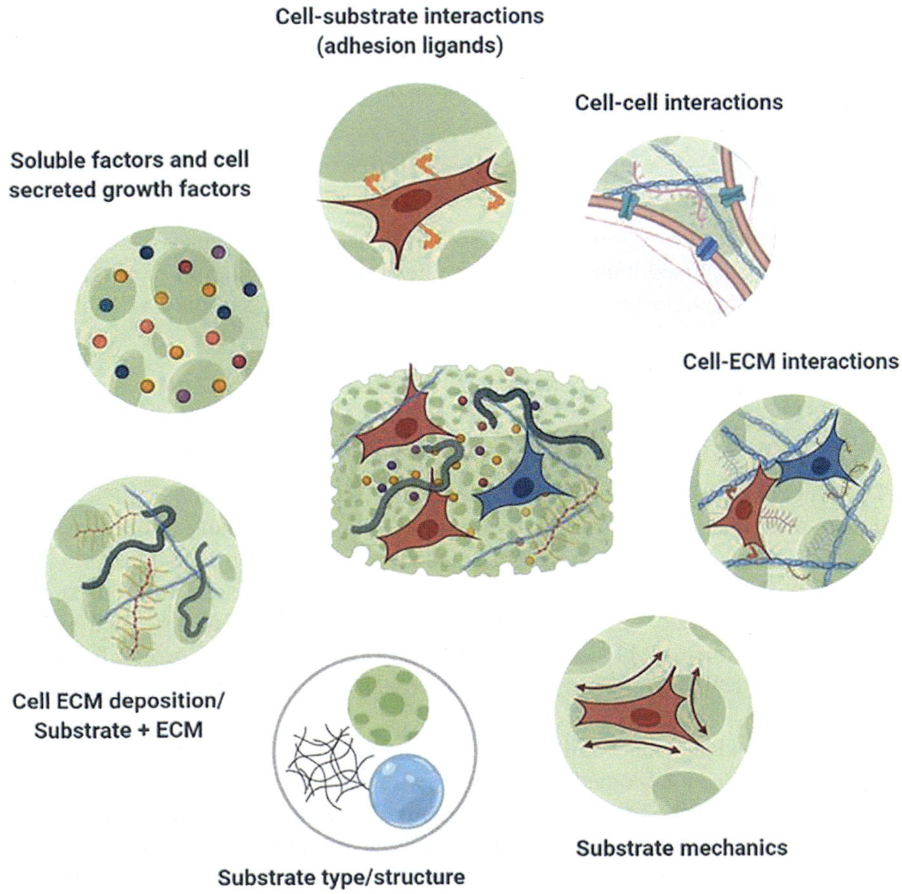

FIGURE 15.3 The figure represents the interactions studied with the help of in vivo models, including ECM-cell interaction, growth factors, cell–cell interaction, and substrate mechanisms.

Since the extracellular matrix is one of the complex networks of proteins existing in the body. Some important extracellular proteins are collagen, glycoproteins, and elastin. Naturally occurring biomaterials that mimic the ECM are significant because they create a natural-like environment for cells, promoting integration and repair, eventually leading to no cell rejection and finally leading to healing. Using biomaterials along with animal models has been in trend for a long time. This goes back to when William Harvey in the 17th century used animals to demonstrate the concept of blood circulation, a crucial step in experimental biology [8]. Building on such early explorations, the importance of animal models evolved remarkably, especially in the mid-20th century. By the 1970s, a pivotal advancement occurred when Dr. W.T. Green employed animal models to explore TE prospects .

Scientist Green was involved in inventing TE constructs using chondrocytes in mice, developing the integration of biological science with engineering frameworks. His research encompasses the capacity of in vivo models to mimic human physiological responses, thus laying the groundwork for modern TE and cultivating the application of these models in biomedical research.

This historical backdrop illustrates the transformative role of in vivo models in bridging fundamental biology and therapeutic applications [5].

In vivo models are categorized based on their immune status and compatibility with the tissue or cells introduced. Each type of model offers different insights because of differentiated morphology and physiology, such as **xenograft models**, where human cells are induced in immunodeficient models to study human-tissue responses, **syngeneic models**, which involve genetically identical organisms, which are useful for studying immune responses, and **allograft models** where tissues are induced between genetically varied individuals of the same species. Other types include **orthotopic models**, which allow implantation in the native site for realistic anatomical studies, and **GEMM (genetically engineered mouse models)**, where specific genes are modified to model human diseases [9].

These diverse model types enable tailored experimentation across a range of immune interactions, biological processes, and disease conditions. One more category is introduced after these four, Disease-Specific models, in which animals are engineered or induced to have certain diseases, such as diabetes or osteoporosis, to test certain treatments in TE.

15.3 Types

Xenograft models: Xenograft models are a useful approach to evaluate the effectiveness and safety of newly engineered tissues and the direct effects of humanized monoclonal antibodies [10]. The xenograft animal models require the use of immunocompromised mice to enable engraftment. Xenografts typically make use of only a few human cell lines that grow quickly and are often sensitive to chemotherapy.

These models help in optimizing different biomaterials and scaffolds by allowing the scientists to filter and test them under conditions that mimic human physiology before moving to human trials. So that this can be quoted, that particular animal xenograft model performs well with a given natural or synthetic biomaterial for effective TE.

Syngeneic models: The syngeneic model involves using genetically similar or nearly genetically similar organisms to observe immune responses and tumor dynamics within a controlled and genetically consistent environment [7]. They provide insights into immune responses. In TE, syngeneic models provide valuable insights into how newly engineered tissues and biomaterials behave in the presence of a fully functional immune system. Both xenograft and syngeneic models simplify tumor complexity, reducing the interactions between neoplastic and stromal cells, including immune, vascular, and fibroblastic compartments [11]. Hence, it helps bridge laboratory research and clinical application, validating the evaluation of both cellular responses and tissue repair processes in newly engineered constructs before human trials.

Genetically engineered mouse models: Genetically engineered mouse models (GEMMs) have emerged as powerful and increasingly influential tools in therapeutic biomedical research, with particular significance in the field of cancer research. One of the major strengths of GEMMs lies in their ability to allow tumors to develop within their native tissue microenvironments, thereby closely replicating the complex biological and pathological processes observed in human disease [4]. By preserving the genetic architecture of cancerous tissues, GEMMs are capable of mimicking not only the molecular and biological characteristics of cancer but also aspects of its clinical progression. This makes them a highly valuable and more physiologically relevant alternative to conventional preclinical models.

Despite their clear advantages, only a limited number of GEMMs have been extensively characterized and validated for preclinical drug evaluation. Nevertheless, these models hold considerable potential to bridge the long-standing gap between basic laboratory research and clinical translation. Within the context of TE, the interaction between biomaterials and in vivo models is critically important for driving progress in regenerative medicine and the development of personalized therapeutic strategies. Biomaterials act as foundational scaffolds, offering both structural support and essential biochemical cues that guide tissue formation by promoting cell adhesion, proliferation, and differentiation [12].

In contrast, in vivo models provide a dynamic and highly complex biological environment that enables comprehensive evaluation of biomaterial performance, including biocompatibility, functional integration, and long-term stability following implantation at the target site. By closely simulating physiological conditions, these animal models generate detailed and clinically relevant insights into how biomaterials interact with native tissues, support vascularization, and withstand mechanical and biological stresses over time.

The integration of natural and synthetic biomaterials with in vivo model systems, therefore, represents a critical step in bridging laboratory-based research with translational applications, ultimately facilitating the development of more effective and personalized regenerative therapies [4]. Importantly, the capacity of in vivo models to capture the complexity of cellular processes such as cell proliferation, signaling pathways, cell–cell communication, and cell–matrix interactions provides an essential framework for advancing our understanding of tissue regeneration. Collectively, these insights pave the way toward a more seamless transition between experimental research and clinical implementation.

15.4 Mimicking the physiological conditions of a living organism

This property of mimicking a similar environment has given a space in research for assertive inferences about the drugs, reliability of a treatment, and exact counter for a certain disorder [13]. In vitro and in silico models fail to mimic the conditions occurring in the cell to provide the exact climate for reference, here enter the in vivo models, giving another level of detailed image of the cell current conditions, including how a cell signals, how cells interact with each other in different processes, including immune responses, wound healing and healing with or without drugs [14].

By providing a physiologically relevant environment, in vivo models are adding value and supporting the successful outcomes in personalized medicine and regenerative therapies. This comprehensive modeling helps us understand the complex environments, hence enabling us to perform tests and design procedures relevant to TE and innovations in treatments by making them easier to perform [15], along with being feasible and available for every class of society by bridging the gaps between research and clinical trials.

Vascularization: Adequate and well-organized vascularization is a fundamental requirement for the success of regenerative medicine, particularly within the field of TE. An understanding of the core principles of vascularization, including the role of angiogenic cells, the design and optimization of biomaterial scaffolds, and the application of advanced biofabrication strategies, is essential for developing functional engineered tissues. In addition, increasing attention has been directed toward the therapeutic applications of vascularized tissue constructs in regenerative medicine, alongside ongoing discussions surrounding emerging challenges and future directions within this rapidly evolving field.

Vascularization not only plays a critical role in supporting the survival of engineered tissues but also significantly enhances their integration with host tissues, functional performance, and long-term stability following implantation. As such, it represents a cornerstone of advanced regenerative strategies aimed at achieving clinically relevant and durable outcomes [8]. Without the establishment of an effective vascular network, engineered tissues are limited in size, functionality, and translational potential.

From a biological perspective, vascularization is a tightly regulated process that ensures tissues receive sufficient oxygen, nutrients, and metabolic support required for survival and homeostasis. In adult organisms, vascularization is typically activated during physiological processes such as tissue repair, regeneration, and wound healing, where the formation of new blood vessels restores circulation to damaged or ischemic regions [16]. Beyond physiological repair, vascularization also plays a pivotal role in pathological conditions, including disease progression and tumor development. Rapidly proliferating tumors often stimulate angiogenesis to satisfy their elevated metabolic demands, thereby facilitating continued growth and survival [17].

A comprehensive understanding of vascularization across both regenerative and pathological contexts is therefore indispensable for TE research. In particular, elucidating these mechanisms is crucial for the development of functional, vascularized tissues and organ constructs within in vivo models, ultimately supporting the translation of engineered systems from experimental settings to clinical applications.

Functional performance of engineered tissues: The functional performance of engineered tissue refers to how effectively the tissue replicates the physiological properties and behaviors of natural tissues when implanted or tested in the body. Key factors that determine functional performance include structural integrity, mechanical strength, biocompatibility, and vascularization. For example, engineered tissue for muscle repair should contract similarly to native muscle, while skin substitutes need to protect underlying tissues and aid in wound healing [18]. The goal in TE is to create constructs that not only survive within the biological environment but also support cell integration, growth, and differentiation, allowing them to perform intended physiological roles effectively over time [19]. Effective testing in an in vivo model is critical to evaluating these aspects, ensuring that the engineered tissues behave consistently under various biological conditions, integrate well with existing tissues, and maintain functionality in the long term.

Observing healing processes: Identification of cell healing process, how immune responses are generated, how cell adhesion occurs, then how cell recovery occurs can be done by using these animal models. What responses cells generate after receiving a healing in the form of TE and tissue repair by being able to study cell interactions and cell-ECM interaction, cell signaling, and the complete cell growth mechanism.

Studying cell interactions, precisely cell-to-cell and cell-to-extracellular matrix (ECM) communication, is a crucial step in TE. By exploring important cell mechanisms like cell signaling, adhesion, and growth, researchers can decode how cells coordinate for immune responses, recover from damage, and regenerate tissue [20]. TE anchorages insights to design biomaterials and scaffolds that support natural healing processes. This understanding also allows us to make precise interventions to mimic or enhance natural healing, providing innovative solutions for tissue repair and regeneration, key areas assessed through in vivo models for translational applications.

By mimicking disease environment: Studying the immune system with different diseases and treatments has provided an edge for new-age research. Introducing new branches like pharmacodynamics and pharmacokinetics would not have been possible without the concrete results seen in animal models, ultimately justifying the experiments and use of in vivo models in the complete research [18]. Biomedical research has evolved and will keep on evolving due to a platform founded by in vivo modeling that helps in disease progression and treatment testing in short durations, ultimately enhancing the therapeutic potential of the drugs and designed treatments.

15.5 Recent trends and innovations

CRISPR-Cas9 Gene Editing: The novel integration of CRISPR-Cas9 and in vivo models has revolutionized research by providing precise genetic modifications and changes that are required to adjust to the treatments. This innovative technology has made the procedure easy to monitor, perform, and less time-consuming (Alijabali et al., 2024). The precision and efficiency associated with this technique are at par as it is associated with the root level cause, that is, the genetic reasons for a particular event to occur, and hence is free from other time-consuming methods where other diagnostic tests are performed [10]. The in vivo models provide a realistic tissue environment that is dynamic and patient-specific, which makes the theranostics easy and doable, where biologically relevant conditions are easy to create to reach the conclusions, eventually filling the voids associated with research conduction and clinical trials [10]. Future trends associated with these techniques, including CRISPR gene editing, include the development of such biomaterials that act as genetic material delivery platforms for genetic manipulation and editing in in vivo models.
3D bioprinting and organoids: For creating Native like functionality and more precise models for testing, scientists are combining 3d bioprinting and organoids technologies to enhance the results based on in vivo models for creating a new area in the field of regenerative therapy and personalized medicine. Such techniques are now used in waste removal by creating bioprinted vascular constructs and also leading to improvement in nutrient absorption, hence supporting in vivo studies in the long term [21]. Advancements in biomaterials, such as the development of smart materials that have responsive properties, can be used for enhancing cell behaviors when integrated with different implants and drugs. The combination of bioprinted tissues and organoid models is explored in diseases such as cancer, fibrosis, and neurodegeneration [22].
In vivo imaging techniques: When we can track the immune responses in real time, giving us a clear monitoring of tissue compatibility and inflammation occurring inside the cell, this can give us detailed cell signaling information and how cells act in the presence of different chemical and physical components [10]. Scaffold visualization, biodegradation, and compatibility can be assessed after the integration with native tissues, providing transparent reactions from the cell microenvironment. Different techniques like fluorescence and bioluminescence imaging using nanoparticles, micro-CT, ultrasound imaging, PAI, SPECT, PET, and MRI help evaluate the perfusion and growth occurring inside the cell [23]. Newly developed and recent trends include multimodal imaging, smart contrast agents, and longitudinal imaging.

15.6 Conclusion and future discussions

Selection of the most applicable in vivo models is essential to perform tissue regeneration, drug development, and theranostics for meticulous modeling of therapeutic efficiency. By progressing in innovational preclinical trials using in vivo animal models that mimic the human cell microenvironment, advances in the field of regenerative medicine and improvement in success rates for potential novel therapeutics in clinical trials can be assured.

Coming to overcome the difficulties and making desirable changes, new models and development have been taken up as a futuristic approach, such as integrating artificial intelligence and machine learning in in vivo models. This will not only refine personalized medicine but also make regenerative therapies more efficient [24]. Similarly, real-time monitoring will enhance the overall procedure and will be able to provide real-time monitoring and will helping us figure out how scaffolds behave inside the body over time with different drug interactions.

Another approach that has been in trend in recent times is humanized models; these models are patient-derived xenografts models or could be bioengineered organoids, which are bound to provide better outcomes in comparison.

Similar hybrid approaches are also considered, like organ-on-chip technology, where these preexisting in vivo models are combined with an organ-on-chip platform, bound to provide results in personalized testing environments. This will allow researchers to reduce animal reliance to test biomaterials and therapies more efficiently while ensuring high translational potential.

Lastly, the most in-trend technique which can be induced is injectable hydrogels, stem cell-loaded scaffolds, and gene editing technologies like CRISPR.

This will provide an edge for next-generation biomaterials where these are designed in such a way that they perform dynamically inside the cell microenvironment and respond effectively to provide treatments and therapies [10]. This synergy between in vivo models and emerging biomedical technologies will escalate our ability to restore damaged muscles, veins, bones, and tissues, ultimately revolutionizing the field of regenerative medicine.

List of abbreviations

PET	positron emission tomography
CRISPR	clustered regularly short interspaced palindromic repeats
CT	computed tomography
MRI	magnetic resonance imaging
PAI	photoacoustic imaging
SPECT	single-photon emission computed tomography
CAS9	CRISPR-associated protein 9
ECM	extracellular Matrix
PLGA	poly(lactic-*co*-glycolic acid)
PGA	polyglycolic acid
GEMM	genetically engineered mouse model

References

[1] S. Park, K.A. Rahaman, Y.C. Kim, H. Jeon, H.S. Han, Fostering tissue engineering and regenerative medicine to treat musculoskeletal disorders in bone and muscle, Bioactive Materials 40 (2024) 345–365.

[2] M.U.A. Khan, M.A. Aslam, M.F.B. Abdullah, W.S. Al-Arjan, G.M. Stojanovic, A. Hasan, Hydrogels: classifications, fundamental properties, applications, and scopes in recent advances in tissue engineering and regenerative medicine–a comprehensive review, Arabian Journal of Chemistry 17 (10) (2024) 105968.

[3] S. Liu, J.M. Yu, Y.C. Gan, X.Z. Qiu, Z.C. Gao, H. Wang, et al., Biomimetic natural biomaterials for tissue engineering and regenerative medicine: new biosynthesis methods, recent advances, and emerging applications, Military Medical Research 10 (1) (2023) 16.

[4] B. Godin, E. Touitou, Transdermal skin delivery: predictions for humans from in vivo, ex vivo, and animal models, Advanced Drug Delivery Reviews 59 (11) (2007) 1152–1161.

[5] C.A. Vacanti, The history of tissue engineering, Journal of Cellular and Molecular Medicine 10 (3) (2007) 569.

[6] C. Su, D. Lin, X. Huang, J. Feng, A. Jin, F. Wang, et al., Developing hydrogels for gene therapy and tissue engineering, Journal of Nanobiotechnology 22 (1) (2024) 182.

[7] S. Ud-Din, A. Bayat, Non-animal models of wound healing in cutaneous repair: in silico, in vitro, ex vivo, and in vivo models of wounds and scars in human skin, Wound Repair and Regeneration 25 (2) (2017) 164–176.

[8] P. Mukherjee, S. Roy, D. Ghosh, S.K. Nandi, Role of animal models in biomedical research: a review, Laboratory Animal Research 38 (1) (2022) 18.

[9] Z. Xu, B. Mao, H. Liu, S. Wang, X. Chen, S. Guo, Integrative multi-omics characterization of 12 syngeneic mouse models, iScience (2025).

[10] A. Gu, J. Li, M.Y. Li, Y. Liu, Patient-derived xenograft model in cancer: establishment and applications, MedComm 6 (2) (2025) e70059.

[11] M. Mahdizade Ari, L. Dadgar, Z. Elahi, R. Ghanavati, B. Taheri, Genetically engineered microorganisms and their impact on human health, International Journal of Clinical Practice 2024 (1) (2024) 6638269.

[12] A. Hotta, S. Schrepfer, A. Nagy, Genetically engineered hypoimmunogenic cell therapy, Nature Reviews Bioengineering 2 (11) (2024) 960–979.

[13] Z. Li, W. Zheng, H. Wang, Y. Cheng, Y. Fang, F. Wu, et al., Application of animal models in cancer research: recent progress and future prospects, Cancer Management and Research (2021) 2455–2475.

[14] M. Boudna, A.D. Campos, P. Vychytilova-Faltejskova, T. Machackova, O. Slaby, K. Souckova, Strategies for labelling of exogenous and endogenous extracellular vesicles and their application for in vitro and in vivo functional studies, Cell Communication and Signaling 22 (1) (2024) 171.

[15] S.A. Mosaddad, A. Hussain, H. Tebyaniyan, Exploring the use of animal models in craniofacial regenerative medicine: a narrative review, Tissue Engineering Part B: Reviews 30 (1) (2024) 29–59.

[16] M.U. Aslam Khan, M.A. Aslam, M.F. Bin Abdullah, G.M. Stojanović, Current perspectives of protein in bone tissue engineering: bone structure, ideal scaffolds, fabrication techniques, applications, scopes, and future advances, ACS Applied Bio Materials 7 (8) (2024) 5082–5106.

[17] A. Pierre, R. Favory, S. Lancel, S. Preau, Can patient-derived in vitro models improve clinical translation in critical care research when used before animal studies? Intensive Care Medicine Experimental 13 (1) (2025) 1–6.

[18] A. Horton, I.T. Schiefer, Pharmacokinetics and pharmacodynamics of nitric oxide mimetic agents, Nitric Oxide 84 (2019) 69–78.

[19] C.M. Moysidou, C. Barberio, R.M. Owens, Advances in engineering human tissue models, Frontiers in Bioengineering and Biotechnology 8 (2021) 620962.

[20] S. Das, S. Sahoo, S. Pattanaik, R.K. Prusty, B. Barik, B.S. Satapathy, et al., Advancing breast cancer research: a comprehensive review of in vitro and in vivo experimental models, Medical Oncology 42 (8) (2025) 316.

[21] Y. Zhang, X. Lv, Y. Wang, X. Chen, J. Zhang, D. Su, Recent advances in self-immobilizing fluorescent probes for in vivo imaging, Smart Molecules 2 (3) (2024) e20240031.

[22] M. Cabral, K. Cheng, D. Zhu, Three-dimensional bioprinting of organoids: past, present, and prospective, Tissue Engineering Part A 30 (11-12) (2024) 314–321.

[23] Y. Zhang, G. Li, J. Wang, F. Zhou, X. Ren, J. Su, Small joint organoids 3D bioprinting: construction strategy and application, Small 20 (8) (2024) 2302506.

[24] R. Bagherpour, G. Bagherpour, P. Mohammadi, Application of artificial intelligence in tissue engineering, Tissue Engineering Part B: Reviews 31 (1) (2025) 31–43.

Chapter 16

Animals as preclinical models for tissue engineering in retina

Shikha Upreti, Neilanjan Maitra, and Madhumita P. Ghosh
Centre for Medical Biotechnology, Amity Institute of Biotechnology, Amity University, Noida, Uttar Pradesh, India

16.1 Introduction

The European Medicines Agency (EMA), the US Federal Food and Drug Administration (FDA), and the International Society for Stem Cell Research (ISSCR) recommend the use of large animal models for translational research [1]. The practice of maintaining cells and tissues in vitro allows for a systematic and standardized analysis of various cellular, biophysical, and biochemical queries in a controlled environment, but these can only offer an abstract insight into the pathophysiology of diseases and disorders [2]. On the other hand, animal models, even with their inherent natural variability, cannot be overlooked as they provide a coherent physiological system that is under neuronal control, and the outcomes achieved can be directed to optimize the experiments concerning their translatability and the welfare of the animals.

Small rodent animals, specifically rats and mice, are generally regarded as valuable for the study of disease mechanisms and fundamental biology, but the translatability of the outcomes from such small animals into clinical applications is often questionable. As there was a considerable difference in bioavailability of pharmaceuticals between primates, pigs, rodents, and humans [1] other animals like horses, sheep, and pig were preferred as models due to their long gestational periods. This approach resulted in sufficient temporal resolution to translate their findings into human parameters. Due to nonavailability of larger animals and restrictions on their use as models for preclinical testing, rodents remained the only choices for biomedical research.

The development of stem cell transplantation of fetal tissue origin gained significance in rodent models of retinal diseases despite the limitation of nonhuman features like small eyes, rod-dominant vision, absence of macula, and short life span [2]. The chicken has the advantage of having a relatively large proportion of cone photoreceptor cells, somewhat similar to that of the human retina [3]. RCS rats have been developed as models of retinitis pigmentosa (RP) and when administered with bone marrow-derived mesenchymal stem cells, resulted in improvement in visual acuity and associated vascular pathology [4]. Human central nervous system-derived stem cells (HuCNS-SC) when transplanted into a rodent model of retinal degeneration, RCS rat, could maintain photoreceptor density, visual acuity, and luminescence sensitivity functional outcomes [5].

16.2 Tissue engineering and its challenges

Tissue engineering encompasses designing biodegradable or nondegradable scaffolds that establish a physical interface with the sheets of cultured cells. The interface of cultured monolayered cells with different biomaterials requires an effective delivery system for site-directed transplantation of cells. Retina, being of neural origin, is difficult to regenerate since most retinal degenerations like glaucoma, age-related macular degeneration (AMD), or diabetic retinopathy (DR) involve irreversible loss of retinal pigment epithelium (RPE), rods, cones, and retinal ganglion cells (RGCs). Since the 1990s the construction of scaffolds that are able to carry the RPCs and deliver in the vicinity of the damaged inner retina has gained impetus and reported success in the recovery of photoreceptors and RPE [6]. The site-specific integration of cells into the damaged tissue and long-term functionality of the tissue are of utmost significance.

Fundamentals of Translational Tissue Engineering. DOI: https://doi.org/10.1016/B978-0-443-33455-9.00008-0

In retinal degenerations, where the RPE cell sheet rests on the blood retinal barrier and supports the photoreceptors, becomes dysfunctional and even an initiative of bolus cell injection at the degeneration site becomes ineffective frequently due to inflammation, and requires mechanical support for maintaining retinal microenvironment through selective passage into and out of the retina. A series of biomaterials have been found adequate for the creation of scaffolds for the delivery of RPCs [7]. Besides being biodegradable, the scaffolds need to be nontoxic, thin enough not to affect the focal length of the eye, strong enough to survive the transplant procedure, yet flexible enough to conform to the curvature of the retina [8–10]. Several studies have shown that bioengineered scaffolds used to deliver normal human retinal pigment epithelium (ARPE-19) cells could maintain the structural and functional integrity of damaged RPE [11,12]. Retinal degeneration, being a frequent ocular problem in old age, and eye injuries requires reversal of the functional RPE layer through the use of various biomaterials as scaffolds [13]. ARPE-19 cells and human corneal epithelial (HCE-T) cells were cultured on poly(ε-caprolactone) (PCL) nanofibers of different diameters, nominally 500 and 1300nm which demonstrated that the response can differ markedly from scaffold architecture [14]. RPE cells undergo dedifferentiation in culture, losing important characteristics after repeated passage. Transplanted hESC-RPE/polyester composites survived surgical delivery in a pig model under optimal growth conditions to achieve a high level of differentiation [15], ability, and physiological function in vitro used to deliver an intact cell sheet into the eye [16].

16.3 Ocular disease models

Retinal degenerative diseases that lead to irreversible blindness include age AMD, glaucoma, DR, and RP. Cell replacement therapy evolved as an effective therapeutics of these diseases, but appropriate animal models work as syndicate to the therapies. The success of stem cell delivery in retinal degeneration relies on well-fabricated scaffolds for survival, differentiation, and delivery of transplanted cells [17]. There is a paucity of suitable animal models for complex ocular diseases like DR and AMD. The inherited optic neuropathies are a group of diseases that are characterized by optic nerve dysfunction, primarily due to alterations in ganglion cell function and ultimately cell death. The clinical phenotypes of such diseases are slightly heterogeneous both within and between various disease states. Although various animal models are available, but only a few are summarised below.

16.4 Retinitis pigmentosa

RP is a genetic disease characterized by inheritance of a mutated gene, which results in loss of photoreceptors. To restore photoreceptors and replace RPE, a number of cellular sources like embryonic stem cells (ESCs), fetal tissue, iPSCs, and adult stem cells have been found to grow and multiply in culture that could be grafted for the repair of a degenerating retina. The success of an efficient cell transplantation method could be tested only in preclinical or animal models. Animal models of retinal degeneration have helped us to analyze the functional and structural changes in the dimensions of the retinal layers during the onset of degenerative diseases. Among the animal models, mouse models of retinal diseases with a primary focus on photoreceptor loss and RPE dysfunction, and synaptic abnormalities helped in the assessment of novel therapeutic modalities. This led to mutant mouse models of photoreceptor dysfunction and AMD [18] through gene knockout techniques. An RP GTPase regulator (RPGR) deficient model of both rod and cone cell degeneration was created to elucidate the pathobiology of the disease. The RPGR mutation led to the disruption of the unidirectional movement of opsin in the photoreceptor cell with resultant cell degeneration [19]. The symptoms of RP in humans could be simulated in a porcine model too [20]. Therapeutic RPE replacement strategies are currently being tested in small clinical trials (phase I) for AMD and in some types of RP (China: NCT02755428, NCT03944239, NCT03046407; United States of America: NCT01345006, NCT01344993, NCT02286089).

16.5 Glaucoma models

Glaucoma and DR involve retinal neuronal death due to increased intraocular pressure (IOP), followed by obstruction of aqueous humor outflow and glutamate neurotoxicity, leading to death of RGCs, which cannot regenerate once the neurons are damaged. DR is also associated with neovascularization and macular edema where anti-VEGF therapy proves to be the only sustainable solution to prevent vision loss. Recent postmortem studies of individuals with advanced retinal degeneration have shown that 88% of the retina's inner nuclear layer and 48% of the retinal ganglion layer remain viable even after significant retinal degeneration [21–24]. This advocates the revival of residual viable neuronal cells in retinal ganglion layer through engineered RGCs on fabricated scaffolds. Other animal models of retinal cell degeneration and

RGC apoptosis have been developed using N-methyl-D-aspartate (NMDA), Cocl2, or optic nerve crush, initiating the caspase cascade and death of RGCs [25].

Interestingly, extremolytes such as ectoine and hydroxyectoine also had a protective effect on the hypoxia-damaged retina in the porcine organ culture model and thus represent a potentially interesting therapeutic approach for glaucoma patients [26].

16.6 Age-related macular degeneration

AMD is a degenerative disease of the macula marked by a genetic mutation, which leads to blindness due to deposits of yellow lipoproteins between the RPE and Bruch's membrane [2]. AMD can be classified into neovascular (wet) and atrophic (dry) AMD. In its early phase, dry (nonexudative) AMD affects RPE and secondarily destroys photoreceptors, whereas wet (neovascular) AMD affects the RPE but often involves degeneration and aberrant growth of the choriocapillaris and deep capillary plexus [6].

RPE resting on the Bruch's membrane plays a very significant role in the visual process and demands immediate attention. Stem cells grown from varying sources, fabricated with scaffolds, were delivered in retinal degeneration models for the repair of RPE [27]. Animals that served as models of AMD include rodents like mice and rats, nonhuman primates, which can recreate histological features of AMD [28]. In a review describing RPE transplantation, subgroup analyses revealed use of young and adolescent animals was better than that of adult animals [29]. In mice, models of retinal degeneration hydrogel-mediated cotransplantation of RPE and photoreceptors restored vision [30].

Co-transplantation of hESC-derived retinal organoids with RPE leads to retinal reconstruction in immunodeficient retinal degenerate RCS rats [31]. In case of cone cell dysfunction, a cone-specific human induced pluripotent stem cell (hiPSC) GFP-reporter line transplanted into recipient mice revealed extensive and polarized incorporation of cones into mouse outer nuclear layer and showed long-term survival up to six months [32].

The retinal remodeling observed in this manner is identical to that visible in genetic models. hESC-RPE monolayer xenograft induced to grow to embryoid body structures, multiplied in laminin-coated plates, and then integrated in the submacular location in surgical nonhuman primates [33].

16.7 Retinal ganglion cells

Transplantation of stem cells or progenitors into the retina to compensate for RGC loss did not show appreciable progress. In rat models, transplanted RGCs can migrate into the ganglion cell layer and are capable of extending neurites into injured adult retinas ex vivo [34]. Recent evidence indicates that transplanted postnatal mouse RGCs acquire the morphology of endogenous RGCs, with axons orienting toward the optic nerve head of the host retina. These engrafted RGCs form synapses with the lateral geniculate nucleus and respond to light [35]. The potential application of hiPSCs-derived retinal organoids as intraocular implants is to replace lost RGCs [36]. Transplantation of hiPSCs onto the inner surface of rhesus monkey retinas using biodegradable polylactic-coglycolic acid (PLGA) scaffolds directed toward an RGC fate resulted in green fluorescent protein-labeled axonal projections toward the host optic nerve [35].

16.8 Mutant models of retinal degeneration

A genetically modified mouse, for example, the optineurin E50K knock-in or the glutamate aspartate transporter (*GLAST*)-deficient mouse was raised to mimic glaucomatous condition, and excitotoxicity was induced by injecting the glutamate analog NMDA intravitreally [37]. The phytochemical, Carvacrol (CAR), is found to be neuroprotective and antiapoptotic against RGC degeneration in a rabbit model of retinal cell degeneration [38].

More interesting results were obtained with saffron, referred to as "repron," in an animal model of RP (*rd*1 and *rd*10 mice), showing a slowing down of rod degeneration due to genetic mutation [39]. Attempts to use genetically engineered cell lines, $Bhlhb4^{-/-}$ and $Islet1^{-/-}$, a transcription factor expressed in rod bipolar cells, which is also required for maintenance and neural maturation of rod bipolar cells, were engaged for rod bipolar cell fate, and the nascent cells succumbed to apoptosis [40]. Allogenically iPSC-RPE transplanted in retinas of RCS rats and mutant models of immune-deficient $NOD/SCID/IL\text{-}2Rgc^{null}$ (NSG) mice rescued visual function as observed through fundus photography and focal electroretinogram [41]. Exposure to multicharacteristic opsin gene therapy to mouse models of RP, *rd*1, and *rd*10 mice attenuates structural retinal degeneration [42]. A mutant model of RP, TgP23H, showed altered rhodopsin expression but rescued cone photoreceptor morphology [43] (Table 16.1).

TABLE 16.1 Scaffolds, cellular therapies, and animal models used in retinal tissue engineering.

Scaffold used	Cells	Animal model	Reference
Polyester composites	hESC-RPE	Pig	[15,16]
Laminin-coated plates	hESC-derived RPE	Nonhuman primates	[44]
Hydrogel-scaffolds	RPE and photoreceptors	Mice	[45]
PLGA scaffolds	hiPSC derived RGCs	Rhesus monkey	[35]
Bioengineered scaffolds	ARPE-19 cells	Rodents	[11,12]
Biodegradable PLGA scaffolds	hiPSCs	Rhesus monkey	[35]
Laminin and biomaterial-based scaffolds	Human stem cell-derived retinal organoids	RCS rats	[31]
Genetically engineered rod bipolar cells	Bhlhb4–/–, Islet1–/– rod bipolar cells	Mouse models	[40]
Laminin-coated plates	iPSC-RPE	RCS rats, NOD/SCID/IL-2RGCnull mice	[41]

hiPSC, Human-induced pluripotent stem cells; *PLGA*, polylactic-co-glycolic acid.

16.9 Conclusion

This chapter is an attempt at developing the reader's insight on some of the latest techniques used for preventing vision loss, while also highlighting their scientific advancements and limitations. It thoroughly examines various therapeutic approaches studied across a range of animal models, from simple rodent models to larger animals and genetically engineered mutants. Earlier studies have utilized rodent models to investigate conditions such as elevated IOP, RGC degeneration, and photoreceptor loss, as well as pathologies like AMD and neovascularization. These findings were subsequently extrapolated to larger animal models, such as horses and pigs, to enhance translatability to human applications. More recently, mutant models with targeted gene mutations have enabled detailed analysis of the structural and functional defects associated with specific genetic abnormalities, providing critical insights into disease mechanisms and potential therapeutic interventions.

References

[1] J.H. Laura, N.J. Koehl, J.P. O'Shea, E.S. Kostewicz, R. Holm, B.T. Griffin, The pig as a preclinical model for predicting oral bioavailability and in vivo performance of pharmaceutical oral dosage forms: a PEARRL review, The Journal of Pharmacy and Pharmacology 71 (4) (2018) 581–602. Available from https://doi.org/10.1111/jphp.12912.

[2] H. Klassen, B.A. Tucker, C.G. Liew, M.L. Cour, H.C. Wang, Stem cells in large animal models of retinal and neurological disease, Stem Cells Internationals: (2012) 460504. Available from https://doi.org/10.1155/2012/460504.

[3] A. Wu, R. Lu, E. Lee, Tissue engineering in age-related macular degeneration: a mini-review, Journal of Biological Engineering 16 (11) (2022). Available from https://doi.org/10.1186/s13036-022-00291-y.

[4] S. Wang, B. Lu, S. Girman, J. Duan, T. McFarland, Q. Zhang, et al., Non-invasive stem cell therapy in a rat model for retinal degeneration and vascular pathology, PLoS One (2010). Available from https://doi.org/10.1371/journal.pone.0009200.

[5] T.J. McGill, B. Cottam, B. Lu, S. Wang, S. Girman, C. Tian, et al., Transplantation of human central nervous system stem cells – neuroprotection in retinal degeneration, The European Journal of Neuroscience 35 (3) (2012) 468–477. Available from https://doi.org/10.1111/j.1460-9568.2011.07970.x.

[6] L.J. Rizzolo, I.O. Nasonkin, R.A. Adelman, Retinal cell transplantation, biomaterials, and in vitro models for developing next-generation therapies of age-related macular degeneration, Stem Cells Translational Medicine. 11 (3) (2022) 269–281. Available from https://doi.org/10.1093/stcltm/szac001.

[7] C.E. White, R.M. Olabisi, Scaffolds for retinal pigment epithelial cell transplantation in age-related macular degeneration, Journal of Tissue Engineering 8 (2017) 2041731417720841. Available from https://doi.org/10.1177/2041731417720841.

[8] J. Ambati, B.J. Fowler, Mechanisms of age-related macular degeneration, Neuron 75 (1) (2012) 26–39. Available from https://doi.org/10.1016/j.neuron.2012.06.018.

[9] N.J.Y. Yeo, E.J.J. Chan, C. Cheung, Choroidal neovascularization: mechanisms of endothelial dysfunction, Frontiers Pharmacology 10 (2019) 1363. Available from https://doi.org/10.3389/fphar.2019.01363.

[10] R.F. Spaide, G.J. Jaffe, D. Sarraf, et al., Consensus nomenclature for reporting neovascular age-related macular degeneration data: consensus on neovascular age-related macular degeneration nomenclature study group, Ophthalmology 127 (5) (2020) 616–636. Available from https://doi.org/10.1016/j.ophtha.2019.11.004.

[11] S. Sapino, G. Chindamo, E. Peira, D. Chirio, F. Foglietta, L. Serpe, et al., Development of ARPE-19-equipped ocular cell model for in vitro investigation on ophthalmic formulations, Pharmaceuticals 15 (10) (2023) 2472. Available from https://doi.org/10.3390/pharmaceutics15102472.

[12] S. Bagewadi, S. Parameswaran, S. Krishnakumar, S. Sethuraman, A. Subramanian, Tissue engineering approaches towards the regeneration of biomimetic scaffolds for age-related macular degeneration, Journal of Material Chemistry B. 9 (30) (2020) 5935–5953. Available from https://doi.org/10.1039/d1tb00976a.

[13] D.S.R. Nair, M.J. Seiler, K.H. Patel, V. Thomas, J.C. Martinez Camarillo, M.S. Humayun, et al., Tissue engineering strategies for retina regeneration, Applied Sciences. 11 (5) (2021) 2154. Available from https://doi.org/10.3390/app11052154.

[14] L. Krishna, S. Nilawar, M. Ponnalagu, M. Subramani, C. Jayadev, R. Shetty, et al., Fiber diameter differentially regulates function of retinal pigment and corneal epithelial cells on nanofibrous tissue scaffolds, ACS Applied Bio Materials 3 (2) (2020) 823–837. Available from https://doi.org/10.1021/acsabm.9b00897.

[15] Ahmado, A. 2015. Optimisation of Substrate and Growth Conditions of Retinal Pigment Epithelial Cells Destined for Transplantation (Doctoral thesis), UCL discovery.

[16] A.A. Zuber, D.E. Robinson, R.D. Short, et al., Development of a surface to increase retinal pigment epithelial cell (ARPE-19) proliferation under reduced serum conditions, Journal of Materials Science: Materials in Medicine 25 (2014) 1367–1373. Available from https://doi.org/10.1007/s10856-014-5163-y.

[17] M. Trese, C.V. Regatieri, M.J. Young, Advances in retinal tissue engineering, Materials 5 (1) (2012) 108–120. Available from https://doi.org/10.3390/ma5010108.

[18] S.E. Brockerhoff, J.B. Hurley, U. JanssenBienhold, S.C. Neuhauss, W. Driever, J.E. Dowling, A behavioral screen for isolating zebrafish mutants with visual system defects, Proceedings of the National Academy of Sciences of the United States of America 92 (23) (1995) 10545–10549. Available from https://doi.org/10.1073/pnas.92.23.10545.

[19] D.H. Hong, B.S. Pawlyk, J. Shang, M.A. Sandberg, E.L. Berson, T. Li, A retinitis pigmentosa GTPase regulator (RPGR)-deficient mouse model for X-linked retinitis pigmentosa (RP3), Proceedings of the National Academy of Sciences of the United States of America 97 (7) (2000) 3649–3654. Available from https://doi.org/10.1073/pnas.060037497.

[20] R. Petters, C. Alexander, K. Wells, et al., Genetically engineered large animal model for studying cone photoreceptor survival and degeneration in retinitis pigmentosa, Nature Biotechnology 15 (1997) 965–970. Available from https://doi.org/10.1038/nbt1097-965.

[21] J.L. Stone, W.E. Barlow, M.S. Humayun, E. de Juan Jr., A.H. Milam, Morphometric analysis of macular photoreceptors and ganglion cells in retinas with retinitis pigmentosa, Archives of Ophthalmology 110 (11) (1992) 1634–1639. Available from https://doi.org/10.1001/archopht.1992.01080230134038.

[22] A. Santos, M.S. Humayun, E. de Juan Jr, R.J. Greenburg, M.J. Marsh, I.B. Klock, et al., Preservation of the inner retina in retinitis pigmentosa. a morphometric analysis, Archives of Ophthalmology 115 (4) (1997) 511–515. Available from https://doi.org/10.1001/archopht.1997.01100150513011.

[23] M.S. Humayun, M. Prince, E. de Juan Jr, Y. Barron, M. Moskowitz, I.B. Klock, et al., Morphometric analysis of the extramacular retina from postmortem eyes with retinitis pigmentosa, Investigative Ophthalmology & Visual Science 40 (1) (1999) 143–148.

[24] M. Pavlidis, T. Stupp, R. Naskar, C. Cengiz, S. Thanos, Retinal ganglion cells resistant to advanced glaucoma: a postmortem study of human retinas with the carbocyanine dye diI, Investigative Ophthalmology & Visual Science 44 (12) (2003) 5196–5205. Available from https://doi.org/10.1167/iovs.03-0614.

[25] M. Niwa, H. Aoki, A. Hirata, H. Tomita, P.G. Green, H. Akira, Retinal cell degeneration in animal models, International Journal of Molecular Sciences. 17 (1) (2016) 110. Available from https://doi.org/10.3390/ijms17010110.

[26] T. Tsai, A.M. Mueller-Buehl, Y. Satgunarajah, S. Kuehn, H.B. Dick, S.C. Joachim, Protective effect of the extremolytes ectoine and hydroxyectoine in a porcine organ culture, Graefe's Arch. Clin. Exp. Ophthalmol. 258 (10) (2020) 2185–2203. Available from https://doi.org/10.1007/s00417-020-04854-x.

[27] Y. Wang, D. Zhang, B. Shen, Y. Zhang, P. Gu, Stem/progenitor cells and biodegradable scaffolds in the treatment of retinal degenerative diseases, Current Stem Cell Research & Therapy 13 (3) (2018) 160–173. Available from https://doi.org/10.2174/1574888x1366617122723.

[28] M.E. Pennesi, M. Neuringer, R.J. Courtney, Animal models of age-related macular degeneration, Molecular Aspects of Medicine 33 (4) (2012) 487–509. Available from https://doi.org/10.1016/j.mam.2012.06.003.

[29] C. Koster, K.E. Wever, E.L. Wagstaff, K.T. van den Hurk, C.R. Hooijmans, A.A. Bergen, A systematic review on transplantation studies of the retinal pigment epithelium in animal models, International Journal of Molecular Sciences. 21 (2020) 2719. Available from https://doi.org/10.3390/ijms21082719.

[30] N. Mitrousis, S. Hacibekiroglu, M.T. Ho, Y. Sauvé, A. Nagy, D. van der Kooy, et al., Hydrogel-mediated co-transplantation of retinal pigmented epithelium and photoreceptors restores vision in an animal model of advanced retinal degeneration, Biomaterials 257 (2020) 120233. Available from https://doi.org/10.1016/j.biomaterials.2020.120233.

[31] B.B. Thomas, B. Lin, J.C. Martinez-Camarillo, D. Zhu, B.T. McLelland, G. Nistor, et al., Co-grafts of human embryonic stem cell derived retina organoids and retinal pigment epithelium for retinal reconstruction in immunodeficient retinal degenerate royal college of surgeons rats, Frontiers Neuroscience 15 (2021) 752958. Available from https://doi.org/10.3389/fnins.2021.752958.

[32] S.J. Gasparini, K. Tessmer, M. Reh, et al., Transplanted human cones incorporate into the retina and function in a murine cone degeneration model, The Journal of Clinical Investigation 132 (12) (2022) e154619. Available from https://doi.org/10.1172/JCI154619.

[33] Z. Liu, T. Ilmarinen, G.S.W. Tan, et al., Submacular integration of hESC-RPE monolayer xenografts in a surgical non-human primate model, Stem Cell Research Therapy 12 (423) (2021). Available from https://doi.org/10.1186/s13287-021-02395-6.

[34] J. Hertz, B. Qu, Y. Hu, R.D. Patel, D.A. Valenzuela, J.L. Goldberg, Survival and integration of developing and progenitor-derived retinal ganglion cells following transplantation, Cell Transplantation 23 (7) (2014) 855–872. Available from https://doi.org/10.3727/096368913X667024.

[35] P. Venugopalan, Y. Wang, T. Nguyen, et al., Transplanted neurons integrate into adult retinas and respond to light, Nature Communications 7 (2016) 10472. Available from https://doi.org/10.1038/ncomms10472.

[36] Z. Luo, B. Xian, K. Li, K. Li, R. Yang, M. Chen, et al., Biodegradable scaffolds facilitate epiretinal transplantation of hiPSC-Derived retinal neurons in nonhuman primates, Acta Biomaterialia 134 (2021) 289–301. Available from https://doi.org/10.1016/j.actbio.2021.07.040.

[37] T. Tsai, S. Reinehr, L. Deppe, A. Strubbe, N. Kluge, H.B. Dick, et al., Glaucoma animal models beyond chronic IOP increase, International Journal of Molecular Sciences. 25 (2) (2024) 906. Available from https://doi.org/10.3390/ijms25020906.

[38] A. Inda, S. Martinez, C. Bessone, et al., Evidence of the protective role of Carvacrol in a retinal degeneration animal model, Experimental Eye Research 244 (2024) 109938. Available from https://doi.org/10.1016/j.exer.2024.109938.

[39] I. Piano, F. Corsi, A. Galante, S. Bisti, C. Gargini, Resilience and neuroprotection: the efficacy of saffron Repron® in counteracting the progression of inherited retinal degeneration, Acta Ophthalmologica 102 (S279) (2024). Available from https://doi.org/10.1111/aos.16331.

[40] D.E. Bramblett, M.E. Pennesi, S.M. Wu, M.J. Tsai, The transcription factor Bhlhb4 is required for rod bipolar cell maturation, Neuron 43 (6) (2004) 779–793. Available from https://doi.org/10.1016/j.neuron.2004.08.032.

[41] H. Zhang, B. Su, L. Jiao, Z.H. Xu, et al., Transplantation of GMP-grade human iPSC-derived retinal pigment epithelial cells in rodent model: the first pre-clinical study for safety and efficacy in China, Annals of Translational Medicine 9 (3) (2021) 245. Available from https://doi.org/10.21037/atm-20-4707.

[42] S. Batabyal, S. Kim, M. Carlson, et al., Multicharacteristic opsin gene therapy attenuates retinal degeneration and restore vision in mouse models of retinitis pigmentosa, Translational Vision Science & Technology 13 (10) (2024) 25. Available from https://doi.org/10.1167/tvst.13.10.25.

[43] P.A. Scott, J.P. Fernandez de Castro, H.J. Kaplan, M.A. McCall, A Pro23His mutation alters prenatal rod photoreceptor morphology in a transgenic swine model of retinitis pigmentosa, Investigative Ophthalmology & Visual Science 55 (4) (2014) 2452–2459. Available from https://doi.org/10.1167/iovs.13-13723.

[44] W.S. Tan, Y. Lai, Y. Chung, S. Adusumalli, X.Y. Lee, K. Tryggvason, et al., Retina-specific laminin-based generation of photoreceptor progenitors from human pluripotent stem cells under xeno-free and chemically defined conditions, Nature Protocols 20 (9) (2025) 2481–2503.

[45] Y. Wei, U. Alexandre, X. Ma, Hydrogels to support transplantation of human embryonic stem cell-derived retinal pigment epithelial cells, Brain Sciences 12 (12) (2022) 1620.

[46] M. Gray, S. Guido, A. Kugadas, Editorial: The use of large animal models to improve pre-clinical translational research, Frontiers in Veterinary Science 9 (2022). Available from https://doi.org/10.3389/fvets.2022.1086912.

[47] Y. Zhu, B. Cao, A. Tolone, J. Yan, G. Christensen, B. Arango-Gonzalez, M. Ueffing, François Paquet-Durand, In vitro Model Systems for Studies Into Retinal Neuroprotection, Frontiers in Neuroscience 16 (2022). Available from https://doi.org/10.3389/fnins.2022.938089.

Chapter 17

Circadian mechanisms and the importance of biological timing in tissue engineering

Amlan Chakraborty[1], **Kamal Dua**[2], **and Hannah J. Durrington**[1,3]

[1]*Division of Immunology, Immunity to Infection and Respiratory Medicine, University of Manchester, Manchester, United Kingdom,* [2]*NICM Health Research Institute & School of Science, Western Sydney University, Sydney, NSW, Australia,* [3]*Wythenshawe Hospital, Manchester University NHS Foundation Trust, Manchester, United Kingdom*

17.1 Introduction

Circadian rhythms are intrinsic, near-24-hour cycles that regulate physiological functions, including metabolism, immune responses, cell cycle, and tissue regeneration, in almost all organisms. These rhythms are controlled by a central clock located in the suprachiasmatic nucleus (SCN) of the hypothalamus and also by peripheral clocks present in virtually every cell and tissue of the body. The synchronization of these clocks with external cues, particularly light, is crucial for maintaining homeostasis and optimal functioning of various biological systems.

At the core of circadian rhythms lie transcription-translation feedback loops governed by key clock genes, including *Circadian Locomotor Output Cycles Kaput* (*CLOCK*), *Brain and Muscle ARNT-Like 1* (*BMAL1*), *Period* (*PER*), and *Cryptochrome* (*CRY*) (Fig. 17.1). These molecular regulators orchestrate cellular processes in alignment with environmental cues, particularly the light-dark cycle. Circadian disruption, where there is a misalignment between internal biological clocks and light-dark signals or feeding cues, such as those caused by shift work or irregular sleep patterns, has been linked to numerous health conditions, including respiratory, cardiovascular, metabolic, immune, and neuronal disorders.

Tissue engineering, which seeks to create biological substitutes to restore or replace damaged tissues, has traditionally focused on biomaterial design and cell biology. In recent years, the understanding of circadian rhythms has expanded to include their role in tissue engineering. As tissue engineering evolves, integrating circadian biology into its approaches could enhance outcomes, especially in treating chronic diseases such as respiratory diseases, cardiovascular disorders, metabolic conditions, immune disorders, and neuronal disorders. This chapter explores the mechanistic underpinnings of circadian rhythms and their application in tissue engineering, highlighting recent research and potential clinical implications.

17.2 Molecular basis of circadian rhythms

17.2.1 Core clock mechanisms

Circadian rhythms are maintained by a complex network of genes and proteins that form transcription-translation feedback loops (TTFL), ensuring the regular oscillation of cellular functions. The primary components include (Fig. 17.1)

- **CLOCK and BMAL1:** These transcription factors form a heterodimer that activates the transcription of target genes, including Period (PER1, PER2, PER3) and Cryptochrome (CRY1, CRY2).
- **PER and CRY proteins:** These proteins accumulate in the cytoplasm, translocate to the nucleus, and inhibit the CLOCK-BMAL1 complex, thus closing the feedback loop and inhibiting their transcription.
- **Additional regulators:** Proteins such as REV-ERBα/β and RORα/β further modulate the rhythmic expression of clock-controlled genes, influencing processes like metabolism, inflammation, and cell cycle progression.

Fundamentals of Translational Tissue Engineering. DOI: https://doi.org/10.1016/B978-0-443-33455-9.00013-4

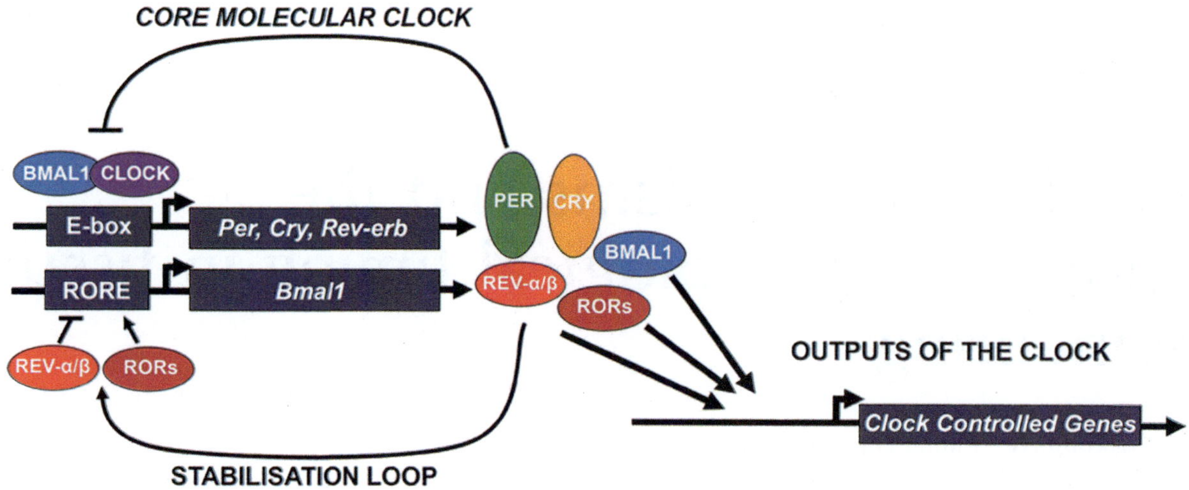

FIGURE 17.1 Schematic of circadian clock feedback loop. The peripheral oscillator operates on a 24-hour cycle through a TTFL. In this loop, the BMAL1/CLOCK heterodimer promotes the transcription of the PER and CRY genes. The resulting PER and CRY proteins inhibit the BMAL1/CLOCK activity, establishing a negative feedback mechanism. As PER and CRY proteins degrade over time, partly due to the action of SIRT1, the inhibition is lifted, allowing the cycle to restart. Additionally, the transcription of BMAL1 is tightly regulated by the opposing actions of REV-ERB and ROR. REV-ERB suppresses BMAL1 transcription, adding an extra layer of feedback control, while ROR promotes BMAL1 transcription, counteracting REV-ERB's inhibitory effect. This regulatory balance involves specific DNA response elements, such as the enhancer box (E-box) for BMAL1/CLOCK binding and the RAR-related orphan receptors response element (RORE) for REV-ERB and ROR interactions. *TTFL*, Transcription-translation feedback loop.
Adapted with permission from P.S. Cunningham, C. Jackson, A. Chakraborty, J. Cain, H.J. Durrington, J. Blaikley, Circadian regulation of pulmonary disease: the importance of timing, Clinical Science 137 (11) (2023) 895–912.

17.2.2 Central and peripheral clocks

The master clock, located in the SCN of the hypothalamus, coordinates peripheral clocks in tissues, aligning physiological processes with the external environment (Fig. 17.2). Peripheral clocks, present in nearly all body cells, are regulated by both SCN signals and local environmental cues such as feeding, temperature, and hormonal changes. This organization allows for tissue-specific regulation, ensuring that functions such as energy metabolism, immune responses, and cellular repair occur at optimal times of the day.

17.3 The role of circadian rhythms in tissue engineering

17.3.1 Circadian rhythms in stem cell biology and regenerative medicine

Stem cells, the cornerstone of tissue engineering, exhibit circadian oscillations that influence their proliferation, differentiation, and regenerative capacity. Circadian regulation of stem cells is mediated by clock genes, which control key pathways involved in cell cycle progression, DNA repair, and apoptosis. Studies have shown that circadian misalignment, such as that caused by shift work or chronic jet lag, impairs stem cell function and diminishes tissue repair capacity [2].

In tissue engineering, timing the delivery of stem cells to align with circadian peaks of regenerative activity could enhance outcomes (Table 17.1). For example, mesenchymal stem cells (MSCs) display time-dependent variations in their secretory profiles, affecting their antiinflammatory and immunomodulatory properties. Delivering MSCs at optimal circadian phases may maximize their therapeutic potential (chronotherapy), particularly in inflammatory conditions such as respiratory and cardiovascular diseases.

17.3.2 Biomaterials and circadian synchronization

Biomaterials used in tissue engineering, such as scaffolds and hydrogels, can be designed to support circadian synchronization within engineered tissues. By incorporating circadian cues, such as oscillatory delivery of growth factors or mechanical stimuli, biomaterials can promote rhythmic gene expression in encapsulated cells, enhancing their functionality (Fig. 17.3). For instance, scaffolds embedded with rhythmic light stimuli have been shown to entrain circadian rhythms in neuronal cultures, suggesting potential applications in repairing circadian disruptions observed in neurodegenerative disorders [3]. Scaffold-based biomaterials embedded with rhythmic light stimuli have demonstrated the

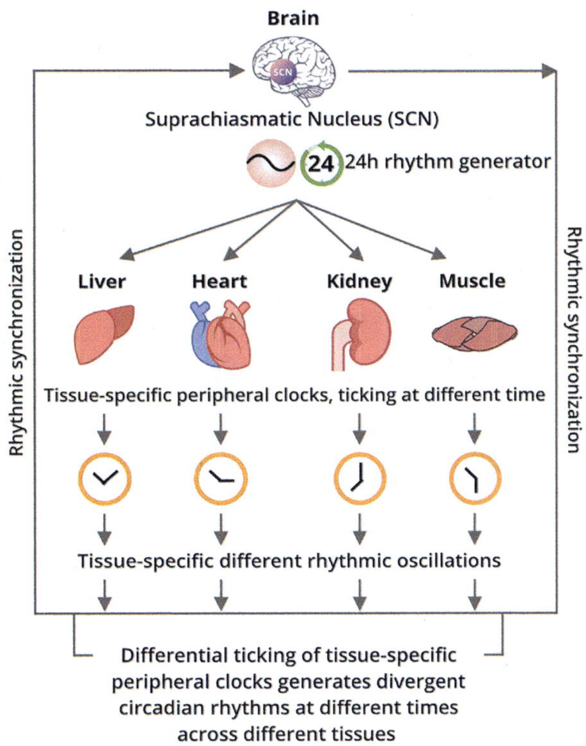

FIGURE 17.2 A representative connectome illustrates the relationship between the master clock and tissue-specific peripheral clocks. A representative connectome illustrates the hierarchical relationship between the central pacemaker, located in the SCN, and peripheral clocks residing in various tissues. The SCN-driven rhythmic signals are disseminated through humoral and neural pathways, ensuring systemic synchrony under normal conditions. Peripheral clocks typically remain in phase with the SCN; however, disruptions such as altered feeding regimens or aberrant light exposure can induce transient misalignment. The extent and rate of resynchronization vary across tissues, reflecting the complexity of circadian homeostasis [1]. *SCN*, Suprachiasmatic nucleus. Adapted with permission from D. Chowdhury, C. Wang, A.P. Lu, H.L. Zhu, Understanding quantitative circadian regulations are crucial towards advancing chronotherapy, Cells 8 (8) (2019).

TABLE 17.1 Circadian influence on stem cell-based tissue engineering.

Stem cell type	Circadian effects	Implications for tissue engineering
Mesenchymal stem cells	Variation in differentiation and proliferation	Optimize timing for cell harvest and therapy
Neural stem cells	Neurogenesis peaks during specific circadian phases	Timing interventions for neuronal repair
Hematopoietic stem cells	Circadian rhythms influence immune modulation	Align cell delivery with immune peaks

capacity to entrain circadian rhythms in neuronal cultures, offering a potential therapeutic avenue for neurodegenerative disorders characterized by circadian disruption. This approach is particularly relevant in tissues capable of direct photoreception, such as the brain and retina. The cited reference appears to be a secondary review; verification is required to ensure the presence of primary empirical evidence supporting this claim.

17.4 Circadian rhythms in disease and tissue engineering applications

17.4.1 Respiratory diseases

Circadian rhythms profoundly affect respiratory function, influencing airway caliber, lung inflammation, and response to environmental stimuli [4](). Asthma, a disease characterized by circadian variation in symptoms, shows peak exacerbations overnight and early morning due to oscillations in airway inflammation and bronchial responsiveness [5–8]. Clock gene

AI-generated figure

FIGURE 17.3 Circadian modulation of biomaterials in tissue engineering. Illustration showing how biomaterials can be designed with embedded rhythmic cues to synchronize cellular functions within engineered tissues.

TABLE 17.2 Circadian modulation in cardiovascular tissue engineering.

Cardiovascular component	Circadian regulation	Implications for tissue engineering
Endothelial cells	Rhythmic expression of vasodilatory genes	Timing of graft implantation to reduce thrombosis
Cardiac myocytes	Circadian variation in contractility	Align patch integration with metabolic activity peaks
Vascular smooth muscle	Regulation of vascular tone	Circadian synchronization for stable graft function

dysregulation has been implicated in the pathogenesis of asthma and chronic obstructive pulmonary disease (COPD), highlighting the need for chronotherapy approaches in respiratory disease management [5,6].

In tissue engineering, aligning the timing of therapeutic interventions with circadian phases of airway reactivity and inflammation could enhance the effectiveness of engineered respiratory tissues. For example, the delivery of stem cell-derived epithelial cells at specific circadian phases may optimize their integration into damaged lung tissue, improving outcomes in COPD and other chronic lung diseases.

17.4.2 Cardiovascular diseases

The cardiovascular system exhibits strong circadian rhythms that regulate heart rate, blood pressure, and vascular tone. Disruption of these rhythms, such as through night-shift work or irregular sleep patterns, is associated with increased risks of myocardial infarction, hypertension, and stroke. Circadian dysregulation affects endothelial function, platelet aggregation, and inflammatory responses, contributing to cardiovascular pathology [9].

Recent advances in cardiovascular tissue engineering emphasize the need to incorporate circadian considerations into the design of cardiac patches, vascular grafts, and bioengineered heart tissues. For example, endothelial cells display circadian oscillations in the expression of genes involved in vasodilation and thrombosis, suggesting that the timing of cell seeding and conditioning may impact the success of engineered cardiovascular tissues. Timing therapeutic interventions to coincide with periods of optimal endothelial function could enhance the integration and performance of bioengineered tissues in cardiovascular repair (Table 17.2).

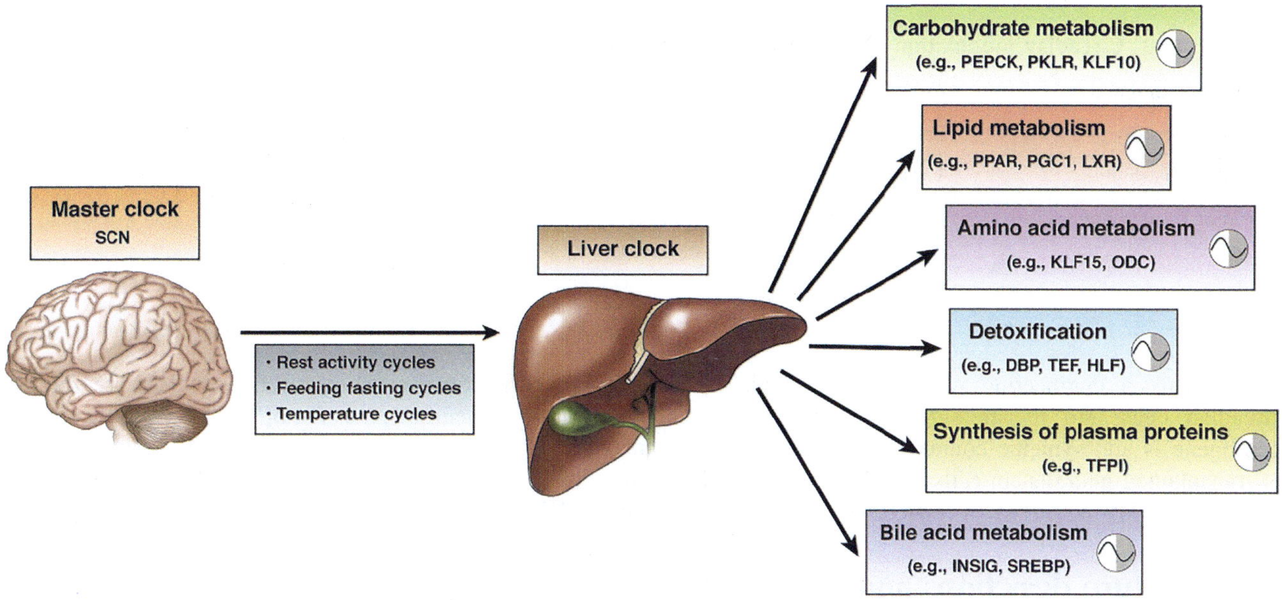

FIGURE 17.4 Circadian timing in metabolic tissue engineering. The hypothalamic master clock coordinates peripheral oscillators, like the liver clock, by regulating rhythmic patterns of activity, feeding behaviors, and temperature fluctuations. In turn, the liver clock orchestrates the periodic expression of key master regulators and enzymes that control critical hepatic metabolic processes [11].
Adapted with permission from H. Reinke, G. Asher, Circadian clock control of liver metabolic functions, Gastroenterology 150 (3) (2016) 574–580.

17.4.3 Metabolic disorders

Metabolic disorders, including obesity, type 2 diabetes, and nonalcoholic fatty liver disease, are closely linked to circadian disruptions. The liver, a central organ in metabolism, exhibits robust circadian rhythms that regulate glucose homeostasis, lipid metabolism, and detoxification processes (Fig. 17.4). Disruption of hepatic clocks, often seen in shift workers or individuals with irregular eating patterns, contributes to metabolic dysregulation and disease progression [10].

Tissue engineering approaches for metabolic disorders, such as the development of bioengineered liver tissues or pancreatic islets, can benefit from circadian optimization. For instance, hepatocyte proliferation and function exhibit time-of-day variations, suggesting that circadian timing of cell harvest, culture, and transplantation could improve tissue viability and function. Additionally, rhythmic delivery of insulin from bioengineered pancreatic tissues could better mimic natural insulin oscillations, improving glycemic control in diabetic patients.

17.4.4 Immune disorders

The immune system is tightly regulated by circadian rhythms, with immune cell trafficking, cytokine production, and pathogen response all showing time-of-day variations. Circadian disruption has been linked to autoimmune diseases, increased susceptibility to infections, and impaired vaccine responses Refs. The clock gene BMAL1, for example, plays a critical role in regulating inflammatory responses and immune cell differentiation, influencing disease outcomes in conditions such as rheumatoid arthritis and systemic lupus erythematosus [12].

In tissue engineering, aligning immune modulation with circadian rhythms can enhance the integration and function of bioengineered tissues. For instance, the timing of immunosuppressive therapies or cytokine delivery could be optimized to coincide with circadian peaks of immune tolerance, reducing the risk of graft rejection. Moreover, designing scaffolds that release immunomodulatory agents in a circadian manner could promote a more favorable immune environment for tissue regeneration.

17.4.5 Neuronal disorders

Circadian rhythms are intimately linked to neuronal function, influencing sleep-wake cycles, mood regulation, and cognitive performance. Disruptions in circadian timing are associated with neurodegenerative diseases such as Alzheimer's and Parkinson's, as well as mood disorders like depression and bipolar disorder. Clock genes regulate key processes in the brain, including neurogenesis, synaptic plasticity, and neurotransmitter release [12].

TABLE 17.3 Circadian modulation in neuronal tissue engineering.

Neuronal function	Circadian influence	Engineering implications
Neurogenesis	Peaks during sleep phases	Timing stem cell delivery for enhanced integration
Synaptic plasticity	Varies with circadian cycle	Synchronizing neurotrophic factor release
Neurotransmitter release	Rhythmic dopamine and serotonin levels	Aligning therapies with neurotransmitter cycles

Neural tissue engineering, which aims to repair or replace damaged neuronal circuits, can benefit from incorporating circadian cues into its strategies. For example, circadian timing of neurotrophic factor delivery could enhance neurogenesis and synaptic connectivity in engineered brain tissues. Additionally, synchronizing the circadian rhythms of transplanted neural stem cells with host tissue could improve their integration and functional outcomes, offering new avenues for treating circadian-related neuronal disorders (Table 17.3).

17.5 Future directions and challenges

17.5.1 Integrating chronobiology into tissue engineering protocols

While the potential benefits of circadian optimization in tissue engineering are clear, several challenges remain. A major hurdle is the need for precise control of circadian cues within engineered tissues, particularly in complex, multicellular constructs. Advances in biomaterials that can dynamically deliver circadian signals, such as light, temperature, or rhythmic mechanical stimulation, will be critical.

Furthermore, personalized chronotherapy approaches, which tailor interventions based on an individual's circadian profile, could revolutionize the field. For example, chronotyping patients to determine their optimal timing for receiving stem cell transplants or other bioengineered therapies could enhance therapeutic efficacy and minimize side effects.

17.5.2 Circadian disruptions in clinical settings

Clinical settings often involve disruptions to natural circadian rhythms, such as irregular lighting, noise, and nocturnal interventions. Addressing these disruptions will be essential for maximizing the success of tissue engineering therapies. Strategies to entrain patients' circadian rhythms before, during, and after tissue engineering procedures could include controlled light exposure, timed feeding schedules, and appropriately timed pharmacological interventions.

17.5.3 Ethical considerations

As the field advances, ethical considerations will also come into play, particularly regarding the use of genetic and epigenetic modifications to enhance circadian function in engineered tissues. Balancing the potential benefits with ethical concerns about altering circadian biology will be crucial as the technology moves toward clinical application.

17.6 Conclusion

Circadian rhythms are fundamental regulators of physiological function, with profound implications for tissue engineering. By understanding and harnessing these rhythms, it is possible to optimize tissue repair and regeneration, particularly in the context of chronic diseases such as respiratory, cardiovascular, metabolic, immune, and neuronal disorders. Future research should continue to explore the intricate relationship between circadian biology and tissue engineering, paving the way for innovative chronotherapy approaches that align with the body's natural timing systems.

AI disclosure

During the preparation of this work, the author(s) used Grammarly and ChatGPT to check grammar and spelling. After using this tool/service, the author(s) reviewed and edited the content as needed and take(s) full responsibility for the content of the publication.

The AI-generated image(s) were reviewed, edited, and verified by the author(s) for scientific accuracy prior to inclusion in this chapter. The author(s) take full responsibility for the content, accuracy, and interpretation of all figures presented, including those generated with AI assistance.

References

[1] D. Chowdhury, C. Wang, A.P. Lu, H.L. Zhu, Understanding quantitative circadian regulations are crucial towards advancing chronotherapy, Cells 8 (8) (2019). Available from https://doi.org/10.3390/cells8080883, https://www.mdpi.com/2073-4409/8/8/883/pdf.

[2] E. Paatela, D. Munson, N. Kikyo, Circadian regulation in tissue regeneration, International Journal of Molecular Sciences 20 (9) (2019) 2263. Available from https://doi.org/10.3390/ijms20092263.

[3] M. Naveed, O.Y. Chao, J.W. Hill, Y.M. Yang, J.P. Huston, R. Cao, Circadian neurogenetics and its implications in neurophysiology, behavior, and chronomedicine, Neuroscience and Biobehavioral Reviews 157 (2024). Available from https://doi.org/10.1016/j.neubiorev.2023.105523, https://www.sciencedirect.com/science/journal/01497634.

[4] H.J. DurHrington, K. Krakowiak, P. Meijer, N. Begley, R. Maidstone, L. Goosey, J.E. Gibbs, J.F. Blaikley, L.G. Gregory, C.M. Lloyd, A.S.I. Loudon, D.W. Ray, Circadian asthma airway responses are gated by REV-ERBα, European Respiratory Journal (2020) 1902407. Available from https://doi.org/10.1183/13993003.02407-2019.

[5] P.S. Cunningham, C. Jackson, A. Chakraborty, J. Cain, H.J. Durrington, J. Blaikley, Circadian regulation of pulmonary disease: the importance of timing, Clinical Science (Lond) 137 (11) (2023) 895–912. Available from https://doi.org/10.1042/CS20220061.

[6] R. Wang, C.S. Murray, S.J. Fowler, A. Simpson, H.J. Durrington, Asthma diagnosis: into the fourth dimension, Thorax 76 (6) (2021) 624–631. Available from https://doi.org/10.1136/thoraxjnl-2020-216421, http://thorax.bmj.com/.

[7] A Chakraborty, J. Cain, R. Maidstone, K. Krakowiak, V. Deugi, n Begley, et al., Airway inflammation and hyperresponsiveness in asthma is regulated by the circadian clock through REV-ERB, European Respiratory 62 (67) (2023) OA4208.

[8] A Chakraborty, J. Cain, R. Maidstone, S124 Single-cell transcriptomics identifies unique pathways regulating airway hyperresponsiveness via circadian regulator rev-erba in airway epithelium, Thorax 79 (2024) A87.

[9] F. Portaluppi, R. Tiseo, M.H. Smolensky, R.C. Hermida, D.E. Ayala, F. Fabbian, Circadian rhythms and cardiovascular health, Sleep Medicine Reviews 16 (2) (2012) 151–166. Available from https://doi.org/10.1016/j.smrv.2011.04.003.

[10] I. Shimizu, Y. Yoshida, T. Minamino, A role for circadian clock in metabolic disease, Hypertension Research 39 (7) (2016) 483–491. Available from https://doi.org/10.1038/hr.2016.12, http://www.nature.com/hr/archive/index.html.

[11] H. Reinke, G. Asher, Circadian clock control of liver metabolic functions, Gastroenterology 150 (3) (2016) 574–580. Available from https://doi.org/10.1053/j.gastro.2015.11.043, http://www.journals.elsevier.com/gastroenterology/.

[12] E. Vieira, G.G. Mirizio, G.R. Barin, R.V. de Andrade, N.F.S. Nimer, L. La Sala, Clock genes, inflammation and the immune system—implications for diabetes, obesity and neurodegenerative diseases, International Journal of Molecular Sciences 21 (24) (2020) 1–8. Available from https://doi.org/10.3390/ijms21249743, https://www.mdpi.com/1422-0067/21/24/9743/pdf.

Chapter 18

Tissue engineering strategies in retina

Shikha Upreti, Neilanjan Maitra, and Madhumita P. Ghosh
Centre for Medical Biotechnology, Amity Institute of Biotechnology, Amity University, Noida, Uttar Pradesh, India

18.1 Introduction

The human retina is an important part of the central nervous system, originating from the frontal portion of the neural tube. It is a transparent light-sensitive tissue composed of the cellular layers of retinal pigment epithelium (RPE) that supports the blood retinal barrier, an outer nuclear layer containing photoreceptors (PRs), and inner nuclear layer lined with the inner retinal ganglion layer [1]. As the retina absorbs light energy transmitted by PRs, the conversion of visual information into chemical and subsequently neural signals is finally relayed to retinal ganglion cells (RGCs) [2]. The information thus received by the brain in the form of an image is then processed via the optic nerve, comprising the RGCs.

Retinal degenerative diseases like age-related macular degeneration (AMD) and retinitis pigmentosa (RP) are a typical outcome of RPE dysfunction, and resultant deterioration of PR turns out to be the most common pathologies [3]. Currently, available therapies are aimed at delaying the progression of the disease, as there exists no permanent cure yet except antiangiogenic treatment for neovascular AMD [4]. The regeneration of the RPE, PRs, and, more recently, RGCs has shown significant recovery with the efficient methods of culturing stem/progenitor cells. Even reports from clinical trials of human embryonic stem cell–derived RPE (hESC-RPE) are encouraging, but the prolonged survival of the cells and their effective integration into the damaged retina is yet to be established. The incidences of frequent teratoma and formation of rosette-like structures at the integration site demand research on more precise methods of subretinal delivery of the cells.

To ensure correct orientation and establish physical interface with the PRs, the use of a preformed monolayer along with a supporting substrate is necessary. In that context, various biomaterials are in use for the construction of scaffolds for the successful delivery of cells. These nanostructured materials are usually of natural, synthetic, or biological membrane origin [5–8]. More recently, scaffold-free technique by blending cell transplantation, biomaterials, and organ fabrication methods has been used to create a multilayered retina through inkjet bioprinting techniques. They used a bioprinting system that could very precisely, without a carrier, use a unique inkjet to deposit PRs on bioprinted RPE [9]. The advantage of 3D bioprinting is enhanced implant adaptation due to the use of biocompatible materials with reduced risk of rejection and irritation. In this chapter, the initial use of stem cells was briefly summarize and the advancements made for the successful delivery of stem cells through the use of biomaterials as a support, and how it aids the physical interaction of cells within the damaged layers, preferably PRs and RPE. (Fig. 18.1).

18.2 Stem cell therapy

The regeneration of the RPE, PRs, Müller, glial, and, more recently, RGCs has been the area of interest for scientists to address the issue of recurrent loss of vision. The three types of pluripotent cells, such as embryonic, induced, and reprogrammed cells, are being explored as the current therapy for retinal replacements. These cells with the ability to multiply indefinitely and mature into the tissue type of their vicinity have gained significance in replacement therapies against retinal degeneration. The ESCs are derived from the blastocyst of the developing embryo, which gives rise to the three germ layers, such as ectoderm, mesoderm, and endoderm [10]. The damage to the inner nuclear layers, amacrine layer, and retinal ganglion layer in the case of glaucoma is an outcome of increased intraocular pressure (IOP) and excitotoxicity caused by the excessive release of glutamate. To date, treatment options include suppressing IOP through intraocular injections of timolol and brimonidine and regeneration of RGCs by injecting iPSCs [11–13].

Fundamentals of Translational Tissue Engineering. DOI: https://doi.org/10.1016/B978-0-443-33455-9.00015-8

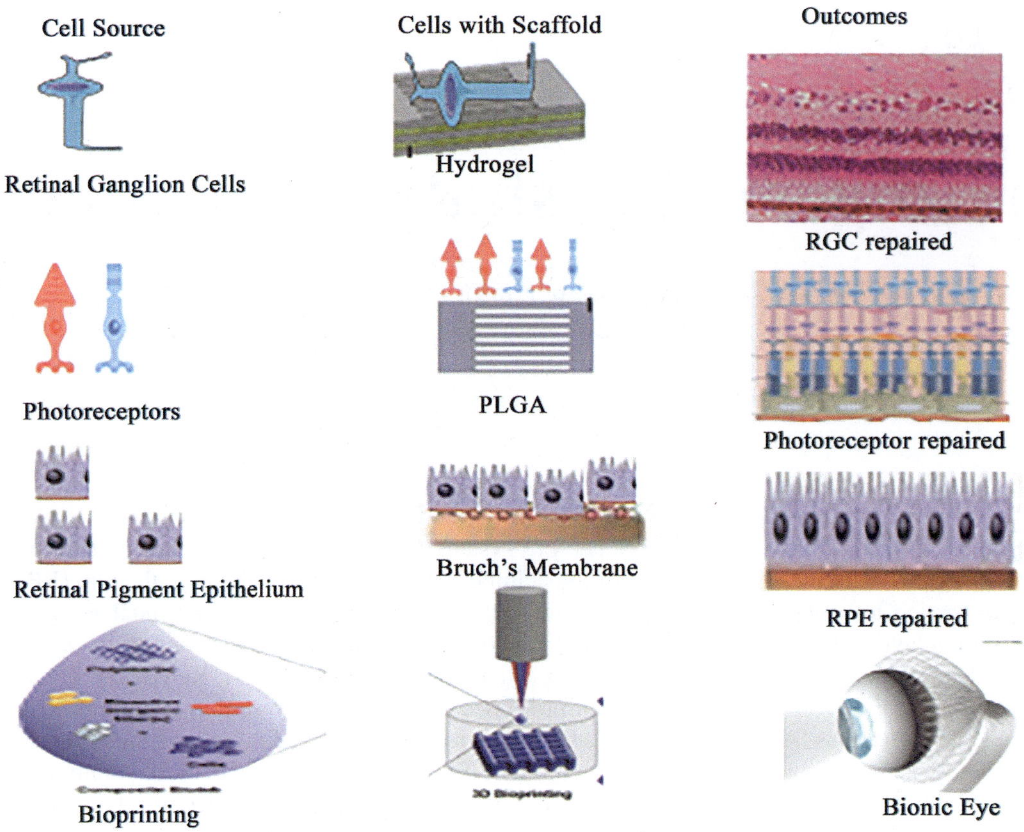

FIGURE 18.1 **Advancements in scaffold-based delivery of different retinal cell types.** Retina is derived from animal sources like rats or pigs and extracted in culture. The culture is grown to a specific cell type stage of either ESCs, iPSCs, or RPCs and combined with bioprinted scaffolds to target degeneration caused by conditions mimicking diseases like AMD, where RPE is damaged. *AMD*, Age-related macular degeneration; *ESCs*, embryonic stem cells; *iPSCs*, induced pluripotent stem cells; *RPCs*, retinal precursor cells.
From Madhumita. P. Ghosh.

Diabetic retinopathy, a microvascular condition due to long-term effects of diabetes, is associated with neovascularization preceded by structural deformations of the neural retina [14,15]. Present therapeutics to address these ocular challenges involve the use of stem cells such as the human embryonic stem cells (hESCs) grown to RPCs, injected within the vitreous and subretinal space, showing recovery of damaged cells within the retina and differentiated into functional RGCs when injected in RGC ablated mice model, thus improving visual acuity [16]. But the introduction of stem cells or RPCs can sometimes lead to immune rejection, even though subretinal space is an immune-privileged site [17]. ESCs were successfully transplanted in mouse and primate models [18,19] with no immune rejection but caused a teratoma because of the heterogeneity and pluripotent nature of the stem cells [20,21], thereby resulting in chromosomal instability [22]. ESCs, iPSCs, and mesenchymal stem cells (MSCs), being heterogeneous, exhibit limited proliferation and differentiation in vitro, impacting their survival, but this can be overcome by the use of a specific RPC^{+} marker like stem cell factor receptor c-Kit [23].

Intravitreally transplanted bone marrow-derived stem cells (BMSCs) and neuroprotective factors were reported to be easily accepted with good effects by a number of preclinical animal studies and clinical studies [24,25]. The ESCs and iPSCs have been used for generating PRs, which are a part of the outer nuclear layer (ONL) and are responsible for color vision [26,27]. Their successful integration and survival in the ONL led to the use of MSCs for the generation of PRs [28]. The loss of RGCs being the primary cause of vision loss necessitated the use of embryonic or MSCs to generate functional RGCs. The MSCs have achieved success to some extent in alleviating IOP, but the RGC-like cells generated have not yet been tested in a clinical setting [29].

18.3 Tissue engineering—scaffolds

Creation of an implantable graft using biodegradable or nonbiodegradable polymer scaffolds that sustain the growth of cells in a monolayer sheet is called tissue engineering (TE). These engineered constructs, together with the extracellular matrix (ECM), can enhance survival, integration, and visual function of the cells [30]. The use of scaffolds as nanobased structures for efficient site-specific delivery of the cells gained importance as it acts as a support to carry the precultured cells maturing into a functional monolayer that can be transplanted as cell-based scaffold complexes beneath the retina.

The nanomaterials work as an interface between the physical structures and the biological environment of the body. So far, the cell-based therapies through TE have been employed for restructuring of PRs, RPCs, the retinal sheet, and RPE [31–36]. Commonly, two approaches are used, which are known as scaffold-free and scaffold-based for the delivery of cells and their physical integration. Scaffold-free approaches, as the name suggests, can produce functional tissue aided by the cells' competence to merge into larger cohesive constructs without an external support system. The scaffold-free approach of delivering the cells directly into the subretinal space, although causing integration into the damaged layer, may result in random localization of grafts along with reflux of cells from the injection site [37], which may lead to fibrosis and retinal detachment. The property of cells to secrete their own ECM on reaching confluency is exploited in the form of cell sheets to develop a scaffold-free TE approach. Compared with free cell suspension techniques, the presence of the ECM promotes faster adhesion and attachment to the host tissue without any pretreatment [38]. The cultured cells are harvested in a thermoresponsive culture dish without using enzymes, which enables reversible cell adhesion to and detachment from the dish by switching the hydrophobicity of its surface [39]. In the scaffold-based approach, the cells are seeded on the material scaffold and delivered in the subretinal space so that the cells are able to integrate into the site, differentiate, and maintain the functionality of the target tissue (Table 18.1).

18.4 Retinal pigment epithelium

RPE cells are found on the Bruch's membrane (BrM), a specialized basement membrane, responsible for supplying nutrients to the retina and retinal vascular endothelial cells. Simulation of the natural environment together with the biocompatibility helped RPE monolayer regeneration from RPE scaffolds [35,40,41]. Collagen types I, III, and IV, gelatin, alginates, laminin, fibronectin, Matrigel, silk fibroin, and vitronectin are some of the natural biopolymers used for subretinal placement of human RPE and normal RPE cells (ARPE-19 cells) [42]. On a microfabricated poly (glycerol sebacate) scaffold about 45 m thick, RPCs adhered strongly with unique mechanical properties when grown for 7 days, a mixture of immature and mature cells, indicating the tendency of differentiation [43–45].

Since the correct functioning of cells in the subretinal space is determined by their organization, integration, and subsequent differentiation into retinal cell types, a lack of an interface or TE scaffold for delivering the cells in the

TABLE 18.1 Types of nanomaterials/scaffolds used with different biological sources using various types of retinal cells.

Scaffold	Cells	Source	References
Nanostructured biomaterials	RPE, PRs, RGCs	Natural, synthetic, or biological membranes	[5,6,8]
3D bioprinting	PRs, RPE, multilayered retina	Bioprinted RPE and PRs	[9]
Biodegradable polymer scaffolds	PRs, RPCs, RPE	Poly (glycerol sebacate), Collagen types I, III, IV, gelatin, silk fibroin, alginates	[43,46,55,58,59]
Electrospun nanofibers	PRs, RPCs	Silk fibroin (SF), poly (L-lactic acid-co-ε-caprolactone) (PLCL), laminin, chitosan	[62–65]
RPE Patch	RPE	Bioengineered scaffolds (synthetic/natural)	[47,51,54]
3D Outer Blood-Retina Barrier Model	Endothelial cells, pericytes, fibroblasts, RPE	Bioprinted structure with BrM-like development	[75,76]
Stem Cell Therapy	hESCs, iPSCs, RPCs, BMSCs	Human embryonic stem cells, bone marrow-derived stem cells, induced pluripotent stem cells	[10,16,17,22,24,29]
Scaffold-Free Techniques	PRs, RPE	ECM-based cell sheets	[38,39]
Bioprinting Techniques	PRs, RPE	Inkjet, laser-assisted, extrusion bioprinting	[72–74]

Source:

subretinal space can lead to cell death [46]. Both transplanted retinal cells and tissues have demonstrated the formation of rosettes within the retina [47], with a study showing that transplantation with a scaffold improved survival of RPE cells and differentiation of RPCs [48,49]. Various reports are now available demonstrating the ability of a bioengineered scaffold, created using an RPE patch using stem cell-derived RPE [50], to maintain the polarity and laminar structure of the transplanted RPE cells [48,51–54].

The performance of the nanofibrous synthetic or biological scaffolds, as compared to their membrane counterparts, for cell survival and transplantation is better and not material dependent [55,56]. Scaffolds can further be classified as classical 3D structured scaffolds with interconnected pores, fibrous hydrogel scaffolds, or a combination of both. They provide a suitable microenvironment for the growth and functionalization of living cells, both in vitro and in vivo. For retinal TE, the scaffold should ideally mimic BrM in terms of permeability and flexibility to support RPE cells [57] and avoid damaging surrounding tissues. Additionally, the scaffold must be mechanically stable to withstand surgical manipulation [58]. When polymers—poly(ε-caprolactone) (PCL), poly (glycerol sebacate) (PGS), and poly(1,8-octanediol-co-citrate) (POC) were combined to create a nanofibrous scaffold, they exhibited promising hydrophilicity, degradability, and biocompatibility. In vitro studies showed faster cell proliferation with the addition of POC, indicating potential for such applications [59–61].

18.5 Retinal precursor cells and photoreceptors

The capacity of PRs to retain their architecture for a length of time after retinal degeneration indicates that the introduction of new progenitors will adjunct the ability of the cells to reconnect the remaining host synaptic pathways to restore vision. This way, when the electric field was applied to form nanofibers of silk fibroin (SF) and poly (L-lactic acid-co-ε-caprolactone) (PLCL) and develop scaffolds (SF: PLCL1:1) for RPCs *in vitro*, it promoted RPC proliferation and enhanced their differentiation toward specific retinal neurons, especially PRs [62]. Similarly, proliferation of RPCs was also enhanced when a hybrid scaffold of cationic chitosan-graft-poly(ε-caprolactone)/polycaprolactone (CS-PCL/PCL, 20/80) was produced via electrospinning [63]. Some studies also reported that combining polycaprolactone (PCL) with laminin or chitosan in electrospun nanofibers could enhance cell adhesion, proliferation, differentiation, and also promote expression of a specific gene related to PRs and bipolar cells [64–66].

18.6 3D printing

Scaffold-free techniques have emerged as promising tools for preserving the retina's complex layered structure, offering new avenues in ocular TE for clinical applications.

Advances in TE now enable the development of *in vitro* models to study retinal diseases. Retinal 3D printing enhances imaging, improves optical coherence tomography (OCT) visualization, and provides detailed insights into retinal anatomy and pathology, benefiting both education and clinical practice [67–69]. Applications of bioprinters and 3D printed materials include drug delivery systems, research models, and retinal prosthetics [70] for degenerative diseases. However, replicating the retina's intricate anatomy remains a key challenge.

18.7 3D bioprinting

The primary techniques utilized in 3D bioprinting include inkjet bioprinting [71], which relies on the deposition of bioink droplets, laser-assisted bioprinting, driven by laser stimulation, and extrusion bioprinting, which employs mechanical force to create a continuous bioink stream. Each technique offers distinct advantages and limitations, making them suitable for different applications in TE [72,73]. The eye, with its immune-privileged status and advanced diagnostic tools, is particularly suited for 3D bioprinting, reducing the likelihood of graft rejection and improving clinical outcomes [48,74].

Recent advancements include the creation of a three-dimensional outer blood-retina barrier (BRB) model. This model was achieved by bioprinting endothelial cells, pericytes, and fibroblasts onto a biodegradable scaffold to form a basal layer, with an RPE monolayer established on top. This design promotes BrM-like structural development through gene expression modulation in choroidal cells [75]. Additionally, RPE spheroids derived from human iPSCs were successfully seeded and bioprinted onto collagen vitrigel (CV) membranes to create engineered RPE sheets. These sheets demonstrated improved cell adhesion and ECM organization, critical for functional tissue integration [76,77].

The integration of 3D printing and bioprinting in ophthalmology continues to drive interdisciplinary collaboration, accelerating the translation of these innovations from experimental models to clinical therapies (Table 18.2).

TABLE 18.2 Clinical trials and current R&D status of the degenerative diseases showing cell damage in retina.

Retinal degenerative diseases	Current cell replacement therapies			Clinical trials	Systemic literature search		
	Scaffold based	Scaffold free	Gene therapy		Scaffold based	Scaffold free	Gene therapy
Photoreceptor degeneration	NCT02464436	NCT03963154	NC-T00999609	I/II I III, I, II	ReNeuron Agency	[78]	[79]
(RPE) detachment	NCT02590692	NC-T01345006,	NC-T01461213	I I II	[78]	[80]	[81]
	Retinal Patch			I/IIa	[82]		
Age-related macular degeneration (AMD)	NCT01691261 UMIN00001192	NCT01625559	NC-T03748784	I, I I, II	[83] [84] [85]	[19]	[86]
	NCT04339764	NCT01518127	NC-T03066258	I/IIa II	NEI Department, NIH	[87]	[86]
		NCT00346060		II		[88]	
	Retinal ImplantNT-501				[89]		
Retinitis pigmentosa	NCT03963154	NCT01531348	NCT0256736 NC-T03326336 NC-T01482195 NC-T03328130	I/II I, I/II	[90] [89]	[91]	[92] [74]
		NCT02464436		I/II		NEI Departme-nt NIH	
		NCT03073733		II		[93]	
		NCT02464436		I/II		[93]	
		NCT01068561		I		[94]	
		NCT01068561		I		[94]	
NPDR	NCT01691261	NCT01518842 NCT01736059	NC-T03846193	I I I/II	[83]	[95] [96]	[97]
Dry AMD	NCT02590692, NCT01469832 NCT04339764	NC-T01344993,	NC-T03144999	I/II I/II I	[36] [78] [98]	[98]	
SMD	NCT0146983	NCT01345006 NCT03178149	NC-T00643747 NC-T01736592	I/II I	[80]	[96] [92]	[84] [93]
RGC Degeneration	NTC 04577300 NTC 02862938 NTC 02330978	NCT01920867 NC-T03011541,	NC-T02161380	II, II, I I I I	[99] [99] [99]	[93] [84]	[92] [100]

Source:

Table 18.1 types of nanomaterials/scaffolds used with different biological sources using various types of retinal cells Table 18.2 illustrates a summary of all therapies that have made use of tissue engineering in vitro (Table 18.2).

18.8 Conclusion

Advancements in stem cell therapy, TE, and 3D bioprinting have significantly advanced the field of retinal regenerative medicine. Stem cell-based approaches, particularly involving ESCs, iPSCs, and MSCs, have shown potential in restoring retinal functionality, though challenges such as immune rejection, teratoma formation, and limited cell integration persist. TE, through scaffold-based and scaffold-free methods, has enhanced the survival, differentiation, and functionality of transplanted cells. Innovations in nanostructured and biodegradable scaffolds provide critical support for cell delivery, promoting integration with host tissue while reducing adverse outcomes.

The introduction of 3D bioprinting further revolutionizes retinal repair by creating multilayered tissue constructs that mimic the intricate architecture of the retina. These models enable precise cellular organization and offer applications ranging from *in vitro* disease modeling to therapeutic implants. Despite promising progress, replicating the retina's complex structure and achieving long-term integration remains a substantial hurdle [101]. Continued interdisciplinary research and technological advancements will be pivotal in translating these approaches from laboratory research to clinical solutions, ultimately improving outcomes for patients with retinal degenerative diseases.

List of abbreviations

RPE	Retinal pigment epithelium
RGC	Retinal ganglion cells
PR	Photoreceptor
RP	Retinitis pigmentosa
hESC	Human embryonic stem cell
ESC	Embryonic stem cell
iPSC	Induced pluripotent stem cell
RPC	Retinal precursor cell
IOP	Intraocular pressure
MSC	Mesenchymal stem cell
ONL	Outer nuclear layer
TE	Tissue engineering
ECM	Extracellular matrix
BrM	Bruch's membrane
PCL	Poly (Є caprolactone)
PGS	Poly (Glycerol sebacate)
POC	Poly (1,8-octanediol-co-citrate)
SF	Silk fibroin
PLCL	Poly (L-lactic acid-co-Є-caprolactone)
CS	Cationic chitosan
OCT	Optical Coherence tomography
CV	Collagen vitrigel
BMSC	Bone marrow-derived stem cells

References

[1] E. Sernagor, S.J. Eglen, R.O.L. Wong, Development of retinal ganglion cell structure and function, Progress in Retinal and Eye Research 20 (2) (2001) 139–174. Available from https://doi.org/10.1016/S1350-9462(00)00024-0.

[2] P. Sterling, How Retinal Circuits Optimize the Transfer of Visual Information, MIT Press, (2003), pp. 234–259. Available from 10.7551/mitpress/7131.003.0022.

[3] B.S. Atiyeh, M. Costagliola, Cultured epithelial autograft (CEA) in burn treatment: three decades later, Burns 33 (4) (2007) 405–413. Available from https://doi.org/10.1016/j.burns.2006.11.002.

[4] Y.M. Paulus, A. Sodhi, Anti-angiogenic therapy for retinal disease, Handbook of Experimental Pharmacology 242 (2017) 271–307. Available from: https://doi.org/10.1007/164_2016_78, http://www.springer.com/series/164.

[5] J.P. Vacanti, R. Langer, Tissue engineering: the design and fabrication of living replacement devices for surgical reconstruction and transplantation, The Lancet 354 (1) (1999) S32. Available from https://doi.org/10.1016/s0140-6736(99)90247-7.

[6] R. Langer, Biomaterials in drug delivery and tissue engineering: one laboratory's experience, Accounts of Chemical Research 33 (2) (2000) 94–101. Available from https://doi.org/10.1021/ar9800993.

[7] J.F. Mano, G.A. Silva, H.S. Azevedo, P.B. Malafaya, R.A. Sousa, S.S. Silva, et al., Natural origin biodegradable systems in tissue engineering and regenerative medicine: present status and some moving trends, Journal of the Royal Society Interface 4 (17) (2007) 999–1030. Available from https://doi.org/10.1098/rsif.2007.0220.

[8] G. Chan, D.J. Mooney, New materials for tissue engineering: towards greater control over the biological response, Trends in Biotechnology 26 (7) (2008) 382–392. Available from https://doi.org/10.1016/j.tibtech.2008.03.011.

[9] E. Masaeli, V. Forster, S. Picaud, F. Karamali, M.H. Nasr-Esfahani, C. Marquette, Tissue engineering of retina through high resolution 3-dimensional inkjet bioprinting, Biofabrication 12 (2) (2020) 25006. Available from https://doi.org/10.1088/1758-5090/ab4a20.

[10] T. Vazin, W.J. Freed, Human embryonic stem cells: derivation, culture, and differentiation: a review, Restorative Neurology and Neuroscience 28 (4) (2010) 589–603. Available from https://doi.org/10.3233/RNN-2010-0543.

[11] E. Taka, C. Karavasili, N. Bouropoulos, T. Moschakis, D.D. Andreadis, C.K. Zacharis, et al., Ocular co-delivery of timolol and brimonidine from a self-assembling peptide hydrogel for the treatment of glaucoma: in vitro and ex vivo evaluation, Pharmaceuticals 13 (6) (2020) 1–13. Available from https://doi.org/10.3390/ph13060126, https://www.mdpi.com/1424-8247/13/6/126/pdf.

[12] Z. Tang, Y. Zhang, Y. Wang, D. Zhang, B. Shen, M. Luo, et al., Progress of stem/progenitor cell-based therapy for retinal degeneration, Journal of Translational Medicine 15 (1) (2017) Available from http://www.translational-medicine.com/home/, https://doi.org/10.1186/s12967-017-1183-y, http://www.translational-medicine.com/home/.

[13] F. Deng, M. Chen, Y. Liu, H. Hu, Y. Xiong, C. Xu, et al., Stage-specific differentiation of iPSCs toward retinal ganglion cell lineage, Molecular Vision 22 (2016) 536.

[14] H.W. van Dijk, P.H.B. Kok, M. Garvin, M. Sonka, J.H. Devries, R.P.J. Michels, et al., Selective loss of inner retinal layer thickness in type 1 diabetic patients with minimal diabetic retinopathy, Investigative Ophthalmology and Visual Science 50 (7) (2009) 3404–3409. Available from https://doi.org/10.1167/iovs.08-3143NetherlandsAvailable from: http://www.iovs.org/cgi/reprint/50/7/3404.

[15] M.S. Divya, V.A. Rasheed, T. Schmidt, S. Lalitha, S. Hattar, J. James, Intraocular injection of ES cell-derived neural progenitors improve visual function in retinal ganglion cell-depleted mouse models. bioRxiv, India bioRxiv, 2017. Available from: https://www.biorxiv.org, https://doi.org/10.1101/114678.

[16] M.J. Young, J. Young Park, Cell and Animal Models used for Retinal Stem Cell Research, Springer Science and Business Media LLC, (2019), pp. 87–122. Available from https://doi.org/10.1007/978-3-319-98080-5_4.

[17] K.P. Kennelly, T.M. Holmes, D.M. Wallace, C. O'Farrelly, D.J. Keegan, Early subretinal allograft rejection is characterized by innate immune activity, Cell Transplantation 26 (6) (2017) 983–1000. Available from https://doi.org/10.3727/096368917X694697, https://doi.org/10.3727/096368917X694697.

[18] H. Terasakib, M. Eirakug, Y. Sasaid, Takahashia, et al., Transplantation of human embryonic stem cell-derived retinal tissue in two primate models of retinal degeneration, Proceedings of the National Academy of the USA 113 (1) (2015) .

[19] Z. Qu, Y. Guan, L. Cui, J. Song, J. Gu, H. Zhao, et al., Transplantation of rat embryonic stem cell-derived retinal progenitor cells preserves the retinal structure and function in rat retinal degeneration, Stem Cell Research & Therapy 6 (1) (2015). Available from https://doi.org/10.1186/s13287-015-0207-x.

[20] H. Hentze, R. Graichen, A. Colman, Cell therapy and the safety of embryonic stem cell-derived grafts, Trends in Biotechnology 25 (1) (2007) 24–32. Available from https://doi.org/10.1016/j.tibtech.2006.10.010.

[21] H. Hentze, P.L. Soong, S.T. Wang, B.W. Phillips, T.C. Putti, N.R. Dunn, Teratoma formation by human embryonic stem cells: evaluation of essential parameters for future safety studies, Stem Cell Research 2 (3) (2009) 198–210. Available from https://doi.org/10.1016/j.scr.2009.02.002.

[22] S.H. Moon, J.S. Kim, S.J. Park, J.J. Lim, H.J. Lee, S.M. Lee, et al., Effect of chromosome instability on the maintenance and differentiation of human embryonic stem cells in vitro and in vivo, Stem Cell Research 6 (1) (2011) 50–59. Available from https://doi.org/10.1016/j.scr.2010.08.006.

[23] X. Chen, S. Wang, H. Xu, J.D. Pereira, K.E. Hatzistergos, D. Saur, et al., Evidence for a retinal progenitor cell in the postnatal and adult mouse, Stem Cell Research 23 (2017) 20–32. Available from https://doi.org/10.1016/j.scr.2017.06.010Available from: http://www.elsevier.com/wps/find/journaldescription.cws_home/711630/description#description.

[24] J.N. Weiss, S. Levy, S.C. Benes, Stem cell ophthalmology treatment study (SCOTS): bone marrow-derived stem cells in the treatment of Leber's hereditary optic neuropathy, Neural Regeneration Research 11 (10) (2016) 1685–1694. Available from https://doi.org/10.4103/1673-5374.193251Available from: http://www.nrronline.org/temp/NeuralRegenRes11101685-7961715_220657.pdf.

[25] E. Lawley, P. Baranov, M. Young, Hybrid vitronectin-mimicking polycaprolactone scaffolds for human retinal progenitor cell differentiation and transplantation, Journal of Biomaterials Applications 29 (6) (2015) 894–902.

[26] U. Schraermeyer, G. Thumann, T. Luther, N. Kociok, S. Arnhold, K. Kruttwig, et al., Subretinally transplanted embryonic stem cells rescue photoreceptor cells from degeneration in the RCS rats, Cell Transplantation 10 (8) (2001) 673–680. Available from https://doi.org/10.3727/000000001783986215.

[27] E.L. West, A. Gonzalez-Cordero, C. Hippert, F. Osakada, J.P. Martinez-Barbera, R.A. Pearson, et al., Defining the integration capacity of embryonic stem cell-derived photoreceptor precursors, Stem Cells 30 (7) (2012) 1424–1435. Available from https://doi.org/10.1002/stem.1123UnitedKingdomAvailable from: http://onlinelibrary.wiley.com/doi/10.1002/stem.1123/pdf.

[28] W. Lu, D. Duan, Z. Ackbarkhan, M. Lu, M.L. Huang, Differentiation of human olfactory mucosa mesenchymal stem cells into photoreceptor cells in vitro, International Journal of Ophthalmology 10 (10) (2017) 1504–1509. Available from https://doi.org/10.18240/ijo.2017.10.04, http://www.ijo.cn/.

[29] C.R. Harrell, C. Fellabaum, A. Arsenijevic, B.S. Markovic, V. Djonov, V. Volarevic, Therapeutic potential of mesenchymal stem cells and their secretome in the treatment of glaucoma, Stem Cells International 2019 (2019). Available from https://doi.org/10.1155/2019/7869130, http://www.hindawi.com/journals/sci/contents/.

[30] T. Ren, Y. van der Merwe, M.B. Steketee, Developing extracellular matrix technology to treat retinal or optic nerve injury, eNeuro 2 (5) (2015) .

[31] A.S.L. Kwan, S. Wang, R.D. Lund, Photoreceptor layer reconstruction in a rodent model of retinal degeneration, Experimental Neurology 159 (1) (1999) 21–33. Available from https://doi.org/10.1006/exnr.1999.7157.

[32] R.E. MacLaren, R.A. Pearson, A. MacNeil, R.H. Douglas, T.E. Salt, M. Akimoto, et al., Retinal repair by transplantation of photoreceptor precursors, Nature 444 (7116) (2006) 203–207. Available from https://doi.org/10.1038/nature05161, http://www.nature.com/nature/index.html.

[33] M.J. Seiler, B.B. Thomas, Z. Chen, R. Wu, S.R. Sadda, R.B. Aramant, Retinal transplants restore visual responses: trans-synaptic tracing from visually responsive sites labels transplant neurons, European Journal of Neuroscience 28 (1) (2008) 208–220. Available from https://doi.org/10.1111/j.1460-9568.2008.06279.x.

[34] S. Wang, B. Lu, S. Girman, T. Holmes, N. Bischoff, R.D. Lund, Morphological and functional rescue in RCS rats after RPE cell line transplantation at a later stage of degeneration, Investigative Ophthalmology and Visual Science 49 (1) (2008) 416–421. Available from https://doi.org/10.1167/iovs.07-0992UnitedStates, http://www.iovs.org/cgi/reprint/49/1/416.

[35] N.C. Hunt, D. Hallam, V. Chichagova, D.H. Steel, M. Lako, The application of biomaterials to tissue engineering neural retina and retinal pigment epithelium, Advanced Healthcare Materials 7 (23) (2018). Available from https://doi.org/10.1002/adhm.201800226, http://onlinelibrary.wiley.com/journal/10.1002/(ISSN)2192-2659.

[36] M.S. Kyaw, Y. Kamano, Y. Yahata, T. Tanaka, N. Sato, F. Toyama, et al., Endodontic regeneration therapy: current strategies and tissue engineering solutions, Cells 14 (6) (2025) 422.

[37] R. Sharma, D. Bose, A. Maminishkis, K. Bharti, Retinal pigment epithelium replacement therapy for age-related macular degeneration: are we there yet? Annual Review of Pharmacology and Toxicology 60 (1) (2020) 553–572. Available from https://doi.org/10.1146/annurev-pharmtox-010919-023245.

[38] J.C. Valdoz, B.C. Johnson, D.J. Jacobs, N.A. Franks, E.L. Dodson, C. Sanders, et al., The ECM: to scaffold, or not to scaffold, that is the question, International Journal of Molecular Sciences 22 (23) (2021). Available from https://doi.org/10.3390/ijms222312690, https://www.mdpi.com/1422-0067/22/23/12690/pdf.

[39] A.K.A.S. Brun-Graeppi, C. Richard, M. Bessodes, D. Scherman, O.-W. Merten, Thermoresponsive surfaces for cell culture and enzyme-free cell detachment, Progress in Polymer Science 35 (11) (2010) 1311–1324. Available from https://doi.org/10.1016/j.progpolymsci.2010.07.007.

[40] Z. Liu, B.H. Parikh, Q.S. Tan, D.S. Wong, K.H. Ong, W. Yu, et al., Surgical transplantation of human RPE stem cell-derived RPE monolayers into non-human primates with immunosuppression, Stem Cell Reports 16 (2) (2021) 237–251.

[41] I.K. Lee, R. Xie, A. Luz-Madrigal, S. Min, J. Zhu, J. Jin, et al., Micromolded honeycomb scaffold design to support the generation of a bilayered RPE and photoreceptor cell construct, Bioactive Materials 30 (2023) 142–153.

[42] H. Liu, L. Jing, J. Sun, D. Huang, An overview of scaffolds for retinal pigment epithelium research, Procedia Manufacturing 53 (2021) 492–499. Available from https://doi.org/10.1016/j.promfg.2021.06.051.

[43] W.L. Neeley, S. Redenti, H. Klassen, S. Tao, T. Desai, M.J. Young, et al., A microfabricated scaffold for retinal progenitor cell grafting, Biomaterials 29 (4) (2008) 418–426. Available from https://doi.org/10.1016/j.biomaterials.2007.10.007.

[44] S. Behtaj, F. Karamali, E. Masaeli, Y.G. Anissimov, M. Rybachuk, Electrospun PGS/PCL, PLLA/PCL, PLGA/PCL and pure PCL scaffolds for retinal progenitor cell cultivation, Biochemical Engineering Journal 166 (2021) 107846.

[45] H.V. Gudiseva, V. Vrathasha, J. He, D. Bungatavula, J.M. O'Brien, V.R. Chavali, Single cell sequencing of induced pluripotent stem cell derived retinal ganglion cells (iPSC-RGC) reveals distinct molecular signatures and RGC subtypes, Genes 12 (12) (2021) 2015.

[46] Y. Edger, S. Tan, Biofabrication of choroid-retina tissue construct for modelling of age-related macular degeneration disease, Neural Regeneration Research (2019).

[47] M.J. Seiler, R.B. Aramant, Cell replacement and visual restoration by retinal sheet transplants, Progress in Retinal and Eye Research 31 (6) (2012) 661–687. Available from https://doi.org/10.1016/j.preteyeres.2012.06.003.

[48] S.R. Hynes, E.B. Lavik, A tissue-engineered approach towards retinal repair: scaffolds for cell transplantation to the subretinal space, Graefe's Archive for Clinical and Experimental Ophthalmology 248 (6) (2010) 763–778. Available from https://doi.org/10.1007/s00417-009-1263-7.

[49] J. Yao, S.L. Tao, M.J. Young, Synthetic polymer scaffolds for stem cell transplantation in retinal tissue engineering, Polymers 3 (2) (2011) 899–914.

[50] R. Sharma, V. Khristov, A. Rising, B.S. Jha, R. Dejene, N. Hotaling, et al., Clinical-grade stem cell–derived retinal pigment epithelium patch rescues retinal degeneration in rodents and pigs, Science Translational Medicine 11 (475) (2019) eaat5580.

[51] B. Diniz, P. Thomas, B. Thomas, R. Ribeiro, Y. Hu, R. Brant, et al., Subretinal implantation of retinal pigment epithelial cells derived from human embryonic stem cells: improved survival when implanted as a monolayer, Investigative Ophthalmology and Visual Science 54 (7) (2013) 5087–5096. Available from https://doi.org/10.1167/iovs.12-11239UnitedStates, http://www.iovs.org/content/54/7/5087.full.pdf+html.

[52] E.L. West, R.A. Pearson, R.E. MacLaren, J.C. Sowden, R.R. Ali, Cell transplantation strategies for retinal repair, Progress in Brain Research 175 (2009) 3–21. Available from https://doi.org/10.1016/S0079-6123(09)17501-5.

[53] J. Hsiung, D. Zhu, D.R. Hinton, Polarized human embryonic stem cell-derived retinal pigment epithelial cell monolayers have higher resistance to oxidative stress-induced cell death than nonpolarized cultures, Stem Cells Translational Medicine 4 (1) (2015) 10–20. Available from https://doi.org/10.5966/sctm.2014-0205.
[54] C.E. White, R.M. Olabisi, Scaffolds for retinal pigment epithelial cell transplantation in age-related macular degeneration, Journal of Tissue Engineering (2017) .
[55] N.A. Hotaling, V. Khristov, Q. Wan, R. Sharma, B.S. Jha, M. Lotfi, et al., Nanofiber scaffold-based tissue-engineered retinal pigment epithelium to treat degenerative eye diseases, Journal of Ocular Pharmacology and Therapeutics 32 (5) (2016) 272–285. Available from https://doi.org/10.1089/jop.2015.0157.
[56] M. Hatami, N.Z. Mehrjardi, S. Kiani, K. Hemmesi, H. Azizi, A. Shahverdi, et al., Human embryonic stem cell-derived neural precursor transplants in collagen scaffolds promote recovery in injured rat spinal cord, Cytotherapy 11 (5) (2009) 618–630.
[57] B. Belgio, F. Boschetti, S. Mantero, Towards an in vitro retinal model to study and develop new therapies for age-related macular degeneration, Bioengineering 8 (2) (2021) 18.
[58] M. Abedin Zadeh, M. Khoder, A.A. Al-Kinani, H.M. Younes, R.G. Alany, Retinal cell regeneration using tissue engineered polymeric scaffolds, Drug Discovery Today 24 (8) (2019) 1669–1678. Available from https://doi.org/10.1016/j.drudis.2019.04.009.
[59] F. Jafari, S.N. Khorasani, F. Alihosseini, D. Semnani, S. Khalili, R.E. Neisiany, Development of an electrospun scaffold for retinal tissue engineering, Polymer Science, Series B 62 (3) (2020) 290–298. Available from https://doi.org/10.1134/s1560090420030069.
[60] P.R. Buenzli, M. Lanaro, C.S. Wong, M.P. McLaughlin, M.C. Allenby, M.A. Woodruff, et al., Cell proliferation and migration explain pore bridging dynamics in 3D printed scaffolds of different pore size, Acta Biomaterialia 114 (2020) 285–295.
[61] C.G. Jeong, S.J. Hollister, A comparison of the influence of material on in vitro cartilage tissue engineering with PCL, PGS, and POC 3D scaffold architecture seeded with chondrocytes, Biomaterials 31 (15) (2010) 4304–4312.
[62] D. Zhang, N. Ni, J. Chen, Q. Yao, B. Shen, Y. Zhang, et al., Electrospun SF/PLCL nanofibrous membrane: a potential scaffold for retinal progenitor cell proliferation and differentiation, Scientific Reports 5 (1) (2015). Available from https://doi.org/10.1038/srep14326.
[63] H. Chen, X. Fan, J. Xia, P. Chen, X. Zhou, J. Huang, et al., Electrospun chitosan-graft-poly (e-caprolactone)/poly (e-caprolactone) nanofibrous scaffolds for retinal tissue engineering, International Journal of Nanomedicine 6 (2011) 453–461.
[64] S. Redenti, S. Tao, J. Yang, P. Gu, H. Klassen, S. Saigal, et al., Retinal tissue engineering using mouse retinal progenitor cells and a novel biodegradable, thin-film poly(e-caprolactone) nanowire scaffold, Journal of Ocular Biology, Diseases, and Informatics 1 (1) (2008) 19–29. Available from https://doi.org/10.1007/s12177-008-9005-3, http://www.springer.com/humana+press/ophthalmology/journal/12177.
[65] M.R. Steedman, S.L. Tao, H. Klassen, T.A. Desai, Enhanced differentiation of retinal progenitor cells using microfabricated topographical cues, Biomedical Microdevices 12 (3) (2010) 363–369. Available from https://doi.org/10.1007/s10544-009-9392-7.
[66] P. Baranov, A. Michaelson, J. Kundu, R.L. Carrier, M. Young, Interphotoreceptor matrix-poly (ϵ-caprolactone) composite scaffolds for human photoreceptor differentiation, Journal of Tissue Engineering 5 (2014) 2041731414554139.
[67] A.C. Sommer, E.Z. Blumenthal, Implementations of 3D printing in ophthalmology, Graefe's Archive for Clinical and Experimental Ophthalmology 257 (9) (2019) 1815–1822. Available from https://doi.org/10.1007/s00417-019-04312-3.
[68] P. Xie, Z. Hu, X. Zhang, X. Li, Z. Gao, D. Yuan, et al., Application of 3-dimensional printing technology to construct an eye model for fundus viewing study, PLoS ONE 9 (11) (2014) e109373. Available from https://doi.org/10.1371/journal.pone.0109373.
[69] K.Y. Wu, A. Tabari, É. Mazerolle, S.D. Tran, Towards precision ophthalmology: the role of 3D printing and bioprinting in oculoplastic surgery, retinal, corneal, and glaucoma treatment, Biomimetics 9 (3) (2024) 145.
[70] U. Garg, N. Jain, S. Kaul, V.K. Rai, M. Pandey, U. Nagaich, et al., The emerging role of 3D-printing in ocular drug delivery: challenges, current status, and future prospects, Journal of Drug Delivery Science and Technology 76 (2022) 103798.
[71] N. Poomathi, S. Singh, C. Prakash, R.V. Patil, P.T. Perumal, V.A. Barathi, et al., Bioprinting in ophthalmology: current advances and future pathways, Rapid Prototyping Journal 25 (3) (2019) 496–514.
[72] R.M. Nerem, Tissue engineering: the hope, the hype, and the future, Tissue Engineering 12 (5) (2006) 1143–1150. Available from https://doi.org/10.1089/ten.2006.12.1143.
[73] H. Gudapati, M. Dey, I. Ozbolat, A comprehensive review on droplet-based bioprinting: past, present and future, Biomaterials 102 (2016) 20–42. Available from https://doi.org/10.1016/j.biomaterials.2016.06.012.
[74] R. Agarwal, A. Sharma, S. Shukla, Recent advances in 3D bioprinting for ocular tissue regeneration, Current Opinion in Ophthalmology 32 (1) (2021) 62–69. Available from https://doi.org/10.1097/ICU.0000000000000735.
[75] M.J. Song, R. Quinn, E. Nguyen, C. Hampton, R. Sharma, T.S. Park, et al., Bioprinted 3D outer retina barrier uncovers RPE-dependent choroidal phenotype in advanced macular degeneration, Nature Methods 20 (1) (2023) 149–161. Available from https://doi.org/10.1038/s41592-022-01701-1, https://www.nature.com/nmeth/.
[76] W. Wang, T. Yang, S. Chen, L. Liang, Y. Wang, Y. Ding, et al., Tissue engineering RPE sheet derived from hiPSC-RPE cell spheroids supplemented with Y-27632 and RepSox, Journal of Biological Engineering 18 (1) (2024). Available from https://doi.org/10.1186/s13036-024-00405-8.
[77] F. Kuwata, H. Ohnishi, N. Yamamoto, T. Takezawa, M. Yamashita, H. Okuyama, et al., Transplantation of human induced pluripotent stem cell-derived airway cells on vitrigel membrane into rat nasal cavity, Tissue Engineering. Part A 28 (13-14) (2022) 586–594.
[78] M. Nishida, Y. Tanaka, Y. Tanaka, S. Amaya, N. Tanaka, H. Uyama, et al., Human iPS cell derived RPE strips for secure delivery of graft cells at a target place with minimal surgical invasion, Scientific Reports 11 (1) (2021). Available from https://doi.org/10.1038/s41598-021-00703-x.

[79] Y. Jiménez-Gómez, D. Alba-Molina, M. Blanco-Blanco, L. Pérez-Fajardo, F. Reyes-Ortega, L. Ortega-Llamas, et al., Novel treatments for age-related macular degeneration: a review of clinical advances in sustained drug delivery systems, Pharmaceutics 14 (7) (2022) 1473. Available from https://doi.org/10.3390/pharmaceutics14071473.
[80] S.D. Schwartz, J.P. Hubschman, G. Heilwell, V. Franco-Cardenas, C.K. Pan, R.M. Ostrick, et al., Embryonic stem cell trials for macular degeneration: apreliminary report, The Lancet 379 (9817) (2012) 713–720.
[81] R.E. MacLaren, M. Groppe, A.R. Barnard, C.L. Cottriall, T. Tolmachova, L. Seymour, et al., Retinal gene therapy in patients with choroideremia: initial findings from a phase 1/2 clinical trial, The Lancet 383 (9923) (2014) 1129–1137.
[82] L. da Cruz, K. Fynes, O. Georgiadis, J. Kerby, Y.H. Luo, A. Ahmado, et al., Phase 1 clinical study of an embryonic stem cell–derived retinal pigment epithelium patch in age-related macular degeneration, Nature Biotechnology 36 (4) (2018) 328–337. Available from https://doi.org/10.1038/nbt.4114.
[83] S.D. Schwartz, C.D. Regillo, B.L. Lam, D. Eliott, P.J. Rosenfeld, N.Z. Gregori, et al., Human embryonic stem cell-derived retinal pigment epithelium in patients with age-related macular degeneration and Stargardt's macular dystrophy: follow-up of two open-label phase 1/2 studies, The Lancet 385 (9967) (2015) 509–516. Available from https://doi.org/10.1016/s0140-6736(14)61376-3.
[84] X. Gu, X. Yu, C. Zhao, P. Duan, T. Zhao, Y. Liu, et al., Efficacy and safety of autologous bone marrow mesenchymal stem cell transplantation in patients with diabetic retinopathy, Cellular Physiology and Biochemistry 49 (1) (2018) 40–52. Available from https://doi.org/10.1159/000492838.
[85] S. Russell, J. Bennett, J.A. Wellman, D.C. Chung, Z.-F. Yu, A. Tillman, et al., Efficacy and safety of voretigene neparvovec (AAV2-hRPE65v2) in patients with RPE65-mediated inherited retinal dystrophy: a randomised, controlled, open-label, phase 3 trial, The Lancet 390 (10097) (2017) 849–860.
[86] M. Mandai, A. Watanabe, Y. Kurimoto, Y. Hirami, C. Morinaga, T. Daimon, et al., Autologous induced stem-cell–derived retinal cells for macular degeneration, New England Journal of Medicine 376 (11) (2017) 1038–1046. Available from https://doi.org/10.1056/nejmoa1608368.
[87] V. Khristov, J. Hartford, Q. Wan, M.R. Lotfi, K. Miyagishima, A. Maminishkis, et al., Optimizing biodegradable scaffolds for developing iPS cell derived RPE tissue, Investigative Ophthalmology & Visual Science 55 (13) (2014). Available from https://doi.org/10.1134/S00062979060700911175–1175, Available from: .
[88] K. Zhang, J.J. Hopkins, J.S. Heier, D.G. Birch, L.S. Halperin, T.A. Albini, et al., Ciliary neurotrophic factor delivered by encapsulated cell intraocular implants for treatment of geographic atrophy in age-related macular degeneration, Proceedings of the National Academy of Sciences of the United States of America 108 (15) (2011) 6241–6245. Available from https://doi.org/10.1073/pnas.1018987108, http://www.pnas.org/content/108/15/6241.full.pdf+html.
[89] D. Terrell, J. Comander, Current stem-cell approaches for the treatment of inherited retinal degenerations, Seminars in Ophthalmology 34 (4) (2019) 287–292. Available from https://doi.org/10.1080/08820538.2019.1620808.
[90] I.K. Sugino, A. Rapista, Q. Sun, J. Wang, C.F. Nunes, N. Cheewatrakoolpong, et al., A method to enhance cell survival on Bruch's membrane in eyes affected by age and age-related macular degeneration, Investigative Ophthalmology & Visual Science 52 (13) (2011) 9598–9609.
[91] A. Chaffiol, M. Provansal, C. Joffrois, K. Blaize, G. Labernede, R. Goulet, et al., In vivo optogenetic stimulation of the primate retina activates the visual cortex after long-term transduction, Molecular Therapy Methods & Clinical Development 24 (2022) 1-0.
[92] A. Tuekprakhon, S. Sangkitporn, A. Trinavarat, A.R. Pawestri, V. Vamvanij, M. Ruangchainikom, et al., Intravitreal autologous mesenchymal stem cell transplantation: a non-randomized phase I clinical trial in patients with retinitis pigmentosa, Stem Cell Research & Therapy (2021). Available from https://doi.org/10.1186/s13287-020-02122-7.
[93] D.S. Rajendran Nair, M.J. Seiler, K.H. Patel, V. Thomas, J.C.M. Camarillo, M.S. Humayun, et al., Tissue engineering strategies for retina regeneration, Applied Sciences (Switzerland) 11 (5) (2021) 1–19. Available from https://doi.org/10.3390/app11052154, https://www.mdpi.com/2076-3417/11/5/2154/pdf.
[94] H. Uyama, M. Mandai, M. Takahashi, Stem-cell-based therapies for retinal degenerative diseases: current challenges in the establishment of new treatment strategies, Development, Growth & Differentiation 63 (1) (2021) 59–71. Available from https://doi.org/10.1111/dgd.12704.
[95] A.H. Kashani, J.S. Lebkowski, F.M. Rahhal, R.L. Avery, H. Salehi-Had, S. Chen, et al., One-year follow-up in a phase 1/2a clinical trial of an allogeneic rpe cell bioengineered implant for advanced dry age-related macular degeneration, Translational Vision Science and Technology 10 (10) (2021). Available from https://doi.org/10.1167/TVST.10.10.13, https://tvst.arvojournals.org/article.aspx?articleid=2777956.
[96] J.W. Bainbridge, A.J. Smith, S.S. Barker, S. Robbie, R. Henderson, K. Balaggan, et al., Effect of gene therapyonvisual func tion in Leber's congenital amaurosis, The New England Journal of Medicine 358 (2008) 2231–2239.
[97] J. Trincao-Marques, et al., Gene and cell therapy for age-related macular degeneration: a review, Survey of Ophthalmology (2024) .
[98] A.H. Kashani, J.S. Lebkowski, F.M. Rahhal, R.L. Avery, H. Salehi-Had, W. Dang, et al., A bioengineered retinal pigment epithelial monolayer for advanced, dry age-related macular degeneration, Science Translational Medicine 10 (435) (2018). Available from https://doi.org/10.1126/scitranslmed.aao4097, http://stm.sciencemag.org/content/scitransmed/10/435/eaao4097.full.pdf.
[99] R.M. Coco-Martin, S. Pastor-Idoate, J.C. Pastor, Cell replacement therapy for retinal and optic nerve diseases: cell sources, clinical trials and challenges, Pharmaceutics 13 (6) (2021) 865.
[100] J. Guy, et al., Gene therapy for Leber hereditary optic neuropathy: low-and medium-dose visual results, Ophthalmology 124 (11) (2017) 1621–1634.
[101] N. Poomathi, S. Singh, C. Prakash, R.V. Patil, P.T. Perumal, V.A. Barathi, K.K. Balasubramanian, S. Ramakrishna, N.U. Maheshwari, Bioprinting in ophthalmology: current advances and future pathways, Rapid Prototyping Journal 25 (2018) 496–514. Available from 10.1108/rpj-06-2018-0144.

Section III

Biosensors in tissue engineering

Chapter 19

Biomaterial-based biosensors

Sabina Khan[1], Ruma Rani[2], Parth Malik[3], and Tapan Kumar Mukherjee[4]

[1]*Kadi Sarva Vishwavidyalaya, Gandhinagar, Gujarat, India,* [2]*Indian Council of Agricultural Research (ICAR)-National Research Centre on Equines, Hisar, Haryana, India,* [3]*Swarrnim School of Sciences, Swarrnim Startup and Innovation University (SSIU), Gandhinagar, Gujarat, India,* [4]*Department of Biotechnology, Amity University, Kolkata, West Bengal, India*

19.1 Introduction

In recent years, tremendous advancements have been reported in materials science, owing to a consistent need for multifunctional, energy-efficient, and dynamic materials having cutting-edge needs across industries, better healthcare, steadfast energy transfer, and catalytic applications. Creditable tributes for this are hereby bestowed to the increasing precision and reliability of characterization techniques, which have been the result of rigorous efforts of multi and cross-disciplinary collaborations, spanning biology, chemistry, engineering, mathematical simulations, and physical theories. The net result of such cumulative advances has provided materials that are intensively eco-friendly, durable and could be tailored to specific needs. The diversity of such multifunctional materials encompasses nanomaterials (NMs), porous, semiconductor, and topological materials with the remarkable ease of functionalization [1]. This chapter intends to update our understanding of biomaterials and their usability engineering for biosensing applications. With due progress, the definition of biomaterials has aligned with their practical applications, encompassing classification with intriguing insights into their origins and integration into scientific endeavors. The backbone aspect of biomaterials' usefulness in biosensing pertains to their exceptional reproducibility in the preparation mechanisms, the possibility of size-shape tunable physicochemical behavior, and the robustness of functionalization abilities. The integration of nanoscience has enriched the biomaterials research, with the well-understood top-down (physicist's choice) and bottom-up (chemist's choice) mechanisms [2]. Recent understanding of self-assembly at the molecular scale has augmented the spontaneity of *vis-à-vis* their 3D arrangements supported by noncovalent interactions with the ease of stabilization modulation [2]. The distinctive recognition of a biomaterial pertains to an entity that has been engineered to interact with biological systems for a medical rationale. The interactions generally occur either for therapeutic purposes (to treat, facilitate, repair, or replace a tissue function of the body) or for diagnostic purposes. Most well-illustrated biomaterial configurations to date are categorized as natural (organic/inorganic compounds from biological resources) or synthetic (polymers, ceramics, or metals), whereby the recent interest in biopolymers has made noteworthy strides via compatible essence (Table 19.1) [3]. Recent interest in exploring biocompatible polymers for diagnostic or therapeutic intent has been in their carbohydrate and protein-derived structures, allowing for the feasibility of molecular interactions and responses analogous to the physiological environment [4]. The noticeable distinctions of these compatible polymeric entities have considerably minimized the concerns behind the uncertain toxicities of NMs manifested by their dimensionally tuned quantum effects.

Keeping the above distinctions in mind, this chapter focuses on the application of biomaterials in biosensing, spanning their self-assembly-mediated preparation and engineering physicochemical properties to accomplish precise detection of analyte molecules. The distinctive outcomes of such engineered biosensors enable the analyte detection across various domains, enabling a precise screening of contaminants, toxins, pathogens, and related entities [5]. Towards the end, we have talked about the significance of NMs-driven quantum effects as assets in biomaterials' biosensing abilities, conferring a nanocomposite-like functional essence to biomaterials.

19.2 Biomaterials

19.2.1 Definition and features

In most simplified language, a biomaterial refers to any substance that is derived or obtained from biological organisms such as plants, animals, bacteria, fungi, or other living entities. A familiar terminology for these materials is the

Fundamentals of Translational Tissue Engineering. DOI: https://doi.org/10.1016/B978-0-443-33455-9.00012-2

TABLE 19.1 Terminological distinction of biomaterials configurations along with a description of select useful and compatible monomers to make biocompatible polymers [3].

Type of biomaterial	Broad recognition/nomenclature	Distinguishing features
Natural	Organic: collagen, chitosan, Alginate	Biocompatible, biodegradable, and often bioactive
	Inorganic	Bioinert, stable, and often used for structural support
Biomolecules	Enzymes	Catalysts for biochemical reactions, biocompatible, and specific to substrates
	Proteins	Structural or functional biomolecules, biocompatible, involved in biological processes
	Antibodies	Immune system proteins that bind to specific antigens, used in diagnostics and therapeutics
	Nucleic Acids	DNA and RNA, carriers of genetic information, used in gene therapy and diagnostics
Synthetic/Man-made	Metals	Strong, durable, can be bioinert or bioactive depending on the metal and surface treatment
	Polymers	Versatile, can be tailored for specific applications, and range from bioinert to bioactive
	Ceramics	Inert, used for bone implants due to similarity with natural minerals
Select polymers having manifold applications of biomaterials (compatibility issues)		
Polysaccharides	Polycaprolactone	Biodegradable, biocompatible, slow degradation rate
	Chitosan	Biocompatible, biodegradable, nontoxic, antifungal, antibacterial, antitumor, potent hemostatic activity
	Alginate	Nontoxic, Biodegradable, stable, enhances collagen type 1 production, activates macrophages to produce cytokines
	Albumin	Biodegradable, biocompatible, used in drug delivery and tissue engineering
	Dextran	Biodegradable, nontoxic, used in drug delivery, blood plasma substitute
	Gelatin	Biodegradable, biocompatible, promotes cell adhesion, growth, used in wound dressings and tissue engineering
	Collagen	Promotes cell growth and proliferation, aids in blood coagulation
	Hyaluronic acid	Nontoxic, biodegradable, antiinflammatory, biocompatible, viscoelastic, induces cell adhesion, nonimmunogenic

"biologically derived materials", wherein the accomplished biomedical need could be mimicking a native biological function or treatment of an injury (via drug delivery, timely diagnosis, or both) [6]. A lot of discussions and debates preceded the discussions for a consensual definition of biomaterials, eventually leading to a consensus conference in 1987 for a definition within biomaterial science. From this conference emerged the definition of biomaterial as a nonviable material utilized in medical devices, intended to interact with biological systems [7]. However, the concept of "nonviable" required reassessment, leading to a revised definition introduced in a materials science dictionary in 1999, which defined biomaterials as 'a substance capable of interfacing with biological systems to assess, treat, augment, or substitute any bodily tissue, organ, or function of the body' [8].

An application-oriented definition of a biomaterial describes it as any substance with significant biocompatibility that facilitates its entry into the human body. Biomaterials could be made naturally (from plants, animals, or microbes) or synthetically in the laboratory (as compatible polymers *via* optimum selection of monomers). Prominent factors affecting the functional performance of biomaterials include their optimum composition, desired physicochemical attributes,

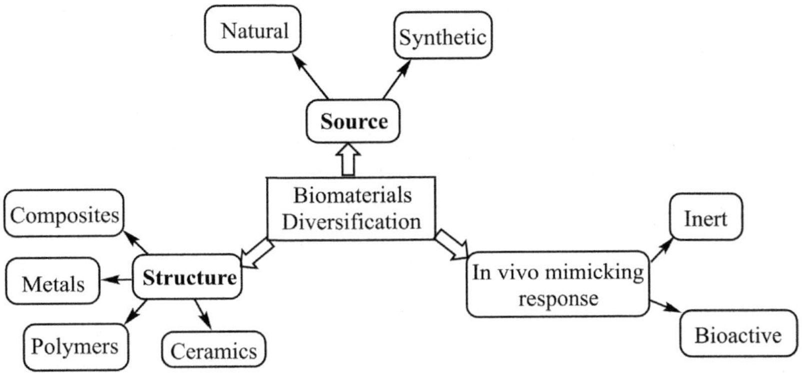

FIGURE 19.1 Diversity prospects of biomaterials pertaining to their source distinctions, structural makeup, and biological usefulness.

surface topography, and the ability to defy infection and engineer the surgical replacement(s). Biomaterials exhibit key characteristics that pique the interest of scientists and industries. Foremost among these is biocompatibility, ensuring they do not elicit toxic effects on biological systems. Biodegradability is equally crucial, denoting the material's resistance to decomposition by microbial action [9]. Fig. 19.1 summarizes the diversity prospects of biomaterials with reference to their preparation distinctions, compositional diversity, and biological usefulness. This chapter focuses mainly on the application of natural and synthetic biomaterials for the purpose of biosensing. It is pertinent to mention here that natural biomaterials used for biosensing primarily include the native biomolecules such as enzymes, proteins, antibodies, nucleic acids, polysaccharides, or their derivatives. The synthetic counterparts, however, require a modification from the man-made efforts, which are majorly intended to reduce the toxicity, improve the interaction precision and efficacy for infinitesimal detection [10]. Readers must note here that our discussion herein is confined to the biosensing applications, and the criteria for drug delivery and tissue engineering purposes are quite distinct and can be searched for in erstwhile specific literature sources.

Current interests in the widespread adoption of biomaterials in biosensing are largely driven by reduced pressure to make synthetic materials, cumbersome handling controls in surface engineering, and remaining environmental risks. With the emerging progress in functionalization and characterization approaches, multiple nonbiological and synthetic materials have emerged as exceptional sensing probes, with uncommon electronic and optical properties [11]. The following paragraphs focus on the classification criterion of biomaterials, fundamentals of biosensing, and the prominent biosensing traits of biomaterials in their natural and engineered forms, followed by a discussion of recent studies thereof.

19.2.2 Biomaterial-driven classification of biosensors

Development of biomaterials requires expertise from multiple disciplines, mainly medical sciences, biophysics, material science and engineering, chemistry, and biology. Prominently being optimized as a research domain of material science and engineering, metals, alloys, polymers, ceramics, and composites are majorly distinct classes of biomaterials on the basis of their specific in vivo functions. Optimization of biomaterials is decided by the principles of medical science, wherein the logical information about the specific kinds of devices needed, along with their physical, mechanical, chemical, and biological traits. This warrants support from biophysics, wherein compliance of functional performance (magnetic, electrical, thermal, and physical properties) with the designated in vivo function is optimized [12]. Finally, chemistry also lends major support to the development of rightful biomaterials, whereby the toxicity or cross-reactivity of biomaterials in the in vivo conditions is minimized. The effect of surface coating and functionalization of biomaterials on their in vivo performance is determined *by* knowing their fundamental constitution and their specific response in the physiological environment. Modern considerations for developing biomaterials have significantly progressed from the yesteryear and inceptive considerations (that relied primarily on the physical support, emphasizing the inertness), whereby, in addition to the inertness ability, the ability to assist the healing of injured body tissues is also being worked upon. Familiar configurations of such entities are designed in multiple forms, such as a complex made up of hybrid composites or polymer-coated metals. Earlier validation of biomaterials did not screen them exhaustively for their interactions in the physiological environment, as the materials being developed were strictly inert. However, with the gradual progress, inclusion of noninert biomaterials has necessitated a need for screening stimulus-based physical and chemical response, involving the dynamics of redox modulation, corrosive perturbation, and replicating the native physiological tissues. Such constraints have consolidated a need for biocompatibility evaluation for the biomaterials, being defined as the characteristic ability of a material to perform a characteristic function within the living environment, chronologically supported by optimal host responses, least inflammatory and toxicity repercussions from local and systemic viewpoints [13]. Thereby, biocompatibility of any biomaterial is a complex function of its constituent makeup, applications, host-specific responses that are in turn affected by gender, food habits, smoking and drinking history, age, and the surviving geography of the host organism. Even though biocompatibility screening can be made using in vitro models, wide variations in the in vitro and physiological conditions continue to jeopardize the assumptions and reliability.

Using biomaterials for biosensing necessitates a prior sound understanding of the sensing working mechanism, wherein a specific quantitative or semiquantitative information is screened using a biological recognition element (BRE). The BRE herein refers to a material that senses or detects the analyte through its biological origin, hereby being relatable to physiologically essential biomolecules such as enzymes, nucleic acids, proteins, antibodies, or polysaccharides. Accordingly, the distinguished terminologies are enzyme, nucleic acid, protein, antibody, or polysaccharide biosensors. Apart from the biomaterial comprising the sensing probe, a biosensor's operation is primarily governed by the electronic transduction domain (actuation mechanism), which converts the biological interactions between the sensing probe and the analyte into an electronic signal. The response of this electronic signal is what we receive as the quantification or output of a biosensor function. The interconversion of a biomaterial (probe)-analyte biological interaction into a corresponding electronic response relies on one or more mechanisms, including physicochemical, optical, piezoelectric, or electrochemical effects. Theoretical conceptualization of an enzyme-based biosensor was first unveiled by Leland Clark in 1962 [14]. Thereafter, Updike and Hicks in 1967 provided the framework for a working enzyme electrode aimed at the detection of glucose, presently recognized as the first global biosensor [15]. The classification of biosensors on a global scale is based either on probe type or on the transduction mechanism [16]. Although both classifications exhibit their uniqueness, the one relying on the transduction mechanism is more fundamental with respect to detection precision, which varies with the specific biomaterial (probe sensitivity). The next section describes the fundamentals and recent developments in probe-based biosensor configurations.

19.3 Functional aspects of biomaterials based biosensor platforms

As described in the previous section, enzymes, proteins, antibodies, nucleic acids, and polysaccharides are the major biomaterials (of natural grade) used as receptor components in a biosensor. These entities prevail in the physiological environment and are thoroughly understood for their explicit functions, distinguished via their unique biochemical activities. For instance, enzymes are characteristically functional on perceiving their specific substrates, while proteins get activated on meeting their receptor (which mediates their phosphorylation). Similarly, antibodies distinctly recognize antigens (or pathogens) and neutralize their toxicity, while nucleic acids are the templates of nucleotides and build up over them via mutual receptivity. This section discusses the prominent aspects of biosensors having these entities as receptor moieties.

19.3.1 Biosensors using enzymes as sensing agents

In 1962, through the dedicated efforts of Clark and Lyons, the first enzyme biosensor was proposed in 1962, functioning via the detection of glucose in the blood plasma using glucose oxidase in the optimized configuration of the electrochemical biosensor. Thereafter, significant strides have been made in the development of enzyme biosensors via the characteristic affinity of enzymes; some prominent enzymes from the human body are urease, peroxidase, amylases, hydrolases, lipases, and others. The working principle of enzymatic biosensors relies on the efficacy of enzymes to bind their specific analyte, thereby specifically detecting the substrate from a mixture of compounds [17]. The distinct recognition of analytes by the sensing probe (enzymes) could be enacted via multiple mechanisms, such as (1) conversion of analyte into a sensor-detectable product, (2) detection of enzyme activation or inhibition via analyte, (3) screening the altered properties of enzymes in response to their combination with substrates.

The long-lasting use of enzyme-based biosensors is due to the reusability benefits of enzymes, which are presently used to a greater extent through immobilization and matrix adsorption. The immobilization of enzymes on functionalized NMs enhances their surface area (SA), whereby activation energy of corresponding reactions is substantially reduced. A pegging challenge in the long-lasting working ability of enzymatic biosensors is the retention of enzyme stability, for which numerous engineered provisions are in use. For instance, many enzymes prevail in the attached or cell membrane-inserted forms, anchored with hydrophobic subunits or lipid chains. The availability of such enzymes could be reengineered via their functionalization on the electrodes, facilitating the native structural traits of enzymes, besides optimizing their electrocatalytic attributes and stability. Several biomimetic mechanisms have been explored to modify the electrode surfaces for a feasible accommodation of membrane-bound enzymes. The prominent assortments herein comprise the formation of self-assembled monolayers of hydrophobic compounds, lipid bilayers, or engineered liposomal assembly. Electrodes of inert and conducting materials such as gold, Au have been demonstrated as active platforms to accomplish high detection specificity and turnover of enzyme catalysis, facilitating selective target detection, signal amplification, prompt response generation, and broad range of operational versatility *via* electrochemical transduction [18,19]. Knowing the fact that a major proportion of redox enzymes reported to date are membrane-bound (remaining conjugated to the external region or transmembrane), many of the demonstrated enzymatic biosensors rely on soluble enzymes. Prominent reasons for this include a high structural complexity with a weaker stability of purified membrane-bound enzyme configurations. This was also demonstrated recently, wherein a frequent dissociation of membrane-bound enzymes' hydrophobic subunits was observed, resulting in impaired activity and stability [20]. To materialize the maximum benefit of membrane-bound enzymes used in electrochemical devices, it is essential to optimize the immobilization assays, which assist in the maintenance of natural enzyme configurations *by* stabilizing the hydrophobic domains. In this reference, the formation of a biomimetic membrane on the electrode surface via the assistance of phospholipid layers is a versatile and reliable strategy for reconstituting the membrane-bound enzymes [21,22]. Deposition of liposomes and formation of hydrophobic compounds monolayers on the electrode surface are some other potential mechanisms having been explored herewith [23,24].

As an illustration, biosensors based on the use of lipases, a group of enzymes associated with the lipase–esterase family, have alkaline sensitivity, including serine at the active site, besides being in the folded regime subsequent to the expression of genetic traits [25,26]. The frequent use of lipases is credited to their wide substrate specificity and good commercial availability. Cellular provisions of lipases could vary from bacterial, fungal, or mammalian, exhibiting a prominent structural resemblance with cholinesterases (acetylcholinesterase and butyrylcholinesterase), exhibiting increasing utility for biosensing due to their neurotoxic sensitivity [27–29]. The major substrate of lipases is triacylglycerol, generating glycerol and carboxylic acids as the products of enzyme-catalyzed reaction. Lipase functions in sustaining metabolic processes, playing a central role in retinol biosynthesis, triacylglycerol degradation, and lipid metabolism. A salient aspect of lipase functioning is the distinction of catalyzed reactions in different organisms, wherein acetylation, hydrolysis of carboxylic esters, and transesterification are the most prominent Fig. 19.2 [30–32]. Lipases are extensively used in conjunction with different analytes, due to (1) being an analyte itself, the activity of a lipase isoform facilitates the control of a biotechnological process and the discrete disease diagnosis, (2) serving as tools to hydrolyze esters, wherein esters serve as analytes, (3) in the development of a lipase inhibitor assay. In all these possibilities, lipases function as recognition agents, with the corresponding screening of substrates and inhibitors. The usefulness of lipases can be estimated from the diversity of substrates on which they act. In a noted study by *Lianghua and colleagues*, lipase from *Penicillium expansum* was observed to act on ibuprofen in the racemic mixture (equivalent proportion of dextro and laevo rotator forms), providing (S)-ibuprofen n-propylester and (R)-form as the products [32]. This reaction could be engineered for preparing optically pure ibuprofen. Adjacent observation was the hydrolytic breakdown of 4-nitrophenyl palmitate into 4-nitrophenol and palmitate, wherein the potent activity of yeast-lipase was noticed Fig. 19.3. Likewise, aqueous hydrolysis of 1,2,3-trihexaicosanoylglycerol to 1,2-dihexaicosanoylglycerol and hexaicosanoate and of tripalmitin and trimyristin to dipalmitin + palmitate and Dimyristin + myristate were noticed using lipase from *Saccharomyces cerevisiae* [33].

Sensing configurations based on lipases, using electrochemical sensors and the principles of voltammetry and impidimetry have been reported with significant promise. Lipase assisted hydrolytic breakdown of an ester into alcohol and organic acid resulted in altered pH, which was used by *Kartal and colleagues* using lipase from fungus *Candida rugosa.* The reaction involved immobilization of lipase on a glass pH electrode followed by facilitated hydrolysis to tributyrin as a substrate [34]. The sensor worked by recording the pH decrement by the glass electrode, caused by the

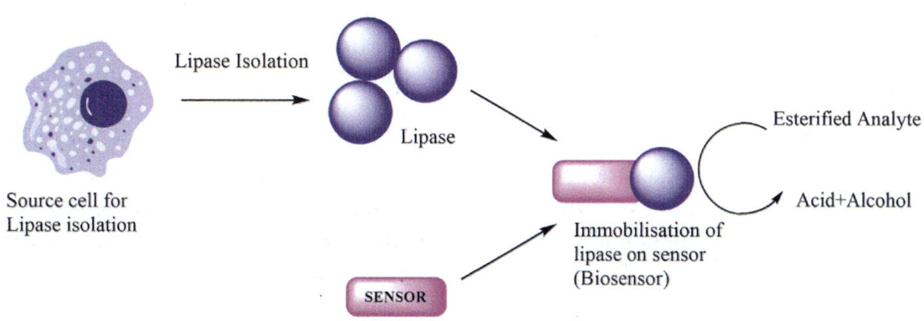

FIGURE 19.2 Pictorial understanding of lipase ability as a biosensing probe. Lipase is an enzyme generated in the pancreas and assists the metabolism of ingested fats in the intestines.

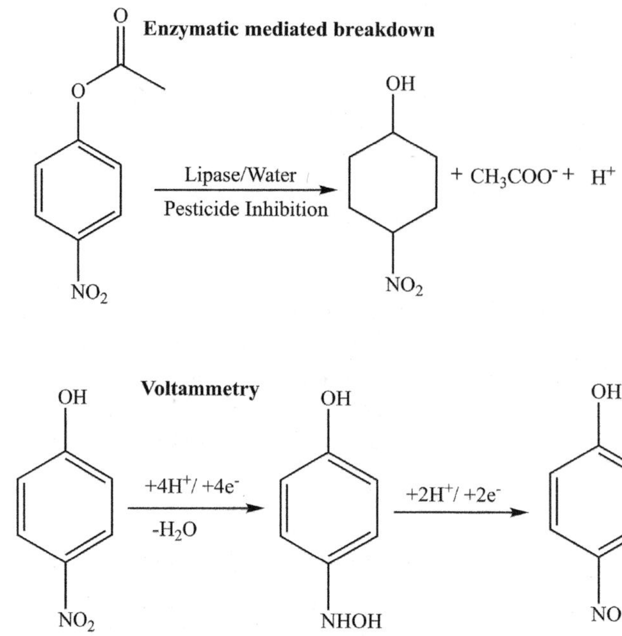

FIGURE 19.3 Voltammetry facilitated screening of lipase inhibitors, such as malathion, parathion, and chlorfenvinphos using p-ntrophenyl acetate as enzyme substrate.

release of butyric acid. The assay was standardized to ascertain the pesticide, methyl-parathion, which worked by inhibiting the lipase and preventing it from acidifying the medium. Corresponding detection threshold for the assay was estimated as 93 μmol•L^{-1} with a linear working range within 64–455 μmol•L^{-1}. Lipase-aided biosensing has also been reported *via* immobilization on the surface of a voltammetric electrode, working *via* the formation of an electroactive molecule. This conceptualization was worked out by *Ma and associates* in their work on methyl-parathion assay, wherein lipase from bacterium *Burkholderia cepacia* was immobilized on amine functionalized zeolitic imidazoles framework nanoparticles (NPs) and the modified NPs were captured into chitosan on a glassy carbon electrode (GCE) [35]. The optimized configuration exhibited suitability for screening the pesticides having p-nitrophenol functionality in their structure (Fig. 19.3). The functionality got hydrolyzed and got electrochemically oxidized in the course of a cyclic voltammetry screening. Ascertaining methyl-parathion using this configuration, the investigators screened a 0.28 μmol•L^{-1} as the detection threshold. A similar configuration was optimized by *Reedy and the group* for screening the pesticides chlorfenvinphos and malathion in an aqueous environment [36]. Contrary to the earlier attempt, pesticides herein served as inhibitors of the enzymes. The assay worked *viaCandida rugosa* lipase catalyzed conversion of p-nitrophenyl acetate to p-nitrophenol and acetate, generated p-nitrophenol was electrochemically oxidized, deciphered as 0.24 V characteristic peak. Screening using lipases has also been attempted using impedance variation tracking. The study monitored pesticide diazinon using lipases from *C. rugosa* and porcine pancreas, coated on Au electrode as a self-assembled monolayer and glutaraldehyde cross-linkage [37]. The assay worked via lipase-mediated diazinon conversion to diethyl phosphorothioic acid and 2-isopropyl-4-methyl-6-hydroxypyrimidine, ascertained via varied medium impedance. The optimized assay exhibited a threshold lipase detection of 10 and 100 nmol•L^{-1} from fungi and porcine pancreas, respectively.

Table 19.2 summarizes the select enzymes that have been screened for their biosensing abilities along with their microbial resources, characteristic nutrient needs in culturing and the working distinctions of pH, temperature, detection sensitivity, etc. It may be noted that one substrate could be screened using more than one enzyme and with varied microbial sources, thereby the selection constraints could be suitably optimized. For instance, both alcohol oxidase and alcohol dehydrogenase identify ethanol but from microbes *Pichia pastoris, Hansenula polymorpha,* and *Saccharomyces Cerevisiae*, having different optimal temperatures and pHs. Similar is the scenario for glutamate oxidase and glutamate dehydrogenase suitability for L-glutamate. The efficacy of enzyme-based biosensors could be substantially improved using compatible NPs, ensuring efficient long-term capping and nonaggregated state prevalence. In general, most enzymes function well within a (6–7.5) pH range. Presently, much interest is focused on the possible reuse of NPs via immobilization on a washable solid surface after screening interactions [38–62].

TABLE 19.2 Select enzyme configurations with their sources, substrate, and biosensing distinctions *vis-à-vis* stimulus detection and optimal working conditions [38–62].

Enzyme	Source	Substrate	Whereabouts of interactions and biosensing attributes
Alcohol oxidase	*Pichia pastoris Hansenulap olymorpha* EC 1.1.3.13	Ethanol	• Alcohol oxidase converts ethanol to acetaldehyde and H_2O_2 • **Target**: Short-chain alcohols like ethanol, ideal for ethanol biosensing • **Optimal conditions:** pH 7.0–8.0 and 25°C–35°C • **Sensing distinctions**: Screens ethanol by detecting H_2O_2, selective and sensitive functioning
Ascorbate oxidase	*Cucurbita sp* EC 1.10.3.3	L-Ascorbic acid	• Ascorbate oxidase oxidizes L-ascorbic acid to dehydroascorbic acid, producing water • **Specificity:** Targets L-ascorbic acid for measuring vitamin C levels • **Measurement**: Detects vitamin C *via* altered oxygen or oxidation products • **Optimal conditions**: pH 5.0–6.0 and 20°C–35°C. • **Sensing distinctions:** High specificity, broad detection range, and reliable for routine analysis
Alcohol dehydrogenase	*Saccharomyces cerevisiae* E.C. 1.1.1.1	Ethanol	• Alcohol dehydrogenase oxidizes ethanol to acetaldehyde using NAD^+ • **Substrate specificity**: Highly specific for ethanol; can act on other alcohols • **Biosensor applications**: Measures ethanol in beverages, fermentation, and clinical samples • **Optimal conditions**: pH 8.5 and 30°C • **Advantages**: Easy to use, cost-effective, and reusable
Cholesterol oxidase	*Streptomyces sp*, Porcine pancreas EC 1.1.3.6	Cholesterol	• **Enzyme function**: Oxidizes cholesterol to cholest-4-en-3-one, forming H_2O_2 • Highly specific for cholesterol, suitable for blood and food analysis • **Measurement**: Quantifies cholesterol via H_2O_2 signal • **Optimal conditions**: pH 7.0–8.0 and 35°C–40°C • **Distinctions:** Rapid response, high specificity, and portability
Choline oxidase	*Alcaligenessp* (EC 1.1.3.17)	Choline acetylcholine	• **Choline oxidase** oxidizes choline to betaine, producing H_2O_2 • **Specificity:** Targets choline and acetylcholine • **Measurement:** Electrochemical screening of choline via H_2O_2 • **Optimal conditions**: pH 7.0–8.0 and 30°C–37°C • **Distinctions:** High specificity, rapid response, and portability for point-of-care testing
Glucose oxidase (GOD)	*Aspergillus niger*, E.C. 1.1.3.4	β-D-Glucose	• GOD oxidizes β-D-Glucose to gluconolactone and H_2O_2 using FAD • **Optimal pH**: 6.0, exhibits an electrochemical behavior at optimal pH, ideal for biosensing • **Interactions**: Stable with both substrate and pH conditions

(*Continued*)

TABLE 19.2 (Continued)

Enzyme	Source	Substrate	Whereabouts of interactions and biosensing attributes
Glutamate oxidase	*Streptomyces sp.* EC 1.4.3.11	L-Glutamate	• **Enzyme:** Glutamate oxidase (EC 1.4.3.11) from *Streptomyces* sp. converts L-glutamate to α-ketoglutarate • **Biosensing**: Targets L-glutamate; detects by oxidizing H_2O_2 at 0.3 V • **Optimization**: Works best at pH 7–9 and 37°C–55°C. • **Advantages:** Ensures L-glutamate selectivity and sensitive detection
Glutamate dehydrogenase	Bovine liver E.C. 1.4.1.2	L-Glutamate	• Glutamatede hydrogenase catalyzes oxidative deamination of L-Glutamate • **Products:** α-Ketoglutarate and ammonia (NH_3) • **Cofactors required:** NAD^+ or $NADP^+$ • **Sensing abilities**: Measures L-glutamate by detecting NADH or NADPH electrochemically • **Uses**: Clinical diagnostics, food quality control, and environmental monitoring
Glucose dehydrogenase	*Pseudomonas sp.* *Escherichia coli* EC 1.1.1.47	Glucose	• **Reaction type**: Glucose oxidation • **Mechanism**: Glucose is oxidized to gluconolactone, reducing **NAD^+**or $NADP^+$ to NADH or NADPH • **Cofactor**: NAD^+ or $NADP^+$ (some use FAD) • **Sensing distinctions**: Measures glucose by detecting NADH/NADPH electrochemically or optically • **Uses:** Medical diagnosis, food quality control, and industrial applications
Lactate oxidase	*Pediococcus* sp. *Aerococcus viridians* EC 1.1.3.2	L-Lactate	• Lactate oxidase oxidizes L-lactate to pyruvate, producing H_2O_2 • **Target**: L-lactate, ideal for diagnostics and sports • **Measurement**: Detects L-lactate via H_2O_2 • **Optimal conditions**: pH 7.0–8.0 and 25°C–37°C • **Sensing behavior**: Highly specific, linear response over a broad range, and robust for continuous use
Lactate dehydrogenase	Rabbit muscle Chicken heart EC 1.1.1.27	L-Lactate	• Enables lactate and pyruvate interconversion • **Product:** L-Lactate is converted to pyruvate • **Mechanism:** Converts L-lactate to pyruvate, reducing NAD^+ to NADH • **Biosensor use**: Measures lactate *via* NADH changes • **Output:** Provides an electrical signal or colorimetric change indicating lactate levels
Laccase	*Trametespubescens* *Paraconiothyrium variable* *Trametes versicolor* (EC 1.1.3.4)	Polyphenols	• Lactate oxidase oxidizes L-lactate to pyruvate, forming H_2O_2 • **Target**: L-lactate, ideal for diagnostics and sports • **Measurement**: Screens L-lactate *via* H_2O_2 formation • **Optimal conditions**: pH 7.0–8.0 and 25°C–37°C • **Sensing distinctions:** Specific, linear response over a broad range, robust for continuous use

(Continued)

TABLE 19.2 (Continued)

Enzyme	Source	Substrate	Whereabouts of interactions and biosensing attributes
Tyrosinase	Mushroom EC 1.14.18.1	Monophenols Dihydroxyphenols Bisphenol A	• Oxidizes monophenols to o-diphenols and o-quinones • **Substrate specificity**: Targets monophenols like tyrosine and phenol, and bisphenol A • **Sensing abilities**: Detects phenolic compounds in food, pharmaceuticals, and the environment • **Optimal conditions**: pH 5.0–7.0 and 25°C–37°C • **Distinctions:** Rapid, low-cost, eco-friendly, and reusable

EC, Enzyme Commission Number; *FAD*, Flavin Adenine Dinucleotide; *NAD*$^+$, Nicotinamide Adenine Dinucleotide; NADH, Nicotinamide Adenine Dinucleotide; *NADP*$^+$, Nicotinamide Adenine Dinucleotide Phosphate; *NADPH*, Nicotinamide Adenine Dinucleotide Phosphate.

19.3.2 Biosensors using proteins as sensing agents

Proteins are essential biological macromolecules with diverse structures and functions, playing pivotal roles in various physiological processes within living organisms. Their remarkable properties, such as specificity, selectivity, and catalytic activity, have led to their extensive utilization in the development of biosensors, innovative devices designed for the detection and quantification of specific analytes in biological, environmental, and clinical samples [63].

Protein-based biosensors utilize individual proteins as recognition elements, which are directly linked to signal transducers, encompassing various types such as small-molecule-binding proteins, antibodies, regulatory proteins, enzymes, or small peptide fragments. The resulting signals can range from electrochemical to fluorescent or bioluminescent, leading to shorter assay times than cell-based systems and reducing inherent variability. These biosensors can be derived from a diverse array of proteins by reconfiguring their recognition and regulatory domains and integrating them into different host cells, typically designed to interact with specific target ligands or metabolites. This interaction elicits structural or conformational changes or influences the behavior of other associated proteins [64].

The development of protein-based biosensors targeting small molecules requires careful consideration of various critical elements, characteristics, and methodologies. Generally, these biosensors consist of a ligand-binding domain, a linker, and a reporter domain responsible for detection and quantification [65]. Upon coupling with the analyte, the biosensor generates a signal within the ligand-binding domain, which is then transmitted through the linker, inducing a change in the reporter domain. This signal can manifest as a conformational alteration of the molecule, a physicochemical transformation, the creation of an intermediate, or a modification in the stability or quantity of the reporter. Additionally, the reporter domain must be detectable through optical, electrochemical, or other means, aside from its potential for change [66].

Protein-based sensors work either via genetically encoded and chemically modified signaling optimizations. In genetically encoded systems, engineered proteins incorporate recognition and transduction elements without additional modification. For instance, ligand-binding proteins fused to reporter proteins such as GFP exhibit fluorescence changes upon ligand binding. Conversely, chemically modified systems require postpurification protein alterations, such as covalently attaching small reporter molecules, enabling the measurement of fluorescence changes following ligand binding. Leveraging the optical modality in biosensor constructs offers several advantages, including direct and real-time detection of numerous target molecules, compact size, and cost-effectiveness [67]. Additionally, engineering a protein-based biosensor to detect a specific molecule can be achieved through various approaches, such as rational design, directed evolution, de novo design, mutagenesis, DNA shuffling, and recombinant DNA techniques. Charge transfer is particularly crucial in the design of protein sensors that employ rational design or direct evolution techniques. These methods typically focus on modifying existing proteins or protein structures to enhance their sensing capabilities. Rational design involves making specific changes to a protein's structure based on known information about its function and properties, while direct evolution utilizes iterative cycles of mutation and selection to optimize protein function. Both approaches aim to improve the interaction between the protein sensor and its target molecules, often by optimizing charge transfer processes for efficient signal transduction. On the other hand, de novo protein design, which involves creating entirely new proteins from scratch, may not always prioritize charge transfer in the same way since it starts with a clean slate and focuses more on achieving desired functionalities [68].

Among the select noteworthy research efforts, *Ambrosio and colleagues* (2020) optimized directed evolution to enhance the responsiveness of the VanR transcription factor to vanillin while reducing its affinity for vanillic acid [69]. Similarly, ***Espinosa-Urgel** and colleagues* engineered the TtgR (a protein that controls the expression of TtgABC efflux pump) transcriptional repressor from *Pseudomonas putida*, coupled with a fluorescent reporter gene, through directed evolution to detect a wide range of antibiotics and organic compounds [70]. Strategic placement of a fluorescent protein within a highly sensitive region to the ligand of interest is essential for achieving a reliable output signal, as highlighted by *Chesterfield and colleagues* (2020). The investigators demonstrated that inserting a fluorescent protein into the flexible loop of a biosensor without altering the ligand-binding site enhanced the sensitivity to local conformational changes and increased the biosensor's signal output [71]. Through another worthy effort, *Yodhida and colleagues* [72] developed an optical biosensor for leucine–isoleucine–valine working through fluorescence modulation driven by intracellular branched chain amino acids (BCAAs exhibiting high sensitivity and specificity). The provision responded effectively to leucine, isoleucine, and valine in mammalian cells. However, the response was affected by changes in pH and the presence of non-BCAAs, glucose, and pyruvate, indicating a complex regulation of intracellular BCAA extents as affected by the cellular conditions. Impaired actions of L-type amino acid transporter 1 (LAT1) altered basal BCAA levels, suggesting a potential application of BCAA imaging in screening anticancer drugs targeting amino acid transporters like LAT1 [72]. Another attempt by *Kurnik and colleagues* [73] reported a biosensor using the SH3 domain from human Fyn kinase, distinguishing between peptide targets with rapid and quantitative response, even in whole blood. This protein-based biosensor leverages protein folding for discrimination between ligands, potentially expanding the chemical space accessible to real-time electrochemical biosensors [73].

In another study, *Janeena and associates* (2022) proposed a method for genetically encoding a redox-active amino acid into various Green Fluorescent Protein (GFP) variants using site-specific incorporation Fig. 19.4. Among these variants, GFP Dopa (Dopamine) exhibited superior biophysical properties. This GFP Dopa variant was then utilized to create an electrochemical sensor for detecting Zn^{+2}. The electrode, fabricated with GFP Dopa/Nafion, demonstrated improved electrochemical response, attributed to increased effective SA and available binding sites for Zn^{+2}. The hydrophobic nature of the polymer-coated GFP Dopa protected the protein from hydrolytic dissolution. This electrochemical sensing system serves as a prototype for diagnosing bladder cancer in biological fluids like urine, boasting enhanced selectivity and sensitivity with a detection limit of 0.129 nM for Zn^{2+} in spiked urine samples, thus offering potential applications in bladder cancer diagnostics [74]. Similarly, *Murugappan and associates* (2022) developed a protein-based electrochemical biosensor platform, exemplified by two amino acid biosensors, Cys-LeuFS and Cys-GlyFS, which utilize protein conformational changes for detecting small molecules. These biosensors, derived from SBP-based optical biosensors,

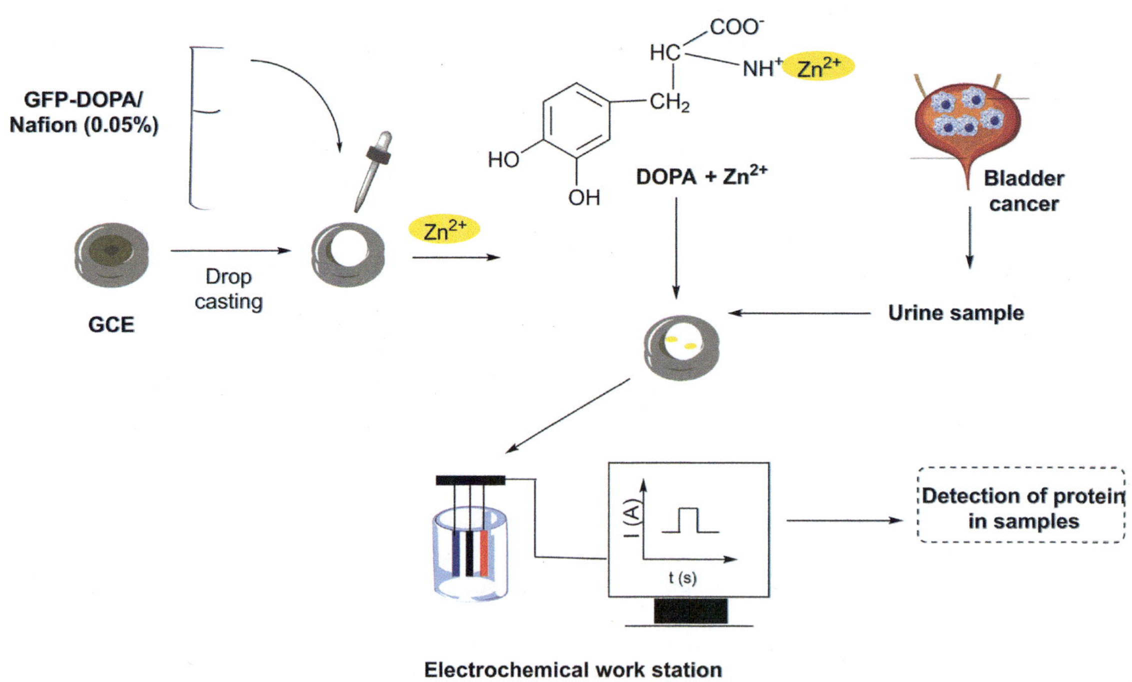

FIGURE 19.4 Schematic working scheme of protein-based metal sensor optimized for selective, real time Zn^{2+} detection in urine samples.

exhibit increased sensitivity and an expanded concentration range when integrated with an electrochemical detection platform. Their modular design allows for future engineering and diversification efforts, leveraging the ligand specificity inherent in this protein family. This integration of dynamic binding-protein-based bioreceptors with sensitive electroanalytical techniques offers promise for medical diagnostics and environmental assessment [75].

The recent past witnessed the emergence of protein biosensors as advanced tools for swift and accurate COVID-19 detection, emerging as significant remedial measures to ease the troublesome aftermaths of the pandemic. In a notable effort in this area, *Alfredo Quijano-Rubio and colleagues* [76] demonstrate the creation of a versatile array of protein-based biosensors using de novo–designed protein switches, in which peptide interactions trigger specific biological responses. These adaptable devices transition from a dormant state to an active luminescent state upon binding analytes, simplifying sensor design and enabling direct assessment in solution. The investigators engineered biosensors capable of screening multiple targets with clinical sensitivity, including Bcl-2, IgG1 Fc domain, HER2 receptor, Botulinum neurotoxin B, cardiac troponin I, and antihepatitis B virus antibody. Employing this platform for diagnostic purposes related to SARS-CoV-2, the scientists developed probes to screen the spike protein and antibodies against the membrane and nucleocapsid proteins. The spike protein sensor, incorporating a specially designed receptor binding domain, achieves an LOD of 15 pM and significantly higher luminescence. This method highlights the adaptability and sensitivity of de novo protein design, facilitating the prompt detection of diverse analytes for generating multistate protein systems [76]. A notable, curious attempt from 2023 introduced a novel miniature surface molecularly imprinted biosensor, optimizing the sequential deposition of Au NPs, reduced graphene oxide (rGO), poly(methylene blue)/poly(ionic liquids), and poly(ionic liquids) onto the tip of an acupuncture needle. Removal of the SARS-CoV-2-S protein template retrieved the molecularly imprinted biosensor for SARS-CoV-2-S protein detection. This biosensor exhibits a linear range from 0.1 to 1000 ng•mL^{-1} and an impressive LOD of 38 pg mL^{-1}, surpassing the reported values for molecularly imprinted biosensors. Additionally, the developed miniature biosensor demonstrates specificity and stability via SARS-CoV-2-S protein detection in clinical serum samples [77].

Hence, in conclusion, protein-based biosensors represent a cutting-edge approach in the realm of sensor technology, offering unparalleled sensitivity, specificity, and versatility for the detection of a wide range of analytes. The integration of engineered proteins into biosensor platforms holds immense potential for revolutionizing healthcare, biotechnology, and environmental science, paving the way for the next-generation diagnosis and monitoring devices.

19.3.3 Biosensors using antibodies as sensing probes

Antibody-based biosensors utilize the precise binding capabilities of antibodies to detect a wide array of substances. These sensors function by combining antibodies, which act as the recognition element, with various transduction mechanisms to generate measurable signals [78]. The core principle of these biosensors is the antigen–antibody interaction, where antibodies specifically bind to target antigens. This binding is then converted into a detectable signal by the transducer, which can operate through optical, electrochemical, piezoelectric, or thermal methods. The processed signal yields a quantitative or qualitative readout [79].

The main components of antibody-based biosensors include the recognition element (antibody), which binds the target antigen with high specificity; the transducer, which converts this interaction into a measurable signal; and the signal processor, which interprets and outputs the data. These biosensors are classified by their transduction mechanisms: optical (using fluorescence, surface plasmon resonance, or chemiluminescence), electrochemical (including amperometric, potentiometric, and conductometric methods), and piezoelectric (such as Quartz Crystal Microbalance) [16]. In their development, antibodies can be monoclonal, providing consistent specificity, or polyclonal, recognizing multiple target sites. Immobilization methods such as physical adsorption, covalent binding, and affinity binding are essential for attaching antibodies to the sensor surface. Signal amplification strategies, such as enzymatic amplification or NP labeling, further enhance the sensors' sensitivity. There are also numerous advantages, such as the ability to attach tags like His or C-myc for easier isolation and immobilization. Additionally, various labels (like green fluorescent protein or enzymes) can be directly fused to the antibody fragment, which simplifies detection. A wide array of antibody formats is also available, including single-chain variable fragments (scFv), antigen-binding fragments (Fab), reengineered immunoglobulin G (IgG), and dimmers [80–82]. These biosensors have diverse applications. In clinical diagnostics, they can screen disease biomarkers and pathogens, exemplified by SARS-CoV-2 detection. In environmental monitoring, these biosensors can screen pollutants and biological warfare agents. In food safety, they detect allergens and foodborne pathogens, ensuring product safety and preventing allergic reactions. In pharmaceutical analysis, the antibody-based biosensors can screen drug texture and quality, study protein interactions, and monitor bioprocess parameters for optimizing yield.

In a noteworthy effort, *Gehring and colleagues* developed an amperometric assay to screen for *Salmonella Typhimurium* cells. The method involved capturing the cells with magnetic beads conjugated to antibodies, then detecting them with an alkaline phosphatase (AP)-labeled goat anti-Salmonella antibody. Once the beads were deposited on graphite ink electrodes, the AP-catalyzed conversion of the substrate para-aminophenyl phosphate (pAPP) to the electroactive para-aminophenol (PaP) was monitored electrochemically. The signal generated was directly proportional to the number of bacterial cells captured, with a sensitivity of 8×10^3 cells•mL^{-1} [83]. Similarly, *Lin and colleagues* recently developed a method to immobilize monoclonal antibodies on screen-printed carbon electrodes (SPCE) for capturing *E. coli O157* from pure cultures. The scientists employed a horseradish peroxidase (HRP)-labeled polyclonal antibody for detection using an indirect sandwich assay. The response current was significantly amplified by functionalization with Au NPs and using FEDC as a mediator, allowing detection of ~ 5•10^3 CFU•mL^{-1} within 1 hour. This assay demonstrated excellent selectivity and specificity, showing minimal cross-reactivity with other food pathogens, such as *L. monocytogenes*, *Salmonella choleraesuis,* and *Vibrio parahaemolyticus*, highlighting the effectiveness of a selective BRE [84]. In another research, *Radke and Alocilja* developed a high-density microelectrode array capable of detecting *E. coli* O157 with a sensitivity range of 1•10^4 to 1•10^7 CFU•mL^{-1}, utilizing a goat anti-IgG polyclonal capture antibody [85].

Another significant effort by *Hnaiein and colleagues* reported an innovative conductimetric immunosensor for screening *Escherichia coli* contamination, wherein the probe comprised a biotinylated polyclonal antibody on streptavidin-coated magnetite NPs, immobilized on a conductimetric electrode *via* glutaraldehyde coupling. Conductimetric readings were obtained using alternating-current (AC) voltage, and the presence of NPs enhanced conductivity, allowing detection down to 0.5 CFU•mL^{-1}. A slight background signal was observed when testing with *S. epidermidis*, attributed to the polyclonal capture antibody, suggesting that monoclonal or recombinant antibodies might be more appropriate for certain assay formats [86].

Surface Plasmon Resonance (SPR) is another key phenomenon used in antibody-based biosensors for real-time, label-free detection of biomolecular interactions. Fig. 19.5 depicts an overview of the process, illustrating the integrated functioning of various components of an SPR-antibody-based biosensor. In these biosensors, antibodies are immobilized on the metal surface. As target antigens bind to these antibodies, the refractive index near the surface changes, causing a shift in SPR angle or wavelength. This shift is proportional to antigen concentration, enabling precise measurement of binding affinity and kinetics without the need for labels. In a study, researchers developed an SPR immunosensor with an amplification strategy using antibody-functionalized gold-coated magnetic NPs (mAuNPs) for detecting the CD5 biomarker. The SPR sensor and mAuNPs were coated with 11-mercaptoundecanoic acid, and antibodies (anti-CD52A for capture and anti-CD52B for detection) were immobilized using a coupling method. After forming the anti-CD52A/CD5 complex on the SPR sensor, AuNPs-anti-CD52B conjugates were used to amplify the signal, achieving a detection limit of 8.31 fM and a quantification limit of 27.70 fM. This approach significantly improved sensitivity compared to direct detection and performed well in human blood serum [87]. Another study introduced a versatile antibody-mediated protocol for capturing and analyzing NPs using a protein A/G-functionalized SPR biosensor. This approach used antistreptavidin antibodies to enable regenerable capture of biotin-containing NPs, such as drug-delivery vesicles, and antibodies against glycophorin A and B to specifically capture red blood cell–derived extracellular vesicles (RBC EVs). The biosensor effectively determined the size and concentration of NP subpopulations from complex biological samples. The mean size of surface-captured RBC EVs matched those measured by established methods like NP-tracking analysis

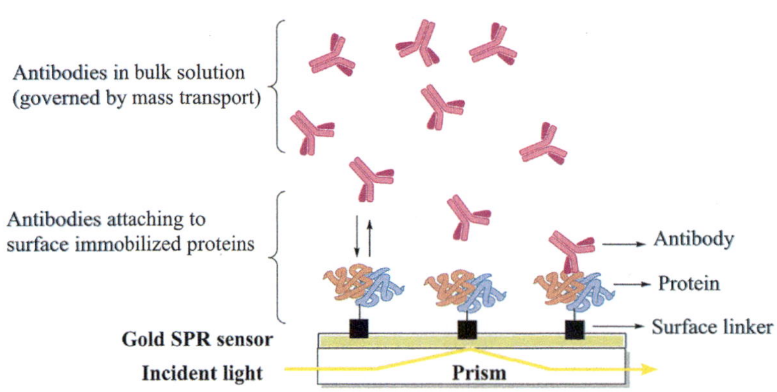

FIGURE 19.5 Working of a Surface Plasmon Resonance (SPR)-antibody-based in an integrated regime. Antibodies from a bulk screen the binding of those bound to immobilized proteins.

and dynamic light scattering. Overall, the Protein A/G-functionalized SPR biosensor offers a flexible alternative to traditional NP-capturing techniques [88].

Table 19.3 summarizes the working aspects of select antibody-based biosensors developed in the last decade, along with the applications and usage precision cum accuracy [89–98]. The diversity of screened stimuli herein elucidates the functionalization potential of antibodies to intercept with manifold antigens in distinct biochemical environments. The precision efficacy is justified by the femto and nano gram LODs for prostate-specific antigen and HER2 protein in breast cancer. Similarly, prompt response generation significantly reduced the response time by 5 minutes for screening of cardiac troponin, an explicit cardiac arrest marker. Efficacy is also indicated by 88.7% and 90.63% sensitivity and specificity for COVID-19 screening. Future advancements in antibody-based biosensors aim to improve sensitivity, specificity, and portability, incorporating advanced materials and microfluidic technologies. However, ensuring stability, reproducibility, and cost-effectiveness remain formidable challenges for wholesome applications. With ongoing technological progress and advances in competitive biological assays, these biosensors are poised to enable rapid, precise, and sensitive detection.

TABLE 19.3 Notable distinctions of biosensors utilizing antibodies as detection probes. Integration compatibility with nanomaterials paved the way for screening multiple analytes in varied biochemical environments [89–98].

Theme of the study	Antigen detected application	Prominent traits of working configuration	Performance aspects
Development of MCLW optical biosensor for screening CRP in diluted real serum sample	A potent infection and inflammation risk predictor, CRP extent should be < 3 $\mu g \cdot mL^{-1}$ in a normal person	• Spin-coated deposition of NC on the MCLW sensor chip • CRP-Ab immobilized via hydrophobic adsorption • No additional treatment for ligand immobilization • Tracking of refractive index changes near the sensor surface	• NC deposition on the sensor chip engineered the largest signal difference for the sample than control • Real-time CRP detection in (0.1–10) $\mu g \cdot mL^{-1}$ range • The outer surface of the metal oxide film minimized the environmental stress • The extensive penetration depth of the evanescent field arising from the sensor surface is a unique aspect
Development of silver nanoprism (AgNPRs) etching-based plasmonic biosensor for screening cancer biomarkers	PSA	• Triangular AgNPRs use PSA immune response to oxidize glucose *via* glucose oxidase • H_2O_2 generated oxidizes triangular AgNPRs into spherical NPs: causing a shift in blue SPR peak • Ultrasensitive, cost-effective, and easy to operate: suitability for point-of-care test	• Quasilinear response to logarithmic PSA extents within 10 $fg \cdot mL^{-1}$ to 100 $pg \cdot mL^{-1}$ • LOD of 4.1 $fg \cdot mL^{-1}$: 5 orders higher than 1.25 $fg \cdot mL^{-1}$ of HRP-ELISA • 16 screened donor serum samples exhibited no significant differences for PSA amounts
Development of ECL-based immunoassay for ultrasensitive ZIKV detection in human biological fluids	ZIKV, an emerging mosquito-transmitted flavivirus	• PSB were loaded with a large number of ECL labels, which were conjugated with anti-ZIKV MAb to generate anti-ZIKV-PSBs • Designed anti-ZIKV-PSBs selectively captured ZIKV in solution, forming ZIKV-anti-ZIKV-PSB complexes • The complexes were screened for ECL intensity on subsequent separation of magnetic beads	• Optimized configuration of anti-ZIKV-PSBs could detect even 1 PFU-ZIKV in 100 μL saline, human plasma or urine • Sensing assembly could be used to screen ZIKV-similar viruses in clinical diagnosis, epidemiological and laboratory research

(Continued)

TABLE 19.3 (Continued)

Theme of the study	Antigen detected application	Prominent traits of working configuration	Performance aspects
Development of a label-free electrical biosensor *via* interfacing graphene with interdigitated capacitor microelectrodes biofunctionalized with E. coli O157:H7 specific antibodies	*Escherichia coli* O157:H7	• Workable probe configuration responded to the varied dielectric behavior, *via* (1) polarization of captured cell surface charges, (2) cell's native bioactivity, (3) cell-wall's electronegativity modulated relaxation time, and (4) charge carrier mobility of graphene altered electrical properties on *E. coli O157:H7* capture over the sensor surface	• Graphene nanoconstructs exhibited superior chemical properties on the sensor surface, like high carrier mobility alongside compatibility with antibodies and bacteria • Sensitive capacitance changes with graphene capacitors were explicit for the *E. coli O157:H7* strain with [10–29,33–99] cells per mL precision
Designing a sensitive electrochemical aptasensor for screening the BC biomarker, HER2	HER2/neu protein Breast cancer biomarker	• HER2 aptamer was first hybridized with FcNS prior to HER2 binding • Released FcNS were hybridized with HRP-labeled DNA/Au nanospheres • $RecJ_f$ exonuclease intervention recycled HER2 *via* degradation of the unbound and bound aptamer connected to FcNS	• Generated FcNS in large amounts was captured by Au electrode via β-cyclodexterin-ferrocene host-guest proximity • HRP catalyzed o-phenylenediamine with H_2O_2, generating an amplified signal • Optimized configuration exhibited a good sensitivity with 4.9 ng•mL^{-1} LOD and high HER2 specificity
Development of an SPR-based sensor chip (using SAM) to assess the mycotoxin contamination in wheat and corn	Aflatoxin B1, ochratoxin A, zearalenone, and deoxynivalenon contaminations	• Antigen immobilization on sensor chip surface via SAM-assisted hydrazone conjugation • Carboxymethyl groups were engineered via NHS-EDC solution proximity to N-hydroxysuccinic acid esters • Amine and other nucleophilic groups on the toxin structure were covalently attached to N-hydroxysuccinimide	• Minimum LOD for aflatoxin B1, ochratoxin A, zearalenone, and deoxynivalenol were 0.59, 1.27. 7.07 and 3.26 ng•mL^{-1}, respectively • Low cross-reactivity of all mycotoxins • Comparative screening with HPLC-MS revealed good agreement
A robust and prompt point-of-care lateral flow immunoassay for screening anti-SARS-CoV-2 IgM and IgG antibodies in human blood	SARS-CoV-2 spike protein COVID-19 diagnosis	• Test strip comprises of plastic backing, sample, conjugate, absorbent pad, and NC membrane • Every component needs a pretreatment, *viz.* NC membrane is attached to a plastic backing layer for cutting and handling • Anti-human IgM, IgG, and anti-rabbit IgG immobilized at test M, G, and control line • Conjugate pad was sprayed with a mixture of Au NPs-COVID-19 recombinant antigen conjugate and Au NPs-rabbit IgG	• Analysis of the assay was done on 379 PCR-confirmed and 128 nondiagnosed cases at 8 clinical sites • Overall testing sensitivity and specificity were 88.66% and 90.63%, respectively • Clinical screening on venous and fingerstick blood samples revealed consistency with higher sensitivity than a single IgM or IgG test

(Continued)

TABLE 19.3 (Continued)

Theme of the study	Antigen detected application	Prominent traits of working configuration	Performance aspects
		• Sample pad was pretreated with BSA (3%w/v) and Tw-20 (0.5%w/v) before use	
Optimization of Cu-MOF and PANI composite to design an impedimetric sensor for screening cardiac marker troponin I	Cardiac troponin I, an important marker for heart attack	• Solvothermally prepared $Cu_3(BTC)_2$-PANI composite was coated as a thin layer on SPE • Electroconductive thin film was conjugated with antitroponin antibodies • Optimization of PANI content revealed 50 mg as inadequate for conducting films, while 150 mg resulted in SA damage, so 100 mg PANI was chosen	• Developed Cu_3(BTC)-PANI assembly enabled troponin I detection over clinically significant range [1–29,33–170] ng•mL^{-1} • Entire analysis was completed in 5 min, exhibiting specificity towards troponin I even in the presence of interfering proteins
Development of two biosensors for IL-6 screening in blood plasma using array SPRi has been optimized	Interleukin-6 (IL-6) Inflammation biomarker	• One of the developed sensors comprises a mouse monoclonal anti-IL-6 Ab as the immobilized receptor using cysteamine linker • Second optimized configuration used galiellactone as receptor, an IL-6 specific inhibitor, immobilized through octadecanethiol as linker • Both configurations were validated *via* screening the IL-6 blood plasma extent variations before and after resection of the ovarian tumor and the endometrial cyst	• Biosensor with Ab as receptor gave a linear analytical response between 3 (LOQ) and 20 pg•mL^{-1}, exhibiting (8–9.8)% ranged precision and [97–107]% recovery • Sensor using galiellalactone as receptor generated a linear analytical response within (1.1(LOQ)-20) pg•mL^{-1}, (3.5–9.3)% precision, and [101–105]% recovery
A textile OECT based biosensor to screen *L. monocytogenes* in food samples	Listeria monocytogenes, screened using culturing, Listeria Precis, PCR, and OECT biosensor that employed PEDOT and PSS doping the organic channel	• AFM screened topographic maps of the Gold Gate • Electrochemical activity on gate electrodes was ascertained and based on extracted DNA amount from samples before hybridization with culture probe, immobilized over the gate gold surface	• Assay accomplished 1.05 ng•μL^{-1} as LOD, for 0.56 pM *L. onocytogenes* (ATCC 7644) • Optimized assembly enabled specific and rapid detection in samples

Ab, Antibody; *AFM*, atomic force microscopy; *BC*, breast cancer; *CR*, C-reactive protein; *ECL*, electrogenerated chemiluminescence; *EDC*, N-ethyl-N′-(3-dimethylaminopropyl)carbodiimide.; *FcNS*, ferrocene labeled DNA/Au nanospheres; *HER2*, human epidermal growth factor receptor-2; *HRP*, horseradish peroxidase; *IL-6*, interleukin-6; *MAb*, monoclonal antibody; *MCLW*, metal cad leaky waveguide; *MOF*, metal organic framework; *MOF*, metal organic framework; *NHS*, N-hydroxysuccinimide; *OECT*, organic electrochemical transistor; *PEDOT*, poly(3,4-ethylenedioxythiophene); *PSA*, prostate-specific antigen; *PSB*, polystyrene beads; *PSS*, polystyrene sulfonate; *SA*, surface area; *SAM*, self-assembled monolayer; *SPE*, screen-printed carbon electrode; *ZIKV*, zika virus.

19.3.4 Nucleic acids probe-based biosensors

A nucleic acid biosensor (NAB) uses oligonucleotides or a definitive stretch of nucleotides having a known base sequence as the sensing probe. Current progress in the biosensors usability is focused on improving the precision, promptness, and accuracy of transduction technology. Research in NABs attracts an increasing number of researchers' attention owing to the sequence variability and specificity. The working framework of biosensors utilizing nucleic acid (NA) sequences as probes involves single-stranded DNA, which can hybridize with its complementary strand, thereby enabling highly efficient detection of complementary DNA or RNA strands [99,100]. Based on the diversity of the applications or different screened entities, two broad categories of NABs are distinguished. The first category uses DNA/RNA as an analyte that is normally perceived *via* hybridization and is famously referred to as a genosensor, while the second category uses the fixed length of DNA or RNA bases to screen specific biological or chemical entities, such as target proteins, antigens, environmental contaminants, or drugs [101]. Functional constituents of NABs include aptamers, DNAzymes, and aptazymes, exhibiting the ability of integration over a lab-on-a-chip assay. Most commonly known nucleic acids include deoxyribonucleic and ribonucleic acids (DNA and RNA), prevailing universally in the living cells as well as viruses. Initially, the role of nucleic acids was believed to transport and facilitate the passage of genetic information. However, from the 1990s, a manifold short sequence of NAs have come to the fore (recognized as aptamers), having been extracted from large genomic libraries, conferring them a high affinity and specificity [102,103]. The first major advance in the functioning optimization of DNA biosensors was accomplished in 1994, wherein the DNA assisted the catalytic efficacy [104]. The used catalytic DNA domains were subsequently referred to as DNA enzymes, deoxyribozymes, or catalytic DNA.

Lately, though, DNAzymes and aptamers have been combined and referred to as allosteric DNAzymes or aptazymes. As of now, the DNAzymes, aptamers, and aptazymes are together referred to as functional DNAs with the roles extending over the Watson–Crick postulated identification of complementary strands. The usefulness of nucleic acids in biosensing stems from their natural occurrence and the template-based compatibility for interacting with diverse stimuli that have been employed for enhancing the working modality of optimized assembly. With the blending of multifunctional NMs, the NABs have been the subject of intense interest in the emerging analytical applications, revolutionizing the development of next-generation biomedical devices. In comparison to conventional biosensors, the attributes of wider detection targets, enhanced working lifetime, and dynamic functioning (modulations of PCPs) propel DNA biosensors as distinct devices for accurate detection. The helical loops in DNA structure, with the topography of aligned nucleotide bases and phosphate backbone, engineer the DNA structures for signal conduction near the biosensor surface for an improved working efficacy. The native morphology of DNA exhibits substantial familiarity for the detection of specific stimuli, spanning nucleic acids, proteins, metal ions, and small biomolecules [105]. Advances in DNA nanotechnology have made the creation of DNA hybridization-based nanotechnology a readily exploitable avenue, via amplification of sensing responses *and* efficacious transduction. On the basis of a select probe design, sensors relying on DNA, aptamers, and DNAzymes are known. Their basic workable modalities are discussed ahead.

19.3.4.1 DNA based biosensors

Also known by the terms DNA sensors or genosensors, sensors based on DNA are highly compatible for screening the individual genomic details or NA-mediated screening of pathogenic invasion [100,106,107]. The application potential of DNA-based biosensors assists in improved access for sequence-specific information, making way for accurate stimulus detection in medical diagnosis, environmental, and food industries [108,109]. In all these domains, the detection is facilitated through explicit DNA hybridization, encompassed via the hybridization ability of NA strands (the fundamental detection framework of DNA sensors). The complementary nature of nucleotide base pairs, along with the compatibility of DNA, makes DNA hybridization a familiar event in diagnostic laboratories, contrary to the direct sequencing method. Measurement sensitivity in these assays is engineered via sequence amplification methods (polymerase chain reaction and others), efficacy of sequence hybridization alongside the extent of background signal, together serving as the sensitivity affecting parameters. A number of previously reported parameters, such as ionic strength, reaction temperature, and a potential DNA computation circuit, are known to modulate specificity herein. Screening of genomic sequences has consolidated a highly sensitive framework for quantitative detection of pathogens and genetic variations. Relying on this principle, the pathogenic microbes could be screened for minimizing environmental and food contamination. Another authentic area includes the feasibility optimization of microRNAs, owing to their regulation of critical gene expression and point-of-care (POC) diagnostics.

19.3.4.2 Biosensors working through aptamers

Aptamers work analogously to the "Lock and Key" mechanism for interacting with their targets, integrating in vitro chosen ligands using exponential enrichment (SELEX), screening the ability of NAs to bind a wide range of analytes other than DNA or RNA with high affinity and specificity [110]. The short stretch of NAs used as probes herein is referred to as "aptamers". Fascinations behind the use of aptamers are attributed to their ability as novel recognition elements, first noticed around 2004 [111,112]. Distinctive prospects of aptasensors comprise their high affinity, modest preparation, and the ability to form Watson–Crick base pair, suggesting the feasibility of targets amongst inorganic ions (Na^+, Mn^{+2}, etc.), organic entities (adenosine triphosphate, ATP, cocaine), large biomolecules (peptides, proteins), and perhaps even the whole organisms (bacteria, cells). To date, studies on these mechanisms have revealed their compatibility with nearly all diverse detection needs and readout assays [113,114]. The progress of aptamers has witnessed significant expansion over the past few years, extending compatibility from select model targets (thrombin, ATP, and cocaine) to the more emergent targets (toxins, drugs, antibiotics, insecticides, biomarkers). Readers are suggested to consult the 2016 review article by *Du and Dong*, featured in ACS-Analytical Chemistry, wherein broader applications extend across clinical diagnosis, monitoring of water, environment, and food quality.

19.3.4.3 DNAzyme-based biosensors

DNAzymes represent a class of catalytic NAs having the ability to cleave specific substrates in the co-existing scenario of metal ions (majorly cations) and amino acids (e.g., L-histidine). The distinction of these entities as biosensing moieties pertains to their utility for immobilizing the metal ions, a prospect lacking in aptamer-based biosensors [113]. To put it in simple terms, DNAzymes comprise certain manually screened DNA strands that are compatible with the environment and exhibit resistance to nuclease degradation, besides having enzymatic activity. These DNA strands possess remarkable potential to function as biometric probes wherein NA cleavage function and peroxide catalysis are the notable abilities in biosensing. The DNAzyme equipped for nucleic acid cleavage comprises a loop-resembling catalytic framework anchored by two substrate-identifying regions [115]. Fig. 19.6A depicts the working principle of DNAzyme optimized for nucleic acid cleavage, wherein substrate-identifying domains could capture the substrate strand as per the Watson–Crick model. Prevalence of the catalytic core is essential for cleavage, a process that gets activated once the external catalytic core is enveloped into the catalytic domain of the DNAzyme. Thereafter, the substrate strand is cut off and separated from the substrate-recognition domains owing to the decreasing melting point [116,117]. It is important to note that during this process, some specific small molecules (metal ions or amino acids) could only function as the catalytic core of the concerned DNAzyme [118,119]. The activation of the cleavage process is acutely specific *vis-à-vis the* small-molecule detection purpose. The most common module of NABs optimizes through a molecular beacon, worked out *via* a single

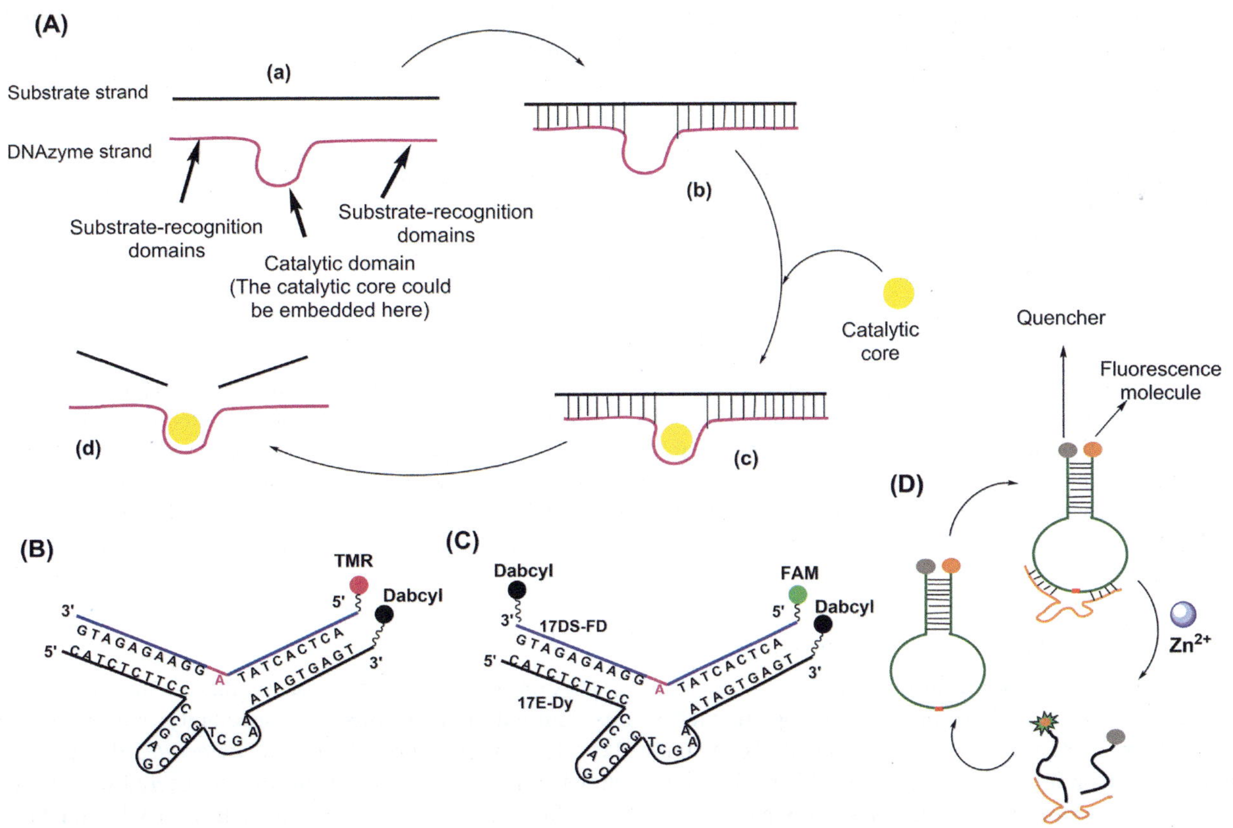

FIGURE 19.6 (A) Working modality of nucleic acid cleavage capable DNAzyme sensing probe, (B) modified DNAzyme facilitated Pb^{+2} sensor, (C) improved configuration of DNAzyme based Pb^{+2} sensor, and (D) DNAzyme facilitated amplified biosensing.

quencher, a double quencher, and a hairpin-shaped substrate. As of now, beacons with single quenchers, double quenchers, and hairpin resembling substrates are the known configurations of molecular beacons (Fig. 19.6B–D). The beacon with a single quencher pertains to a configuration that is engineered with a fluorescent entity and a quencher apiece. In a notable effort illustrating single quencher possessing molecular beacons, *Li and colleagues* reported a significantly diligent DNAzyme biosensor optimized for Pb^{+2} screening [120]. The scientists herewith optimized the working configuration of the fluorescence sensor *via* conjugating the 5' end of substrate DNA with fluorophore, 6-carboxylic tetramethylrhodamine (TMR), and the 3' terminal with fluorescence quencher, 4-(40-dimethylaminophenylazo) benzoic acid. The Pb^{+2} sensing was accomplished *via* DNAzyme's cleavage, leading to separation of fluorescent moiety and the quencher. Consequently, there arises a discontinuity in the FRET between TMR and Dabcyl, resulting in increased fluorescent intensity. The optimized version herein exhibited ~80-fold higher response for Pb^{2+} than other divalent ions, albeit one constraint was the low temperature explicit functional state. This was because of the fact that when working temperature exceeded the melting temperature, select DNAzymes release the substrate strand without cleavage, generating high background noise and weakening the method's selectivity. To address this bottleneck, a beacon with one fluorescent molecule and double quenchers was demonstrated. The configuration reported by *Liu and colleagues*, worked through a fluorescent molecule that is quenched not only by the quencher on DNAzyme but also by the quencher on substrate strand (Rh-17DS) [121]. This smart modification enabled the quenching of fluorescent molecules even when the substrate strand remained separated from the DNAzyme, unusually, since the Dabcyl moiety on the substrate strand could also quench the FAM. Analysis revealed 60% increment in the relative fluorescence intensity of the workable configuration.

Another configuration of a beacon is also known, using hairpin morphology and a stem-loop of DNA hairpin, which acts as a DNAzyme substrate. In a notable effort of such diversity, *Zhao and accomplices* optimized the working of a DNAzyme-based amplified biosensor, working through a fluorescent quench pair engineered at the DNA hairpin's terminal [122]. Having the cleavage sites on the hairpin loop, the DNAzyme cuts the loop *via* Zn^{2+} scissoring controls, releasing NA fragments in conjugation with fluorescent quenching molecules to amplify the fluorescence intensity. This strategy did not require the modification of DNAzyme, which prevented the suppression of DNAzyme's catalytic performance. The very first configuration of DNAzyme exhibiting catalytic activity towards peroxides was postulated by *Travascioand colleagues* in 1998 [123]. The optimized configuration worked *via* a combination of guanine-rich DNA sequences with hemin to form HRP-DNAzyme that is capable of catalyzing ABTS2-, 2,20-azino-bis (3-ethylbenzothiazoline-6-sulfonic acid) to colored products. Of note, HRP-DNAzymes are used as reporter tags of DNAzyme and DNA-aptamer biosensors owing to their native intense fluorescence ability that does not significantly alter the corresponding activity of these sensors. In a notable effort, *Willner and colleagues* reported an aptamer-HRP-DNAzyme hairpin resembling a probe to detect AMP and lysozyme [124]. Analysis revealed the functioning *via* analyte hybridization with a DNA-aptamer sequence, untying the DNAzyme from the stem to form a G-quadruplex similar morphology. The HRP-DNAzyme could also screen aniline polymerization (polyaniline) using higher SPR-signal enhancement. Encouraged by this study, *Li and colleagues* worked on an SPR biosensor using HRP-DNAzyme mediated signal amplification [125]. The study screened "bleomycin" with an LOD as low as 0.35 pM.

19.3.4.4 Biosensors working through DNA hybridization

Prompt and diligent detection of specific biomarkers is of utmost significance in biochemical analysis, particularly in the event of pandemics, like the COVID-19 situation [126,127]. The conventional assays like polymerase chain reaction, loop-mediated thermal amplification, DNA chip-based microarray, and ELISA mandate conduct by skilled professionals due to the complementary aspects and long incubation time. Owing to this, prompt and steadfast methods in airports, railway stations, etc., are needed. The approach optimized *via* DNA hybridization amplification provides amicable options with its high detection speed, sensitivity, and stability. The broad approaches based on DNA hybridization are discussed in subsequent paragraphs.

19.3.4.4.1 DNA hairpin-based biosensor probes

DNA hairpin implies the hairpin morphology acquired by ssDNA on folding proximity of its singular nucleotide stretch. According to Watson–Crick base pair complementarity, this configuration exhibits high specificity and can modestly convert hybridization into a physical signal. To mention the select latest efforts, *Fang and associates* demonstrated a molecular beacon for surface-immobilized DNA hybridization, wherein a generated hairpin was engineered with TMR and dimethylaminoazobenzene aminoexal-3-acryinido (DBCAL) as fluorescence-quenching pairs. Both TMR and DBCAL were separated on the hairpin opening *via* target intervention, resulting in enhanced fluorescent signal with nanomolar sensitivity [128]. In another worthy effort, *Fan and associates* optimized an electrochemical DNA biosensor using the DNA hairpin for identifying the sequence-specific DNA [129]. The optimized configuration comprised a DNA hairpin engineered with MB and a ferrocene molecule, immobilized over the working electrode *via* gold and hydro sulfonyl covalent reaction. With the opening of the hairpin on the interception of sequence-specific DNA, MB became isolated from the sensor surface, resulting in decreased intensity of the electrochemical signal. The sensor precision was good enough as 10 pM, and the equipment was not bulky as the fluorescence method with no significant alteration from the sample's light transmittance. The inverse correlation of response current with the target molecule's extent was indeed a concern, due to which the sensor response to a great extent was anticipated as being suppressed in the background noise. Suggested remedies by researchers for this referred to hairpin morphology use *via* "signal gain mode" or introduction of a "reference probe" [130].

19.3.4.4.2 Signal integrated sensing probes

Despite a high sensitivity, moderate expenditure, and chemical stability, biosensor configurations using DNA probes could be improved for LOD due to surface density constraints. Optimization efforts decipher that a denser surface distribution of DNA biometric constituents could accomplish the feasibility of a target and sensing surface probe combination [131]. An impairing factor in DNA sensors is the steric hindrance resulting from the closer gathering of probes over sensor surface, hindering the target-probe proximity. Cyclic amplification assays using hairpins, which work *via* hybridization chain reaction (HCR) and catalytic hairpin assembly (CHA), are well-suited for enhanced response. The HCR assay was optimized through a single nucleic acid strand with alternate hybridization of sticky-terminal DNA hairpins, forming a long double helix. In a noted effort, *Hou and colleagues* designed an HCR-based electrochemical sensor for highly sensitive microRNA screening *via* signal attenuated mode [132]. The optimized configuration herein displayed good sensitivity with an LOD of 1 pM. While HCR may not necessarily require the target for activation (besides no autostoppage provision), the CHA needs the involvement of original ssDNA as the activation step in every reaction stage. In a diligent effort, *Duan and colleagues* optimized the functioning of a fiber optic-driven biosensor, using CHA and nanocomposites' aided signal amplification for screening 18 food samples [133]. The working design exhibited an LOD of 12 pM, with the current efforts being focused on amalgamating CHA with advanced modules of biomedical inspection. A noted attempt from *Wu and colleagues* integrated the CHA approach with rolling circle amplification for screening the telomerase actions, accomplishing high in vitro sensitivity [134]. The RCA modulation enabled telomerase activity and target amplification, while CHA provision facilitated the transformation of the enhanced RCA product to fluorescence enhancement-triggered screening.

19.3.4.4.3 Detection principles for nucleic acid–based biosensors

Over the past few years, biosensing advances have been accomplished using enzymes, antibodies, and oligonucleotides with explicit recognition distinctions [135,136]. Contrary to the enzyme or antibody-based biosensors, those using DNA probes are anchored by distinguished benefits of high thermal tolerance, easy modification, and adequate surface regeneration *via* exploitable functional chemistry [137]. Incentives for NAB manifest *via* referable DNA libraries through which DNA probes with varied target affinity are available [138]. Herein, a discussion on optimized probe configurations with some recent studies is described.

To date, the most common detection optimization for DNA-aptamer workability has been to functionalize the probe with a reporter molecule (ferrocene, methylene blue, MB) and an immobilizing moiety (alkane thiol, alkane amino, streptavidin, and hydrazoate) at the 5' and 3' termini of the DNA strand, respectively. Changes in DNA-aptamer design could be quantified by quantifying the electrochemical variations on the electrode surface. In a notable effort herein, *Liu and associates* developed a 34-mer interferon binding aptamer for interferon-gamma (IFN-γ) detection [139]. The optimized probe configuration was immobilized on an electrode surface using the covalent interaction of gold and alkyl mercaptan. In the event of missing a specific target, the DNA aptamer spontaneously folded to acquire secondary spatial loop morphology to prevent the access of reporting molecules and the sensing electrode surface. The amalgamation of target and DNA aptamer altered the aptamer's conformation through an increase in the separation of the reporting molecule and the electrode surface. Consequently, variations in electronic transfer efficacy between the reporting molecules and electrode surface varied, resulting in 0.06 nm as LOD with the 10 nM extension of the linear range. On similar lines, *Chen and colleagues* proposed another electrochemical configuration working through 34-mer IFN-binding aptamer engineered using MB and disulfide, *vis-à-vis* optimized semiconducting steps [140]. The optimized configuration detected IFN-γ within (1–500) ng•mL^{-1} with an LOD of 1.3 ng•mL^{-1}.

Another strategy to use a DNA aptamer could be through the use of conjugated polymers, frequently employed as reporter tags *vis-à-vis* fluorescent and colorimetric biosensors owing to the remarkable optoelectronic attributes [141]. With the ability to absorb over the conjugated polymers *via* electrostatic forces, confirmation variations of DNA aptamer (on target binding) enable fine-tuning of the conjugated polymer's structure, whereby changes in absorption and emission wavelengths are perceived [142,143]. The optimized provision has been validated to screen human thrombin with a threshold of $2 \cdot 10^{-15}$ M [144]. Despite significant advantages of functional DNA aptamer sensors, the complexity arises due to cumbersome modification procedure required to firmly immobilize the aptamers over the sensor surface. Potent attempts to overcome this concern have worked on firm adsorption of single DNA strands on graphene oxide (GO, exhibiting splendid photoelectric and carrier transport ability), surface *via* base stacking and hydrogen bonding [145]. For instance, in a notable effort, *Yu and colleagues* proposed a modest configuration of electrochemically reduced GO (ERGO) based aptamer sensor having ERGO deposited over a glassy carbon electrode. Likewise, the guanine-rich DNA aptamer conjugated with MB was physically adsorbed over the ERGO *via* π-π stacking proximity [146]. The Use of optimized configuration to screen Pb^{+2} catalyzed MB-conjugated aptamer to fold as G-quadruplex without the involvement of electrode surface, weakening the electrochemical signal of sensor surface. The configuration deciphered a linear range from 10^{-5} to 10^{-9} M with an LOD of 0.51 fM. To improve the workable sensitivity and range of DNA aptamer biosensors (otherwise kept restrictive following Langmuir Isothermal Adsorption), aptamers with distinct detection feasibilities can be combined to accomplish a wider detection. Such a provision is well-suited for the detection of multiple viruses and inflammation biomarkers, whose detection range may prevail over higher orders. Nevertheless, owing to unaffected maximum sensing efficacy, the sensitivity of such sensors is substantially reduced. The aptamers choice could be streamlined in a manner to optimize the functional confirmation, enabling a shift in the aptamer detection range. In this way, the working sensitivity of sensing is not likely to reduce, making the operation feasible to screen tumor biomarkers [147,148].

Despite significant improvement in functional performance *vis-à-vis* responding time and distinguishing accuracy, caution must be exercised *vis-à-vis* operational cost through reuse provisions. A commonly practiced approach is to flush the sensor surface with running buffer swiftly for a long duration to facilitate recovery. Though feasible, a major lacuna of such recovery is the possible drift in sensor baseline subsequent to the surface flushing. Besides, the rapid-flowing conditions could tamper with the sensor surface, specifically when aptamers are immobilized *via* electrostatic forces. Recovery could also be made *via* a denaturant or a surfactant, but it poses a vulnerable impact on sensor bioactivity. In order to overcome such concerns, another provision, DNA aptamer substitution, has been proposed to accomplish the sensor surface recovery more thoroughly, posing no risk to the surface distribution of DNA aptamer [149].

19.3.5 Biosensors using polysaccharides and their functionalized assemblies as probes

Polysaccharides are a class of complex carbohydrates formed by the polymerization of monosaccharide units. The advantages and inclinations for polysaccharides in being used as biosensor probes stem from their presence in all living organisms, exhibiting crucial roles in various physiological processes [150]. The structural diversity and functional versatility of these natural materials make them feasible for sustained biological interactions with multiple analytes such as antigens, toxins, environmental carcinogens, proteins, and several others. The most researched and prominent sources of polysaccharides are plants, animals, algae, and microorganisms. Polysaccharides are composed of simpler compounds linked as short or long chain assemblies (mono or di- saccharides) *via* glycosidic bonds [151]. Being biocompatible and responsive to biological stimuli, polysaccharides are adaptable to complex biochemical environments while also exhibiting mechanical strength, rigidity, and durability [152]. These biomacromolecules are distinguished by their multiple functional groups, such as -OH, -COOH, $-NH_2$—endowing them with capabilities for selective adsorption, numerous active sites, and fractionated interactions. Such abilities confer them an aqueous proximity, enhancing the mobility of polymer chains for graduated variations *via* entropic modulation [153].

The varied preparation routes could manifest in distinct functions of polysaccharides through dissimilar alkyl chain lengths and functional group conformations, influencing their properties and broadening the scope of applications. Additionally, polysaccharides can carry a positive, negative, or neutral charge that can be adjusted by pH changes, further altering their ionic proximity [154]. Depending on the similarity or dissimilarity of constituent monosaccharides and the nature of linkages, the polysaccharides can be either homo or hetero natured. This structural variability allows polysaccharides to participate in diverse biological processes, making them suitable for various biosensing applications, through their stacking, hydrogen bonding, van der Waals, and other noncovalent interactions Fig. 19.7. Homo-polysaccharides, made up of similar smaller sugars, include cellulose (used in paper and textiles), starch (a plant energy reserve and food industry thickener), inulin (a dietary fiber used in food and pharmaceuticals), and glycogen (animal energy reserve). Hetero-polysaccharides, containing dissimilar monosaccharides, include alginate (in biomedical applications), hyaluronic acid (in cosmetics and medicine), chitosan (in drug delivery and water purification), and heparin (an anticoagulant in medical settings) [155].

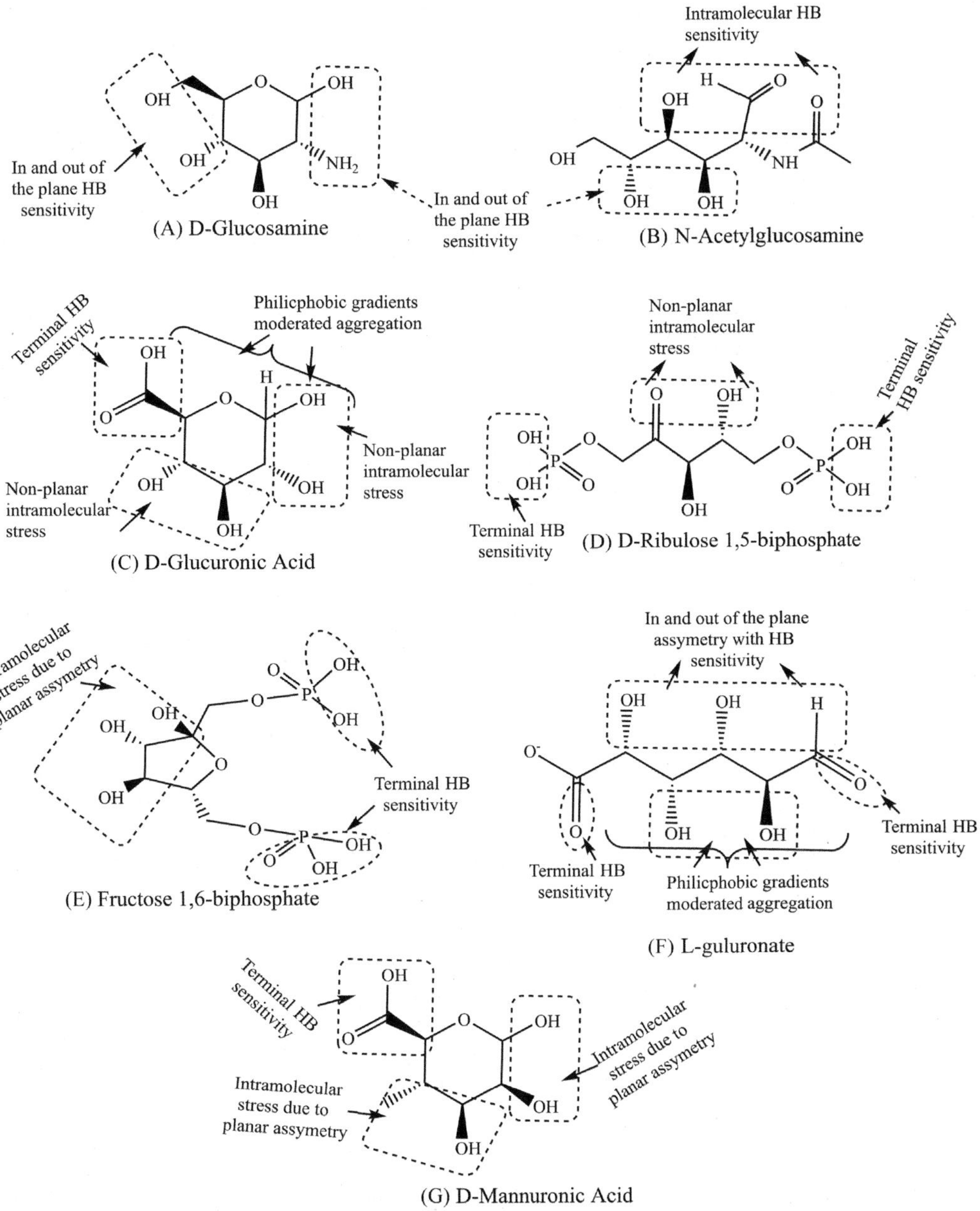

FIGURE 19.7 Molecular structures of select monosaccharides depicting their dispersion stabilizing abilities *via* their hydrophobic and HB driven moderated interactions. *HB*, Hydrogen bonding.

19.3.5.1 Distinct attributes of polysaccharides for biosensing

Polysaccharides, with their unique properties and functionalities, have gained attention as promising materials for biosensor development in light of their biocompatibility, biodegradability, and abundance in nature [156–158]. Polysaccharides can be used as immobilization matrices for biomolecules enhancing the sensing stability and sensitivity. The prominent unique attributes of polysaccharides for their biosensing use are as ahead.

- **Biocompatibility**: Polysaccharides are naturally occurring biopolymers that are generally nontoxic and pose minimal risks of accidental cross-reactivity in the medical and environmental applications. These abilities could screen some compounds which may be present in toxins and contaminant in combined form in the screened analytes.
- **Functional Diversity**: The presence of various functional groups (e.g., –OH, –COOH, $-NH_2$, $-CO-NH_2$) allows polysaccharides to interact with a wide range of biomolecules, facilitating their immobilization and high SA-driven greater access on sensor surfaces.

- **Structural Versatility**: Polysaccharides can form various structures, such as films, hydrogels, and NPs, providing flexibility in sensor design and fabrication. The plant origin of polysaccharides' constitutional units, monosaccharides makes them suitable as aggregation stabilizing agents. The functional ability to fractionate the strong hydrophobicity in drugs, dyes, and other strong textured environmental contaminants could be readily accomplished *via* segregated –H, -OH, -NH_2 and localized π-conjugation activities in monosaccharides. These abilities enable a precise sensing ability *via* fractionating the stronger analyte CF.
- **Mechanical Strength**: Many polysaccharides exhibit good mechanical properties, such as long-range stearic stabilization with a higher SA, enabling adsorptive abilities. Such attributes could be assets in immobilizing the enzymes and other analytes, enabling their reuse.

The specific whereabouts regarding the polysaccharides integrated contribution in varying combinations with multiple analyte-based sensing are discussed ahead.

19.3.5.2 Polysaccharides in enzyme-based biosensors

Enzyme-based biosensors rely on the catalytic activity of enzymes to detect specific analytes. Polysaccharides serve as excellent support materials for enzyme immobilization due to their ability to provide a conducive environment for enzyme activity.

Chitosan, with its amino groups, offers numerous functionalization possibilities and forms stable films, making it an ideal matrix for enzyme immobilization. The GOx immobilization on chitosan film benefits from strong electrostatic interactions between chitosan's –NH_2 groups and GOx's –COOH groups, enhancing the enzyme's stability and activity. The film-forming capability of chitosan ensures a robust and uniform surface, improving sensor sensitivity and response time. In a noted 2022 effort, *Kim and colleagues* prepared a composite of GO, cobalt, and chitosan that demonstrated strong stability and efficient electron transfer through the bio-electromechanical working framework. The versatile functionalization ability of GO apart from the layered morphology and stacking interactions argued well for bridging cobalt and chitosan in the structural assembly *vis-à-vis* Au thin-film electrode enabled real-time D-glucose screening within 1.0–15.0 mM extents. The sensor maintained consistent detection responses even after 14 days [159]. Alginate's gel-forming abilities make it suitable for encapsulating enzymes within a hydrogel matrix. An alginate-based biosensor for lactate detection has been reported *via* entrapping lactate oxidase (LOx) within an alginate gel. The gel matrix provides a protective environment for the enzyme, maintaining its activity over extended periods. The porous structure of the alginate gel allows for efficient substrate diffusion, ensuring rapid and accurate detection of lactate levels. In a noted attempt, enzymes such as HRP and LOx were effectively immobilized within alginate hydrogels to enable electrochemiluminescence (ECL) aided detection through selective reaction with H_2O_2 and lactic acid. Initially embedded in $CaCO_3$ microspheres and coated with a nanoshell, the enzymes were dispersed in alginate, applied onto GCEs *via* drop-casting and cross-linked using Ca^{2+}. The biosensors demonstrated consistent performance, with Michaelis-Menten constants of 7.71 ± 0.62 μM and 8.41 ± 0.43 μM for HRP and LOx, respectively along with LODs of 5.38 ± 0.05 μM for H_2O_2 and 0.50 ± 0.03 μM for lactic acid. The probe configurations maintained functionality till 28 days [160].

19.3.5.3 Polysaccharides in immunosensors

Immunosensors utilize the specific binding interactions between antibodies and antigens to detect target analytes. Polysaccharides can be used to immobilize antibodies on sensor surfaces, enhancing the specificity and sensitivity of immunosensors. The hyaluronic acid's ability to bind to specific cell surface receptors makes it well-suited for immobilizing antibodies in immunosensors. For example, a hyaluronic acid–based immunosensor for cancer biomarkers works through a hyaluronic acid matrix as a biocompatible surface for antibody immobilization, preserving their binding activity. This results in highly sensitive and specific detection of marker proteins, facilitating early disease diagnosis and monitoring. In a noted effort, *Mobed and associates* developed a label-free electrochemical immunosensor using modified indium tin oxide-polyethylene terephthalate electrodes for screening hyaluronic acid in real samples. The optimized probe works via a disposable ITO-coated PET electrode, accomplishing Au NPs aided immobilization of hyaluronic acid antibodies. Encapsulation in KCC1-NH-CS_2 enhanced sensor performance, with excellent reproducibility, selectivity, and stability, and a 0.078 to 160 ng mL^{-1} detection range, besides a low limit of quantification (0.078 ng mL^{-1}) in human plasma [161].

19.3.5.4 Polysaccharides usefulness in nucleic acid biosensors

The NABs use complementary base pairing to detect specific DNA or RNA sequences. Polysaccharides, like chitosan, are ideal for immobilizing nucleic acid probes due to their stability and ability to form films with a cationic sensitivity at acidic pH. The complementary nature of nucleotide base pairs augments the hybridization efficiency and sensitivity, following conjugation with analyte and probe *vis-à-vis* segregated polysaccharide assortment. Chitosan-grafted zirconia NPs have been used to immobilize ssDNA probes for detecting *M. tuberculosis*, improving biocompatibility, stability, and electron transfer. The nanoscale ZrO_2 colloids move toward the cathode, facilitating the DNA/CHIT-NanoZrO_2/ITO bioelectrode-driven DNA detection to as low as 0.00078 μM extents [162].

19.3.5.5 Polysaccharides augmented sensitivity in electrochemical biosensors

Electrochemical biosensors ascertain the electrical signals generated by the interaction between the target analyte and the BRE. Polysaccharides can be used to modify electrode surfaces, enabling improved sensitivity and selectivity. With its high SA, spaced functional groups, and porous structure, cellulose is ideal for modifying electrode surfaces. One study herein reported a cellulose-based biosensor for screening COVID-SARS, with natural cellulose fibers from *Gluconacetobacter hansenii* being a biodegradable substrate for COVID-19 diagnosis. The label-free potentiometric test uses bacterial cellulose as substrate deposited over a GO-engineered carbon-based electrode and human angiotensin-converting enzyme-2 (ACE2) as the receptor. The probe screens SARS-CoV-2 along with its variants sensitively and specifically from nasopharyngeal/oropharyngeal samples within 10 minutes, costing $3.50 per unit [163].

19.3.5.6 Polysaccharides as functional components in optical biosensors

Optical biosensors detect changes in light response behavior (e.g., intensity, wavelength, transmittance, reflection, or refraction) resulting from the interaction between the target analyte and BRE. Polysaccharides can be used to immobilize biomolecules on optical sensor surfaces, improving the sensitivity and stability of sensing. Chitosan's transparent film-forming abilities are highly suited for pathogen screening *via* perceived changes in light for the immobilized antibodies and pathogens' interactions. The stable chitosan film enhances sensitivity and specificity in antibody immobilization. Apart from this, a novel point-of-care assay has been optimized as an immuno-dot-blot modulation for diagnosing cystic echinococcosis. The methodology involves forming a sandwich complex between a Au nanoprobe (chitosan-Au NPs-protein A) and hydatid cyst antigen (Ag B), detecting anti-Ag B antibodies. Protein A is conjugated to chitosan-Au NPs *via* glutaraldehyde while Ag B prevails over a nitrocellulose membrane. Positive signals are visible to the naked eye, correlating with antibody, serum and Ag B concentration [164]. An alginate-based optical biosensor is also reported for glucose detection through alginate gel matrix guarded immobilized GOx, maintaining its activity over time. The probe facilitates efficient glucose diffusion to the enzyme, enabling quick and accurate detection. A research study investigated implantable glucose biosensors using polyelectrolyte multilayer (PEM) microspheres, encapsulating GOx in calcium alginate microspheres with an oxygen-quenched ruthenium framework. The multilayer nanothin films stabilized the enzyme with controlled substrate diffusion, enabling stable encapsulation with a reversible response [165].

19.4 Amalgamation of nanomaterials for enhanced sensing precision

With a pressing momentum towards NMs use, the precision and accuracy for biosensing have been significantly improved. The nanoscale dimensions argue well for quantum confinement-driven energy constriction of charge carriers, to manifest the shape and size-modulated physicochemical properties. The prominent NMs that have been explored significantly for biosensing include carbon nanotubes (CNTs), quantum dots (QDs), and metal and metal oxide NPs. While rod morphology enables the immobilization of multiple analytes simultaneously, the QDs could be coupled or conjugated with fluorophore and other optically responsive materials. The functionalization of NMs with varying probes is the first way to use NMs for biosensing, whereby the analyte sensing of probe is perceived by the sensitive energy levels that are recorded *via* transducer. Another way to use NMs for biosensing is *via* NPs adsorption or immobilization on a solid surface. The recognition of analyte by the NPs alters their native surface plasmon resonance peak through which their aggregated or monodispersed prevalence is inferred. The corresponding changes are recorded through blue (increasing energy, finer dispersion) or red shift (decreasing energy, aggregated state) in the UV–Vis spectra of used NPs (as probes). Another feasibility of using NMs as probes could be *via* their thin-layer deposition over the surface of electrodes, whereby the changes in current or potential difference before and after analyte recognition are recorded. The feasible modifications in this regard include the preparation of electrodes using Au or graphene, with the obvious reasons being the inert nature (for Au) and electrically conducting sensitivity (with latter). Readers suggested referring to more specific literature sources in this regard, wherein classification on the basis of specific NMs and the transduction mechanisms is optimized [166–168]. A common advantage of using the manifold NMs morphologies correlates with their closely spaced energy levels and dimensionally controlled quantum confinement. Secondly, their multifunctional abilities make it possible to screen manifold analytes within a short duration. A fine illustration of such distinctions is realized in the anticipated lab-on-a-chip assay, whereby surface engineering of a versatile high SA substrate enables prompt conduct of manifold interactions.

Mutually supplementing roles of NMs in additive response are being optimized with advanced computational software-assisted combinatorial validation. For instance, the use of Au and Ag NPs is rigorously explored for disease diagnosis and simultaneous drug delivery [169,170]. However, if one functionalizes the surface of CNTs or graphene with Au or Ag NPs, the sensing response could be obtained in further reduced time, even though the assembly cost of the device would increase. In this way, the size and shape modulated PCPs of NMs manifest the constriction of their constitutional energy levels to perceive the interacting stimulus promptly. Such attributes could not be accomplished in bulk materials due to much less exposure of functional SA, taking longer to screen the samples. The use of plant extracts as stabilizing and reducing agents in the preparation of metal NPs makes their assembly compatible with the least possibility of inadvertent cross-response. Additive outcomes for multiple cancer screening has been explored through functionalizing GO thin sheets with Au or Ag NPs whereby surface versatility and prompt coupling of electronic excitations and deexcitations are replicated and recorded. The functionalization of biocompatible NMs with proteins, antibodies, and nucleic acids has been in robust use and optimization. A further supplementing aspect is the use of nanoscale architecture of certain mono and disaccharides, such as gelatin and alginate NPs, which, on functionalized GO and its derivatives, could significantly augment the precision and accuracy of sensing. Thus, NMs use in biosensing anticipates well for future needs and facilitates quick response in much less time with reduced material and energy inputs, enabling a higher room for needful cautions.

19.5 Summary and future prospects

The biocompatible essence of enzymes, proteins, antibodies, nucleic acids, and polysaccharides presents a robust workable platform to synchronize and acclimatize the functional performance of biosensors. The best working conditions for accurate detection in a biosensor necessitate an analyte–probe proximity *via* noncovalent interactions so that the random chances of cross-reactivity are minimized. With the feasibility of predictive computational techniques, a lot could be accomplished through combinatorial optimization of varied biomaterials. For instance, antibodies could be functionalized with gelatin or alginate NPs to form a workable screening platform; similarly, amino acids could also be functionalized with antibodies for explicit analyte detection with accuracy. The realm of NMs sums up the current trend of feasible biosensing, wherein functionalization and derivatization could multiply the application prospects. Blessed by the characteristic SPR of metallic NPs, the sensing phenomenon could be improved with respect to screening time through the use of NPs in varied morphologies and functionalized configurations. Extracting proteins from animals, microbes, plants, and humans and screening their comparative efficacy for biosensing could be a potential area for environmental and diagnostic applications. The feasibility could be extended to NABs with the manifested base pair complement association to screen even the minutest extents. The future holds well for the biosensors development and applications, considering the diversity of environmental contaminants and the emerging resistance phenomenon in pathogens. A notable caution that could advance the understanding for improved sensitivity should be to preserve the databases of probe configurations *vis-à-vis* analyte screening precision so that the configurations with novel prospects and anticipated outcomes could be finalized. Similarly, comparative studies could be consolidated using different plant parts as NPs stabilizing and capping agents. Anticipations do provide a clue for synergistic response of more than one kind of NPs but analysis and optimizations on workable stoichiometries are essential to work out. To summarize, we have discussed the feasibility and workability of biomaterials as sensing probes through a discussion of their interaction diversity. Readers must understand that nanoscale quantum confinement is an explicit function of dimensions and this attribute varies with size and in no case, remains totally missing. Size-controlled energy content and the modulation in physicochemical properties are the key aspects to optimize the NMs' usability in biosensing *vis-à-vis* SA-driven proximity.

19.6 Conclusions

We have discussed the structure-function significance of enzymes (from microbes), proteins, antibodies, nucleic acids, and polysaccharides as natural assets for enhanced biosensing precision. The spaced functional groups with distributed hydrophilic–hydrophobic sensitivity make these molecules well-suited screening agents for environmental contaminants, microbial pathogens, and other toxic industrial effluents. Manifold improvements could be accomplished *via* the combined use of these moieties for enhanced screening precision with lower detection time and reduced time limits. For example, functionalizing metallic NPs with antibodies or proteins could improve the detection threshold compared to using proteins or antibodies alone as probes. Although much has been accomplished in biosensing, room for explicit response still prevails and could be optimized *via* computational chemistry-driven augmented probe-analyte structural compatibility.

Reactivity of analytes could be an interfering source in their reproducible detection and it could be addressed *via* predictive assays optimized through prompt interactions and combinative abilities through stoichiometry control. The role of stabilizing agent (plant extract constituents in NPs) warrants a thorough consideration for still improving the detection precision and feasible reuse ability through immobilization on the sensing surface. A close monitoring of each study with a specific analyte is the need of the hour for coherent efforts aimed at continued betterment, aimed at higher sensitivity of probes in reproducible settings.

References

[1] G.A. Naikoo, F. Arshad, M. Almas, I.U. Hassan, M.Z. Pedram, A.A.A. Aljabali, et al., 2D materials, synthesis, characterization and toxicity: a critical review, Chemico-Biological Interactions 365 (2022). Available from https://doi.org/10.1016/j.cbi.2022.110081, www.elsevier.com/locate/chembioint.

[2] G.A. Ozin, K. Hou, B.V. Lotsch, L. Cademartiri, D.P. Puzzo, F. Scotognella, et al., Nanofabrication by self-assembly, Materials Today 12 (5) (2009) 12–23. Available from https://doi.org/10.1016/S1369-7021(09)70156-7.

[3] M.Z.A. Mahmud, M.H. Mobarak, N. Hossain, Emerging trends in biomaterials for sustainable food packaging: a comprehensive review, Heliyon 10 (1) (2024). Available from https://doi.org/10.1016/j.heliyon.2024.e24122, https://www.sciencedirect.com/science/journal/24058440.

[4] P.C. Ray, H. Yu, P.P. Fu, Toxicity and environmental risks of nanomaterials: challenges and future needs, Journal of Environmental Science and Health - Part C Environmental Carcinogenesis and Ecotoxicology Reviews 27 (1) (2009) 1–35. Available from https://doi.org/10.1080/10590500802708267.

[5] M.H. Bakri, A.C. Özarslan, A. Erarslan, Y. Basaran Elalmis, F. Ciftci, Biomedical applications of wearable biosensors, Next Materials 3 (2024) 100084. Available from https://doi.org/10.1016/j.nxmate.2023.100084.

[6] D.F. Williams, On the nature of biomaterials, Biomaterials 30 (30) (2009) 5897–5909. Available from https://doi.org/10.1016/j.biomaterials.2009.07.027.

[7] Williams, D.F., 1987. Definitions in Biomaterials.

[8] D.F. Williams, The Williams Dictionary of Biomaterials, Liverpool University Press, (1999).

[9] B.D. Ratner, A.S. Hoffman, F.J. Schoen, J.E. Lemons, Biomaterials Science: An Introduction to Materials, third ed. Elsevier Inc, United States, (2013), pp. 1–1555. Available from http://doi.org/10.1016/B978-0-08-087780-8.00148-0, http://www.sciencedirect.com/science/book/9780123746269.

[10] N. Huebsch, D.J. Mooney, Inspiration and application in the evolution of biomaterials, Nature 462 (7272) (2009) 426–432. Available from https://doi.org/10.1038/nature08601.

[11] A. Verma, B.C. Yadav, Comprehensive review on two dimensional nanomaterials for optical biosensors: present progress and outlook, Sustainable Materials and Technologies 40 (2024) e00900. Available from https://doi.org/10.1016/j.susmat.2024.e00900.

[12] B.D. Ratner, A.S. Hoffman, F.J. Schoen, J.E. Lemons, Introduction - biomaterials science: an evolving, multidisciplinary endeavor, Biomaterials Science: An Introduction to Materials: Third Edition, Elsevier Inc, United States, (2013), p. xxv. Available from https://doi.org/10.1016/B978-0-08-087780-8.00153-4, http://www.sciencedirect.com/science/book/9780123746269.

[13] B. Huzum, B. Puha, R. Necoara, S. Gheorghevici, G. Puha, A. Filip, et al., Biocompatibility assessment of biomaterials used in orthopedic devices: an overview (Review), Experimental and Therapeutic Medicine 22 (5) (2021). Available from https://doi.org/10.3892/etm.2021.10750.

[14] J.D. Newman, S.J. Setford, Enzymatic biosensors, Molecular Biotechnology 32 (3) (2006) 249–268. Available from https://doi.org/10.1385/MB:32:3:249.

[15] S.A. Pullano, M. Greco, M.G. Bianco, D. Foti, A. Brunetti, A.S. Fiorillo, Glucose biosensors in clinical practice: principles, limits and perspectives of currently used devices, Theranostics 12 (2) (2022) 493–511. Available from https://doi.org/10.7150/thno.64035, https://www.thno.org/v12p0493.

[16] E.O. Polat, M.M. Cetin, A.F. Tabak, E.B. Güven, B.Ö. Uysal, T. Arsan, et al., Transducer technologies for biosensors and their wearable applications, Biosensors 12 (6) (2022). Available from https://doi.org/10.3390/bios12060385, https://www.mdpi.com/2079-6374/12/6/385/pdf?version=1654160906.

[17] J. Wang, Nanomaterial-based electrochemical biosensors, Analyst 130 (4) (2005) 421–426. Available from https://doi.org/10.1039/b414248a, http://pubs.rsc.org/en/journals/journal/an.

[18] A. Heller, B. Feldman, Electrochemical glucose sensors and their applications in diabetes management, Chemical Reviews 108 (7) (2008) 2482–2505. Available from https://doi.org/10.1021/cr068069y.

[19] L.J.C. Jeuken, Electrodes for integral membrane enzymes, Natural Product Reports 26 (10) (2009) 1234–1240. Available from https://doi.org/10.1039/b903252e.

[20] T. Laftsoglou, L.J.C. Jeuken, Supramolecular electrode assemblies for bioelectrochemistry, Chemical Communications 53 (27) (2017) 3801–3809. Available from https://doi.org/10.1039/c7cc01154g, http://pubs.rsc.org/en/journals/journal/cc.

[21] P.O. Saboe, E. Conte, M. Farell, G.C. Bazan, M. Kumar, Biomimetic and bioinspired approaches for wiring enzymes to electrode interfaces, Energy and Environmental Science 10 (1) (2017) 14–42. Available from https://doi.org/10.1039/c6ee02801b, http://pubs.rsc.org/en/journals/journal/ee.

[22] D. Marchal, J. Pantigny, J.M. Laval, J. Moiroux, C. Bourdillon, Rate constants in two dimensions of electron transfer between Pyruvate oxidase, a membrane enzyme, and ubiquinone (coenzyme Q8), its water-insoluble electron carrier, Biochemistry 40 (5) (2001) 1248–1256. Available from https://doi.org/10.1021/bi002325y.

[23] P. Infossi, E. Lojou, J.P. Chauvin, G. Herbette, M. Brugna, M.T. Giudici-Orticoni, Aquifex aeolicus membrane hydrogenase for hydrogen biooxidation: role of lipids and physiological partners in enzyme stability and activity, International Journal of Hydrogen Energy 35 (19) (2010) 10778–10789. Available from https://doi.org/10.1016/j.ijhydene.2010.02.054, http://www.journals.elsevier.com/international-journal-of-hydrogen-energy/.

[24] C.C. Akoh, G.C. Lee, Y.C. Liaw, T.H. Huang, J.F. Shaw, GDSL family of serine esterases/lipases, Progress in Lipid Research 43 (6) (2004) 534–552. Available from https://doi.org/10.1016/j.plipres.2004.09.002.

[25] M.H. Doolittle, M. Péterfy, Mechanisms of lipase maturation, Clinical Lipidology 5 (1) (2010) 117–130. Available from https://doi.org/10.2217/clp.09.84.

[26] M. Pohanka, Cholinesterases, a target of pharmacology and toxicology, Biomedical Papers 155 (3) (2011) 219–230. Available from https://doi.org/10.5507/bp.2011.036, http://biomed.papers.upol.cz/pdfs/bio/2011/03/02.pdf.

[27] M. Pohanka, Acetylcholinesterase inhibitors: a patent review (2008 present), Expert Opinion on Therapeutic Patents 22 (8) (2012) 871–886. Available from https://doi.org/10.1517/13543776.2012.701620.

[28] M. Pohanka, Biosensors containing acetylcholinesterase and butyrylcholinesterase as recognition tools for detection of various compounds, Chemical Papers 69 (1) (2015) 4–16. Available from https://doi.org/10.2478/s11696-014-0542-x, http://rd.springer.com/journal/volumesAndIssues/11696.

[29] R.N. Patel, Stereoselective biotransformations in synthesis of some pharmaceutical intermediates, Advances in Applied Microbiology 43 (1997) 91–140. Available from https://doi.org/10.1016/s0065-2164(08)70224-7, http://www.sciencedirect.com/science/journal/00652164.

[30] H. Matsumae, M. Furui, T. Shibatani, T. Tosa, Production of optically active 3-phenylglycidic acid ester by the lipase from *Serratia marcescens* on a hollow-fiber membrane reactor, Journal of Fermentation and Bioengineering 78 (1) (1994) 59–63. Available from https://doi.org/10.1016/0922-338X(94)90179-1.

[31] A. Acosta, M. Filice, G. Fernandez-Lorente, J.M. Palomo, J.M. Guisan, Kinetically controlled synthesis of monoglyceryl esters from chiral and prochiral acids methyl esters catalyzed by immobilized *Rhizomucor miehei* lipase, Bioresource Technology 102 (2) (2011) 507–512. Available from https://doi.org/10.1016/j.biortech.2010.08.095.

[32] T. Lianghua, X. Liming, S. Min, G. Huaying, Purification and application of a lipase from *Penicillium expansum* PED-03, Applied Biochemistry and Biotechnology 142 (2) (2007) 194–199. Available from https://doi.org/10.1007/s12010-007-0043-2.

[33] K. Athenstaedt, G. Daum, Tgl4p and Tgl5p, two triacylglycerol lipases of the yeast *Saccharomyces cerevisiae* are localized to lipid particles, Journal of Biological Chemistry 280 (45) (2005) 37301–37309. Available from https://doi.org/10.1074/jbc.M507261200.

[34] F. Kartal, A. Kilinç, S. Timur, Lipase biosensor for tributyrin and pesticide detection, International Journal of Environmental Analytical Chemistry 87 (10-11) (2007) 715–722. Available from https://doi.org/10.1080/03067310701327741.

[35] B. Ma, L.Z. Cheong, X. Weng, C.P. Tan, C. Shen, Lipase@ZIF-8 nanoparticles-based biosensor for direct and sensitive detection of methyl parathion, Electrochimica Acta 283 (2018) 509–516. Available from https://doi.org/10.1016/j.electacta.2018.06.176, http://www.journals.elsevier.com/electrochimica-acta/.

[36] K. Gangadhara Reddy, G. Madhavi, B.E. Kumara Swamy, Mobilized lipase enzymatic biosensor for the determination of Chlorfenvinphos and Malathion in contaminated water samples: a voltammetric study, Journal of Molecular Liquids 198 (2014) 181–186. Available from https://doi.org/10.1016/j.molliq.2014.06.019.

[37] N. Zehani, S.V. Dzyadevych, R. Kherrat, N.J. Jaffrezic-Renault, Sensitive impedimetric biosensor for direct detection of diazinon based on lipases, Frontiers in Chemistry 2 (2014). Available from https://doi.org/10.3389/fchem.2014.00044, http://journal.frontiersin.org/article/10.3389/fchem.2014.00044/full.

[38] M. Gamella, S. Campuzano, J. Manso, G.G. de Rivera, F. López-Colino, A.J. Reviejo, et al., A novel non-invasive electrochemical biosensing device for in situ determination of the alcohol content in blood by monitoring ethanol in sweat, Analytica Chimica Acta 806 (2014) 1–7. Available from https://doi.org/10.1016/j.aca.2013.09.020.

[39] O. Secchi, M. Zinellu, Y. Spissu, M. Pirisinu, G. Bazzu, R. Migheli, et al., Further in-vitro characterization of an implantable biosensor for ethanol monitoring in the brain, Sensors (Switzerland) 13 (7) (2013) 9522–9535. Available from https://doi.org/10.3390/s130709522, http://www.mdpi.com/1424-8220/13/7/9522/pdf.

[40] Y. Wen, J. Xu, M. Liu, D. Li, H. He, Amperometric vitamin C biosensor based on the immobilization of ascorbate oxidase into the biocompatible sandwich-type composite film, Applied Biochemistry and Biotechnology 167 (7) (2012) 2023–2038. Available from https://doi.org/10.1007/s12010-012-9711-y, http://www.springer.com/humana+press/journal/12010.

[41] M. Liu, Y. Wen, J. Xu, H. He, D. Li, R. Yue, et al., An amperometric biosensor based on ascorbate oxidase immobilized in poly(3,4-ethylenedioxythiophene)/multi-walled carbon nanotubes composite films for the determination of L-ascorbic acid, Analytical Sciences 27 (5) (2011) 477–482. Available from https://doi.org/10.2116/analsci.27.477, http://www.jstage.jst.go.jp/article/analsci/27/5/477/_pdf.

[42] G. Wu, M.H. Zaman, L. Li, H. Lu, L. Deng, A sensitive NADH and ethanol biosensor based on graphene-Au nanorods nanocomposites, Talanta 134 (2013) 1–6. Available from https://doi.org/10.1016/j.talanta.2014.10.061.

[43] K. Lata, V. Dhull, V. Hooda, Fabrication and optimization of ChE/ChO/HRP-AuNPs/c-MWCNTs based silver electrode for determining total cholesterol in Serum, Biochemistry Research International 2016 (2016). Available from https://doi.org/10.1155/2016/1545206, http://www.hindawi.com/journals/bcri/.

[44] V. Aggarwal, J. Malik, A. Prashant, P.K. Jaiwal, C.S. Pundir, Amperometric determination of serum total cholesterol with nanoparticles of cholesterol esterase and cholesterol oxidase, Analytical Biochemistry 500 (2016) 6–11. Available from https://doi.org/10.1016/j.ab.2016.01.019.

[45] M.M. Rahman, A.M. Asiri, Selective choline biosensors based on choline oxidase co-immobilized into self-assembled monolayers on micro-chips at low potential, Analytical Methods 7 (22) (2015) 9426–9434. Available from https://doi.org/10.1039/c5ay02456k, http://pubs.rsc.org/en/journals/journal/ay.

[46] A.T. Tunç, E. Aynacı Koyuncu, F. Arslan, Development of an acetylcholinesterase–choline oxidase based biosensor for acetylcholine determination, Artificial Cells, Nanomedicine and Biotechnology 44 (7) (2016) 1659–1664. Available from https://doi.org/10.3109/21691401.2015.1080167, http://www.tandfonline.com/loi/ianb20#.VmugQbfovcs.

[47] M.F. Hossain, J.Y. Park, Plain to point network reduced graphene oxide-activated carbon composites decorated with platinum nanoparticles for urine glucose detection, Scientific Reports 6 (2016). Available from https://doi.org/10.1038/srep21009, www.nature.com/srep/index.html.

[48] Ö. Sağlam, B. Kızılkaya, H. Uysal, Y. Dilgin, Biosensing of glucose in flow injection analysis system based on glucose oxidase-quantum dot modified pencil graphite electrode, Talanta 147 (2016) 315–321. Available from https://doi.org/10.1016/j.talanta.2015.09.050.

[49] C.P. McMahon, G. Rocchitta, P.A. Serra, S.M. Kirwan, J.P. Lowry, R.D. O'Neill, Control of the oxygen dependence of an implantable polymer/enzyme composite biosensor for glutamate, Analytical Chemistry 78 (7) (2006) 2352–2359. Available from https://doi.org/10.1021/ac0518194.

[50] O. Soldatkin, A. Nazarova, N. Krisanova, A. Borysov, D. Kucherenko, I. Kucherenko, et al., Monitoring of the velocity of high-affinity glutamate uptake by isolated brain nerve terminals using amperometric glutamate biosensor, Talanta 135 (2015) 67–74. Available from https://doi.org/10.1016/j.talanta.2014.12.031.

[51] S.Y. Keskin, C.S. Keskin, Quantitative determination of glycine in aqueous solution using glutamate dehydrogenase-immobilized glyoxal agarose beads, Applied Biochemistry and Biotechnology 172 (1) (2014) 289–297. Available from https://doi.org/10.1007/s12010-013-0543-1.

[52] A. Gholizadeh, S. Shahrokhian, A. Iraji zad, S. Mohajerzadeh, M. Vosoughi, S. Darbari, et al., Mediator-less highly sensitive voltammetric detection of glutamate using glutamate dehydrogenase/vertically aligned CNTs grown on silicon substrate, Biosensors and Bioelectronics 31 (1) (2012) 110–115. Available from https://doi.org/10.1016/j.bios.2011.10.002.

[53] Y. Lin, P. Yu, J. Hao, Y. Wang, T. Ohsaka, L. Mao, Continuous and simultaneous electrochemical measurements of glucose, lactate, and ascorbate in rat brain following brain ischemia, Analytical Chemistry 86 (8) (2014) 3895–3901. Available from https://doi.org/10.1021/ac4042087, http://pubs.acs.org/journal/ancham.

[54] D.M. Kim, M.Y. Kim, S.S. Reddy, J. Cho, C.H. Cho, S. Jung, et al., Electron-transfer mediator for a NAD-glucose dehydrogenase-based glucose sensor, Analytical Chemistry 85 (23) (2013) 11643–11649. Available from https://doi.org/10.1021/ac403217t.

[55] G. Rocchitta, O. Secchi, M.D. Alvau, D. Farina, G. Bazzu, G. Calia, et al., Simultaneous telemetric monitoring of brain glucose and lactate and motion in freely moving rats, Analytical Chemistry 85 (21) (2013) 10282–10288. Available from https://doi.org/10.1021/ac402071w.

[56] N. Hernández-Ibáñez, L. García-Cruz, V. Montiel, C.W. Foster, C.E. Banks, J. Iniesta, Electrochemical lactate biosensor based upon chitosan/carbon nanotubes modified screen-printed graphite electrodes for the determination of lactate in embryonic cell cultures, Biosensors and Bioelectronics 77 (2016) 1168–1174. Available from https://doi.org/10.1016/j.bios.2015.11.005, www.elsevier.com/locate/bios.

[57] S.N.A. Jenie, B. Prieto-Simon, N.H. Voelcker, Development of L-lactate dehydrogenase biosensor based on porous silicon resonant microcavities as fluorescence enhancers, Biosensors and Bioelectronics 74 (2015) 637–643. Available from https://doi.org/10.1016/j.bios.2015.07.025, www.elsevier.com/locate/bios.

[58] S. Azzouzi, L. Rotariu, A.M. Benito, W.K. Maser, M. Ben Ali, C. Bala, A novel amperometric biosensor based on gold nanoparticles anchored on reduced graphene oxide for sensitive detection of L-lactate tumor biomarker, Biosensors and Bioelectronics 69 (2015) 280–286. Available from https://doi.org/10.1016/j.bios.2015.03.012, www.elsevier.com/locate/bios.

[59] J.C. Gonzalez-Rivera, J.F. Osma, Fabrication of an amperometric flow-injection microfluidic biosensor based on laccase for in situ determination of phenolic compounds, BioMed Research International 2015 (2015). Available from https://doi.org/10.1155/2015/845261, http://www.hindawi.com/journals/biomed/.

[60] S. Sadeghi, E. Fooladi, M. Malekaneh, A new amperometric biosensor based on Fe3O4/polyaniline/laccase/chitosan biocomposite-modified carbon paste electrode for determination of catechol in tea leaves, Applied Biochemistry and Biotechnology 175 (3) (2015) 1603–1616. Available from https://doi.org/10.1007/s12010-014-1380-6, http://www.springer.com/humana+press/journal/12010.

[61] F. Campanhã Vicentini, L.L.C. Garcia, L.C.S. Figueiredo-Filho, B.C. Janegitz, O. Fatibello-Filho, A biosensor based on gold nanoparticles, dihexadecylphosphate, and tyrosinase for the determination of catechol in natural water, Enzyme and Microbial Technology 84 (2016) 17–23. Available from https://doi.org/10.1016/j.enzmictec.2015.12.004.

[62] J. Kochana, K. Wapiennik, J. Kozak, P. Knihnicki, A. Pollap, M. Woźniakiewicz, et al., Tyrosinase-based biosensor for determination of bisphenol A in a flow-batch system, Talanta 144 (2015) 163–170. Available from https://doi.org/10.1016/j.talanta.2015.05.078.

[63] B. Alberts, A. Johnson, J. Lewis. Molecular Biology of the Cell, fourth Ed. Garland Science Protein Function, 2002.

[64] O. Smutok, E. Katz, Biosensors: electrochemical devices—general concepts and performance, Biosensors 13 (1) (2023) 44. Available from https://doi.org/10.3390/bios13010044.

[65] J. Feng, B.W. Jester, C.E. Tinberg, D.J. Mandell, M.S. Antunes, R. Chari, et al., A general strategy to construct small molecule biosensors in eukaryotes, eLife 4 (2015) 2015 https://doi.org/10.7554/eLife.10606, http://elifesciences.org/content/elife/4/e10606.full.pdf.

[66] V. Naresh, N. Lee, A review on biosensors and recent development of nanostructured materials-enabled biosensors, Sensors 21 (4) (2021) 1109. Available from https://doi.org/10.3390/s21041109.

[67] M. Sadoine, Y. Ishikawa, T.J. Kleist, M.M. Wudick, M. Nakamura, G. Grossmann, et al., Designs, applications, and limitations of genetically encoded fluorescent sensors to explore plant biology, Plant Physiology 187 (2) (2021) 485–503. Available from https://doi.org/10.1093/plphys/kiab353, https://academic.oup.com/plphys/article/187/2/485/6359373?login=true.

[68] P. Chaisupa, R.C. Wright, State-of-the-art in engineering small molecule biosensors and their applications in metabolic engineering, SLAS Technology 29 (2) (2024) 100113. Available from https://doi.org/10.1016/j.slast.2023.10.005.

[69] V. D'Ambrosio, S. Pramanik, K. Goroncy, T. Jakočiūnas, D. Schönauer, M.D. Davari, et al., Directed evolution of VanR biosensor specificity in yeast, Biotechnology Notes 1 (2020) 9–15. Available from https://doi.org/10.1016/j.biotno.2020.01.002.

[70] M. Espinosa-Urgel, L. Serrano, J.L. Ramos, A.M. Fernández-Escamilla, Engineering biological approaches for detection of toxic compounds: a new microbial biosensor based on the *Pseudomonas putida* TtgR repressor, Molecular Biotechnology 57 (6) (2015) 558–564. Available from https://doi.org/10.1007/s12033-015-9849-2, http://www.springer.com/humana+press/journal/12033.

[71] R.J. Chesterfield, J.H. Whitfield, B. Pouvreau, D. Cao, K. Alexandrov, C.A. Beveridge, et al., Rational design of novel fluorescent enzyme biosensors for direct detection of strigolactones, ACS Synthetic Biology 9 (8) (2020) 2107–2118. Available from https://doi.org/10.1021/acssynbio.0c00192, http://pubs.acs.org/journal/asbcd6.

[72] T. Yoshida, H. Nakajima, S. Takahashi, A. Kakizuka, H. Imamura, OLIVe: a genetically encoded fluorescent biosensor for quantitative imaging of branched-chain amino acid levels inside single living cells, ACS Sensors 4 (12) (2019) 3333–3342. Available from https://doi.org/10.1021/acssensors.9b02067, http://pubs.acs.org/journal/ascefj.

[73] M. Kurnik, E.Z. Pang, K.W. Plaxco, An electrochemical biosensor architecture based on protein folding supports direct real-time measurements in whole blood, Angewandte Chemie - International Edition 59 (42) (2020) 18442–18445. Available from https://doi.org/10.1002/anie.202007256, http://onlinelibrary.wiley.com/journal/10.1002/(ISSN)1521-3773.

[74] A. Janeena, S. Prem, P. Mohandass, S. Ashokraj, J. Narayanan, S. Easwaramoorthi, et al., Selection and screening of genetically encoded fluorescent protein as a sensor for cancer theranostics, Biosensors and Bioelectronics: X 10 (2022) 100129. Available from https://doi.org/10.1016/j.biosx.2022.100129.

[75] K. Murugappan, U. Sundaramoorthy, A.M. Damry, D.R. Nisbet, C.J. Jackson, A. Tricoli, Electrodetection of small molecules by conformation-mediated signal enhancement, JACS Au 2 (11) (2022) 2481–2490. Available from https://doi.org/10.1021/jacsau.2c00291Australia, https://pubs.acs.org/page/jaaucr/about.html.

[76] A. Quijano-Rubio, H.W. Yeh, J. Park, H. Lee, R.A. Langan, S.E. Boyken, et al., De novo design of modular and tunable protein biosensors, Nature 591 (7850) (2021) 482–487. Available from https://doi.org/10.1038/s41586-021-03258-z, http://www.nature.com/nature/index.html.

[77] X. Yang, Z.-Z. Yin, G. Zheng, M. Zhou, H. Zhang, J. Li, et al., Molecularly imprinted miniature electrochemical biosensor for SARS-CoV-2 spike protein based on Au nanoparticles and reduced graphene oxide modified acupuncture needle, Bioelectrochemistry (Amsterdam, Netherlands) 151 (2023) 108375. Available from https://doi.org/10.1016/j.bioelechem.2023.108375.

[78] S. Sharma, H. Byrne, R.J. O'Kennedy, Antibodies and antibody-derived analytical biosensors, Essays in Biochemistry 60 (1) (2016) 9–18. Available from https://doi.org/10.1042/EBC20150002, http://essays.biochemistry.org/content/ppebio/60/1/9.full.pdf.

[79] C. Wang, M. Liu, Z. Wang, S. Li, Y. Deng, N. He, Point-of-care diagnostics for infectious diseases: From methods to devices, Nano Today 37 (2021) 101092. Available from https://doi.org/10.1016/j.nantod.2021.101092.

[80] W.J.J. Finlay, N.C. DeVore, E.N. Dobrovolskaia, A. Gam, C.S. Goodyear, J.E. Slater, Exploiting the avian immunoglobulin system to simplify the generation of recombinant antibodies to allergenic proteins, Clinical & Experimental Allergy 35 (8) (2005) 1040–1048. Available from https://doi.org/10.1111/j.1365-2222.2005.02307.x.

[81] P. Leonard, P. Säfsten, S. Hearty, B. McDonnell, W. Finlay, R. O'Kennedy, High throughput ranking of recombinant avian scFv antibody fragments from crude lysates using the Biacore A100, Journal of Immunological Methods 323 (2) (2007) 172–179. Available from https://doi.org/10.1016/j.jim.2007.04.010.

[82] Y.C. Lee, S.J.C. Leu, C.J. Hu, N.Y. Shih, I.J. Huang, H.H. Wu, et al., Chicken single-chain variable fragments against the SARS-CoV spike protein, Journal of Virological Methods 146 (1-2) (2007) 104–111. Available from https://doi.org/10.1016/j.jviromet.2007.06.010.

[83] A.G. Gehring, C.G. Crawford, R.S. Mazenko, L.J. Van Houten, J.D. Brewster, Enzyme-linked immunomagnetic electrochemical detection of *Salmonella typhimurium*, Journal of Immunological Methods 195 (1-2) (1996) 15–25. Available from https://doi.org/10.1016/0022-1759(96)00076-2, www.elsevier.com/locate/jim.

[84] Y.H. Lin, S.H. Chen, Y.C. Chuang, Y.C. Lu, T.Y. Shen, C.A. Chang, et al., Disposable amperometric immunosensing strips fabricated by Au nanoparticles-modified screen-printed carbon electrodes for the detection of foodborne pathogen Escherichia coli O157:H7, Biosensors and Bioelectronics 23 (12) (2008) 1832–1837. Available from https://doi.org/10.1016/j.bios.2008.02.030.

[85] S.M. Radke, E.C. Alocilja, A high density microelectrode array biosensor for detection of E. coli O157:H7, Biosensors and Bioelectronics 20 (8) (2005) 1662–1667. Available from https://doi.org/10.1016/j.bios.2004.07.021, www.elsevier.com/locate/bios.

[86] M. Hnaiein, W.M. Hassen, A. Abdelghani, C. Fournier-Wirth, J. Coste, F. Bessueille, et al., A conductometric immunosensor based on functionalized magnetite nanoparticles for *E. coli* detection, Electrochemistry Communications 10 (8) (2008) 1152–1154. Available from https://doi.org/10.1016/j.elecom.2008.04.009.

[87] A. Kausaite-Minkstimiene, A. Popov, A. Ramanaviciene, Surface plasmon resonance immunosensor with antibody-functionalized magneto-plasmonic nanoparticles for ultrasensitive quantification of the CD5 biomarker, ACS Applied Materials and Interfaces 14 (18) (2022) 20720–20728. Available from https://doi.org/10.1021/acsami.2c02936, http://pubs.acs.org/journal/aamick.

[88] P. Parkkila, K. Härkönen, P. Ilvonen, S. Laitinen, T. Viitala, Protein A/G-based surface plasmon resonance biosensor for regenerable antibody-mediated capture and analysis of nanoparticles, Colloids and Surfaces A: Physicochemical and Engineering Aspects 654 (2022) 130015. Available from https://doi.org/10.1016/j.colsurfa.2022.130015.

[89] B.B. Kim, W.J. Im, J.Y. Byun, H.M. Kim, M.G. Kim, Y.B. Shin, Label-free CRP detection using optical biosensor with one-step immobilization of antibody on nitrocellulose membrane, Sensors and Actuators, B: Chemical 190 (2014) 243–248. Available from https://doi.org/10.1016/j.snb.2013.08.078.

[90] J. Liang, C. Yao, X. Li, Z. Wu, C. Huang, Q. Fu, et al., Silver nanoprism etching-based plasmonic ELISA for the high sensitive detection of prostate-specific antigen, Elsevier Ltd, China Biosensors and Bioelectronics 69 (2015) 128–134. Available from https://doi.org/10.1016/j.bios.2015.02.026, www.elsevier.com/locate/bios.

[91] D. Acharya, P. Bastola, L. Le, A.M. Paul, E. Fernandez, M.S. Diamond, et al., An ultrasensitive electrogenerated chemiluminescence-based immunoassay for specific detection of Zika virus, Scientific Reports 6 (2016). Available from https://doi.org/10.1038/srep32227, www.nature.com/srep/index.html.

[92] A. Pandey, Y. Gurbuz, V. Ozguz, J.H. Niazi, A. Qureshi, Graphene-interfaced electrical biosensor for label-free and sensitive detection of foodborne pathogenic *E. coli* O157:H7, Biosensors and Bioelectronics 91 (2017) 225–231. Available from https://doi.org/10.1016/j.bios.2016.12.041, www.elsevier.com/locate/bios.

[93] S. Yang, M. You, F. Zhang, Q. Wang, P. He, A sensitive electrochemical aptasensing platform based on exonuclease recycling amplification and host-guest recognition for detection of breast cancer biomarker HER2, Sensors and Actuators, B: Chemical 258 (2018) 796–802. Available from https://doi.org/10.1016/j.snb.2017.11.119.

[94] T. Wei, P. Ren, L. Huang, Z. Ouyang, Z. Wang, X. Kong, et al., Simultaneous detection of aflatoxin B1, ochratoxin A, zearalenone and deoxynivalenol in corn and wheat using surface plasmon resonance, Food Chemistry 300 (2019) 125176. Available from https://doi.org/10.1016/j.foodchem.2019.125176.

[95] A. Gupta, S.K. Sharma, V. Pachauri, S. Ingebrandt, S. Singh, A.L. Sharma, et al., Sensitive impedimetric detection of troponin I with metal-organic framework composite electrode, RSC Advances 11 (2021). Available from https://doi.org/10.1039/D0RA06665F.

[96] Z. Li, Y. Yi, X. Luo, N. Xiong, Y. Liu, S. Li, et al., Development and clinical application of a rapid IgM-IgG combined antibody test for SARS-CoV-2 infection diagnosis, Journal of Medical Virology 92 (9) (2020) 1518–1524. Available from https://doi.org/10.1002/jmv.25727, http://onlinelibrary.wiley.com/journal/10.1002/(ISSN)1096-9071.

[97] B. Szymanska, Z. Lukaszewski, L. Oldak, B. Zelazowska-Rutkowska, K. Hermanowicz-Szamatowicz, E. Gorodkiewicz, Two biosensors for the determination of interleukin-6 in blood plasma by array SPRi, Biosensors 12 (6) (2022) 412. Available from https://doi.org/10.3390/bios12060412.

[98] P. Vizzini, E. Beltrame, N. Coppedè, F. Vurro, F. Andreatta, E. Torelli, et al., Detection of *Listeria monocytogenes* in foods with a textile organic electrochemical transistor biosensor, Applied Microbiology and Biotechnology 107 (11) (2023) 3789–3800. Available from https://doi.org/10.1007/s00253-023-12543-y, https://www.springer.com/journal/253.

[99] A.P.F. Turner, Biosensors: sense and sensibility, Chemical Society Reviews 42 (8) (2013) 3184–3196. Available from https://doi.org/10.1039/c3cs35528d.

[100] F. TELES, L. FONSECA, Trends in DNA biosensors, Talanta 77 (2) (2008) 606–623. Available from https://doi.org/10.1016/j.talanta.2008.07.024.

[101] I. Palchetti, M. Mascini, Nucleic acid biosensors for environmental pollution monitoring, Analyst 133 (7) (2008) 846–854. Available from https://doi.org/10.1039/b802920m, http://pubs.rsc.org/en/journals/journal/an.

[102] A.D. Ellington, J.W. Szostak, In vitro selection of RNA molecules that bind specific ligands, Nature 346 (6287) (1990) 818–822. Available from https://doi.org/10.1038/346818a0.

[103] C. Tuerk, L. Gold, Systematic evolution of ligands by exponential enrichment: RNA ligands to bacteriophage T4 DNA polymerase, Science (New York, N.Y.) 249 (4968) (1990) 505–510. Available from https://doi.org/10.1126/science.2200121.

[104] R.R. Breaker, G.F. Joyce, A DNA enzyme that cleaves RNA, Chemistry and Biology 1 (4) (1994) 223–229. Available from https://doi.org/10.1016/1074-5521(94)90014-0.

[105] Y. Du, S. Dong, Nucleic acid biosensors: recent advances and perspectives, Analytical Chemistry 89 (1) (2017) 189–215. Available from https://doi.org/10.1021/acs.analchem.6b04190, http://pubs.acs.org/journal/ancham.

[106] D. Li, S. Song, C. Fan, Target-responsive structural switching for nucleic acid-based sensors, Accounts of Chemical Research 43 (5) (2010) 631–641. Available from https://doi.org/10.1021/ar900245u.

[107] T.G. Drummond, M.G. Hill, J.K. Barton, Electrochemical DNA sensors, Nature Biotechnology 21 (10) (2003) 1192–1199. Available from https://doi.org/10.1038/nbt873.

[108] C.I.L. Justino, T.A. Rocha-Santos, A.C. Duarte, Review of analytical figures of merit of sensors and biosensors in clinical applications, TrAC - Trends in Analytical Chemistry 29 (10) (2010) 1172–1183. Available from https://doi.org/10.1016/j.trac.2010.07.008, www.elsevier.com/locate/trac.

[109] M. Pedrero, S. Campuzano, J.M. Pingarrón, Electroanalytical sensors and devices for multiplexed detection of foodborne pathogen microorganisms, Sensors 9 (7) (2009) 5503–5520. Available from https://doi.org/10.3390/s90705503Spain, http://www.mdpi.com/1424-8220/9/7/5503/pdf.

[110] Y. Du, B. Li, E. Wang, "Fitting" makes "sensing" simple: label-free detection strategies based on nucleic acid aptamers, Accounts of Chemical Research 46 (2) (2013) 203–213. Available from https://doi.org/10.1021/ar300011g.

[111] M. Famulok, G. Mayer, Aptamer modules as sensors and detectors, Accounts of Chemical Research 44 (12) (2011) 1349–1358. Available from https://doi.org/10.1021/ar2000293.

[112] Y.C. Lim, A.Z. Kouzani, W. Duan, Aptasensors: a review, Journal of Biomedical Nanotechnology 6 (2) (2010) 93–105. Available from https://doi.org/10.1166/jbn.2010.1103.

[113] J. Liu, Z. Cao, Y. Lu, Functional nucleic acid sensors, Chemical Reviews 109 (5) (2009) 1948–1998. Available from https://doi.org/10.1021/cr030183i, http://pubs.acs.org/doi/pdfplus/10.1021/cr030183i.

[114] N. Navani, Y. Li, Nucleic acid aptamers and enzymes as sensors, Current Opinion in Chemical Biology 10 (3) (2006) 272–281. Available from https://doi.org/10.1016/j.cbpa.2006.04.003.

[115] S.W. Santoro, G.F. Joyce, A general purpose RNA-cleaving DNA enzyme, Proceedings of the National Academy of Sciences of the United States of America 94 (9) (1997) 4262–4266. Available from https://doi.org/10.1073/pnas.94.9.4262.

[116] A. Panjkovich, F. Melo, Comparison of different melting temperature calculation methods for short DNA sequences, Bioinformatics (Oxford, England) 21 (6) (2005) 711–722. Available from https://doi.org/10.1093/bioinformatics/bti066, http://bioinformatics.oxfordjournals.org/.

[117] P. Wu, S.I. Nakano, N. Sugimoto, Temperature dependence of thermodynamic properties for DNA/DNA and RNA/DNA duplex formation, European Journal of Biochemistry 269 (12) (2002) 2821–2830. Available from https://doi.org/10.1046/j.1432-1033.2002.02970.x.

[118] H.K. Kim, I. Rasnik, J. Liu, T. Ha, Y. Lu, Dissecting metal ion-dependent folding and catalysis of a single DNAzyme, Nature Chemical Biology 3 (12) (2007) 763–768. Available from https://doi.org/10.1038/nchembio.2007.45, http://www.nature.com/nchembio.

[119] X.B. Zhang, R.M. Kong, Y. Lu, Metal ion sensors based on DNAzymes and related DNA molecules, Annual Review of Analytical Chemistry 4 (2011) 105–128. Available from https://doi.org/10.1146/annurev.anchem.111808.073617.

[120] J. Li, Y. Lu, A highly sensitive and selective catalytic DNA biosensor for lead ions [9], Journal of the American Chemical Society 122 (42) (2000) 10466–10467. Available from https://doi.org/10.1021/ja0021316.

[121] J. Liu, Y. Lu, Improving fluorescent DNAzyme biosensors by combining inter- and intramolecular quenchers, Analytical Chemistry 75 (23) (2003) 6666–6672. Available from https://doi.org/10.1021/ac034924r.

[122] X.H. Zhao, L. Gong, X.B. Zhang, B. Yang, T. Fu, R. Hu, et al., Versatile DNAzyme-based amplified biosensing platforms for nucleic acid, protein, and enzyme activity detection, Analytical Chemistry 85 (7) (2013) 3614–3620. Available from https://doi.org/10.1021/ac303457u.

[123] P. Travascio, Y. Li, D. Sen, DNA-enhanced peroxidase activity of a DNA aptamer-hemin complex, Chemistry and Biology 5 (9) (1998) 505–517. Available from https://doi.org/10.1016/S1074-5521(98)90006-0, http://www.cell.com.

[124] C. Teller, S. Shimron, I. Willner, Aptamer-DNAzyme hairpins for amplified biosensing, Analytical Chemistry 81 (21) (2009) 9114–9119. Available from https://doi.org/10.1021/ac901773bIsrael, http://pubs.acs.org/doi/pdfplus/10.1021/ac901773b.

[125] H. Li, J. Chang, T. Hou, F. Li, HRP-mimicking DNAzyme-catalyzed in situ generation of polyaniline to assist signal amplification for ultrasensitive surface plasmon resonance biosensing, Analytical Chemistry 89 (1) (2017) 673–680. Available from https://doi.org/10.1021/acs.analchem.6b02988, http://pubs.acs.org/journal/ancham.

[126] A.D. Whetton, G.W. Preston, S. Abubeker, N. Geifman, Proteomics and informatics for understanding phases and identifying biomarkers in COVID-19 disease, Journal of Proteome Research 19 (11) (2020) 4219–4232. Available from https://doi.org/10.1021/acs.jproteome.0c00326, http://pubs.acs.org/journal/jprobs.

[127] S.T. Dekosky, P.M. Kochanek, A.B. Valadka, R.S.B. Clark, S.H.Y. Chou, A.K. Au, et al., Blood biomarkers for detection of brain injury in COVID-19 patients, Journal of Neurotrauma 38 (1) (2021) 1–43. Available from https://doi.org/10.1089/neu.2020.7332, www.liebertonline.com/neu.

[128] X. Fang, X. Liu, S. Schuster, W. Tan, Designing a novel molecular beacon for surface-immobilized DNA hybridization studies, Journal of the American Chemical Society 121 (12) (1999) 2921–2922. Available from https://doi.org/10.1021/ja9837809, http://pubs.acs.org/journal/jacsat.

[129] C. Fan, K.W. Plaxco, A.J. Heeger, Electrochemical interrogation of conformational changes as a reagentless method for the sequence-specific detection of DNA, Proceedings of the National Academy of Sciences of the United States of America 100 (16) (2003) 9134–9137. Available from https://doi.org/10.1073/pnas.1633515100.

[130] E. Xiong, Z. Li, X. Zhang, J. Zhou, X. Yan, Y. Liu, et al., Triple-helix molecular switch electrochemical ratiometric biosensor for ultrasensitive detection of nucleic acids, Analytical Chemistry 89 (17) (2017) 8830–8835. Available from https://doi.org/10.1021/acs.analchem.7b01251, http://pubs.acs.org/journal/ancham.

[131] A.W. Peterson, R.J. Heaton, R.M. Georgiadis, The effect of surface probe density on DNA hybridization, Nucleic Acids Research 29 (24) (2001) 5163–5168. Available from https://doi.org/10.1093/nar/29.24.5163.

[132] T. Hou, W. Li, X. Liu, F. Li, Label-free and enzyme-free homogeneous electrochemical biosensing strategy based on hybridization chain reaction: a facile, sensitive, and highly specific microRNA assay, Analytical Chemistry 87 (22) (2015) 11368–11374. Available from https://doi.org/10.1021/acs.analchem.5b02790, http://pubs.acs.org/journal/ancham.

[133] Z. Chen, S. Chengjun, L. Zewei, L. Kunping, Y. Xijian, Z. Haimin, et al., Fiber optic biosensor for detection of genetically modified food based on catalytic hairpin assembly reaction and nanocomposites assisted signal amplification, Sensors and Actuators, B: Chemical 254 (2018) 956–965. Available from https://doi.org/10.1016/j.snb.2017.07.174.

[134] Y. Liu, S. Li, L. Zhang, Q. Zhao, N. Li, Y. Wu, Catalytic hairpin assembly-assisted rolling circle amplification for high-sensitive telomerase activity detection, ACS Omega 5 (20) (2020) 11836–11841. Available from https://doi.org/10.1021/acsomega.0c01459, pubs.acs.org/journal/acsodf.

[135] F. Long, A. Zhu, H. Shi, Recent advances in optical biosensors for environmental monitoring and early warning, Sensors 13 (10) (2013) 13928–13948. Available from https://doi.org/10.3390/s131013928.

[136] Q. Hu, J. Wu, L. Chen, X. Lou, F. Xia, Recent development of DNA-modified AIEgen probes for biomedical application, Chemical Research in Chinese Universities 37 (1) (2021) 66–72. Available from https://doi.org/10.1007/s40242-021-0388-z, http://www.springer.com/chemistry/journal/40242.

[137] J. Wang, Survey and summary: from DNA biosensors to gene chips, Nucleic Acids Research 28 (16) (2000) 3011–3016. Available from https://doi.org/10.1093/nar/28.16.3011.

[138] J. Canoura, H. Yu, O. Alkhamis, D. Roncancio, R. Farhana, Y. Xiao, Accelerating post-SELEX aptamer engineering using exonuclease digestion, Journal of the American Chemical Society 143 (2) (2021) 805–816. Available from https://doi.org/10.1021/jacs.0c09559.

[139] Y. Liu, N. Tuleouva, E. Ramanculov, A. Revzin, Aptamer-based electrochemical biosensor for interferon gamma detection, Analytical Chemistry 82 (19) (2010) 8131–8136. Available from https://doi.org/10.1021/ac101409t.

[140] Y. Chen, T.S. Pui, P. Kongsuphol, K.C. Tang, S.K. Arya, Aptamer-based array electrodes for quantitative interferon-γ detection, Biosensors and Bioelectronics 53 (2014) 257–262. Available from https://doi.org/10.1016/j.bios.2013.09.046.

[141] H.M. Pan, S. Gonuguntla, S. Li, D. Trau, 3.33 Conjugated polymers for biosensor devices, Comprehensive Biomaterials II, Elsevier, Singapore, (2017), pp. 716–754. Available from http://doi.org/10.1016/B978-0-12-803581-8.10144-4, http://doi.org/10.1016/B978-0-12-803581-8.10144-4.

[142] A.V. Kabanov, P.L. Felgner, L.W. Seymour, Self-assembling complexes for gene delivery, Laboratory and Clinical Trials (1998) 197–218. Available from https://doi.org/10.1136/mp.55.3.207-b.

[143] H.A. Ho, M. Boissinot, M.G. Bergeron, G. Corbeil, K. Doré, D. Boudreau, et al., Colorimetric and fluorometric detection of nucleic acids using cationic polythiophene derivatives, Angewandte Chemie - International Edition 41 (9) (2002) 1548–1551. Available from https://doi.org/10.1002/1521-3773(20020503)41:9<1548::AID-ANIE1548>3.0.CO;2-I.

[144] H.A. Ho, M. Leclerc, Optical sensors based on hybrid aptamer/conjugated polymer complexes, Journal of the American Chemical Society 126 (5) (2004) 1384–1387. Available from https://doi.org/10.1021/ja037289f.

[145] J. Liu, Adsorption of DNA onto gold nanoparticles and graphene oxide: surface science and applications, Physical Chemistry Chemical Physics 14 (30) (2012) 10485–10496. Available from https://doi.org/10.1039/c2cp41186e.

[146] S.H. Yu, C.S. Lee, T.H. Kim, Electrochemical detection of ultratrace lead ion through attaching and detaching DNA aptamer from electrochemically reduced graphene oxide electrode, Nanomaterials 9 (6) (2019). Available from https://doi.org/10.3390/nano9060817, https://www.mdpi.com/2079-4991/9/6/817/pdf.

[147] D. Kang, A. Vallée-Bélisle, A. Porchetta, K.W. Plaxco, F. Ricci, Re-engineering electrochemical biosensors to narrow or extend their useful dynamic range, Angewandte Chemie - International Edition 51 (27) (2012) 6717–6721. Available from https://doi.org/10.1002/anie.201202204.

[148] H. Satış, H.S. Özger, P. Aysert Yıldız, K. Hızel, Ö. Gulbahar, G. Erbaş, et al., Prognostic value of interleukin-18 and its association with other inflammatory markers and disease severity in COVID-19, Cytokine 137 (2021). Available from https://doi.org/10.1016/j.cyto.2020.155302, https://www.sciencedirect.com/journal/cytokine/issues.

[149] Y. Lu, X. Li, L. Zhang, P. Yu, L. Su, L. Mao, Aptamer-based electrochemical sensors with aptamer–complementary DNA oligonucleotides as probe, Analytical Chemistry 80 (6) (2008) 1883–1890. Available from https://doi.org/10.1021/ac7018014.

[150] S. Ullah, A.A. Khalil, F. Shaukat, Y. Song, Sources, extraction and biomedical properties of polysaccharides, Foods 8 (8) (2019). Available from https://doi.org/10.3390/foods8080304, https://www.mdpi.com/2304-8158/8/8/304/pdf.

[151] A.S.A. Mohammed, M. Naveed, N. Jost, Polysaccharides: classification, chemical properties, and future perspective applications in fields of pharmacology and biological medicine (a review of current applications and upcoming potentialities), Journal of Polymers and the Environment 29 (8) (2021) 2359–2371. Available from https://doi.org/10.1007/s10924-021-02052-2, http://www.kluweronline.com/issn/1566-2543/.

[152] N. Jabeen, M. Atif, Polysaccharides based biopolymers for biomedical applications: a review, Polymers for Advanced Technologies 35 (1) (2024). Available from https://doi.org/10.1002/pat.6203, http://onlinelibrary.wiley.com/journal/10.1002/(ISSN)1099-1581.

[153] M. Bilal, H.M.N. Iqbal, Naturally-derived biopolymers: potential platforms for enzyme immobilization, International Journal of Biological Macromolecules 130 (2019) 462–482. Available from https://doi.org/10.1016/j.ijbiomac.2019.02.152, www.elsevier.com/locate/ijbiomac.

[154] M. Luo, X. Zhang, J. Wu, J. Zhao, Modifications of polysaccharide-based biomaterials under structure-property relationship for biomedical applications, Carbohydrate Polymers 266 (2021) 118097. Available from https://doi.org/10.1016/j.carbpol.2021.118097.

[155] Z. Behrooznia, J. Nourmohammadi, Polysaccharide-based materials as an eco-friendly alternative in biomedical, environmental, and food packaging, Giant 19 (2024) 100301. Available from https://doi.org/10.1016/j.giant.2024.100301.

[156] M. Rinaudo, Chitin and chitosan: properties and applications, Progress in Polymer Science 31 (7) (2006) 603–632. Available from https://doi.org/10.1016/j.progpolymsci.2006.06.001.

[157] X. Guo, Y. Wang, Y. Qin, P. Shen, Q. Peng, Structures, properties and application of alginic acid: a review, International Journal of Biological Macromolecules 162 (2020) 618–628. Available from https://doi.org/10.1016/j.ijbiomac.2020.06.180, www.elsevier.com/locate/ijbiomac.

[158] S. Sharmeen, M.S. Rahman, M.M. Islam, M.S. Islam, M. Shahruzzaman, A.K. Mallik, et al., Application of polysaccharides in enzyme immobilization, Functional Polysaccharides for Biomedical Applications, Elsevier, Bangladesh, (2019), pp. 357–395. Available from http://doi.org/10.1016/B978-0-08-102555-0.00011-X, http://www.sciencedirect.com/science/book/9780081025550.

[159] D.S. Kim, X. Yang, J.H. Lee, H.Y. Yoo, C. Park, S.W. Kim, et al., Development of GO/Co/chitosan-based nano-biosensor for real-time detection of D-glucose, Biosensors 12 (7) (2022). Available from https://doi.org/10.3390/bios12070464, https://www.mdpi.com/2079-6374/12/7/464/pdf?version=1656336861.

[160] L.S. Ferraraccio, D. Di Lisa, L. Pastorino, P. Bertoncello, Enzymes encapsulated within alginate hydrogels: bioelectrocatalysis and electrochemiluminescence applications, Analytical Chemistry 94 (46) (2022) 16122–16131. Available from https://doi.org/10.1021/acs.analchem.2c03389, http://pubs.acs.org/journal/ancham.

[161] A. Mobed, F. Kohansal, S. Dolati, M. Hasanzadeh, A novel immuno-device based on the specific binding of AuNP-supported CTAB with biotinylated antibody of hyaluronic acid toward an early-stage recognition of a biomarker: a bioanalytical assay in real samples using disposal biosensor technology, RSC Advances 12 (44) (2022) 28473–28488. Available from https://doi.org/10.1039/d2ra04984h, http://pubs.rsc.org/en/journals/journal/ra.

[162] M. Das, C. Dhand, G. Sumana, A.K. Srivastava, R. Nagarajan, L. Nain, et al., Electrophoretic fabrication of chitosan-zirconium-oxide nanobiocomposite platform for nucleic acid detection, Biomacromolecules 12 (3) (2011) 540–547. Available from https://doi.org/10.1021/bm1013074.

[163] L.F. de Lima, A.L. Ferreira, I. Ranjan, R.G. Collman, W.R. de Araujo, C. de la Fuente-Nunez, A bacterial cellulose-based and low-cost electrochemical biosensor for ultrasensitive detection of SARS-CoV-2, Cell Reports Physical Science 4 (8) (2023). Available from https://doi.org/10.1016/j.xcrp.2023.101476, https://www.cell.com/cell-reports-physical-science/aims.

[164] H. Safarpour, H. Majdi, A. Masjedi, A.S. Pagheh, M.D.L. Pereira, S.M. Rodrigues Oliveira, et al., Development of optical biosensor using protein A-conjugated chitosan–gold nanoparticles for diagnosis of cystic echinococcosis, Biosensors 11 (5) (2021) 134. Available from https://doi.org/10.3390/bios11050134.

[165] J.Q. Brown, R. Srivastava, M.J. McShane, Encapsulation of glucose oxidase and an oxygen-quenched fluorophore in polyelectrolyte-coated calcium alginate microspheres as optical glucose sensor systems, Biosensors and Bioelectronics 21 (1) (2005) 212–216. Available from https://doi.org/10.1016/j.bios.2004.08.020.

[166] P. Malik, V. Katyal, V. Malik, A. Asatkar, G. Inwati, T.K. Mukherjee, Nanobiosensors: concept and variations, ISRN Nanomaterials (2013).

[167] P. Malik, R. Gupta, V. Malik, R.K. Ameta, Emerging nanomaterials for improved biosensing, Measurement: Sensors 16 (2021). Available from https://doi.org/10.1016/j.measen.2021.100050, www.journals.elsevier.com/measurement-sensors.

[168] P. Malik, A. Sangwan, Nanotechnology: a tool for improving efficiency of bio-energy, Computers & Applied Sciences 1 (2012) 37–49.

[169] P. Malik, T.K. Mukherjee, Recent advances in gold and silver nanoparticle based therapies for lung and breast cancers, International Journal of Pharmaceutics 553 (1-2) (2018) 483–509. Available from https://doi.org/10.1016/j.ijpharm.2018.10.048, www.elsevier.com/locate/ijpharm.

[170] P. Malik, R.K. Ameta. Recent Progress in Au nanoparticle treatment of lung cancers in Biomedical Engineering and its Applications in Healthcare. Springer Nature Publications, pp. 978–981.

Chapter 20

Functional Biomaterials for Biosensing Technologies

Ramesh Kumar[1], Muskan[1], Teena Gakhar[2], and Suman Nagpal[3]

[1]*Department of Biotechnology, Indira Gandhi University Meerpur, Rewari, Haryana, India,* [2]*University School of Information and Communication Technology, GGSIPU, New Delhi, India,* [3]*Environmental Sciences, Indira Gandhi University, Rewari, Haryana, India*

20.1 Introduction

We are regularly confronted with the necessity to identify and monitor a diverse array of substances, including pathogenic organisms and high-value biochemical compounds, for various applications in healthcare, security, environmental, and industrial fields. Biosensors have emerged as an indispensable part of healthcare and environmental monitoring due to growing demands. As the demand for biosensors continues to grow, the development of suitable biomaterials has become a crucial aspect of the research and development in this field, which is designed to specifically and sensibly detect biological molecules [1]. The biosensor stands at the intersection of biomaterials science and sensor technology, representing among the most active and promising areas in modern Biomedical Engineering [2].

The advancement in materials science, biotechnology, and engineering has developed biomaterials from rudimentary implantable substances (e.g., metals) to sophisticated biosensing platforms. Biomaterials, the engineered substances designed to interact with biosensors, have revolutionized the landscape of biosensor development over the past several decades. Such materials can be biopsied in the wild or synthesized in laboratories with diverse strategies, including metals, polymers, ceramics, or composite materials. They can be customized for particular biosensing applications, such as the realms of medical diagnostics or environmental monitoring [3] (Fig. 20.1).

Biomaterials play multiple roles in biosensors. Their contributions include

- **Immobilization matrices:** Imparting stable support for biological recognition elements.
- **Signal transduction:** Functioning as mediators of biological signals and translating these signals into quantifiable readouts.
- **Increasing sensitivity:** Enhancing signals to identify low concentrations of analytes.

Therefore by integrating biomaterials, biosensors can be made more useful (in terms of portability, cost-effectiveness, and ease of use) while maintaining high performance making them indispensable in diverse fields.

20.2 Classification of biomaterials

Biomaterials are classified into two categories, that is, natural and synthetic, based on their origin and properties, where each offers unique advantages for biosensor applications.

- Natural biomaterials: Biocompatible and biodegradable solutions

Natural biomaterials come from biological sources like plants, animals, or other living organisms and have unique properties that make them highly attractive for biosensor applications [4]. Such components contain proteins, polysaccharides, and various organic materials that are usually biocompatible and biodegradable [5]. Some of the commonly used natural biomaterials in biosensors include

Fundamentals of Translational Tissue Engineering. DOI: https://doi.org/10.1016/B978-0-443-33455-9.00010-9

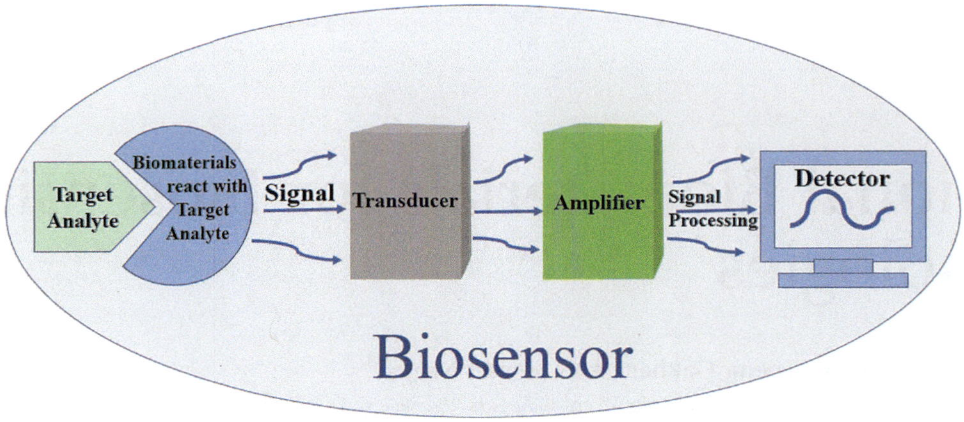

FIGURE 20.1 **Structure and operating principle of biosensor.** Working principle of biosensors using biomaterials for target analyte detection.

- **Proteins**: Proteins are commonly used in biosensors due to their ability to provide structural stability and compatibility [6]. Often used proteins include collagen, fibrin, and silk fibroin that provide structural stability and biocompatibility. Enzymes (mainly proteins) are natural catalysts that have an important role in biosensors as they increase the speed of reaction, thus enabling rapid and specific detection of biomolecules.
- **Polysaccharides**: Chitosan, alginate, and cellulose are other materials supported by their hydrophilicity, which supports the gel formation and immobilization of biomolecules on biosensors. As an example, chitosan is commonly applied in biosensors because of its film-forming capabilities and biocompatible characteristics.
- **Nucleic acids:** As DNA and RNA can hybridize with complementary sequences, they are used as recognition elements for detecting specific genetic sequences or pathogens [7]. Biosensors based on DNA are pivotal for disease diagnostics and environmental monitoring.
- **Synthetic biomaterials: Engineered for stability and functionality**

Synthetic biomaterials are designed and produced in the laboratory, facilitating predefined properties such as mechanical strength, stability, and functionality. Biosensors that use synthetic materials include

- **Polymers:** Synthetic polymers such as polyethylene glycol (PEG) and polyvinyl alcohol provide flexibility, durability, and chemical resistance. Conductive polymers such as polyaniline and polypyrrole have been employed in biosensors as signal amplifiers owing to their conductivity. PEG is a polymer used commonly for enhancing biocompatibility and reducing fouling.
- **Metal nanoparticles:** Metals such as gold, platinum, and silver are used in biosensors because of their exceptional catalytic and conductivity properties. In particular, gold nanoparticles (AuNPs) are widespread among all kinds of sensors due to their high surface area and the ability of this kind of nanoparticle to be functionalized with biomolecules as well as enhancing biosensor sensitivity and stability.
- **Ceramics:** Zirconia and alumina ceramics are the most stable, biocompatible, and wear-resistive biosensors, which can last for a long time period. Due to their structural characteristics and thermal stability, they can be utilized in harsh-space applications such as industrial implementations and environmental monitoring sensors.
- **Composites:** Composites are two or more materials that have been combined to achieve desirable material properties that cannot be achieved at a reasonable cost by any single material. For example, the cooccurrence of polymers with nanomaterials (polymer-metal composites, etc.) can enhance conductivity and biocompatibility [1] (Fig. 20.2).

20.3 Biosensor comparison and selection for applications

Advancements in biosensors require both natural and synthetic biomaterials. The performance, sensitivity, and durability of biosensors depend largely on the type of biomaterials selection. This table presents a comparative analysis to help choose suitable biomaterials for different applications considering factors (Table 20.1).

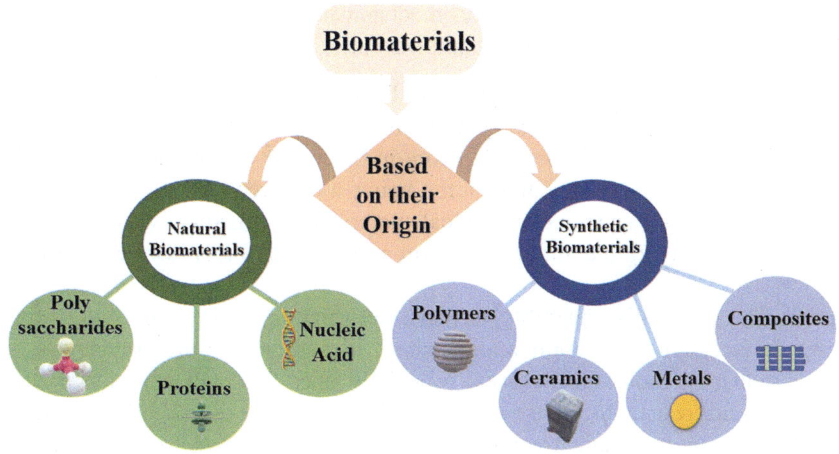

FIGURE 20.2 **Classification of biomaterials.** Overview of biomaterials classified based on their origin and composition.

TABLE 20.1 Comparison of natural and synthetic biomaterials.

Aspect	Natural biomaterials	Synthetic biomaterials
Source	Derived from living organisms (e.g., collagen, silk, chitosan)	Artificially developed materials (e.g., polymers, metals, ceramics)
Biocompatibility	High, since they are modeled after native biological structures	Moderate to high, difference in design and coating being a determining factor
Biodegradability	Primary biodegradable	Possibly biodegradable or nonbiodegradable
Mechanical strength	Tunable but generally lower strength	Strong, resistant enough for structural applications
Immunogenicity	Unlikely to provoke immune reactions	Result in inflammatory or immune responses
Customization	Limited, as properties rely on natural source	Very customizable in terms of properties and shapes
Production	Difficult to extract and usually expensive	More simple and scalable manufacturing
Examples	Collagen, chitosan, alginate	Polylactic acid (PLA), titanium, hydroxyapatite
Applications	Tissue engineering, wound healing, drug delivery	Orthopedic implants, dental restorations, drug delivery systems

20.4 Comparison of natural and synthetic biomaterials

Synthetic biomaterials and natural biomaterials have their respective pros and cons, so the selection of biomaterials is application-dependent. Natural materials have superior biocompatibility and biodegradability compared to their synthetic counterparts, making them the perfect candidates for medical applications where a reduced immune response is critical. On the other hand, synthetic materials offer better durability and can be customized for long-term and industrial applications. Advances in composite biomaterials are filling in these gaps by providing customized solutions that combine advantages of both types of biomaterials (Table 20.1).

20.5 Core properties for biomaterial selection

The selection of biomaterials for biosensors is a critical step in ensuring their performance, accuracy, and durability. Key material characteristics affecting performance, ease of use, and longevity in vitro or in vivo impact the value/redemption of biosensors. Such characteristics are biocompatibility, mechanical strength, chemical stability, and sensitivity [3]. A brief description of each of the features that matter to biosensors:

- **Biocompatibility**: The design and development of effective biosensors require careful consideration of various material properties to ensure accurate performance especially in the medical field where they interface with sensitive biological materials. So, biocompatibility is one of the major factors in biosensor design as they should not produce unwanted tissue responses or interfere with the biological systems being monitored (De) [8]. For biosensors, which are often interfaced with blood, cells, or other sensitive biological materials in a medical context, high biocompatibility is an absolute necessity. Maximized negligible interference or a biocompatible biosensor ensures reliable and reading accuracy [9].
- **Mechanical properties**: The material's capability to resist deformation under many constrains such as compressing, stretching, or bending. It is essential in applications where biosensors require robustness, such as for wearables, implantable sensors, or industrial monitors. Sensors, especially ones that will experience some form of mechanical stress, require materials that are resistant to these exact forces so that they do not break or deform and can perform as expected over time. Due to their mechanical robustness, metals and ceramics are typically utilized in the structural components of biosensors.
- **Chemical stability:** When applied for bio detection, biosensors are harbored in the variable chemical environments like biological fluids (blood or saliva) or beyond body polluted environment. The stability in a chemical sense allows the biosensor to retain its integrity and deliver a constant performance in these ambient environments.
- **Sensitivity and specificity:** The nature of biosensor analytes must be sensitive and selective to detect biological molecules, pathogens, or used chemicals even in complex mixtures as they form an essential portion for analyzing samples efficiently. Due to their specific nature for target molecules, biomaterials such as enzymes, antibodies, and aptamers are widely utilized as bioreceptors in biosensors [10]. They additionally include conductive polymers and nanomaterials that are designed to intensify the stimulus signal output [11]. Especially conductive polymers, including polyaniline (PANI), polypyrrole (PPy), poly(3,4-ethylenedioxythiophene) (PEDOT), and polythiophene, are beneficial to the electrochemical biosensors since they can conduct electrical currents and well combine with biological molecules [12].
- **Real-time detection:** Biosensors are based on biomaterials that react with the target analyte in real-time and therefore can give an immediate response as feedback from the sensor. Biocomponent molecules like antibodies and aptamers are bound to their suitable target in a short time, and conductive materials and nanomaterials favor fast signal transduction; thus response times for biosensors remain instantaneous [13]. Durable conductive polymers such as PEDOT and PPy are used in flexible biosensors where the physical movement poses a challenge to the performance of the device over time [14] (Fig. 20.3).

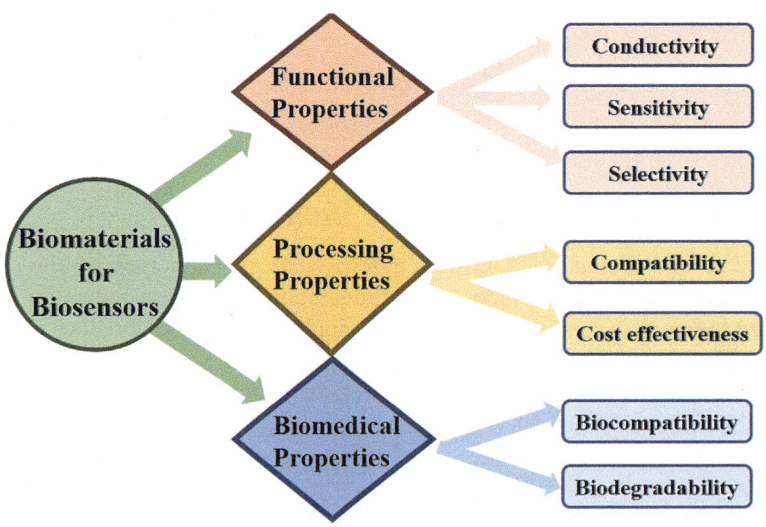

FIGURE 20.3 **Core properties for the selection of biomaterial.** Overview of essential properties of biomaterials categorized into functional, processing, and biomedical properties.

20.6 Various biomaterials used in biosensors

The performance and specificity of biosensors are heavily influenced by the biomaterials employed. Below is a categorized overview of the most common biomaterials used:

- **Biological recognition elements**
- **Enzymes**: Enzymes are biological catalysts, accelerating reactions to produce measurable outputs like color changes, pH shifts, or conductivity alterations. Examples include the cholesterol biosensor (cholesterol oxidase) and the lactate sensor (lactate oxidase).

Pros: Enzymes have high substrate specificity, readily allowing selective detection. This generates fast, measurable responses that are perfect for real-time tracking.

- **Antibodies:** These are proteins made by the immune system that specifically bind to antigens. Due to the high specificity and the inherent affinities of antibodies to a given target, such as proteins, toxins, or pathogens, they are often used as recognition elements in biosensors.

COVID-19: Antibodies have been used in COVID biosensors to detect viral proteins.

Pros: Antibody-based biosensors offer high specificity, allowing precise discrimination between molecules that differ by only a small functional group. They can be designed to specifically bind many different things; therefore they are versatile in the applications for biosensing.

- **RNA/DNA aptamer:** Aptamers are short, single-stranded DNA or RNA sequences that fold up to evoke 3D geometries enabling specificity for molecular targets such as proteins, small molecules, and even whole cells [15]. Aptamers have a similar role as antibodies but are chemically synthesized with some advantages [16]. Aptamers are used in biosensors for cancer diagnosis, toxin detection, or viral RNA identification.

Pros: Aptamers can be synthesized to bind virtually any molecule and are typically stable and cheaper than antibodies. They are also straightforward to modify for improved efficiency in a range of climate situations.

- **Peptides**: Peptides are short chains of amino acids that can either imitate binding properties of proteins or create a specific recognition site for target molecules [17]. Peptide-based biosensors can be used to detect toxins, pathogens, and biomarkers in medical diagnosis and environmental monitoring [18]. These peptides may be tuned for biosensors to detect proteins, hormones, or other analytes and are generally far more stable than biomolecules found in nature especially when subjected to diverse or extreme conditions.

Pros: Advantages of peptides include stability, ease of synthesis, and modifiability [17]. These synthetic structures can be tailored to bind specific targets and are more stable than proteins in extreme conditions leading to various applications for biosensors.

- **Structural and functional elements**
- **Proteins**—Proteins may also serve as recognition elements in biosensors besides enzymes and antibodies, such as receptors or binding proteins [1]. That is because these proteins are often involved in sensing specific molecules and have sites for this that arise naturally. Glucose-binding proteins in glucose sensors and ion channels or transport proteins in biosensors detecting neurotransmitters or ions involved in neuronal studies [18]. Others have been engineered to detect pH or metabolic indicators.

Pros: Proteins can provide specificity without significant modification. Certain proteins are extremely stable and can be compatible with multiple sensor designs, making protein-based biosensors attractive in this regard.

- **Cell-based materials:** Cell-based biosensors utilize whole cells (e.g., bacteria, yeast, and mammalian cells) as sensing elements for various target biological or chemical species. Cells are capable of producing a measurable reaction, such as fluorescence or bioluminescence, or a change in metabolic activity, on exposure to specific analytes. Cell-based biosensors are most commonly employed in toxicity screening, environmental monitoring, and drugs discovery. For example, the detection of pollutants using biosensors with bacterial cells and drug interaction studies using mammalian cells.

Pros: Cell-based sensors are widely capable of detecting multiple analytes with functional responses to diverse stimuli such as complex mixtures. Moreover, they can replicate in vivo scenarios thus making them ideal for applications that have realistic biological interactions.

20.7 Advancements in biomaterial technology that could enhance biosensor capabilities

Development of new generation biosensors is a growing trend due to advancements in biomaterial technology possessing the most prominent and specific features with excellent stability and sensitivity. This is improving the efficiency of biosensors in medical diagnostics, environmental monitoring, food safety, and other applications. In other words, these are some of the most interesting areas of development in biomaterials, which will likely improve biosensors' capabilities:

1. **Nanomaterials with engineered features/functions**
   - **Surface functionalization:** Nanomaterials such as gold nanoparticles, carbon nanotubes (CNTs), and graphene are being surface functionalized with different chemical groups or biomolecules to enhance their specificity towards target analytes [10]. Graphene oxide functionalized for target biomolecule detection shows high specificity, indicating potential application in the early stage diagnosis of diseases [19].
   - **Two-dimensional (2D) materials:** The new 2D nanomaterials (aka. TMDs, MXenes) provide a high surface area coupled with good conduction properties, adjustable electronic characteristics. Such materials will improve the signal transduction in biosensors, typically in situations where the detection of low levels of analyte is necessary.
   - **Hybrid nanocomposites:** The association of nanomaterials such as graphene with metallic nanoparticles, polymers, or enzymes gives place to hybrid structures that increase the sensitivity, specificity, and stability. For example, the composites of graphene and gold nanoparticles can offer practical use in electrochemical sensors due to their high nature for low-concentration biomarkers facilitated through their ability to rapidly transfer electrons.
   - **Miniaturization of biosensors:** Nanomaterials are also crucial for the miniaturization of biosensors so that they lose their functions but are preferable to use in mobile, wearable, or infiltrating devices. Nanomaterials enable the miniaturization of highly sensitive sensing components with high surface area and conductivity (Ofek) [20], allowing more enhancement in detectivity while maintaining low-power operation with compact hardware. CNTs and graphene, for example, form ultrathin and stretchable electrodes that can be incorporated into compact biosensor structures without compromising their electrical conductivity and detection sensitivity.
2. **Synthetic bioreceptors**
   - **Molecularly imprinted polymers (MIPs):** These are synthetic polymers that have had molecular imprints added to them for target analytes, enabling the recognition and binding of specific molecules. Molecularly imprinted polymers (MIPs) are the best synthetic and bio-mimic polymer to mimic natural recognition elements like antibody but are more stable, cheaper, and resistant to harsh conditions. This is highly helpful in developing biosensors. They have very important applications when it comes to environmental and industrial monitoring (De) [8].
3. **Degradable and biocompatible materials**
   - **Disposable polymers:** Various novel biodegradable polymers such as polylactic acid (PLA), polycaprolactone (PCL), and poly(lactic-co-glycolic acid) (PLGA) for biosensors intended to be used in implantable devices [1]. These materials slowly perish in the body, and interference surgery is not required, so side effects are less likely to occur.
   - **Hydrogels for stabilization of biomolecules:** Due to their well-porous and water-retentive property, hydrogels are the most preferable material for encapsulating enzymes, antibodies, and cells in biosensors. The advancement of hydrogel technology has made it possible to achieve stabilizing capacity for immobilized biomolecules to enable long-term and dependable biosensing applications in clinical diagnostics, food safety, and environmental monitoring.
4. **Self-healing and responsive biomaterials**
   - **Self-healing polymers:** An exciting development in biosensors for wearable or implantable applications is self-healing polymers that can recover from mechanical wound damage. Such materials prolong the lifetime of biosensors, thus allowing for continuous health monitoring where repeated replacement of sensors is not appropriate.
   - **Stimuli-responsive materials**: A responsive material that reacts in a certain way with external stimuli based on the kind of tendency it may have (pH, temperature, or even light). For example, pH-sensitive polymers can change their conformation in response to changes in acidity, allowing for the real-time detection of dynamic processes with biological fluids or environmental samples. These tunable materials facilitate biosensor sensitivity and signal amplification when required.
5. **3D printing and microfabrication for customizable biosensors**
   - **3D-printed biomaterials**: 3D-printed biomaterials can be designed with complex, tunable architectures to enhance biosensor function (e.g., binding of target molecules) as polymer, hydrogel, and composite. This method is especially useful to develop user-oriented biosensors for dedicated applications like wearables adjusted towards the particular anatomical arrangement of a person [21,22].

- **Microfluidic integration:** Advanced microfabrication techniques are enabling the combination of biosensors with microfluidic channels for the manipulation and processing of small sample volumes. It allows for relatively small lab-on-a-chip devices capable of multiplexed detection of biomarkers and more complex analyses integrated in one chip, which increases the compactness and functionality of biosensors.

6. **Signal amplification and detection approaches**
   - **Enzyme cascades and nano-catalysts:** Enzyme cascades and nano-catalysts (e.g., platinum or palladium nanoparticles) are improving signal amplification by speeding up reactions to generate detectable signals amiability for rapid detection of rare markers during late-stage cancer diagnostics [23]. These innovations are clearly very useful when one needs to make biosensors several folds more sensitive to low concentrations of an analyte.
   - **Surface plasmon resonance (SPR)**: Improvements in SPR and localized SPR (LSPR) methods using gold nanoparticles or nanorods allow the biosensors to detect molecular interactions with high sensitivity. Improving optics biosensor performances with those methods could permit detection of interactions at the nanoscales, useful for protein–protein interaction study, and DNA hybridization (Fig. 20.4).

Applications:

- **Early disease detection:**
  The sensitivity and selectivity of the biosensors have been improved by Biomaterials which can be used for the early detection of the disease. Noble nanomaterials such as gold nanoparticles and quantum dots enhance signal amplification to the extent that they can detect trace amounts of disease biomarkers, such as those for cancer, cardiovascular disorders, and infectious diseases, in blood or tissue samples at concentrations as low as picograms per milliliter. For example, DNA aptamer-based biosensors can detect genetic mutations or the viral RNA sequences pertinent to certain diseases, facilitating early intervention. Furthermore, to be stable with the aid of biomaterials, biosensor components are suitable for real-time monitoring and rapid diagnostics for clinical and point-of-care.
  CNTs and graphene-based biosensors are used in the detection of biomarkers troponin and myoglobin that are released into circulation during heart attacks. Identifying these biomarkers before the onset of clinical symptoms provides an opportunity for rapid diagnosis and effective medical management, helping to save lives from cardiovascular diseases.
- **Pollution detection by biosensor:**
- Detections of pollutants, including heavy metals, pesticides, organic compounds, and toxins, are important for environmental protection and human health.
- **Detection of gaseous pollutants**: Gaseous pollutants like nitrogen dioxide (NO_2), sulfur dioxide (SO_2), carbon monoxide, and volatile organic compounds (VOCs) are frequent in overlap urban and industrial regions, so advanced biosensors for detecting these gases have been developed. These sensors use materials such as graphene and metal oxide nanoparticles, whose electrical resistance changes as the concentration of a specific gas changes. Thus with these sensors, real-time quantification of pollution levels and air quality analysis can be done.
- **Heavy metal detection**: Biosensors are specifically designed to be able to detect heavy metals such as lead, mercury, cadmium, and arsenic from water sources, preventing contamination of drinking water [24]. The use of conductive nanomaterials such as gold nanoparticles and CNTs has increased the sensitivity of these sensors in trace-level detection.

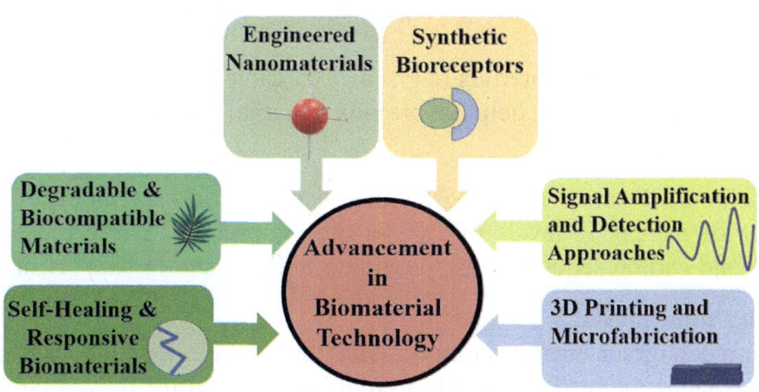

FIGURE 20.4 **Advancements in biomaterial technology enhance biosensor capabilities.** Recent advancements in biomaterial technology, including signal amplification, 3D printing, and microfabrication techniques for biomedical applications.

Particularly in regions subjected to mining or industrial processes, these biosensors can offer real-time insights into the safety of water systems [25].

- **Water quality monitoring**
- **Pesticide and herbicide detection**: Enzyme-based biosensors, such as immobilized acetylcholinesterase on nanomaterials, can have trace levels of pesticides and herbicides monitored even from agricultural runoff. They form a cornerstone of our ability to monitor and manage chronic pollution in rivers, lakes, and groundwater, safeguarding ecosystem health as well as drinking water.
- **Detection of microbial contaminants**: Biosensors that incorporate aptamers or antibodies (bio-recognition particulate matter) can detect specific microbial pathogens, E. coli, etc. in water. These could be useful in monitoring bacterial contamination in water continuously and in real-time during flood scenarios, post waste-water discharge scenario or after a natural disaster.
- **Soil: soil health and agricultural monitoring** Soil nutrient biosensing for detecting nitrogen, phosphorus, and potassium, which are key nutrients needed by the plants [26]. Nanomaterial-based sensors such as graphene oxide sensor provide accurate sensitivity towards a specific ion, which are used at specific places and thus the indirect detection of nutrient content in soil, enabling farmers to monitor the quality of soil and use fertilizers in optimal amounts so that yield may be improved with the needful preservation of the environment [27].
- **Soil pollutant detection:** Biosensors can be used to pinpoint the presence of pollutants in soil such as pesticides, heavy metals, and organic contaminants. It uses nanomaterials such as CNTs and gold nanoparticles to build biosensors with a high sensitivity, making them perfect for low-concentration contaminant detection, which is able to affect the growth of crops and enter the food chain [28,29].
- **Microbes and pathogen detection:** If microorganisms in the soil are pathogenic bacteria or fungi that affect crops, advanced biosensors can assess them when certain pathological characteristics of cells develop. These sensors contribute to sustainable agriculture by preventing disease and enabling targeted interventions.
- **Drug delivery systems** Biomaterials are an important element in smart drug delivery systems as they aid in the controlled release of therapeutic agents. Such systems are initiated by physiological triggers, including pH, temperature, or particular enzymes so that the drugs are released at the right place and the right time. Hydrogel-based biosensors are another example that delivers cancer drugs at the sites of tumors and in that way reduces unwanted effects. Furthermore, biodegradable biomaterials provide opportunities for drug delivery systems that do not require surgical removal since they break down naturally after their task is achieved.
- **Environmental monitoring**
- **Identifying algal blooms and their toxins**: Harmful algal blooms (HABs) produce toxins that can pollute drinking water sources and pose a risk to aquatic species. Aptamer and antibody-based biosensors that are specific to algal toxins are able to detect these harmful events quickly allowing for prompt management actions to protect drinking water resources [13].
- **Geolocation environmental household monitoring**: Innovative biosensors have been developed that can be carried on the person or attached to drones and buoys to monitor environmental properties over vast distances. Novel portable biosensors, which improve the electrochemical sensitivity of detection through nanomaterials, allow for on-site advanced testing for pollutants with fast data without laboratory analyses.
- **Internet of Things (IoT) integration**: Numerous environmental biosensors have integrated IoT, which supports real-time transmission and remote monitoring of data. Sensors connected to the Internet of Things can send data on environmental indicators to centralized platforms, giving authorities real-time insights into air quality, water safety, and environmental health to patentes.
- **Greenhouse gas (GHG) detection**: Nanomaterials (e.g., metal-organic frameworks (MOFs) and CNTs) based sensors can detect low levels of GHGs and could help to provide reliable measurements of emissions from agricultural, industrial, and natural sources [30].

Advanced biosensors based on pH-sensitive electrochemical nanomaterials for monitoring ocean acidification, one of the significant consequences of increasing levels of CO_2 in the atmosphere. These biosensors can be used in marine environments to detect CO_2 and carbon sequestration, providing needed data in support of climate mitigation [31].

- **Detection in real-time and multiplexed:**
  Biosensor based on quantum dots-graphene nanocomposite has been developed for the simultaneous detection of several disease biomarkers in one reaction (assay). The ability to multiplex is vital for full-spectrum diagnostics since it allows for a wide health evaluation from only one sample (Fig. 20.5).

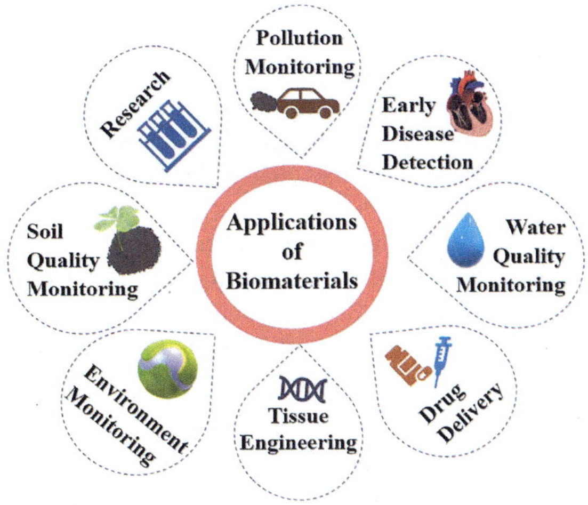

FIGURE 20.5 **Application of Biomaterials in various fields.** Diverse applications in environmental monitoring, healthcare, and tissue engineering.

20.8 Challenges

Despite these advantages, the use of biomaterials-based biosensors faces some hurdles in their development and implementation, affecting performance, robustness, and scalability (da Silva) [32]. Among them are the biocompatibility problem, the stability of materials used, and high manufacturing costs.

- **Biocompatibility issues**
- **Immune reactions and inflammation**: Biocompatibility is important for medically applied biosensors—especially those used as implantable sensors. Some biomaterials, particularly synthetic ones, are recognized as foreign bodies by the immune system after implantation and induce multiple inflammatory responses that often result in rejection or are encapsulated by adjacent tissue. This reaction can disrupt sensor functionality, reducing the overall lifetime of biosensors, but most importantly, it is extremely problematic for micro-engineered sensors because pneumatic contacts make this reaction-dependent monitoring very expensive in terms of manufacturing and implementation cost [33].
- **Protein fouling**: Biological fluids like blood or saliva samples can adhere to the sensor surface due to the adhesion of proteins on the surfaces and making biofouling. The presence of these proteins may mask the active sensing region, resulting in reduced sensitivity and erroneous measurements. Finding and optimizing coatings or other treatments that mitigate biofouling, but are also biocompatible, is fraught with difficulty, adding further complexity to sensor design.
- **Cytotoxicity of specific nanomaterials**: At high concentrations, there may be cytotoxicity associated with the presence of some nanomaterials (e.g., metal nanoparticles or certain carbon-based materials). These substances may build up in tissues or organs over time and pose a risk to health. The integration of high biosensing sensitivity and conductivity, on one hand, and safety, on the other hand, presents a remarkable challenge in designing medical biosensors.
- **Material stability**
- **Destruction of biomolecules**: A vast majority of biosensors utilize biomolecules (for instance, enzymes, antibodies, or aptamers) for recognition elements. These molecules are typically sensitive to environmental factors such as temperature, pH, and humidity, which leads them to denature or degrade with time. The degradation shortens the wear and tear sensor life span and performance, which limits their use in long-term monitoring.
- **Environmental sensitivity of polymers**: A few synthetic polymers applied in biosensors undergo changes in temperature, humidity, or UV light (bulb exposure), which lead to morphological and functional character alteration [34]. As an example, conductive polymers might undergo a loss of their electrical conductivity or mechanical flexibility in the presence of negative environmental conditions. Although it is possible to design stable materials that are responsive in a range of conditions, there still exists a limitation since they must be designed for utility across different environments—important for sensors for outdoor or wearable applications [35].
- **Limited reusability and shelf life**: In point-of-care diagnostics, biosensors are often intended for single use. But reusable biosensors would be cheaper and greener. It is particularly problematic to develop stable biomaterials that are still functional after several uses of the device or have an extended shelf life, because biological recognition elements can lose their functionality within days [36].

20.9 Additional challenges

- **Sensor miniaturization:** Durability is a major issue as biosensors are miniaturized and made portable. Architectures for miniaturized sensors often rely on sensitive components, which are more likely to be mechanically compromised. Mechanical stability over time is one of the main challenges in ensuring reliable performance of small and portable sensors, particularly when it comes to wearable devices that undergo movement upon use or bast empirical weather conditions.
- **Cost of next-generation biomaterials:** Many high-performance biomaterials, including engineered nanomaterials, aptamers, and some polymers, are expensive to manufacture at scale. Gold nanoparticles and fractionalized carbon nanomaterials require complex manufacturing processes that further increase the cost. These costs must be reduced to enable scale-up and accessibility, particularly for use in resource-limited settings, while maintaining a level of performance benefit expected from ideal biosensors (Fig. 20.6).

20.10 Future prospects in biomaterials-based biosensors

Biosensors on biomaterials are in a very rapid development, progress of new materials science and nanotechnology, and bioengineering creates great new opportunities for biosensor applications.

1. **Biosensors of the next-generation-wearable and implant**:
   - **Expansion of health monitoring**: Future biosensors will probably target continuous health monitoring but with a larger spectrum, such as glucose, lactate, electrolyte levels, and hormone levels. You will see these sensors migrated to wearables (smartwatch, patch, or even implant devices).
   - **Biodegradable and biocompatible materials**: Creation of fully biocompatible biomaterials for implantable biosensors is desirable as it minimizes potential inflammation or rejection (when the body recognizes something foreign to it) reaction, avoiding the need for removal surgery.
   - **Smart drug delivery systems**: Implantable biosensors with feedback capability could make way for smart drug delivery systems in which the sensor may recognize fluctuations of a given biomarker and respond by expunging an appropriate amount of the drug. This could change everything for diseases such as diabetes, cancer, and even hormonal issues.
2. **Component with AI and Internet of Things integration**
   - **Real-time data analytics**: AI can be integrated with biosensors that are used to analyze real-time data, which will offer insights into health trends and effectively predict potential health issues before they reach a critical level. It becomes specifically meaningful in the context of pay-as-you-go healthcare [37,38].

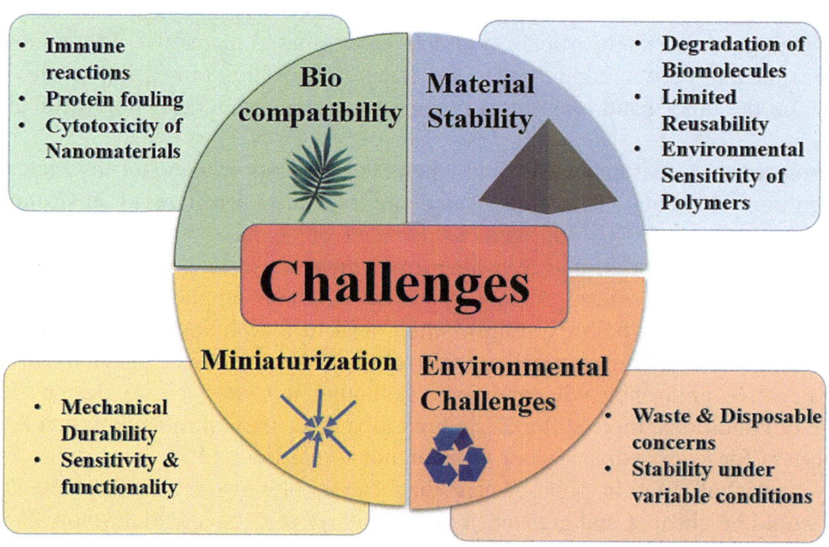

FIGURE 20.6 **Challenges associated with biomaterial technology.** Common challenges faced in biomaterial design, including biocompatibility and environmental sensitivity.

- **Connected healthcare systems**: With the use of IoT technology, biosensors can send data wirelessly to healthcare providers for remote monitoring. Utilizing biosensors for observation is beneficial when caring for older patients, chronic disease management, and recovery of surgery since constant monitoring is useful.
- **Environmental monitoring networks**: Connected biosensors could also form networks for distributed environmental monitoring, detecting pollutants or tracking the health of ecosystems at scale and driving faster responses to environmental threats [39,40]

3. **Point-of-care testing and resource-limited settings:**
   - **Portable, inexpensive sensors**—Low-cost sensors that are simple to use and deploy for near-patient testing will enable diagnostics in resource-limited locations. For example, paper-based biosensors are low-cost, single-use, and can be fashioned for multiplexed assays.
   - **Multiplexed sensing**: Multiplexed biosensors are useful for point-of-care diagnostics because they can simultaneously detect multiple biomarkers, something we might need in situations where a comprehensive health profile is needed quickly. Multiplexed biosensors could, for example, allow quick diagnosis of coinfections or provide multiple health biomarker measures from a single sample.

20.11 Conclusion

The introduction of biomaterials in biosensing has drastically changed the paradigm in biological diagnostics, environmental sensing, and process monitoring in various industries. Biomaterials have been used to enhance sensitivity, specificity, and stability in biosensors to make them priceless tools in in situ determination of analytes in various contexts. Advances in biomaterials science through the introduction of nanomaterials, synthetic bioreceptors, and biodegradable polymers have further extended functionality and variety in biosensors. Despite this unprecedented development, several problems have yet to be addressed, including material stability, material biocompatibility, and the cost of production. Solutions to these problems invoke novel paradigms in the shape of functionalized surfaces, self-healing materials, and fusion between biomaterials and nanotechnology. Furthermore, the incorporation of newer technologies in the shape of artificial intelligence, IoT (Internet of Things), and microfluidics is opening pathways towards next-gen biosensors with remote, real-time multiplexed detection capabilities. The direction in biomaterial-based biosensors in the future should be to engineer economical yet sustainable platforms to aid in personalized medicine, environmental sustainability, and point-of-care testing. Addressing current limitations through an interdisciplinary perspective, biomaterial-based biosensors have immense potential to revolutionize healthcare, environmental monitoring, and industrial processes to drive an improved healthier and sustainable future.

References

[1] Jian Xu, H. Lee, Anti-biofouling strategies for long-term continuous use of implantable biosensors, Chemosensors 8 (3) (2020) 66. Available from https://doi.org/10.3390/chemosensors8030066.

[2] A. Fatima, I. Younas, M.W. Ali, An overview on recent advances in biosensor technology and its future application, Archives Of Pharmacy Practice 13 (2022) 5–10. Available from https://doi.org/10.51847/ltogi43jil.

[3] N. Wongkaew, M. Simsek, C. Griesche, A.J. Baeumner, Functional nanomaterials and nanostructures enhancing electrochemical biosensors and lab-on-a-chip performances: recent progress, applications, and future perspective, Chemical Reviews 119 (2018) 120–194. Available from https://doi.org/10.1021/acs.chemrev.8b00172.

[4] A.B. Allen, L.B. Priddy, M.-T.A. Li, R.E. Guldberg, Functional augmentation of naturally-derived materials for tissue regeneration, Annals of Biomedical Engineering 43 (3) (2015) 555–567. Available from https://doi.org/10.1007/s10439-014-1192-4.

[5] M.S. Birajdar, H. Joo, W.-G. Koh, H. Park, Natural bio-based monomers for biomedical applications: a review, Biomaterials Research 25 (1) (2021). Available from https://doi.org/10.1186/s40824-021-00208-8.

[6] F. Luzi, D. Puglia, L. Torre, Biomass, biopolymer-based materials, and bioenergy, (2019), pp. 179–201. Available from https://doi.org/10.1016/b978-0-08-102426-3.00010-2.

[7] M. Puertas-Bartolomé, A. Mora-Boza, L. García-Fernández, Emerging biofabrication techniques: a review on natural polymers for biomedical applications, Polymers 13 (2021) 1209. Available from https://doi.org/10.3390/polym13081209.

[8] L. Stefano, I. De, S. Rea, M. Longobardi, Giocondo, Giardina, Organic–inorganic interfaces for a new generation of hybrid biosensors, Biosensors: Emerging Materials and Applications (2011).

[9] S. Itescu, J.-H. Ankersmit, A.A. Kocher, M.D. Schuster, Immunobiology of left ventricular assist devices, Progress in Cardiovascular Diseases 43 (1) (2000) 67–80. Available from https://doi.org/10.1053/pcad.2000.7191.

[10] Y. Wang, Y. Guo, J. Lu, Y. Sun, X. Yu, S.C.B. Gopinath, et al., Nanodetection of head and neck cancer on titanium oxide sensing surface, Nanoscale Research Letters 15 (1) (2020) Available from https://doi.org/10.1186/s11671-020-3262-x, http://www.springer.com/materials/nanotechnology/journal/11671.

[11] D. Grieshaber, R. MacKenzie, J. Vörös, E. Reimhult, Electrochemical biosensors - sensor principles and architectures, Sensors (Basel, Switzerland) 8 (3) (2008) 1400–1458. Available from https://doi.org/10.3390/s8031400.

[12] C. Zhu, G. Yang, H. Li, D. Du, Y. Lin, Electrochemical sensors and biosensors based on nanomaterials and nanostructures, Analytical Chemistry 87 (1) (2015) 230–249. Available from https://doi.org/10.1021/ac5039863.

[13] A. Hendricks, C.M. Mackie, E. Luy, C. Sonnichsen, J. Smith, I. Grundke, et al., Compact and automated eDNA sampler for in situ monitoring of marine environments, Scientific Reports 13 (1) (2023). Available from https://doi.org/10.1038/s41598-023-32310-3.

[14] G. Yang, K.L. Kampstra, M.R. Abidian, High Performance Conducting Polymer Nanofiber Biosensors for Detection of Biomolecules, Advanced Materials 26 (2014) 4954–4960. Available from https://doi.org/10.1002/adma.201400753.

[15] F.D. Guerra, M.F. Attia, D.C. Whitehead, F. Alexis, Nanotechnology for environmental remediation: materials and applications, Molecules 23 (7) (2018) 1760. Available from https://doi.org/10.3390/molecules23071760.

[16] J. McCarthy, R. Weissleder, Multifunctional magnetic nanoparticles for targeted imaging and therapy☆, Advanced Drug Delivery Reviews 60 (2008) 1241–1251. Available from https://doi.org/10.1016/j.addr.2008.03.014.

[17] M.S.H. Akash, K. Rehman, M. Tariq, S. Chen, Development of therapeutic proteins: advances and challenges, Turkish Journal of Biology 39 (3) (2015) 343–358. Available from https://doi.org/10.3906/biy-1411-8.

[18] B. Strehlitz, N. Nikolaus, R. Stoltenburg, Protein detection with aptamer biosensors, Sensors 8 (2008) 4296–4307. Available from https://doi.org/10.3390/s8074296.

[19] A.M. Smith, L.M. Gilbertson, Rational ligand design to improve agrochemical delivery efficiency and advance agriculture sustainability, ACS Sustainable Chemistry & Engineering 6 (11) (2018) 13599–13610. Available from https://doi.org/10.1021/acssuschemeng.8b03457.

[20] O. Almog, H. Rakefet, Y. Ben-Yoav, T. Sverdlov, S. Shmilovich, Y. Krylov, Shacham-diamand, integrated polypyrrole flexible conductors for biochips and MEMS applications, Journal of Atomic and Molecular Physics 2012 (1) (2012).

[21] S.R. Paital, N.B. Dahotre, Calcium phosphate coatings for bio-implant applications: materials, performance factors, and methodologies, Materials Science and Engineering: R: Reports 66 (1–3) (2009) 1–70. Available from https://doi.org/10.1016/j.mser.2009.05.001.

[22] J.J. Ghidoni, C.W. Hall, Some requirements and problems in tissue interfacing, Journal of Biomedical Materials Research 5 (6) (1971) 17–20. Available from https://doi.org/10.1002/jbm.820050605.

[23] C.S. Lee, R.E.A. Gwyther, M. Freeley, D. Jones, M. Palma, Fabrication and functionalisation of nanocarbon-based field-effect transistor biosensors, Available from http://onlinelibrary.wiley.com/journal/10.1002/(ISSN)1439–7633.

[24] K.-H. Tseng, M.-Y. Chung, L.-H. Chen, P.-Y. Chang, Green smart campus monitoring and detection using LoRa, Sensors (Basel, Switzerland) 21 (19) (2021) 6582. Available from https://doi.org/10.3390/s21196582.

[25] O.A. Sadik, A. Mulchandani, Chemical and biological sensors: meeting the challenges of environmental monitoring, ACS Symposium Series 762 (2000) 1–6. Available from https://doi.org/10.1021/bk-2000-0762.ch001, https://pubs.acs.org/series/symposium.

[26] P. Sharma, V. Pandey, M.M.M. Sharma, A. Patra, B. Singh, S. Mehta, et al., A review on biosensors and nanosensors application in agroecosystems, Nanoscale Research Letters 16 (1) (2021). Available from https://doi.org/10.1186/s11671-021-03593-0.

[27] A. Priyam, P. Prasad Singh, Understanding the safety aspect of biologically synthesized agriculturally useful nanomaterials for humans and environment, Frontiers in Nanoscience and Nanotechnology 4 (2018). Available from https://doi.org/10.15761/fnn.1000176.

[28] D.P. Roberts, N.M. Short, J. Sill, D.K. Lakshman, X. Hu, M. Buser, Precision agriculture and geospatial techniques for sustainable disease control, Indian Phytopathology 74 (2) (2021) 287–305. Available from https://doi.org/10.1007/s42360-021-00334-2.

[29] V. Kumar, K. Arora, Trends in nano-inspired biosensors for plants, Materials Science for Energy Technologies 3 (2020) 255–273. Available from https://doi.org/10.1016/j.mset.2019.10.004.

[30] S.L. Ullo, G.R. Sinha, Advances in smart environment monitoring systems using IoT and sensors, Sensors (Basel, Switzerland) 20 (11) (2020) 3113. Available from https://doi.org/10.3390/s20113113.

[31] Alshamsi A. Anwar Y. Almulla M. Aldohoori M. Hamad N. Awad M. 2017 6 28 2017/06/28 2017 International Conference on Electrical and Computing Technologies and Applications, ICECTA 2017 doi: https://doi.org/10.1109/ICECTA.2017.8251998 9781538608722 1 4 Institute of Electrical and Electronics Engineers Inc. United Arab Emirates Monitoring pollution: Applying IoT to create a smart environment 2018-.

[32] S. Neves, M.M. Pereira, M. Da, González-García, P. Hernández-Santos, Fanjul-Bolado, Future trends in the market for electrochemical biosensing. " current opinion in, Electrochemistry. 10 (2018).

[33] D. Yogev, T. Goldberg, A. Arami, S. Tejman-Yarden, T.E. Winkler, B.M. Maoz, Current state of the art and future directions for implantable sensors in medical technology: clinical needs and engineering challenges, APL Bioengineering 7 (3) (2023). Available from https://doi.org/10.1063/5.0152290.

[34] J.H. Boss, I. Shajrawi, D.G. Mendes, Histological patterns of the tissue reaction to polymers in orthopaedic surgery, Clinical Materials 13 (1993) 11–17. Available from https://doi.org/10.1016/0267-6605(93)90083-J.

[35] S. Seetharaman, D. Sankaranarayanan, M. Gupta, Magnesium-based temporary implants: potential, current status, applications, and challenges, Journal of Functional Biomaterials 14 (2023) 324. Available from https://doi.org/10.3390/jfb14060324.

[36] R. Nißler, O. Bader, M. Dohmen, S.G. Walter, C. Noll, G. Selvaggio, et al., Remote near infrared identification of pathogens with multiplexed nanosensors, Nature Communications 11 (1) (2020). Available from https://doi.org/10.1038/s41467-020-19718-5.

[37] A. Das, M.P. Sarma, K.K. Sarma, N. Mastorakis, N. Mastorakis, V. Mladenov, et al., Design of an IoT based real time environment monitoring system using legacy sensors, MATEC Web of Conferences 210 (2018) 03008. Available from https://doi.org/10.1051/matecconf/201821003008.

[38] B.C. Kavitha, R. Vallikannu R. India IoT based intelligent industry monitoring system, in: 6th International Conference on Signal Processing and Integrated Networks, SPIN. Institute of Electrical and Electronics Engineers Inc, 2019. doi: https://doi.org/10.1109/SPIN.2019.8711597.

[39] J. Asharf, N. Moustafa, H. Khurshid, E. Debie, W. Haider, A. Wahab, A review of intrusion detection systems using machine and deep learning in internet of things: challenges, solutions and future directions, Electronics 9 (7) (2020) 1177. Available from https://doi.org/10.3390/electronics9071177.

[40] A. Kamilaris, F.O.O. Ostermann, Geospatial analysis and the internet of things, ISPRS International Journal of Geo-Information. 7 (7) (2018) 269. Available from https://doi.org/10.3390/ijgi7070269.

Chapter 21

Electroconductive hydrogels in tissue and regenerative engineering

Anthony Guiseppi-Elie[1,2,3], Robabeh Mazhabi[4], and Dyer Narinesingh[5]

[1]Department of Biomedical Engineering, Bioelectronics, Biosensors and Biochips (C3B®), Texas A&M University, College Station, TX, United States, [2]ABTECH Scientific, Inc., Biotechnology Research Park, Richmond, VA, United States, [3]Department of Cardiovascular Sciences, Houston Methodist Research Institute, Houston, TX, United States, [4]West China School of Basic Medical Sciences & Forensic Medicine, Sichuan University, Chengdu, P.R. China, [5]Department of Chemistry, University of the West indies, St. Augustine, Trinidad and Tobago

21.1 Introduction

Electroconductive hydrogels (ECHs) are polymer composites comprising one and two-dimensional inherently conductive electroactive polymers (CEPs) within cross-linked, highly hydratable hydrogels. First introduced in 1994 by Guiseppi–Elie [1–4] to address the limited switching speeds of organic transistors fashioned from inherently conductive polymers (CPs) [1], since then publications featuring the term "ECHs" have increased tremendously [2,3]. This growth aligns with their newfound applications in biomedical engineering, particularly research, clinical interest, and bench-to-bedside commercial development in wearable bioelectronics, tissue and regenerative engineering, and advanced therapeutics. ECHs comprise the electrical and electrochemical properties of the discontinuous inclusion and the high hydration and biocompatibility of the continuous hydrogel matrix that maintain their three-dimensional structure in the swollen state. Both constituents, being individually stimuli-responsive, are combined into an evolving class of multiresponsive, engineered polymeric materials that have one or more actuated properties that may be endogenously or exogenously altered. Among these are electrical conductivity, permeability, swellability, elastic modulus, and optical properties, each of which may be engineered in a controlled fashion and made responsive (single or multi) to external stimuli such as electric fields, light, temperature, pH, ionic strength, moisture, external stress, or chemical/biochemical compounds. Accordingly, ECHs are finding ever-increasing applications in diverse fields such as actuators and sensors, wearable electronics, protective clothing, energy storage devices [4], electro-stimulated drug delivery systems [5], tissue engineering (TE) [6], and regenerative engineering (RE) [7].

Hydrogels are widely explored in TE/RE [8,9] and can be tuned to mimic biochemical, mechanical, and topographical cues intended to mimic the native extracellular matrix (ECM) [10]. However, their insulating nature often limits their potential for the modulation of electrically sensitive cells and tissues such as cardiac and neural tissues [6]. The synthesis of CEPs within hydrogels confers an additional range of properties. With the advent of microengineering fabrication techniques such as 3D printing, electrospinning, and other lithography-based approaches, it is possible to exert precise control over the composition, geometry, and spatial arrangement of cells and biomolecules within ECHs. Shown to be permissive for neuronal cells, RMS-13, C2C12 murine myoblasts, and HUVECs and to support the electrode-to-tissue interface in neurostimulation, ECHs have been fabricated to produce functionally relevant engineering structures [11–13]. ECHs have also been employed as stimuli-responsive materials in chemical and bioanalytical biosensors [14], e.g., for subcutaneous glucose [15] or galactose [16] monitoring, in electronic NOSES and TONGUES, electronic tattoos, in implantable and wearable bioelectronics [17], explored in photothermal therapy treatment of breast cancer [18] and as guidance structures for nerve regeneration [19].

The hydrogel components of ECHs may be natural, synthetic, or hybrid combinations. The natural hydrogels include agarose, hyaluronic acid, alginate, chitosan, fibrin, collagen, gelatin, dextran, heparan, and cellulose [20]. Common synthetic hydrogels include poly(acrylates/methacrylates), polyethyleneglycol, and polyvinyl alcohol [21]. Hybrid hydrogels, that seek to combine the natural with the synthetic, include, e.g., gelatin methacryloyl (GelMA) hydrogels

Fundamentals of Translational Tissue Engineering. DOI: https://doi.org/10.1016/B978-0-443-33455-9.00025-0

[22,23]. Within these hydrogels, the crosslinking between the polymer chains can be physical, chemical, or a combination thereof that appreciably influences their properties, an important factor in their applications in TE/RE [24]. Advantages associated with the use of natural hydrogels in biomedical applications include biocompatibility, biodegradability, and antibiotic/antifungal properties. However, their stability and strength are usually much lower than those of synthetic hydrogels [2].

The conductive components of ECHs include CEPs such as polythiophene (PTh) and its derivatives, including poly(3,4-ethylenedioxythiophene) (PEDOT) [25], polyanilines (PAn or PANI) [26], polypyrroles (PPy) [27,28], and their derivatives. CEPs are highly conjugated polymers [29,30] that facilitate the transfer of π-electrons along and between adjacent chains. CEPs possess high conductivity equal to insulators, semiconductors, and low-end metals and are widely explored in the field of electroactive TE [31,32].

Methods of synthesis of ECHs include chemical oxidative, electrochemical, hyphenated chemical–electrochemical, metal-catalyzed, and enzyme-catalyzed polymerization. Additionally, compounding of separately synthesized and stabilized nanoparticles of the CEP within the hydrogel matrix is possible. Chemical oxidative polymerization is the most widely used, especially in the large-scale synthesis of CEPs and ECHs. The nature and concentration of the reagents, the reaction conditions, and the solvent(s) used all strongly influence the polymerization rates and the resultant electrical conductivity, morphology, and yield of the synthesized polymers [24,29,30,33–36]. Electropolymerization methods include potentiostatic, potentiodynamic, or amperometric, which initiate the electropolymerization of oxidizable monomer partitioned within the hydrogel [2]. In the hyphenated chemical–electrochemical coupling synthetic approach, the chemical oxidative polymerization step produces the isolated CEP domains, which, upon passage of modest charge, result in a highly conductive and redox-active ECH. Alternatively, the reactive monomer, which may be a combination of both the hydrogel and CEP precursors, may be combined, along with photo or thermally labile free radical initiators, into a single cocktail or prepolymer cocktail [37]. In the enzymatic method, the enzymes act in the presence of an oxidizing agent (e.g., H_2O_2 or O_2) to produce radicals or reactive intermediates that initiate polymerization. Whilst laboratory-based chemical oxidative polymerization synthesis of CEPs/ECHs is the norm, tissue-based intracellular oxidative polymerization is a rapidly emerging technique that can mimic cell behavior and can be triggered by intracellular reactive oxygen species (ROS) without the external stimuli. However, the technique remains challenging because of the complexity of the cellular microenvironment [38,39]. In the compounding synthetic technique, the ECHs may be produced from stabilized dispersions of presynthesized nanoparticles of CEPs mixed into hydrogel monomer cocktails prior to crosslinking [10,40].

ECHs may be fashioned into a wide range of nano- or microforms using a range of polymer processing techniques [41]. These include electrospinning, 3-D printing, microlithographic fabrication, and microspheres (microgels) formation. Electrospinning, a simple, rapid, cost-effective nanofabrication technique, uses electrostatic forces to produce fibers from a polymer solution or dope. It facilitates the production of 1-D nanofibers, tailoring their desired physicochemical properties for intended biomedical applications [40,42–44] with structural similarities to the natural ECM [45–48]. 3-D printing of ECHs is a rational biofabrication strategy [17,49] that may employ one of several digital model-based layer-by-layer deposition techniques, including Fused Deposition Modeling (FDM), Direct Ink Writing (DIW), Inject 3-D Printing, and Stereolithography (SLA). Presently, SLA provides the best resolution printing of ECHs. 3-D printed ECH and scaffolds reported for use in biomedical applications include PEDOT-PSS(DIW), PEGDA/PEDOT-PSS(SLA), GelMA/PEDOT-PSS(DIW), PAA/PPy-Chi(DIW), GelMA/PANI(SLA), and p(HEMA-*co*-EGMA/PEDOT-PSS(SLA)) [50–53]. 3D printed ECHs function as scaffolds for growing cells to create artificial tissues and organs or may be cell-laden prior to printing [54]. Microlithographic fabrication includes a range of additive and subtractive techniques that exploit the photopolymerizability of ECH precursor cocktails (dope) and photomasks. Like 3D printing, challenges are ensuring microfabricated ECHs have suitable mechanical strength, modulus, and biocompatibility. Key biotechnical properties are achieved by precise control over crosslinking methods (chemical, physical, or photochemical) that ensure acceptable resolution while ensuring cell viability during and after the microfabrication process [51]. EC-microgels (microspheres) by any of the aforementioned techniques [55]. The results are dispersions of conductive, colloidal particles made of cross-linked hydrophilic or amphiphilic polymer chains and their CEP counterparts [56,57]. EC-microgels may be formed via water-in-oil emulsions or via microfluidic techniques. They are typically less than 10 μm in diameter and are attractive for wound repair and drug delivery [58]. Following synthesis and fabrication, ECH microforms may be conditioned to achieve targeted biotechnical properties. These include: (1) solvent conditioning to extract unreacted monomer and achieve equilibrium hydration under the tonic conditions of deployment, (2) reactive conditioning to promote chemical conjugations of bioactive moieties to the preformed construct, and (3) thermal annealing, to negate time-temperature dependent changes in properties following deployment.

21.2 Electroconductive hydrogels in tissue and regenerative engineering

TE aims to combine natural, synthetic, or hybrid biomaterials of appropriate physical and chemical properties with cells, biomolecular cues, and exogenously applied forces (mechanical, electrical, and magnetic) to create bioactive 3D constructs that mimic and restore damaged native tissue [59,60]. Such scaffolds present a highly interconnected porous framework to support the diffusion of nutrients and gases and for the migration, adhesion, and growth of cells to replace the damaged tissue [61]. RE integrates TE, stem cell biology, and developmental biology to create complex replacement tissue constructs [62,63]. An engineered three-dimensional microenvironment supports cellular differentiation into functional tissue.

ECHs offer many advantages over CEPs for biomedical applications in TE and RE, although several issues should be considered:

1. The balance of electrical conductivity and tissue-like elastic modulus as the incorporation of large amounts of conductive components increases not only electrical conductivity but also mechanical stiffness and brittleness [64,65].
2. The postfabrication conditioning should be incorporated in parts design [9].
3. The biocompatibility and utility of ECHs have not been well demonstrated for biomedical applications, leaving them at a premature stage of clinical deployment.

Fig. 21.1 illustrates the breadth of applicability of ECHs and provides specific examples of ongoing work in these areas.

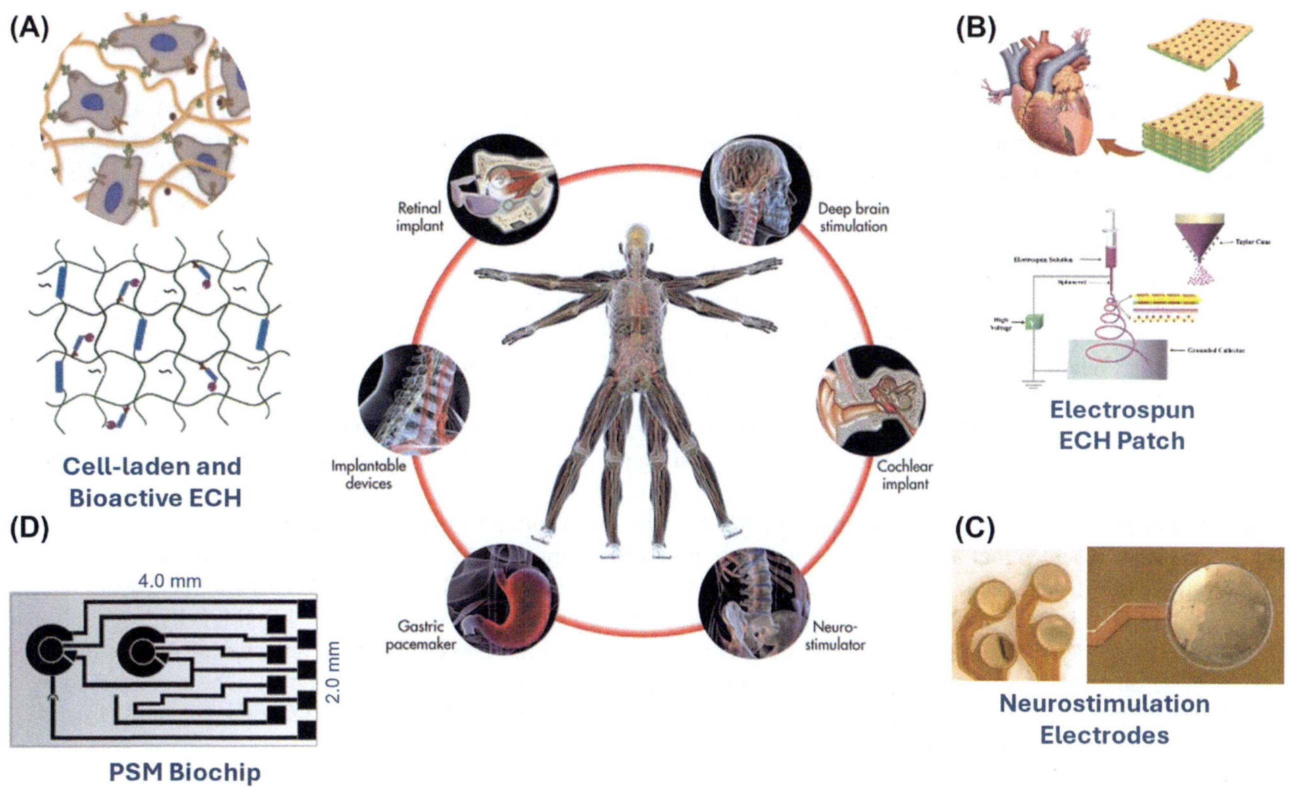

FIGURE 21.1 Biotechnical applications and broad applicability of ECHs in TE and RE: (A) Cell-laden and/or molecularly engineered SMART materials for drug and factor delivery. (B) Electrospun ECHs of PPy for electrical cardiac patches to correct traumatic defects in electrical tissue from cardiac infarction or axomy. (C) Biointerfaces for direct communication to nerves in neuromodulation, and (D) for biosensing applications, such as in the development of a minimally invasive PSM Biochip. *ECH*, Electroconductive hydrogel; *PSM*, physiological status monitoring; *RE*, regenerative engineering; *TE*, tissue engineering.
Reproduced with permission from A [66], C [37], D [67,68].

21.2.1 ABIO–BIO interfaces in neurostimulation devices

Biointerface engineering refers to the purposeful manipulation of that interphasial region between an abiotic material (e.g., metal, semiconductor, glass, or synthetic polymer) and a material of biological origin, to achieve a desired technical objective [69]. Chronically indwelling neurostimulation electrodes are needed to stimulate the vagus nerve of the chronically depressed [42,43], to stimulate the spinal cord of paraplegics [44] and in the chronically implanted electrodes of brain–computer interfaces (BCI) used to record electrical impulses, stimulate neurons with electrical impulses, and/or translate brain activity into actuatable commands. Typical electrodes are plagued by fibrous encapsulation [40], leading to increased interfacial impedance and an escalating power penalty [70]. Such fibrous encapsulation is the result of the inflammatory cascade [71] triggered by the foreign body response [72]. The goal of biointerface engineering in this context is to reduce interfacial impedance as well as mitigate the foreign body response [73].

In the work of Bhat et al. [37], using the electrodeposition of an ECH on sputter-deposited gold-on-polyimide electrodes, various charge densities (0–100 mC/cm^2) were explored to address interfacial impedance and each electrode was coated with a phosphoryl choline-containing bioactive hydrogel to address biocompatibility at the ABIO–BIO interface. SEM analysis and electroactive area calculations identified films fabricated with a charge density of 50 $mC\ cm^{-2}$ as being well suited for neurostimulation electrodes. Charge transfer resistance demonstrated a strong inverse correlation (−0.83) with the charge density of electrodeposition. On average, the addition of polypyrrole and hydrogel to neurostimulation electrodes decreased charge transfer resistance by 82 %. These results support the use of interfacial engineering techniques to mitigate high interfacial impedance and combat the foreign body response to epidurally implanted neurostimulation electrodes. One challenge is the robustness of these modifying polymers to interfacial failure, which is addressed through specific immobilization strategies that involve covalent attachment of the ECH to the metallic electrode [74].

The data in Fig. 21.2A reflect that biomimetic poly(HEMA)-based hydrogels support the attachment and spindle-like morphology of RMS-13 human muscle fibroblasts but do so in a cross-linker concentration-dependent manner. These hydrogels show increasing stiffness with increased mole% of the cross-linker TEGDA and this correlates with the free:bound water ratio within the hydrogel but is anti-correlated with the cell viability (Fig. 21.2B) [75]. Lightly cross-linked hydrogels accommodate invagination of the RMS-13 spindles that produce physical rather than receptor-mediated attachment. The formation of hybrid bioactive and inherently conductive constructs of composites formed from polyaniline-polyacrylamidomethylpropane sulfonic acid (PAn-PAAMPSA) nanomaterials (0.00–10.0 wt.%) within poly [(2-hydroxy ethyl methacrylate-*co*-N-acrylamide)-*co*-polyethyleneglycol methacrylate)] [p(HEMA-*co*-HMMA-*co*-PEGMA)] hydrogels were achieved using microlithographic fabrication and 3-D printing. Hybrid constructs formed by combining a nonconductive base (0.00 wt.% PAn-PAAMPSA) and ECHs (varying wt% PAn-PAAMPSA) using these two production techniques were directly compared. Both base and ECHs supported the attachment and growth of NIH/3T3 fibroblast cells. When the base hydrogel was rendered bioactive by the inclusion of collagen (>200 g mL^{-1}), it also supported the attachment, but not the differentiation, of PC-12 neural progenitor cells. Fig. 21.2C (i) and (ii), show the attachment and growth of NIH/3T3 cells was well supported on both nonconductive, base (0.00 wt.% PAn-PAAMPSA), and electroconductive (0.10 wt.% PAn-PAAMPSA) hydrogels. Although the hydrogels did not contain any bioactive reagent, such as protein or peptide to support receptor-mediated attachment, the fibroblasts could still attach to these materials and proliferate [53]. The attachment of PC-12 cells was limited to hydrogels containing less than 200 g mL^{-1} collagen, which indicated that at those concentrations, motifs that supported receptor-mediated attachment did not present on the surface of the gel at adequate density and were not bioavailable to the cells. Fig. 21.2C (iii) and (iv) show the bright field micrograph of PC-12 cells attached to the collagen-containing p(HEMA-*co*-HMMA-*co*-PEGMA) hydrogel (200 and 500 $\mu g\ mL^{-1}$ collagen). The cells maintained their regular chromaffin morphology without any sign of neurite outgrowth [53]. Neurite outgrowth of PC-12 cells could be achieved by the addition of the CEP component as was confirmed by the use of p(HEMA-*co*-PEGMA) hydrogels containing 0.1 and 0.5 wt.% PEDOT:PSS (Fig. 21.2D (i), (ii), and (iii)) [51].

21.2.2 Wound repair and healing

The care and rehabilitation of acute and chronic wounds and injured organs have a significant social and economic impact on patients and global health [76,77]. Tissue transplantation is a general strategy to treat these patients, but this approach is largely limited by donor shortage. Scaffold development for hosting progenitor and stem cells for TE/RE necessitates engineering biomaterials that mimic many roles of ECMs, including their topographical, mechanical, and electrical properties [78,79]. Hydrogels have been widely used in medical applications and are considered safe in many situations [80–83]. However, most hydrogels are electrical insulators and cannot establish effective electrical coupling between the

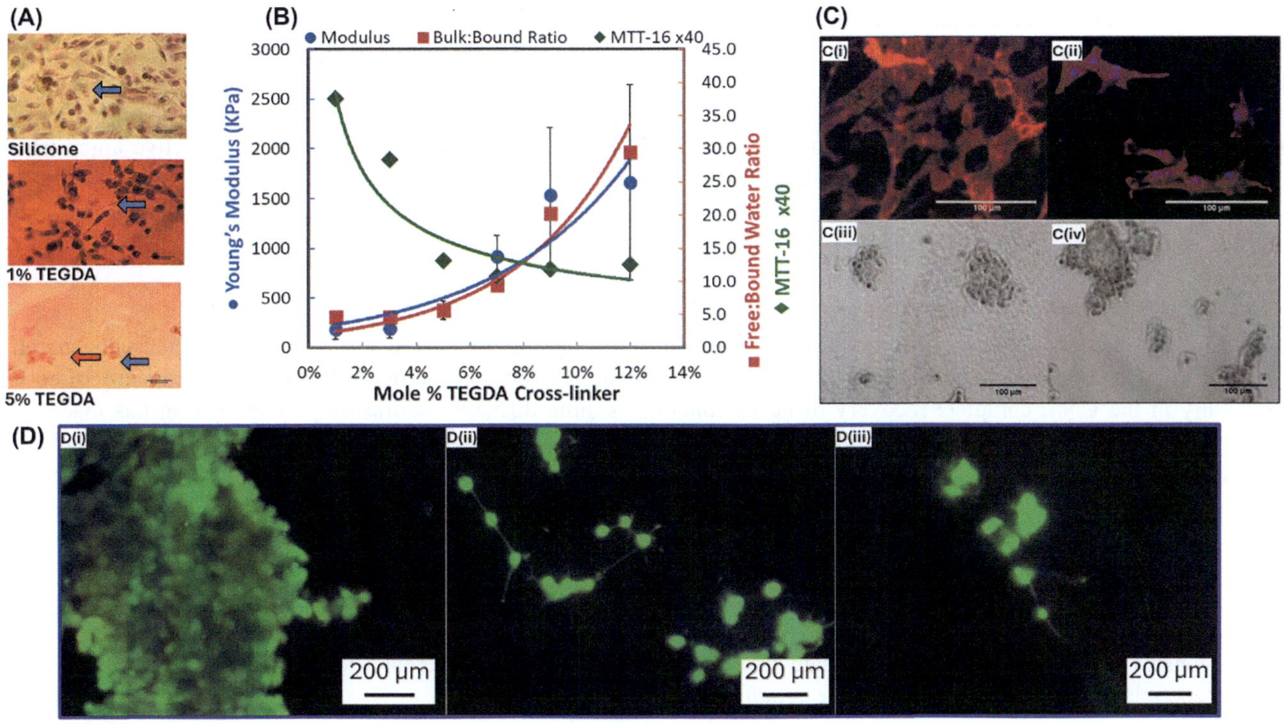

FIGURE 21.2 Cellular interactions with hydrogels and ECHs. (A) Attachment and morphology of RMS-13 human muscle fibroblasts on silicone (control) and on the base, biomimetic poly(HEMA)-based hydrogel with 1 and 5 mole% of TEGDA cross-linker on day 8 following H&E staining, scale 50 μm. Blue arrows show attached cells with spindle-shaped morphology on silicone and on low cross-linker hydrogels, red arrows show rounded (not well attached) cells on stiffer hydrogels [75]. (B) Trends in the free:bound water ratio (■), the elastic modulus (●), and the MTT cell viability (♦) as a function of varying mole% of the cross-linker TEGDA in base, biomimetic poly(HEMA)-based hydrogels [75]. (C) Attachment and morphology of NIH/3T3 mouse fibroblast cells on p(HEMA-*co*-HMMA-*co*-PEGMA) hydrogels showing behavior on the nonconductive, base hydrogel (0.0 wt.% PAn-PAAMPSA) [C(i)] and the ECH (0.1 wt.% PAn-PAAMPSA) [C(ii)] [53]. (D) Attachment and neurite-like morphology of differentiated PC-12 cells cultured on 3-D printed hydrogel disks with 0.0 wrt.% (pristine) [D(i)] showing exclusively chromaffin-like morphology, 0.1 wt.% PEDOT:PSS [D(ii)], and 0.5 wt.% PEDOT:PSS [D(iii)] showing evidence of neurite outgrowth [51]. *ECH*, Electroconductive hydrogel; *H&E*, hematoxylin and eosin. Reproduced with permission from (A) [75], (B) [75], (C) [53], and (D) [51].

scaffold and native tissue; as a result, this limits their therapeutic efficacy in native electroconductive tissues [49,84]. In this case, ECHs are indicated.

The physical and optoelectronic properties of ECHs, including bulk mixed electronic/ionic conduction, enable the fabrication of multifunctional biomaterials that passively affect cellular response and modulate electric field, charge injection, or drug delivery, allowing these materials to actively affect tissue regeneration [2,85]. Implanting an ECH of suitable biocompatibility and appropriate conductivity can effectively reconstruct physiological functions related to electrical conduction, such as ventricular contraction via ECH-based, synthetic Purkinje fibers [55,79,86], nerve conduction [78,87–89], transepithelial potential (TEP) [90], and skeletal muscle contraction [79,91,92]. At the cellular level, ECH-based scaffolds could mediate the adhesion, proliferation, migration, and differentiation of different cell types, including cardiomyocytes (CMs), adipose-derived stem cells (ADSCs), neural stem cells (NSCs), myoblasts, and human mesenchymal stem cells (MSCs) [78,86,88,93–96]. The conductive microenvironment provided by ECHs could enhance electrical communications among endogenous or exogenous cells and so promote damaged tissue regeneration [97]. In addition, electrical stimulation has been widely used in TE to induce physiologically relevant functions, such as cardiac function, neurite outgrowth, and wound healing [98–100]. In many cases, ECH-based devices are promising for electrical stimulation of living tissues to promote tissue repair and texture of synthetic meats [101] and fish [102].

21.2.2.1 Electroconductive hydrogel for myocardium repair

Cardiovascular disease is a disease with a high mortality rate. Among various cardiovascular diseases, myocardial infarction (MI) is the most fatal [103]. When MI occurs, it is generally because the coronary artery is partially or completely blocked by atheromatous plaque, resulting in irreversible fibrosis or necrosis of cardiac cells due to hypoxia and scar tissue formation at the injury site, leading to impaired electrical signal propagation and cardiac contraction. Current clinical treatment for MI includes coronary intervention, thrombolytic therapy, and pharmacological therapy, but these therapies cannot save the necrotic myocardial cells [104]. The unique properties of ECHs make them particularly suitable as scaffolds for cardiac tissue engineering [105]. Therefore implantation of ECHs is an effective approach for cardiac tissue repair, not only providing mechanical support for weakened myocardial tissue but also potentially engineered to restore electrical impulse propagation via Purkinje-like cardiac fibers to the MI area, thereby preventing cardiac arrhythmia. Furthermore, ECHs could serve as platforms for local delivery of therapeutic drugs [106].

21.2.2.2 Electroconductive hydrogel for nerve repair

Nerve injury, such as spinal cord injury (SCI), traumatic brain injury, peripheral nerve injury, and neurodegenerative disorder, may lead to loss of sensory or motor function in patients. Because of the limited regenerative capacity of nerves, especially in the CNS, complete recovery of nerve function is quite difficult. Neural tissue engineering has made great progress in nerve damage repair and nerve function regeneration [107]. Among the various scaffold materials, hydrogels are well-suited for nerve repair because of their biocompatibility, high water content, and similar mechanical properties to natural neural tissue. In addition, hydrogels can deliver factors [108], drugs/cells to the target sites [109,110]. However, electrical signals are essential for neuronal cells to survive, differentiate, and activate neural circuits. Electrical activity plays an essential role in the early development of the nervous system by regulating the expression profiles of genes related to voltage-gated ion channels. Studies have shown that ECHs can be used as an electrical stimulator and greatly promote neural cell proliferation, elongation, orientation, and neuronal differentiation, making it promising for neural tissue repairs [111–113]. ECHs could facilitate proliferation, migration, and differentiation of neural cells toward neurons, being beneficial for two categories of nerve repair [114].

21.2.2.3 Electroconductive hydrogels for spinal cord repair

SCI is a serious and disabling condition that causes loss of sensory and motor function in the lower limbs of patients. According to the WHO, there are up to 500,000 people globally who suffer from SCI each year [115]. SCI leads to neuronal necrosis and axonal disruption, and due to the limited regenerative capacity of the CNS, its clinical treatment effect is affected. Furthermore, glial scar formation (gliosis) at the injury site acts as the major barrier to the regeneration of axons, resulting in the disruption of electrical nerve signal transmission [69,116]. Implantation of ECHs could fill the irregular gaps and provide structural support, as well as restore electrical signal propagation pathways for the spinal cord. In addition, ECHs could facilitate proliferation, migration, and differentiation of neural cells toward neurons, thereby being beneficial to SCI repair [117].

21.2.2.3.1 Electroconductive hydrogels for peripheral nerve injury repair

PNI is a common clinical condition caused by trauma, tumor, or medically induced injury. Compared to the CNS, the PNS has a definite capacity for regeneration and repair, but the reconstruction of major nerve defects (>4.0 cm) remains a major challenge [118]. Clinically, for short, small-defect gaps, direct suture is possible with no tension on the nerve; for long defect distances, nerve tissue grafts or nerve conduit bridging is required. Autologous nerve grafts are currently the gold standard for nerve repair but have many associated challenges, including a shortage of nerve graft donors, potential risk of neuroma, and the possibility of infection [119]. Neural guide conduits (NGCs) are increasingly considered a potential alternative to nerve autografts for the treatment of peripheral nerve injury [120–122]. An ideal material for NGCs should have the following properties: suitable elasticity, good biocompatibility, biodegradability, high porosity, support directed nerve regeneration, and electroconductivity [123,124]. Therefore, the conductive NGCs based on ECHs have great potential for the treatment of peripheral nerve injury [125].

21.2.2.4 Electroconductive hydrogel for skin repair

The integrity of the skin, the largest organ of the human body, plays an important role in maintaining the health of the body because it is in direct contact with the external environment and prevents the invasion of pathogens and microorganisms. However, the skin is extremely vulnerable to skin defects caused by trauma, burns, ulcers, and surgical trauma. Like myocardium and nerve, skin is also sensitive to electrical signals and has a conductivity from 2.6 to 1×10^{-4} mS cm^{-1}. Epithelial tissue transports ions on the epidermis to form a transepithelial potential (TEP) and functions as an epidermal battery. In intact skin, TEP, maintained by the epidermis, is 10–60 mV, equivalent to the total polarization distribution of ion channel charges in epithelial cells [126]. When the skin is broken, the balance of the transmembrane potential is disrupted, resulting in a decrease in the potential at the wound. Recent studies have shown that external or endogenous ES may induce cell migration and proliferation to the wound along the electrical gradient until the wound heals and the TEP of the original epidermal battery is reestablished. As such, based on their properties, ECH dressings can provide a conductive pathway for endogenous bioelectricity as well as guided cell adhesion, migration, and orientation, thus favoring wound healing [127,128].

21.2.2.5 Electroconductive hydrogel for muscle repair

The skeletal muscle tissue accounts for 40% of body weight and is usually damaged by a tumor, traumatic injury, or muscular dystrophy. Compared to nerve or myocardium, skeletal muscle tissue has an inherent regeneration ability unless severely damaged. Similar to other electroactive tissues, the function of skeletal muscle depends on electrical signal conduction and ECHs can promote the recovery of impaired skeletal muscle functions and tissue regeneration [129]. The ECHs can maintain a suitable microenvironment for growth of myoblasts and myotube formation, and as such can significantly enhance proliferation and myogenic differentiation of C2C12 myoblasts. Therefore, ECHs have great potential for skeletal muscle injury repairs [96,130].

21.2.2.5.1 Electro-stimulated drug release

On-demand drug delivery is one of the main research directions of the drug delivery system (DDS). At present, with the development of various stimuli-responsive materials and technologies, it is possible to regulate drug release using external or internal stimuli. Among them, electrical stimulation DDS has great potential due to its easy combination with a sensor or microchip and precise time and space-controlled release ability. Electro-stimulated drug release is a method of releasing drugs from a material that responds to electrical stimuli. It uses electroactive biomaterials that can change their physical and chemical properties in response to electrical signals. Despite excellent properties, hydrogels generally suffer from insufficient electric field responsivity, critical for electrical-stimuli systems [131–134]. ECHs have been widely studied in order to eliminate such limitations [135]. Electro-stimulated drug release devices (EDRDs) are engineered devices that produce a programmed drug release profile influenced by the application of an impressed voltage or current [136,137], which benefit from a high loading capacity and low voltage actuation. In all cases, these polymeric materials, which are both electronically and ionically conductive, provide a noncytotoxic interface between the device and native living tissue or cell culture medium [138].

CEPs possess substantial porosity and delocalized charge centers to allow counter-ion diffusion and electromigration inside the polymeric electrode body in response to oxidation or reduction [139–142]. However, the continued development and exploitation of CEPs as drug release platforms suffer from the following major limitations: (1) Passive loss of loaded drug molecules by diffusion, (2) loss of loaded drug molecules as a result of anion exchange with the bathing environment, (3) poor oxidative stability of the pristine CEP, particularly under repeated cycling and in the presence of weakly oxidizing agents, (4) a type of "membrane fatigue" that is the result of the fast switching of the CEP and the generally slower diffusion of the drug molecule that gives rise to an accumulation of salts within the polymer and hence a reduction in its drug eluting capability, (5) inherently low drug loading levels where in the equilibrium pay load is governed by the density of redox-active sites on the CEP component, (6) poor control and unpredictability of the binding and release kinetics and (7) low diffusion coefficients of the released drug through dense semicrystalline polymer fibrils resulting in poor release kinetics and necessitating the application of slow cycling conditions that seek to match ON–OFF rates with attendant transport properties. Several of these limitations are addressed by the development of electro-responsive ECH [115,143,144]. These limitations are also addressed by the development of ECHs that offer the use of lower electrode potentials and faster switching [145]. The electro-stimulated retention of Ca^{2+} ions from an ECH formulated from a conetwork of a poly(HEMA)-based hydrogel and polyaniline has been demonstrated by Wilson et al. [97]. The release profile of passively loaded Ca^{2+} ions was measured using an ion-specific Ca^{2+} electrode from the ECH. The ECH is seen to release up to 65% of its loaded Ca^{2+} ions under open circuit (OC) conditions (or in the absence of an applied potential) and to do so with considerable time lag. However, the ECH instantaneously releases the loaded Ca^{2+} ions upon application of an oxidizing (+200 mV vs. Ag/AgCl) or reducing (−200mV vs. Ag/AgCl) potential. The amount released under electro-stimulated conditions is now ca. 25%, reflecting the preferential ingress of Cl^- ions from the bathing solution.

21.3 Conclusions and perspectives

This chapter introduced ECS, summarized their classifications, their various methods of synthesis and fabrication, and provided well-documented examples of their application in TE and RE. Recent advances in in vivo wound injury repair experiments demonstrate that ECHs are highly promising for repairing damage in electroactive organs or tissues. As discussed above, ECHs can promote the communication of electrical signals between cells, as well as mediate the adhesion, proliferation, migration, and differentiation of cells

ECHs represent a unique avenue for combining the responsive properties of both component materials within an aqueous milieu hospitable to biological molecules such as peptide sequences, enzymes, antibodies, and DNA. The combination of hydrogels and CEPs allows both materials to retain their unique responsive properties. In addition, ECHs enable a new class of devices with low interfacial impedance, suitable for neural prosthetic devices such as deep brain stimulation electrodes, low-voltage actuation for electrically stimulated drug release devices, and in vivo biocompatibility in implantable biosensors [2].

Finally, it is recommended that follow-up in vivo studies be performed to confirm whether these ECH candidates have the potential to promote nerve tissue regeneration in relevant medical conditions, such as for the treatment of the CNS and PNS traumatic injuries that produce critical size defects [114]. While outside the scope of this chapter, it is important to highlight that ECHs have great potential to improve the electrode-tissue integration and stability for BCI technologies [146]. Alzheimer's, CNS, and CVD remain among the major causes of death worldwide. As such, the future of neural and cardiac TE holds great potential. ECHs represent a unique platform for treating these diseases and improving the quality of human life. The combination of several approaches, such as 3D bioprinting, bioreactors, and stem cell engineering with ECHs, could further advance the development of cardiac and neural tissues or their corresponding organs in their entirety. The long-term bench-to-bedside success of TE/RE will strongly depend on impact-driven interdisciplinary collaboration among engineers, biologists, material scientists, and associated medical specialties, among others. However, despite the numerous hurdles to overcome, advances are rapidly being realized and subsequent clinical translatability will broaden.

References

[1] G.G.W. Anthony Guiseppi-Elie, T. Matsue, Chemical and biological sensors based on electrically conducting polymers, Handbook of Conducting Polymers, Marcel Dekker, (1998), p. 963 vol. 34.

[2] A. Guiseppi-Elie, Electroconductive hydrogels: synthesis, characterization and biomedical applications, Biomaterials 31 (10) (2010) 2701–2716. Available from https://doi.org/10.1016/j.biomaterials.2009.12.052.

[3] N.A. Peppas, R.M. Ottenbrite, K. Park, T. Okano, Biomedical Applications of Hydrogels Handbook, Springer, (2010).

[4] W. Zhang, P. Feng, J. Chen, Z. Sun, B. Zhao, Electrically conductive hydrogels for flexible energy storage systems, Progress in Polymer Science 88 (2019) 220–240. Available from https://doi.org/10.1016/j.progpolymsci.2018.09.001, http://www.sciencedirect.com/science/journal/00796700.

[5] M. Bansal, A. Dravid, Z. Aqrawe, J. Montgomery, Z. Wu, D. Svirskis, Conducting polymer hydrogels for electrically responsive drug delivery, Journal of Controlled Release 328 (2020) 192–209. Available from https://doi.org/10.1016/j.jconrel.2020.08.051, www.elsevier.com/locate/jconrel.

[6] R.A. Green, R.T. Hassarati, J.A. Goding, S. Baek, N.H. Lovell, P.J. Martens, et al., Conductive hydrogels: mechanically robust hybrids for use as biomaterials, Macromolecular Bioscience 12 (4) (2012) 494–501. Available from https://doi.org/10.1002/mabi.201100490.

[7] S. Liu, X. Li, L. Han, Recent developments in stimuli-responsive hydrogels for biomedical applications, Biosurface and Biotribology 8 (4) (2022) 290–306. Available from https://doi.org/10.1049/bsb2.12050, https://digital-library.theiet.org/content/journals/bsbt/5/1;jsessionid=2u2fkm73f1rsv.x-iet-live-01.

[8] A.C. Jen, M.C. Wake, A.G. Mikos, Review: hydrogels for cell immobilization, Biotechnology and Bioengineering 50 (4) (1996) 357–364. Available from https://doi.org/10.1002/(SICI)1097-0290, www.interscience.wiley.com/jpages/0006-3592.

[9] E.M. Ahmed, Hydrogel: preparation, characterization, and applications: a review, Journal of Advanced Research 6 (2) (2015) 105–121. Available from https://doi.org/10.1016/j.jare.2013.07.006, http://www.elsevier.com/wps/find/journaldescription.cws_home/722881/description#description.

[10] X. Guan, M. Avci-Adali, E. Alarçin, H. Cheng, S.S. Kashaf, Y. Li, et al., Development of hydrogels for regenerative engineering, Biotechnology Journal 12 (5) (2017) Available from https://doi.org/10.1002/biot.201600394, http://onlinelibrary.wiley.com/journal/10.1002/(ISSN)1860-7314.

[11] S. Agarwala, J.M. Lee, W.L. Ng, M. Layani, W.Y. Yeong, S. Magdassi, A novel 3D bioprinted flexible and biocompatible hydrogel bioelectronic platform, Biosensors and Bioelectronics 102 (2018) 365–371. Available from https://doi.org/10.1016/j.bios.2017.11.039, www.elsevier.com/locate/bios.

[12] M. Li, Y. Guo, Y. Wei, A. Macdiarmid, P. Lelkes, Electrospinning polyaniline-contained gelatin nanofibers for tissue engineering applications, Biomaterials 27 (13) (2006) 2705–2715. Available from https://doi.org/10.1016/j.biomaterials.2005.11.037.

[13] D. Kim, J. Kim, Y. Ko, K. Shim, J.H. Kim, J. You, A facile approach for constructing conductive polymer patterns for application in electrochromic devices and flexible microelectrodes, ACS Applied Materials and Interfaces 8 (48) (2016) 33175–33182. Available from https://doi.org/10.1021/acsami.6b10103, http://pubs.acs.org/journal/aamick.

[14] G. Rocchitta, A. Spanu, S. Babudieri, G. Latte, G. Madeddu, G. Galleri, et al., Enzyme biosensors for biomedical applications: strategies for safeguarding analytical performances in biological fluids, Sensors 16 (6) (2016) 780. Available from https://doi.org/10.3390/s16060780.

[15] C.N. Kotanen, O. Karunwi, A. Guiseppi-Elie, Biofabrication using pyrrole electropolymerization for the immobilization of glucose oxidase and lactate oxidase on implanted microfabricated biotransducers, Bioengineering 1 (1) (2014) 85–110. Available from https://doi.org/10.3390/bioengineering1010085, https://www.mdpi.com/2306-5354/1/1/85/pdf.

[16] S.I. Brahim, D. Maharajh, D. Narinesingh, A. Guiseppi-Elie, Design and characterization of a galactose biosensor using a novel polypyrrole-hydrogel composite membrane, Analytical Letters 35 (5) (2002) 797–812. Available from https://doi.org/10.1081/AL-120004070.

[17] S.S. Athukorala, T.S. Tran, R. Balu, V.K. Truong, J. Chapman, N.K. Dutta, et al., 3D printable electrically conductive hydrogel scaffolds for biomedical applications: a review, Polymers 13 (3) (2021) 1–21. Available from https://doi.org/10.3390/polym13030474, https://www.mdpi.com/2073-4360/13/3/474.

[18] Q. Yang, P. Li, H. Ran, J. Wan, H. Chen, H. Chen, et al., Polypyrrole-coated phase-change liquid perfluorocarbon nanoparticles for the visualized photothermal-chemotherapy of breast cancer, Acta Biomaterialia 90 (2019) 337–349. Available from https://doi.org/10.1016/j.actbio.2019.03.056, http://www.journals.elsevier.com/acta-biomaterialia.

[19] S. Shrestha, B.K. Shrestha, J.I. Kim, S. Won Ko, C.H. Park, C.S. Kim, Electrodeless coating polypyrrole on chitosan grafted polyurethane with functionalized multiwall carbon nanotubes electrospun scaffold for nerve tissue engineering, Carbon 136 (2018) 430–443. Available from https://doi.org/10.1016/j.carbon.2018.04.064, http://www.journals.elsevier.com/carbon/.

[20] Z. Shi, X. Gao, M.W. Ullah, S. Li, Q. Wang, G. Yang, Electroconductive natural polymer-based hydrogels, Biomaterials 111 (2016) 40–54. Available from https://doi.org/10.1016/j.biomaterials.2016.09.020, http://www.journals.elsevier.com/biomaterials/.

[21] B.W. Walker, R. Portillo Lara, E. Mogadam, C. Hsiang Yu, W. Kimball, N. Annabi, Rational design of microfabricated electroconductive hydrogels for biomedical applications, Progress in Polymer Science 92 (2019) 135–157. Available from https://doi.org/10.1016/j.progpolymsci.2019.02.007, http://www.sciencedirect.com/science/journal/00796700.

[22] M. Zhu, Y. Wang, G. Ferracci, J. Zheng, N.J. Cho, B.H. Lee, Gelatin methacryloyl and its hydrogels with an exceptional degree of controllability and batch-to-batch consistency, Scientific Reports 9 (1) (2019) Available from https://doi.org/10.1038/s41598-019-42186-x, www.nature.com/srep/index.html.

[23] K. Yue, G. Trujillo-de Santiago, M.M. Alvarez, A. Tamayol, N. Annabi, A. Khademhosseini, Synthesis, properties, and biomedical applications of gelatin methacryloyl (GelMA) hydrogels, Biomaterials 73 (2015) 254–271. Available from https://doi.org/10.1016/j.biomaterials.2015.08.045, http://www.journals.elsevier.com/biomaterials/.

[24] M. Omastová, M. Trchová, J. Kovářová, J. Stejskal, Synthesis and structural study of polypyrroles prepared in the presence of surfactants, Synthetic Metals 138 (3) (2003) 447–455. Available from https://doi.org/10.1016/S0379-6779(02)00498-8.

[25] S.M. Mousavi, S.A. Hashemi, S. Bahrani, K. Yousefi, G. Behbudi, A. Babapoor, et al., Recent advancements in polythiophene-based materials and their biomedical, geno sensor and DNA detection, International Journal of Molecular Sciences 22 (13) (2021) 6850. Available from https://doi.org/10.3390/ijms22136850.

[26] C.D. Pina, E. Falletta, Advances in polyaniline for biomedical applications, Current Medicinal Chemistry 29 (2) (2022) 329–357. Available from https://doi.org/10.2174/0929867328666210419135519.

[27] E.N. Zare, T. Agarwal, A. Zarepour, F. Pinelli, A. Zarrabi, F. Rossi, et al., Electroconductive multi-functional polypyrrole composites for biomedical applications, Applied Materials Today 24 (2021) 101117. Available from https://doi.org/10.1016/j.apmt.2021.101117.

[28] S. Cui, J. Mao, M. Rouabhia, S. Elkoun, Z. Zhang, A biocompatible polypyrrole membrane for biomedical applications, RSC Advances 11 (28) (2021) 16996–17006. Available from https://doi.org/10.1039/d1ra01338f, http://pubs.rsc.org/en/journals/journal/ra.

[29] Y. Kudoh, Properties of polypyrrole prepared by chemical polymerization using aqueous solution containing Fe2(SO4)3 and anionic surfactant, Synthetic Metals 79 (1) (1996) 17–22. Available from https://doi.org/10.1016/0379-6779(96)80124-X, http://www.journals.elsevier.com/synthetic-metals/.

[30] K. Tsuchiya, K. Ogino, Catalytic oxidative polymerization of thiophene derivatives, Polymer Journal 45 (3) (2013) 281–286. Available from https://doi.org/10.1038/pj.2012.146.

[31] J. Tropp, J. Rivnay, Design of biodegradable and biocompatible conjugated polymers for bioelectronics, Journal of Materials Chemistry C 9 (39) (2021) 13543–13556. Available from https://doi.org/10.1039/d1tc03600a, http://pubs.rsc.org/en/journals/journal/tc.

[32] T. Nezakati, A. Seifalian, A. Tan, A.M. Seifalian, Conductive polymers: opportunities and challenges in biomedical applications, Chemical Reviews 118 (14) (2018) 6766–6843. Available from https://doi.org/10.1021/acs.chemrev.6b00275, http://pubs.acs.org/journal/chreay.

[33] A. Yussuf, M. Al-Saleh, S. Al-Enezi, G. Abraham, Synthesis and characterization of conductive polypyrrole: the influence of the oxidants and monomer on the electrical, thermal, and morphological properties, International Journal of Polymer Science 2018 (2018) Available from https://doi.org/10.1155/2018/4191747, http://www.hindawi.com/journals/ijps/.

[34] Z. Deljoo Kojabad, S.A. Shojaosadati, Chemical synthesis of polypyrrole nanostructures: Optimization and applications for neural microelectrodes, Materials and Design 96 (2016) 378–384. Available from https://doi.org/10.1016/j.matdes.2016.02.045.

[35] D.D. Hebert, M.A. Naley, C.C. Cunningham, D.J. Sharp, E.E. Murphy, V. Stanton, et al., Enabling conducting polymer applications: methods for achieving high molecular weight in chemical oxidative polymerization in alkyl- and ether- substituted thiophenes. MDPI, United States, Materials 14 (20) (2021) Available from https://doi.org/10.3390/ma14206146, https://www.mdpi.com/1996-1944/14/20/6146/pdf.

[36] T.A.P. Hai, R. Sugimoto, Effect of molar ratio of oxidizer/3-hexylthiophene monomer in chemical oxidative polymerization of poly(3-hexylthiophene), Journal of Molecular Structure 1146 (2017) 660–668. Available from https://doi.org/10.1016/j.molstruc.2017.06.019, https://www.journals.elsevier.com/journal-of-molecular-structure.

[37] A. Bhat, A.R. Graham, H. Trivedi, M.K. Hogan, P.J. Horner, A. Guiseppi-Elie, Engineering the ABIO-BIO interface of neurostimulation electrodes using polypyrrole and bioactive hydrogels, Pure and Applied Chemistry 92 (2020) 897–907. Available from https://doi.org/10.1515/pac-2019-1107, http://www.degruyter.com/view/j/pac.

[38] Z. Zhou, K. Maxeiner, D.Y.W. Ng, T. Weil, Polymer chemistry in living cells, Accounts of Chemical Research 55 (20) (2022) 2998–3009. Available from https://doi.org/10.1021/acs.accounts.2c00420, http://pubs.acs.org/journal/achre4.

[39] Y. Dai, T. Li, Z. Zhang, Y. Tan, S. Pan, L. Zhang, et al., Oxidative polymerization in living cells, Journal of the American Chemical Society 143 (28) (2021) 10709–10717. Available from https://doi.org/10.1021/jacs.1c04821, http://pubs.acs.org/journal/jacsat.

[40] I. Moreno-Duarte, L.R. Morse, M. Alam, M. Bikson, R. Zafonte, F. Fregni, Targeted therapies using electrical and magnetic neural stimulation for the treatment of chronic pain in spinal cord injury, Neuroimage 85 (2014) 1003–1013. Available from https://doi.org/10.1016/j.neuroimage.2013.05.097, http://www.elsevier.com/inca/publications/store/6/2/2/9/2/5/index.htt.

[41] B. Maji, B. Barik, P. Dash, Methods for design and fabrication of nanosensors, Nanosensors for Smart Manufacturing, Elsevier, India, (2021), pp. 3–18. Available from https://doi.org/10.1016/B978-0-12-823358-0.00002-2, https://www.elsevier.com/books/nanosensors-for-smart-manufacturing/thomas/978-0-12-823358-0.

[42] K.W. Hatton, J.T. McLarney, T. Pittman, B.G. Fahy, Vagal nerve stimulation: overview and implications for anesthesiologists, Anesthesia and Analgesia 103 (5) (2006) 1241–1249. Available from https://doi.org/10.1213/01.ane.0000244532.71743.c6, http://journals.lww.com/anesthesia-analgesia/toc/publishahead.

[43] N. Khodaparast, M.P. Kilgard, R. Casavant, A. Ruiz, I. Qureshi, P.D. Ganzer, et al., Vagus nerve stimulation during rehabilitative training improves forelimb recovery after chronic ischemic stroke in rats, Neurorehabilitation and Neural Repair 30 (7) (2016) 676–684. Available from https://doi.org/10.1177/1545968315616494.

[44] E.C. Meyers, B.R. Solorzano, J. James, P.D. Ganzer, E.S. Lai, R.L. Rennaker, et al., Vagus nerve stimulation enhances stable plasticity and generalization of stroke recovery, Stroke; a Journal of Cerebral Circulation 49 (3) (2018) 710–717. Available from https://doi.org/10.1161/STROKEAHA.117.019202, http://stroke.ahajournals.org/.

[45] G.G. de Lima, S. Lyons, D.M. Devine, M.J.D. Nugent, Electrospinning of Hydrogels for Biomedical Applications. Springer Science and Business Media LLC, 2018, pp. 219–258. doi: 10.1007/978-981-10-6077-9_9.

[46] A. De Mori, M.P. Fernández, G. Blunn, G. Tozzi, M. Roldo, 3D printing and electrospinning of composite hydrogels for cartilage and bone tissue engineering, Polymers 10 (3) (2018) Available from https://doi.org/10.3390/polym10030285, http://www.mdpi.com/2073-4360/10/3/285/pdf.

[47] M.Z.A. Zulkifli, D. Nordin, N. Shaari, S.K. Kamarudin, Overview of electrospinning for tissue engineering applications, Polymers 15 (11) (2023) Available from https://doi.org/10.3390/polym15112418, http://www.mdpi.com/journal/polymers.

[48] M. Rahmati, D.K. Mills, A.M. Urbanska, M.R. Saeb, J.R. Venugopal, S. Ramakrishna, et al., Electrospinning for tissue engineering applications, Progress in Materials Science 117 (2021) 100721. Available from https://doi.org/10.1016/j.pmatsci.2020.100721.

[49] B. Guo, P.X. Ma, Conducting polymers for tissue engineering, Biomacromolecules 19 (6) (2018) 1764–1782. Available from https://doi.org/10.1021/acs.biomac.8b00276, http://pubs.acs.org/journal/bomaf6.

[50] H. Yuk, B. Lu, S. Lin, K. Qu, J. Xu, J. Luo, et al., 3D printing of conducting polymers, Nature Communications 11 (1) (2020) Available from https://doi.org/10.1038/s41467-020-15316-7, http://www.nature.com/ncomms/index.html.

[51] J.R. Aggas, S. Abasi, J.F. Phipps, D.A. Podstawczyk, A. Guiseppi-Elie, Microfabricated and 3-D printed electroconductive hydrogels of PEDOT:PSS and their application in bioelectronics, Biosensors and Bioelectronics 168 (2020) Available from https://doi.org/10.1016/j.bios.2020.112568, www.elsevier.com/locate/bios.

[52] B. Oh, S. Baek, K.S. Nam, C. Sung, C. Yang, Y.S. Lim, et al., 3D printable and biocompatible PEDOT:PSS-ionic liquid colloids with high conductivity for rapid on-demand fabrication of 3D bioelectronics, Nature Communications 15 (1) (2024) Available from https://doi.org/10.1038/s41467-024-50264-6, https://www.nature.com/ncomms/.

[53] J.R. Aggas, S. Abasi, B. Smith, M. Zimmerman, M. Deprest, A. Guiseppi-Elie, Microfabricated and 3-D printed soft bioelectronic constructs from PAn-PAAMPSA-containing hydrogels, Bioengineering 5 (4) (2018) Available from https://doi.org/10.3390/bioengineering5040087, https://www.mdpi.com/2306-5354/5/4/87/pdf.

[54] H. Rastin, B. Zhang, J. Bi, K. Hassan, T.T. Tung, D. Losic, 3D printing of cell-laden electroconductive bioinks for tissue engineering applications, Journal of Materials Chemistry B 8 (27) (2020) 5862–5876. Available from https://doi.org/10.1039/d0tb00627k, http://pubs.rsc.org/en/journals/journal/tb.

[55] A. Mihic, Z. Cui, J. Wu, G. Vlacic, Y. Miyagi, S.H. Li, et al., A conductive polymer hydrogel supports cell electrical signaling and improves cardiac function after implantation into myocardial infarct, Circulation 132 (8) (2015) 772–784. Available from https://doi.org/10.1161/CIRCULATIONAHA.114.014937, http://circ.ahajournals.org.

[56] A. Casella, J. Lowen, K.H. Griffin, N. Shimamoto, D.H. Ramos-Rodriguez, A. Panitch, et al., Conductive microgel annealed scaffolds enhance myogenic potential of myoblastic cells, Advanced Healthcare Materials 13 (25) (2024) Available from https://doi.org/10.1002/adhm.202302500, http://onlinelibrary.wiley.com/journal/10.1002/(ISSN)2192-2659.

[57] S. Shang, J. Liu, Y. He, P. Zhu, Smart conducting PNIPAM-co-AAc microgels with controllable phase transition and stimuli responsibility, Materials Letters 272 (2020) 127862. Available from https://doi.org/10.1016/j.matlet.2020.127862.

[58] G. Aguirre, P. Marcasuzaa, L. Billon, Soft self-assembled mechanoelectrical transducer films from conductive microgel waterborne dispersions, ACS Applied Materials and Interfaces 15 (23) (2023) 28310–28320. Available from https://doi.org/10.1021/acsami.3c04622, http://pubs.acs.org/journal/aamick.

[59] J.F. Mano, G.A. Silva, H.S. Azevedo, P.B. Malafaya, R.A. Sousa, S.S. Silva, et al., Natural origin biodegradable systems in tissue engineering and regenerative medicine: present status and some moving trends, Journal of The Royal Society Interface 4 (17) (2007) 999–1030. Available from https://doi.org/10.1098/rsif.2007.0220.

[60] J.P. Vacanti, C.A. Vacanti, The history and scope of tissue engineering, Principles of Tissue Engineering, fourth ed. Elsevier Inc., United States, (2013), pp. 3–8. Available from https://doi.org/10.1016/B978-0-12-398358-9.00001-X, http://www.sciencedirect.com/science/book/9780123983589.

[61] I.M. El-Sherbiny, M.H. Yacoub, Hydrogel scaffolds for tissue engineering: progress and challenges, Global Cardiology Science and Practice 2013 (3) (2013) 38. Available from https://doi.org/10.5339/gcsp.2013.38.

[62] D. Goldenberg, C. McLaughlin, S.V. Koduru, D.J. Ravnic, Regenerative engineering: current applications and future perspectives, Frontiers in Surgery 8 (2021) Available from https://doi.org/10.3389/fsurg.2021.731031, journal.frontiersin.org/journal/surgery.

[63] C.J. Esdaille, K.S. Washington, C.T. Laurencin, Regenerative engineering: a review of recent advances and future directions, Regenerative Medicine 16 (5) (2021) 495–512. Available from https://doi.org/10.2217/rme-2021-0016, http://www.futuremedicine.com/loi/rme.

[64] J. Yi, G. Choe, J. Park, J.Y. Lee, Graphene oxide-incorporated hydrogels for biomedical applications, Polymer Journal 52 (8) (2020) 823–837. Available from https://doi.org/10.1038/s41428-020-0350-9, http://www.nature.com/pj/journal/v42/n1/index.html.

[65] Y. Shi, C. Ma, L. Peng, G. Yu, Conductive "smart" hybrid hydrogels with PNIPAM and nanostructured conductive polymers, Advanced Functional Materials 25 (8) (2015) 1219–1225. Available from https://doi.org/10.1002/adfm.201404247, http://onlinelibrary.wiley.com/journal/10.1002/(ISSN)1616-3028.

[66] J.R. Aggas, A. Guiseppi-Elie, Responsive polymers in the fabrication of enzyme-based biosensors, Biomaterials Science: An Introduction to Materials in Medicine, Elsevier, United States, (2020), pp. 1267–1286. Available from https://doi.org/10.1016/B978-0-12-816137-1.00079-9, https://www.sciencedirect.com/book/9780128161371.

[67] J.R. Aggas, A. Guiseppi-Elie, Modeling a penta-analyte biochip for physiological status monitoring in the triage of hemorrhagic trauma and for allograft stratification. in: Proceedings of the 1st International Online Conference on Bioengineering MDPI SciForum, 2024.

[68] A. Guiseppi-Elie, An implantable biochip to influence patient outcomes following trauma-induced hemorrhage, Analytical and Bioanalytical Chemistry 399 (1) (2011) 403–419. Available from https://doi.org/10.1007/s00216-010-4271-x.

[69] X. Li, D. Liu, Z. Xiao, Y. Zhao, S. Han, B. Chen, et al., Scaffold-facilitated locomotor improvement post complete spinal cord injury: motor axon regeneration versus endogenous neuronal relay formation, Biomaterials 197 (2019) 20–31. Available from https://doi.org/10.1016/j.biomaterials.2019.01.012, http://www.journals.elsevier.com/biomaterials/.

[70] S. Meijs, M. Fjorback, C. Jensen, S. Sørensen, K. Rechendorff, N.J.M. Rijkhoff, Influence of fibrous encapsulation on electro-chemical properties of TiN electrodes, Medical Engineering and Physics 38 (5) (2016) 468–476. Available from https://doi.org/10.1016/j.medengphy.2016.02.010, www.elsevier.com/locate/medengphy.

[71] M. Kastellorizios, N. Tipnis, D.J. Burgess, Foreign body reaction to subcutaneous implants, Advances in Experimental Medicine and Biology 865 (2015) 93–108. Available from https://doi.org/10.1007/978-3-319-18603-0_6, http://www.springer.com/series/5584.

[72] J.M. Anderson, A. Rodriguez, D.T. Chang, Foreign body reaction to biomaterials, Seminars in Immunology 20 (2) (2008) 86–100. Available from https://doi.org/10.1016/j.smim.2007.11.004.

[73] R.A. Green, N.H. Lovell, G.G. Wallace, L.A. Poole-Warren, Conducting polymers for neural interfaces: challenges in developing an effective long-term implant, Biomaterials 29 (24-25) (2008) 3393–3399. Available from https://doi.org/10.1016/j.biomaterials.2008.04.047.

[74] A. Guiseppi-Elie, A.M. Wilson, J.M. Tour, T.W. Brockmann, P. Zhang, D.L. Allara, Specific immobilization of electropolymerized polypyrrole thin films onto interdigitated microsensor electrode arrays, Langmuir: The ACS Journal of Surfaces and Colloids 11 (5) (1995) 1768–1776. Available from https://doi.org/10.1021/la00005a055.

[75] A. Guiseppi-Elie, C. Dong, C.Z. Dinu, Crosslink density of a biomimetic poly(HEMA)-based hydrogel influences growth and proliferation of attachment dependent RMS 13 cells, Journal of Materials Chemistry 22 (37) (2012) 19529–19539. Available from https://doi.org/10.1039/c2jm32516k.

[76] V. Gounden, M. Singh, Hydrogels and wound healing: current and future prospects, Gels 10 (1) (2024) Available from https://doi.org/10.3390/gels10010043, www.mdpi.com/journal/gels.

[77] L. Edgar, T. Pu, B. Porter, J.M. Aziz, C. La Pointe, A. Asthana, et al., Regenerative medicine, organ bioengineering and transplantation, British Journal of Surgery 107 (7) (2020) 793–800. Available from https://doi.org/10.1002/bjs.11686.

[78] S. Abasi, J.R. Aggas, A. Guiseppi-Elie, Permissive electroconductive nanocomposites for neuronal progenitor cells. In: International IEEE/EMBS Conference on Neural Engineering, NER IEEE Computer Society United States, 2019. doi: 10.1109/NER.2019.8716893 5 875–878. http://ieeexplore.ieee.org/xpl/conhome.jsp?punumber=1001963 2019-

[79] C. Mancino, T. Hendrickson, L.V. Whitney, F. Paradiso, S. Abasi, E. Tasciotti, et al., Electrospun electroconductive constructs of aligned fibers for cardiac tissue engineering, Nanomedicine: Nanotechnology, Biology and Medicine 44 (2022) 102567. Available from https://doi.org/10.1016/j.nano.2022.102567.

[80] A.M.J.B.Ao.H.H. Wilson, E. Hydrogels, A.M. Wilson, G. Justin, A. Guiseppi-Elie, Abstract methods for the synthesis of electroconductive hydrogels (ECH), as polymer blends and as polymer conetworks via chemical oxidation, electrochemical, or a combination of chemical oxidation followed by electrochemical synthesis techniques, are described, 2010.

[81] K.Y. Lee, D.J. Mooney, Hydrogels for tissue engineering, Chemical Reviews 101 (7) (2001) 1869–1879. Available from https://doi.org/10.1021/cr000108x.

[82] N. Bhattarai, J. Gunn, M. Zhang, Chitosan-based hydrogels for controlled, localized drug delivery, Advanced Drug Delivery Reviews 62 (1) (2010) 83–99. Available from https://doi.org/10.1016/j.addr.2009.07.019.

[83] F.Y. Hsieh, H.H. Lin, Sh Hsu, 3D bioprinting of neural stem cell-laden thermoresponsive biodegradable polyurethane hydrogel and potential in central nervous system repair, Biomaterials 71 (2015) 48–57. Available from https://doi.org/10.1016/j.biomaterials.2015.08.028, http://www.journals.elsevier.com/biomaterials/.

[84] A. Casella, A. Panitch, J.K. Leach, Endogenous electric signaling as a blueprint for conductive materials in tissue engineering, Bioelectricity 3 (1) (2021) 27–41. Available from https://doi.org/10.1089/bioe.2020.0027, home.liebertpub.com/publications/bioelectricity/647/.

[85] Z.J. Rogers, M.P. Zeevi, R. Koppes, S.A. Bencherif, Electroconductive hydrogels for tissue engineering: current status and future perspectives, Bioelectricity 2 (3) (2020) 279–292. Available from https://doi.org/10.1089/bioe.2020.0025, home.liebertpub.com/publications/bioelectricity/647/.

[86] R. Bao, B. Tan, S. Liang, N. Zhang, W. Wang, W. Liu, A π-π conjugation-containing soft and conductive injectable polymer hydrogel highly efficiently rebuilds cardiac function after myocardial infarction, Biomaterials 122 (2017) 63–71. Available from https://doi.org/10.1016/j.biomaterials.2017.01.012, http://www.journals.elsevier.com/biomaterials/.

[87] L. Zhou, L. Fan, X. Yi, Z. Zhou, C. Liu, R. Fu, et al., Soft conducting polymer hydrogels cross-linked and doped by tannic acid for spinal cord injury repair, ACS Nano 12 (11) (2018) 10957–10967. Available from https://doi.org/10.1021/acsnano.8b04609, http://pubs.acs.org/journal/ancac3.

[88] G. Justin, A. Guiseppi-Elie, Electroconductive blends of poly(HEMA-co-PEGMA-co-HMMAco-SPMA) and poly(Py-co-PyBA): in vitro biocompatibility, Journal of Bioactive and Compatible Polymers 25 (2) (2010) 121–140. Available from https://doi.org/10.1177/0883911509350660.

[89] P. Zarrintaj, E. Zangene, S. Manouchehri, L.M. Amirabad, N. Baheiraei, M.R. Hadjighasem, et al., Conductive biomaterials as nerve conduits: recent advances and future challenges, Applied Materials Today 20 (2020) 100784. Available from https://doi.org/10.1016/j.apmt.2020.100784.

[90] Y. Liang, L. Qiao, B. Qiao, B. Guo, Conductive hydrogels for tissue repair, Chemical Science 14 (12) (2023) 3091–3116. Available from https://doi.org/10.1039/d3sc00145h, http://pubs.rsc.org/en/journals/journal/sc.

[91] B. Guo, J. Qu, X. Zhao, M. Zhang, Degradable conductive self-healing hydrogels based on dextran-graft-tetraaniline and N-carboxyethyl chitosan as injectable carriers for myoblast cell therapy and muscle regeneration, Acta Biomaterialia 84 (2019) 180–193. Available from https://doi.org/10.1016/j.actbio.2018.12.008, http://www.journals.elsevier.com/acta-biomaterialia.

[92] P. Sikorski, Electroconductive scaffolds for tissue engineering applications, Biomaterials Science 8 (20) (2020) 5583–5588. Available from https://doi.org/10.1039/d0bm01176b, http://pubs.rsc.org/en/journals/journal/bm.

[93] C.N. Kotanen, A.N. Wilson, C. Dong, C.Z. Dinu, G.A. Justin, A. Guiseppi-Elie, The effect of the physicochemical properties of bioactive electroconductive hydrogels on the growth and proliferation of attachment dependent cells, Biomaterials 34 (27) (2013) 6318–6327. Available from https://doi.org/10.1016/j.biomaterials.2013.05.022.

[94] S.R. Shin, S.M. Jung, M. Zalabany, K. Kim, P. Zorlutuna, S.B. Kim, et al., Carbon-nanotube-embedded hydrogel sheets for engineering cardiac constructs and bioactuators, ACS Nano 7 (3) (2013) 2369–2380. Available from https://doi.org/10.1021/nn305559j.

[95] Y. Wu, L. Wang, B. Guo, P.X. Ma, Interwoven aligned conductive nanofiber yarn/hydrogel composite scaffolds for engineered 3D cardiac anisotropy, ACS Nano 11 (6) (2017) 5646–5659. Available from https://doi.org/10.1021/acsnano.7b01062, http://pubs.acs.org/journal/ancac3.

[96] H.Y. Gong, J. Park, W. Kim, J. Kim, J.Y. Lee, W.G. Koh, A novel conductive and micropatterned PEG-based hydrogel enabling the topographical and electrical stimulation of myoblasts, ACS Applied Materials and Interfaces 11 (51) (2019) 47695–47706. Available from https://doi.org/10.1021/acsami.9b16005, http://pubs.acs.org/journal/aamick.

[97] K. Hosoyama, M. Ahumada, K. Goel, M. Ruel, E.J. Suuronen, E.I. Alarcon, Electroconductive materials as biomimetic platforms for tissue regeneration, Biotechnology Advances 37 (3) (2019) 444–458. Available from https://doi.org/10.1016/j.biotechadv.2019.02.011, www.elsevier.com/inca/publications/store/5/2/5/4/5/5/index.htt.

[98] N. Tandon, C. Cannizzaro, P.H.G. Chao, R. Maidhof, A. Marsano, H.T.H. Au, et al., Electrical stimulation systems for cardiac tissue engineering, Nature Protocols 4 (2) (2009) 155–173. Available from https://doi.org/10.1038/nprot.2008.183.

[99] B.C. Thompson, R.T. Richardson, S.E. Moulton, A.J. Evans, S. O'Leary, G.M. Clark, et al., Conducting polymers, dual neurotrophins and pulsed electrical stimulation - dramatic effects on neurite outgrowth, Journal of Controlled Release 141 (2) (2010) 161–167. Available from https://doi.org/10.1016/j.jconrel.2009.09.016.

[100] M. Radisic, H. Park, H. Shing, T. Consi, F.J. Schoen, R. Langer, et al., Functional assembly of engineered myocardium by electrical stimulation of cardiac myocytes cultured on scaffolds, Proceedings of the National Academy of Sciences of the United States of America. 101 (52) (2004) 18129–18134. Available from https://doi.org/10.1073/pnas.0407817101.

[101] M. Sajid Arshad, M. Javed, M. Sohaib, F. Saeed, A. Imran, Z. Amjad, et al., Tissue engineering approaches to develop cultured meat from cells: a mini review, Cogent Food & Agriculture 3 (1) (2017) 1320814. Available from https://doi.org/10.1080/23311932.2017.1320814.

[102] H. Lou, H. Lu, S. Zhang, Y. Shi, E. Xu, D. Liu, et al., Highly aligned myotubes formation of piscine satellite cells in 3D fibrin hydrogels of cultured meat, International Journal of Biological Macromolecules 282 (Pt 2) (2024) 136879. Available from https://doi.org/10.1016/j.ijbiomac.2024.136879.

[103] G.W. Reed, J.E. Rossi, C.P. Cannon, Acute myocardial infarction, The Lancet 389 (10065) (2017) 197–210. Available from https://doi.org/10.1016/S0140-6736(16)30677-8, http://www.journals.elsevier.com/the-lancet/.

[104] L.M. Ptaszek, M. Mansour, J.N. Ruskin, K.R. Chien, Towards regenerative therapy for cardiac disease, The Lancet 379 (9819) (2012) 933–942. Available from https://doi.org/10.1016/S0140-6736(12)60075-0, http://www.journals.elsevier.com/the-lancet/.

[105] H. Esmaeili, A. Patino-Guerrero, M. Hasany, M.O. Ansari, A. Memic, A. Dolatshahi-Pirouz, et al., Electroconductive biomaterials for cardiac tissue engineering, Acta Biomaterialia 139 (2022) 118–140. Available from https://doi.org/10.1016/j.actbio.2021.08.031, http://www.journals.elsevier.com/acta-biomaterialia.

[106] L. Wang, J. Jiang, W. Hua, A. Darabi, X. Song, C. Song, et al., Mussel-inspired conductive cryogel as cardiac tissue patch to repair myocardial infarction by migration of conductive nanoparticles, Advanced Functional Materials 26 (24) (2016) 4293–4305. Available from https://doi.org/10.1002/adfm.201505372, http://onlinelibrary.wiley.com/journal/10.1002/(ISSN)1616-3028.

[107] G.A.A. Saracino, D. Cigognini, D. Silva, A. Caprini, F. Gelain, Nanomaterials design and tests for neural tissue engineering, Chemical Society Reviews 42 (1) (2013) 225–262. Available from https://doi.org/10.1039/c2cs35065c.

[108] A.O. Boztas, A. Guiseppi-Elie, Immobilization and release of the redox mediator ferrocene monocarboxylic acid from within cross-linked p (HEMA-co-PEGMA-co-HMMA) hydrogels, Biomacromolecules 10 (8) (2009) 2135–2143. Available from https://doi.org/10.1021/bm900299b, http://pubs.acs.org/doi/pdfplus/10.1021/bm900299b.

[109] J. George, C.C. Hsu, L.T.B. Nguyen, H. Ye, Z. Cui, Neural tissue engineering with structured hydrogels in CNS models and therapies, Biotechnology Advances 42 (2020) Available from https://doi.org/10.1016/j.biotechadv.2019.03.009, www.elsevier.com/inca/publications/store/5/2/5/4/5/5/index.htt.

[110] Y. Zhao, S. Song, X. Ren, J. Zhang, Q. Lin, Y. Zhao, Supramolecular adhesive hydrogels for tissue engineering applications, Chemical Reviews 122 (6) (2022) 5604–5640. Available from https://doi.org/10.1021/acs.chemrev.1c00815, http://pubs.acs.org/journal/chreay.

[111] D.N. Heo, S.J. Lee, R. Timsina, X. Qiu, N.J. Castro, L.G. Zhang, Development of 3D printable conductive hydrogel with crystallized PEDOT:PSS for neural tissue engineering, Materials Science and Engineering C 99 (2019) 582–590. Available from https://doi.org/10.1016/j.msec.2019.02.008.

[112] X. Yao, Y. Qian, C. Fan, Electroactive nanomaterials in the peripheral nerve regeneration, Journal of Materials Chemistry B 9 (35) (2021) 6958–6972. Available from https://doi.org/10.1039/d1tb00686j, http://pubs.rsc.org/en/journals/journal/tb.

[113] A.N. Koppes, K.W. Keating, A.L. McGregor, R.A. Koppes, K.R. Kearns, A.M. Ziemba, et al., Robust neurite extension following exogenous electrical stimulation within single walled carbon nanotube-composite hydrogels, Acta Biomaterialia 39 (2016) 34–43. Available from https://doi.org/10.1016/j.actbio.2016.05.014.

[114] Z. Zhang, M. Ma, Strategies to enhance the ability of nerve guidance conduits to promote directional nerve growth, Biomedical Engineering Online 23 (1) (2024) Available from https://doi.org/10.1186/s12938-024-01233-z, https://biomedical-engineering-online.biomedcentral.com/.

[115] S. Murdan, Electro-responsive drug delivery from hydrogels, Journal of Controlled Release 92 (1-2) (2003) 1–17. Available from https://doi.org/10.1016/S0168-3659(03)00303-1, www.elsevier.com/locate/jconrel.

[116] Q. Zhang, B. Shi, J. Ding, L. Yan, J.P. Thawani, C. Fu, et al., Polymer scaffolds facilitate spinal cord injury repair, Acta Biomaterialia 88 (2019) 57–77. Available from https://doi.org/10.1016/j.actbio.2019.01.056, http://www.journals.elsevier.com/acta-biomaterialia.

[117] C. Rinoldi, M. Lanzi, R. Fiorelli, P. Nakielski, K. Zembrzycki, T. Kowalewski, et al., Three-dimensional printable conductive semi-interpenetrating polymer network hydrogel for neural tissue applications, Biomacromolecules 22 (7) (2021) 3084–3098. Available from https://doi.org/10.1021/acs.biomac.1c00524, http://pubs.acs.org/journal/bomaf6.

[118] X. Navarro, M. Vivó, A. Valero-Cabré, Neural plasticity after peripheral nerve injury and regeneration, Progress in Neurobiology 82 (4) (2007) 163–201. Available from https://doi.org/10.1016/j.pneurobio.2007.06.005.

[119] R. Sullivan, T. Dailey, K. Duncan, N. Abel, C. Borlongan, Peripheral nerve injury: stem cell therapy and peripheral nerve transfer, International Journal of Molecular Sciences 17 (12) (2016) Available from https://doi.org/10.3390/ijms17122101, http://www.mdpi.com/1422-0067/17/12/2101/pdf.

[120] S. Vijayavenkataraman, Nerve guide conduits for peripheral nerve injury repair: a review on design, materials and fabrication methods, Acta Biomaterialia 106 (2020) 54–69. Available from https://doi.org/10.1016/j.actbio.2020.02.003, http://www.journals.elsevier.com/acta-biomaterialia.

[121] W.A. Lackington, A.J. Ryan, F.J. O'Brien, Advances in nerve guidance conduit-based therapeutics for peripheral nerve repair, ACS Biomaterials Science and Engineering 3 (7) (2017) 1221–1235. Available from https://doi.org/10.1021/acsbiomaterials.6b00500, http://pubs.acs.org/journal/abseba.

[122] N.U. Kang, S.J. Lee, S.J. Gwak, Fabrication techniques of nerve guidance conduits for nerve regeneration, Yonsei Medical Journal 63 (2) (2022) 114–123. Available from https://doi.org/10.3349/ymj.2022.63.2.114, http://eymj.org/Synapse/Data/PDFData/0069YMJ/ymj-63-114.pdf.

[123] R.V. Bellamkonda, Peripheral nerve regeneration: an opinion on channels, scaffolds and anisotropy, Biomaterials 27 (19) (2006) 3515–3518. Available from https://doi.org/10.1016/j.biomaterials.2006.02.030.

[124] H. Jiang, Y. Qian, C. Fan, Y. Ouyang, Polymeric guide conduits for peripheral nerve tissue engineering, Frontiers in Bioengineering and Biotechnology 8 (2020) Available from https://doi.org/10.3389/fbioe.2020.582646, http://journal.frontiersin.org/journal/bioengineering-and-biotechnology#archive.

[125] F. Zhang, M. Zhang, S. Liu, C. Li, Z. Ding, T. Wan, et al., Application of hybrid electrically conductive hydrogels promotes peripheral nerve regeneration, Gels 8 (1) (2022) Available from https://doi.org/10.3390/gels8010041, https://www.mdpi.com/2310-2861/8/1/41/pdf.

[126] C. Martin-Granados, C.D. McCaig, Harnessing the electric spark of life to cure skin wounds, Advances in Wound Care 3 (2) (2014) 127–138. Available from https://doi.org/10.1089/wound.2013.0451.

[127] R. Yu, H. Zhang, B. Guo, Conductive biomaterials as bioactive wound dressing for wound healing and skin tissue engineering, Nano-Micro Letters. 14 (1) (2022) Available from https://doi.org/10.1007/s40820-021-00751-y, http://www.springer.com/engineering/journal/40820.

[128] L. Nie, Q. Wei, J. Li, Y. Deng, X. He, X. Gao, et al., Fabrication and desired properties of conductive hydrogel dressings for wound healing, RSC Advances 13 (13) (2023) 8502–8522. Available from https://doi.org/10.1039/d2ra07195a, http://pubs.rsc.org/en/journals/journal/ra.

[129] R. Dong, P.X. Ma, B. Guo, Conductive biomaterials for muscle tissue engineering, Biomaterials 229 (2020) 119584. Available from https://doi.org/10.1016/j.biomaterials.2019.119584.

[130] H. Jo, M. Sim, S. Kim, S. Yang, Y. Yoo, J.-H. Park, et al., Electrically conductive graphene/polyacrylamide hydrogels produced by mild chemical reduction for enhanced myoblast growth and differentiation, Acta Biomaterialia 48 (2017) 100–109. Available from https://doi.org/10.1016/j.actbio.2016.10.035.

[131] S. Ganguly, D. Ray, P. Das, P.P. Maity, S. Mondal, V.K. Aswal, et al., Mechanically robust dual responsive water dispersible-graphene based conductive elastomeric hydrogel for tunable pulsatile drug release, Ultrasonics Sonochemistry 42 (2018) 212–227. Available from https://doi.org/10.1016/j.ultsonch.2017.11.028, www.elsevier.com/inca/publications/store/5/2/5/4/5/1.

[132] Y. Yun, H. Wu, J. Gao, W. Dai, L. Deng, O. Lv, et al., Facile synthesis of Ca2+-crosslinked sodium alginate/graphene oxide hybrids as electro- and pH-responsive drug carrier, Materials Science and Engineering C 108 (2020) Available from https://doi.org/10.1016/j.msec.2019.110380, https://www.journals.elsevier.com/materials-science-and-engineering-c.

[133] H. Ge, Y. Kong, D. Shou, L. Deng, Communication—three-dimensional electro- and pH-responsive polypyrrole/alginate hybrid for dual-controlled drug delivery, Journal of the Electrochemical Society 163 (2016).

[134] R. Wang, D. Shou, O. Lv, Y. Kong, L. Deng, J. Shen, pH-Controlled drug delivery with hybrid aerogel of chitosan, carboxymethyl cellulose and graphene oxide as the carrier, International Journal of Biological Macromolecules 103 (2017) 248–253. Available from https://doi.org/10.1016/j.ijbiomac.2017.05.064, www.elsevier.com/locate/ijbiomac.

[135] J. Xu, Y.L. Tsai, S.H. Hsu, Design strategies of conductive hydrogel for biomedical applications, Molecules (Basel, Switzerland) 25 (22) (2020) Available from https://doi.org/10.3390/MOLECULES25225296, https://www.mdpi.com/1420-3049/25/22/5296/pdf.

[136] M. Staples, K. Daniel, M.J. Cima, R. Langer, Application of micro- and nano-electromechanical devices to drug delivery, Pharmaceutical Research 23 (5) (2006) 847–863. Available from https://doi.org/10.1007/s11095-006-9906-4.

[137] D.A. LaVan, T. McGuire, R. Langer, Small-scale systems for in vivo drug delivery, Nature Biotechnology 21 (10) (2003) 1184–1191. Available from https://doi.org/10.1038/nbt876.

[138] J.M. Fonner, L. Forciniti, H. Nguyen, J.D. Byrne, Y.-F. Kou, J. Syeda-Nawaz, et al., Biocompatibility implications of polypyrrole synthesis techniques, Biomedical Materials 3 (3) (2008) 034124. Available from https://doi.org/10.1088/1748-6041/3/3/034124.

[139] A.G. MacDiarmid, "Synthetic metals": a novel role for organic polymers, Current Applied Physics 1 (4-5) (2001) 269–279. Available from https://doi.org/10.1016/S1567-1739(01)00051-7.

[140] R. Patil, Y. Harima, K. Yamashita, K. Komaguchi, Y. Itagaki, M. Shiotani, Charge carriers in polyaniline film: a correlation between mobility and in-situ ESR measurements, Journal of Electroanalytical Chemistry 518 (1) (2002) 13–19. Available from https://doi.org/10.1016/s0022-0728(01)00689-1.

[141] A.J. Epstein, J. Joo, R.S. Kohlman, G. Du, A.G. MacDiarmid, E.J. Oh, et al., Inhomogeneous disorder and the modified Drude metallic state of conducting polymers, Synthetic Metals 65 (2-3) (1994) 149–157. Available from https://doi.org/10.1016/0379-6779(94)90176-7.

[142] S. Roth, H. Bleier, W. Pukacki, Charge transport in conducting polymers, Faraday Discussions of the Chemical Society 88 (1989) 223–233. Available from https://doi.org/10.1039/DC9898800223.

[143] R. Tomer, D. Dimitrijevic, A.T. Florence, Electrically controlled release of macromolecules from cross-linked hyaluronic acid hydrogels, Journal of Controlled Release 33 (3) (1995) 405–413. Available from https://doi.org/10.1016/0168-3659(94)00115-B.

[144] I. Stassen, T. Sloboda, G. Hambitzer, Membrane with controllable permeability for drugs, Synthetic Metals 71 (1-3) (1995) 2243–2244. Available from https://doi.org/10.1016/0379-6779(94)03241-w.

[145] L.M. Lira, S.I. Córdoba De Torresi, Conducting polymer-hydrogel composites for electrochemical release devices: synthesis and characterization of semi-interpenetrating polyaniline- polyacrylamide networks, Electrochemistry Communications 7 (7) (2005) 717–723. Available from https://doi.org/10.1016/j.elecom.2005.04.027.

[146] H. Yuk, B. Lu, X. Zhao, Hydrogel bioelectronics, Chemical Society Reviews 48 (6) (2019) 1642–1667. Available from https://doi.org/10.1039/c8cs00595h, http://pubs.rsc.org/en/journals/journal/cs.

Chapter 22

Advanced biosensors developed using biomaterials at macro, micro, and nanoscale for tissue engineering applications

Sushmita Kushwaha[1,2,*], Shakra Nazli[1,2,*], and Raja Gopal Rayavarapu[1,2]

[1]*Nanomaterial Toxicology Laboratory, Food, Drug & Chemical, Environment and Systems Toxicology (FEST) Division, Drug and Chemical Toxicology Group, CSIR-Indian Institute of Toxicology Research (CSIR-IITR), Lucknow, Uttar Pradesh, India,* [2]*Academy of Scientific and Innovative Research (AcSIR), Ghaziabad, Uttar Pradesh, India*

22.1 Introduction

Tissue engineering is an interdisciplinary field that combines biological sciences, physical sciences, and engineering to create substitutes aimed at restoring, enhancing, or replacing damaged tissues or organs, bone repair, and skin regeneration. There are two primary approaches to tissue engineering: (1) the cell-based approach, which relies on the use of living cells to regenerate tissue. They may be of autografts, allografts, or xenografts types. (2) Scaffold-based approach, which provides a 3D structure that mimics the tissue's characteristics and functions. These can be natural (e.g., collagen) or synthetic [e.g., polylactic acid(PLA)], which are designed to be biocompatible and customizable for different tissue types. Functionality and viability of engineered tissues depend upon the biosensors as they play a crucial role in real-time monitoring, controlled tissue development process [1,2].

A biosensor is a device that incorporates a biological component with an instrument which can convert biochemical reactions into a measurable signal. Recently, it has gained huge recognition, mainly in nanotechnology, tissue engineering, and clinical applications [3,4]. Clarks and Lyson designed a device in the 1960s to measure glucose using immobilized glucose oxidase on an electrode [5]. Tissue engineering and regenerative medicines are growing as reconstructive therapies for tissue reconstruction and restoring damaged tissues and organs and lost functions [6]. Microfluidic chips mainly play a crucial role in tissue engineering. They can monitor minuscule biological signals within small tissue constructs with high specificity by utilizing ultrasensitive optical, electrochemical, and acoustic sensing technologies. Biosensors are also used for monitoring of biomarkers of certain diseases [4]. There were many issues while using the synthetic-based biosensors, such as saturation, nonspecific binding, strong affinities, and difficulty in selection. Biomaterial-based biosensors solved all these problems by showing higher efficacy, particular binding, selectivity, and biocompatibility [7]. The bioreceptor components of biosensors have been extensively engineered to perform diverse functions for detecting toxic substances, proteins, genes, cells, or tissue activity. This gives biosensors a distinct advantage and allows them to provide timely information on health-related questions like early disease detection and treatment-making in clinical applications [7,8]. Several fabrication techniques have been developed since the advent of the first biosensors, including photolithography, SPR (Surface Plasmon Resonance), 3D printing, and nanotechnology, for applications in tissue engineering and medical diagnostics [7,9–11]. Micro- and nanosensors have been developed for clinical applications to detect biological signals in the microenvironment. They can measure pH, oxygen level, and glucose concentration. For example, in novel immunofluorescent assay or exosome

* Equal First authors

Fundamentals of Translational Tissue Engineering. DOI: https://doi.org/10.1016/B978-0-443-33455-9.00032-8

detection systems, magnetic nanobeads are utilized [12]. A highly fluorescent nanocrystal, quantum dots, can detect specific proteins and DNA [13]. Notably, biosensors featuring various micro- and nanostructured surfaces have been effectively employed in vivo studies. This chapter provides an overview of biosensors, the types of biomaterials used in their fabrication, and the applications of these devices in tissue engineering.

22.2 Overview of biosensors

Biosensors are instruments that translate biological or chemical reactions into signals indicating a particular analyte's concentration. Biomolecules like an antibody, aptamers, enzymes, nucleic acids, or cells are bioreceptors capable of detecting or identifying target analytes as shown in the schematic representation in Fig. 22.1. These biomolecules make these biosensors highly selective due to the specific interaction between bioreceptors and analytes, known as biorecognition [14,15]. The elements typically found in a biosensor are:

Analyte: The analyte is a particular substance that the biosensor intends to detect or quantify, which differs according www.flaticon.comto the biosensor's application. For example, glucose is used for diabetes monitoring.

Bioreceptor: Bioreceptors are used to bind and recognize analytes. It provides selectivity, ensuring the biosensor interacts with the specific substance. Depending on the sensor's application, bioreceptors can include antibodies, enzymes, DNA/RNA, aptamers, cells, or proteins.

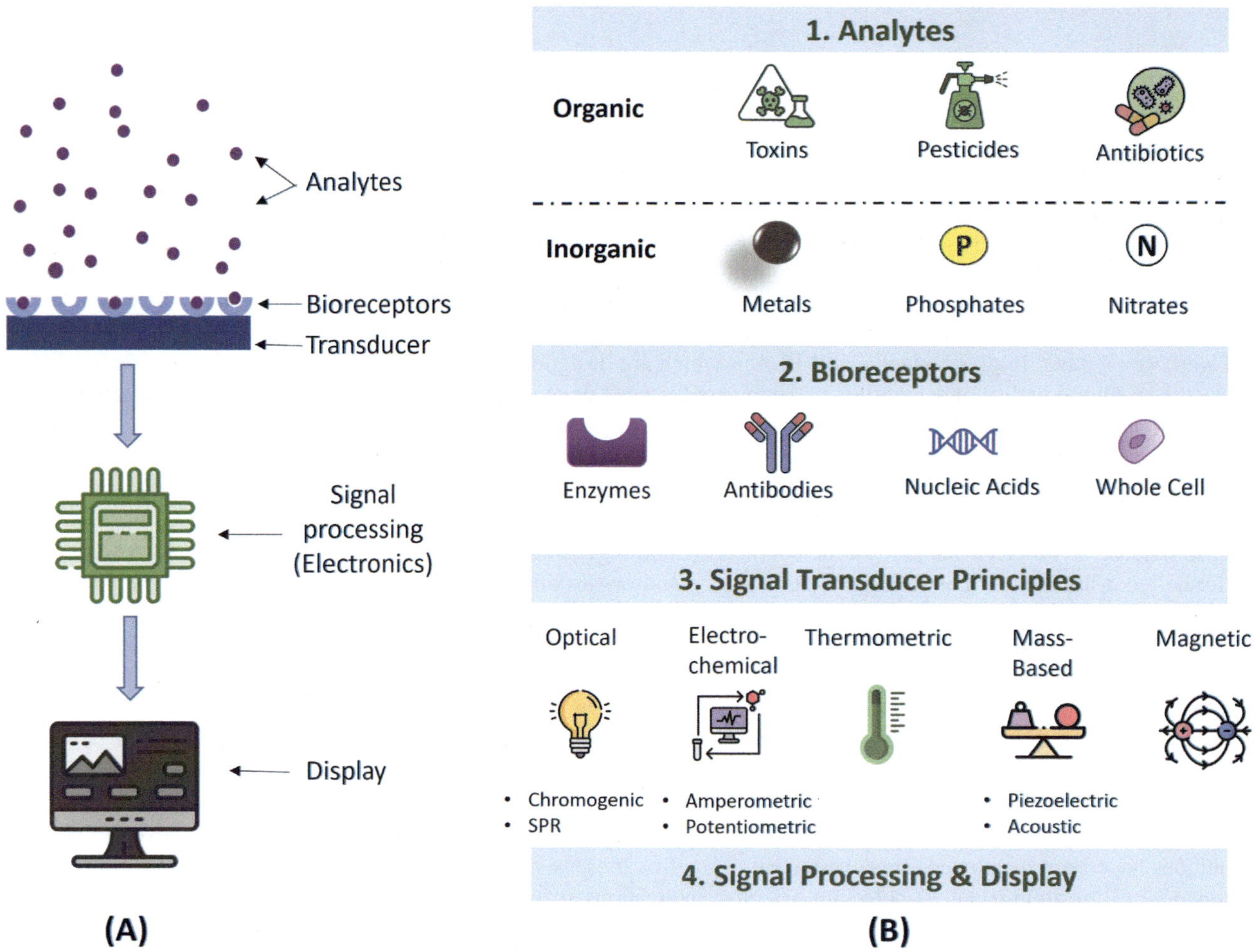

FIGURE 22.1 (A) Schematic representation of a biosensor. (B1) Different types of Organic and inorganic analytes, (B2) different forms of bioreceptors, i.e., enzymes, antibodies, nucleic acids, whole cells, (B3) different types of transducers showing optical, electrochemical, thermometric, mass-based, magnetic, and (B4) signal processing and display.
Icon from www.flaticon.com.

Transducer: Bioreceptor–analyte interaction transduced in a quantifiable electrical, optical, thermal, or mechanical signal by transducers.

Electronics: It processes the signal produced by the transducer and amplifies filters, and converts it, usually changing it from analog to digital for straightforward interpretation.

Display: The display shows the final output in a clear format that is easy to read. After the electronics process the signal, the display presents the data in numbers, graphs, images, or text, typically using simple interfaces like liquid-crystal display (LCD) screens or digital panels [14,15].

22.2.1 Characteristics of biosensors

For high functionality, better performance, and good commercial applications, some static and dynamic requisites are necessary for a biosensor, which are:

- **Selectivity:** Selectivity means detecting a particular analyte in a sample mixture of impurities. Biosensor selectivity depends on three critical aspects: the analyte, the transducer, and the pH of the solution.
- **Sensitivity:** Sensitivity is defined by the smallest detectable amount of analyte, identified with a few steps and at deficient concentrations, such as nanogram mL^{-1} or femtogram mL^{-1}.
- **Linearity:** A sensor's linearity is evaluated by how closely its calibration curve aligns with a straight line. A higher degree of linearity corresponds to higher detection of substrate concentrations.
- **Reproducibility:** Reproducibility means precision, where repeated measurements of the same sample give similar outputs and accuracy, which means the average reading is close to the actual value.
- **Detection limit and response time:** The detection limit of biosensors means the lowest analyte concentration that can be reliably sensed at a sufficiently low signal-to-noise (S/N) ratio. Response time indicates the time required to obtain 95% of the outcomes [14–16].

22.2.2 Classification of biosensors

Due to their diverse and multidisciplinary nature, the classifications of biosensors according to a wide range of criteria are:

- **Based on bioreceptor type:**
  A bioreceptor is a molecule, such as an enzyme, DNA probe, or antigen/antibody, that specifically binds and identifies a particular analyte.
  Bioreceptors in enzyme-based biosensors are enzymes that can be used for an extended period because enzymes are not consumed. The life of a biosensor depends on the enzyme stability. For example, glucose oxidase in glucose biosensors is used for glucose detection in blood [17,18].
  Antibodies are attached to the transducer and work as bioreceptors in antibody-based biosensors. For example, protein-based biosensors for HIV detection [19].
  In cell-based biosensors, living cells detect the changes in the intracellular and extracellular microenvironment and physiological parameters. It generates the response after the interaction between stimuli and the cell as in case of early oral cancer detection [18].

- **Based on transducer type:**
  Different biosensors based on transducers, such as electrochemical biosensors, which use electrodes as transducers. After target–analyte interaction, either electron transfers or there is a change in potential, which leads to current or voltage generation.
  The optical transducer provides real-time, label-free parallel detection of various analytes, which is used at a large scale in research applications and medical diagnostics [4].
  Piezoelectric/Acoustic biosensors have piezoelectric crystals which undergo elastic deformation on electrical stimulus and generate an electrical potential when mechanical force is applied [18].

- **Based on analyte type:**
  Analyte-based biosensors detect biological or chemical substances known as analytes, such as glucose, DNA, proteins, toxins, or pathogens.
  A metabolite biosensor is a biosensor based on analytes and it is used in the detection of natural products and the overproduction of natural products, but its own quantities and varieties restrict it [20].
  A glucose biosensor is also an analyte-based biosensor for detecting glucose [17].

22.2.3 Biomaterials in biosensor development

Some critical functional features of a biomaterial used in biosensor development differ in sensors based on measurement techniques like transparency for optical devices or electrical conductivity for electrochemical techniques. These properties ensure the sensor's sensitivity, selectivity, stability, and responsiveness [21,22]. In tissue engineering applications, the properties of ideal biomaterials are that they are bioactive, biocompatible, biodegradable, and sterilizable. Based on their function and nature, different biomaterials are used in biosensors, such as natural biomaterials, synthetic biomaterials, and inorganic biomaterials. But due to high biocompatibility, effective biodegradability, and minimal toxicity, natural polymers like chitosan, gelatin, collagen, cellulose, and alginates usages are more than synthetic polymers such as polylactic-co-glycolic acid (PLGA), polycaprolactone (PCL), PLA, fibronectin, and polyurethane [23] (Table 22.1).

22.3 Advancements in biosensors

Biosensors are essential in tissue engineering, as they play an important role in monitoring, controlling, and optimizing the conditions necessary for the successful growth and development of engineered tissues. Various fabrication techniques, as shown, are used to develop biosensors to support these functions, each contributing to enhancing the precision, efficiency, and safety of tissue regeneration. By providing real-time data and feedback, these biosensors help refine tissue engineering processes, ultimately leading to improved outcomes in clinical applications.

22.3.1 Nanomaterial-based biosensors

Different nanomaterials like carbon nanotubes, graphene sheets, nanocrystals, and metal nanoparticles, due to their greater surface area and excellent electrical properties, have significantly enhanced the sensitivity and efficiency of detection of biosensors [12]. Almost every nanomaterial is biofunctionalized with molecules such as antibodies, peptides, and nucleic acids, using approaches such as covalent bonding, electrostatic forces, polymer entrapment, or Van der Waals forces, without altering the intrinsic properties of the components [13,39]. Incorporating nanomaterials within electrochemical biosensors gives exosomes a better sensing limit. Nanomaterials enhance sensor performance by speeding up signal transmission and amplifying the detection of biological interactions. Zhou et al. developed a micro gold chip with multiple channels for exosome analysis. The expression of an epithelial cell adhesion molecule and an antigen on the prostate membrane was measured using silver and copper nanoparticle oxidation [39,40]. Fluorescence quenching properties of nanocrystals called quantum dots are used in optical transducers because fluorescence reappears when the quencher is removed. This approach is commonly used for optical sensing of DNA and oligonucleotides [41,42].

22.3.2 Photolithography in biosensors

Photolithography transfers geometric patterns from a photomask onto a silicon or glass surface. It involves multiple steps: pattern designing and printing onto a mask, then coating the substrate with photoresist and transferring the pattern to the substrate through UV exposure, followed by pouring of polymers to develop the mold, and finally baking it [12,43–45]. This technique is mainly used for proteins and cell patterning in tissue engineering and biosensor applications. For example, in a study by Mir et al., the fabrication of a protein array on polymer-coated silicon film, AZ5214, was done using protein photolithography. This allows the precise patterning of protein spots with diameters as small as 2 micrometers (μm). This is significant for application in detection devices and advancing biochip fabrication, with the possibility of producing microarrays detecting different proteins with dimensions as small as 50–100 nm [43]. An electrochemical 20 μm bandwidth, interdigitated microarray can detect DNA, enzymes, and antibodies.

22.3.3 Biosensor based on surface plasmon resonance

Biosensors based on SPR have appeared as an appropriate technique for detecting molecules without any labeling and real-time monitoring [44,45]. In SPR instruments, polarized light is directed through a high-refractive index prism onto a metal surface (like silver, gold or copper) at a particular resonance angle, which excites the surface plasmons. When biomolecules bind near the surface, this changes the local refractive index and prevents the formation of plasmons and the reflection of light back to the detector (Fig. 22.2A). Change in the resonance angle or reflected light intensity quantifies the molecular binding, and the limit of detection limit can be as low as are 10 pg mm^{-1} [44]. The SPR method is mainly used to calculate kinetics, thermodynamics, and the affinity constant of receptor–target interactions. Banères-Roquet et al. studied the DNA-RNA, antigen–antibody interaction and also the interaction of drugs with albumin [46]. Liu et al. [47] cultured the cancer cells on the roof of a PDMS (Polydimethylsiloxane) SPR flow chamber to measure the real-time

TABLE 22.1 Biomaterials/nanoparticle and bioreceptors used in the development of the biosensors.

Bioreceptor type	Target analyte	Biomaterial/ nanoparticle used	Biosensor application	Example	References
Enzyme	Glucose	Glucose oxidase, SWNTs, Chitosan, Nafion	Detection of glucose levels in human serum	Glucose oxidase (micro-biosensor)	[17]
Antibody on the cantilever surface	HIV-specific antigens	Metallic nanoparticles	Detection of HIV infection	Anti-GP120 monoclonal antibodies	[19]
Enzyme	Glucose	Chitosan/CNT composite	Biosensors for glucose monitoring	Glucose oxidase	[24]
Enzyme	Catechol	Poly (l-lysine) grafted Zein	Detection of catechol	Laccase	[25]
Antibody-based receptor	HIV-specific antigens	Piezoelectric materials	Detection of HIV	Recombinant gp41 and synthetic p24 peptides	[26]
Nucleic acid	Thrombin	Gold nanoparticles	Thrombin detection	Aptamer	[27]
DNA aptamer-based bioreceptor	Thrombin	Magnetic nanoparticles	Thrombin detection	Aptamer	[28]
Cell-based receptor	Odorant molecules	Collagen-micropillars	Detection of odorant molecules	Encapsulated live cells	[29]
Cell receptor	Odorant molecules	Collagen-pillar (cells expressing olfactory receptors)	Detection of volatile compounds	GPCR	[30]
DNA/RNA probe	DNA sequence	Polypyrrole-based substrate	Genetic mutation detection	Composite nanofiber	[31]
Lectin	Glycoproteins	Graphene-based nanomaterials	Glycoprotein detection in cancer biomarkers	Concanavalin A	[32]
Peptide nucleic acid probes	Circulating tumor DNA	AuNPs	Early cancer diagnosis	Peptide nucleic acid	[33]
Peptide nucleic acids probes	microRNAs	Nanoporous metal-organic frameworks	Cancer diagnosis	Peptide nucleic acid	[34]
Microbial cell	Environmental toxins	Alginate microbeads	Detection and removal of cadmium	*Trametes versicolor*	[35]
Microalgae	Herbicides	Alginate beads, silica hydrogels	Water quality monitoring	*Chlamydomonas reinhardtii*	[36]
lipopolysaccharide	Bacteria	Chitosan-stabilized silver nanoparticles	Bacterial pathogen detection	*Escherichia coli*	[37]
Protein	Biotin	Liquid crystals	Biotin detection in biochemical assays	Streptavidin	[38]

GPCR, G-protein coupled receptor; *SWNTs*, single-walled carbon nanotubes.

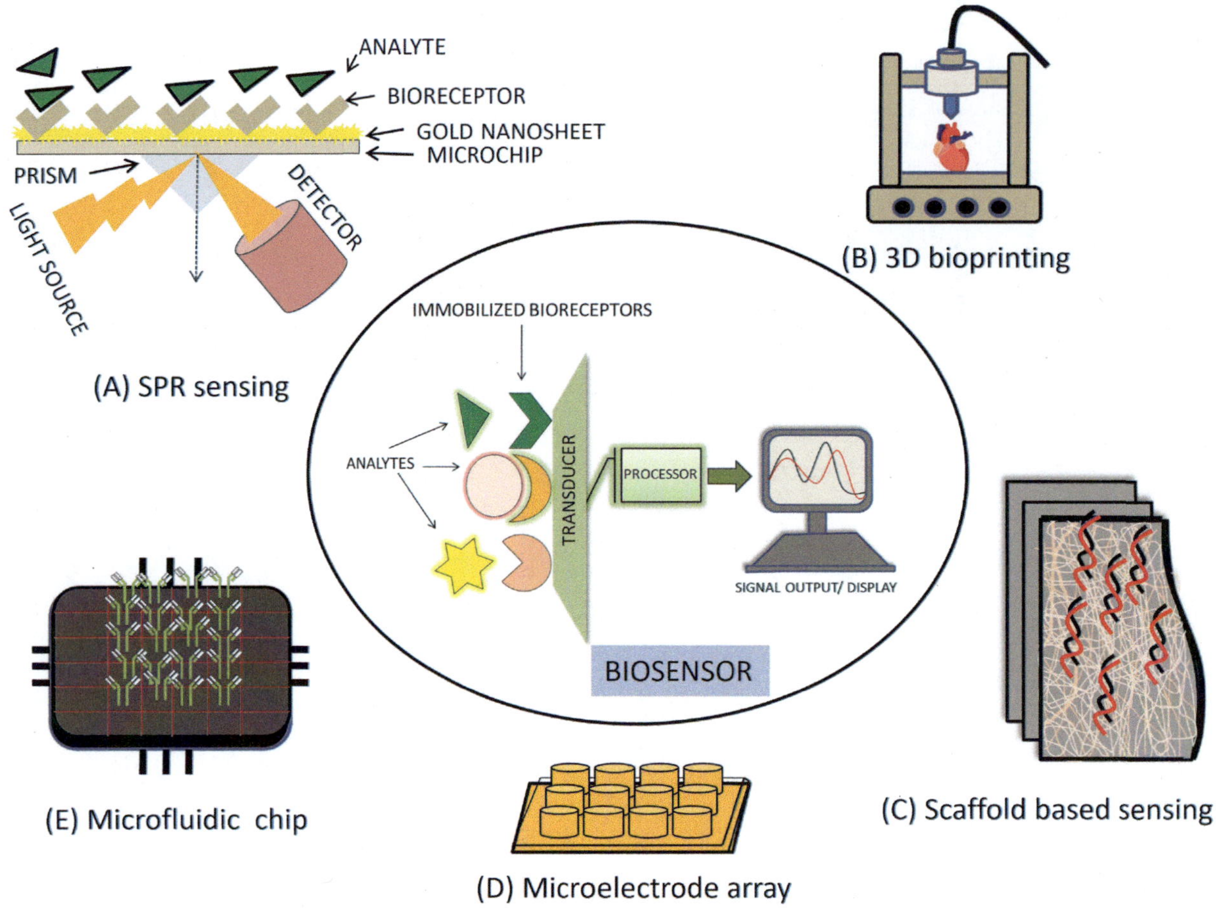

FIGURE 22.2 Some of the fabrication techniques used in advancement of biosensors: (A) Surface plasmon resonance-based biosensors, light is passing through prism and reflecting off the metal-coated or biopolymer-supported chip. (B) 3D bioprinting technique for biosensors development using bioinks composed of cells, growth factors, and other biomaterials. (C) Scaffold of cellulose, fibronectin, and silk fibroin for developing biocompatible biosensors. (D) Microelectrode array containing biosensors developed on cell/tissue chip for sensing neural activities. (E) Microfluidic chip functionalized with biomolecules and facilitated with micro channels.

secretion of VEGF (vascular endothelial growth factor). Human ovarian carcinoma (SKOV-3) cells were adhered to the chamber's ceiling using gelatin before placing onto the sensor. The secretion of VEGF was detected by immobilized monoclonal anti-VEGF antibodies [45,47].

22.3.4 Microelectromechanics system based biosensors

Microelectromechanics system (MEMS)-based biosensors combine the principles of MEMS technology with biosensing elements to detect biological analytes. Due to their compact size, electronic functionality, and capacity for fast operation, MEMS provide significant advantages over other sensors. MEMS consist of components ranging from 1 to 100μm in size [48]. Through the microfabrication technique, mechanical components of MEMS like sensors, actuators, and electronics are integrated onto the same substrate, typically on a piece of silicon (Fig. 22.2D) [49]. A MEMS-based biosensor, a microcantilever, operates by converting biological recognition events into nanomechanical motion. In this system, the target is detected when it interconnects with a chemically sensitive cantilever's surface, which is directly proportional to some properties like displacement, resistance, or resonant frequency. A 200-μm-long MEMS microcantilever performed a label-free detection of C-reactive protein. The specific binding between CRP and anti-CRP, which causes the microcantilever deflection, is measured using a position-sensitive detector [50]. Membrane-based NEMS/MEMS biosensors provide a sustainable, cost-efficient, ultra-miniaturized, and user-friendly solution for analytical techniques across various applications. In capacitive-based biosensors, a biomodified cantilever or membrane acting as the sensing element interacts with the adsorbed target molecules, and the sensor deforms because of changes in electrostatic properties, steric interactions, hydration, etc. The important advantages include label-free detection, high sensitivity, fast response, and simultaneous sensing with sensor arrays, and easy probe preparation and testing process [51].

22.3.5 3D printing, microfluidics, and lab-on-chip-based biosensors

The process of transferring a 3D computer-aided design into physical form by patterning using a print head, nozzle, or other tools is called 3D Printing (Fig. 22.2B) [52]. In 1986, Charles W. Hull reported stereolithography as a tool to fabricate 3D structures for the first time. It has developed into a versatile fabrication technology, providing distinct benefits for biomedical applications such as diagnostic tools, scaffolds for 3D-printed implants (Fig. 22.2C), prosthetics, and tissue engineering [53]. An innovative graphene and polylactic acid (PLA) based 3D-printed enzymatic electrode using extrusion technology was developed by Marzo et al. [54]. In this biosensor, the horseradish peroxidase enzyme was used to detect the hydrogen peroxide, and a gold nanoparticle was added to increase the electron transfer. The limit of detection and quantification was 11.1 and 37 μM, respectively [54,55]. Lehman et al. [56] developed a 3D-printed plasmonic biosensor using 2-hydroxymethylmethacrylate (HEMA) biomaterial embedded with gold nanoparticles for noninvasive, in situ live cell measurements. This biosensor provides scaffolds for cell adhesion, growth, and proliferation, and analyzes the cell culture [55,56].

Microfluidic devices are an up-and-coming solution for miniaturized fluid systems because they can precisely manage small volumes of samples and reagents in a controlled environment (Fig. 22.2E). With the help of 3D design software, 3D printing provides an efficient method for producing geometrically intricate microfluidic devices, bypassing the difficulties of traditional manufacturing techniques [53]. Zhou et al. developed bone stromal cells containing 3D-printed gelatin-based methacrylate hydrogel to investigate their interactions with breast cancer cells [57]. An electrochemical impedance spectroscopy (EIS) based "liver-on-chip" microfluidic device was developed by Shin et al. to detect liver albumin and glutathione-S-transferase-alpha [58].

22.4 The applications of biosensors in tissue engineering

Biosensors have emerged as an important instrument in the tissue engineering field, enabling precise monitoring and control of biological processes critical for tissue growth and regeneration with the help of three-dimensional cell cultures developed on chips called "organs-on-chip" models and microfluidic devices. In tissue engineering, biosensors are applied to monitor cell viability, metabolic activity, and the microenvironment, such as glucose, H_2O_2, adenosine concentration, etc., which provide crucial feedback for optimizing scaffold designs and biomaterial interactions. To monitor the cellular behavior and signals, tissue engineering depends on the interaction of the cell with its neighboring cells or the base material. Here we further explore biosensors' critical roles in different tissue engineering applications [4,59].

22.4.1 Detection of small molecules

Glucose: The role of biosensors is crucial in tissue engineering applications like diabetes, glucose detection, and regulation in engineered tissue constructs for cell metabolism and culture. Various glucose biosensors, like electrochemical biosensors, are either coupled with enzymes or can be nonenzymatic. The enzymatic one, for example, a glucose oxidase–based sensor, is most frequently used among these two, especially in cell culture. The latest developments in nano biosensors have become very useful in detecting blood glucose levels in diabetic patients. Nano biosensors have features like enhanced diffusion, an increase in surface-to-volume ratio, and a meager detection limit (LOD), which improves bioreceptor sensitivity and selectivity in the particular sensor [60–62]. Enzyme-based biosensors have fabulous detection limits but limitations like high cost, complex enzyme immobilization procedures, constraints with reproducibility, and chemical/thermal instability. Nonenzymatic sensors have a low price, high reproducibility, and stability, which can be used to overcome the problems of the sensors based on enzymes [4].

Hydrogen Peroxide: The presence of H_2O_2 at abnormal levels indicates stress and hypoxia and can harm the biological system. So, careful measurement of H_2O_2 is crucial for clinical applications and tissue engineering. Some biosensors play an essential role in its detection, such as graphene-based H_2O_2 biosensors. The presence of nanomaterials in these biosensors makes them highly useful because they are smaller than cells, which is why there is minimal physical disturbance during the measurement. Because of that, it has high stability and reproducibility and gives a rapid amperometric response time of <2 s and a good linearity with a detection limit of 3×10^{-6} M [4].

Signaling molecules: Biosensors can detect various signaling molecules (such as cytokines, chemokines, local interferon-gamma (IFN-γ), nucleotides), and these signaling molecules significantly contribute to maintaining cell behavior and environment. Changes in the level of these molecules indicate a change in the normal physiology of the tissue/cell. In many physiological and pathological events, there is a role of nucleotide signaling like adenosine diphosphate (ADP) and adenosine triphosphate (ATP), present in the extracellular matrix, and there are biosensors to measure [4]. (Table 22.2).

TABLE 22.2 Biosensors in tissue engineering applications and fabrication techniques used in its development.

Biosensors	Application	Technique used in biosensor	References
Electromechanical biosensors	Investigation of a 3D cardiac model and drug testing	MEA and AFM	[63]
Surface plasmon resonance-based biosensor	Screening of drug–target protein interactions	Dextran hydrogel-coated SPR imaging	[64]
Integrated biosensor for sepsis	Sepsis diagnosis	Magneto-electrochemical sensing strategy	[65]
Flexible impedimetric biosensors	Detection of VEGF	Photolithography and electrochemical impedance spectroscopy	[66]
Electrochemical biosensors	Detection of Aβ 42 protein in Alzheimer's disease diagnosis	N-doped graphene sheet-based immunomagnetic sensing	[67]
Fibronectin dot array biosensor	Cell adhesion kinetics and integrin profiling	Photolithography and fluorescence microscopy	[68]
Microelectronic cardiac patch (microECP)	Online monitoring of the engineered tissue function	Microelectrode array recording system integrated with an electrospun nanofibre scaffold and cardiac cells	[69]
Impedance-based sensing chip	Monitoring of cell growth and its cytotoxicity	Fibronectin-coated gold microarray electrodes	[70]
Microfluidic chip	Monitoring of cell migration, stimulation of cytokine, and an anticancer drug	Electrical impedance spectroscopy	[71]
Quantum dot–based cellulose biosensor (CdTe)	pH and glucose levels detection	Wet spinning and transmission electron microscopy	[72]

AFM, Atomic force microscopy; *MEA*, microelectrode array; *VEGF*, vascular endothelial growth factor.

22.4.2 Monitoring tissue growth, drug effects, and stimulation in diseased models

Biosensors are used for real-time monitoring of tissue growth during tissue engineering of specific organ models under "organ-on-chip." Cell-based biosensors using microelectrode array technology are very frequently used for neurobiological research, for example, to study various properties of neural systems such as network formation and functioning, neural behavior upon drug treatment, neurite growth, electrical stimulation, and memory function [71]. Various types of microelectrode arrays have 10 to 100 electrodes [73]. Alzheimer's is a neurodegenerative disease which is distinguished by the aggregation of β-amyloid and affects a large number of the aged population. In a study by Kim et al., a multielectrode array biosensor based on density is utilized to monitor exosomes' impact on neural network formation and intracerebral communication and its application in Alzheimer's disease [74]. A noninvasive screening with a magnetic N-doped graphene-based gold electrode biosensor using Aβ 1–28 (Aβab) antibodies for the detection of Aβ42 protein was done by Shan et al. for the diagnosis of Alzheimer's disease. These biosensors can be reconstructed quickly due to the superparamagnetic property of Aβab-MNG, can be reused, and have good reproducibility with a detection limit of 5 picogram /mL−1 [67].

Cardiovascular diseases like coronary artery disease, cardiac attacks, and cardiac failure are leading causes of death globally. Damaged heart tissue can be repaired or replaced using tissue engineering. Lab-on-a-chip for cardiac tissue engineering is essential for controlled monitoring of tissue growth. In a research by Zhang et al., endothelialized myocardial tissues were engineered on the microfluidic chip with the help of 3D bioprinting for the screening of pharmaceutical compounds for their cardiovascular toxicity [75]. The cardiac contractility can be monitored with a biosensor developed by lithography of polymethylisiloxane, named flexible micropost array [76]. In the study of Caluori et al., cardiomyocytes (CM) were placed on the combined platform of microelectrode array (MEA) and atomic force microscopy for real-time recording of the beating force of the same. This model was developed for the diseased CMs' drug response testing and Duchenne Muscular Dystrophy characterization [63].

Liver on-chip model can be used for monitoring the physiology and regeneration of the liver tissue, drugs and hepatotoxicity analysis, and modeling liver diseases. In vitro drug analysis was done using a hepatic lobule chip; during the analysis, cells were exposed to isoniazid, acetaminophen (APAP), and rifampin to measure the CYP1A1/2 and UGT activities for evaluating the enzymatic response of the cells [77]. Huh et al. developed a lung-on-chip microdevice to study the development of pulmonary edema in cancer patients treated with interleukin-2 (IL-2) [78].

22.4.3 Biosensors used in controlled drug delivery

There are many adverse effects related to conventional treatment methods, which have led to the establishment of a controlled drug transport system. So many studies have been done on biosensor-integrated drug delivery systems to study different chronic diseases [79].

Applications in cancer treatment: The microenvironment of cancer cells differs from the typical cell environment. It is mildly acidic, high in glutathione concentration, and has increased levels of hyaluronidase [80]. In cancer treatment, stimulus-based biosensors operate efficiently because they release their payload in response to environmental changes (e.g., tumor-specific microenvironments). Glucose oxidase (GOx) induces the oxidation of beta-D-glucose into gluconic

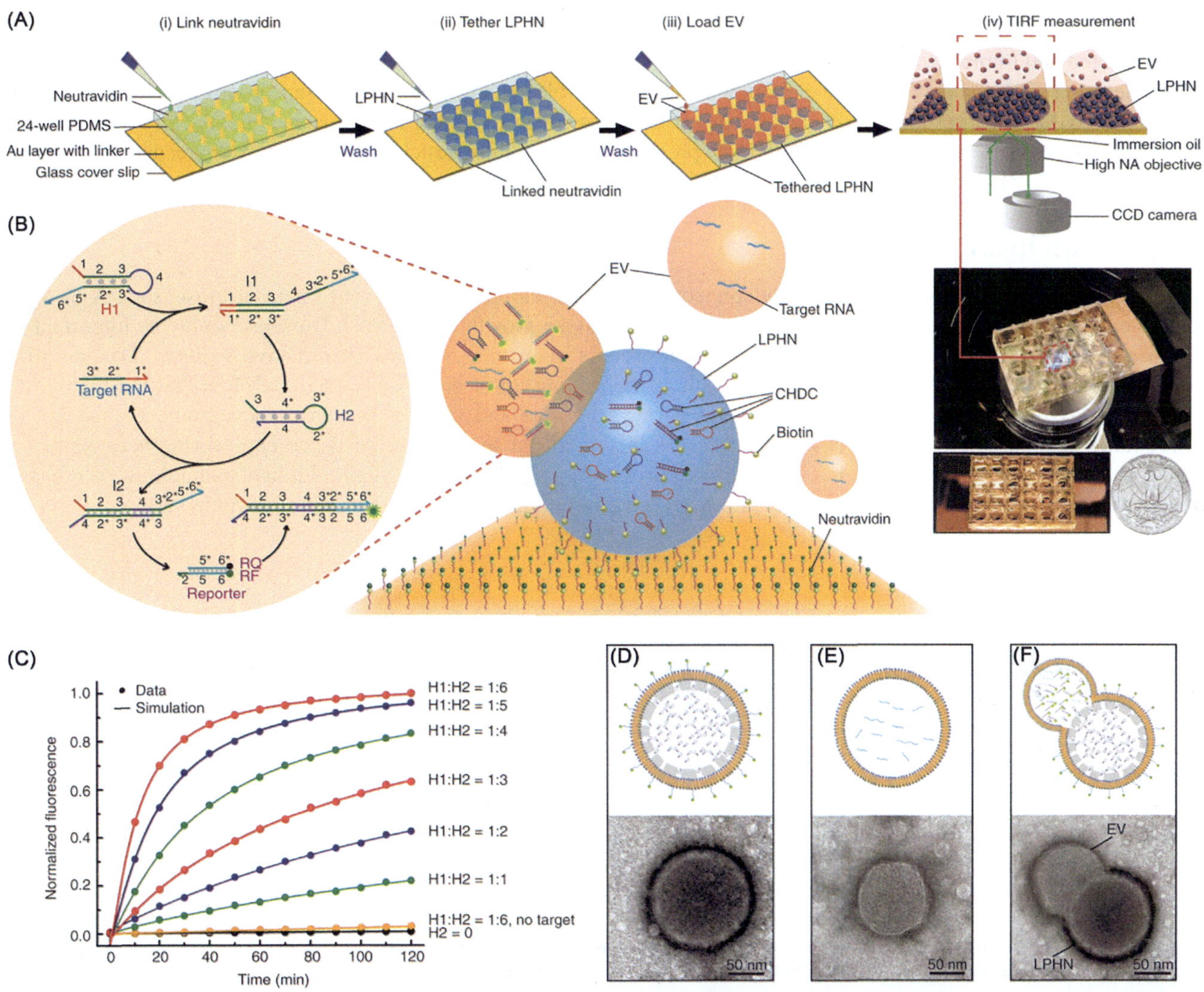

FIGURE 22.3 Hairpin-shaped DNA-lipid-based biochip. (A) Lipid-based assay chip; which is functionalized with neutravidin along with hybrid lipid-based nanoparticles, its reaction with pancreatic cancer–derived extracellular vesicles, and the fluorescence measurement. (B) Binding of hairpin DNA with target mRNA. (C) Fluorescence results of binding. (D–F) TEM images of components of biochip-based reactions.
From J. Hu, Y. Sheng, K.J. Kwak, J. Shi, B. Yu, L. James Lee (2017). A signal-amplifiable biochip quantifies extracellular vesicle-associated RNAs for early cancer detection. Nature Communications 8 (1) (2017). https://doi.org/10.1038/s41467-017-01942-1.

acid and H_2O_2 in the presence of oxygen. GOx helps monitor glucose levels through oxygen usage, pH alterations, or H_2O_2 production, which facilitates drug release in a Biosensor-integrated drug delivery system [81].

Applications in cardiovascular treatment: A drug delivery system based on stimulus response is critical in targeted treatment at the disease site and reduces the effect on nontargeted sites. In the case of cardiovascular diseases (CVD), several stimuli-responsive drug delivery methods are reported, which include pH-, redox-, hypoxia-, and enzyme-responsive particles.

pH is related to disease conditions, so it is used in drug delivery systems (i.e., CVD) in specific microenvironments or tissues like myocardial infarction, ischemia/reperfusion, and other CVDs, which generate many reactive oxygen species (ROS) under hypoxic conditions. Nanoparticles receptive towards ROS made from materials like poly (ethylene glycol) and poly (propylene sulfide) encapsulated the effective plant antioxidant ginsenoside Rg3 (Rg3). When injected into the injured rat's myocardium, Rg3 is released in the presence of ROS [81].

22.4.4 Biosensors in cancer detection and bioimaging

Biosensors, because of their noninvasive nature, the ability to provide near real-time feedback, and high levels of accuracy and reliability, can be used as sensing and imaging devices in disease diagnostics and management. Due to their precise labeling function, biosensors are used for in vivo targeting and imaging. VEGF- and Prostrate-specific antigen-labeled dual biomarker chip was developed by Pan et al. for the detection of prostate cancers and their circulating tumors. The biomarkers were thiolated and immobilized on a silicon chip deposited with Au nanorod. The targeted molecules are combined with the biomarkers and analyzed by UV–Visible absorbance [82]. A visible signal-amplifiable biosensor was developed by Hu and co-workers to detect cancer in its early stages. The biochip was developed using hairpin-shaped DNA embedded in a lipid-based nano-polymer to capture the target mRNA expressed on extracellular vesicles derived from pancreatic cancer cells, as shown in (Fig. 22.3). This binding generates the enhanced signal through fluorescence [83].

22.5 Conclusion

The success of tissue engineering is directly connected to the development and advancement of biosensors. A biosensor integrated with tissue engineering is dynamic as it monitors tissue growth in real time. Biosensors play an important role because of the difficulties involved in the translation process of in vitro systems into in vivo systems. Despite all the developments, specific challenges still prevail, such as the scaling up of biosensors, their stability, and the high cost of production. To cut down the cost of tissue engineering, the cost and time of biosensor development must be low. Introducing nanomaterials into biosensing devices has increased the success rate of tissue engineering because of their high sensitivity, selectivity, and accuracy. Therefore current trends in biosensors align with nanotechnology, as it enables the development of biosensors at low cost, with high sensitivity and accuracy, and simple instrumentation.

Acknowledgment

We thank the esteemed editors for the invitation to contribute a chapter in the book titled: "Fundamentals of Translational Tissue Engineering". We acknowledge the CSIR-Indian Institute of Toxicology Research (CSIR-IITR), Lucknow, as a platform for overwhelming support. Sushmita Kushwaha and Shakra Nazli thank UGC for providing the fellowship. The publication committee of the CSIR-IITR has approved the book chapter, and the manuscript number for communication is IITR/SEC/MS/2024/71.

References

[1] E. Lavik, R. Langer, Tissue engineering: current state and perspectives, Applied Microbiology and Biotechnology 65 (1) (2004) 1–8. Available from https://doi.org/10.1007/s00253-004-1580-z, link.springer.de/link/service/journals/00253/index.htm.

[2] A. Shafiee, A. Atala, Tissue engineering: toward a new era of medicine, Annual Review of Medicine 68 (1) (2017) 29–40. Available from https://doi.org/10.1146/annurev-med-102715-092331.

[3] B.D. Malhotra, R. Singhal, A. Chaubey, S.K. Sharma, A. Kumar, Recent trends in biosensors, Current Applied Physics 5 (2) (2005) 92–97. Available from https://doi.org/10.1016/j.cap.2004.06.021, http://www.elsevier.com/.

[4] A. Hasan, M. Nurunnabi, M. Morshed, A. Paul, A. Polini, T. Kuila, et al., Recent advances in application of biosensors in tissue engineering, BioMed Research International 2014 (2014). Available from https://doi.org/10.1155/2014/307519, http://www.hindawi.com/journals/biomed/.

[5] L.C. Clark, C. Lyons, Electrode systems for continuous monitoring in cardiovascular surgery, Annals of the New York Academy of Sciences 102 (1) (1962) 29–45. Available from https://doi.org/10.1111/j.1749-6632.1962.tb13623.x.

[6] R. Feiner, S. Fleischer, A. Shapira, O. Kalish, T. Dvir, Multifunctional degradable electronic scaffolds for cardiac tissue engineering, Journal of Controlled Release 281 (2018) 189–195. Available from https://doi.org/10.1016/j.jconrel.2018.05.023, www.elsevier.com/locate/jconrel.
[7] W. Zhang, X. Li, R. Zou, H. Wu, H. Shi, S. Yu, et al., Multifunctional glucose biosensors from Fe3O4 nanoparticles modified chitosan/graphene nanocomposites, Scientific Reports 5 (2015). Available from https://doi.org/10.1038/srep11129, www.nature.com/srep/index.html.
[8] B.N.G. Giepmans, S.R. Adams, M.H. Ellisman, R.Y. Tsien, The fluorescent toolbox for assessing protein location and function, Science (New York, N.Y.) 312 (5771) (2006) 217–224. Available from https://doi.org/10.1126/science.1124618.
[9] S. Vigneshvar, C.C. Sudhakumari, B. Senthilkumaran, H. Prakash, Recent advances in biosensor technology for potential applications – an overview, Frontiers in Bioengineering and Biotechnology 4 (2016). Available from https://doi.org/10.3389/fbioe.2016.00011.
[10] W. Zhang, C. Chen, D. Yang, G. Dong, S. Jia, B. Zhao, et al., Optical biosensors based on nitrogen-doped graphene functionalized with magnetic nanoparticles, Advanced Materials Interfaces 3 (20) (2016). Available from https://doi.org/10.1002/admi.201600590, http://onlinelibrary.wiley.com/journal/10.1002/(ISSN)2196-7350.
[11] K.T.M. Tran, T.D. Nguyen, Lithography-based methods to manufacture biomaterials at small scales, Journal of Science: Advanced Materials and Devices 2 (1) (2017) 1–14. Available from https://doi.org/10.1016/j.jsamd.2016.12.001, www.journals.elsevier.com/journal-of-science-advanced-materials-and-devices.
[12] Y. Kabe, S. Sakamoto, M. Hatakeyama, Y. Yamaguchi, M. Suematsu, M. Itonaga, et al., Application of high-performance magnetic nanobeads to biological sensing devices, Analytical and Bioanalytical Chemistry 411 (9) (2019) 1825–1837. Available from https://doi.org/10.1007/s00216-018-1548-y, link.springer.de/link/service/journals/00216/index.htm.
[13] D. Wu, Y. Liu, Y. Wang, L. Hu, H. Ma, G. Wang, et al., Label-free electrochemiluminescent immunosensor for detection of prostate specific antigen based on aminated graphene quantum dots and carboxyl graphene quantum dots, Scientific Reports 6 (2016). Available from https://doi.org/10.1038/srep20511, www.nature.com/srep/index.html.
[14] N. Bhalla, P. Jolly, N. Formisano, P. Estrela, Introduction to biosensors, Essays in Biochemistry 60 (1) (2016) 1–8. Available from https://doi.org/10.1042/EBC20150001, http://essays.biochemistry.org/content/ppebio/60/1/1.full.pdf.
[15] A.A. Ensafi, An introduction to sensors and biosensors, Electrochemical Biosensors, Elsevier, Iran, (2019), pp. 1–10. Available from http://doi.org/10.1016/B978-0-12-816491-4.00001-2, http://www.sciencedirect.com/science/book/9780128164914.
[16] V. Naresh, N. Lee, A review on biosensors and recent development of nanostructured materials-enabled biosensors, Sensors 21 (4) (2021) 1109. Available from https://doi.org/10.3390/s21041109.
[17] X. Wang, Y. Zhang, C. Cheng, R. Dong, J. Hao, Glucose in human serum determined by capillary electrophoresis with glucose micro-biosensor, Analyst 136 (8) (2011) 1753–1759. Available from https://doi.org/10.1039/c0an00348d, http://pubs.rsc.org/en/journals/journal/an.
[18] H. Liu, J. Ge, L. Yang, E. Ma, Advanced biomaterials for biosensor and theranostics, Biomaterials in Translational Medicine, Elsevier, China, (2019), pp. 213–255. Available from http://doi.org/10.1016/B978-0-12-813477-1.00010-4, https://www.sciencedirect.com/book/9780128134771.
[19] S. Nandi, A. Mondal, A. Roberts, S. Gandhi, Biosensor platforms for rapid HIV detection, Advances in Clinical Chemistry 98 (2020) 1–34. Available from https://doi.org/10.1016/bs.acc.2020.02.001, http://www.elsevier.com/wps/find/bookdescription.cws_home/705213/description#description.
[20] S. Shi, Y. Xie, G. Wang, Y. Luo, Metabolite-based biosensors for natural product discovery and overproduction, Current Opinion in Biotechnology 75 (2022) 102699. Available from https://doi.org/10.1016/j.copbio.2022.102699.
[21] J.A.C. Pérez, J.E. Sosa-Hernández, S.M. Hussain, M. Bilal, R. Parra-Saldivar, H.M.N. Iqbal, Bioinspired biomaterials and enzyme-based biosensors for point-of-care applications with reference to cancer and bio-imaging, Biocatalysis and Agricultural Biotechnology 17 (2019) 168–176. Available from https://doi.org/10.1016/j.bcab.2018.11.015, http://www.journals.elsevier.com/biocatalysis-and-agricultural-biotechnology/.
[22] P. Pavlickova, E.M. Schneider, H. Hug, Advances in recombinant antibody microarrays, Clinica Chimica Acta 343 (1-2) (2004) 17–35. Available from https://doi.org/10.1016/j.cccn.2004.01.009.
[23] A.E. Eldeeb, S. Salah, N.A. Elkasabgy, Biomaterials for tissue engineering applications and current updates in the field: a comprehensive review, AAPS PharmSciTech 23 (7) (2022). Available from https://doi.org/10.1208/s12249-022-02419-1, https://www.springer.com/journal/12249.
[24] W. Suginta, P. Khunkaewla, A. Schulte, Electrochemical biosensor applications of polysaccharides chitin and chitosan, Chemical Reviews 113 (7) (2013) 5458–5479. Available from https://doi.org/10.1021/cr300325r.
[25] H.S. Sofi, R. Ashraf, A.H. Khan, M.A. Beigh, S. Majeed, F.A. Sheikh, Reconstructing nanofibers from natural polymers using surface functionalization approaches for applications in tissue engineering, drug delivery and biosensing devices, Materials Science and Engineering C 94 (2019) 1102–1124. Available from https://doi.org/10.1016/j.msec.2018.10.069.
[26] H. Nawaz, M. Tahir, S. Anwar, M.I. Majeed, N. Rashid, Detection of HIV virus using biosensor, Nanobiosensors: From Design to Applications, wiley, Pakistan, (2020), pp. 149–169. Available from http://doi.org/10.1002/9783527345137.ch7, https://www.wiley.com/en-gb/Nanobiosensors:+From+Design+to+Applications-p-9783527345106.
[27] T. Fan, Y. Du, Y. Yao, J. Wu, S. Meng, J. Luo, et al., Rolling circle amplification triggered poly adenine-gold nanoparticles production for label-free electrochemical detection of thrombin, Sensors and Actuators, B: Chemical 266 (2018) 9–18. Available from https://doi.org/10.1016/j.snb.2018.03.112.
[28] H. Sun, N. Wang, L. Zhang, H. Meng, Z. Li, Aptamer-based sensors for thrombin detection application, Chemosensors 10 (7) (2022) 255. Available from https://doi.org/10.3390/chemosensors10070255.
[29] Y. Hirata, Y. Morimoto, S. Takeuchi, Japan cell-laden micropillars detect gaseous odorants on a liquid-air interface 2018. In: Proceedings of the IEEE International Conference on Micro Electro Mechanical Systems (MEMS), pp. 304–307. Institute of Electrical and Electronics Engineers Inc, 2018. https://doi.org/10.1109/MEMSYS.2018.8346546.

[30] H. Deng, T. Nakamoto, Biosensors for odor detection: a review, Biosensors 13 (12) (2023) 1000. Available from https://doi.org/10.3390/bios13121000.

[31] X. Zhang, X. Tan, P. Wang, J. Qin, Application of polypyrrole-based electrochemical biosensor for the early diagnosis of colorectal cancer, Nanomaterials 13 (4) (2023) 674. Available from https://doi.org/10.3390/nano13040674.

[32] J. Filip, S. Zavahir, L. Klukova, J. Tkac, P. Kasak, Immobilization of concanavalin A lectin on a reduced graphene oxide-thionine surface by glutaraldehyde crosslinking for the construction of an impedimetric biosensor, Journal of Electroanalytical Chemistry 794 (2017) 156–163. Available from https://doi.org/10.1016/j.jelechem.2017.04.019.

[33] A.H. Nguyen, S.J. Sim, Nanoplasmonic biosensor: detection and amplification of dual bio-signatures of circulating tumor DNA, Biosensors and Bioelectronics 67 (2015) 443–449. Available from https://doi.org/10.1016/j.bios.2014.09.003, www.elsevier.com/locate/bios.

[34] R. D'Agata, M.C. Giuffrida, G. Spoto, Peptide nucleic acid-based biosensors for cancer diagnosis, Molecules (Basel, Switzerland) 22 (11) (2017). Available from https://doi.org/10.3390/molecules22111951, http://www.mdpi.com/1420-3049/22/11/1951/pdf.

[35] M.Y. Arıca, Y. Kaçar, Ö. Genç, Entrapment of white-rot fungus Trametes versicolor in Ca-alginate beads: preparation and biosorption kinetic analysis for cadmium removal from an aqueous solution, Bioresource Technology 80 (2) (2001) 121–129. Available from https://doi.org/10.1016/s0960-8524(01)00084-0.

[36] Y. Ferro, M. Perullini, M. Jobbagy, S. Bilmes, C. Durrieu, Development of a biosensor for environmental monitoring based on microalgae immobilized in silica hydrogels, Sensors 12 (12) (2012) 16879–16891. Available from https://doi.org/10.3390/s121216879.

[37] M. Imran, C.J. Ehrhardt, M.F. Bertino, M.R. Shah, V.K. Yadavalli, Chitosan stabilized silver nanoparticles for the electrochemical detection of lipopolysaccharide: a facile biosensing approach for gram-negative bacteria, Micromachines 11 (4) (2020). Available from https://doi.org/10.3390/MI11040413, https://www.mdpi.com/2072-666X/11/4/413.

[38] H. Wang, T. Xu, Z. Wang, Y. Liu, H. Chen, J. Jiang, et al., Highly sensitive and label-free detection of biotin using a liquid crystal-based optofluidic biosensor, Biomedical Optics Express 14 (7) (2023) 3763–3774. Available from https://doi.org/10.1364/BOE.494783, https://opg.optica.org/directpdfaccess/799b2ee1-e7cd-482b-a2b1e73294018ec8_532291/boe-14-7-3763.pdf?da=1&id=532291&seq=0&mobile=no.

[39] Y.G. Zhou, R.M. Mohamadi, M. Poudineh, L. Kermanshah, S. Ahmed, T.S. Safaei, et al., Interrogating circulating microsomes and exosomes using metal nanoparticles, Small (Weinheim an der Bergstrasse, Germany) 12 (6) (2016) 727–732. Available from https://doi.org/10.1002/smll.201502365, http://onlinelibrary.wiley.com/journal/10.1002/(ISSN)1613-6829.

[40] L. Zhang, C. Gu, J. Wen, G. Liu, H. Liu, L. Li, Recent advances in nanomaterial-based biosensors for the detection of exosomes, Analytical and Bioanalytical Chemistry 413 (1) (2021) 83–102. Available from https://doi.org/10.1007/s00216-020-03000-0, https://link.springer.com/journal/216/volumes-and-issues.

[41] M. Holzinger, A.L. Goff, S. Cosnier, Nanomaterials for biosensing applications: a review, Frontiers in Chemistry 2 (2014). Available from https://doi.org/10.3389/fchem.2014.00063, http://journal.frontiersin.org/article/10.3389/fchem.2014.00063/full.

[42] R. Freeman, J. Girsh, I. Willner, Nucleic acid/quantum dots (QDs) hybrid systems for optical and photoelectrochemical sensing, ACS Applied Materials and Interfaces 5 (8) (2013) 2815–2834. Available from https://doi.org/10.1021/am303189h.

[43] P.S. Petrou, M. Chatzichristidi, A.M. Douvas, P. Argitis, K. Misiakos, S.E. Kakabakos, A biomolecule friendly photolithographic process for fabrication of protein microarrays on polymeric films coated on silicon chips, Biosensors and Bioelectronics 22 (9-10) (2007) 1994–2002. Available from https://doi.org/10.1016/j.bios.2006.08.036.

[44] P.N. Abadian, C.P. Kelley, E.D. Goluch, Cellular analysis and detection using surface plasmon resonance techniques, Analytical Chemistry 86 (6) (2014) 2799–2812. Available from https://doi.org/10.1021/ac500135s, http://pubs.acs.org/journal/ancham.

[45] B. Karki, A. Uniyal, A. Pal, V. Srivastava, Advances in surface plasmon resonance-based biosensor technologies for cancer cell detection, International Journal of Optics 2022 (2022). Available from https://doi.org/10.1155/2022/1476254, http://www.hindawi.com/journals/ijo/.

[46] F. Banères-Roquet, M. Gualtieri, P. Villain-Guillot, M. Pugnière, J.P. Leonetti, Use of a surface plasmon resonance method to investigate antibiotic and plasma protein interactions, Antimicrobial Agents and Chemotherapy 53 (4) (2009) 1528–1531. Available from https://doi.org/10.1128/AAC.00971-08, http://aac.asm.org/cgi/reprint/53/4/1528.

[47] W. Liu, Y. Chen, M. Yan, Surface plasmon resonance imaging of limited glycoprotein samples, Analyst 133 (9) (2008) 1268–1273. Available from https://doi.org/10.1039/b804235g, http://pubs.rsc.org/en/journals/journal/an.

[48] P.G. Gopinath, V.R. Anitha, S.A. Mastani, Microcantilever based biosensor for disease detection applications, Journal of Medical and Bioengineering 4 (4) (2015) 307–311. Available from https://doi.org/10.12720/jomb.4.4.307-311.

[49] B. Derkus, Applying the miniaturization technologies for biosensor design, Biosensors and Bioelectronics 79 (2016) 901–913. Available from https://doi.org/10.1016/j.bios.2016.01.033.

[50] C.H. Chen, R.Z. Hwang, L.S. Huang, S.M. Lin, H.C. Chen, Y.C. Yang, et al., A wireless Bio-MEMS sensor for C-reactive protein detection based on nanomechanics, IEEE Transactions on Biomedical Engineering 56 (2) (2009) 462–470. Available from https://doi.org/10.1109/TBME.2008.2003262.

[51] H. Stapf, F. Selbmann, Y. Joseph, P. Rahimi, Membrane-based NEMS/MEMS biosensors, ACS Applied Electronic Materials 6 (4) (2024) 2120–2133. Available from https://doi.org/10.1021/acsaelm.3c01732, pubs.acs.org/journal/aaembp.

[52] R. Amin, S. Knowlton, A. Hart, B. Yenilmez, F. Ghaderinezhad, S. Katebifar, et al., 3D-printed microfluidic devices, Biofabrication 8 (2) (2016) 022001. Available from https://doi.org/10.1088/1758-5090/8/2/022001.

[53] M. Sharafeldin, A. Jones, J.F. Rusling, 3D-printed biosensor arrays for medical diagnostics, Micromachines 9 (8) (2018). Available from https://doi.org/10.3390/mi9080394, http://www.mdpi.com/2072-666X/9/8/394/pdf.

[54] A.M. López Marzo, C.C. Mayorga-Martinez, M. Pumera, 3D-printed graphene direct electron transfer enzyme biosensors, Biosensors and Bioelectronics 151 (2020) 111980. Available from https://doi.org/10.1016/j.bios.2019.111980.

[55] G. Remaggi, A. Zaccarelli, L. Elviri, 3D printing technologies in biosensors production: recent developments, Chemosensors 10 (2) (2022). Available from https://doi.org/10.3390/chemosensors10020065, https://www.mdpi.com/journal/chemosensors.

[56] S.E. Lehman, J.M. McCracken, L.A. Miller, S. Jayalath, R.G. Nuzzo, Biocompliant composite Au/pHEMA plasmonic scaffolds for 3D cell culture and noninvasive sensing of cellular metabolites, Advanced Healthcare Materials 10 (4) (2021). Available from https://doi.org/10.1002/adhm.202001040, http://onlinelibrary.wiley.com/journal/10.1002/(ISSN)2192-2659.

[57] X. Zhou, W. Zhu, M. Nowicki, S. Miao, H. Cui, B. Holmes, et al., 3D bioprinting a cell-laden bone matrix for breast cancer metastasis study, ACS Applied Materials and Interfaces 8 (44) (2016) 30017–30026. Available from https://doi.org/10.1021/acsami.6b10673, http://pubs.acs.org/journal/aamick.

[58] S.R. Shin, T. Kilic, Y.S. Zhang, H. Avci, N. Hu, D. Kim, et al., Label-free and regenerative electrochemical microfluidic biosensors for continual monitoring of cell secretomes, Advanced Science 4 (5) (2017). Available from https://doi.org/10.1002/advs.201600522, http://onlinelibrary.wiley.com/journal/10.1002/(ISSN)2198-3844.

[59] Y.-C.E. Li, I.-C. Lee, The current trends of biosensors in tissue engineering, Biosensors 10 (8) (2020) 88. Available from https://doi.org/10.3390/bios10080088.

[60] K.K. Jain, Applications of nanobiotechnology in clinical diagnostics, Clinical Chemistry 53 (11) (2007) 2002–2009. Available from https://doi.org/10.1373/clinchem.2007.090795.

[61] V. Scognamiglio, Nanotechnology in glucose monitoring: advances and challenges in the last 10 years, Biosensors and Bioelectronics 47 (2013) 12–25. Available from https://doi.org/10.1016/j.bios.2013.02.043.

[62] M. Gajendiran, J. Choi, S.J. Kim, K. Kim, H. Shin, H.J. Koo, et al., Conductive biomaterials for tissue engineering applications, Journal of Industrial and Engineering Chemistry 51 (2017) 12–26. Available from https://doi.org/10.1016/j.jiec.2017.02.031, http://www.sciencedirect.com/science/journal/1226086X.

[63] G. Caluori, J. Pribyl, M. Pesl, S. Jelinkova, V. Rotrekl, P. Skladal, et al., Non-invasive electromechanical cell-based biosensors for improved investigation of 3D cardiac models, Republic Biosensors and Bioelectronics 124-125 (2019) 129–135. Available from https://doi.org/10.1016/j.bios.2018.10.021, www.elsevier.com/locate/bios.

[64] S. Li, M. Yang, W. Zhou, T.G. Johnston, R. Wang, J. Zhu, Dextran hydrogel coated surface plasmon resonance imaging (SPRi) sensor for sensitive and label-free detection of small molecule drugs, Applied Surface Science 355 (2015) 570–576. Available from https://doi.org/10.1016/j.apsusc.2015.05.020, http://www.journals.elsevier.com/applied-surface-science/.

[65] J. Min, M. Nothing, B. Coble, H. Zheng, J. Park, H. Im, et al., Integrated biosensor for rapid and point-of-care sepsis diagnosis, ACS Nano 12 (4) (2018) 3378–3384. Available from https://doi.org/10.1021/acsnano.7b08965, http://pubs.acs.org/journal/ancac3.

[66] M. Xu, V.K. Yadavalli, Flexible biosensors for the impedimetric detection of protein targets using silk-conductive polymer biocomposites, ACS Sensors 4 (4) (2019) 1040–1047. Available from https://doi.org/10.1021/acssensors.9b00230, http://pubs.acs.org/journal/ascefj.

[67] S.S. Li, C.W. Lin, K.C. Wei, C.Y. Huang, P.H. Hsu, H.L. Liu, et al., Non-invasive screening for early Alzheimer's disease diagnosis by a sensitively immunomagnetic biosensor, Scientific Reports 6 (2016). Available from https://doi.org/10.1038/srep25155, www.nature.com/srep/index.html.

[68] A. Gonzalez-Pujana, E. Santos-Vizcaino, M. García-Hernando, B. Hernaez-Estrada, M. M. de Pancorbo, F. Benito-Lopez, et al., Extracellular matrix protein microarray-based biosensor with single cell resolution: integrin profiling and characterization of cell-biomaterial interactions, Sensors and Actuators B: Chemical 299 (2019) 126954. Available from https://doi.org/10.1016/j.snb.2019.126954.

[69] R. Feiner, L. Engel, S. Fleischer, M. Malki, I. Gal, A. Shapira, et al., Engineered hybrid cardiac patches with multifunctional electronics for online monitoring and regulation of tissue function, Nature Materials 15 (6) (2016) 679–685. Available from https://doi.org/10.1038/nmat4590, http://www.nature.com/nmat/.

[70] C. Xiao, J.H.T. Luong, On-line monitoring of cell growth and cytotoxicity using electric cell-substrate impedance sensing (ECIS), Biotechnology Progress 19 (3) (2003) 1000–1005. Available from https://doi.org/10.1021/bp025733x.

[71] L. Liu, X. Xiao, K.F. Lei, C.H. Huang, Quantitative impedimetric monitoring of cell migration under the stimulation of cytokine or anti-cancer drug in a microfluidic chip, Biomicrofluidics 9 (3) (2015). Available from https://doi.org/10.1063/1.4922488, http://scitation.aip.org/content/aip/journal/bmf.

[72] J. Yao, P. Ji, B. Wang, H. Wang, S. Chen, Color-tunable luminescent macrofibers based on CdTe QDs-loaded bacterial cellulose nanofibers for pH and glucose sensing, Sensors and Actuators, B: Chemical 254 (2018) 110–119. Available from https://doi.org/10.1016/j.snb.2017.07.071.

[73] T. Nyberg, A. Shimada, K. Torimitsu, Ion conducting polymer microelectrodes for interfacing with neural networks, Journal of Neuroscience Methods 160 (1) (2007) 16–25. Available from https://doi.org/10.1016/j.jneumeth.2006.08.008.

[74] S. Kim, J. Jeon, D. Ganbat, T. Kim, K. Shin, S. Hong, et al., Alteration of neural network and hippocampal slice activation through exosomes derived from 5XFAD nasal lavage fluid, International Journal of Molecular Sciences 24 (18) (2023) 14064. Available from https://doi.org/10.3390/ijms241814064.

[75] Y.S. Zhang, A. Arneri, S. Bersini, S.R. Shin, K. Zhu, Z. Goli-Malekabadi, et al., Bioprinting 3D microfibrous scaffolds for engineering endothelialized myocardium and heart-on-a-chip, Biomaterials 110 (2016) 45–59. Available from https://doi.org/10.1016/j.biomaterials.2016.09.003, http://www.journals.elsevier.com/biomaterials/.

[76] K.M. Beussman, M.L. Rodriguez, A. Leonard, N. Taparia, C.R. Thompson, N.J. Sniadecki, Micropost arrays for measuring stem cell-derived cardiomyocyte contractility, Methods (San Diego, Calif.) 94 (2016) 43–50. Available from https://doi.org/10.1016/j.ymeth.2015.09.005, http://www.elsevier.com/inca/publications/store/6/2/2/9/1/4/index.htt.

[77] C. Ma, L. Zhao, E.M. Zhou, J. Xu, S. Shen, J. Wang, On-chip construction of liver lobule-like microtissue and its application for adverse drug reaction assay, Analytical Chemistry 88 (3) (2016) 1719–1727. Available from https://doi.org/10.1021/acs.analchem.5b03869, http://pubs.acs.org/journal/ancham.

[78] D. Huh, D.C. Leslie, B.D. Matthews, J.P. Fraser, S. Jurek, G.A. Hamilton, et al., A human disease model of drug toxicity-induced pulmonary edema in a lung-on-a-chip microdevice, Science Translational Medicine 4 (159) (2012). Available from https://doi.org/10.1126/scitranslmed.3004249, http://stm.sciencemag.org/content/4/159/159ra147.full.pdf.

[79] D. Olvera, M.G. Monaghan, Electroactive material-based biosensors for detection and drug delivery, Advanced Drug Delivery Reviews 170 (2021) 396–424. Available from https://doi.org/10.1016/j.addr.2020.09.011, www.elsevier.com/locate/drugdeliv.

[80] Y. Lu, A.A. Aimetti, R. Langer, Z. Gu, Bioresponsive materials, Nature Reviews Materials 2 (1) (2016). Available from https://doi.org/10.1038/natrevmats.2016.75, www.nature.com/natrevmats/.

[81] I. Cicha, R. Priefer, P. Severino, E.B. Souto, S. Jain, Biosensor-integrated drug delivery systems as new materials for biomedical applications, Biomolecules 12 (9) (2022) 1198. Available from https://doi.org/10.3390/biom12091198.

[82] L.H. Pan, S.T. Pang, P.Y. Fang, C.K. Chuang, H.W. Yang, Label-free biochips for accurate detection of prostate cancer in the clinic: dual biomarkers and circulating tumor cells, Theranostics 7 (17) (2017) 4289–4300. Available from https://doi.org/10.7150/thno.21092, http://www.thno.org/v07p4289.pdf.

[83] J. Hu, Y. Sheng, K.J. Kwak, J. Shi, B. Yu, L.J. Lee, A signal-amplifiable biochip quantifies extracellular vesicle-associated RNAs for early cancer detection, Nature Communications 8 (1) (2017). Available from https://doi.org/10.1038/s41467-017-01942-1, http://www.nature.com/ncomms/index.html.

Section IV

Clinical aspects of tissue engineering

Chapter 23

Patents on biomaterials

Nishant Tyagi[1,*] and Shivangi Gupta[2,*]
[1]*Voxvess Biotech Lab Private Limited, New Delhi, India,* [2]*Eicher Shroff Center for Stem Cell Research, Delhi, India*

23.1 Introduction

Over the past years, biomaterials have made remarkable progress in biomedical science, acting as the building block for development from tissue regeneration to intelligent medical devices with the ability to modify their properties to meet certain criteria and specific characteristics. In tissue engineering (TE), these materials are designed to mimic the natural extracellular matrix (ECM), providing structural and biochemical cues essential for cell growth, differentiation, and the repair of damaged tissues such as bone, cartilage, and skin [1].

Beyond this, biomaterials have a number of applications, from targeted drug delivery systems to long-term medical implants and biosensors that can monitor health in real time. The rapid increase in the development of biomaterials is due to continuous innovation, which leads to the evolution of advanced polymers, ceramics, and nanocomposites. These materials are engineered to regulate mechanical strength, biocompatibility, and controlled degradation to match the requirements of uses such as cardiac stents and orthopedic implants [2]. Simultaneously, the emergence of "smart" biomaterials like pH-sensitive hydrogels, temperature-responsive has enabled personalized and targeted therapies [3]. Development of techniques like 3D bioprinting has allowed for the fabrication of complex tissue constructs, while emerging 4D bioprinting technologies, where materials can dynamically change properties postimplantation, promise greater adaptability and precision in clinical outcomes [4].

Among these developments, the protection of intellectual property via patents has become a key driver of advancement. Patents not only safeguard new biomaterials and their synthesis routes but also encourage investment in research and commercialization. In 2023 more than 30,000 biomaterial-related patents were filed globally. Notably, India saw a 40% increase in filings since 2020, led by institutions such as the Indian Institute of Science (IISc), which secured a patent for a silk fibroin-based corneal substitute. While discrepancies remain, with United States and China leading in patent grant, India faces challenges, including slow approval process timelines, limited industry-academia linkage, and ambiguous interpretations of the Biological Diversity Act [5].

This chapter examines the relationship between biomaterials innovation and the global patent landscape. It analyzes and highlights challenges such as regulatory hurdles and ethical dilemmas, and showcases landmark case studies, including delayed approvals of biodegradable stents and collaborative advances in 4D bioprinting. It aims to serve as a guide for researchers, entrepreneurs, and decision makers navigating the intersection of biomaterials and intellectual property in TE

23.2 Patentability and regulatory aspects of biomaterials

23.2.1 Definition and types of patents in biomaterials

Patents are legal documents that grant inventors the exclusive right to make, use, sell, or license their innovations for a defined period, typically 20 years. In the context of biomaterials, patents are vital tools that protect intellectual property by covering novel compositions, methods of fabrication, and biomedical applications. These include innovations in medical implants, drug delivery systems, regenerative therapies, and bioengineered scaffolds (Bhapkar, R. A., & Pimpalshende). Biomaterial-related patents generally fall into four main categories, which are listed in Table 23.1.

*. Authors contributed equally to the study.

Fundamentals of Translational Tissue Engineering. DOI: https://doi.org/10.1016/B978-0-443-33455-9.00026-2

TABLE 23.1 Type of patents in tissue engineering and its scope and relevance in biomaterials.

Patent type	Scope	Significance in biomaterials
Utility patents	Protect functional inventions such as novel materials or biomedical devices	Most common in biomaterials, e.g., drug eluting hydrogels, 3D-printed scaffolds
Design patents	Covers esthetic or physical appearance of product	Important for patient-specific implants, e.g., shape of custom biodegradable stents
Plant patents	Apply to genetically engineered plant varieties	Emerging relevance, e.g., neem-based antimicrobial biomaterials
Process patents	Protect manufacturing or synthesis methods	Key for scalable production, e.g., laser-assisted collagen crosslinking methods

TABLE 23.2 Regional patentability guidelines.

Region	Patentability guidelines	Implications for biomaterials
The United States (USPTO)	Permits patenting of naturally derived substances if they are "marked differently" from their natural forms (e.g., genetically modified algae for wound healing)	Allows broader protection, especially for biologically engineered materials
European Union (EPO)	Applies strict standards under Directive 98/44/EC, especially regarding human-derived biological material	Limits patents involving human embryonic stem cells and bioprinting
India (IPO)	Requires compliance with the Biological Diversity Act, 2002; mandates deposition of biological samples in recognized repositories	Delays common due to clearance from the National Biodiversity Authority (NBA); e.g., a 2022 AIIMS patent for a zinc-magnesium stent faced a 3-year pendency for compliance

Globally, over 30,000 biomaterial-related patents were filed in 2023, with utility and process patents dominating the landscape. While China led in raw patent numbers with over 12,000 filings, India's share remained relatively low (~5% of total biomaterials filings), indicating significant room for growth, particularly in advanced sectors such as bioelectronics and nanotherapeutics [6].

23.3 Patentability criteria for biomaterials

To qualify for patent protection, biomaterials must meet four universally accepted criteria of patentability [7]

1. **Novelty**: The material or process must be new and not previously disclosed in any public domain.
2. **Inventive step (nonobviousness)**: The invention should involve a significant advancement beyond existing knowledge or techniques.
3. **Industrial applicability (utility)**: The biomaterial must have a practical application in a relevant industry, particularly in healthcare or biotechnology.
4. **Patentable subject matter**: The invention must fall within the scope of what is legally allowed for patenting, as defined by regional regulations.

These criteria are interpreted differently across jurisdictions, leading to regional variations that significantly impact biomaterial-related patents. Some of them are discussed in Table 23.2.

23.4 Nonpatentable Inventions under Indian Patent Law

Under Sections 3 and 4 of the Indian Patent Act, certain categories of inventions are excluded from patent protection, which include

1. **Section 3(b)**: Inventions that pose a risk to human, animal, or plant life or the environment are not patentable.
2. **Section 3(c)**: Naturally occurring substances, living organisms, or materials already existing in nature cannot be patented.
3. **Section 3(d)**: New forms of known substances that do not demonstrate enhanced efficacy, as well as discoveries of new properties or uses of known substances, are not eligible for patent protection.
4. **Section 3(e)**: Mixtures of known ingredients that do not produce a synergistic effect are excluded from patentability.
5. **Section 3(h)**: Agricultural or horticultural methods, such as farming practices, are not considered inventions under the law.
6. **Section 3(i)**: Any process for the treatment of humans or animals (such as surgical, therapeutic, or diagnostic methods) is not patentable.
7. **Section 3(p)**: Inventions that are based on traditional knowledge or that duplicate known traditional practices cannot be patented.

These clauses ensure that the patent system in India supports genuine innovation while respecting ethical considerations and indigenous resources [8].

23.5 Role of patents in advancing tissue engineering and biotechnology

Patents play a critical role in driving innovation and protecting intellectual property in the field of TE and biotechnology. By providing exclusive rights to inventors and institutions, patents create a secure environment for researchers and entrepreneurs to invest time, capital, and expertise into developing novel biomaterials, advanced therapeutic strategies, and scalable biomedical technologies.

In recent years, patents have contributed significantly to innovation in TE by enabling the development and protection of technologies such as biofunctionalized scaffolds, cell-laden matrices, and advanced delivery platforms. These inventions not only enhance biological performance, such as improving stem cell adhesion or controlled degradation, but also attract translational partnerships and investment [9].

India has also emerged as a promising contributor, particularly in the field of natural biomaterials and regenerative implants. However, patent processing often faces regulatory delays due to biodiversity laws and unclear administrative protocols. Despite this, India's academic institutions continue to file meaningful patents with strong translational potential.

Globally, there has been a substantial increase in filings within domains like 4D bioprinting and dynamic biomaterials. The emergence of adaptive bioinks and programmable tissue scaffolds has driven a 15% increase in 4D printing-related patents between 2022 and 2024, reflecting a shift toward more responsive and personalized therapies [10].

As shown in Table 23.3, patents in TE span diverse applications, from smart scaffolds to adaptive bioinks. While global institutions dominate in high-tech filings, Indian contributions reflect strong potential despite regulatory delays. These examples underscore the importance of streamlined patent systems and collaborative innovation to translate biomaterials research into real-world therapies.

23.6 Applications beyond tissue engineering

While TE remains a central focus of biomaterials innovation, its utility extends far beyond scaffold fabrication and organ regeneration. Patents over the past decade reflect a rapid expansion into adjacent fields such as targeted drug delivery, orthopedic implants, and biosensor development, each benefiting from tailored material properties and advanced functionalization strategies.

In drug delivery, biomaterial-based nanocarriers have enabled the controlled, site-specific release of therapeutics, particularly in oncology. Over 5000 patents now focus on lipid and polymer-based nanoparticles. A notable example from India is NanoHeal's 2024 turmeric curcumin nano formulation, designed for cancer therapy. Despite its innovation, the patent faced delays under Section 3(d) of the Indian Patents Act, which requires proof of "enhanced efficacy" for known compounds.

Orthopedic applications also reflect growing patent activity. More than 8500 patents globally relate to high-performance materials like polyether ether ketone (PEEK), titanium alloys, and bioactive ceramics. In 2023 SCTIMST (India)

TABLE 23.3 Summarizes key patents from recent years that exemplify the diverse contributions of both global and Indian research institutions.

Assignee/Institution	Patented innovation	Application area	Year	Remarks
MIT (USA)	Biofunctionalized scaffolds	Bone regeneration via stem cells	2022	Improves stem cell adhesion and osteogenic differentiation
IISc (India)	Silk fibroin-based corneal substitute	Corneal tissue engineering	2023	NBA approval caused an 18-month delay in patent processing
MIT–IIT Delhi (Joint)	Temperature-responsive bioinks (4D bioprinting)	Smart implants	2024	Enables postimplantation adaptability
AIIMS (India)	Zinc-magnesium biodegradable stent	Cardiovascular scaffold	2022	Delayed under Section 3(c) for lack of "inventive step"
Organovo (USA)	3D-bioprinted liver tissue	Drug testing, transplant models	2023	Used in pharmaceutical validation pipelines

developed a low-cost titanium alloy knee implant, an important step toward making orthopedic solutions more accessible. However, Indian startups continue to face intellectual property (IP) conflicts and international filing barriers that slow their path to market [11].

Patent trends also reflect the evolving application landscape. As illustrated in Fig. 23.1, biomaterials are increasingly being utilized in areas such as disease diagnostics (23%), orthopedic devices (16%), and cardiovascular grafts (19%). These sectors now surpass tissue regeneration in global patent share, underscoring the expansion of biomaterials into mainstream medical technologies.

Smart biomaterials are also transforming wearable and diagnostic technologies. Biosensors, often based on polymeric or hydrogel platforms, account for a significant portion of biomaterial-related patents. Since 2017, filings in this category have grown by over 80%, driven by applications such as glucose monitoring contact lenses and sweat-based diagnostic patches. A Bengaluru-based startup developed one such biosensor, but commercialization was hindered by unclear regulatory guidelines with approval delayed for over three years [12].

These applications highlight the growing relevance of biomaterials in precision medicine, rehabilitation, and real-time monitoring fields that demand innovation beyond structural repair. Patent systems must evolve to not only protect such innovations but also streamline their pathway from laboratory to clinic.

23.7 Challenges in patent commercialization

Despite the rise in biomaterial-related patent filings globally, translating these innovations into market-ready products remains a complex and uneven process especially in developing countries. Patent commercialization is often slowed by multilayered regulatory approvals, restrictive legal frameworks, and funding disparities between nations.

In India, one of the most significant challenges is the prolonged patent examination process. The average time for patent approval exceeds five years, in contrast to approximately 2.5 years in the United States. This delay discourages startups and academic institutions from pursuing timely IP protection, often pushing them to file internationally where systems are faster and more industry aligned. A notable example is a Bengaluru-based startup developing a 3D-printed trabecular bone scaffold, which has awaited approval for over 4 years due to unclear biocompatibility testing standards.

Adhering to biodiversity laws presents another major challenge. The Biological Diversity Act mandates prior approval from the National Biodiversity Authority (NBA) for patents involving biological materials sourced from India. This has led to delays of up to two years in projects using neem-based antimicrobial dressings and silkworm-derived silk fibroin, as seen in the 2023 IISc corneal substitute patent. While these regulations aim to protect indigenous resources, their unclear interpretation and slow processing often discourage innovation.

The absence of consistent funding hinders progress toward effective commercialization. In 2023 India allocated \$120 million to biomaterials research and development through the Department of Biotechnology (DBT), compared to the \$4 billion invested by the US National Institutes of Health (NIH). There is a persistent shortage of financial resources for

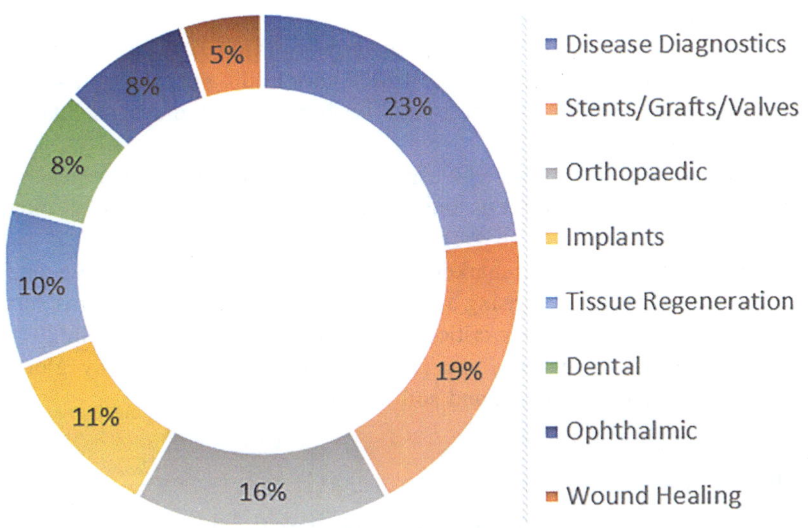

FIGURE 23.1 Global distribution of biomaterial-related patents by application area.

clinical translation, validation trials, and international patent filings, all of which are essential steps to advance lab-scale biomaterials into global markets.

Together, these factors contribute to a situation where only around 20% of biomaterial-related patents in India reach commercial deployment, compared to over 60% in the United States. Bridging this gap will require regulatory reform, increased funding for translational research, and stronger collaboration between academia, industry, and IP authorities [13].

23.8 Global biomaterials patent landscape: a country-wise overview

23.8.1 United States

The United States leads the global biomaterials patent landscape, both in volume and translational success. Between 2015 and 2024, over 12,500 biomaterial-related patents were filed, with an average grant rate of approximately 70%. This dominance is driven by strong public investment, with the National Institutes of Health (NIH) alone allocating over $4 billion toward biomaterials research in 2023 [14].

A key strength of the US system lies in its supportive intellectual property infrastructure, including expedited examination pathways and well-established industry-academia partnerships. Leading institutions such as MIT and Stanford routinely collaborate with startups and large biotech firms to translate patented innovations into FDA-approved technologies. For instance, MIT's portfolio includes over 150 patents related to AI-driven scaffolds and programmable biomaterials, many of which have advanced to clinical or commercial use [15].

Regulatory flexibility is another differentiator. The USPTO permits patents on stem cell-based innovations, provided they do not involve the destruction of human embryos. This has enabled rapid growth in regenerative medicine, with approximately 30% of US TE patents involving stem cells [16].

Overall, the United States offers a mature ecosystem that supports not just invention, but commercialization, making it a model for emerging economies seeking to strengthen their own biomaterials innovation pipelines.

23.8.2 China

China has become a major contributor in the global biomaterials patent landscape, driven by strong government-led initiatives and aggressive innovation strategies. Between 2015 and 2024, China filed over 10,200 biomaterial-related patents, placing it just behind the United States in total volume. This output reflects a focused national agenda, particularly under the "Made in China 2025" policy, which prioritizes biotechnology, regenerative medicine, and advanced materials.

The patent grant rate in China stands at approximately 65%, reflecting an efficient system with favorable timelines for both domestic and foreign applicants. In 2023 more than 200 biomaterials startups received funding through provincial and national incubator schemes. Public R&D investment was estimated at over $2.5 billion in 2023, according to WIPO and industry reports.

Key institutions driving innovation include Tsinghua University, which has filed over 200 patents related to 3D bioprinting, including a 2024 patent for vascularized liver tissue. Other leading contributors include the Chinese Academy of Sciences and Zhejiang University, particularly in areas like bioactive ceramics and nanocomposite drug delivery systems.

Regulatory policies in China are generally pro-patent, especially for biomaterials and regenerative technologies. Unlike the EU, China does not impose ethical restrictions on patents involving stem cells or bioprinted tissues. This has accelerated innovation in areas that are more tightly regulated elsewhere.

In terms of commercialization, China's IP to market pipeline is rapidly maturing, supported by industry-academia linkages and government subsidies for patent licensing and clinical trials. However, concerns persist around patent quality and global competitiveness with some innovations criticized for incremental novelty or limited translational depth.

Still, China's combination of volume, funding, and policy backing makes it a key global driver of biomaterials innovation, particularly in bioelectronics, implants, and soft TE [17].

23.8.3 European Union

The European Union holds a significant position in the global biomaterials patent landscape, with over 8,000 patents filed between 2015 and 2024, and approximately 6,200 filings in 2023 alone. The average grant rate stands at around 60%, reflecting a moderately efficient patent system supported by high-quality applications.

Funding for biomaterials research in the EU is largely driven by programs such as Horizon Europe and the European Research Council (ERC), which have collectively invested over €1.5 billion in biomaterials and TE in the past five years. Leading research institutions include the Max Planck Institutes, ETH Zurich, and the University of Oxford, all of which contribute to developments in regenerative materials, ECM-based therapies, and biodegradable devices.

However, the EU's regulatory environment is among the strictest globally, particularly in areas involving human biological materials. The Directive 98/44/EC prohibits patenting of inventions that involve the destruction of human embryos, restricting stem cell-based and bioprinting-related patents. This ethical alignment with European norms tends to slow down the pace of translational research.

Despite these constraints, the EU excels in material quality and sustainability, with strong contributions in algae-based wound dressings, bioceramic, and bioresorbable implants. Commercialization is generally cautious but steady, supported by strong collaboration between academia, small biotech firms, and med-tech manufacturers.

Overall, while the EU does not lead in patent volume, it contributes significantly to niche, high-value innovations, particularly those emphasizing ethical compliance, material biodegradability, and clinical safety.

23.8.4 India

India's biomaterials innovation landscape has expanded rapidly, though it still lags behind global leaders in both patent volume and commercialization. From 2015 to 2024, approximately 1500 biomaterials patents were filed, with 1800 filings in 2023, reflecting a 45% year on year growth. Despite this progress, the country's grant rate remains low at ~45%, although it has improved from 35% in 2020.

India's biomaterials research is largely driven by publicly funded institutions such as IISc, AIIMS, and the IITs, supported by schemes like BIRAC, Startup India, and the Department of Biotechnology (DBT), which allocated around $120 million in 2023. Notable innovations include IISc's silk fibroin-based corneal substitute and AIIMS's zinc-magnesium biodegradable stent, though both faced multiyear delays due to regulatory and compliance issues.

The biggest hurdle in India's IP ecosystem is its legal and regulatory framework. Sections 3(b) and 3(c) of the Indian Patents Act block patents deemed "contrary to morality" or those involving natural substances without a sufficient inventive step. Additionally, the Biological Diversity Act, 2002 mandates approval from the National Biodiversity Authority (NBA) for any patent involving native biological materials, causing delays of up to 2 years [18].

To improve its standing, India will need to implement faster patent examination processes, streamline biodiversity compliance, and increase funding for clinical translation, particularly in bioelectronics and regenerative implants.

23.8.5 Japan

Japan continues to be a key player in the global biomaterials domain, especially in orthopedics, dental materials, and nano-enabled coatings. With approximately 3,000 patents filed between 2015 and 2024 and a 55% grant rate, Japan emphasizes high-quality, clinically relevant innovations over sheer volume.

Funding is robust, provided by agencies such as the Japan Agency for Medical Research and Development (AMED) and Japan Science and Technology Agency (JST). Private sector involvement is strong as well, with major medical device manufacturers collaborating with academic centers such as Osaka University, University of Tokyo, and RIKEN. These institutions have pioneered innovations in ceramic-based implants, antimicrobial coatings, and nanostructured biomaterials for spinal and dental applications [19].

Regulatory conditions in Japan are balanced and innovation-friendly. Unlike the EU, Japan permits patents involving stem cells and regenerative medicine under ethical oversight, allowing researchers to explore a wide range of biologically active materials.

Commercialization pathways are efficient, supported by clear patent laws, fast-track innovation schemes, and strong tech transfer offices in universities. Japan has a particularly strong record in bringing orthopedic and dental biomaterials to market, many of which are now internationally licensed.

While Japan may not lead in emerging domains like 4D bioprinting or AI-driven biomaterials, its strength lies in precision manufacturing, nanotechnology, and regulatory predictability making it a stable hub for biomaterials innovation in Asia [20] (Fig. 23.2).

The global biomaterials patent landscape is dynamic and rapidly evolving, with each major region contributing in distinct ways. The United States and China dominate in volume and translation, backed by massive public funding and streamlined IP frameworks. The European Union emphasizes ethical compliance and material safety, though often at the cost of regulatory speed. Japan's strengths lie in precision, nanotechnology, and clinical implementation, while India, despite strong academic innovation, continues to struggle with regulatory bottlenecks, limited funding, and commercialization delays. By evaluating these national ecosystems through a common framework, this chapter has highlighted not

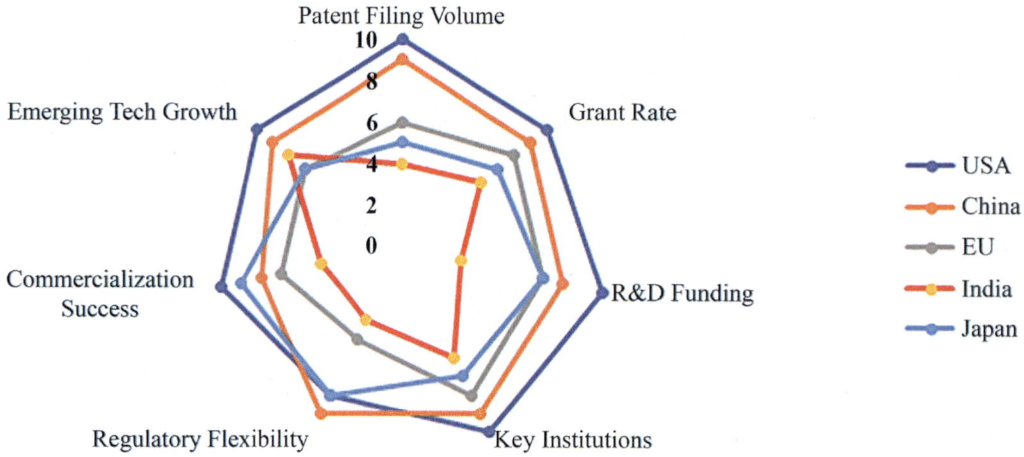

FIGURE 23.2 Comparative Radar Chart of Global Biomaterials Innovation Performance (2015–24).

only the trends and innovations shaping biomaterials research but also the structural gaps that must be addressed to ensure global accessibility and impact. As biomaterials increasingly intersect with smart systems, AI, and 4D fabrication, the need for adaptive, ethical, and supportive patent systems becomes more urgent than ever.

23.9 Conclusion

Biomaterials have emerged as foundational tools in the advancement of modern healthcare, particularly within TE, regenerative medicine, and intelligent therapeutic systems. This chapter has explored how patent systems, when effectively structured and ethically balanced, play a pivotal role in protecting innovation, fostering commercialization, and advancing global research. From utility and process patents to challenges in biodiversity compliance, the complexity of the biomaterials patent landscape reflects both the promise and barriers that define this rapidly evolving field.

The comparative analysis of five major global ecosystems, that is, the United States, China, European Union, India, and Japan, reveals that innovation strength is not solely determined by the number of patents filed, but by the ecosystem's ability to convert those patents into real-world applications. While the United States and China dominate in filing volume and commercialization, the EU leads in ethical frameworks and material sustainability. Japan offers a model of clinical precision and regulatory balance, and India, despite institutional excellence, continues to face regulatory delays, limited funding, and underdeveloped commercialization pathways.

The future of biomaterials innovation will be shaped by advances in 4D bioprinting, AI-integrated design, biodegradable smart implants, and personalized biosensors. To fully harness this potential, nations must invest not only in research and patent protection but also in building integrated frameworks that support ethical compliance, industry-academia collaboration, and international licensing. Ultimately, accelerating innovation in biomaterials requires patent systems that are not only protective but also progressive, adaptable, and globally inclusive.

AI disclosure

During the preparation of this work, the author(s) used Grammarly and ChatGPT to check grammar and spelling. After using this tool/service, the author(s) reviewed and edited the content as needed and take(s) full responsibility for the content of the publication.

References

[1] R. Agrawal, A. Kumar, M.K.A. Mohammed, S. Singh, Biomaterial types, properties, medical applications, and other factors: a recent review, Journal of Zhejiang University: Science A 24 (11) (2023) 1027–1042. Available from https://doi.org/10.1631/jzus.A2200403, https://www.springer.com/journal/11582.

[2] R. Song, M. Murphy, C. Li, K. Ting, C. Soo, Z. Zheng, Current development of biodegradable polymeric materials for biomedical applications, Drug Design, Development and Therapy 12 (2018) 3117–3145. Available from https://doi.org/10.2147/DDDT.S165440, https://doi.org/10.2147/DDDT.S165440.

[3] S. Amukarimi, S. Ramakrishna, M. Mozafari, Smart biomaterials—a proposed definition and overview of the field, Current Opinion in Biomedical Engineering 19 (2021) 100311. Available from https://doi.org/10.1016/j.cobme.2021.100311.

[4] Z.U. Arif, M.Y. Khalid, R. Noroozi, M. Hossain, H.T.H. Shi, A. Tariq, et al., Additive manufacturing of sustainable biomaterials for biomedical applications, Asian Journal of Pharmaceutical Sciences 18 (3) (2023). Available from https://doi.org/10.1016/j.ajps.2023.100812, http://www.journals.elsevier.com/asian-journal-of-pharmaceutical-sciences/recent-articles/.

[5] D. Hernández-Melchor, E. López-Bayghen, A. Padilla-Viveros, The patent landscape in the field of stem cell therapy: closing the gap between research and clinic, F1000Research 11 (2022). Available from https://doi.org/10.12688/f1000research.123799.3.

[6] Global Innovation Index 2024, 2024. Available from: https://www.wipo.int/web-publications/global-innovation-index-2024/en/.

[7] S.M. Thomas, Intellectual property rights in biotechnology, Expert Opinion on Therapeutic Patents 6 (9) (1996) 845–854. Available from https://doi.org/10.1517/13543776.6.9.845, http://www.tandfonline.com/loi/ietp20.

[8] K.K. Singh, Biotechnology and intellectual property rights: legal and social implications, Biotechnology and Intellectual Property Rights: Legal and Social Implications, Springer India, India, (2015), pp. 1–254. Available from http://doi.org/10.1007/978-81-322-2059-6, http://doi.org/10.1007/978-81-322-2059-6.

[9] D. Cao, J. Ding, Recent advances in regenerative biomaterials, Regenerative Biomaterials 9 (2022). Available from https://doi.org/10.1093/rb/rbac098.

[10] S. Liu, J.M. Yu, Y.C. Gan, X.Z. Qiu, Z.C. Gao, H. Wang, et al., Biomimetic natural biomaterials for tissue engineering and regenerative medicine: new biosynthesis methods, recent advances, and emerging applications, Military Medical Research 10 (1) (2023). Available from https://doi.org/10.1186/s40779-023-00448-w, https://mmrjournal.biomedcentral.com/.

[11] K. Dzobo, N.E. Thomford, D.A. Senthebane, H. Shipanga, A. Rowe, C. Dandara, et al., Advances in regenerative medicine and tissue engineering: innovation and transformation of medicine, Stem Cells International 2018 (2018). Available from https://doi.org/10.1155/2018/2495848, http://www.hindawi.com/journals/sci/contents/.

[12] S. Bhat, A. Kumar, Biomaterials and bioengineering tomorrow's healthcare, Biomatter 3 (3) (2013). Available from https://doi.org/10.4161/biom.24717.

[13] S.S. Tiwari, S. Raman, P. Martin, Regenerative medicine in India: trends and challenges in innovation and regulation, Regenerative Medicine 12 (7) (2017) 875–885. Available from https://doi.org/10.2217/rme-2017-0094, http://www.futuremedicine.com/loi/rme.

[14] NIBIB Budget Overview FY2023, 2023. Available from: https://www.nibib.nih.gov/about-nibib/overview/budget.

[15] MIT researchers create scaffold to support growing tissues, 2006. Available from: https://news.mit.edu/2006/scaffold.

[16] Intellectual Property and Stem Cell Research, 2021. Available from: https://www.uspto.gov/ip-policy/ip-and-healthcare/stem-cell-research.

[17] A. Fatimi, Exploring the patent landscape and innovation of hydrogel-based bioinks used for 3D bioprinting, Recent Advances in Drug Delivery and Formulation 16 (2) (2022) 145–163. Available from https://doi.org/10.2174/2667387816666220429095834.

[18] E. Differding, Biotechnology in India: an analysis of 'biotechnology industry research assistance council' (BIRAC)-supported projects, Chembiochem: a European Journal of Chemical Biology 24 (21) (2023). Available from https://doi.org/10.1002/cbic.202300302.

[19] Y. Yamada, M. Haga, Y. Matsuoka, Funding trends in Japan agency for medical research and development (AMED): focus on psychiatry, JMA journal 8 (2) (2025) 385–394. Available from https://doi.org/10.31662/jmaj.2024-0391.

[20] A. Kojima, H. Hanaoka, Y. Uyama, Characteristics of multi-regional clinical trials conducted in Asia, focusing on japan's participation and small/medium companies-sponsored trials, Therapeutic Innovation & Regulatory Science (2025). Available from https://doi.org/10.1007/s43441-025-00837-4.

Chapter 24

Challenges in clinical translation of laboratory discoveries of tissue engineered products

Rashi Mathur[1], Sweta Singh[2], Devansh Dahiya[1], Priyam Mangal[1], Raunak Varshney[1], and Sachin Sony[1]

[1]*Applied Chemistry and Nanoscience Department, Institute of Nuclear Medicine and Allied Sciences, DRDO, Brig S.K. Mazumdar Marg, New Delhi, Delhi, India,* [2]*Clinical Research and Medical Management Department, Institute of Nuclear Medicine and Allied Sciences, DRDO, Brig S.K. Mazumdar Marg, New Delhi, Delhi, India*

24.1 Introduction

TEPs are a revolutionary combination of biology, materials science, and engineering. By mimicking the natural form and activity of a human tissue using cells, scaffolds, and biologically active compounds, they aim to develop functional tissues and organs for therapeutic application [1–5]. Tissue engineering's main objective is to overcome the limitations of traditional transplant methods, such as organ shortages and the risk of rejection. TEPs have a wide range of applications, which include skin grafts for burn victims, tailored organs for transplant, and regenerative therapy for injured tissues [6,7]. Advancements in bioprinting, stem cell research, and biomaterials have accelerated the research field, resulting in new therapies for diseases such as cardiovascular disease, orthopedic injuries, and degenerative disorders. Advancements in research have enabled TEPs to transform regenerative medicine, enhance therapeutic outcomes, and improve quality of life.

TEPs had been traditionally developed from the integration of science, engineering, and medical requirements. At first, scientists focused mostly on understanding biological functions and the development of scaffolds able to mimic the extracellular matrix. By the late 20th century, the development of basic tissue structures was facilitated by advancements in cell culture techniques, specifically the ability to reproduce human cells in vitro [8,9]. Initial developments in skin substitutes focused on essential therapeutic requirements, such as burn therapy. In biomaterials research development, complex scaffolds have been designed to facilitate the incorporation of cells, growth factors, and mechanical properties relevant to specific tissues. The development of regulatory bodies, especially from organizations such as the FDA, provided pathways for the clinical study and approval of these products [10,11]. Tissue engineering became an essential component of advanced regenerative medicine as a result of multidisciplinary cooperation, driving concepts like gene-edited structures, stem cell synthesis, and 3D bioprinting. This development represents not only technical advancement but also a better knowledge of tissue biology and the ability of the body to regenerate [4,12,13].

Several clinically available TEPs have made significant advances in regenerative medicine [14–16]. *Apligraf*, a dermal substitute used for the treatment of chronic wounds such as diabetic ulcers, is developed by the combination of human skin cells and a bovine collagen matrix [17,18]. Another product with the name *TransCyte* is a temporary wound care that helps in burn injury and surgical wounds [19,20]. It contains live human fibroblasts on a silicone membrane. The other class of material is bone grafts like *OsteoSet*, a biodegradable calcium sulfate product used in orthopedic procedures to facilitate bone healing and regeneration [21,22]. *Infuse*, a widely used bone graft, integrates synthetic human BMP-2 with a collagen matrix, facilitating bone growth in spinal fusions and many orthopedic applications. Another major concern is cartilage regeneration [23,24]. *Matrix autologous chondrocyte implantation (MACI) ()* is the process of extracting a patient's native chondrocytes, expanding them in a lab, and implanting them on a scaffold to repair damaged cartilage in joints [25,26]. Other applications of this kind of material are vascular grafts known as *Caval Patch*, commonly used for the replacement or repair of damaged blood vessels, usually obtained from human or animal tissue [27,28]. In dental

Fundamentals of Translational Tissue Engineering. DOI: https://doi.org/10.1016/B978-0-443-33455-9.00034-1

applications, *OsteoGrow*, a product used in dental bone regeneration, incorporating growth factors and scaffolds to enhance healing, is frequently used in dental implants [29,30]. These products represent different uses of tissue engineering, offering new approaches to challenging medical issues and enhancing medical outcomes. As research progresses, the variety and effectiveness of TEPs are expected to expand further.

The timeline offers a comprehensive overview of the advancement of tissue engineering, covering major academic, scientific, and technological milestones from 1930 to 2010. from fundamental concepts to practical bioprinting techniques, the timeline highlights major developments in the advancement of tissue engineering. It starts in 1930 when Alexis Carrel and Charles Lindbergh developed a perfusion pump, therefore laying the foundation for artificial organ development. The first synthetic skin substitute had been identified by 1962, which helped engineered skin product manufacturing—most famously Howard Green's skin constructions—to develop in 1980 [31–33]. Officially first used in 1987, the term "tissue engineering" was used by 1988 when Epicel, the first cell-based product, was used clinically. Products like Carticel launched modified skin and cartilage into the market in 1990. Using biomaterials in 1996 marked ongoing progress; notable FDA approvals came from Apligraf in 1998 and Dermagraft in 2001 (Table 24.1). With Thomas Boland's inkjet bioprinting method in 2004, a historic first in bioprinting allowed tissue formation free from a scaffold. With the commercialization of the Novogen MMX 3D bioprinter in 2009 and concluding in 2010 with a greater spectrum of complex, commercially viable bioprinted tissue products, this technology developed further. From theoretical foundations to practical, market-ready medicines, this development highlights a profound journey in regenerative medicine [34–37].

however, clinical translation of tissue engineering refers to the process of turning experimental results in tissue engineering into practical therapy and medical equipment that can be used in patient care. It is a challenging, multistep process that requires addressing manufacturing, ethical, scientific, and regulatory barriers. Replicating the complex structures and functions of human tissues and organs in vitro, while ensuring their appropriate function in vivo, is a considerable challenge [38–43]. TEPs must undergo rigorous evaluation to comply with safety and effectiveness guidelines. Additionally, developing TEPs at a scale acceptable for clinical application while ensuring quality and consistency poses significant challenges. An additional challenge to this research and development could include costs, which areoften considerable in tissue engineering. Although these challenges exist, developments in research, materials science, and a better understanding of biological frameworks are facilitating the clinical translation process.

Therefore, translating lab results into medical applications is essential for innovative therapeutic methods, as they can significantly increase patient outcomes and quality of life. The rapid advancement of healthcare technologies is made possible by bridgingthe gap between clinical practice and research. By translating experimental results into practical approaches, researchers may target multiple injuries for which there are no viable therapies, which could reduce the rate of

TABLE 24.1 Timeline for the evolution of the engineered tissue.

Year/decade	Key milestone
1930	Alexis Carrel and Charles Lindbergh laid the foundation for artificial organ development, publishing *The Culture of New Organs* and creating an artificial perfusion pump
1962	First recorded instance of a synthetic skin substitute
1980	First successful manufacturing of engineered skin products (e.g., Howard Green's skin construct)
1987	The term tissue engineering was introduced during a session of the US National Science Foundation in Washington, DC
1988	First cell-based tissue-engineered medical application (Epicel, developed by Genzyme Tissue Repair Corporation)
1990	Market introduction of tissue-engineered skin and cartilage products, such as Carticel
1996	Biomaterials introduced in tissue regeneration
1998	FDA approval of Apligraf, the first skin construct with dermis and epidermis
2001	FDA approval of Dermagraft, a skin substitute with dermal fibroblasts in a polymeric scaffold
2004	Invention of inkjet bioprinting technology, leading to the first bioprinted 3D tissue without a matrix (patented by Thomas Boland)
2009	Commercialization of the Novogen MMX 3D bioprinter, advancing bioprinting technology
2010	Expansion of bioprinted products, increasing complexity, and commercial availability

morbidity and mortality caused by an array of diseases. The multidisciplinary strategy employed here can increase innovation and accelerate the development of comprehensive solutions, establishing a clear route from bench to bedside and, as a result, attracting prospective funding sources. Understanding the translation process assists researchers in navigating the complex regulatory framework, accelerating the approval process, and encouraging commercial entrance [44–48]. Successful translation improves the significance of research investment and promotes constant innovation in the area of study. In summary, the translation of laboratory findings into clinical applications is critical not only for the potential of cutting-edge treatments but also for promoting collaboration, creativity, and a comprehensive understanding of regulatory systems, all of which are important for the advancement of healthcare. This chapter covers the fundamental elements and procedures that are relevant to the therapeutic application of tissue engineering.

24.2 Challenges in clinical translation of tissue-engineered products

The pathway to the clinical translation of TEPs provides several challenges that must be addressed. Some of the major problems in the process are shown in Fig. 24.1.

24.2.1 Regulatory hurdles

24.2.1.1 Compliance with regulatory agencies (Food and Drug Administration, European Medicines Agency)

Frameworks developed for ensuring safety and efficacy significantly impact the clinical translation of TEPs [49,50]. Considering the regulatory framework for TEPs is necessary to ensure safety, effectiveness, and quality. Countries have different regulatory pathways due to their unique healthcare systems and approaches to innovation.

TEPs can fall under categories such as drugs, devices, or biologics, depending on their composition and intended use [51–53]. Compliance with a broad range of standards and guidelines is required due to this level of complexity. The

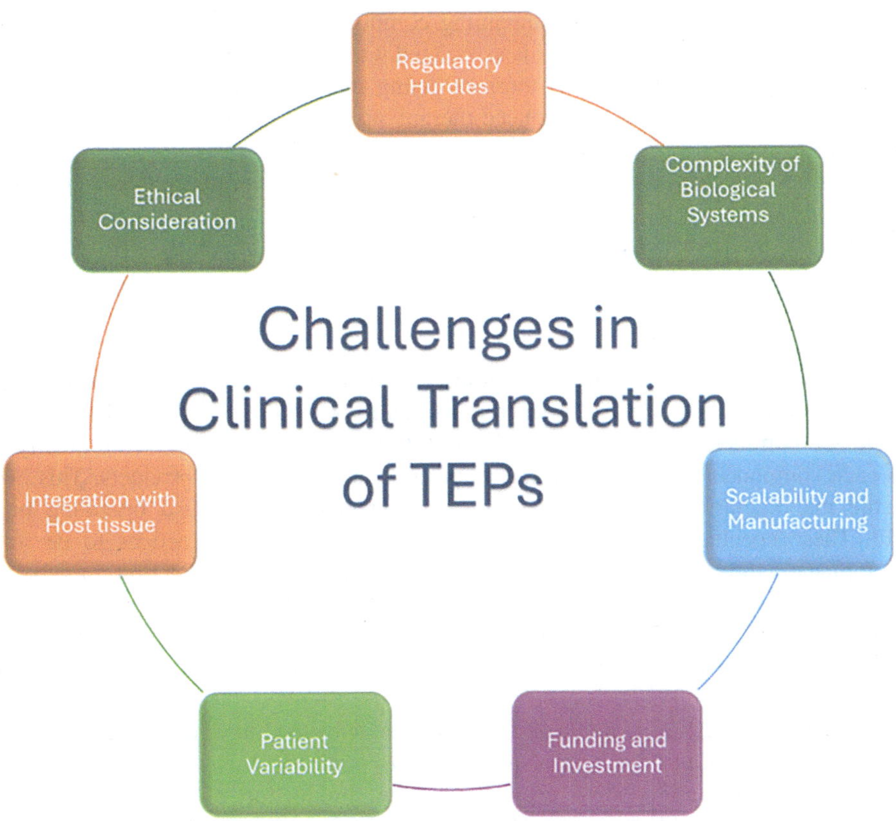

FIGURE 24.1 Overview of the challenges in clinical translation of tissue-engineered products.

regulatory systems for TEPs are complex and differ significantly across various areas [54–56]. This includes a discussion of national and international regulatory structures utilized for TEPs. The United States Food and Drug Administration (FDA) controls TEPs mostly under the Federal Food, Drug, and Cosmetic Act (FDCA) [57–59]. Products are categorized as medicines, biologics, or devices based on their attributes. The details of some of the regulatory bodies and frameworks are given in Table 24.2.

The regulatory framework for TEPs is currently being developed, requiring explicit guidelines and standardized protocols for their evaluation. Insufficient regulatory capacity and oversight infrastructure could inhibit the approval process and innovation within the sector. Several initiatives are in progress to align regulations with international standards, potentially facilitating access for Indian companies to global markets [81–83]. The Indian government is continuously promoting research and development in regenerative medicine, perhaps resulting in more robust regulatory frameworks in the future. The regulation of TEPs offers challenging circumstances that integrate innovation with safety and efficacy across several areas [84,85]. Compiling a detailed Investigational New Drug (IND) application or a Biologics License Application (BLA) needs extensive data on product safety, efficacy, and manufacturing methods [86]. The review process may be lengthy since requests for further data or clarification might extend deadlines. Developing comprehensive preclinical and clinical research protocols is required to establish safety and efficacy. This involves identifying goals for research, selecting suitable animal models, and developing a human study that complies with ethical norms. Regulatory bodies frequently demand comprehensive data ahead of trial approval [87–89].

While countries such as the United States and the European Union established regulatory systems, developing countries like India are currently constructing their regulatory guidelines, which present both problems and opportunities for advancement in regenerative medicine. To successfully navigate this evolving environment, the ongoing cooperation among governmental agencies, industry stakeholders, and researchers will be critical [90,91]. Compliance with Good Manufacturing Practices (GMP) is needed for the manufacturing of TEPs, which involves rigorous quality control protocols during production to ensure product quality, safety, and effectiveness. This compliance is a resource-intensive activity and needs detailed documentation, significant financial and human resources. As a result, startups and smaller enterprises may encounter challenges in obtaining sufficient capital to successfully proceed through regulatory systems [92–94]. After approval and implementation of TEPs, regulatory bodies frequently require continuous monitoring and reporting of harmful situations, which includes maintaining compliance to postmarket research studies regarding long-term safety and efficacy [95,96]. The ethical problems associated with stem cell application and other biological materials could complicate regulatory frameworks, such as those affecting matrix-induced autologous chondrocyte implantation (MACI) and Holoclar (the first stem cell-basedadvanced therapy approved in Europe to restore the surface of the eye in adults suffering from moderate-to-severe limbal stem-cell deficiency [LSCD] 97). Equally critical for compliance and public acceptance is to ensure ethical sourcing and usage of raw material. Overcoming these regulatory barriers is critical for effective translation of TEPs from the laboratory to clinical environments, and it requires careful planning, teamwork, and a solid understanding of the regulatory framework [98–100].

24.2.1.2 Understanding good manufacturing practices

Good manufacturing practices (GMPs) are important guidelines and regulations that ensure products, including TEPs, remain consistently developed and regulated in order to meet quality standards. Understanding GMP is necessary for the effective development and commercialization of TEPs [101,102]. GMP highlights an overall quality management system (QMS) that covers each component of manufacturing, from raw materials to final products. This ensures that quality is implemented into the manufacturing process at every stage. Developing standard operating procedures (SOPs) is necessary for GMP. These specified protocols define accurate processes and practices to be followed to, maintain consistency and compliance throughout manufacturing. GMP requires that production facilities and equipment be developed to reduce the risks of contamination, mix-ups, and errors. This includes proper layout, cleanliness, and maintenance protocols. Individuals working on the production process should have proper training and qualifications. GMPs need regular training courses to verify that employees are proficient in standard procedures and regulatory standards [103–106].

All raw materials must be obtained from authorized vendors and subjected to rigorous verification and documentation. This is to verify the identity, quality, and purity of resources before proceeding with their use in manufacturing [107,108]. Manufacturing operations require validation to ensure consistent production of goods that comply with established standards. This requires doing thorough testing and keeping documentation during the production process [109,110]. Good Manufacturing Practice requires the development of quality control protocols, including systematic testing of

TABLE 24.2 Regulatory bodies with their classifications.

Region	Regulatory body	Regulatory framework	Classification and requirements	Examples
United States (U.S.)	FDA (Food and Drug Administration)	Federal Food, Drug, and Cosmetic Act (FDCA)	TEPs classified as drugs, biologics, or devices based on characteristics Class I: Low-risk, generally exempt from premarket notification. Class II: Moderate risk, requires 510(k) premarket notification. Class III: High risk, requires premarket approval (PMA) with rigorous clinical testing [60,61]. Human Cell and Tissue Products: Regulated under	SYMVESS, Apligraf, Integra Artificial Skin, STRATAGRAFT, Collagen nerve cuff, etc [62].
			Section 361 of the Public Health Service Act with fewer requirements if minimally manipulated [63–65].	
European Union (EU)	EMA (European Medicines Agency)	Advanced Therapy Medicinal Products (ATMP) Regulation (EC) No 1394/2007	Covers somatic cell therapies, gene therapies, and TEPs. Classification based on composition and intended use, leading to centralized or national authorization. Requires comprehensive assessment of quality, safety, and efficacy [66,67].	ChondroCelect, Provenge, Zalmoxis, Alofisel, Ebvallo [68].
Japan	PMDA (Pharmaceuticals and Medical Devices Agency)	Regenerative Medicine Products Regulation (2014)	TEPs regulated as regenerative medicine products. 2014 framework for accelerated approval based on early clinical data [69,70].	Abecma, Alofisel, Breyanzi, Delytact, HeartSheet, JACC, etc [71].
Canada	Health Canada	Food and Drugs Act	TEPs categorized similarly to drugs and medical devices based on characteristics [72,73].	Brineura, Contrave, Mezera, Ozempic, Rayaldee, etc [74].
India	CDSCO (Central Drugs Standard	Drugs and Cosmetics Act; Draft	TEPs classified as medical devices if human tissues are manipulated. Clinical trials must	Cholederm, APCEDEN,
	Control Organization)	Guidelines for TEPs (2021)	comply with Good Clinical Practice (GCP) guidelines [75–77]. Approval process includes ethics committee review and riskbased evidence requirements. Postmarketing surveillance and manufacturing regulations outlined in Draft Guidelines (2021) [78,79].	Stempeucel, etc [80].

products throughout different phases of production. This ensures that any violations of quality requirements are immediately discovered and corrected [111,112].

Maintaining thorough documentation for GMP compliance is important. These documents must consist of records of production processes, testing results, and training, and any adjustments from established procedures. Complete documentation is required for transparency as well as accountability [113,114]. Every change to the production process, instrumentation, or chemicals must be carefully monitored and documented. Change control procedures are established to verify that changes do not have a negative impact on product quality [115,116]. Compliance with GMP criteria is critical for product quality and is required for regulatory approval. Regulatory bodies, including the FDA and EMA, rigorously monitor compliance with GMP, with noncompliance sometimes causing by substantial fines or delays in product clearance. Understanding and executing GMP is important for TEP developers to verify that their products comply with safety and effectiveness criteria, which allows a smoother transfer from laboratory to clinical applications [117–119].

24.2.1.3 Clinical trial design considerations

Designing clinical trials for TEPs requires systematic planning and research to ensure that the studies properly demonstrate safety and effectiveness. Clearly describe the primary and secondary goals associated with the experiment. Primary objectives usually focus on the principal endpoint (e.g., safety or efficacy), whereas secondary objectives may evaluate additional outcomes, such as quality of life or long-term consequences. Identify a suitable target population that reflects the individuals who will ultimately benefit from the TEP. Carefully consider the criteria for inclusion and exclusion, to ensure that the research population is representative and that the results can be applied generally [120–123].

Randomization is a useful method for reducing bias and ensuring that groups can be compared to one another. A control group (either placebo or conventional therapy) is important to evaluate the comparative effectiveness of the TEP. Select relevant and measurable endpoints that align with the trial objectives. Common objectives for TEPs may consist of clinical data (e.g., healing rates), functional evaluations, and patient-reported outcomes. Perform a power analysis and determine the correct sample size to recognize significant differences between groups [124,125]. This analysis has to consider expected effect sizes, variability, and the level of statistical significance.

Establish the trial duration depending on the estimated time for results to come out. Long-term observation may be required in order to evaluate the long-term effects and safety of TEPs. Verify that the study design meets the requirements established by regulatory organizations like the FDA or EMA [126,127]. This may require interactions with regulatory authorities throughout the planning process. Develop strong collection approaches that ensure precise and reliable data. This involves the selection of appropriate clinical data management systems and the development of standard operating procedures for data management [128,129].

Implement a detailed safety monitoring strategy, including the development of an independent Data Safety Monitoring Board (DSMB) to analyze adverse events and determine the risk-to-benefit ratio throughout the study. Investigate ethical concerns related to informed consent, specifically in the context of vulnerable people. Ensure that this research protocol receives approval and review from an Institutional Review Board (IRB) or Ethics Committee [130–132]. Develop a detailed statistical analysis strategy defining the data analysis methods and covering methods for managing missing data and controlling for confounding variables. Consider incorporating an interim analysis to review data at certain time points during the study. This can allow early research cancellation for safety or effectiveness, or changes to the trial protocol if necessary [133,134].

Designing a clinical investigation for TEPs needs a critical evaluation of these variables to ensure that the trial is scientifically valid and capable of producing accurate data for regulatory submissions.

24.2.2 Manufacturing challenges

24.2.2.1 Scalability issues

Ensuring consistent quality and characteristics in different batches of TEPs is a challenge due to several interlinked factors, like various donors or cell lines demonstrating variation in growth rates, differentiation capabilities, and response to stimuli, which affects the uniformity of products [135,136]. In addition, errors in the procurement or manufacturing of biomaterials, such as scaffolds, can result in variations in mechanical characteristics, porosity, and degradation rates. Even small modifications in the fabrication process between batches might lead to significant variations in performance. Environmental factors, including temperature, pH, and oxygen levels, can affect cellular activity and product properties during culture. Developing a three-dimensional (3D) architecture that mimics native tissues is very challenging, often resulting in variations in cell distribution and functionality, exacerbatedby the complexity of vascularization in larger constructs, which can vary significantly between different batches [137–141]. The manufacturing process includes several stages—cell isolation, expansion, scaffold synthesis, and tissue maturation—each offering opportunities for variability. As manufacturing is scaled up, it becomes more challenging to regulate variables such as nutrition supply and waste removal. Larger production areas are becoming more vulnerable to contamination, which impacts batch quality and can lead to different results between batches when combined with a lack of standard production and testing protocols. Different regulatory requirements could affect the development of uniform manufacturing processes, specifically if operations move across different jurisdictions. Addressing these issues is critical for enhancing the consistency and reliability of TEPs, and the failure to do so may result in unpredictable clinical results, inhibiting regulatory approval and market acceptance [142–145].

24.2.2.2 *Ensuring reproducibility*

Achieving reproducibility in TEPs requires strategic methods that focus on uniformity, quality control, and process management. It is important to establish a standard protocol validated by complete appropriate documentation. Standard protocols should be established for all stages of TEPS production, including cell sourcing, culture conditions, scaffold fabrication, product formulationmaturation, and detailed records of all procedures, materials, and equipment. This documentation is critical for tracking and reproducibility. Real-time surveillance of critical parameters such as temperature, pH, and nutrient concentrations during cell culture and product development is mandatory for maintaining uniform conditions [146–148]. In addition, robust testing processes must be established for final goods, including mechanical, biological, and chemical evaluations to verify compliance with established guidelines. The utilization of well-defined biomaterials with uniform characteristics, validated by detailed physical and chemical characterization, drastically improves repeatability [149,150]. Working with reliable vendors and requiring documentation of material characteristics and fabrication methods is important. To reduce contamination risk, strong aseptic protocols and a regulated environment must be implemented [151–155]. Continuous monitoring of environmental factors, including air quality and humidity, in manufacturing areas is essential for maintaining uniformity. The utilization of standard cell lines or primary cells from reliable sources decreases biological variability.

24.2.3 Biological and clinical considerations

Biocompatibility is an essential component of TEPs that significantly influences their therapeutic use. Ensuring the biocompatibility of TEPs promotes effective integration, reduces side effects, and improves overall patient outcomes, which results in enhancing the probability of regulatory approval and clinical acceptance. As research develops in biomaterials and tissue engineering, continuous endeavors to enhance biocompatibility will remain critical to the efficacy of these novel treatments [156–158]. The regulatory agencies demand rigorous biocompatibility studies (e.g., ISO 10993 standards) for materials used in TEPs. Establishing biocompatibility is important for getting approval.

Biocompatible materials increase interactions with surrounding tissue, facilitating cell adhesion, proliferation, and differentiation. The successful use of TEPs in clinical environments requires effective integration. These compounds decrease inflammatory responses, therefore minimizing problems such as long-term inflammation or tissue rejection. Biocompatibility maintains the stability and function of implants throughout time, influencing patient outcomes to enhance healing and overall treatment experiences [159–161]. Advancements in biomaterials research, including smart materials and bioactive scaffolds, increases performance via biocompatibility. It is essential that degradation products from resorbable materials be nontoxic and facilitate tissue regrowth [162,163]. Prolonged biocompatibility is critical for maintaining the mechanical and biological functional characteristics of TEPs. Evaluating biocompatibility can facilitate the development of tailored TEPs, modification in therapies to meet unique patient requirements, and improve overall results [164].

Similarly, clinical considerations of TEPs focus on their safety, effectiveness, integration into host tissue, and long-term functions. These items must exhibit biocompatibility to prevent immunological rejection and inflammation. Professionals must determine the risk of infection, tumor development, or abnormal tissue proliferation, particularly when employing stem cells or genetically altered elements. Surgical implantation procedures, patient-specific characteristics (including age, underlying diseases, and immunological state), and postoperative observation are required factors in determining therapeutic results. Tissue-engineered products (TEPs) require mechanical and structural properties that are suitable for their specified objective, whether for skin, cartilage, bone, or organ repair. Maintaining uniform quality, traceability of biological resources, and compliance with GMP is important for secure clinical application [31,135,165–169].

The complex nature of tissue engineering methods and their effects has given rise to several ethical problems. The use of human tissues can cause questions about permission, whereas the application of animal tissues might provoke debate on animal rights. The use of embryonic stem cells will provoke significant ethical discussions over their acquisition and application. Another concern is the consent and transparency that ensure donors of biological materials understand the intended use of their tissues and provide consent without pressure [98,170–172]. The engineering of tissues or organs possessing human-like properties may provoke theoretical queries on identity and humanity. Ethical challenges may arise about the hazards associated with TEPs, particularly when the long-term effects remain unknown. The possibility that advanced tissue engineering may become too costly has generated concern about fair treatment for persons from different economic statuses. A complex approach that respects personal rights, social principles, and scientific advancement is essential to tackle these ethical issues [173–175].

24.3 Innovations driving the field

24.3.1 Advances in biomaterials

The development of responsive materials that respond to environmental stimuli—such as pH levels, temperature variations, and certain biomolecules—improves the regulated release of growth factors and medicines, facilitating tissue regeneration. Many biomaterials have self-healing capabilities that enable them to maintain structural integrity over time [176,177]. 3D printing allows the fabrication of customized scaffolds that mimic the structure of natural tissues. Novel bioinks enhance the incorporation of live cells during manufacturing. Bioactive glasses and ceramics adhere to bone and stimulate osteogenesis, enhancing orthopedic repair [178,179]. Newman's natural polymer-derived compounds, such as collagen and gelatin, increase cell attachment and replicate the extracellular matrix. Decellularized tissues maintain their natural architecture to enhance functioning [180]. Nano-coatings enhance surface characteristics, whereas nanoparticles enable targeted medication administration, enhancing healing and minimizing adverse effects. Composite materials can improve mechanical and biological performance, and more accurately mimic natural tissue properties [181,182]. Layered architectures provide diverse functional features, whereas organ-on-a-chip technologies enable the examination of biomaterial interactions in a regulated setting. Improvements in vascular networks within TEPs facilitate nutrition transport and waste elimination. Further developments in biomaterials are important for the advancement of tissue engineering, resulting in more efficient TEPs and enhanced patient outcomes in regenerative medicine [183–186].

24.4 Stem cell-based technologies

Stem cell technologies are important for the development of TEPs by providing the cellular components required for tissue regeneration. Embryonic stem cells (ESCs) and induced pluripotent stem cells (iPSCs) contain the capability to develop into any cell type, but mesenchymal stem cells (MSCs) derived from sources such as bone marrow and adipose tissue are esteemed for their versatility and immunomodulatory characteristics. Recent advancements facilitate the accurate differentiation of stem cells into particular cell types required for tissue applications. The selection of biomaterials affects stem cell destiny and improves integration into synthetic tissues [187–189]. Scaffolds provide a structure for cell adherence and proliferation, whereas bioreactors and dynamic cultures replicate physiological environments, improving cellular development. The controlled release of growth factors from biomaterials enhances stem cell viability and differentiation, promoting the development of functional tissues for cartilage, bone, and cardiac engineering [190,191]. Mesenchymal stem cells mitigate inflammatory reactions, enhancing the reception of TEPs by the host. The use of autologous stem cells reduces the risk of rejection, while improvements in induced pluripotent stem cell (iPSC) technology facilitate the creation of patient-specific TEPs [192,193]. Also, 3D bioprinting improves the precise organizing of stem cells within scaffolds, resulting in specific tissue structures.

24.5 Bioreactor systems

Bioreactor systems offer controlled conditions for cellular and tissue development and are essential for the development of TEPs. As technology advances, these systems will improve the scalability, tailoring, and effectiveness of TEPs in regenerative medicine [194,195]. Bioreactors provide ideal conditions for cellular proliferation with precise control of temperature, pH, oxygen concentrations, and food supply. They enhance result reproducibility by reducing variability and provide mechanical forces that replicate physiological circumstances, promoting cell differentiation and tissue development. These techniques provide uniform cell seeding on scaffolds and promote nutritional supply via continuous blood supply, which increases cell viability [196,197]. These bioreactors can also enhance vascularization in tissue structures, which is important for the sustained success of TEPs postimplantation. Different types of bioreactor setups, like batch and continuous systems, can be tailored for specific tissue types, such as cartilage and bone. Some of the models possess different compartments for cultivating various cell types, whilst others use bioprinting methods to fabricate complex, multilayered tissues. Advanced bioreactors offer real-time monitoring and adjustments through sensors,

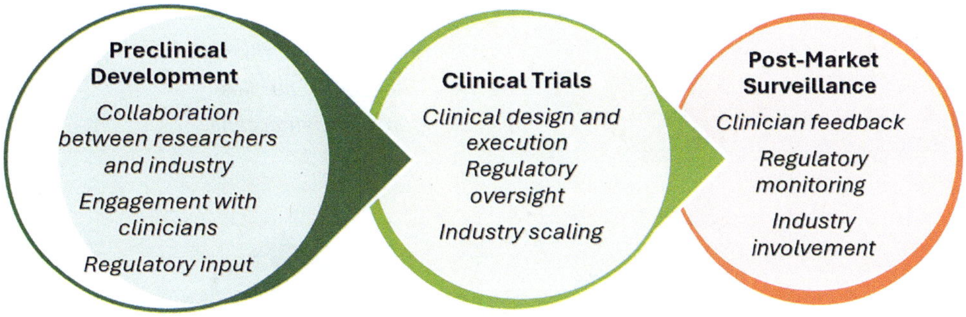

FIGURE 24.2 Key stages of collaborations between various agencies.

which results in enhancing the optimization of culture conditions. Hence, they often act as platforms for preclinical evaluation of TEPs, enabling researchers to examine mechanical and biological properties in regulated environmenical trials. They not onlnts [198–201].

24.6 Interdisciplinary collaboration

The clinical application and translation of TEPs will require interdisciplinary collaboration among researchers, clinicians, and regulatory bodies [202]. Academicians and researchers typically lead this development, focusing on key areas like biocompatibility, scaffold design, and tissue regeneration methods, generally testing new biomaterials with animal models and exploring innovative techniques such as stem cells and gene editing. To make the movement of the products from research to clinics, researchers must work closely with clinicians to ensure that these products meet real human health needs. Engaging regulatory bodies early is also crucial for understanding the requirements for clinical studies, while collaboration with industry partners helps scale up production for trials [203–205]. Clinicians provide essential insights into patient needs and treatment challenges, offering feedback on the functionality of products and the design of clinical trials. They not only gather data on patient outcomes but also assist with recruitment and monitoring of patients during studies. Regulatory bodies like the US FDA, EMA, and CDSCO in India ensure that TEPs comply with safety, efficacy, and quality standards. They establish guidelines for product approval, review of clinical study data, and collaborate with researchers to define regulatory pathways [206–208]. Industry partners, including biotech firms and CROs, play a critical role in transforming research into commercially viable products. They manage product development, ensure compliance with GMP, and oversee clinical trial operations, ensuring that the products are both safe and effective for widespread clinical use [209,210]. Open communication is thus important among researchers, clinicians, and industry for adapting products based on real-world data and ongoing advancements.

Key stages of collaboration: Below are discussed the various stages (Fig. 24.2) which are the pillars for collaboration and form the foundation for a successful translational strategy.

24.7 Preclinical development

- *Collaboration between researchers and industry*: Researchers develop TEPs, while industry partners help refine manufacturing processes.
- *Engagement with clinicians*: Clinicians provide input into patient-specific needs and anatomical considerations.
- *Regulatory input*: Regulatory agencies help define the preclinical testing requirements and the path forward for clinical trials.

24.8 Clinical trials

- *Clinical design and execution:* Clinicians lead the design and execution of clinical trials, working with industry for operational support.
- *Regulatory oversight*: Regulatory agencies assess the safety and efficacy data from clinical trials.
- *Industry scaling*: If successful, industry partners help scale the product for largescale trials and eventual commercialization.

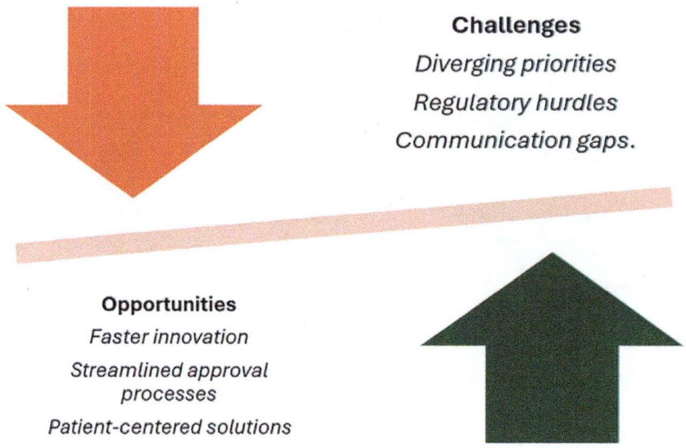

FIGURE 24.3 Balancing opportunities and challenges.

24.9 Postmarket surveillance

- *Clinician feedback*: Ongoing clinician feedback helps refine the product and its usage in patient care.
- *Regulatory monitoring*: Regulatory bodies monitor postmarket data for any signs of long-term safety issues.
- *Industry involvement*: Industry partners continue to manufacture and distribute the product, ensuring compliance with regulatory standards.

24.10 Challenges and opportunities in interdisciplinary collaboration

Diverging priorities: Researchers may focus on innovation, while industry stakeholders may prioritize marketability and cost-effectiveness, and clinicians may focus on patient outcomes (Fig. 24.3).

- *Regulatory hurdles*: Regulatory agencies may be slow to adopt new guidelines for TEPs, delaying clinical trials and approvals.
- *Communication gaps*: Lack of understanding between disciplines can lead to inefficiencies or misunderstandings about product requirements.

- **Opportunities**:
- *Faster innovation*: Collaborating early and frequently can lead to faster innovation and more targeted product development.
- *Streamlined approval processes*: Regulatory bodies can provide earlier guidance, helping reduce time to market.
- *Patient-centered solutions*: Combining clinical experience with scientific innovation can lead to more effective treatments that are better aligned with patient needs.

24.11 Conclusion

To summarize we can say that while the field of TEPs is at an exciting crossroads and has great potential to transform healthcare, there are many obstacles to be crossed. The major ones are the regulatory hurdles that put these products under complex frameworks requiring rigorous safety and efficacy validation. Ethical concerns, such as sourcing cells and tissues, logistical issues, like scaling production for widespread use, also pose significant barriers. The high costs of development and the need for multidisciplinary collaboration can additionally slow progress. On the other hand, tissue engineering holds the promise of addressing unmet clinical needs, such as repairing or replacing damaged tissues and organs. Many innovative solutions are emerging through advances in bioprinting, biomaterials, and stem cell research. In the near future, this successful translation could bring a paradigm shift in regenerative medicine, improving patient outcomes and quality of life. Thus, seamless collaboration across multiple sectors is required for the successful clinical translation of TEPs. Researchers, clinicians, regulatory bodies, and industry stakeholders must work together from the beginning of product development through clinical trials and postmarket surveillance. Clear communication, shared goals, and mutual understanding are key to overcoming challenges and accelerating the path from tissue engineering labs to clinics.

References

[1] K. Parveenns, ReviewArticle NewwEraainnHealthhCare::TissueeEngineeringg, Internet Vol. 1 (2006). Available from www.pubstemcell.com.

[2] F.M. Chen, X. Liu, Advancing biomaterials of human origin for tissue engineering, Progress in Polymer Science (2016) 86–168. Available from https://doi.org/10.1016/j.progpolymsci.2015.02.004.

[3] D.K. Shanmugam, S.C. Anitha, V. Souresh, Y. Madhavan, S. Sampath, D. Catakapatri Venugopal, et al., Current advancements in the development of bionic organs using regenerative medicine and 3D tissue engineering, Materials Technology (2023). Available from https://doi.org/10.1080/10667857.2023.2242732.

[4] K. Dzobo, N.E. Thomford, D.A. Senthebane, H. Shipanga, A. Rowe, C. Dandara, et al., Advances in regenerative medicine and tissue engineering: Innovation and transformation of medicine, Stem Cells Int 2018 (2018). Available from https://doi.org/10.1155/2018/2495848. PubMed PMID: 30154861.

[5] F. Han, J. Wang, L. Ding, Y. Hu, W. Li, Z. Yuan, et al., Tissue engineering and regenerative medicine: achievements, future, and sustainability in Asia, Frontiers in Bioengineering and Biotechnology (2020). Available from https://doi.org/10.3389/fbioe.2020.00083.

[6] G. Cangelosi, S. Mancin, D. Bei, E. Clementi, P. Pantanetti, G. Caggianelli, et al., Multidisciplinary management and autologous skin grafting in a patient with severe burns: a case study, Medicina (Lithuania) 60 (8) (2024). Available from https://doi.org/10.3390/medicina60081201.

[7] S. Ji, S. Xiao, Z. Xia, Consensus on the treatment of second-degree burn wounds (2024 edition), Burns Trauma 12 (2024). Available from https://doi.org/10.1093/burnst/tkad061.

[8] T. Ramos, L. Moroni, Tissue Engineering and regenerative medicine 2019: the role of biofabrication - a year in review, Tissue Engineering - Part C: Methods (2020) 91–106. Available from https://doi.org/10.1089/ten.tec.2019.0344.

[9] E. Knight, S. Przyborski, Advances in 3D cell culture technologies enabling tissue-like structures to be created in vitro, Journal of Anatomy (2015) 746–756. Available from https://doi.org/10.1111/joa.12257. PubMed PMID: 25411113.

[10] R. Stone, S. Natesan, C.J. Kowalczewski, L.H. Mangum, N.E. Clay, R.M. Clohessy, et al., Advancements in regenerative strategies through the continuum of burn care, Frontiers in Pharmacology (2018). Available from https://doi.org/10.3389/fphar.2018.00672.

[11] J.J. Darrow, J. Avorn, A.S. Kesselheim, FDA approval and regulation of pharmaceuticals, 1983-2018, JAMA: The Journal of the American Medical Association 323 (2) (2020) 164–176. Available from https://doi.org/10.1001/jama.2019.20288. PubMed PMID: 31935033.

[12] Y. Zhao, A. Aili, Z. Jia, T. Wen, A. Muheremu, Novel tissue engineering scaffolds in the treatment of spinal cord injury—a bibliometric study, Bioengineering 12 (4) (2025) 347. Available from https://doi.org/10.3390/bioengineering12040347.

[13] E.S. Gomez-Hinostroza, N. Gurdo, M.V.G. Alvan Vargas, P.I. Nikel, M.E. Guazzaroni, L.P. Guaman, et al., Current landscape and future directions of synthetic biology in South America, Frontiers in Bioengineering and Biotechnology (2023). Available from https://doi.org/10.3389/fbioe.2023.1069628.

[14] C.R. Meretsky, A. Polychronis, D. Liovas, A.T. Schiuma, Advances in tissue engineering and its future in regenerative medicine compared to traditional reconstructive techniques: a comparative analysis, Cureus (2024). Available from https://doi.org/10.7759/cureus.68872. Sep 7.

[15] M. Lele, S. Kapur, S. Hargett, N.M. Sureshbabu, A.K. Gaharwar, APPLIED SCIENCES AND ENGINEERING Global trends in clinical trials involving engineered biomaterials, Sci. Adv [Internet]. (2024) Available from https://www.science.org.

[16] F. De Chiara, A. Ferret-Miñana, J.M. Fernández-Costa, J. Ramón-Azcón, The tissue engineering revolution: from bench research to clinical reality, Biomedicines (2024). Available from https://doi.org/10.3390/biomedicines12020453.

[17] Curran M.P., Plosker G.L., Armstrong D., Veves A. Bilayered Bioengineered Skin Substitute (Apligraf ®1) A Review of its Use in the Treatment of Venous Leg Ulcers and Diabetic Foot Ulcers.

[18] L. Zaulyanov, R.S. Kirsner, A review of a bi-layered living cell treatment (Apligraf) in the treatment of venous leg ulcers and diabetic foot ulcers, Clinical Interventions in Aging (2007) 93–98. Available from https://doi.org/10.2147/ciia.2007.2.1.93. PubMed PMID: 18044080.

[19] B. ter Horst, G. Chouhan, N.S. Moiemen, L.M. Grover, Advances in keratinocyte delivery in burn wound care, Advanced Drug Delivery Reviews (2018) 18–32. Available from https://doi.org/10.1016/j.addr.2017.06.012. PubMed PMID: 28668483.

[20] J.R. Saffle, Closure of the excised burn wound: temporary skin substitutes, Clinics in Plastic Surgery (2009) 627–641. Available from https://doi.org/10.1016/j.cps.2009.05.005. PubMed PMID: 19793557.

[21] A.W. Barone, S. Andreana, R. Dziak, Current use of calcium sulfate bone grafts, Medical Research Archives [Internet] (2020) Available from http://journals.ke-i.org/index.php/mra.

[22] T. Kurien, R.G. Pearson, B.E. Scammell, Bone graft substitutes currently available in orthopaedic practice THE EVIDENCE, FOR THEIR USE 95 (5) (2013). Available from https://doi.org/10.1302/0301-620X.95B5.

[23] C.E. Gillman, A.C. Jayasuriya, FDA-approved bone grafts and bone graft substitute devices in bone regeneration, Materials Science and Engineering C (2021). Available from https://doi.org/10.1016/j.msec.2021.112466. PubMed PMID: 34702541.

[24] V. Campana, G. Milano, E. Pagano, M. Barba, C. Cicione, G. Salonna, et al., Bone substitutes in orthopaedic surgery: from basic science to clinical practice, Journal of Materials Science. Materials in Medicine 25 (10) (2014) 2445–2461. Available from https://doi.org/10.1007/s10856-014-5240-2. PubMed PMID: 24865980.

[25] R.L. Davies, N.J. Kuiper, Regenerative medicine: a review of the evolution of autologous chondrocyte implantation (aci) therapy, Bioengineering (2019). Available from https://doi.org/10.3390/bioengineering6010022.

[26] H. Mistry, M. Connock, J. Pink, D. Shyangdan, C. Clar, P. Royle, et al., Autologous chondrocyte implantation in the knee: Systematic review and economic evaluation, Health Technology Assessment (2017) V–160. Available from https://doi.org/10.3310/hta21060. PubMed PMID: 28244303.

[27] L.E. Niklason, J.H. Lawson, Bioengineered human blood vessels, Science (New York, N.Y.) (2020). Available from https://doi.org/10.1126/science.aaw8682. PubMed PMID: 33033191.

[28] P. Gupta, B.B. Mandal, Tissue-engineered vascular grafts: emerging trends and technologies, Advanced Functional Materials (2021). Available from https://doi.org/10.1002/adfm.202100027.

[29] A. Ho-Shui-Ling, J. Bolander, L.E. Rustom, A.W. Johnson, F.P. Luyten, C. Picart, Bone regeneration strategies: Engineered scaffolds, bioactive molecules and stem cells current stage and future perspectives, Biomaterials (2018) 143–162. Available from https://doi.org/10.1016/j.biomaterials.2018.07.017. PubMed PMID: 30036727.

[30] G. Asensio, B. Vázquez-Lasa, L. Rojo, Achievements in the topographic design of commercial titanium dental implants: Towards anti-peri-implantitis surfaces, Journal of Clinical Medicine (2019). Available from https://doi.org/10.3390/jcm8111982.

[31] E.K. Sims, A.L.J. Carr, R.A. Oram, L.A. DiMeglio, C. Evans-Molina, 100 years of insulin: celebrating the past, present and future of diabetes therapy, Nature Medicine (2021) 1154–1164. Available from https://doi.org/10.1038/s41591-021-01418-2. PubMed PMID: 34267380.

[32] M.M. Heravi, V. Zadsirjan, Prescribed drugs containing nitrogen heterocycles: an overview, RSC Advances (2020) 44247–44311. Available from https://doi.org/10.1039/d0ra09198g.

[33] K. Schürrle, History, current state, and emerging applications of industrial biotechnology, in: Magnus Fröhling, M. Hiete (Eds.), Sustainability and Life Cycle Assessment in Industrial Biotechnology [Internet], Springer International Publishing, Cham, (2020), pp. 13–51. Available from https://doi.org/10.1007/10_2018_81https://doi.org/10.1007/10_2018_81.

[34] M. Ayran, B.and U.S. Bulut, Bioprinting, in: O. Gunduz, C. Egles, PRA, FD and UCB (Eds.), Biomaterials and Tissue Engineering [Internet], Springer International Publishing, Cham, (2023), pp. 357–384. Available from https://doi.org/10.1007/978-3-031-35832-6_13, https://doi.org/10.1007/978-3-031-35832-6_13.

[35] N. Shankhwar, P. Saha, M. Zurek-Mortka, S. Thomas, A. Unal, Chapter 24 - Tissue engineered products—Translational avenues, in: C.P. Sharma, T. Chandy, V. Thomas, F.G. Thankam (Eds.), Tissue Engineering [Internet], Academic Press, (2022), pp. 575–625. Available from https://doi.org/10.1016/B978-0-12-824064-9.00007-1, https://www.sciencedirect.com/science/article/pii/B9780128240649000071.

[36] Q. Ramadan, M. Zourob, 3D bioprinting at the frontier of regenerative medicine, pharmaceutical, and food industries, Frontiers in Medical Technology (2020). Available from https://doi.org/10.3389/fmedt.2020.607648. PubMed PMID: 35047890.

[37] A. Fatimi, O.V. Okoro, D. Podstawczyk, J. Siminska-Stanny, A. Shavandi, Natural hydrogel-based bio-inks for 3D bioprinting in tissue engineering: a review, Gels (2022). Available from https://doi.org/10.3390/gels8030179.

[38] B. Landesmann, A. Paini, JRC summer school on non-animal approaches in science, May 2021, Alternatives to Laboratory Animals: ATLA 49 (6) (2021) 235–300. Available from https://doi.org/10.1177/02611929211065919. PubMed PMID: 35081789.

[39] Y. Dong, X. Zhou, Y. Ding, Y. Luo, H. Zhao, Advances in tumor microenvironment: Applications and challenges of 3D bioprinting, Biochemical and Biophysical Research Communications (2024). Available from https://doi.org/10.1016/j.bbrc.2024.150339. PubMed PMID: 39032359.

[40] M. Raasch, E. Fritsche, A. Kurtz, M. Bauer, A.S. Mosig, Microphysiological systems meet hiPSC technology – new tools for disease modeling of liver infections in basic research and drug development, Advanced Drug Delivery Reviews (2019) 51–67. Available from https://doi.org/10.1016/j.addr.2018.06.008. PubMed PMID: 29908880.

[41] S. Ramesh, A. Deep, A. Tamayol, A. Kamaraj, C. Mahajan, S. Madihally, Advancing 3D bioprinting through machine learning and artificial intelligence, Bioprinting (2024). Available from https://doi.org/10.1016/j.bprint.2024.e00331.

[42] E. Jagtiani, M. Yeolekar, S. Naik, V. Patravale, In vitro blood brain barrier models: An overview, Journal of Controlled Release (2022) 13–30. Available from https://doi.org/10.1016/j.jconrel.2022.01.011. PubMed PMID: 35026351.

[43] H.S. Santana, M.S.A. Palma, M.G.M. Lopes, J. Souza, G.A.S. Lima, O.P. Taranto, et al., Microfluidic devices and 3D printing for synthesis and screening of drugs and tissue engineering, Industrial & Engineering Chemistry Research 59 (9) (2020) 3794–3810. Available from https://doi.org/10.1021/acs.iecr.9b03787.

[44] D. Hartl, V. de Luca, A. Kostikova, J. Laramie, S. Kennedy, E. Ferrero, et al., Translational precision medicine: an industry perspective, Journal of Translational Medicine (2021). Available from https://doi.org/10.1186/s12967-021-02910-6. PubMed PMID: 34090480.

[45] C. Acquadro, K. Conway, A. Hareendran, N. Aaronson, Literature review of methods to translate health-related quality of life questionnaires for use in multinational clinical trials, Value in Health 11 (3) (2008) 509–521. Available from https://doi.org/10.1111/j.1524-4733.2007.00292.x. PubMed PMID: 18179659.

[46] M.J. Khoury, M. Gwinn, P.W. Yoon, N. Dowling, C.A. Moore, L. Bradley, The continuum of translation research in genomic medicine: How can we accelerate the appropriate integration of human genome discoveries into health care and disease prevention? Genetics in Medicine (2007) 665–674. Available from https://doi.org/10.1097/GIM.0b013e31815699d0. PubMed PMID: 18073579.

[47] S. Nattel, P.T. Sager, J. Hüser, J. Heijman, D. Dobrev, Why translation from basic discoveries to clinical applications is so difficult for atrial fibrillation and possible approaches to improving it, Cardiovascular Research 117 (7) (2021) 1616–1631. Available from https://doi.org/10.1093/cvr/cvab093. PubMed PMID: 33769493.

[48] J.D. Oliver, S. Jia, L.R. Halpern, E.M. Graham, E.C. Turner, J.S. Colombo, et al., Innovative molecular and cellular therapeutics in cleft palate tissue engineering, Tissue Engineering - Part B: Reviews (2021) 215–237. Available from https://doi.org/10.1089/ten.teb.2020.0181. PubMed PMID: 32873216.

[49] M.M. Godoi, E.M. Reis, J. Koepp, J. Ferreira, Perspective from developers: Tissue-engineered products for skin wound healing, International Journal of Pharmaceutics (2024). Available from https://doi.org/10.1016/j.ijpharm.2024.124319. PubMed PMID: 38866084.

[50] F. Xu, C. Dawson, M. Lamb, E. Mueller, E. Stefanek, M. Akbari, et al., Hydrogels for tissue engineering: addressing key design needs toward clinical translation, Frontiers in Bioengineering and Biotechnology (2022). Available from https://doi.org/10.3389/fbioe.2022.849831.

[51] C. Pimenta, V. Bettiol, T. Alencar-Silva, O.L. Franco, R. Pogue, J.L. Carvalho, et al., Advanced therapies and regulatory framework in different areas of the globe: past, present, and future, Clinical Therapeutics (2021) e103–e138. Available from https://doi.org/10.1016/j.clinthera.2021.02.006. PubMed PMID: 33892966.

[52] D.K. Gupta, A. Tiwari, Y. Yadav, P. Soni, M. Joshi, Ensuring safety and efficacy in combination products: regulatory challenges and best practices, Frontiers in Medical Technology (2024). Available from https://doi.org/10.3389/fmedt.2024.1377443.

[53] E. Olesti, Y. Nuevo, M. Bachiller, E. Guillen, J. Bascuas, S. Varea, et al., Academic challenges on advanced therapy medicinal products' development: a regulatory perspective, Cytotherapy (2024) 221–230. Available from https://doi.org/10.1016/j.jcyt.2023.12.005. PubMed PMID: 38260921.

[54] P. Sebbagh, A. Cannone, G. Gremion, V. Gremeaux, W. Raffoul, N. Hirt-Burri, et al., Current status of PRP manufacturing requirements & European regulatory frameworks: practical tools for the appropriate implementation of PRP therapies in musculoskeletal regenerative medicine, Bioengineering 10 (3) (2023). Available from https://doi.org/10.3390/bioengineering10030292.

[55] J.M. Granjeiro, P.G.de M. Borchio, I.P.B. Ribeiro, K.B.S. Paiva, Bioengineering breakthroughs: The impact of stem cell models on advanced therapy medicinal product development, World J Stem Cells 16 (10) (2024) 860–872. Available from https://doi.org/10.4252/wjsc.v16.i10.860.

[56] C.K. Wang, T.H. Tsai, C.H. Lee, Regulation of exosomes as biologic medicines: Regulatory challenges faced in exosome development and manufacturing processes, Clinical and Translational Science (2024). Available from https://doi.org/10.1111/cts.13904. PubMed PMID: 39115257.

[57] F.B. Palumbo, L.H. Rosebush, L.M. Zeta, Navigating through a complex and inconsistent regulatory framework: section 503b of the federal food drug and cosmetic act outsourcing facilities engaged in clinical investigation, Ther Innov Regul Sci 50 (3) (2016) 270–278. Available from https://doi.org/10.1177/2168479015618695.

[58] W.H. Kitchens, FDA regulatory approval process for medical products, Advances in the Study of Entrepreneurship, Innovation, and Economic Growth, Emerald Group Publishing Ltd, (2016), pp. 201–229. Available from https://doi.org/10.1108/S1048-473620160000026007.

[59] Study of the laws that regulate IPR and Health.

[60] G.F. Safety and Efficacy of Orthopaedic Surgical Devices Under the Safety and Efficacy of Orthopaedic Surgical Devices Under the FDA's Updated Premarket Notification Program FDA's Updated Premarket Notification Program [Internet]. Available from: https://scholarship.law.missouri.edu/mlrhttps://scholarship.law.missouri.edu/mlr/vol84/iss3/8.

[61] G.A.V. Norman, translational toolbox drugs, devices, and the FDA: part 2 an overview of approval processes: FDA approval of medical devices, Basic Trans Science (2016).

[62] M. Streit, L.R. Braathen, Apligraf – a living human skin equivalent for the treatment of chronic wounds, The International Journal of Artificial Organs 23 (12) (2000) 831–833. Available from https://doi.org/10.1177/039139880002301208.

[63] Lewis M.S. Halted Innovation: The Expansion of Federal Jurisdiction over Medicine and the Human Body [Internet]. 2018. Available from: https://perma.cc/E6.

[64] W.H. Fang, C.T. Vangsness, Governmental regulations and increasing food and drug administration oversight of regenerative medicine products: what's new in 2020? Arthroscopy - Journal of Arthroscopic and Related Surgery 36 (10) (2020) 2765–2770. Available from https://doi.org/10.1016/j.arthro.2020.05.015. PubMed PMID: 32442713.

[65] E.D. Quinley, J. Schwartz, Quality and regulatory issues in cellular therapy, Transfusion Medicine and Hemostasis, Elsevier, (2019), pp. 543–551. Available from https://doi.org/10.1016/b978-0-12-813726-0.00089-1.

[66] Kaur K., Harjai R. Classification of cellular therapy and gene therapy (CGT) products: How regulators are working with industry [Internet]. Available from: www.fda.gov.

[67] K. Joyce, Z. Buljovcic, G. Rosic, M. Kaszkin-Bettag, A. Pandit, Issues with tissues: trends in tissue-engineered products in clinical trials in the European union, Tissue Engineering - Part B: Reviews (2023) 78–88. Available from https://doi.org/10.1089/ten.teb.2022.0094. PubMed PMID: 36062927.

[68] S. Bellino, A. La Salvia, M.F. Cometa, R. Botta, Cell-based medicinal products approved in the European Union: current evidence and perspectives, Frontiers in Pharmacology (2023). Available from https://doi.org/10.3389/fphar.2023.1200808.

[69] R. Kurauchi, H. Kasai, T. Ito, Characteristics of medical products comprising human cells, genes, or tissues developed in Japan and the European union compared via public assessment reports, Front Bioeng Biotechnol 8 (2020). Available from https://doi.org/10.3389/fbioe.2020.606606. Dec 23.

[70] T. Qiu, E. Hanna, M. Dabbous, B. Borislav, M. Toumi, Regenerative medicine regulatory policies: a systematic review and international comparison, Health Policy (Amsterdam, Netherlands) (2020) 701–713. Available from https://doi.org/10.1016/j.healthpol.2020.05.004. PubMed PMID: 32499078.

[71] K. Patel, Review. Int J of Pharm Sci. 3 (2025) 1290–1311. Available from https://doi.org/10.5281/zenodo.14878652.

[72] A.S. Kesselheim, B. Wang, J.M. Franklin, J.J. Darrow, Trends in utilization of FDA expedited drug development and approval programs, 1987-2014: Cohort study, BMJ (Online) (2015). Available from https://doi.org/10.1136/bmj.h4633. PubMed PMID: 26400751.

[73] Joshi D., Sharma I., Gupta S., Gurjeet Singh T., Dhiman S., Prashar A., et al. A global comparison of implementation and effectiveness of materiovigilance program: overview of regulations [Internet]. doi:10.1007/s11356-021-16345-5/Published.

[74] Canada [Internet]. [cited 2025 Apr 6]. Available from: https://www.canada.ca/en/health-canada/services/publications/drugs-health-products/drug-medical-device-highlights-2018.html.

[75] Verma R., Shelar Y., Venkatesh M.P., Bansode M. Overview of Current Regulatory Practices, Challenges and Guidance for Regulation of Stem Cells and Stem Cellderived Products in India. Vol. 6. 2019.

[76] S. Muthu, M. Jeyaraman, M. Basha Kotner, N. Jeyaraman, R.L. Rajendran, S. Sharma, et al., Evolution of mesenchymal stem cell therapy as an advanced therapeutic medicinal product (ATMP)—an Indian perspective, Bioengineering (2022). Available from https://doi.org/10.3390/bioengineering9030111.

[77] D. Lu, F. Dou, F. Qi, J. Gao, Tailored to fit: China optimizes policies and regulations regarding drug registration and review to promote innovation in traditional Chinese medicine, Drug Discov Ther (2024). Available from https://doi.org/10.5582/ddt.2024.01055Aug 31.

[78] M.P. Sekar, H. Budharaju, A. Zennifer, S. Sethuraman, N. Vermeulen, D. Sundaramurthi, et al., Current standards and ethical landscape of engineered tissues—3D bioprinting perspective, Journal of Tissue Engineering (2021). Available from https://doi.org/10.1177/20417314211027677.

[79] F. Fregni, M.A. Nitsche, C.K. Loo, A.R. Brunoni, P. Marangolo, J. Leite, et al., Regulatory considerations for the clinical and research use of transcranial direct current stimulation (tDCS): Review and recommendations from an expert panel, Clinical Research and Regulatory Affairs (2015) 22–35. Available from https://doi.org/10.3109/10601333.2015.980944. PubMed PMID: 25983531.

[80] A. Gupta, StemOneTM/Stempeucel®: CDSCO approved, adult human bone marrow-derived, cultured, pooled, allogenic mesenchymal stem cells for knee osteoarthritis, Biomedicines (2023). Available from https://doi.org/10.3390/biomedicines11112894.

[81] C. Marquis, M. Raynard, Institutional strategies in emerging markets, Academy of Management Annals 9 (1) (2015) 291–335. Available from https://doi.org/10.1080/19416520.2015.1014661.

[82] H.J. Allen, Regulatory innovation and permission to fail, SSRN Electronic Journal (2022). Available from https://doi.org/10.2139/ssrn.4239079.

[83] C.P. Efunniyi, A.O. Abhulimen, A.N. Obiki-Osafiele, O.S. Osundare, E.E. Agu, I.A. Adeniran, Strengthening corporate governance and financial compliance: Enhancing accountability and transparency, Finance & Accounting Research Journal 6 (8) (2024) 1597–1616. Available from https://doi.org/10.51594/farj.v6i8.1509.

[84] N. Beheshtizadeh, M. Gharibshahian, Z. Pazhouhnia, M. Rostami, A.R. Zangi, R. Maleki, et al., Commercialization and regulation of regenerative medicine products: Promises, advances and challenges, Biomedicine and Pharmacotherapy (2022). Available from https://doi.org/10.1016/j.biopha.2022.113431. PubMed PMID: 36076549.

[85] Y. Walter, Managing the race to the moon: global policy and governance in Artificial Intelligence regulation—a contemporary overview and an analysis of socioeconomic consequences, Discover Artificial Intelligence 4 (1) (2024). Available from https://doi.org/10.1007/s44163-024-00109-4.

[86] B. Arondekar, M.S. Duh, R.H. Bhak, M. DerSarkissian, L. Huynh, K. Wang, et al., Real-world evidence in support of oncology product registration: a systematic review of new drug application and biologics license application approvals from 2015-2020, Clinical Cancer Research: an Official Journal of the American Association for Cancer Research 28 (1) (2022) 27–35. Available from https://doi.org/10.1158/1078-0432.CCR-21-2639. PubMed PMID: 34667027.

[87] A. Chorniy, J. Bailey, A. Civan, M. Maloney, Regulatory review time and pharmaceutical research and development, Health Economics (United Kingdom) 30 (1) (2021) 113–128. Available from https://doi.org/10.1002/hec.4180. PubMed PMID: 33078483.

[88] C. Samara, A. Garcia, C. Henry, L. Vallotton, L. Cariolato, J. Desmeules, et al., Safety surveillance during drug development: comparative evaluation of existing regulations, Advances in Therapy (2023) 2147–2185. Available from https://doi.org/10.1007/s12325-023-02492-3. PubMed PMID: 37020083.

[89] J. Shen, B. Swift, R. Mamelok, S. Pine, J. Sinclair, M. Attar, Design and conduct considerations for first-in-human trials, Clinical and Translational Science 12 (1) (2019) 6–19. Available from https://doi.org/10.1111/cts.12582. PubMed PMID: 3004804630048046.

[90] A. Rosemann, F. Vasen, G. Bortz, Global diversification in medicine regulation: insights from regenerative stem cell medicine, Sci Cult (Lond) 28 (2) (2019) 223–249. Available from https://doi.org/10.1080/09505431.2018.1556253.

[91] S. Kabbashi, I.A. Roomaney, M. Chetty, Bridging the gap between omics research and dental practice, BDJ Open (2024). Available from https://doi.org/10.1038/s41405-024-00199-3.

[92] E. Bicudo, I. Brass, Institutional and infrastructure challenges for hospitals producing advanced therapies in the UK: the concept of 'point-of-care manufacturing readiness, Regenerative Med 17 (10) (2022) 719–737. Available from https://doi.org/10.2217/rme-2022-0064. PubMed PMID: 36065826.

[93] Chandra, Saha, Bangabandhu, Sheikh, Mujibur, Rahman, et al., Artificial intelligence in pharmaceutical manufacturing: enhancing quality control and decision making, Rivista Italiana di Filosofia Analitica Junior [Internet] 14 (2) (2023) 2037–4445. Available from https://www.researchgate.net/publication/375330771.

[94] H.O. Bello, C. Idemudia, T.V. Iyelolu., Navigating financial compliance in small and medium-sized enterprises (SMEs): overcoming challenges and implementing effective solutions, World Journal of Advanced Research and Reviews 23 (1) (2024) 042–050. Available from https://doi.org/10.30574/wjarr.2024.23.1.1984.

[95] A. Lavertu, B. Vora, K.M. Giacomini, R. Altman, S. Rensi, A New Era in Pharmacovigilance: toward real-world data and digital monitoring, Clinical Pharmacology and Therapeutics (2021) 1197–1202. Available from https://doi.org/10.1002/cpt.2172. PubMed PMID: 33492663.

[96] K.D Alvis, P. Gunasena, T. Hapuarachchi, K. Gunawardana, S. Gunathilake, S. Bandara, et al., Designing and conducting clinical research: methodological approaches, Uva Clinical Anaesthesia, (2024).

[97] F. Migliorini, J. Eschweiler, C. Götze, A. Driessen, M. Tingart, N. Maffulli, Matrix-induced autologous chondrocyte implantation (mACI) versus autologous matrix-induced chondrogenesis (AMIC) for chondral defects of the knee: A systematic review, British Medical Bulletin (2022) 47–59. Available from https://doi.org/10.1093/bmb/ldac004. PubMed PMID: 35175354.

[98] A.F.J. De Kanter, K.R. Jongsma, M.C. Verhaar, A.L. Bredenoord, The ethical implications of tissue engineering for regenerative purposes: a systematic review, Tissue Engineering - Part B: Reviews (2023) 167–187. Available from https://doi.org/10.1089/ten.teb.2022.0033. PubMed PMID: 36112697.

[99] A.D. Ogbu, N.L. Eyo-Udo, M.A. Adeyinka, W. Ozowe, A.H. Ikevuje., A conceptual procurement model for sustainability and climate change mitigation in the oil, gas, and energy sectors, World Journal of Advanced Research and Reviews 20 (3) (2023) 1935–1952. Available from https://doi.org/10.30574/wjarr.2023.20.3.2304.

[100] A.A. Seyhan, Lost in translation: the valley of death across preclinical and clinical divide – identification of problems and overcoming obstacles, Transl Med Commun 4 (1) (2019). Available from https://doi.org/10.1186/s41231-019-0050-7.

[101] W.T. Seet, M.A. Mat Afandi, M.F. Ishak, M.N.F. Hassan, N. Ahmat, M.H. Ng, et al., Quality management overview for the production of a tissue-engineered human skin substitute in Malaysia, Stem Cell Research and Therapy (2023). Available from https://doi.org/10.1186/s13287-023-03536-9. PubMed PMID: 37858277.

[102] Mateus J., Sandle T., Racine T., Kalergis A.M. Current GMP standards for the production of vaccines and antibodies: An overview.

[103] C.O.R. Okpala, M. Korzeniowska, Understanding the relevance of quality management in agro-food product industry: from ethical considerations to assuring food hygiene quality safety standards and its associated processes, Food Reviews International (2023) 1879–1952. Available from https://doi.org/10.1080/87559129.2021.1938600.

[104] Deshmukh A.S., Dighe P.R., Mahajan V.R., Kunde V.D., Mhaske G.S., Awate S.S. Comprehensive Analysis of Quality Management in Pharmaceutical Manufacturing Process. In: Advanced Concepts in Pharmaceutical Research Vol. 2. B P International (a part of SCIENCEDOMAIN International); 2023. p. 12–23. doi:10.9734/bpi/acpr/v2/6663e.

[105] N.Z. Noor Hasnan, R.K. Basha, N.A.M. Amin, S.H.M. Ramli, J.Y.H. Tang, N.A. Aziz, Analysis of the most frequent nonconformance aspects related to Good Manufacturing Practices (GMP) among small and medium enterprises (SMEs) in the food industry and their main factors, Food Control (2022). Available from https://doi.org/10.1016/j.foodcont.2022.109205.

[106] R. Ahmed, The role of cGMP in the manufacturing unit of a pharmaceutical industry, Radinka Journal of Health Science 2 (2) (2024) Available from https://www.rjupublisher.com/ojs/index.php/RJHS.

[107] Z. Chen, C. Han, M. Gao, S.Y. Kandukuri, K. Zhou, A review on qualification and certification for metal additive manufacturing, Virtual and Physical Prototyping (2022) 382–405. Available from https://doi.org/10.1080/17452759.2021.2018938.

[108] J. Whitley, C. Zwolinski, C. Denis, M. Maughan, L. Hayles, D. Clarke, et al., Development of mRNA manufacturing for vaccines and therapeutics: mRNA platform requirements and development of a scalable production process to support early phase clinical trials, Translational Research (2022) 38–55. Available from https://doi.org/10.1016/j.trsl.2021.11.009. PubMed PMID: 34871810.

[109] H. Wang, Y. Chen, L. Wang, Q. Liu, S. Yang, C. Wang, Advancing herbal medicine: enhancing product quality and safety through robust quality control practices, Frontiers in Pharmacology (2023). Available from https://doi.org/10.3389/fphar.2023.1265178.

[110] Kambala G. Adopting Cloud Services In Enterprise Application Development: A Framework For Decision-Making [Internet]. 2025. Available from: https://www.researchgate.net/publication/389045503.

[111] A.L. Alparslan, Basic good manufacturing practices (GMP), special conditions and inspection processes in the Covid19 (SARS-CoV-2) pandemic, Journal of Research in Pharmacy (2022) 1527–1538. Available from https://doi.org/10.29228/jrp.247.

[112] B.G. Gouveia, P. Rijo, T.S. Gonçalo, C.P. Reis, Good manufacturing practices for medicinal products for human use, Journal of Pharmacy and Bioallied Sciences (2015) 87–96. Available from https://doi.org/10.4103/0975-7406.154424.

[113] J.A. Kirwan, H. Gika, R.D. Beger, D. Bearden, W.B. Dunn, R. Goodacre, et al., Quality assurance and quality control reporting in untargeted metabolic phenotyping: mQACC recommendations for analytical quality management, Metabolomics: Official Journal of the Metabolomic Society (2022). Available from https://doi.org/10.1007/s11306-022-01926-3. PubMed PMID: 36029375.

[114] M.C. Vaghela, S. Rathi, R.L. Shirole, J. Verma, Shaheen, S. Panigrahi, et al., Leveraging AI and machine learning in six-sigma documentation for pharmaceutical quality assurance, Zhongguo ying yong sheng li xue za zhi=Zhongguo yingyong shenglixue zazhi=Chinese journal of applied physiology (2024) e20240005. Available from https://doi.org/10.62958/j.cjap.2024.005. PubMed PMID: 39019923.

[115] F. Leal, A.E. Chis, S. Caton, H. González-Vélez, J.M. García-Gómez, M. Durá, et al., Smart pharmaceutical manufacturing: ensuring end-to-end traceability and data integrity in medicine production, Big Data Research 24 (2021). Available from https://doi.org/10.1016/j.bdr.2020.100172. May 15.

[116] B.K. Singh, M. Delgado-Baquerizo, E. Egidi, E. Guirado, J.E. Leach, H. Liu, et al., Climate change impacts on plant pathogens, food security and paths forward, Nature Reviews. Microbiology (2023) 640–656. Available from https://doi.org/10.1038/s41579-023-00900-7. PubMed PMID: 37131070.

[117] C. Aguilar-Gallardo, A. Bonora-Centelles, Integrating artificial intelligence for academic advanced therapy medicinal products: challenges and opportunities, Applied Sciences (Switzerland) (2024). Available from https://doi.org/10.3390/app14031303.

[118] Good Design Practices for GMP Pharmaceutical Facilities.

[119] Raja N., Kumar Boopathi L. The Crucial Role of Regulatory Affairs: Navigating Compliance and Innovation in the Pharmaceutical Industry.

[120] E. Terpos, J. Mikhael, R. Hajek, A. Chari, S. Zweegman, H.C. Lee, et al., Management of patients with multiple myeloma beyond the clinical-trial setting: understanding the balance between efficacy, safety and tolerability, and quality of life, Blood Cancer Journal (2021). Available from https://doi.org/10.1038/s41408-021-00432-4. PubMed PMID: 33602913.

[121] C. Le Berre, A. Ricciuto, L. Peyrin-Biroulet, D. Turner, Evolving short- and long-term goals of management of inflammatory bowel diseases: getting it right, making it last, Gastroenterology 162 (5) (2022) 1424–1438. Available from https://doi.org/10.1053/j.gastro.2021.09.076. PubMed PMID: 34995529.

[122] Baltes S., Ralph P. Sampling in Software Engineering Research: A Critical Review and Guidelines [Internet]. 2020 Feb 18. Available from: http://arxiv.org/abs/2002.07764.

[123] A.A.H. de Hond, A.M. Leeuwenberg, L. Hooft, I.M.J. Kant, S.W.J. Nijman, H.J.A. van Os, et al., Guidelines and quality criteria for artificial intelligence-based prediction models in healthcare: a scoping review, npj Digital Medicine (2022). Available from https://doi.org/10.1038/s41746-021-00549-7. PubMed PMID: 35013569.

[124] H.B. Burch, P. Perros, T. Bednarczuk, D.S. Cooper, P.J. Dolman, A.M. Leung, et al., Management of thyroid eye disease: a Consensus Statement by the American Thyroid Association and the European Thyroid Association, Eur Thyroid J 11 (6) (2022). Available from https://doi.org/10.1530/ETJ-22-0189. PubMed PMID: 36479875.

[125] P. Zaratin, S. Samadzadeh, M. Seferoğlu, V. Ricigliano, J. dos Santos Silva, A. Tunc, et al., The global patient-reported outcomes for multiple sclerosis initiative: bridging the gap between clinical research and care – updates at the 2023 plenary event, Frontiers in Neurology (2024). Available from https://doi.org/10.3389/fneur.2024.1407257.

[126] E.V. Carraça, B. Rodrigues, S. Franco, I. Nobre, F. Jerónimo, V. Ilharco, et al., Promoting physical activity through supervised vs motivational behavior change interventions in breast cancer survivors on aromatase inhibitors (PAC-WOMAN): protocol for a 3-arm pragmatic randomized controlled trial, BMC Cancer 23 (1) (2023). Available from https://doi.org/10.1186/s12885-023-11137-1. PubMed PMID: 37407950.

[127] G. Wang, L. Chen, F. Luo, J. Luo, J. Xu, Superiority of kinematic alignment over mechanical alignment in total knee arthroplasty during medium- to long-term follow-up: A meta-analysis and trial sequential analysis, Knee Surgery, Sports Traumatology, Arthroscopy 32 (5) (2024) 1240–1252. Available from https://doi.org/10.1002/ksa.12093. PubMed PMID: 38488220.

[128] D. Bhati, M.S. Deogade, D. Kanyal, Improving patient outcomes through effective hospital administration: a comprehensive review, Cureus (2023). Available from https://doi.org/10.7759/cureus.47731. PubMed PMID: 38021686.

[129] Y. Himeur, M. Elnour, F. Fadli, N. Meskin, I. Petri, Y. Rezgui, et al., AI-big data analytics for building automation and management systems: a survey, actual challenges and future perspectives, Artif Intell Rev 56 (6) (2023) 4929–5021. Available from https://doi.org/10.1007/s10462-022-10286-2.

[130] N.A. Dilts, F.E. Harrell, C.J. Lindsell, S. Nwosu, T.G. Stewart, M.S. Shotwell, et al., Securely sharing DSMB reports to speed decision making from multiple, concurrent, independent studies of similar treatments in COVID-19, J Clin Transl Sci 6 (1) (2022). Available from https://doi.org/10.1017/cts.2022.387.

[131] A.J. Huang, L.C. Walter, K. Yaffe, E. Vittinghoff, E. Kornblith, M. Schembri, et al., Treating Incontinence for underlying mental and physical health (TRIUMPH): a study protocol for a multicenter, double-blinded, randomized, 3-arm trial to evaluate the multisystem effects of pharmacologic treatment strategies for urgency-predominant urinary incontinence in ambulatory older women, Trials 24 (1) (2023). Available from https://doi.org/10.1186/s13063-023-07279-z. PubMed PMID: 37085880.

[132] D. Onakomaiya, J. Pan, T. Roberts, H. Tan, S. Nadkarni, M. Godina, et al., Challenges and recommendations to improve institutional review boards' review of community-engaged research proposals: A scoping review, Journal of Clinical and Translational Science (2023). Available from https://doi.org/10.1017/cts.2023.516.

[133] J.R. Carpenter, M. Smuk, Missing data: a statistical framework for practice, Biometrical Journal (2021) 915–947. Available from https://doi.org/10.1002/bimj.202000196. PubMed PMID: 33624862.

[134] J.D. Ciolino, A.M. Kaizer, L.B. Bonner, Guidance on interim analysis methods in clinical trials, Journal of Clinical and Translational Science (2023). Available from https://doi.org/10.1017/cts.2023.552. PubMed PMID: 37313374.

[135] X. Guo, Y. Ma, Y. Min, J. Sun, X. Shi, G. Gao, et al., Progress and prospect of technical and regulatory challenges on tissue-engineered cartilage as therapeutic combination product, Bioactive Materials (2023) 501–518. Available from https://doi.org/10.1016/j.bioactmat.2022.06.015.

[136] S. Schepanski, G.B. Ngoumou, C. Buss, G. Seifert, Assessing in-vitro models for microglial development and fetal programming: a critical review, Frontiers in Immunology (2025). Available from https://doi.org/10.3389/fimmu.2025.1538920.

[137] M.N. Collins, G. Ren, K. Young, S. Pina, R.L. Reis, J.M. Oliveira, Scaffold fabrication technologies and structure/function properties in bone tissue engineering, Advanced Functional Materials (2021). Available from https://doi.org/10.1002/adfm.202010609.

[138] J.W. Jang, K.E. Min, C. Kim, J. Shin, J. Lee, S. Yi, Review: scaffold characteristics, fabrication methods, and biomaterials for the bone tissue engineering, International Journal of Precision Engineering and Manufacturing (2023) 511–529. Available from https://doi.org/10.1007/s12541-022-00755-7.

[139] R. Alva, G.L. Gardner, P. Liang, J.A. Stuart, Supraphysiological Oxygen Levels in Mammalian Cell Culture: Current State and Future Perspectives, Cells (2022). Available from https://doi.org/10.3390/cells11193123. PubMed PMID: 36231085.

[140] A. Agarwal, I.M. Rosas, C. Anagnostopoulou, R. Cannarella, F. Boitrelle, L.V. Munoz, et al., Oxidative stress and assisted reproduction: a comprehensive review of its pathophysiological role and strategies for optimizing embryo culture environment, Antioxidants (2022). Available from https://doi.org/10.3390/antiox11030477.

[141] F. Carotenuto, S. Politi, A. Ul Haq, F. De Matteis, E. Tamburri, M.L. Terranova, et al., From soft to hard biomimetic materials: tuning micro/nano-architecture of scaffolds for tissue regeneration, Micromachines (Basel) 13 (5) (2022). Available from https://doi.org/10.3390/mi13050780.

[142] O. Habanjar, M. Diab-Assaf, F. Caldefie-Chezet, L. Delort, 3D cell culture systems: Tumor application, advantages, and disadvantages, International Journal of Molecular Sciences (2021). Available from https://doi.org/10.3390/ijms222212200. PubMed PMID: 34830082.

[143] P. Benner, L. Meier, A. Pfeffer, K. Krüger, J.E. Oropeza Vargas, D. Weuster-Botz, Lab-scale photobioreactor systems: principles, applications, and scalability, Bioprocess and Biosystems Engineering (2022) 791–813. Available from https://doi.org/10.1007/s00449-022-02711-1. PubMed PMID: 35303143.

[144] Serhan M., Sprowls M., Jackemeyer D., Long M., Perez I.D., Maret W., et al. Total iron measurement in human serum with a smartphone. In: AIChE Annual Meeting, Conference Proceedings. American Institute of Chemical Engineers; 2019. doi:10.1039/x0xx00000x.

[145] L. Falconer, K. Cutajar, A. Krupandan, E. Capuzzo, R.A. Corner, T. Ellis, et al., Planning and licensing for marine aquaculture, Reviews in Aquaculture (2023) 1374–1404. Available from https://doi.org/10.1111/raq.12783.

[146] Simon Jr C.G., Romantseva E. Measurement needs for biofabrication of tissue engineered medical products workshop report [Internet]. 2024 Apr. Available from: https://nvlpubs.nist.gov/nistpubs/SpecialPublications/NIST.SP.1500-23.pdf doi:10.6028/NIST.SP.1500-23.

[147] D. Pamies, M. Leist, S. Coecke, G. Bowe, D.G. Allen, G. Gstraunthaler, et al., Guidance document on good cell and tissue culture practice 2.0 (GCCP 2.0), ALTEX: Alternativen zu Tierexperimenten 39 (1) (2022) 30–70. Available from https://doi.org/10.14573/altex.2111011. PubMed PMID: 34882777.

[148] Y. Shastak, W. Pelletier, A. Kuntz, Insights into analytical precision: understanding the factors influencing accurate vitamin a determination in various samples, Analytica (2024) 54–73. Available from https://doi.org/10.3390/analytica5010004.

[149] P. Jayaraman, R. Lim, J. Ng, M.C. Vemuri, Acceleration of translational mesenchymal stromal cell therapy through consistent quality GMP manufacturing, Frontiers in Cell and Developmental Biology (2021). Available from https://doi.org/10.3389/fcell.2021.648472.

[150] C. Hu, L. Wang, S. Liu, X. Sheng, L. Yin, Recent development of implantable chemical sensors utilizing flexible and biodegradable materials for biomedical applications, ACS Nanotechnology (2024) 3969–3995. Available from https://doi.org/10.1021/acsnano.3c11832. PubMed PMID: 3827167938271679.

[151] A. Tanzini, M. Ruggeri, E. Bianchi, C. Valentino, B. Vigani, F. Ferrari, et al., Robotics and Aseptic Processing in View of Regulatory Requirements, Pharmaceutics (2023). Available from https://doi.org/10.3390/pharmaceutics15061581.

[152] K.J. Ong, J. Johnston, I. Datar, V. Sewalt, D. Holmes, J.A. Shatkin, Food safety considerations and research priorities for the cultured meat and seafood industry, Compr Rev Food Sci Food Saf 20 (6) (2021) 5421–5448. Available from https://doi.org/10.1111/1541-4337.12853. PubMed PMID: 34633147.

[153] B. Elmadhoun, R. Alsaidalani, F. Burczynski, Quality risk management in the final operational stage of sterile pharmaceutical manufacturing: a case study highlighting the management of sustainable related risks in product sterilization, inspection, labeling, packaging, and storage processes, Sustainability 17 (4) (2025) 1670. Available from https://doi.org/10.3390/su17041670.

[154] G. Byrne, O. Damm, L. Monostori, R. Teti, F. van Houten, K. Wegener, et al., Towards high performance living manufacturing systems - A new convergence between biology and engineering, CIRP J Manuf Sci Technol 34 (2021) 6–21. Available from https://doi.org/10.1016/j.cirpj.2020.10.009Aug 1.

[155] U. Dirnagl, G.N. Duda, D.W. Grainger, P. Reinke, R. Roubenoff, Reproducibility, relevance and reliability as barriers to efficient and credible biomedical technology translation, Advanced Drug Delivery Reviews (2022). Available from https://doi.org/10.1016/j.addr.2022.114118. PubMed PMID: 35066104.

[156] D. Peneda Pacheco, N. Suárez Vargas, S. Visentin, P. Petrini, From tissue engineering to engineering tissues: the role and application of: the vitro models, Biomaterials Science (2021) 70–83. Available from https://doi.org/10.1039/d0bm01097a. PubMed PMID: 33191420.

[157] D.F. Williams, Challenges with the development of biomaterials for sustainable tissue engineering, Front Bioeng Biotechnol 7 (MAY) (2019). Available from https://doi.org/10.3389/fbioe.2019.00127. PubMed PMID: 31214584.

[158] L. Ranakoti, B. Gangil, P. Bhandari, T. Singh, S. Sharma, J. Singh, et al., Promising Role of polylactic acid as an ingenious biomaterial in scaffolds, drug delivery, tissue engineering, and medical implants: research developments, and prospective applications, Molecules (Basel, Switzerland) (2023). Available from https://doi.org/10.3390/molecules28020485. PubMed PMID: 36677545.

[159] K.M. Percival, V. Paul, G.A. Husseini, Recent advancements in bone tissue engineering: integrating smart scaffold technologies and bio-responsive systems for enhanced regeneration, International Journal of Molecular Sciences (2024). Available from https://doi.org/10.3390/ijms25116012. PubMed PMID: 38892199.

[160] N. Kutner, K.R. Kunduru, L. Rizik, S. Farah, Recent Advances for improving functionality, biocompatibility, and longevity of implantable medical devices and deliverable drug delivery systems, Advanced Functional Materials (2021). Available from https://doi.org/10.1002/adfm.202010929.

[161] W. Al-Zyoud, D. Haddadin, S.A. Hasan, H. Jaradat, O. Kanoun, Biocompatibility testing for implants: a novel tool for selection and characterization, Materials (2023). Available from https://doi.org/10.3390/ma16216881.

[162] A.R. Aiman, S. Vigneswari, N.A. Amran, V. Murugaiyah, A.A. Amirul, S. Ramakrishna, Advancing regenerative medicine through the development of scaffold, cell biology, biomaterials and strategies of smart material, Regen Eng Transl Med 8 (2) (2022) 298–320. Available from https://doi.org/10.1007/s40883-021-00227-w.

[163] S. Vach Agocsova, M. Culenova, I. Birova, L. Omanikova, B. Moncmanova, L. Danisovic, et al., Resorbable biomaterials used for 3D scaffolds in tissue engineering: a review, Materials (2023). Available from https://doi.org/10.3390/ma16124267.

[164] Y. Lou, H. Wang, G. Ye, Y. Li, C. Liu, M. Yu, et al., Periosteal Tissue Engineering: Current Developments and Perspectives, Advanced Healthcare Materials (2021). Available from https://doi.org/10.1002/adhm.202100215. PubMed PMID: 33938636.

[165] R. Whitaker, B. Hernaez-Estrada, R.M. Hernandez, E. Santos-Vizcaino, K.L. Spiller, Immunomodulatory biomaterials for tissue repair, Chemical Reviews (2021) 11305–11335. Available from https://doi.org/10.1021/acs.chemrev.0c00895. PubMed PMID: 34415742.

[166] P. Trucillo, Biomaterials for drug delivery and human applications, Materials (2024). Available from https://doi.org/10.3390/ma17020456.

[167] Y. Luo, Toward fully automated personalized orthopedic treatments: innovations and interdisciplinary gaps, Bioengineering (2024). Available from https://doi.org/10.3390/bioengineering11080817.

[168] S.E. Jang, L. Qiu, L.L. Chan, E.K. Tan, L. Zeng, Current status of stem cell-derived therapies for parkinson's disease: from cell assessment and imaging modalities to clinical trials, Frontiers in Neuroscience (2020). Available from https://doi.org/10.3389/fnins.2020.558532.

[169] A. Laurent, P. Abdel-Sayed, A. Ducrot, N. Hirt-Burri, C. Scaletta, S. Jaccoud, et al., Development of standardized fetal progenitor cell therapy for cartilage regenerative medicine: Industrial transposition and preliminary safety in xenogeneic transplantation, Biomolecules 11 (2) (2021) 1–35. Available from https://doi.org/10.3390/biom11020250. PubMed PMID: 33572428.

[170] H.B. Baker, J.P. McQuilling, N.M.P. King, Ethical considerations in tissue engineering research: Case studies in translation, Methods (San Diego, Calif.) (2016) 135–144. Available from https://doi.org/10.1016/j.ymeth.2015.08.010. PubMed PMID: 26282436.

[171] I.A. Charitos, A. Ballini, S. Cantore, M. Boccellino, M. Di Domenico, E. Borsani, et al., Stem Cells: a historical review about biological, religious, and ethical issues, Stem Cells International (2021). Available from https://doi.org/10.1155/2021/9978837.

[172] R. Jahns, J. Geiger, I. Schlünder, D. Strech, M. Brumhard, S.G. Von Kielmansegg, Broad donor consent for human biobanks in Germany and Europe: A strategy to facilitate cross-border sharing and exchange of human biological materials and related data, Journal of Laboratory Medicine 43 (6) (2019) 291–299. Available from https://doi.org/10.1515/labmed-2017-0064.

[173] Sesiro D. An Evaluation of Biotechnological Enhancement in the Light of an African Perspective of Personhood: The Metaphysical Aspects of Botho and Human Nature.

[174] S. Canali, V. Schiaffonati, A. Aliverti, Challenges and recommendations for wearable devices in digital health: data quality, interoperability, health equity, fairness, PLOS Digital Health (2022). Available from https://doi.org/10.1371/journal.pdig.0000104.

[175] B. Senyange, J. Wesana, G. Van Huylenbroeck, X. Gellynck, H. De Steur, Tissue engineering in the agri-food industry: current status, socio-economic overview and regulatory compliance, Current Opinion in Biotechnology 91 (2025) 103228. Available from https://doi.org/10.1016/j.copbio.2024.103228.

[176] H. Wei, J. Cui, K. Lin, J. Xie, X. Wang, Recent advances in smart stimuli-responsive biomaterials for bone therapeutics and regeneration, Bone Research (2022). Available from https://doi.org/10.1038/s41413-021-00180-y.

[177] R. Dallaev, Advances in materials with self-healing properties: a brief review, Materials (2024). Available from https://doi.org/10.3390/ma17102464.

[178] Y. Zhu, D. Joralmon, W. Shan, Y. Chen, J. Rong, H. Zhao, et al., 3D printing biomimetic materials and structures for biomedical applications, Bio-Design and Manufacturing. Springer (2021) 405–428. Available from https://doi.org/10.1007/s42242-020-00117-0.

[179] L. He, J. Yin, X. Gao, Additive manufacturing of bioactive glass and its polymer composites as bone tissue engineering scaffolds: a review, Bioengineering (2023). Available from https://doi.org/10.3390/bioengineering10060672.

[180] J. Fan, K. Abedi-Dorcheh, A.S. Vaziri, F. Kazemi-Aghdam, S. Rafieyan, M. Sohrabinejad, et al., A review of recent advances in natural polymer-based scaffolds for musculoskeletal tissue engineering, Polymers (2022). Available from https://doi.org/10.3390/polym14102097.

[181] M. Akbari Edgahi, S.M. Naghib, A. Emamian, H. Ramezanpour, F. Haghiralsadat, D. Tofighi, A practical review over surface modification, nanopatterns, emerging materials, drug delivery systems, and their biophysiochemical properties for dental implants: Recent progresses and advances, Nanotechnology Reviews (2022) 637–679. Available from https://doi.org/10.1515/ntrev-2022-0037.

[182] A. Johnston, A. Callanan, Recent methods for modifying mechanical properties of tissue-engineered scaffolds for clinical applications, Biomimetics (2023). Available from https://doi.org/10.3390/biomimetics8020205.

[183] L.A. Osório, E. Silva, R.E. Mackay, A review of biomaterials and scaffold fabrication for organ-on-a-chip (OOAC) systems, Bioengineering (2021). Available from https://doi.org/10.3390/bioengineering8080113.

[184] B.K. Nahak, A. Mishra, S. Preetam, A. Tiwari, Advances in organ-on-a-chip materials and devices, ACS Applied Bio Materials (2022) 3576–3607. Available from https://doi.org/10.1021/acsabm.2c00041. PubMed PMID: 35839513.

[185] Margolis E.A., Rolle M.W., Putnam A.J., Putnam A. Manufacturing the multiscale vascular hierarchy: progress towards solving tissue engineering's grand challenge. 2023.

[186] S. Najman, S. Stojanović, J. Živković, J. Najdanović, M. Radenković, P. Vasiljević, et al., Applications of biomaterials in regenerative medicine and tissue engineering – concepts and perspective, Contemporary Material 14 (1) (2023). Available from https://doi.org/10.7251/comen2301001n.

[187] A.K. Shukla, G. Gao, B.S. Kim, Applications of 3D bioprinting technology in induced pluripotent stem cells-based tissue engineering, Micromachines (2022). Available from https://doi.org/10.3390/mi13020155.

[188] K. Czerwiec, M. Zawrzykraj, M. Deptuła, A. Skoniecka, A. Tymińska, J. Zieliński, et al., Adipose-derived mesenchymal stromal cells in basic research and clinical applications, International Journal of Molecular Sciences (2023). Available from https://doi.org/10.3390/ijms24043888. PubMed PMID: 36835295.

[189] X. Wan, Z. Liu, L. Li, Manipulation of Stem cells fates: the master and multifaceted roles of biophysical cues of biomaterials, Advanced Functional Materials (2021). Available from https://doi.org/10.1002/adfm.202010626.

[190] X. Huang, Z. Huang, W. Gao, W. Gao, R. He, Y. Li, et al., Current advances in 3D dynamic cell culture systems, Gels (2022). Available from https://doi.org/10.3390/gels8120829.

[191] B. Adhikari, M.A. Stager, M.D. Krebs, Cell-instructive biomaterials in tissue engineering and regenerative medicine, Journal of Biomedical Materials Research - Part A (2023) 660–681. Available from https://doi.org/10.1002/jbm.a.37510. PubMed PMID: 36779265.

[192] Kaligosi D. ARTIFICIAL ORGANS-BIOMATERIALS NATIONAL AND KAPODISTRIAN et al. DEPARTMENT OF PHARMACY INTERDISCIPLINARY POSTGRADUATE STUDIES PROGRAM "NANOMEDICINE"" Master Thesis Title: Artificial Organs-Biomaterials EXAMINATION COMMITTEE."

[193] D.E. Ciolacu, R. Nicu, F. Ciolacu, Natural polymers in heart valve tissue engineering: strategies, advances and challenges, Biomedicines (2022). Available from https://doi.org/10.3390/biomedicines10051095.

[194] A. Rivera-Ordaz, V. Peli, P. Manzini, M. Barilani, L. Lazzari, Critical analysis of cGMP large-scale expansion process in bioreactors of human induced pluripotent stem cells in the framework of quality by design, BioDrugs: Clinical Immunotherapeutics, Biopharmaceuticals and Gene Therapy (2021) 693–714. Available from https://doi.org/10.1007/s40259-021-00503-9. PubMed PMID: 34727354.

[195] L.M. Delgado, Characterization of the biological response to scaffolds, In (2023) 317–330. Available from https://doi.org/10.1007/978-3-031-35832-6_11.

[196] F. Palladino, P.R.F. Marcelino, A.E. Schlogl, Á.H.M. José, R.de C.L.B. Rodrigues, D.L. Fabrino, et al., Bioreactors: applications and innovations for a sustainable and healthy future—a critical review, Applied Sciences (Switzerland) (2024). Available from https://doi.org/10.3390/app14209346.

[197] Y. Li, I.Y. Wong, M. Guo, Reciprocity of cell mechanics with extracellular stimuli: emerging opportunities for translational medicine, Small (Weinheim an der Bergstrasse, Germany) (2022). Available from https://doi.org/10.1002/smll.202107305. PubMed PMID: 35319155.

[198] C. Blume, X. Kraus, S. Heene, S. Loewner, N. Stanislawski, F. Cholewa, et al., Vascular implants – new aspects for in situ tissue engineering, Engineering in Life Sciences (2022) 344–360. Available from https://doi.org/10.1002/elsc.202100100.

[199] P. Kazimierczak, A. Przekora, Bioengineered living bone grafts—a concise review on bioreactors and production techniques in vitro, International Journal of Molecular Sciences (2022). Available from https://doi.org/10.3390/ijms23031765. PubMed PMID: 35163687.

[200] S.J. Reyes, Y. Durocher, P.L. Pham, O. Henry, Modern sensor tools and techniques for monitoring, controlling, and improving cell culture processes, Processes (2022). Available from https://doi.org/10.3390/pr10020189.

[201] K.R. Kim, W. Yeo, Advances in sensor developments for cell culture monitoring, BMEMat 1 (4) (2023). Available from https://doi.org/10.1002/bmm2.12047.

[202] M. Laubach, S. Whyte, H.F. Chan, T. Frankenbach-Désor, S. Mayer-Wagner, F. Hildebrand, et al., Lost in translation: the lack of agreement between surgeons and scientists regarding biomaterials research and innovation for treating bone defects, BMC Medicine 22 (1) (2024) 517. Available from https://doi.org/10.1186/s12916-024-03734-z. PubMed PMID: 39506708.

[203] X. Han, A. Alu, H. Liu, Y. Shi, X. Wei, L. Cai, et al., Biomaterial-assisted biotherapy: A brief review of biomaterials used in drug delivery, vaccine development, gene therapy, and stem cell therapy, Bioactive Materials (2022) 29–48. Available from https://doi.org/10.1016/j.bioactmat.2022.01.011.

[204] A.K. Dubey, E. Mostafavi, Biomaterials-mediated CRISPR/Cas9 delivery: recent challenges and opportunities in gene therapy, Frontiers in Chemistry (2023). Available from https://doi.org/10.3389/fchem.2023.1259435.

[205] A.E. Lottes, K.J. Cavanaugh, Y.Y.F. Chan, V.J. Devlin, C.J. Goergen, R. Jean, et al., Navigating the regulatory pathway for medical devices—a conversation with the FDA, clinicians, researchers, and industry experts, Journal of Cardiovascular Translational Research (2022) 927–943. Available from https://doi.org/10.1007/s12265-022-10232-1. PubMed PMID: 35288821.

[206] A. Spreafico, A.R. Hansen, A.R.A. Razak, P.L. Bedard, L.L. Siu, The future of clinical trial design in oncology, Cancer Discovery (2021) 822–837. Available from https://doi.org/10.1158/2159-8290.CD-20-1301. PubMed PMID: 33811119.

[207] V. Kandi, S. Vadakedath, P.S. Addanki, V. Godishala, V.B. Pinnelli, Clinical trials: the role of regulatory agencies, pharmacovigilance laws, guidelines, risk management, patenting, and publicizing results, Borneo Journal of Pharmacy 6 (1) (2023) 93–109. Available from https://doi.org/10.33084/bjop.v6i1.3263.

[208] E. Baumfeld Andre, R. Reynolds, P. Caubel, L. Azoulay, N.A. Dreyer, Trial designs using real-world data: The changing landscape of the regulatory approval process, Pharmacoepidemiology and Drug Safety (2020) 1201–1212. Available from https://doi.org/10.1002/pds.4932. PubMed PMID: 31823482.

[209] Wu Q., He Q. DIY laboratories and business innovation ecosystems: the case of pharmaceutical industry.

[210] Good design practices for GMP pharmaceutical facilities.

Chapter 25

Clinical trials documentation in nanomaterials and scaffolds

Apoorv Gupta
Reviana Innovations Private Limited, Delhi, India

25.1 Introduction

Nanotechnology is a multidisciplinary area that includes nanoscale materials to address some of healthcare's most serious issues, including medicine delivery and tissue regeneration, etc. [1]. Nanotechnology, which manipulates matter at the molecular and atomic levels, allows for the production of materials and technologies with unique features that are beyond the constraints of conventional techniques [2].

Nanomaterials, including nanoparticles, nanotubes, and dendrimers, are distinguished by their large surface area, customizable chemical characteristics, and capacity to interact with biological systems at cellular and molecular levels [3]. These materials are commonly used in drug delivery systems, where they increase drug solubility, stability, and bioavailability while allowing for tailored distribution to specific tissues or cells [4]. For example, lipid nanoparticles (LNPs) have attracted substantial interest owing to their efficacy in delivering mRNA vaccinations during the COVID-19 pandemic, illustrating the potential of nanotechnology in dealing with global health emergencies [5]

Meanwhile, scaffolds play an important role in tissue engineering and regenerative medicine. These three-dimensional structures resemble the extracellular matrix, forming a supporting framework for cell adhesion, proliferation, and differentiation [6]. Scaffolds are critical for mending and rebuilding injured tissues, with applications, including cartilage repair, tendon and ligament regeneration, and even organ engineering. Advances in scaffold technology, such as the combination of bioactive chemicals with stem cells, have shown promise in clinical trials, allowing for the creation of more effective regenerative medicines [7,8]

Despite its potential, clinical translation of nanomaterials and scaffolds presents major obstacles. Regulatory challenges, safety concerns, and manufacturing scalability are significant impediments to wider implementation [9]. Clinical trials are critical for addressing these hurdles because they provide the required data to assess the safety, effectiveness, and real-world application of these materials [10]. However, the documenting and reporting of these trials require painstaking attention to detail, as the unique features of nanomaterials and scaffolds create difficulties not normally seen in traditional drug research [11].

This chapter delves into the intricacies of clinical trial documentation in nanomaterials and scaffolds, highlighting their applications, challenges, and future directions. The following sections explore the use of nanomaterials in drug delivery systems, the role of scaffolds in regenerative medicine, the obstacles to clinical translation, and emerging trends in the field.

25.2 Nanomaterials in drug delivery systems

Nanomaterials have revolutionized drug delivery by enabling precise control over drug release, targeting, and pharmacokinetics. Their applications span a wide range of therapeutic areas, including cancer, infectious diseases, and neurological disorders [12]

- **Lipid nanoparticles (LNPs) for mRNA therapeutics**: LNPs have emerged as an important method for delivering nucleic acid-based therapies like mRNA vaccinations. These nanoparticles shield mRNA from enzymatic destruction,

Fundamentals of Translational Tissue Engineering. DOI: https://doi.org/10.1016/B978-0-443-33455-9.00030-4

improve cellular uptake, and promote proper translation into target proteins [13]. During the COVID-19 pandemic, LNPs played a crucial role in the quick development and deployment of mRNA vaccines, demonstrating their ability to solve global health concerns [14].

- **siRNA conjugates in gene silencing**: The distribution of short interfering RNA (siRNA) to specific tissues has long been a challenge due to RNA molecules' instability in biological environments [15]. GalNAc-siRNA conjugates addressed this issue by targeting the asialoglycoprotein receptor in hepatocytes, enabling efficient gene silencing in the liver [16]. These conjugates have entered clinical studies to treat genetic and metabolic illnesses.
- **Overcoming the blood–brain barrier (BBB)**: Nanomaterials have also shown promise in delivering treatments across the blood–brain barrier (BBB), which is a significant barrier to treating central nervous system (CNS) illnesses [17]. Strategies such as receptor-mediated transport and cell-penetrating peptides have been used to improve nanoparticle distribution in the brain [18]. Curcumin-loaded nanoparticles have exhibited neuroprotective benefits in preclinical models of neurodegenerative disorders [19].

25.3 Scaffolds in tissue engineering and regenerative medicine

Scaffolds are indispensable tools in regenerative medicine, providing a structural and biochemical environment conducive to tissue repair and regeneration [20].

- **Cartilage repair and regeneration**: Articular cartilage abnormalities, which are commonly caused by injury or osteoarthritis, are difficult to cure due to cartilage's poor regenerating potential [21]. Biomimetic scaffolds consisting of collagen or hyaluronic acid have been created to aid with cartilage healing [22]. These scaffolds offer mechanical support and bioactive signals that promote chondrocyte proliferation and matrix production [22].
- **Tendon and ligament repair:** Fibrous scar tissue development frequently impairs tendon and ligament recovery [23]. Scaffolds containing bioactive substances, such as growth hormones, or seeded with stem cells, can help the regeneration process [24]. For example, scaffolds built to distribute bioactive chemicals with great spatiotemporal accuracy have demonstrated the potential to repair the mechanical characteristics of tendons and ligaments [25].
- **Integration with stem cells:** Stem cell-loaded scaffolds are a cutting-edge technique in tissue engineering [26]. These scaffolds increase tissue regeneration results by creating a supportive milieu for stem cell survival, proliferation, and differentiation [27]. These techniques are now being evaluated in clinical studies for their efficacy in treating musculoskeletal injuries and degenerative disorders [28].

25.4 Systematic review

A search strategy constructed to identify literature related to nanomedicine clinical studies from PubMed and Google Scholar systematically. Further, a PRISMA flowchart is used to screen the literature (Fig. 25.1. Covidence was used to screen the studies. Regarding Inclusion criteria, all human studies that are reviewed publications with English as the only preferred text language are included in the study. The study's full text should be available. The retracted text was also excluded from the final list.

25.4.1 Search strategy

1. (("Nanostructures"[Mesh]) OR "Tissue Scaffolds"[Mesh]) AND ("clinical trial"[Publication Type] OR "clinical trials as topic"[MeSH Terms] OR "clinical trial"[All Fields]) for PubMed
2. ("Nanostructures" or "Tissue Scaffolds") and "clinical trial" for Google Scholar

The combined search strategy yielded 1964 studies from 1994 to November 1, 2024.

25.4.2 PRISMA workflow

1962 publications from PubMed and 367 from Google Scholar were identified using the above-mentioned search strategy, which comes down to 1964 articles after duplication removal. Afterward, 1964 studies were subjected to title/abstract screening, where articles that were in a non-English language, nonhuman studies, retracted articles, and preprints were excluded. This led to the retention of a total of 1745 articles, which were further assessed for their final eligibility. Those articles with full-text or non-primary studies were excluded from the final eligibility list. Finally, 638 lists of articles were used to write the literature for the book.

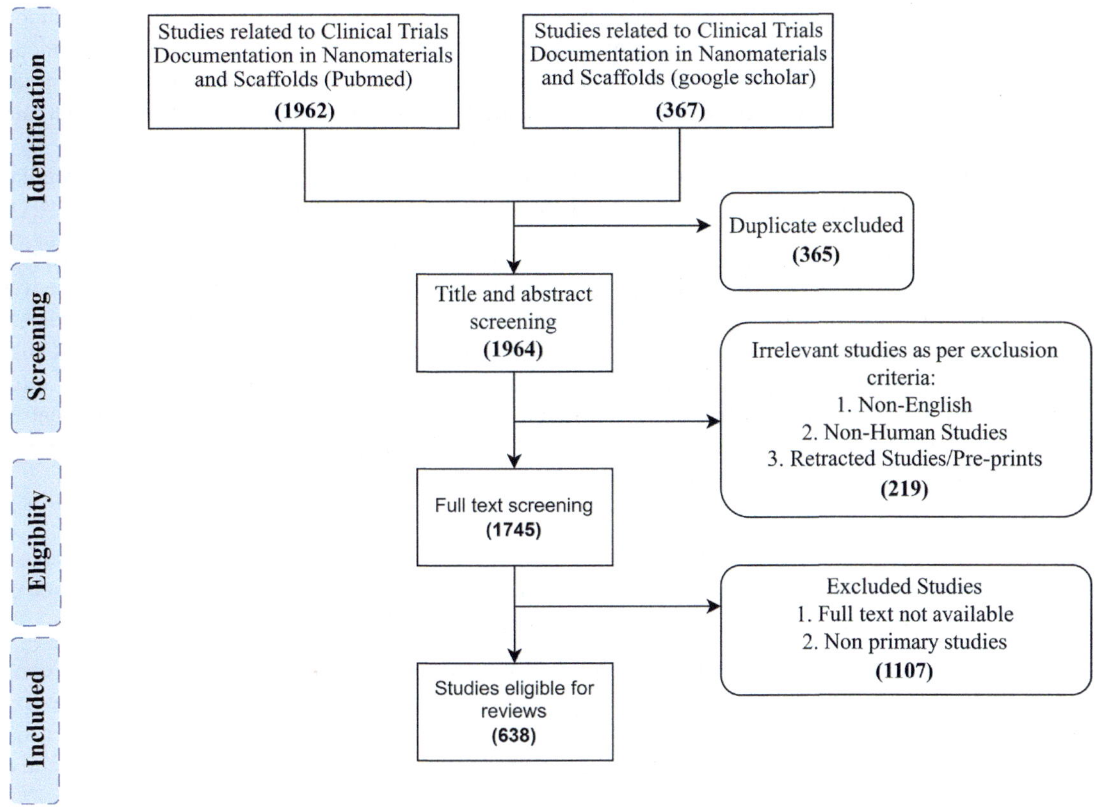

FIGURE 25.1 Prisma flowchart. Consort-flow diagram to identify nanomedicine clinical studies for text mining of bio-entities.

25.5 Challenges in nanomaterial-based clinical trials

The advancement of nanomaterials in clinical settings, particularly for drug delivery and regenerative medicine, offers unprecedented opportunities [29]. However, transitioning nanomaterials from the lab to clinical application entails a myriad of challenges [30]. These challenges span safety concerns, regulatory hurdles, manufacturing, scalability, and the lack of standardized protocols [31]. This section delves into these issues with detailed insights, referencing examples from the research provided.

25.5.1 Safety and toxicity concerns

Nanomaterials interact with biological systems in complicated ways because of their tiny size, large surface area, and distinct physicochemical features [32]. While these characteristics make nanoparticles extremely successful in therapeutic applications, they also raise questions regarding their biocompatibility and long-term safety [33]. Metallic and nanoscale aluminum compounds, for example, have been linked to oxidative damage and intrinsic apoptosis because of their potential to produce reactive oxygen species (ROS) [34]. This emphasizes the importance of rigorous preclinical investigations to prevent any side effects. Furthermore, the fast clearance or build-up of nanoparticles in nontarget organs raises additional safety concerns. For example, curcumin-loaded nanoparticles have shown potential in treating CNS illnesses by crossing the blood–brain barrier (BBB) [35]. However, curcumin's solubility, stability, and toxicity continue to restrict its therapeutic use [36].

25.5.2 Regulatory challenges

The regulatory landscape for nanomedicine remains underdeveloped [37]. Existing frameworks, designed for traditional pharmaceuticals, often fail to address the unique properties and behaviors of nanomaterials [38]. Regulatory challenges include

- **Characterization standards**: Nanomaterials can vary widely in size, shape, surface charge, and composition. This variability complicates the establishment of standardized methods for characterizing nanomaterials in clinical trials [39].
- **Multifunctionality**: Nanomaterials often have multifunctional roles, such as simultaneous imaging and therapy (theranostics) [40]. The dual roles necessitate new evaluation criteria and approval pathways that are currently lacking [41].
- **Long-term monitoring**: Given the potential for slow degradation and prolonged systemic circulation, regulatory bodies require long-term monitoring protocols to assess nanomaterials' safety over extended periods [42].

For example, the development of N-acetylgalactosamine (GalNAc)-siRNA conjugates for RNA interference (RNAi) therapy has highlighted the need for precise guidelines on delivery systems targeting specific tissues like the liver [43].The development of N-acetylgalactosamine (GalNAc)-siRNA conjugates has significantly advanced RNA interference (RNAi) therapy, particularly for targeting the liver [43]. GalNAc-siRNA conjugates leverage the high affinity of GalNAc for the asialoglycoprotein receptor on hepatocytes, ensuring rapid endocytosis and effective delivery of siRNA into liver cells [44]. This targeted delivery mechanism enhances the efficiency of RNAi responses by ensuring sufficient cytoplasmic entry of siRNAs to induce robust and selective gene silencing [45]. Despite their potential, the complexity of siRNA delivery systems underscores the need for precise regulatory guidelines [46].

Specific challenges include

- **Standardization of conjugation protocols**: Ensuring reproducibility and safety across production batches.
- **Optimization of endosomal escape mechanisms**: While delivery to hepatocytes is effective, the exact mechanisms enabling siRNA escape from endosomes into the cytoplasm remain unclear [47].

Evaluation of long-term safety: The prolonged systemic presence of siRNA conjugates necessitates comprehensive monitoring to avoid off-target effects or immunogenicity. Clinical trials, including phase III studies, are exploring the broad therapeutic applicability of GalNAc-siRNA conjugates for conditions like genetic and metabolic liver diseases [48]. These conjugates exemplify a pathway forward in RNAi delivery systems, offering insights that could extend to targeting other tissue types.

25.5.3 Scalability and manufacturing

Producing nanomaterials at a commercial scale while maintaining quality and consistency is a significant challenge. Nanoparticles, such as those used in lipid nanoparticle (LNP)-based delivery systems, require precise formulation to achieve uniform size and stability [49]. Variations in batch production can impact efficacy and safety, creating obstacles for clinical translation [50].

Another challenge lies in the encapsulation of sensitive biological agents, such as mRNA and siRNA, into nanoparticles. Encapsulation efficiency, stability during storage, and shelf-life optimization are critical considerations that require innovative manufacturing techniques.

25.5.4 Clinical trial design complexity

The design of clinical trials for nanomaterials is inherently complex due to their unique pharmacokinetic and pharmacodynamic profiles [51]. Challenges include

- **Biodistribution studies**: Understanding how nanomaterials distribute in the body and accumulate in specific tissues is critical. For instance, the biodistribution of nanoparticles crossing the blood–brain barrier (BBB) needs extensive evaluation to ensure that the therapeutic agents reach the brain without affecting peripheral tissues [52].
- **Patient selection**: Variations in individual responses to nanomedicine can complicate trial outcomes [53]. Personalized approaches based on genetic and physiological differences are necessary but add complexity to trial designs.
- **Combination therapies**: Nanomaterials are often used in combination with other therapies, such as chemotherapy or radiation [54]. Designing trials to evaluate these combinations' synergistic effects poses logistical and analytical challenges.

25.5.5 Ethical and environmental considerations

The ethical implications of introducing nanomaterials into humans, particularly in vulnerable populations, require careful consideration. Potential concerns include

- **Unintended effects**: Long-term effects of nanomaterials on human health and the environment remain largely unknown [55]. For example, the interaction of nanoparticles with the immune system can sometimes lead to adverse immune reactions [42].
- **Environmental impact**: The widespread use of nanomaterials in medicine raises questions about their environmental degradation and potential toxicity [56]. Current research on plant-derived extracellular vesicles (EVs) highlights the potential for biocompatible and sustainable alternatives, which could alleviate some environmental concerns.

25.5.6 Data Interpretation and Standardization

Interpreting data from nanomaterial-based trials requires an interdisciplinary approach that integrates material science, biology, and clinical medicine. However, the lack of standardized protocols for reporting outcomes creates challenges in comparing and validating results across studies. This issue is particularly evident in studies involving stimulus-responsive nanomaterials, where variations in environmental conditions can significantly affect performance [1,57].

25.6 Case study

Scenario: A team develops a novel scaffold using nanostructured materials for repairing articular cartilage in joints. The scaffold combines biocompatible nanomaterials and a cellular matrix to promote tissue regeneration. Listed below are the regulatory challenges that can arise during the clinical process of the material:

25.6.1 Regulatory challenges

1 **Material safety and biocompatibility**: are critical aspects in the clinical translation of nanomaterials and scaffolds because these materials directly interact with biological systems. Biocompatibility ensures that the material performs its intended function without causing toxicity, inflammation, immune reactions, or long-term tissue damage. Safety assessment includes evaluation of cytotoxicity, hemocompatibility, biodegradation, and systemic toxicity following international guidelines such as ISO 10993 and FDA recommendations. For example, gold nanoparticles used in drug delivery and tissue-engineered collagen scaffolds for wound healing are extensively tested for cell viability, inflammatory response, and degradation behavior before clinical application. Proper characterization of particle size, surface charge, purity, and degradation products is essential because small changes in nanomaterial properties can significantly affect biological responses and patient safety.

2 **Clinical translation**: of nanomaterials and scaffolds refers to the process of converting laboratory-based discoveries into safe and effective products for human use. Despite promising preclinical results, several regulatory challenges hinder successful clinical translation, including a lack of standardized characterization methods, variability in manufacturing, scalability issues, long-term toxicity concerns, and a limited understanding of nanoparticle interactions with biological systems. Regulatory agencies such as the FDA and EMA require extensive documentation related to material safety, biocompatibility, pharmacokinetics, biodegradation, and Good Manufacturing Practice (GMP) compliance before approval for clinical trials. For example, nanoparticle-based drug delivery systems such as liposomal doxorubicin underwent rigorous evaluation for toxicity, stability, and reproducibility before clinical approval. Similarly, scaffold-based regenerative products must demonstrate consistent mechanical properties, sterility, and tissue compatibility to ensure patient safety and therapeutic efficacy during clinical applications.

3 **Manufacturing and scalability**: are important regulatory challenges in the clinical development of nanomaterials and scaffolds because materials produced successfully at laboratory scale may behave differently during large-scale industrial production. Regulatory agencies require strict batch-to-batch reproducibility, sterility, stability, and compliance with Good Manufacturing Practices (GMP) to ensure product safety and therapeutic consistency. Even small changes in nanoparticle size, surface charge, or scaffold porosity during manufacturing can alter biological activity, toxicity, and clinical performance. For example, liposomal doxorubicin, a nanoparticle-based anticancer drug, requires highly standardized manufacturing processes because variations in liposome size and drug encapsulation efficiency could affect drug release, circulation time, and toxicity in patients. Similarly, collagen or polymer-based scaffolds used in tissue engineering must maintain consistent mechanical strength, biodegradation rate, and pore structure across all production batches to ensure reliable tissue regeneration and successful regulatory approval for clinical use.

4 **Multidisciplinary oversight**: is an important regulatory challenge in the clinical translation of nanomaterials and scaffolds because these products involve expertise from multiple fields, including nanotechnology, materials science, biology, medicine, toxicology, engineering, and regulatory science. Effective oversight is necessary to ensure that the

material is safe, reproducible, clinically effective, and ethically acceptable throughout preclinical and clinical development. Regulatory agencies often require collaboration among clinicians, biomaterial scientists, pharmacologists, toxicologists, and manufacturing experts to evaluate the complex interactions of nanomaterials with biological systems. For example, in the development of biodegradable polymeric scaffolds for bone tissue engineering, materials scientists are responsible for scaffold fabrication and mechanical strength, biologists evaluate cell compatibility and tissue regeneration, clinicians assess surgical applicability and patient outcomes, while toxicologists and regulatory experts monitor biodegradation products and long-term safety. This multidisciplinary coordination helps ensure proper clinical trial documentation, regulatory compliance, and successful translation of scaffold-based therapies into clinical practice.

5 **Long-term monitoring**:

- **Nanomaterial toxicity**: Regulatory agencies require extensive testing to ensure that the nanomaterials in the scaffold do not provoke harmful immune responses or toxicity. Numerous studies highlight the immunogenicity concerns of nanoparticles and vesicles, like lipid nanoparticles (LNPs) and extracellular vesicles, which require detailed characterization and validation for safety.
- **Standardization**: Due to the high variability in nanomaterial properties (size, surface charge, functionalization), ensuring consistent safety profiles across batches can be difficult.
- **Efficacy testing**: Long-term efficacy data for nanomaterial-based scaffolds are often lacking, as highlighted in studies discussing the limited ability of certain nanostructures to deliver consistent therapeutic benefits over extended periods.
- **Preclinical to clinical transition**: Moving from animal models to human clinical trials is challenging because animal models may not accurately predict human responses to nanostructured scaffolds. Not many nanomaterials have crossed these hurdles, and it is a major challenge in its clinical development.
- **Reproducibility**: Scaling up the production of nanomaterials for scaffold applications while maintaining reproducibility is a significant hurdle.
- **Regulatory approvals**: Complying with international manufacturing standards (e.g., Good Manufacturing Practices) involves rigorous oversight, which can delay clinical deployment.
- **Regulatory frameworks**: Nanostructures for biological use often straddle multiple regulatory domains (e.g., drugs, devices, and biologics). This ambiguity complicates the approval process, as described for lipid nanoparticle-based systems in cancer and infectious diseases.

Postmarket surveillance: Even after approval, nanostructures in scaffolds must be monitored for unforeseen complications, such as degradation products causing inflammation or migration of nanomaterials to unintended sites in the body.

25.6.2 Broader implications

This example reflects the need for:

- **Harmonized regulations**: Unified guidelines across regions can streamline the development pipeline.
- **Robust data standards**: Standardizing toxicity and efficacy testing for nanomaterials can build regulator and public trust.
- **Advanced monitoring tools**: Innovative imaging and biodistribution analysis tools, as mentioned in the context of blood–brain barrier-crossing nanomaterials, could be adapted to scaffolds

25.7 Emerging trends and future directions

The field of nanomaterials and scaffolds is rapidly evolving, with several emerging trends poised to shape its future.

25.7.1 Green nanotechnology

Green nanotechnology emphasizes the use of environmentally friendly methods for synthesizing nanomaterials. For instance, the development of gold nanoparticle-based Nano Swarna Bhasma represents a sustainable approach to cancer therapy.

25.7.2 Stimuli-responsive nanomaterials

These materials can respond to specific environmental cues, such as pH or temperature, enabling controlled drug release. Stimuli-responsive systems hold great promise for improving the precision and efficacy of nanomedicine.

25.7.3 Plant-derived extracellular vesicles

Plant-derived extracellular vesicles are gaining attention as biocompatible and cost-effective nanocarriers. Their potential for delivering therapeutic agents is currently being explored in clinical trials.

25.8 Conclusion

Nanomaterials and scaffolds have ushered in a new era of innovation in medicine, offering unparalleled opportunities for drug delivery and tissue regeneration. However, their clinical translation requires overcoming significant challenges related to safety, regulation, and scalability. By addressing these obstacles through robust clinical trial documentation and interdisciplinary collaboration, these advanced materials can fulfill their potential to transform healthcare.

References

[1] S. Sharifi, N.N. Mahmoud, E. Voke, M.P. Landry, M. Mahmoudi, Importance of standardizing analytical characterization methodology for improved reliability of the nanomedicine literature, Nano-Micro Letters 14 (1) (2022).
[2] S. Sharifi, N.N. Mahmoud, E. Voke, M.P. Landry, M. Mahmoudi, Importance of standardizing analytical characterization methodology for improved reliability of the nanomedicine literature, Nano-Micro Letters 14 (2022) 173. Available from https://doi.org/10.1007/s40820-022-00922-5.
[3] I. Khan, K. Saeed, I. Khan, Nanoparticles: properties, applications and toxicities, Arabian Journal of Chemistry 12 (2019) 908–931. Available from https://doi.org/10.1016/j.arabjc.2017.05.011.
[4] E. Blanco, H. Shen, M. Ferrari, Principles of nanoparticle design for overcoming biological barriers, Nature Biotechnology 33 (2015) 941–951. Available from https://doi.org/10.1038/nbt.3330.
[5] J.K. Patra, G. Das, L.F. Fraceto, et al., Nano based drug delivery systems: recent developments and future prospects, Journal of Nanobiotechnology 16 (2018) 71. Available from https://doi.org/10.1186/s12951-018-0392-8.
[6] X. Hou, T. Zaks, R. Langer, Y. Dong, Lipid nanoparticles for mRNA delivery, Nature Reviews Materials 6 (2021) 1078–1094. Available from https://doi.org/10.1038/s41578-021-00358-0.
[7] K.A. Athanasiou, et al., Self-assembling process in tissue engineering, Annual Review of Biomedical Engineering 15 (2013) 115–136. Available from https://doi.org/10.1146/annurev-bioeng-071812-152423.
[8] M. Tilton, E.T. Camilleri, M.D. Astudillo Potes, et al., Visible light-induced 3D bioprinted injectable scaffold for minimally invasive tissue regeneration, Biomaterials Advances 146 (2023) 213278 https://doi.org/10.1016/j.bioadv.2023.213278.
[9] J.H. Zeng, S.W. Liu, L. Xiong, et al., Clinical studies on scaffolds for bone defect repair: a systematic review, Journal of Orthopaedic Surgery and Research 13 (2018) 33. Available from https://doi.org/10.1186/s13018-018-0731-6.
[10] D.A. Dri, F. Rinaldi, M. Carafa, C. Marianecci, Nanomedicines in clinical trials: regulatory challenges, Drug Delivery and Translational Research 13 (2023) 1640–1656. Available from https://doi.org/10.1007/s13346-023-01234-5.
[11] R.M. Califf, Real-world evidence and clinical trials, New England Journal of Medicine 378 (2018) 216–218. Available from https://doi.org/10.1056/NEJMp1714236.
[12] V.P. Torchilin, Multifunctional nanocarriers, Nature Reviews Drug Discovery 13 (2014) 813–827. Available from https://doi.org/10.1038/nrd4333.
[13] Y. Zhao, L. Huang, Lipid nanoparticles for gene delivery, Advances in Genetics 88 (2014) 13–36. Available from https://doi.org/10.1016/B978-0-12-800148-6.00002-8.
[14] U. Sahin, K. Karikó, Ö. Türeci, mRNA-based therapeutics, Nature Reviews Drug Discovery 13 (2014) 759–780. Available from https://doi.org/10.1038/nrd4278.
[15] C.W. Pouton, C.J.H. Porter, Formulation of lipid-based delivery systems, Advanced Drug Delivery Reviews 60 (2008) 625–637. Available from https://doi.org/10.1016/j.addr.2007.10.010.
[16] J.K. Nair, et al., GalNAc-siRNA conjugates, Journal of the American Chemical Society 136 (2014) 16958–16961. Available from https://doi.org/10.1021/ja505986a.
[17] C. Saraiva, C. Praça, R. Ferreira, et al., Nanoparticle-mediated blood–brain barrier transport, Journal of Controlled Release 235 (2016) 34–47. Available from https://doi.org/10.1016/j.jconrel.2016.05.044.
[18] G. Guidotti, L. Brambilla, D. Rossi, Cell-penetrating peptides, Trends in Pharmacological Sciences 38 (2017) 406–424. Available from https://doi.org/10.1016/j.tips.2017.01.003.
[19] P. Anand, A.B. Kunnumakkara, R.A. Newman, B.B. Aggarwal, Bioavailability of curcumin, Molecular Pharmaceutics 4 (2007) 807–818. Available from https://doi.org/10.1021/mp700113r.
[20] F.J. O'Brien, Biomaterials and scaffolds, Materials Today 14 (2011) 88–95. Available from https://doi.org/10.1016/S1369-7021(11)70058-X.
[21] K.L. Christman, Biomaterials for tissue repair, Science 363 (2019) 340–341. Available from https://doi.org/10.1126/science.aau8275.
[22] M. Tilton, E.T. Camilleri, M.D. Astudillo Potes, et al., Injectable scaffold systems for tissue regeneration, Biomaterials Advances 146 (2023) https://doi.org/10.1016/j.bioadv.2023.213278.

[23] J.H. Zeng, S.W. Liu, L. Xiong, et al., Scaffold applications in bone repair, Journal of Orthopaedic Surgery and Research 13 (2018) 33. Available from https://doi.org/10.1186/s13018-018-0731-6.

[24] F.B. Stillaert, C. Di Bartolo, J.A. Hunt, et al., Human clinical experience with adipose precursor cells seeded on scaffolds, Biomaterials 29 (2008) 3953–3959 https://doi.org/10.1016/j.biomaterials.2008.06.019.

[25] S. Dall'Oca, F. Papacchini, I. Radovic, et al., Repair potential of nano-hybrid composites, Journal of Oral Science 50 (2008) 361–367. Available from https://doi.org/10.2334/josnusd.50.361.

[26] D.K. Lee, T. Kee, Z. Liang, et al., Nanodiamond-based biomaterials for clinical applications, Proceedings of the National Academy of Sciences of the USA 114 (2017) E9445–E9454. Available from https://doi.org/10.1073/pnas.1712431114.

[27] M.S. Roberts, S.E. Cross, M.A. Pellett, Skin transport and safety of nanoparticles, Journal of Investigative Dermatology 138 (2018) 216–223. Available from https://doi.org/10.1016/j.jid.2017.08.027.

[28] S. Quirbach, S. Trattnig, S. Marlovits, et al., MRI evaluation after cartilage transplantation, Skeletal Radiology 38 (2009) 831–840. Available from https://doi.org/10.1007/s00256-009-0703-1.

[29] E. Hatami, M. Jaggi, S.C. Chauhan, M.M. Yallapu, Gambogic acid nanomedicine in cancer therapy, Biochimica et Biophysica Acta - Reviews on Cancer 1874 (2020) 188407. Available from https://doi.org/10.1016/j.bbcan.2020.188407.

[30] A. Roberti, A.F. Valdes, R. Torrecillas, et al., Epigenetics in cancer and nanomedicine, Clinical Epigenetics 11 (2019) 81. Available from https://doi.org/10.1186/s13148-019-0677-4.

[31] D.A. Dri, F. Rinaldi, M. Carafa, C. Marianecci, Clinical translation challenges of nanomedicines, Drug Delivery and Translational Research 13 (2023) 1640–1656. Available from https://doi.org/10.1007/s13346-023-01234-5.

[32] K. Derakhshandeh, M. Soheili, S. Dadashzadeh, et al., PLGA-PEG nanoparticles for drug delivery, International Journal of Nanomedicine 5 (2010) 1089–1100. Available from https://doi.org/10.2147/IJN.S14547.

[33] G.J. Weiss, J. Chao, J.D. Neidhart, et al., First-in-human trial of CRLX101, Investigational New Drugs 31 (2013) 986–1003. Available from https://doi.org/10.1007/s10637-013-9911-4.

[34] U. Roy, J. Rodríguez, P. Barber, et al., HIV nanotherapeutics from bench to clinic, Nanomedicine 10 (2015) 359–373. Available from https://doi.org/10.2217/nnm.14.178.

[35] M.R. McDevitt, G. Sgouros, S. Sofou, Targeted alpha-particle therapy, Annual Review of Biomedical Engineering 20 (2018) 73–93. Available from https://doi.org/10.1146/annurev-bioeng-062117-121257.

[36] Y. Zhao, L. Huang, Lipid nanoparticle systems in gene therapy, Advances in Genetics 88 (2014) 13–36. Available from https://doi.org/10.1016/B978-0-12-800148-6.00002-8.

[37] C.M. Hartshorn, L.M. Russell, P. Grodzinski, The national cancer institute alliance for nanotechnology, WIREs Nanomedicine and Nanobiotechnology 11 (2019) e1570. Available from https://doi.org/10.1002/wnan.1570.

[38] F. Chiodo, M. Marradi, B. Tefsen, et al., Glyconanoparticles for detection systems, PLoS ONE 8 (2013) e73027. Available from https://doi.org/10.1371/journal.pone.0073027.

[39] J.W. Mouton, A. van Peer, K. de Beule, et al., Pharmacokinetics of itraconazole formulations, Antimicrobial Agents Chemotherapy 50 (2006) 4096–4102 https://doi.org/10.1128/AAC.00709-06.

[40] M. Kanai, et al., A phase I study of nanoparticle curcumin (Theracurmin), Cancer Chemotherapy Pharmacology 68 (2011) 65–70. Available from https://doi.org/10.1007/s00280-010-1431-8.

[41] X. Yan-Zhi, W. J-J, Y.P. Chen, et al., Zein-based scaffolds for tissue engineering, International Journal of Oral Science 2 (2010) 142–148. Available from https://doi.org/10.4248/IJOS10056.

[42] M. Chen, X.T. Chen, L.Y. Zhang, et al., Gold nanoshells for multimodal imaging and therapy, Journal of Nanobiotechnology 21 (2023) 138. Available from https://doi.org/10.1186/s12951-023-01869-1.

[43] H.K. Patra, R. Imani, J.R. Jangamreddy, et al., On/off-switchable nanoarchitecture, Scientific Reports 5 (2015) 14571. Available from https://doi.org/10.1038/srep14571.

[44] S. Mishra, Challenges in interventional cardiology trials, Indian Heart Journal 70 (2018) 1–3. Available from https://doi.org/10.1016/j.ihj.2017.12.001.

[45] D.K. Lee, et al., Nanodiamond biomaterials in translational medicine, Proceedings of the National Academy of Sciences USA 114 (2017) E9445–E9454. Available from https://doi.org/10.1073/pnas.1712431114.

[46] M.S. Roberts, et al., Safety considerations in nanoparticle exposure, Journal of Investigative Dermatology 138 (2018) 216–223. Available from https://doi.org/10.1016/j.jid.2017.08.027.

[47] S. Quirbach, et al., Cartilage imaging and scaffold evaluation, Skeletal Radiology 38 (2009) 831–840. Available from https://doi.org/10.1007/s00256-009-0703-1.

[48] F.B. Stillaert, et al., Clinical scaffold applications in regenerative medicine, Biomaterials 29 (2008) 3953–3959. Available from https://doi.org/10.1016/j.biomaterials.2008.06.019.

[49] S. Dall'Oca, et al., Dental nanocomposite repair systems, Journal of Oral Science 50 (2008) 361–367. Available from https://doi.org/10.2334/josnusd.50.361.

[50] K. Derakhshandeh, et al., Nanoparticle drug delivery systems, International Journal of Nanomedicine 5 (2010) 1089–1100. Available from https://doi.org/10.2147/IJN.S14547.

[51] G.J. Weiss, et al., Nanopharmaceutical clinical development, Investigational New Drugs 31 (2013) 986–1003. Available from https://doi.org/10.1007/s10637-013-9911-4.

[52] U. Roy, et al., Translational nanomedicine in infectious diseases, Nanomedicine 10 (2015) 359–373. Available from https://doi.org/10.2217/nnm.14.178.
[53] M.R. McDevitt, et al., Targeted radionuclide therapy, Annual Review of Biomedical Engineering 20 (2018) 73–93. Available from https://doi.org/10.1146/annurev-bioeng-062117-121257.
[54] Y. Zhao, L. Huang, Gene delivery platforms using nanoparticles, Advances in Genetics 88 (2014) 13–36. Available from https://doi.org/10.1016/B978-0-12-800148-6.00002-8.
[55] C.M. Hartshorn, et al., Nanotechnology initiatives in oncology, WIREs Nanomedicine and Nanobiotechnology 11 (2019) e1570. Available from https://doi.org/10.1002/wnan.1570.
[56] F. Chiodo, et al., Glyconanotechnology applications, PLoS ONE 8 (2013) e73027. Available from https://doi.org/10.1371/journal.pone.0073027.
[57] J.W. Mouton, et al., Antifungal nanoformulations, Antimicrobial Agents Chemotherapy 50 (2006) 4096–4102. Available from https://doi.org/10.1128/AAC.00709-06.

Chapter 26

Clinical trial guidelines for tissue engineering

Krishnajit Malakar[1], Anoushka Khanna[2], Bhupendra Singh Butola[2], and Navneet Sharma[1]

[1]*Department of Pharmaceutical Regulatory Affairs, Amity Institute of Pharmacy, Amity University Noida, Noida, Uttar Pradesh, India,* [2]*Department of Textile and Fibre Engineering, Indian Institute of Technology Delhi, New Delhi, India*

26.1 Introduction

Tissue engineering is a multidisciplinary domain that amalgamates ideas from biology, engineering, and clinical medicine to develop viable biological replacements that can restore, maintain, or enhance tissue and organ functionality. In recent decades, progress in tissue engineering has allowed innovative uses in regenerative medicine, organ transplantation, and customized therapeutics. Researchers have developed unique methods for treating numerous medical problems, including organ failure, musculoskeletal deformities, and neurological illnesses, by employing a mix of live cells, biomaterial scaffolds, and biochemical signaling molecules.

Notwithstanding its great promise, the transition of tissue-engineered products (TEPs) from the laboratory to clinical use continues to pose a considerable difficulty. The intricacy of these products, stemming from biological diversity, scaffold integration, and host immunological response, requires the establishment of rigorous regulatory frameworks to guarantee their safety, effectiveness, and long-term survival. Regulatory agencies, including the US Food and Drug Administration (FDA), the European Medicines Agency (EMA), and Japan's Pharmaceuticals and Medical Devices Agency (PMDA) are crucial in the assessment and authorization of these sophisticated medicines. Their guidelines delineate explicit criteria for preclinical testing, clinical trial design, and postmarket surveillance, guaranteeing that TEPs adhere to the highest standards prior to patient availability.

A crucial element of tissue engineering clinical trials is risk assessment, encompassing the evaluation of immunogenicity, biocompatibility, and possible long-term deleterious effects. Tissue-engineered constructions frequently include genetically modified cells or biomaterials with intricate degradation patterns, necessitating comprehensive preclinical investigations to evaluate their interactions with host tissues. Additionally, postmarket surveillance systems must be established to assess long-term patient outcomes, identify any delayed problems, and enhance treatment regimens based on empirical evidence.

A growing aspect affecting clinical trials in tissue engineering is the use of artificial intelligence (AI). AI-driven solutions possess the capacity to enhance several facets of tissue engineering, encompassing patient screening, predictive modeling of cellular activity, scaffold design, and automated 3D bioprinting. Machine learning algorithms can evaluate extensive datasets to choose the most appropriate candidates for clinical trials, minimizing variability and enhancing trial success rates. Moreover, AI-assisted automation in tissue production improves repeatability and accuracy, resulting in more dependable therapeutic outcomes. The increasing dependence on AI in medical applications prompts worries about transparency, regulatory control, and ethical implications. Regulatory bodies must formulate explicit rules for Good Machine Learning Practice (GMLP) to guarantee that AI algorithms employed in clinical trials are interpretable, auditable, and devoid of biases.

The international framework for tissue engineering regulation is always advancing, with a growing focus on standardizing regulatory criteria across many countries. Discrepancies in approval processes among the FDA, EMA, and PMDA provide obstacles for multinational clinical trials and market entrance plans. Cooperative initiatives among regulatory bodies, research institutions, and industry participants are crucial to optimize approval procedures, improve patient safety, and facilitate worldwide access to novel regenerative treatments.

As the discipline progresses, regulatory science must stay flexible to integrate emerging technology and changing clinical requirements. Innovations like adaptive clinical trial designs, digital health technologies, and biomarker-driven

Fundamentals of Translational Tissue Engineering. DOI: https://doi.org/10.1016/B978-0-443-33455-9.00029-8

patient stratification provide novel opportunities for enhancing trial efficiency and precision. Moreover, ethical issues must be highlighted to provide fair access to tissue-engineered medicines, especially for marginalized groups.

26.2 Definition and multidisciplinary nature

Tissue engineering is a multidisciplinary domain that amalgamates concepts from biology, engineering, and clinical medicine to create biological replacements that can restore, maintain, or enhance tissue functionality. In their seminal work, Langer and Vacanti defined tissue engineering as "the application of the principles of engineering and life sciences toward the development of biological substitutes that restore, maintain, or improve tissue function"[1]. This definition underscores the integration of three critical components:

- Cells, originating from autologous, allogenic, or stem cell origins, function as the fundamental units. Their capacity to proliferate, differentiate, and secrete extracellular matrix constituents is crucial for tissue regeneration.
- Biomaterials (scaffolds): Scaffolds offer structural reinforcement and replicate the natural extracellular matrix. The design of these materials, encompassing their mechanical properties, degradation rates, and surface features, is essential for directing cell activity and facilitating integration with host tissues [2]
- Biochemical factors: Growth factors and signaling molecules direct cellular functions, including proliferation, differentiation, and organization into functional tissue. These molecular signals are crucial for the maturation and stability of synthetic tissues.

The effective coordination of these elements necessitates a multidisciplinary approach. Cell biology elucidates the mechanics of cellular activity and development, whereas materials science is essential for creating scaffolds that satisfy certain mechanical and biocompatibility standards. Furthermore, clinical insights guarantee that the created constructions are safe, effective, and appropriate for human use.

26.3 Implications for clinical trial guidelines

The intricate characteristics of TEPs present several issues that need to be resolved in clinical trial guidelines.

26.3.1 Comprehensive product characterization

TEPs are composite entities, including live cells, biomaterial scaffolds, and bioactive chemicals. Niklason et al. [3] showed that functional arteries cultivated in vitro need thorough assessment of mechanical integrity, biocompatibility, and long-term performance to guarantee clinical success. This requires the creation of sophisticated analytical methods and established procedures for product characterization.

26.3.2 Regulatory adaptation

Traditional regulatory frameworks for medicines frequently fail to address the complexity inherent in TEPs. Regulatory authorities must modify criteria to evaluate not just immediate safety and efficacy but also long-term performance, possible immunogenicity, and integration of these goods. The study by [4] on tissue-engineered bladders underscores the necessity for customized regulatory frameworks that tackle the distinct issues associated with implanting live constructions in patients.

26.3.3 Interdisciplinary collaboration

The effective implementation of TEPs in clinical practice depends on the collaboration of biologists, engineers, doctors, and regulatory specialists. This interdisciplinary collaboration guarantees that clinical trials are structured to thoroughly assess the many elements of TEPs, encompassing scaffold efficacy and cellular responses in physiological environments.

26.4 Applications

Tissue engineering has swiftly become a revolutionary method in contemporary medicine, providing novel answers to a wide range of clinical issues. Its uses surpass conventional treatments, influencing regenerative medicine, organ transplantation, illness modeling, and individualized therapeutics.

26.4.1 Regenerative medicine and organ replacement

Engineered tissues provide substitutes for donor organs and traditional grafts, mitigating significant shortages and diminishing difficulties related to immune rejection. Tissue-engineered bladders have been effectively implanted to restore urine function in individuals with congenital anomalies [4]. Furthermore, artificial vascular grafts and cartilage constructions have shown encouraging outcomes in the healing of compromised blood vessels and skeletal frameworks [5]. These advancements highlight the capacity of tissue-engineered items to substitute or mend damaged tissues with constructions that closely resemble their native equivalents both physiologically and mechanically.

26.4.2 Disease modeling and drug testing

Tissue-engineered models function as advanced in vitro platforms for investigating disease processes and evaluating novel pharmaceuticals, in addition to their direct therapeutic uses. By emulating the intricate architecture and cellular interactions of genuine tissues, these models yield more precise predictions of human responses than conventional cell cultures. Engineered liver and cardiac tissues have been utilized to mimic disease development and assess medication toxicity, hence improving the translational potential of preclinical research [6]

26.4.3 Personalized and precision medicine

Tissue engineering facilitates personalized medicine by allowing the development of patient-specific tissue structures. Utilizing autologous cells in conjunction with bespoke scaffolds and regulated biochemical signals, physicians may create personalized tissues that reduce the likelihood of immune rejection and enhance integration. This tailored methodology not only improves therapeutic effectiveness but also provides a framework for personalized disease modeling and pharmacological testing, ultimately resulting in more accurate treatment options [7].

26.4.4 Emerging trends and future directions

Recent technical advancements, including 3D bioprinting and the utilization of decellularized extracellular matrices, are broadening the applications of tissue engineering [8]. These improvements facilitate the exact construction of intricate tissue structures and enhance the replication of the natural microenvironment. As research in this domain progresses, tissue engineering is set to transform the management of organ failure, musculoskeletal deformities, neurological illnesses, and more.

26.5 Need for robust regulatory guidelines

26.5.1 Challenges due to complexity

The intricacy of manufactured tissues stems from several interconnected aspects, such as biological diversity, scaffold integration, and host response. The unpredictability of tissue-engineered constructions postimplantation is influenced by variability in cellular activity, variations in scaffold composition, and the dynamic characteristics of the host's immunological response. Minor variances in cell source or scaffold qualities might result in substantial variations in tissue regeneration and mechanical functioning [9].

This intricate intricacy requires regulatory authorities to implement stringent and consistent testing processes. Conventional pharmaceutical assessment techniques are unable to tackle the distinct issues presented by these hybrid medicines, which integrate living cells with synthetic or natively sourced scaffolding and bioactive compounds. Consequently, innovative regulatory frameworks must encompass.

26.5.2 Comprehensive characterization

Sophisticated analytical methods, including high-resolution imaging and molecular tests, are crucial for evaluating scaffold integration, cellular dynamics, and tissue remodeling throughout time. These strategies ensure that the designed construct preserves its intended form and function under physiological settings.

26.5.3 Multilayered testing protocols

Assessment should encompass in vitro tests, preclinical animal investigations, and extended clinical follow-ups. This stratified methodology encompasses both prompt and prolonged responses, incorporating possible inflammatory reactions, scaffold degradation, or unforeseen host-tissue interactions.

26.5.4 Standardization and quality control

Due to the intrinsic biological diversity in TEPs, regulatory rules must enforce rigorous quality control protocols during the production process. This standardization is essential to reduce batch-to-batch variability and guarantee uniform product performance.

Recent research studies have highlighted these problems. Buma et al. [10] showed that minor alterations in scaffold composition or cell source might significantly affect the functional results of engineered tissues, underscoring the necessity for rigorous testing methodologies. Xi et al. [11] emphasized that traversing the regulatory framework for tissue engineering necessitates customized strategies that include both the biotic and abiotic elements of these products.

26.6 Balancing innovation with patient safety

Tissue-engineered goods promise transformational therapies, providing remedies for hitherto intractable illnesses. Nonetheless, the innovation included in these sophisticated medicines also presents significant hazards, including immunogenicity and tumorigenicity. Regulatory regimes must achieve a delicate equilibrium between promoting innovation and imposing rigorous safety standards to safeguard patients.

The dynamic nature of tissue engineering, characterized by the incorporation of live cells, intricate scaffolding, and powerful biochemical forces, has catalyzed swift technical progress. These solutions can fulfill unaddressed clinical requirements and transform therapy methodologies. Conversely, their biological intricacy presents considerable safety problems. Variability in cellular sources, scaffold materials, and growth factor administration may result in erratic immune responses or potentially facilitate tumorigenic processes [12].

To alleviate these dangers, regulatory bodies are progressively endorsing comprehensive preclinical testing techniques and extended clinical surveillance. Sophisticated in vitro and in vivo models are being created to replicate human physiological reactions, guaranteeing the early detection of any harmful effects during the product development process. Comprehensive immunological tests and tumorigenicity investigations have become vital elements of the review process, designed to identify any anomalous cellular activity prior to clinical use [9].

Additionally, adaptive regulatory mechanisms are being developed to facilitate the gradual approval of novel medicines. This methodology allows items to commence clinical trials with meticulously established parameters and phased evaluations, guaranteeing ongoing safety assessments as the technology progresses. These frameworks foster innovation while upholding a stringent dedication to patient safety [13].

26.7 Illustrations of historical achievements and setbacks

Successful clinical studies, including those with bioengineered bladders, demonstrate the revolutionary capacity of tissue engineering in regenerative medicine. The clinical success of tissue-engineered autologous bladders in spina bifida patients has shown that engineered constructions can restore normal organ function and markedly enhance quality of life [4]. These advancements have facilitated more developments, promoting ongoing research into synthetic organs and tissues.

Conversely, significant setbacks have also occurred. Inconsistent outcomes in stem cell-derived cartilage regeneration have prompted questions about reproducibility, scaffold integration, and the long-term functioning of the engineered tissue [14]. These failures highlight the difficulties presented by biological diversity and the intricate interactions among cells, scaffolds, and the host environment. They emphasize the essential requirement for uniform regulatory supervision and stringent preclinical testing methodologies to guarantee the safety and efficacy of TEPs.

The divergent results successful implementations that instill confidence and failures that warn against hasty clinical application demonstrate the duality of innovation in tissue engineering. They emphasize the necessity of formulating and implementing extensive regulatory frameworks that reconcile the pursuit of technical progress with the obligation of ensuring patient safety. Continuous endeavors to optimize manufacturing processes, improve quality control measures, and standardize assessment techniques are crucial for transitioning these new medicines from the laboratory to clinical use.

26.8 Regulatory framework for clinical trials in tissue engineering

26.8.1 Overview of major regulatory bodies

The regulation of clinical trials in tissue engineering is managed by many major regulatory authorities. These authorities guarantee that TEPs comply with rigorous safety, effectiveness, and quality criteria prior to commercial release. Their advice is essential for reconciling innovation with patient safety, particularly because of the intricacy and biological diversity intrinsic to these sophisticated medicines.

26.8.1.1 Principal organizations

Regulatory agencies such as the US Food and Drug Administration (FDA), the EMA, and Japan's Pharmaceuticals and Medical Devices Agency (PMDA) play a pivotal role in overseeing clinical trials for TEPs. These organizations have developed specialized frameworks to address the unique challenges posed by these hybrid products, which combine living cells with biomaterials and bioactive molecules.

26.8.1.1.1 US Food and Drug Administration

The FDA is essential in guaranteeing that TEPs are created and assessed according to rigorous safety and effectiveness criteria. The FDA, via its Center for Biologics Evaluation and Research (CBER), offers extensive advice for the development and clinical assessment of these novel treatments [15]

26.8.1.1.2 Stringent preclinical investigations

Prior to a TEP's entry into clinical trials, the FDA mandates a comprehensive preclinical evaluation. This entails comprehensive in vitro and in vivo investigations to evaluate the safety, biological efficacy, and functional performance of the product. These preclinical investigations aim to replicate the human physiological milieu, yielding essential data on factors such as tissue integration, scaffold degradation, and possible adverse consequences like immunological rejection or tumorigenicity.

26.8.1.1.3 Standardized manufacturing processes

The FDA underscores the necessity for consistent and repeatable production procedures. This necessitates rigorous compliance with Good Manufacturing Practices (GMP) to guarantee product uniformity and quality. Manufacturers must authenticate each phase of the production process from cell isolation and expansion to scaffold construction and final product assembly to reduce batch-to-batch variability. Stringent quality control protocols and process validations are crucial to preserve the integrity of the TEP and to guarantee its reliable performance upon patient administration.

26.8.1.1.4 Long-term postmarket surveillance

Following the approval and market introduction of a TEP, the FDA requires ongoing postmarket monitoring to assess its safety and effectiveness over an extended period. This monitoring is essential for identifying any delayed detrimental consequences, such as persistent inflammation, scaffold failure, or unforeseen immunological reactions. Ongoing monitoring and reporting systems facilitate the early identification of possible dangers, enabling prompt interventions or revisions to the product as required.

The FDA's regulatory framework seeks to promote innovation in tissue engineering while ensuring patient safety via the integration of stringent preclinical evaluations, standardized manufacturing procedures, and comprehensive postmarket monitoring. This holistic strategy not only promotes the advancement of sophisticated regeneration medicines but also fosters public confidence in these nascent technologies.

26.8.1.1.5 European Medicines Agency

The EMA is pivotal in regulating tissue-engineered goods in Europe, chiefly via its Committee for Advanced Therapies (CAT). This specialist group offers scientific counsel and formulates guidelines specifically designed to address the distinct issues associated with new treatments, including tissue engineering.

26.8.1.1.6 Emphasize quality, safety, and effectiveness

The EMA's recommendations aim to guarantee that TEPs adhere to the utmost standards of quality, safety, and effectiveness. These recommendations mandate producers to furnish detailed data on product characterization, encompassing the consistency and repeatability of biological components. Comprehensive quality evaluations include the origin and behavior of cells, the characteristics of scaffold materials, and the regulated release of bioactive compounds.

26.8.1.1.7 Robust clinical trial designs

The EMA underscores the necessity for resilient clinical trial designs that accommodate the intrinsic unpredictability of biological elements. Trials must be carefully designed to assess both the immediate therapeutic results and the long-term functionality and integration of the altered tissue inside the host. This entails observing possible negative effects, including immunological responses or unforeseen tissue reactions, over a prolonged duration. The EMA's methodology guarantees that any product variability does not jeopardize patient safety or treatment effectiveness.

26.8.1.1.8 Standardization and Regulatory Oversight

The EMA establishes a uniform regulatory framework for tissue-engineered goods by imposing rigorous standards for manufacturing procedures, quality control, and clinical assessment. This framework is essential for attaining uniform clinical results and for enabling the approval and market introduction of novel treatments. The CAT collaborates with industry stakeholders to revise and enhance these standards as new technologies and scientific knowledge develop.

The EMA, via its Committee for Advanced Therapies, establishes a stringent regulatory framework that prioritizes the quality, safety, and efficacy of TEPs. The recommendations guarantee the implementation of rigorous clinical trial designs to manage the unpredictability intrinsic to these complex products, therefore safeguarding patient health and promoting innovation in regenerative medicine.

26.8.1.1.9 Pharmaceuticals and Medical Devices Agency-Japan

In Japan, the PMDA is essential for regulating novel medicines, such as tissue-engineered and regenerative medicine products. The PMDA collaborates with regulatory stakeholders to optimize the approval process, therefore expediting the clinical translation of new medicines while upholding stringent safety and effectiveness criteria.

26.8.1.1.10 Optimized Approval Procedures

The PMDA has instituted specific frameworks, like the Conditional Early Approval System, facilitating accelerated evaluations of regenerative medicine products. This strategy aims to grant patients prompt access to innovative medicines by permitting market entry of goods based on early clinical evidence, contingent upon thorough postmarket surveillance. This method guarantees that, despite the expedited approval procedure, comprehensive data on safety and long-term efficacy are collected to protect patient welfare.

26.8.1.1.11 Collaborative Regulatory Framework

The PMDA engages in active collaboration with industry specialists, university researchers, and global regulatory authorities to consistently revise and align its standards. The PMDA's framework integrates the latest scientific and clinical advancements to tackle the distinct issues associated with TEPs, such as unpredictability in biological components and intricate host reactions. This collaborative method enhances regulatory processes and guarantees that novel therapeutics are assessed comprehensively and adaptively.

26.8.1.1.12 Prioritize Comprehensive Safety and Efficacy

Despite its efficient processes, the PMDA has a robust commitment to ensuring that all regenerative medicine products undergo thorough testing. Comprehensive quality evaluations, encompassing preclinical research and regulated clinical trials, are required to confirm the product's safety, reliability, and therapeutic effectiveness. The simultaneous emphasis on rapid evaluation and thorough safety assessment illustrates the PMDA's dedication to both innovation and patient safeguarding.

These regulatory authorities establish the requirements for clinical trials and promote innovation by offering clear paths for the development of novel medicines. The cooperative endeavors among these organizations and industry stakeholders are crucial for guaranteeing that advancements in tissue engineering result in safe and successful therapeutic applications.

26.8.1.2 Applicable Guidelines

Tissue-engineered goods must adhere to an extensive regulatory system that guarantees their safety, effectiveness, and quality. Essential directives in this field encompass Good Clinical Practice (GCP), the International Council for Harmonization [16] recommendations, and the ISO 10993 standards for the biological assessment of medical devices.

26.8.1.2.1 Good clinical practice

GCP denotes a worldwide ethical and scientific standard for the design, execution, documentation, and reporting of clinical studies involving human participants. It guarantees the prioritization of participants' rights, safety, and well-being, while ensuring the credibility and reliability of the resultant data. Compliance with Good Clinical Practice is essential for preserving public confidence and guaranteeing that studies are executed with utmost integrity.

26.8.1.2.2 International Council for Harmonization Guidelines

The ICH rules provide a standardized framework for pharmaceutical development in Europe, Japan, and the United States. The ICH E6 guideline on GCP offers comprehensive suggestions to guarantee that clinical trials are conducted ethically and are scientifically robust. These principles are essential for promoting the mutual recognition of clinical data among regulatory bodies, therefore expediting the licensing process for novel medicines (ICH E6 (R1) 2016).

26.8.1.2.3 ISO 10993 Standards

The ISO 10993 family of standards pertains to the biological assessment of medical devices, encompassing those utilized in tissue engineering. These standards delineate techniques for evaluating many biological endpoints, including cytotoxicity, sensitization, and irritation. Adherence to [17] guarantees that the biomaterials used in tissue-engineered structures do not provoke negative responses in patients, thus ensuring their long-term safety and efficacy.

These rules together establish a comprehensive regulatory framework that directs the creation, testing, and approval of tissue-engineered goods. By complying with GCP, ICH, and ISO 10993 standards, researchers may guarantee that their breakthrough medicines satisfy worldwide safety and quality criteria, hence enhancing trust among doctors and patients.

26.8.1.3 Geographic variations

The regulatory frameworks for tissue-engineered goods varied markedly around the globe, illustrating variations in healthcare systems, risk tolerance, and governmental agendas. For instance, the US Food and Drug Administration (FDA) employs a risk-based methodology for regulating these items. This technique adjusts the level of regulatory oversight according to the possible dangers linked to a product's components and intended application, thereby providing flexibility for novel therapies while upholding rigorous safety requirements.

The EMA utilizes a centralized approval process only for Advanced Therapy Medicinal Products (ATMPs). This unified method guarantees a standardized assessment of TEPs throughout all EU member states, facilitating the approval procedure and enhancing uniformity in clinical standards. The EMA's strategy is to promote the advancement of innovative regenerative treatments by standardizing regulatory criteria and fostering collaborative evaluations of intricate goods [18].

These geographical disparities highlight the necessity for developers to comprehend and maneuver through various regulatory frameworks when introducing tissue-engineered items to the market. Global harmonization initiatives are under progress; nonetheless, the current diversity requires customized approaches for clinical trial design, product characterization, and postmarket surveillance in various countries.

26.8.2 Clinical Trial Authorization and approval process

26.8.2.1 Investigational new drug applications

Tissue-engineered goods must secure regulatory clearance via processes such as Investigational New Drug (IND) applications in the United States or similar frameworks like the Clinical Trial Application (CTA) in the European Union before being tested in human subjects.

26.8.2.1.1 IND Applications (FDA)

In the United States, the IND application is a pivotal process regulated by the FDA. This application needs comprehensive preclinical data, encompassing pharmacological, toxicological, and manufacturing details, to establish that the proposed TEP is sufficiently safe for preliminary human testing. The FDA's risk-based methodology guarantees that the degree of examination corresponds to the possible hazards linked to the product. The IND proposal must have comprehensive clinical protocols delineating the research design, patient demographics, outcomes, and monitoring strategies. This stringent procedure protects patient safety and guarantees that only items with a robust scientific basis advance to clinical trials.

26.8.2.1.2 Equivalent Frameworks in the EU-Clinical Trial Application

In Europe, tissue-engineered goods are often authorized via the Clinical Trial Application (CTA) procedure. The CTA fulfills a function analogous to the IND by mandating extensive preclinical data and meticulous clinical trial methodologies. The EMA's unified assessment procedure for Advanced Therapy Medicinal goods (ATMPs) guarantees uniform evaluation of goods across all member states. This coordinated method enables efficient approval while upholding rigorous safety and effectiveness criteria.

26.8.2.1.3 Key considerations

Data requirements: Both IND and CTA procedures need comprehensive preclinical evidence, encompassing animal studies and in vitro testing, to evaluate the product's biological activity, possible toxicity, and overall safety profile.

The applications must specify manufacturing methods that comply with Good Manufacturing Practice (GMP) standards, guaranteeing product consistency and quality.

Clinical protocols: Comprehensive clinical trial designs are essential, since they delineate the objectives, methodology, patient monitoring, and risk management techniques to be utilized throughout the experiment.

Regulatory organizations decrease risks and foster the development of safe and effective therapeutics by ensuring that TEPs undergo a comprehensive evaluation procedure prior to clinical testing. This thorough preclinical and clinical assessment is fundamental to the translational process, connecting new laboratory research with practical patient therapies.

26.8.2.2 Regulatory review timelines

The timescales for obtaining clinical trial authorizations varied markedly among areas, indicating variations in regulatory processes and risk evaluations. In the United States, upon submission of an Investigational New Drug (IND) application, the FDA commences a 30-day review period to assess the preclinical data, manufacturing methods, and planned clinical protocols. This brief statutory review period aims to accelerate the progression of promising medicines into clinical trials, contingent upon the fulfillment of all safety and quality standards.

In Europe, the EMA's Committee for Advanced Therapies (CAT) oversees the evaluation of applications for tissue-engineered goods, often classified as Advanced Therapy Medicinal goods (ATMPs). The examination procedure under the EMA is often more protracted than the FDA's schedule. The extended review duration arises from the need to meticulously evaluate intricate biological data, manage variability in manufacturing processes, and guarantee that the clinical trial design adequately considers the inherent uncertainties of these novel medicines. Frequent iterations of inquiries and explanations with candidates are typical, resulting in an extended total evaluation duration [18].

The geographical variations in regulatory review times highlight the necessity for developers to establish customized submission procedures. Comprehending the distinct standards and review procedures in each country is essential for strategizing clinical trial timeframes and guaranteeing the effective submission of all required data. Ultimately, although faster reviews might hasten the accessibility of innovative medicines, they must be reconciled with thorough assessments to guarantee patient safety and product effectiveness.

26.8.2.3 First-in-human trials considerations

First-in-human trials represent a pivotal stage in the transition of TEPs from the laboratory to clinical use. These preliminary clinical trials require substantial preclinical data to warrant the exposure of human patients to these novel medicines. Comprehensive evaluations of biocompatibility and immunogenicity are among the most essential preclinical needs.

26.8.2.3.1 Biocompatibility assessments

Prior to human testing of a TEP, it is essential to establish that its components such as scaffolds and cells integrate harmoniously with host tissues without provoking unfavorable responses. Comprehensive in vitro and in vivo investigations are performed to assess the interaction of the created construct with the adjacent biological environment. These studies evaluate parameters like cell survival, scaffold degradation, and tissue integration, offering predicted insights into the product's performance postimplantation.

26.8.2.3.2 Immunogenicity assessments

Considering that several TEPs include allogenic or stem cell-derived components, evaluating immunogenicity is essential. These assessments ascertain if the product elicits an adverse immunological response, perhaps resulting in rejection or inflammatory consequences. Thorough immunological testing, frequently including animal models and advanced in vitro tests, is crucial to anticipate and reduce possible dangers prior to human trials.

26.8.2.3.3 Supplementary safety data

In addition to biocompatibility and immunogenicity, first-in-human trials need comprehensive data from pharmacokinetic, pharmacodynamic, and toxicity investigations. This extensive dataset offers a whole range of information on the product's performance, safety characteristics, and any bad effects. Regulatory bodies depend on this preclinical research to determine the ethical and scientific justification for commencing human trials.

By aggregating and meticulously assessing this comprehensive preclinical data, researchers may formulate first-in-human studies with clearly delineated risk mitigation techniques. This prudent strategy safeguards patient safety while establishing the scientific groundwork essential for the effective clinical use of tissue-engineered therapeutics.

26.8.3 Ethical and legal considerations

26.8.3.1 Ethical approval process

Institutional Review Boards (IRBs) or Ethics Committees (ECs) are essential for safeguarding the rights, safety, and welfare of human participants in clinical research. The ethical approval procedure for tissue-engineered goods is notably stringent due to the innovative nature and intricacy of these therapies.

Prior to commencing any clinical study, researchers are required to submit a detailed protocol to an Institutional Review Board (IRB) or Ethics Committee (EC). This protocol encompasses comprehensive explanations of the research design, methodologies for acquiring informed consent, risk mitigation techniques, and significant preclinical data that substantiate the product's safety and potential efficacy. Comprehensive preclinical evidence encompassing biocompatibility, immunogenicity, and possible tumorigenicity is essential to validate human exposure and is rigorously evaluated to confirm that the prospective benefits above the hazards [19]

The evaluation procedure entails a comprehensive analysis of

- Risk-benefit analysis: Institutional Review Boards and Ethics Committees assess whether the expected advantages for participants and subsequent patients warrant the associated hazards, particularly considering the experimental characteristics of TEPs.
- Informed consent procedures: Guaranteeing that participants get explicit, thorough information on the study, encompassing its experimental components and any ambiguities related to the therapy.
- Safety monitoring plans: Comprehensive procedures for the continuous oversight of trial participants to swiftly detect and mitigate any harmful effects during the research.

Moreover, ethical review adheres to globally acknowledged ethical norms, including the Declaration of Helsinki and recommendations from the Council for International Organizations of Medical Sciences (CIOMS), which underscore the safeguarding of human subjects in research. These frameworks need that research procedures be formulated to mitigate risks and enhance possible benefits, while maintaining transparency and accountability during the study [20].

By requiring stringent preclinical data and ethical supervision, IRBs/ECs are essential in fostering public confidence and guaranteeing that novel tissue-engineered medicines are implemented in clinical practice with optimal patient safety and ethical integrity.

26.8.3.2 Patient rights and informed consent

In the field of tissue engineering, protecting patient rights is essential. Patients must be thoroughly apprised of the risks, advantages, and uncertainties related to TEPs, with special emphasis on long-term safety issues. This procedure is not just a regulatory obligation but also an ethical need that preserves the autonomy and dignity of each participant.

26.8.3.2.1 Comprehensive disclosure

Informed consent for tissue-engineered treatments necessitates that patients are provided with thorough information about the intervention's nature. This encompasses a detailed elucidation of the manufacturing process of the engineered product, the biological variability intrinsic to the construct, and the possibility of unanticipated long-term consequences. Due to the innovative and intricate characteristics of these treatments, patients must be informed of both the immediate advantages and the possible hazards, such as immunogenicity, scaffold degradation concerns, or the likelihood of delayed adverse effects.

26.8.3.2.2 Long-term safety considerations

TEPs may possess ramifications that continue well beyond the original treatment duration. Consequently, it is imperative for patients to comprehend the long-term monitoring necessities and the potential for side effects or difficulties to arise years posttherapy administration. Clarity about long-term data if accessible and the existing uncertainty is crucial for informed decision-making [21].

26.8.3.2.3 Patient autonomy and decision-making

The informed consent procedure must be executed in a manner that honors patient autonomy. This entails guaranteeing that the patient has adequate time to contemplate the facts presented, pose inquiries, and deliberate the possible results with family members or trusted advisers. Effective communication is essential, and permission documents must be drafted in comprehensible language devoid of superfluous technical jargon.

26.8.3.2.4 Ethical oversight

Institutional Review Boards (IRBs) or Ethics Committees (ECs) are essential in supervising the informed consent process, guaranteeing that all materials adhere to ethical norms and that participants' rights are safeguarded. These entities evaluate consent forms and processes to ensure that the given information is accurate, comprehensive, and presented in a manner that promotes genuinely informed decision-making [20].

By guaranteeing that patients are thoroughly informed and that their rights are meticulously safeguarded, the ethical framework for clinical trials in tissue engineering not only enhances safety and transparency but also cultivates trust in these novel medicines.

26.8.3.3 Considerations for vulnerable populations

Special ethical concerns are essential when conducting clinical studies with vulnerable groups, including pediatric and geriatric patients. These populations frequently exhibit distinct physiological traits and heightened vulnerability to negative outcomes, requiring more safety assessments and intensified regulatory oversight.

26.8.3.3.1 Enhanced safety evaluations

Vulnerable patients may exhibit distinct responses to tissue-engineered treatments owing to differences in immune function, metabolism, and the age or deterioration of organs. Pediatric patients are in a state of development, rendering them especially vulnerable to possible toxicities or unanticipated developmental effects, whereas geriatric patients may present with many comorbidities and reduced physiological resistance. Therefore clinical trial approaches for these populations must include customized preclinical data and enhanced safety monitoring to reduce hazards.

26.8.3.3.2 Regulatory scrutiny and tailored protocols

Regulatory bodies and ethical review committees mandate that studies involving at-risk groups have additional protections. This may need the establishment of more stringent inclusion criteria, enhanced monitoring techniques, and prolonged follow-up durations to identify any delayed detrimental effects. The informed consent approach is modified to guarantee that guardians or legally authorized representatives comprehend the specific hazards and advantages associated with certain patient populations [22].

26.8.3.3.3 Ethical and legal responsibilities

Ethical frameworks, such as the Declaration of Helsinki [23] and CIOMS in Collaboration with [24] recommendations, emphasize the duty to offer additional protections for vulnerable populations. These rules mandate that the prospective advantages of the therapy outweigh the dangers and that all participants, or their representatives, are well informed about the study's nature. Institutional Review Boards (IRBs) or Ethics Committees (ECs) are essential in supervising these studies to guarantee the implementation of supplementary ethical protections.

Through the implementation of customized safety assessments, modification of clinical procedures, and enforcement of stringent regulatory monitoring, the clinical research community may safeguard at-risk individuals while promoting novel tissue-engineered medicines. Such steps are vital to guarantee that the advantages of these novel medicines are attained without jeopardizing patient safety.

26.9 Clinical trial design for tissue-engineered products

26.9.1 Patient inclusion and exclusion criteria

26.9.1.1 Establishing criteria

Formulating stringent inclusion and exclusion criteria is fundamental to the design of clinical trials for TEPs. These criteria must be scientifically substantiated to guarantee that the patient group is adequately homogenous, thereby facilitating repeatable and interpretable results.

26.9.1.1.1 Scientific justification

Inclusion criteria must be established based on explicitly stated clinical and biological factors that are directly associated with the mechanism of action of the TEP. For example, studies may delineate age ranges, illness stages, or biomarker thresholds that correspond with the anticipated therapy response. This selective approach reduces variability in baseline parameters, enabling researchers to more reliably ascribe observed effects to the intervention rather than to extraneous variables [25]

26.9.1.1.2 Ensuring patient homogeneity

A uniform patient population is crucial to minimize confounding factors that may mask the effectiveness and safety indicators of the product. By meticulously choosing volunteers with analogous clinical profiles such as equivalent illness severity or similar treatment histories researchers might attain more uniform outcomes. This homogeneity enhances the rigor of statistical analysis, hence simplifying the detection of genuine treatment effects and the evaluation of repeatability.

26.9.1.1.3 Reproducible results

The primary objective of well-specified inclusion and exclusion criteria is to provide reproducible clinical results. When patients are chosen according to rigorous, scientifically validated criteria, the variability in treatment response is reduced. This uniformity is essential for substantiating the therapeutic potential of the TEP and is a pivotal aspect in obtaining regulatory approval and clinical acceptability.

Furthermore, thorough exclusion criteria are essential to eliminate individuals with potential confounding conditions such as substantial comorbidities or contraindications that may jeopardize safety or skew effectiveness findings. Collectively, these requirements guarantee that the research group is ideally positioned to assess the intervention's genuine benefits.

26.9.1.2 Addressing heterogeneity

Variability in patient responses is a fundamental barrier in clinical studies, especially for TEPs. To guarantee dependable and comprehensible results, it is crucial to consider the variability in patient responses. An effective approach to managing this variability is the implementation of stratification techniques.

26.9.1.2.1 Stratification techniques

Stratification entails segmenting the research population into subgroups according to established characteristic such as age, illness severity, genetic markers, or previous treatment history that are recognized or presumed to affect the therapeutic response. By examining results within these more uniform groupings, researchers may more precisely evaluate the effectiveness and safety of the intervention. This method reduces the confusing influences of variability, resulting in more dependable and repeatable outcomes [26].

26.9.1.2.2 Improving trial reliability

When heterogeneity is adequately managed via stratification, the statistical power of the trial is augmented, revealing the genuine effects of the TEP more clearly. It facilitates more tailored insights on which patient demographics are most likely to gain from the therapy, thereby endorsing the advancement of precision medicine methodologies. Stratification strategies systematically account for variability, therefore reducing bias and enhancing the reliability of clinical trial results.

26.9.1.2.3 Implementation considerations

Effective stratification necessitates meticulous preparation throughout the trial design phase. Researchers must ascertain pertinent stratification parameters grounded in substantial preclinical evidence and previous clinical experience. The analytical strategy must incorporate subgroup analyses to assess varying responses across the strata. This meticulous methodology enhances the dependability of trial results and aids regulatory assessment by showcasing a comprehensive grasp of the variables affecting patient reactions.

26.9.1.3 Use of biomarkers

Biomarker-based selection is an emerging method that holds tremendous potential for boosting patient enrollment in clinical trials of TEPs. By identifying individuals who are most likely to benefit from the medication, biomarkers not only shorten the enrollment process but also increase the overall efficiency and reliability of trial outcomes.

26.9.1.3.1 Enhancing patient recruitment

Biomarkers are quantitative indications of biological processes or reactions that may be utilized to identify patients with certain molecular or cellular profiles. In tissue engineering, the use of biomarkers allows researchers to choose individuals whose biological traits correspond with the mechanism of action of the produced product. For instance, a biomarker may suggest a patient's likelihood to respond favorably to a scaffold meant to encourage cartilage regeneration. By preselecting such individuals, trials can enroll a more homogeneous sample that is likely to display a consistent treatment response [27].

26.9.1.3.2 Improving therapeutic outcomes

The use of biomarker-based selection can lead to better tailored medications. When patients are categorized based on particular biomarkers, it becomes simpler to connect clinical results with underlying molecular causes. This not only raises the possibility of establishing the product's efficacy but also assists in the identification of subgroups that may benefit the most. Consequently, the trial's statistical power is augmented, allowing for a more precise evaluation of the therapeutic efficacy of the TEP.

26.9.1.3.3 Reducing variability and enhancing precision

Employing biomarkers for patient selection reduces interpatient variability, a significant difficulty in clinical trial design. This stratification enhances the precision and reproducibility of results by ensuring that individuals possess important biological features. Moreover, it endorses the tenets of personalized medicine, enabling therapies to be customized to an individual's own genetic or molecular profile, thereby optimizing efficacy and minimizing the likelihood of adverse effects.

26.9.1.3.4 Implementation considerations

The effective utilization of biomarkers necessitates comprehensive validation studies to ascertain their predictive value and repeatability. Researchers must incorporate biomarker assays into the trial design at the outset and guarantee adherence to defined methods for sample collection, processing, and analysis. This degree of rigor is crucial to guarantee that the biomarker data are dependable and can be successfully utilized to inform patient selection.

26.9.2 Risk assessment and long-term monitoring

26.9.2.1 Immunogenicity and biocompatibility

Thorough preclinical testing is crucial to assess the immunogenicity and biocompatibility of TEPs. These evaluations are essential to confirm that the altered structures do not elicit detrimental immune responses and can assimilate successfully with host tissues over an extended period.

26.9.2.1.1 Immunogenicity evaluation

Any TEP may provoke an immune response owing to the presence of allogenic or autologous cells that may exhibit altered behavior in a modified environment. Preclinical investigations must also include extensive immunological tests to identify immune cell activation, cytokine release profiles, and other inflammatory indicators. These assessments assist in forecasting the potential for the implanted product to induce rejection or chronic inflammation, which may jeopardize both therapeutic efficacy and patient safety [2]. Techniques such as flow cytometry, enzyme-linked immunosorbent assays (ELISAs), and animal models are commonly utilized to assess these reactions.

26.9.2.1.2 Biocompatibility assessment

Biocompatibility testing evaluates the interaction between the tissue-engineered construct and the host's biological systems. The scaffold materials and biological components must not elicit toxicity, inflammation, or other harmful consequences upon implantation. Assessments encompass histological analyses, cytotoxicity tests, and in vivo implantation investigations, adhering to standards like ISO 10993. These evaluations offer insights into tissue integration, scaffold breakdown, and the comprehensive healing process. They also direct essential alterations in the design and composition of the build to improve compatibility and mitigate the likelihood of unwanted responses.

26.9.2.1.3 Integrated preclinical testing

Merging immunogenicity and biocompatibility evaluations into a cohesive preclinical testing method is essential for reducing possible hazards. In vitro tests and in vivo animal models are utilized to establish a thorough assessment of the safety of the TEP. This meticulous methodology guarantees that any possible complications whether arising from an unforeseen immunological reaction or a material-related incompatibility are recognized and resolved prior to the therapy progressing to human trials.

By methodically tackling these essential factors, researchers may refine the design of TEPs, improving their safety, effectiveness, and overall clinical success.

26.9.2.2 Long-term follow-up

Long-term follow-up is a critical component in the clinical management of patients receiving TEPs. Due to the novel nature of these therapies, delayed adverse eventssuch as late-onset immune responses, scaffold degradation complications, or unforeseen functional impairmentsmay not become apparent during initial clinical trials. Therefore regulatory agencies mandate robust postmarket surveillance programs to continuously monitor patient outcomes over extended periods.

26.9.2.2.1 Ongoing surveillance strategies

Postmarket monitoring includes many methodologies aimed at collecting and evaluating data about long-term safety and efficacy. This includes the creation of patient registries, regular safety update reports, and methodical gathering of adverse event data. Such surveillance initiatives enable the prompt identification of any delayed adverse occurrences, thereby permitting timely interventions and required product adjustments.

26.9.2.2.2 Regulatory mandates and guidelines

Regulatory systems in areas such as the United States and Europe mandate that tissue-engineered goods undergo long-term monitoring as a prerequisite for approval. The FDA and EMA possess explicit standards delineating the duration and form of postmarket surveillance for sophisticated treatments. These rules guarantee that postapproval, a product's performance is perpetually assessed to protect patient health [28]. The prolonged follow-up yields essential safety data and enhances comprehension of the long-term effectiveness and durability of these novel medicines.

26.9.2.2.3 Importance for patient safety and product improvement

Long-term follow-up statistics are essential for doctors and product developers. They provide insights into the durability of the therapeutic impact and assist in identifying possible areas for enhancement in product design and manufacturing procedures. Furthermore, ongoing monitoring enhances patient confidence by showcasing a dedication to safety and proactive risk management throughout the product's lifespan.

By integrating comprehensive postmarket surveillance into the clinical framework, healthcare providers and regulatory agencies can ensure that tissue-engineered therapies remain safe and effective long after their initial deployment.

26.9.2.3 Postmarket surveillance

Postmarket surveillance is an essential stage in the lifespan of tissue-engineered goods, functioning as an ongoing system for assessing product safety and efficacy beyond regulatory clearance. After a product's market introduction, postapproval studies are essential for detecting late-emerging concerns that may not have been apparent during the initial clinical trials. These investigations facilitate the timely detection, assessment, and management of any adverse events or unexpected problems.

26.9.2.3.1 Comprehensive safety monitoring

Postmarket surveillance initiatives include various activities such as patient registries, observational studies, and regular safety update reports. Through the systematic collection and analysis of empirical data over prolonged durations, these projects yield significant insights into the enduring performance of TEPs. Ongoing surveillance is crucial for identifying infrequent or postponed adverse effects, including immunological reactions or scaffold degradation problems, which may only manifest years postimplantation of the product.

26.9.2.3.2 Regulatory framework and obligations

Regulatory bodies, including the FDA and EMA, require postmarket surveillance as a prerequisite for the approval of sophisticated medicines. These stipulations guarantee that manufacturers are held responsible for the enduring safety of their goods. By following stringent postapproval research standards, corporations may exhibit a sustained dedication to patient safety and facilitate the continuous enhancement of their products informed by real-world data. This regulatory supervision is essential for preserving public confidence and ensuring that any subsequently identified concerns are swiftly addressed [29].

26.9.2.3.3 Benefits for clinical practice and product development

The data produced by postmarket surveillance improve patient safety and offers insights that can propel product innovation and enhancement. Understanding the performance of TEPs across various patient demographics allows researchers to discern factors affecting effectiveness and safety. This feedback loop facilitates ongoing enhancement in product design, manufacturing processes, and clinical procedures, ultimately resulting in improved treatment outcomes.

26.10 Regulatory guidelines for artificial intelligence in tissue engineering clinical trials

26.10.1 Role of artificial intelligence in tissue engineering trials

26.10.1.1 Artificial intelligence-assisted technologies

AI has emerged as an indispensable asset in tissue engineering clinical trials, improving several aspects of therapy development and assessment. AI-driven technologies facilitate the creation and optimization of scaffolds by analyzing extensive datasets about material characteristics and biological interactions. This computational skill allows researchers to develop scaffolds that more accurately replicate the native extracellular matrix, thus enhancing tissue integration and functioning.

Besides scaffold design, AI facilitates cell selection using predictive analytics and machine learning algorithms to pinpoint cell populations with the greatest regeneration potential. This method reduces variability in cell function and improves the consistency of trial results. Furthermore, AI is progressively included in 3D bioprinting methodologies, where it refines printing settings to attain superior accuracy and resolution in tissue constructions. By automating these essential procedures, AI diminishes human error and expedites the manufacture of TEPs.

The integration of AI in these domains enhances clinical trial processes and fosters more rigorous, data-informed decision-making. This eventually improves trial efficiency, resulting in more dependable and scalable regenerative therapeutics [30].

26.10.1.2 Artificial intelligence-driven patient stratification

Machine learning models are increasingly utilized to enhance patient selection in clinical trials for TEPs. Through the analysis of intricate, multidimensional datasets comprising genetic profiles, biomarker levels, and clinical histories AI algorithms may discern certain patient groupings that are most likely to have a positive response to a particular medication. This classification improves trial success rates by minimizing unpredictability and concentrating on those with the greatest treatment potential.

These sophisticated computational techniques provide a more individualized approach to trial design, guaranteeing that treatment effects are not obscured by varied patient responses. AI-driven stratification enhances precision medicine techniques by combining data from imaging, genomics, and electronic health records, hence improving the efficiency and outcomes of clinical trials [31].

This method enhances the statistical power of trials by selecting a more homogeneous study group and informs subsequent product development by identifying critical criteria that contribute to treatment effectiveness. The utilization of machine learning in patient classification signifies a pivotal advancement toward more precise and efficacious regenerative treatments.

26.10.1.3 Regulatory concerns

Transparency in AI decision-making is essential for obtaining regulatory clearance of AI-assisted technologies in clinical trials for tissue engineering. Regulatory bodies mandate that algorithms employed in patient stratification, scaffold design, or other AI-driven processes must be transparent and subject to interpretation and scrutiny, rather than functioning as "black boxes." This openness enables stakeholders to comprehend decision-making processes, validate outcome repeatability, and detect any biases or flaws in the procedure.

Developers must furnish comprehensive documentation of the AI model's design, training data, validation techniques, and performance metrics to comply with these criteria. Explainable AI (XAI) methodologies are progressively utilized to elucidate the decision-making process, allowing regulators to examine the generation of specified outcomes. This degree of transparency is crucial for guaranteeing patient safety and cultivating confidence among professionals, patients, and regulatory authorities.

Topol [32] asserts that transparency in AI is essential for the appropriate incorporation of machine learning into healthcare, highlighting that unambiguous, interpretable models are crucial for clinical acceptance and regulatory success. Moreover, continuous cooperation between technology developers and regulatory bodies is essential for revising standards to align with the fast progress in AI. This partnership facilitates the establishment of standardized assessment methodologies and best practices, ensuring that AI applications are both safe and successful.

By resolving these regulatory issues, developers may bolster the legitimacy of AI-assisted tissue engineering products and expedite the clearance process, hence hastening the availability of new medicines to patients.

26.10.2 Regulatory framework for artificial intelligence-driven approaches

26.10.2.1 Artificial intelligence regulatory bodies

The swift incorporation of AI technology in clinical trials for tissue-engineered goods has prompted major regulatory bodies to formulate specific guidelines to tackle the distinct issues posed by these applications. Regulatory agencies, including the US Food and Drug Administration (FDA), the European Medicines Agency (EMA), and Health Canada have developed frameworks to guarantee that AI-driven methodologies adhere to rigorous requirements for safety, efficacy, and transparency.

26.10.2.1.1 Food and Drug Administration guidelines

The FDA has actively established regulatory requirements for AI uses in healthcare. Their requirements, as detailed in recent papers [33], underscore the necessity for rigorous validation, transparency, and ongoing monitoring of AI algorithms. These standards mandate that AI systems employed for patient selection, trial administration, and data analysis undergo thorough evaluation for accuracy, dependability, and potential biases prior to their integration into clinical trials.

26.10.2.1.2 EMA guidelines

The EMA has established a regulatory framework in Europe that includes comprehensive considerations for advanced therapy medical products (ATMPs), including AI. The EMA's approach mandates thorough evaluations of the clinical significance and replicability of AI results. The EMA seeks to standardize validation and comprehensive review processes to unify regulatory norms throughout EU member states, therefore ensuring the safe and effective application of AI in clinical research.

26.10.2.1.3 Health Canada

Health Canada has contributed to the development of regulatory framework by providing recommendations on the application of AI in clinical studies. Their paradigm emphasizes that AI-driven choices must be underpinned by substantial scientific evidence and be entirely interpretable. Health Canada's strategy requires thorough preclinical and clinical assessment of AI technologies to guarantee they consistently improve trial results while upholding stringent safety criteria.

These regulatory recommendations emphasize the necessity of openness, stringent validation, and ongoing monitoring in the use of AI-driven methodologies in tissue engineering clinical trials. They not only enhance patient safety but also foster innovation by establishing clear channels for the incorporation of innovative technology into clinical research.

26.10.2.2 Good machine learning practice

The concepts of Good Machine Learning Practice (GMLP) are fundamental for incorporating AI into clinical trials in tissue engineering. These principles guarantee that AI algorithms are designed, verified, and implemented in a repeatable and explicable way. GMLP underscores the significance of a transparent machine learning pipeline, encompassing meticulous data collection and preprocessing, as well as model training, validation, and deployment. These methods ensure that the algorithms function consistently across various datasets and environments while also offering interpretable information for examination by regulators and physicians. The EMA's recommendations [34] emphasize that repeatability and explainability are essential for fostering trust in AI-assisted decision-making, facilitating ongoing monitoring and iterative enhancements while reducing any biases or mistakes.

26.10.2.3 Transparency in artificial intelligence algorithms

For AI-assisted methodologies in tissue engineering clinical trials to attain regulatory approval, it is essential that the foundational AI models are interpretable and auditable. Transparency in these algorithms signifies that the decision-making process is explicit, comprehensible, and subject to independent verification by regulatory authorities, doctors, and other stakeholders.

26.10.2.3.1 Interpretable models

Interpretable AI models enable users to comprehend the influence of certain inputs on the result. This clarity is essential in healthcare environments, as decisions immediately influence patient care. By elucidating the reasoning for each prediction, doctors may evaluate whether the AI's suggestions conform to recognized medical principles and ascertain that no latent biases or inaccuracies jeopardize patient safety [35]. Moreover, interpretable models enhance troubleshooting and iterative enhancements, resulting in more robust and dependable clinical instruments.

26.10.2.3.2 Auditability

Auditability denotes the ability to meticulously scrutinize the operations of the AI model, encompassing data pretreatment, feature selection, and algorithm training. This thorough examination allows regulators and independent auditors to verify that the AI system continually satisfies performance criteria and complies with ethical norms. The ability to audit an AI model guarantees that its operations are transparent, repeatable, and devoid of confounding variables that might compromise clinical trial results. The emphasis on auditability fosters trust in AI systems, facilitating regulatory authorities' endorsement of its deployment in sensitive healthcare contexts.

Transparency, achieved by interpretability and auditability, is essential for regulatory approval. Kiseleva et al. [35] assert that these attributes are crucial for validating the dependability of AI models in clinical decision-making, hence enhancing confidence among regulators, doctors, and patients in tissue-engineered medicines.

26.11 Global harmonization and future perspectives

26.11.1 Necessity for standardization

Aligning worldwide regulatory frameworks for tissue-engineered goods is essential to expedite their clearance and maintain uniform quality and safety requirements across various areas. Global standardization can optimize clinical trial procedures and diminish redundant efforts that frequently hinder the commercialization of novel regenerative medicines.

By harmonizing regulatory requirements such as Good Clinical Practice, manufacturing methods, and postmarket surveillanceacross countries, producers may effectively traverse numerous markets using a unified dataset. This synchronized strategy eliminates duplication and administrative encumbrances while promoting international cooperation, resulting in more comprehensive safety and effectiveness assessments. Oberweis et al. [36] assert that standardized regulatory frameworks facilitate the exchange of best practices and data openness, hence accelerating the approval process for tissue-engineered goods.

Global standardization aids patients by guaranteeing access to medicines that adhere to rigorous, generally acknowledged safety criteria, irrespective of geographical location. Efforts by entities such as the International Council for Harmonization (ICH) and the World Health Organization (WHO) bolster the establishment of uniform regulatory guidelines for global adoption, thereby expediting the transition of innovative therapies from research to clinical application.

26.11.2 Innovations in regulatory approaches

Recent advancements in regulatory science are progressively utilizing adaptive trial designs and digital health technologies to enhance the efficiency and accuracy of clinical assessments for tissue-engineered goods. Adaptive trial designs permit alterations to research procedures informed by interim data analysis, facilitating more flexible, efficient, and patient-centered studies. This method aids in optimizing dosage regimens, revising inclusion criteria, and prematurely ending unproductive arms, therefore decreasing total development time and costs.

Simultaneously, digital health tools such as wearable sensors, telemedicine platforms, and real-time data analytics enable ongoing monitoring of patient outcomes and treatment compliance. These technologies offer high-resolution, longitudinal data that improve the safety and effectiveness evaluations of tissue-engineered therapeutics while facilitating remote patient monitoring and decentralized clinical trials. Collectively, these developments signify a substantial advancement in regulatory methodologies, fostering a more dynamic, data-driven framework capable of adapting to the intricacies of contemporary medicinal development [37].

By integrating adaptive trial methodology with digital health technology, regulatory agencies may more effectively mitigate uncertainty and enhance decision-making during the clinical trial process. This advancement in regulatory science expedites market access while guaranteeing the safety and efficacy of innovative treatments, eventually helping patients globally.

26.11.3 Ethical and societal considerations

Ethical frameworks for tissue-engineered goods should encompass not just safety and efficacy but also larger social concerns, especially inequalities in access to innovative medicines. The emergence of novel therapies raises concerns that their elevated costs and intricate production methods may restrict access to those with adequate means or in certain locations, thus increasing existing healthcare disparities. Cheng et al. [38] assert that ethical frameworks must actively confront these inequities, guaranteeing equitable distribution of the advantages of advanced regenerative medicines across all populations.

26.11.3.1 Addressing access disparities

Ethical issues in this context entail formulating policies that encourage low pricing, stimulate scalable production, and facilitate subsidized access for disadvantaged areas. Regulatory authorities, industry participants, and healthcare policymakers must work to establish mechanisms such as tiered pricing, public–private partnerships, and government financing initiatives that facilitate the connection between technical innovation and social advantage.

26.11.3.2 Societal impact and justice

In addition to access, ethical frameworks must evaluate the enduring social consequences of modern medicines. This entails guaranteeing that underprivileged or historically underserved communities are not further disadvantaged by the fast progression of regenerative medicine. By integrating social justice ideas into regulatory and financing choices, stakeholders may advance a more inclusive healthcare system that emphasizes both scientific advancement and equity [20].

26.11.3.3 Collaborative efforts for equitable innovation

Continuous discourse among ethicists, doctors, regulators, and patient advocacy organizations is essential. This partnership aids in identifying possible biases in trial designs, funding distributions, and allocation tactics, eventually shaping recommendations that seek to optimize the social advantages of TEPs. Such conversations may facilitate the establishment of international norms and regulations that enhance worldwide access to breakthrough medications.

26.12 Conclusion

This chapter offers a thorough analysis of the regulatory framework for TEPs, emphasizing the fundamental principles that dictate their clinical development and approval. Tissue engineering is an interdisciplinary domain that integrates cell biology, biomaterials science, and biochemical signaling to develop functioning biological replacements for impaired tissues and organs. This topic has introduced new opportunities in regenerative medicine, although it also poses distinct regulatory concerns owing to the intricacy and variety of created tissues. A fundamental element of this discourse is the stringent regulatory framework imposed by international organizations such as the FDA, EMA, and PMDA. These regulatory agencies guarantee that TEPs undergo comprehensive preclinical assessment, meticulously structured clinical trials, and rigorous postmarket monitoring to validate their safety and effectiveness. The chapter delineates the regulatory approval procedure, encompassing Investigational New Drug (IND) applications in the United States and Clinical Trial Applications (CTA) in the EU, underscoring the significance of risk assessment and longitudinal monitoring.

A major hurdle in clinical trials for tissue engineering is patient heterogeneity, which can profoundly affect trial outcomes. The utilization of biomarkers and AI-driven patient classification is becoming more common to solve this issue. The selection based on biomarkers enables researchers to pinpoint individuals who are most likely to respond positively to a certain medicine, hence enhancing the efficiency and success rates of clinical trials. Likewise, AI-driven machine learning algorithms enhance patient recruitment, evaluate clinical data, and improve trial design. Nonetheless, these technological developments also elicit regulatory apprehensions, especially about transparency, interpretability, and ethical adherence.

The incorporation of AI into tissue engineering has transformed scaffold design, 3D bioprinting, and automated tissue synthesis. AI algorithms can forecast ideal scaffold characteristics, refine biomaterial choices, and augment the accuracy of bioprinting methods, resulting in more uniform and effective tissue structures. Notwithstanding these benefits, regulatory bodies must verify that AI-driven methodologies comply with Good Machine Learning Practice (GMLP) standards, hence ensuring repeatability and explainability in clinical decision-making. In addition to AI and patient selection tactics, the chapter emphasizes the significance of postmarket surveillance in evaluating the long-term safety and efficacy of TEPs. Due to the involvement of live cells and bioactive materials, the behavior of these products within the human body may change over time, requiring ongoing monitoring and subsequent investigations. Regulatory authorities need postmarket monitoring programs to monitor any delayed adverse effects and confirm ongoing therapeutic advantages.

The discourse underscores the necessity for global harmonization of regulatory systems to expedite the licensing and dissemination of tissue-engineered medicines internationally. Harmonizing rules across various regulatory authorities can expedite the development process, diminish unnecessary testing mandates, and enhance international cooperation. Initiatives by entities like the International Council for Harmonization (ICH) and the World Health Organization (WHO) are crucial in advancing regulatory convergence and optimizing approval processes.

Collaboration among regulators, researchers, and doctors is essential for advancing tissue engineering. This partnership enables the conversion of breakthrough research into safe and effective therapeutics by amalgamating varied skills and viewpoints. Regulators guarantee that clinical practices comply with safety standards and ethical requirements, while researchers provide enhanced scientific understanding and technology breakthroughs. Clinicians, at the forefront of patient care, offer essential insights into practical issues and therapeutic results. This collaboration not only expedites product development and regulatory approval but also promotes ongoing enhancement in trial design and treatment effectiveness. Collaborative initiatives facilitate the exchange of best practices and data, perhaps resulting in standardized standards and enhanced worldwide regulatory harmonization. As the area advances, the amalgamation of concepts from other fields is crucial to tackle intricate issues, including enhancing patient selection, augmenting biomaterial efficacy, and alleviating potential bad effects. Ultimately, multidisciplinary collaboration is essential for transitioning TEPs from experimental research to standard clinical practice, so guaranteeing that new medicines reach a wider patient demographic.

The improvement of tissue engineering and regenerative medicine depends on the progression of regulatory science to align with swift technical developments. Innovative AI-driven methodologies and the advancement of more complex TEPs are revolutionizing the therapeutic domain. Regulatory frameworks must consequently develop, embracing more dynamic and adaptable tactics to meet these advancements.

As AI and machine learning become essential for enhancing clinical trial designs, patient stratification, and product production, regulators must formulate new criteria to guarantee that these algorithms are transparent, repeatable, and devoid of bias. This adaptation entails the integration of Good Machine Learning Practice (GMLP) concepts and the establishment of standards for explainable AI in healthcare decision-making. Adopting these technologies can optimize trial procedures, enhance patient results, and expedite product approvals.

Cutting-edge tissue-engineered goods are expanding the limits of conventional therapy. These items frequently exhibit intricate, multicomponent designs that necessitate sophisticated techniques for evaluating safety, effectiveness, and long-term durability. Regulatory agencies must revise current criteria to address the distinct problems presented by these innovative devices, including concerns over scaffold integration, immunological response, and long-term functioning. This may entail adaptive trial designs, improved postmarket surveillance, and international cooperation to standardize regulations worldwide.

In the future, the incorporation of digital health technologies, real-world evidence, and adaptive clinical trial methodology will be essential for regulatory success. Ongoing communication among regulators, business, and academics is crucial for the development and enhancement of these frameworks. By cultivating a collaborative atmosphere, regulatory agencies may guarantee that guidelines adapt to new scientific findings and technology advancements, therefore facilitating the safe and effective implementation of innovative medicines in clinical settings.

The future of regulatory science in tissue engineering depends on its capacity to adapt to developing AI-driven approaches and innovative therapeutic devices. This progressive strategy will be essential in surmounting existing constraints, accelerating the approval procedure, and guaranteeing that novel medicines provide optimal advantages to patients globally.

References

[1] R. Langer, J.P. Vacanti, Tissue engineering, Science (New York, N.Y.) 260 (5110) (2023) 920–926. Available from https://doi.org/10.1126/science.8493529.

[2] D.F. Williams, On the nature of biomaterials, Biomaterials 30 (30) (2019) 5897–5909. Available from https://doi.org/10.1016/j.biomaterials.2009.07.027.

[3] L.E. Niklason, J. Gao, W.M. Abbott, K.K. Hirschi, S. Houser, R. Marini, et al., Functional arteries grown in vitro, Science (New York, N.Y.) 284 (5413) (2019) 489–493. Available from https://doi.org/10.1126/science.284.5413.489.

[4] A. Atala, S.B. Bauer, S. Soker, J.J. Yoo, A.B. Retik, Tissue-engineered autologous bladders for patients needing cystoplasty, The Lancet 367 (9518) (2023) 1241–1246. Available from https://doi.org/10.1016/S0140-6736(06)68438-9.

[5] A.S. Mao, D.J. Mooney, Regenerative medicine: current therapies and future directions, Proceedings of the National Academy of Sciences 112 (47) (2019) 14452–14459. Available from https://doi.org/10.1073/pnas.1508520112.

[6] J. Vanderburgh, J.A. Sterling, S.A. Guelcher, 3D printing of tissue engineered constructs for in vitro modeling of disease progression and drug screening, Annals of Biomedical Engineering 45 (1) (2017) 164–179. Available from https://doi.org/10.1007/s10439-016-1640-4.

[7] J.L. Olson, A. Atala, J.J. Yoo, Tissue engineering: current strategies and future directions, Chonnam Medical Journal 47 (1) (2021) 1. Available from https://doi.org/10.4068/cmj.2011.47.1.1.

[8] S. Badylak, D. Freytes, T. Gilbert, Extracellular matrix as a biological scaffold material: structure and function, Acta Biomaterialia 5 (1) (2019) 1–13. Available from https://doi.org/10.1016/j.actbio.2008.09.013.

[9] Y. Ikada, Challenges in tissue engineering, Journal of the Royal Society Interface 3 (10) (2023) 589–601. Available from https://doi.org/10.1098/rsif.2006.0124.

[10] P. Buma, N.N. Ramrattan, T.G. van Tienen, R.P.H. Veth, Tissue engineering of the meniscus, Biomaterials 25 (9) (2024) 1523–1532. Available from https://doi.org/10.1016/S0142-9612(03)00499-X.

[11] T. Xi, L. Chen, C. Wang, Regulation challenge of tissue engineering and regenerative medicine in China, Burns & Trauma 1 (2) (2023) 56. Available from https://doi.org/10.4103/2321-3868.118927.

[12] P. Marks, S. Gottlieb, Balancing safety and innovation for cell-based regenerative medicine, New England Journal of Medicine 378 (10) (2018) 954–959. Available from https://doi.org/10.1056/NEJMsr1715626.

[13] C. Pimenta, V. Bettiol, T. Alencar-Silva, O.L. Franco, R. Pogue, J.L. Carvalho, et al., Advanced therapies and regulatory framework in different areas of the globe: past, present, and future, Clinical Therapeutics 43 (5) (2021) e103–e138. Available from https://doi.org/10.1016/j.clinthera.2021.02.006.

[14] J. Skoracka, K. Bajewska, M. Kulawik, W. Suchorska, K. Kulcenty, Advances in cartilage tissue regeneration: a review of stem cell therapies, tissue engineering, biomaterials, and clinical trials, EXCLI Journal 23 (2024) 1170. Available from https://doi.org/10.17179/EXCLI2024-7088.

[15] U.S. Food and Drug Administration. https://www.fda.gov/, n.d. (accessed 03.03.25).

[16] International Council for Harmonisation of Technical Requirements for Pharmaceuticals for Human Use (ICH), ICH harmonised guideline integrated addendum to ICH E6(R1): guideline for good clinical practice E6(R2), 2016.

[17] Use of International Standard ISO 10993-1, "Biological evaluation of medical devices-Part 1: Evaluation and testing within a risk management process" Guidance for Industry and Food and Drug Administration Staff Preface Public Comment. https://www.fda.gov/vaccines-blood-biologics/guidance-compliance-regulatory-information-, n.d. (accessed 03.03.25).

[18] Medicines Agency, E., Guideline on quality, non-clinical and clinical requirements for investigational advanced therapy medicinal products in clinical trials, 2025. www.ema.europa.eu/contact.

[19] H.B. Baker, J.P. McQuilling, N.M.P. King, Ethical considerations in tissue engineering research: case studies in translation, Methods (San Diego, Calif.) 99 (2016) 135–144. Available from https://doi.org/10.1016/J.YMETH.2015.08.010.

[20] E.J. Emanuel, D. Wendler, C. Grady, What makes clinical research ethical? JAMA: the Journal of the American Medical Association 283 (20) (2023) 2701–2711. Available from https://doi.org/10.1001/JAMA.283.20.2701.

[21] S.E. Harvey, D. Elbourne, J. Ashcroft, C.M. Jones, K. Rowan, Informed consent in clinical trials in critical care: experience from the PAC-Man Study, Intensive Care Medicine 32 (12) (2023) 2020–2025. Available from https://doi.org/10.1007/S00134-006-0358-4.

[22] A.F.J. De Kanter, K.R. Jongsma, M.C. Verhaar, A.L. Bredenoord, The ethical implications of tissue engineering for regenerative purposes: a systematic review, Tissue Engineering. Part B, Reviews 29 (2) (2023) 167. Available from https://doi.org/10.1089/TEN.TEB.2022.0033.

[23] WMA Declaration of Helsinki – Ethical Principles for Medical Research Involving Human Participants – WMA – The World Medical Association, n.d. Retrieved March 3, 2025, from https://www.wma.net/policies-post/wma-declaration-of-helsinki/.

[24] International Ethical Guidelines for Health-Related Research Involving Humans Prepared by the Council for International Organizations of Medical Sciences (CIOMS) in collaboration with the World Health Organization (WHO), 2016. www.cioms.ch.

[25] L.J. Bonassar, C.A. Vacanti, Tissue engineering: the first decade and beyond, Journal of Cellular Biochemistry 72 (S30-31) (2019) 297–303 https://doi.org/10.1002/(sici)1097-4644(1998)72:30/31+<297::aid-jcb36>3.0.co;2-6.

[26] P.W. Lavori, T.A. Louis, J.C. Bailar, M. Polansky, Designs for experiments — parallel comparisons of treatment, New England Journal of Medicine 309 (21) (2023) 1291–1299. Available from https://doi.org/10.1056/nejm198311243092105.

[27] L. Levstek, L. Janžič, A. Ihan, A.N. Kopitar, Biomarkers for prediction of CAR T therapy outcomes: current and future perspectives, Frontiers in Immunology 15 (2024) 1378944. Available from https://doi.org/10.3389/FIMMU.2024.1378944.

[28] F. Han, J. Wang, L. Ding, Y. Hu, W. Li, Z. Yuan, et al., Tissue engineering and regenerative medicine: achievements, future, and sustainability in Asia, Frontiers in Bioengineering and Biotechnology 8 (2020) 83. Available from https://doi.org/10.3389/FBIOE.2020.00083.

[29] A. Atala, Recent developments in tissue engineering and regenerative medicine, Current Opinion in Pediatrics 18 (2) (2023) 167–171 https://doi.org/10.1097/01.MOP.0000193294.94646.BE.

[30] Z. Zhang, X. Zhou, Y. Fang, Z. Xiong, T. Zhang, AI-driven 3D bioprinting for regenerative medicine: from bench to bedside, Bioactive Materials 45 (2025) 201–230. Available from https://doi.org/10.1016/J.BIOACTMAT.2024.11.021.

[31] A. Esteva, A. Robicquet, B. Ramsundar, V. Kuleshov, M. DePristo, K. Chou, et al., A guide to deep learning in healthcare, Nature Medicine 25 (1) (2019) 24–29. Available from https://doi.org/10.1038/S41591-018-0316-Z.

[32] E.J. Topol, Deep Medicine How Artificial Intelligence Can Make Healthcare Human Again. Basic Books. - References - Scientific Research Publishing, 2019. Retrieved March 4, 2025, Available from https://www.scirp.org/reference/referencespapers?referenceid=3758241.

[33] Artificial Intelligence for Drug Development FDA, 2021. Retrieved March 4, 2025, from https://www.fda.gov/about-fda/center-drug-evaluation-and-research-cder/artificial-intelligence-drug-development.

[34] Artificial intelligence workplan to guide use of AI in medicines regulation European Medicines Agency (EMA), 2021. Retrieved March 4, 2025, Available from https://www.ema.europa.eu/en/news/artificial-intelligence-workplan-guide-use-ai-medicines-regulation.

[35] A. Kiseleva, D. Kotzinos, P. De Hert, Transparency of AI in healthcare as a multilayered system of accountabilities: between legal requirements and technical limitations, Frontiers in Artificial Intelligence 5 (2022) 879603 https://doi.org/10.3389/FRAI.2022.879603/FULL.

[36] C.V. Oberweis, J.A. Marchal, E. López-Ruiz, P. Gálvez-Martín, A worldwide overview of regulatory frameworks for tissue-based products, Tissue Engineering. Part B, Reviews 26 (2) (2020) 181–196. Available from https://doi.org/10.1089/TEN.TEB.2019.0315.

[37] M. Lee, H. Ly, C.C. Möller, M.S. Ringel, Innovation in regulatory science is meeting evolution of clinical evidence generation, Clinical Pharmacology and Therapeutics 105 (4) (2019) 886. Available from https://doi.org/10.1002/CPT.1354.

[38] T. Cheng, S. D'Amico, M. Luo, A.S. Lestoquoy, L. Yinusa-Nyahkoon, L.D. Laird, et al., Health disparities in access to nonpharmacologic therapies in an urban community, Journal of Alternative and Complementary Medicine (New York, N.Y.) 25 (1) (2019) 48–60. Available from https://doi.org/10.1089/ACM.2018.0217.

Chapter 27

Early detection of Parkinson's disease using multimodal information

Shantanu Ghosh[1] and Sumantra Dutta Roy[2]
[1]*Amity Centre for Cognitive Science, Amity University Uttar Pradesh, Noida, Uttar Pradesh, India,* [2]*Department of Electrical Engineering, Indian Institute of Technology Delhi, New Delhi, India*

27.1 Introduction

Parkinson's disease (PD) is a neurodegenerative movement disorder that affects about 8.5 million individuals worldwide [1]. PD is increasing faster than any other neurological disorder, with numbers doubling in the last quarter of a century [2] and is being considered a public health issue in populations above 65 years [3,4]. While clinical symptoms in the advanced stages of the disorder are easily evident, the symptoms are often incompletely rendered or subtle, presenting the diagnosis difficult [5]. The problem arises when affected individuals lose their dopaminergic neurons in the substantia nigra (SNg) and that dopamine concentrations of considerably depleted in the striatum (Str) of patients affected by the condition. Some non-motor comorbidities are also presented that include sleep dysregulation, including REM sleep, incontinence, and mild cognitive impairment. The therapy and management at present involve the relief provided from the symptoms. Beginning treatment at a later stage provides little assistance; dopaminergic neurons are already lost extensively at this point and are considered to be irreversible. The only way is to manage the patient, and at this point, the care of the patient also takes an economic burden on the healthcare system.

About a tenth of the cases have been linked to a genetic mutation of LRRK2, the gene encoding leucine-rich repeat kinase-2, a gene that is also implicated in neurotoxicity and has been tested in a mouse model in vivo [6]. The recent discovery of the role of α-synuclein in dopaminergic neurodegeneration due to severe toxicity provides an insight into the cause of PD etiology [7,8] and strategies to understand its place in the cascade of molecular events for therapeutics [9]. The diagnosis of early PD is tricky and challenging even to the experienced clinician; however, imaging modalities can substantially change the odds. Furthermore, there are decision support tools that could serve as predictors of prognosis and treatment outcomes and can be valuable assets for the clinician to decide on the further course of treatment and use of personalized treatment regimes.

Over the past several decades, machine learning (ML) techniques have matured into a set of best practices that aim at knowledge discovery patterns to inform the diagnosis, prognosis, and response to treatment regimen, as well as genetic blueprints. This knowledge is harvested for affecting the input for understanding the best management of diseases as a decision support system. These systems constitute the end-use forms for a class of techniques that have revolutionized healthcare research using artificial intelligence (AI). With the advent of computer architecture to store and manipulate large volumes of data, scaling up has been possible, and the most important area of development is ML, which uses large volumes of data in pretrained computational models for learning patterns in the data sets. Image-based diagnostic tools have been the major beneficiary of such a computational framework, wherein researchers have used radiological or pathological image data for classification of patients from healthy controls using complex computational neural networks. These methods all used a type of ML known as deep learning (DL), neural networks with more than three layers that are a staple in the field of AI and are designed to generally mimic the activity of neurons in the human brain. Traditionally, the relationship between features (the various pieces of information that are input into the model) and the importance of each feature is unknown when training neural networks. As a result, the weight for each neuron in each layer is assigned at random. However, because the model must "learn" these relationships from scratch, training times may be prolonged. The ability to pretrain neural networks, which is commonly done on large-scale data sets, by initializing networks with weights from previously trained models, thus making them suitable for complex, multidimensional problems.

Fundamentals of Translational Tissue Engineering. DOI: https://doi.org/10.1016/B978-0-443-33455-9.00033-X

27.2 Early Parkinson's disease detection based on preclinical motor and nonmotor, imaging, and biological features

The premotor, or prodromal phase of PD etiology, can last anything between 5 and 20 years, and precedes the onset of neurodegeneration and manifestation of classic clinical motor symptoms. Recent approaches to develop AI-based prediction model that correlates well with the MDS-UPDRS simply by analyzing the patients' nocturnal breathing at home have been quite successful [10]. Another study used only the MDS-UPDRS as input to ML classification, which yielded encouraging results [11] and has the potential to become an effective screening tool to identify at-risk populations. Some models rely on biological features that become apparent clinically, such as the genome-wide association study (GWAS) that implicates the *SNCA* gene locus as a modifier of the age of onset of sporadic PD [12].

Nonmotor symptoms develop in PD due to the factors that go beyond the dopaminergic signaling system [13]. Many of the common non-motor conditions are, but not limited to, hyposmia or olfactory dysfunction, urinary incontinence, REM sleep behavior disorder (RBD) [14], and have been the subject of Convolutional Neural Network (CNN) and Random Forest (RF) algorithms to test the viability of wearable sensors as data gatherers for training symptom detection models using gait, bradykinesia and tremor data [15]. Early models for detecting PD used Partial Least Squares (PLS), Support Vector Machines (SVM), and Bayesian classifiers on DaTScan images [16,17].

Several neuroimaging modalities lend themselves well to sense the changes in PD diagnosis, progression, and monitoring and can be useful input to AI-based diagnostic assessments.

27.2.1 Structural MRI

Substantia nigra (SNg) has been found to show up as hypointense regions in T2w images, as it maps the iron content in various brain areas [16], with increased iron, increased R2 relaxation rates, and R2* is a useful biomarker. Conventional clinical imaging of the structural abnormalities in PD at 1.5 T and 3 T (both T1w and T2w) is not sufficient; higher gradients at 7 T may be able to differentiate healthy individuals from PD patients. The structure of nigrosome 1 may be seen on 7 T MRI scans using T2*w gradient-recalled echo imaging. While most MRIs in early stage PD assessment are performed to rule out alternative pathologies, which inter alia, include tumors, vascular lesions, inflammation, and atypical Parkinsonism, gradient-recalled echo imaging and susceptibility-weighted imaging (SWI) have also been shown to reveal the detailed anatomic structure of the SNg [18].

27.2.2 Diffusion tensor imaging (DTI), Diffusion-weighted magnetic resonance imaging (DWI) and Diffusion spectrum imaging (DSI)

Diffusion tensor imaging (DTI) is a novel technique capable of detecting microstructural abnormalities in white matter that appear normal on standard T1- and T2-weighted imaging [19,20]. Mean diffusivity (MD) and fractional anisotropy (FA) are measures of directionality and diffusion parameters of water molecules calculated from the diffusion tensor imaging protocols. FA is high in healthy white matter, as water moves parallel to the fiber bundles. DTI abnormalities present as increased MD and/or reduced anisotropy. For differential PD diagnosis, DTI and Diffusion-weighted magnetic resonance imaging (DWI) are more sensitive than sMRI. While DWI scans are seen as a differential diagnosis information in most clinically probable patients with heterodegenerative Parkinsonisms, including MSA and PSP, the water diffusion rate is found to be normal in PD [21].

27.2.3 Functional MRI

Functional MRI (fMRI) is based on blood oxygenation level-dependent (BOLD) contrast using the T2*w echo-planar imaging, and relies on tying the local hemodynamics to corresponding neural activity. It has been suggested that ROI-based BOLD imaging to evaluate de novo PD in basal ganglia and thalamus as well as resting-state fMRI (rs-fMRI) to look into normal and atypical functional connectivity analysis [22] and could possibly be used as inputs for AI systems to distinguish healthy controls from PD.

27.2.4 Magnetic resonance spectroscopy

Some research studies have shown that both 1 H and 31 P MRS at 3 T may be useful in differentiating early and advanced PD from healthy controls by focusing on the difference in beta adenosine triphosphate (β-ATP) in the contralateral midbrain, while measurements of β-ATP as well as phosphocreatine were able to distinguish PD and HC groups. However, MRS is of relatively limited utility because the neurochemicals are present in extremely low concentrations and require substantial human expert manipulation for correcting local eddy currents, and so are sensitive to the level of competence and vary from person to person. Furthermore, the brain structures themselves are tiny and positioned deep within the brain, making imaging difficult. These restrictions, however, may be solved with higher-field clinical scanners.

27.2.5 Positron emission tomography

Techniques for brain imaging offer an effective means of *in vivo* examination of the pathogenesis and pathophysiology of PD processes. A nuclear medicine technology called positron emission tomography (PET) imaging enables the measurement of crucial physiological parameters like glucose metabolism and neuroreceptor binding in PD. Loss of dopaminergic neurons and the development of Lewy bodies, ubiquitinated protein deposits, are the pathological hallmarks of PD; for example, 18F-DOPA PET has been used for quite some time to determine the stages of dopaminergic degeneration in PD. It has been demonstrated that the decline of dopamine function starts in the dorsal-caudal putamen contralateral to the clinically affected side [23] and that the rate of degeneration in the caudate nucleus is slower than that of the putamen [24] in early PD. It has been demonstrated that striatal 18F-DOPA binding and motor disability (as measured by the Unified Parkinson's Disease Rating Scale [UPDRS]) are inversely correlated [25–27].

27.2.6 Single photon emission computed tomography

SPECT imaging using 123I-FP-CIT (also known as 123-Isoflupane) as a radiotracer has been shown to be sensitive in a number of imaging studies of PD, especially depicting the presynaptic dopaminergic degeneration in the striatum (caudate and putamen) even in the early stages of PD and hence is emerging as a valuable diagnostic tool [28] as it is highly correlated with PD progression [28,29]. Any condition that results in the loss of presynaptic dopamine neurons will manifest as abnormal in comparison to healthy controls. Thus SPECT cannot distinguish between PD, progressive supranuclear palsy (PSP), multiple system atrophy (MSA), and other neurodegenerative diseases that affect dopamine neurons. This could be one of the most significant issues that AI-based classification will have to address.

Researchers have reported that PD significantly reduces dopamine transporter levels in the striatum, which are linked to disease progression and clinical scores [30,31]. SPECT imaging has been used in a few studies to explore the cortico-basal ganglia thalamocortical (CBGT) circuit (which includes the associative cortex, limbic cortex, sensorimotor cortex, caudate, putamen, thalamus, and pallidum). The specific/striatal binding ratio (SBR) is the ratio of the concentrations of a target brain region's specific binding radioactivity to the nonspecific binding radioactivity of a reference region devoid of dopaminergic transporter (DaT) uptake. In a recent study, SBR values extracted from SPECT imaging and structural connectivity of DTI for PD-related ROIs were used to evaluate differences between healthy controls and PD. This study used a partial least-squares regression framework; MDS-UPDRS scores were predicted using multimodal imaging characteristics [32].

27.3 A general framework for artificial intelligence-based Parkinson's disease diagnosis and severity analysis

Most models of AI for the detection and progression of PD will have the following structure for their use:

The Parkinson's Progression Marker's Initiative (PPMI) is a large-scale public database for PD progression biomarkers [33]. The imaging protocols have been standardized and contain different data sets for the same individuals over a considerable period of time, for example, UPDRS, 123I-Isoflupane, and MDS-UPDRS survey data, among others. It can thus form the basis of a training core of the clinical decision support system for PD. Previously published studies involving the classification of data obtained from the PPMI database using various ML algorithms like Random Forest, Support Vector Machine, Logistic regression, and segmentation of images and classification through surface fitting [11,34–36]. The basic plan for conducting ML is given in Fig. 27.1.

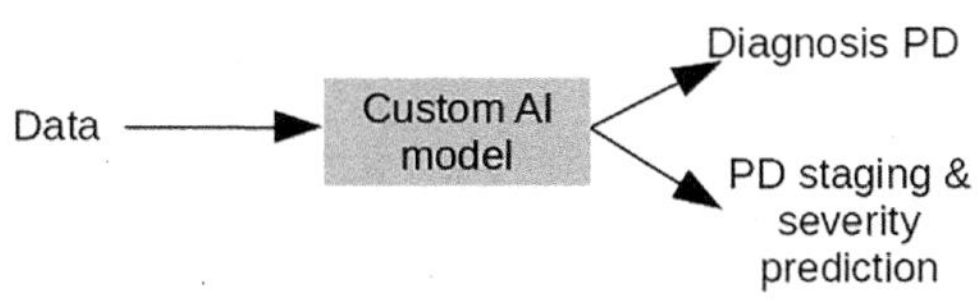

FIGURE 27.1 Overview of the AI model for PD diagnosis and disease severity prediction. *AI*, Artificial intelligence; *PD*, Parkinson's disease.

Briefly, a radial basis function (RBF) kernel may be used, given by $K(x_i, x) = \exp(-y \parallel x_i - x \parallel)$, $\gamma > 0$ for the SVM, thus mapping the inputs into a feature space with infinite dimensions. Furthermore, the parameters C for the SVM and γ for the kernel may be selected using a user-defined input, n-fold cross-validation. Once this algorithm achieves reasonably good classification, it can be used for tunable AI applications for generating the core of a pretrained model, such as the one given in Fig. 27.2, within a transfer learning framework using Convolutional Autoencoder (CA) (adapted from [37]).

Moreover, the inclusion of SBR from DaTScan and other imaging information may indeed provide better classification and prognosis. An AI/ML-based model for PD imaging and diagnosis is proposed for disease severity prediction as in Fig. 27.3.

27.4 Some recent deep learning models for consideration

The process of transforming knowledge from numerous source classes is known as transfer learning. Examples of transfer learning models are Densenet, Inception [38], and ResNet. By acquiring the necessary knowledge from the source classes, transfer learning's effectiveness is enhanced [39]. All ML algorithms (both supervised and unsupervised) need large data to train and process the model, but transfer learning algorithms perform better even on small data sets. In transfer learning, the existing knowledge can be used to fine-tune the model parameters by tuning the weights in the neural layers, based on our applications; hence, several transfer learning strategies are based on relational knowledge, feature representation, parameter characterization and tuning, and instance learning (Fig. 27.4).

Transfer learning is a ML technique where a model created for one job is utilized as the foundation for a model for another task. Densenet is a transfer learning model that consists of layer implementation. In Densenet implementation, each layer is connected to every other layer. In each layer L, the number of fully connected layers is L(L-1)/2. After the basic convolution block, Densenet is built with two main implementations, that is, a transition layer and a dense layer block. Each dense layer consists of predefined normalization such as batch normalization, Relu activation function, and an $n*n$ convolution layer. Finally, the hyperparameter tuning joins the fully connected layers and consists of a dense 4096 Relu activation function. In the final layer, a different type of activation function is preferred, for example, softmax is

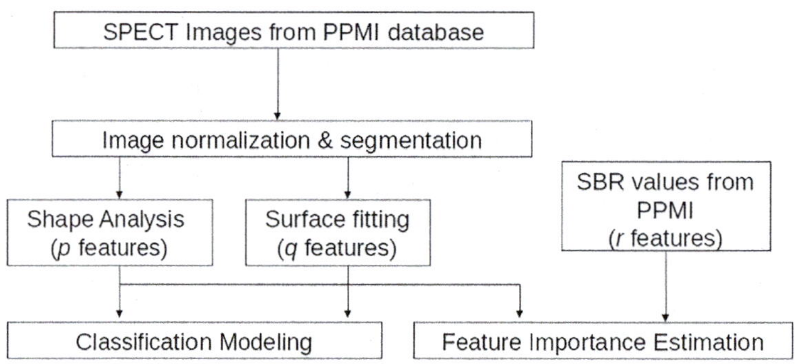

FIGURE 27.2 Architecture of ML-based classification. *ML*, Machine learning.

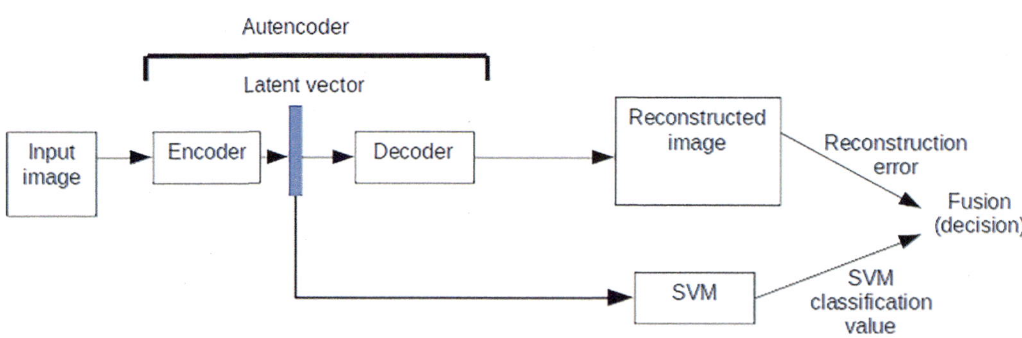

FIGURE 27.3 Decision-based tunable classification.

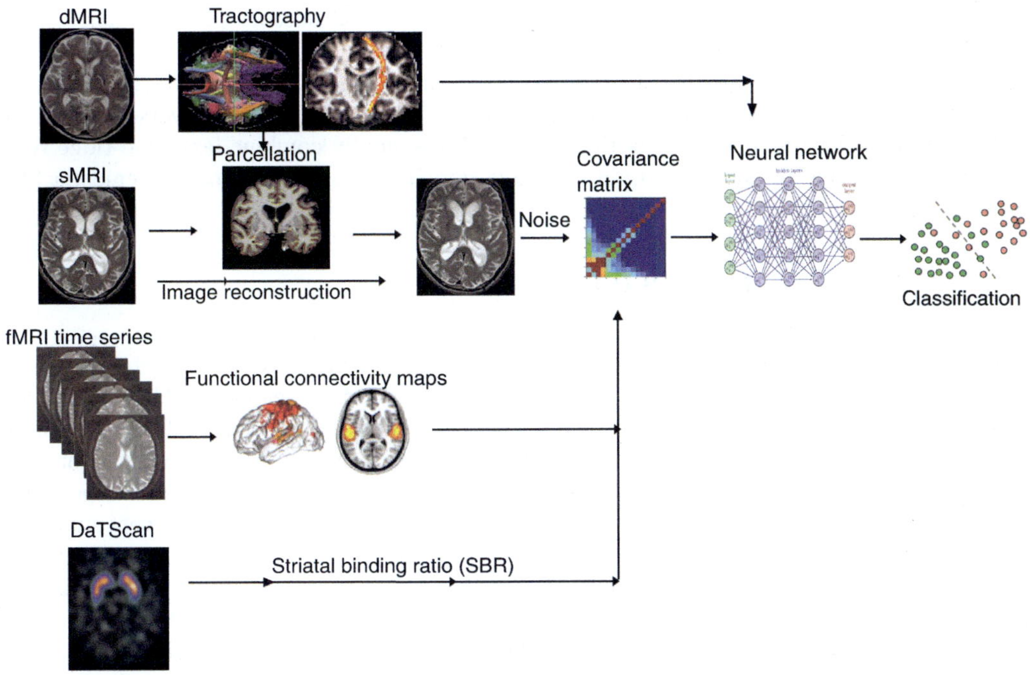

FIGURE 27.4 Recommended multimodal imaging and AI/ML-based model for PD diagnosis, staging, and severity. *AI*, Artificial intelligence; *ML*, machine learning; *PD*, Parkinson's disease.

taken as the activation function for binary classification. When developing transfer learning models, adding more convolution layers to the model will result in overfitting. To address this issue, inception introduced multiple filters of varying sizes. As a result, the model becomes wider rather than deeper, for example, Inception-V3 has 42 layers and steps such as factorized convolutions, smaller convolutions, asymmetric convolutions, auxiliary classifier, and grid size reduction.

Resnet is a residual network that was created to address the issue of vanishing gradients. Resnet is made up of a basic VGG19 connection, to which a short connection is added, and which has 3*3 filters, batch normalization, Relu as an activation function, and global average pooling. After these, one may add dense layers and dropout functions, and because binary classification is used, a final layer can be added with softmax as an activation function.

The input for the transfer learning model VGG16 is a 224*224 RGB-based image. Here, either a 1*1 or 3*3 filter is employed. Typical VGG16 model architectures use 16 convolutional layers and 3 fully connected layers. Usually, there are 5 to 10 pooling layers in this convolutional layer, but not all models do. Finally, fully connected layers are added to the model for hyperparameter tuning, and the final layer's dense 4096 Relu is typically chosen as the activation function for binary classification.

Point- and depth-wise convolutions make up the MobileNet architecture. Depth- and point-wise convolutions are both included in each depth-wise convolution. A MobileNet has 28 layers in total. The input values are 224*224*3. One can add fully connected layers and dense 4096 layers for hyperparameter tuning. In the final layer of softmax, binary classification is frequently performed using the Relu activation function.

AdaBoost (adaptive boosting) [40] is an ensemble of individual (and possibly weak) classifiers trained in sequence, such that the overall classifier is a strong one. The training is based on building on previous misclassifications. The method is the ML technique behind the Viola-Jones face detector module [41], the ubiquitous face detector function available on most smartphones.

XGBoost [42] stands for "eXtreme Gradient Boosting." It involves gradient-boosted decision trees, leading to a higher accuracy over a single tree, through penalization of those "unsuitable" for the task at hand.

In addition to DL models, linear regression models are also useful for the use of AI in medicine. All linear models posit that there is a direct and linear relationship between the underlying behavior/symptom and the variable being predicted. Even though linear models have drawbacks, especially in complex situations, their linearity makes it simple to monitor how a feature affects a prediction. In medicine, a clear understanding is frequently beneficial to examine the transparency of the underlying relationships between factors. For example, linear models have become the industry standard for understanding the impact of genomics on drug efficacy for the past several decades [43]. Similarly, AI can use large, diverse data sets to address numerous stages of drug development, such as high-throughput screening, identification of novel targets, or candidate pharmacological agents, by positioning existing compounds for novel indications, or understanding the risk factors associated with candidate compounds based on predicted adverse events using such linear causal models [44]. Some hybrid models like elastic nets (which essentially combine lasso and ridge regressions) have emerged as the model of choice in predicting mild cognitive impairment and Alzheimer's dementia [45], attributable mainly to their interpretability.

27.5 Conclusion and future directions

Building explainable, highly predictive, and robust ML models is crucial to the success of AI research in neurology and neurosciences, but reproducibility cannot be overemphasized. The interpretability and overall validity of predictive models must be ensured, and this requires code sharing and the release of public data sets like PPMI and Human Connectome. Complex Python ML libraries, like PyTorch [46] and scikit-learn [47], have revolutionized AI research for every interested researcher, and also made these models accessible to a wide range of researchers for use and understanding. Recent advances using AI-enabled detection and assessment of PD rely not only on data that can be easily gathered at a personal level [48] but also on utilizing the available biomarkers for early prediction and diagnosis [49]. In another recent study [50], researchers could streamline personalized treatment protocols by cross-referencing genetic variants of noncoding regions of brain-specific DNA and gene-expression libraries to identify potential sites for pharmacological intervention, followed by an analysis of protein and interactome data sets that led to the identification of potential drugs already available for repurposing in targeted therapies. In sum, it is evident that AI has made giant strides in every sphere of medicine, including PD diagnosis, and, in addition to a model's predictability function, is poised to leverage data and model diversity to advance our understanding of how features interact and impact the interpretability of these models.

References

[1] World Health Organization, Available from https://www.who.int/news-room/fact-sheets/detail/parkinson-disease, 2022 (Accessed 13.11.22).

[2] GBD 2016 Neurology Collaborators, Global, regional, and national burden of neurological disorders, 1990-2016: a systematic analysis for the global burden of disease study 2016, Lancet Neurol 18 (2019) 459–480. Available from https://doi.org/10.1016/S1474-4422(18)30499-X.

[3] World Health Organization, Parkinson Disease: A Public Health Approach. Technical Brief, World Health Organization, Geneva, (2022).

[4] B. Brakedal, L. Toker, K. Haugarvoll, et al., A nationwide study of the incidence, prevalence and mortality of Parkinson's disease in the Norwegian population, npj Parkinson's Disease 8 (2022) 19. Available from https://doi.org/10.1038/s41531-022-00280-4.

[5] A.H. Rajput, B. Rozdilsky, A. Rajput, Accuracy of clinical diagnosis in parkinsonism—a prospective study, Canadian Journal of Neurological Sciences 18 (3) (1991) 275–278.

[6] B.D. Lee, J.H. Shin, J. VanKempen, L. Petrucelli, A.B. West, H.S. Ko, et al., Inhibitors of leucine-rich repeat kinase-2 protect against models of Parkinson's disease, Nature Medicine 16 (9) (2010) 998–1000. Available from https://doi.org/10.1038/nm.2199.

[7] J.A. Obeso, M.C. Rodriguez-Oroz, C.G. Goetz, C. Marin, J.H. Kordower, M. Rodriguez, E.C. Hirsch, M. Farrer, A.H. Schapira, G. Halliday, Missing pieces in the Parkinson's disease puzzle, Nature Medicine 16 (6) (2010) 653–661. Available from https://doi.org/10.1038/nm.2165.

[8] S. Bido, S. Muggeo, L. Massimino, et al., Microglia-specific overexpression of α-synuclein leads to severe dopaminergic neurodegeneration by phagocytic exhaustion and oxidative toxicity, Nature Communications 12 (2021) 6237. Available from https://doi.org/10.1038/s41467-021-26519-x.

[9] L.M.A. Oliveira, T. Gasser, R. Edwards, et al., Alpha-synuclein research: defining strategic moves in the battle against Parkinson's disease, npj Parkinsons Disease 7 (2021) 65. Available from https://doi.org/10.1038/s41531-021-00203-9.

[10] Y. Yang, Y. Yuan, G. Zhang, et al., Artificial intelligence-enabled detection and assessment of Parkinson's disease using nocturnal breathing signals, Nature Medicine 28 (2022) 2207–2215. Available from https://doi.org/10.1038/s41591-022-01932-x.

[11] R. Prashanth, S. Dutta Roy, P.K. Mandal, S. Ghosh, High-accuracy detection of early Parkinson's disease through multimodal features and machine learning, International Journal of Medical Informatics 90 (2016) 13–21. Available from https://doi.org/10.1016/j.ijmedinf.2016.03.001.

[12] K. Brockmann, et al., SNCA: major genetic modifier of age at onset of Parkinson's disease, Movement Disorders 28 (2013) 1217–1221.

[13] A.H.V. Schapira, K.R. Chaudhuri, P. Jenner, Non-motor features of Parkinson disease, Nature Reviews Neuroscience 18 (7) (2017) 435–450. Available from https://doi.org/10.1038/nrn.2017.62.

[14] K. Stiasny-Kolster, Y. Doerr, J.C. Möller, H. Höffken, T.M. Behr, W.H. Oertel, G. Mayer, Combination of 'idiopathic' REM sleep behaviour disorder and olfactory dysfunction as possible indicator for α-synucleinopathy demonstrated by dopamine transporter FP-CIT-SPECT, Brain 128 (1) (2005) 126–137. Available from https://doi.org/10.1093/brain/awh322.

[15] L. Lonini, A. Dai, N. Shawen, T. Simuni, C. Poon, L. Shimanovich, M. Daeschler, R. Ghaffari, J.A. Rogers, A. Jayaraman, Wearable sensors for Parkinson's disease: which data are worth collecting for training symptom detection models, NPJ Digital Medicine 1 (2018) 64. Available from https://doi.org/10.1038/s41746-018-0071-z.

[16] F. Segovia, J.M. Górriz, J. Ramírez, I. Álvarez, J.M. Jiménez-Hoyuela, S.J. Ortega, Improved Parkinsonism diagnosis using a partial least squares based approach, Medical Physics 39 (7Part1) (2012) 4395–4403. Available from https://doi.org/10.1118/1.4730289.

[17] F. Segovia, I.A. Illán, J.M. Górriz, J. Ramírez, A. Rominger, J. Levin, Distinguishing Parkinson's disease from atypical parkinsonian syndromes using PET data and a computer system based on support vector machines and Bayesian networks, Frontiers in Computational Neuroscience 9 (2015) 137. Available from https://doi.org/10.3389/fncom.2015.00137.

[18] Y.J. Bae, J.-M. Kim, C.-H. Sohl, et al., Imaging the substantia nigra in Parkinson disease and other parkinsonian syndromes, Radiology 300 (2) (2021) 260–278. Available from https://doi.org/10.1148/radiol.2021203341.

[19] P.J. Basser, C. Pierpaoli, Microstructural and physiological features of tissues elucidated by quantitative-diffusion-tensor MRI, Journal of Magnetic Resonance, Series B 111 (1996) 209–219.

[20] C. Beaulieu, The basis of anisotropic water diffusion in the nervous system – a technical review, NMR in Biomedicine 15 (2002) 435–455.

[21] S. Lehéricy, E. Bardinet, C. Poupon, M. Vidailhet, C. François, 7 Tesla magnetic resonance imaging: a closer look at substantia nigra anatomy in Parkinson's disease, Movement Disorders 29 (13) (2014) 1574–1581. Available from https://doi.org/10.1002/mds.26043.

[22] K. Ray Chaudhuri, D.G. Healy, A.H.V. Schapira, Non-motor symptoms of Parkinson's disease: diagnosis and management, The Lancet Neurology 5 (3) (2006) 235–245. Available from https://doi.org/10.1016/S1474-4422(06)70373-8.

[23] F.J. Vingerhoets, M. Schulzer, D.B. Calne, B.J. Snow, Which clinical sign of Parkinson's disease best reflects the nigrostriatal lesion? Annals of Neurology 41 (1997) 58–64.

[24] E. Broussolle, C. Dentresangle, P. Landais, L. Garcia-Larrea, P. Pollak, B. Croisile, O. Hibert, F. Bonnefoi, G. Galy, J.C. Froment, D. Comar, The relation of putamen and caudate nucleus 18FDopa uptake to motor and cognitive performances in Parkinson's disease, Journal of the Neurological Sciences 166 (1999) 141–151.

[25] J.M. Fearnley, A.J. Lees, Ageing and Parkinson's disease: substantia nigra regional selectivity, Brain 114 (1991) 2283–2301.

[26] A. Brück, S. Aalto, E. Rauhala, J. Bergman, R. Marttila, J.O. Rinne, A follow-up study on 6-[18F]fluoro-L-dopa uptake in early Parkinson's disease shows nonlinear progression in the putamen, Movement Disorders 24 (2009) 1009–1015.

[27] N. Pavese, M. Rivero-Bosch, S.J. Lewis, A.L. Whone, D.J. Brooks, Progression of monoaminergic dysfunction in Parkinson's disease: a longitudinal (18) F-dopa PET studies, Neuroimage 6 (2011) 1463–1468.

[28] T.C. Booth, et al., The role of functional dopamine-transporter SPECT imaging in parkinsonian syndromes, part 2, American Journal of Neuroradiology 36 (2015) 236–244.

[29] N. Bajaj, R.A. Hauser, I.D. Grachev, Clinical utility of dopamine transporter single photon emission CT (DaT-SPECT) with (123I) isoflupane in diagnosis of parkinsonian syndromes, Journal of Neurology, Neurosurgery & Psychiatry 84 (2013) 1288–1295.

[30] J. Spiegel, et al., Striatal FP-CIT uptake differs in the subtypes of early Parkinson's disease, Journal of Neural Transmission 114 (2007) 331–335.

[31] H.T.S. Benamer, et al., Correlation of Parkinson's disease severity and duration with 123I-FP-CIT SPECT striatal uptake, Movement Disorders 15 (2000) 692–698.

[32] S.J. Son, M. Kim, H. Park, Imaging analysis of Parkinson's disease patients using SPECT and tractography, Scientific Reports 6 (2016) 38070. Available from https://doi.org/10.1038/srep38070.

[33] Parkinson's Progression Markers Initiative, The Parkinson's Progression Markers Initiative (PPMI) database (www.ppmi-info.org/access-data-specimens/download-data). For up-to-date information on the study, visit www.ppmi-info.org.

[34] R. Prashanth, S.D. Roy, P.K. Mandal, S. Ghosh, High accuracy classification of Parkinson's Disease through shape analysis and surface fitting in 123I-isoflupane SPECT imaging, IEEE Journal of Biomedical and Health Informatics 21 (3) (2017) 794–802. Available from https://doi.org/10.1109/JBHI.2016.2547901.

[35] R. Prashanth, S. Dutta Roy, P.K. Mandal, S. Ghosh, Automatic classification and prediction models for early Parkinson's disease diagnosis from SPECT imaging, Expert Systems with Applications 41 (7) (2014) 3333–3342. Available from https://doi.org/10.1016/j.eswa.2013.11.031.

[36] R. Prashanth, S. Dutta Roy, Novel and improved stage estimation in Parkinson's disease using clinical scales and machine learning, Neurocomputing 305 (2018) 78–103.

[37] A. Tellaeche Iglesias, M.Á. Campos Anaya, G. Pajares Martinsanz, I. Pastor-López, On combining convolutional autoencoders and support vector machines for fault detection in industrial textures, Sensors 21 (10) (2021) 3339. Available from https://doi.org/10.3390/s21103339.

[38] C. Szegedy, et al. Rethinking the inception architecture for computer vision, in: Proceedings of the IEEE Conference on Computer Vision and Pattern Recognition (CVPR), 2016, pp. 2818–26.

[39] A. Farahani, B. Pourshojae, K. Rasheed, H.R. Arabnia, A concise review of transfer learning, in: IEEE-International Conference on Computational Science and Computational Intelligence (CSCI), 2020, pp. 344–351.

[40] Y. Freund, R.E. Schapire, A decision-theoretic generalization of on-line learning and an application to boosting, Lecture Notes in Computer Science, Springer, Berlin Heidelberg, (1995), pp. 23–37.

[41] P. Viola, M.J. Jones, Robust real-time face detection, International Journal of Computer Vision 57 (2) (2004) 137–154.

[42] T. Chen, C. Guestrin, XGBoost: a scalable tree boosting system, in: B. Krishnapuram, M. Shah, A.J. Smola, C.C. Aggarwal, D. Shen, R. Rastogi, (Eds.), Proc. 22nd ACM SIGKDD International Conference on Knowledge Discovery and Data Mining, San Francisco, CA, USA, 13-17 August, 2016, ACM, 2016, pp. 785–794.

[43] P. Geeleher, et al., Clinical drug response can be predicted using baseline gene expression levels and in vitro drug sensitivity in cell lines, Genome Biology 15 (2014) R47.

[44] A. Gottlieb, et al., PREDICT: a method for inferring novel drug indications with application to personalized medicine, Molecular Systems Biology 7 (2011) 496.

[45] L. Shen, et al., Identifying neuroimaging and proteomic biomarkers for MCI and AD via the elastic net, in: Proceedings of the International Workshop on Multimodal Brain Image Analysis, 2011, pp. 27–34.
[46] A. Paszke, et al., Automatic differentiation in pytorch, in: NeurIPS 2017, Available from: https://pytorch.org/, 2017.
[47] F. Pedregosa, et al., Scikit-learn: machine learning in Python, Journal of Machine Learning Research 12 (2011) 2825–2830 source available at https://scikit-learn.org/stable/.
[48] Y. Yang, Y. Yuan, G. Zhang, et al., Artificial intelligence-enabled detection and assessment of Parkinson's disease using nocturnal breathing signals, Nature Medicine 28 (2022) 2207–2215. Available from https://doi.org/10.1038/s41591-022-01932-x.
[49] A. Reddy, R.P. Reddy, A.K. Roghani, et al., Artificial intelligence in Parkinson's disease: early detection and diagnostic advancements, Ageing Research Reviews 99 (2024) 102410. Available from https://doi.org/10.1016/j.arr.2024.102410.
[50] L. Dou, Z. Xu, J. Xu, et al., A network-based systems genetics framework identifies pathobiology and drug repurposing in Parkinson's disease, npj Parkinson's diseases 11 (1) (2025). Available from https://doi.org/10.1038/s41531-025-00870-y.

Chapter 28

Regulatory aspects of tissue engineering

Simran Yadav[1], Ashrit Nair[2], Vikesh Kumar Shukla[1], and Navneet Sharma[1]
[1]*Department of Pharmaceutical Regulatory Affairs, Amity Institute of Pharmacy, Amity University Noida, Noida, Uttar Pradesh, India,* [2]*Department of Textile and Fibre Engineering, Indian Institute of Technology, Delhi, India*

28.1 Introduction

A branch of regenerative medicines Tissue Eng. incorporates using engineering techniques, medical knowledge, and physiological principles to create new tissues that replicate the functions and properties of natural tissue [1]. These engineered tissues serve as biological substitutes, intended for implantation to restore, repair, or improve the function of organs in the body [1].

Tissue Eng. aims to create functional structures that help in restoring, maintaining, or enhancing damaged tissues or entire organs. Although FDA has approved engineered tissue products such as cartilage and artificial skin, their application remains limited currently [2,3].

The process of Tissue Eng. involves several critical steps to create new tissue. It begins with scaffolding, where a scaffold is constructed from various materials, such as proteins or plastics, to shape the new tissue [4]. Next, cells are extracted from a tissue source such as a donor organ and grown in culture. These cells are then seeded onto the scaffold that is subsequently incubated in a medium containing growth factor to stimulate cell growth and division [5]. Finally, the engineered tissue is implanted into the body. This field utilizes diverse materials, including human tissues, animal tissues, synthetic materials, or a combination of these to achieve its aims [6].

Regulatory compliance in tissue engineering is essential for ensuring patient safety, enabling market access, supporting international trade, and fostering public trust [7]. Regulatory agencies such as the FDA and EMA oversee the approval processes to safeguard public health by assessing the safety, effectiveness, and quality of tissue-engineered products (TEPs) [8]. Compliance is crucial for companies to legally market their TEPs noncompliance can lead to delays or denial of approval, which can affect revenue and innovation [9].

The regulatory environment for TEPs is intricate with varying requirements based on product type and jurisdiction, which can create confusion for developers, particularly those from academia or small businesses with limited regulatory expertise [10]. Furthermore, the lack of harmonization across regions means that many areas lack clear guidelines for TEPs hindering global development efforts. The rapid evolution of tissue engineering technologies often surpasses current regulations, resulting in gaps that complicate the approval process [11]. Navigating this pathway requires substantial resources, including time and financial investment, as companies must compile comprehensive documentation and conduct clinical trials [12]. Additionally, postmarket surveillance is required to ensure ongoing safety and efficacy after approval, placing further pressure on companies already burdened with initial compliance efforts [13].

In conclusion, while regulatory compliance is crucial for the safe and effective commercialization of TEPs, it presents significant challenges that stakeholders must address carefully to support innovation in this promising field [14].

28.2 Regulatory framework of tissue engineering

TEPs are regulated by different regulatory bodies in different jurisdictions. The regulations depend on the nation where the TEPs are intended to be marked [15]. A customized regulatory framework is essential for incorporating tissue engineering and regenerative medicine into clinical practice [16].

Fundamentals of Translational Tissue Engineering. DOI: https://doi.org/10.1016/B978-0-443-33455-9.00031-6

28.2.1 United States regulatory framework

In the United States, TEPs are regulated through various pathways established by the US FDA, including Biologics License Applications (BLA), 510(k) clearances for Class I and Class II devices, premarket approvals for Class III devices, or designations for human cells, tissues, and cellular and tissue-based products (HCT/Ps) [17]. The regulation of HCT/Ps is managed by the Center for Biologics Evaluation and Research (CBER) under 21 CFR 1271 (refer to Table 28.1) [18]. These products include tissues such as bone, skin, corneas, ligaments, tendons, dura mater, heart valves, hematopoietic stem/progenitor cells from peripheral and cord blood, oocytes, and semen [18].

The Office of Tissues and Advanced Therapies (OTAT), previously part of the FDA's CBER, was responsible for regulating a variety of cellular and genetic therapies [19]. To address the growing number of cell and gene therapy applications, OTAT was reorganized into the Office of Therapeutic Products (OTP) (Table 28.1).

28.2.1.1 The Office of Therapeutic Products' objectives are to [19]

1. Oversee and coordinate various programs while ensuring flexibility to accommodate current and future staff expansion.
2. Promote a more balanced distribution of workload.
3. Meet industry needs and regulatory obligations effectively.
4. Enhance proficiency in specialized areas.

28.2.2 European Union regulations

The European Union (EU) employs various regulatory approaches for tissue engineering products focusing primarily on the tissue engineering process rather than the product itself [20].

European Tissues and Cells Directive

The EU Cell and Tissue Directives 2004/23/EC (EUCTDs) were introduced in 2004 and aim to maintain high standards of safety and quality for human cells and tissues. These guidelines (Directive) cover the donation, collection, testing, processing, preservation, storage, and distribution of human tissues and cells [21].

This includes guidelines for procurement, quality control, Good Manufacturing Practice (GMP) compliance, and inspections of production facilities. The EU also follows a two-stage regulatory strategy designed to balance patient access with safety and quality [22].

28.2.3 Indian regulations

India regulates TEPs under the 1989 Rules for the manufacturing, importing, exporting, storing, and use of Hazardous Microorganisms, Genetically Engineered Organisms, or Cells, which were established as part of the Environment (Protection) Act, 1986.

The key regulatory bodies and guidelines in India responsible for overseeing and regulating TEPs include those established to ensure compliance with scientific, ethical, and safety standards for the development, approval, and commercialization of these advanced therapeutic innovations (Table 28.2; Fig. 28.1.)

TABLE 28.1 Title 21 CFR 1271.

Sub part	Interpretations
Subpart A	General Provisions
Subpart B	Procedures for Registration and Listing
Subpart C	Donor Eligibility
Subpart D	Current Good Tissue Practice
Subpart E	Additional Requirements for Establishments Described in § 1271.10
Subpart F	Inspection and Enforcement of Establishments Described in § 1271.10

TABLE 28.2 Key regulatory bodies and guidelines in India.

RB	Interpretation	References
ICMR	The ICMR has developed the National Guidelines for Stem Cell Research (2017), which categorizes stem cell research into permissible, restrictive, and prohibited areas. These guidelines mandate that any clinical use of stem cells must occur within approved clinical trials.	[23]
DBT	The DBT plays a crucial role in establishing standards for tissue culture and has implemented the National Certification System for Tissue Culture Raised Plants (NCS-TCP) to ensure compliance with safety protocols for tissue culture practices	[24]
CDSCO	The CDSCO oversees the approval process for stem cell therapies and tissue-engineered products. It requires that all clinical trials involving stem cells be registered and conducted under Good Manufacturing Practice (GMP) conditions	[25]
NAC-SCRT	National Apex Committee for Stem Cell Research and Therapy (NAC-SCRT). This committee is responsible for monitoring and overseeing stem cell research activities in India, ensuring adherence to ethical standards and regulatory compliance.	[26]

28.3 Key considerations for tissue-engineered products

Human-derived cells and tissues are not addressed by the current regulations, particularly in the Medical Devices Directive as it was initially assumed that existing laws governing tissue banking, blood products, and cell banking would be sufficient to manage them [27].

As a result, TEPs cannot undergo assessment for a CE (Conformité Europèene), which is a French phrase that translates to "European Conformity" mark, which would have otherwise offered a unified, streamlined pathway to market entry within the EU [28].

In response to the rise of TEPs and the lack of uniform regulations, the European Commission (EC) initiated efforts to address these gaps [29].

SANCO proposed a "Tissue Banking Directive," while ENTERPRISE explored harmonizing regulations for human-derived or TEPs [29].

A consultation in 2002 revealed the need for a distinct regulatory framework, with debates on whether it should be mutual recognition-based, centralized under existing agencies, or handled by a new body [29].

Industry representatives, such as EUCOMED, advocated for a single approval process and clearer distinctions between regulating final TEPs and cell sourcing, while raising concerns about potential overlaps and ethical restrictions.

As discussions progress, the EC appears to favor a flexible, risk-based approach over rigid regulations to accommodate the innovation and diversity in this field.

In India, the regulatory framework for TEPs is evolving, with oversight primarily by the CDSCO under the New Drugs and Clinical Trials Rules, 2019, often classifying them as biological products or drugs based on their composition. Additional guidance comes from the National Guidelines for Stem Cell Research (NGSCR) issued by ICMR and DBT, addressing ethical, scientific, and safety considerations, while the Medical Devices Rules, 2017, may apply to scaffold-based TEPs.

In the United States, TEPs are classified as advanced therapeutic products and are governed by a comprehensive regulatory framework to ensure their safety, efficacy, and quality. These products are regulated by the US FDA under different centers depending on their nature, including the CBER for biologics, the CDRH for scaffold-based TEPs, and the CDER for pharmacological products. The 21st Century Cures Act facilitates innovation in regenerative medicine through expedited pathways like the Regenerative Medicine Advanced Therapy (RMAT) designation, while combination products are overseen by the Office of Combination Products (OCP).

In Europe, TEPs are regulated as advanced therapy medicinal products (ATMPs) under the EMA to ensure their safety, efficacy, and quality. These products are classified under Regulation (EC) No. 1394/2007 and include somatic-cell therapy, gene therapy, and TEPs, with oversight provided by the EMA's Committee for Advanced Therapies (CAT). While most TEPs require centralized EMA approval, some may fall under the hospital exemption rule, allowing their use for specific patients within hospitals without centralized authorization, provided they meet safety standards.

Compliance with GMP is mandatory for every region to ensure consistent quality and sterility with stringent requirements for source material traceability and scaffold standards.

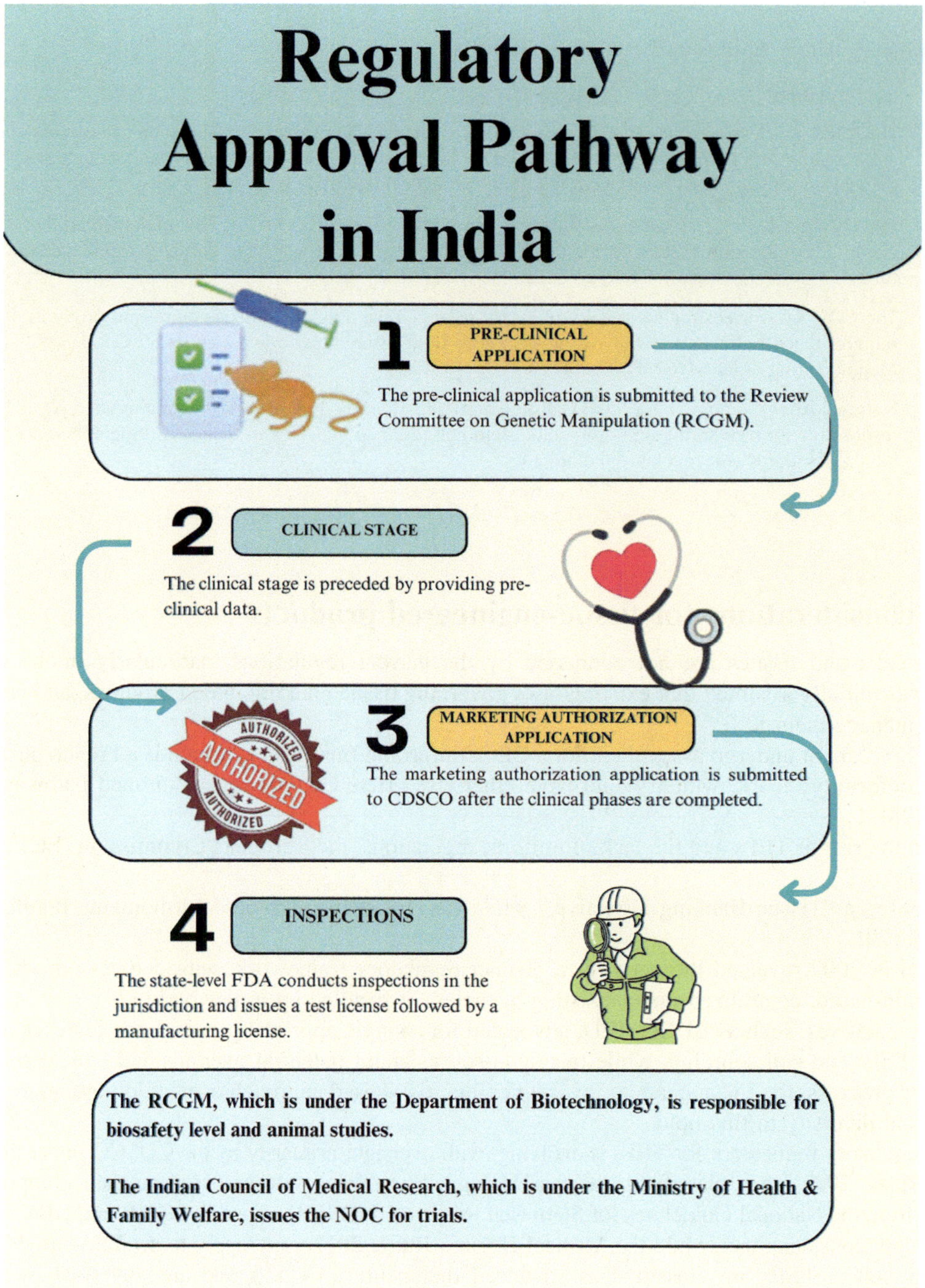

FIGURE 28.1 A flowchart depicting the regulatory approval pathway for TEPs in India. Streamlined steps: regulatory approval pathway for therapeutics in India.

28.4 Other approval pathways

28.4.1 IND approval

RMAT designation for tissue engineering products in the United States can be requested through an Investigational New Drug (IND) application. The request may be made as an adjustment to an already-existing IND or at the time the IND is submitted. Within sixty days, the sponsor will receive notification of the RMAT designation from the OTAT [30].

28.4.2 HCT/P designation

Products that are meant for homologous use and have undergone minimal manipulation might be eligible for regulation under HCT/P (human cells, tissues, and cellular and tissue-based products) criteria without needing premarket approval. However, a product must go through a more stringent process, such as Biologics License Application (BLA) or Premarket Approval (PMA) if it is meant for nonhomologous usage or is subjected to substantial alteration [31].

28.4.3 510 (k) pathway

Certain medical devices that can show significant equivalency to current products fall under this approach. It is frequently used for less complicated or risky TEPs [32].

28.5 Quality control, ethical, and intellectual property considerations in tissue-engineered products

The development and commercialization of TEPs require careful attention to quality control, ethical considerations, and intellectual property rights. Each of these areas plays a crucial role in ensuring that TEPs are safe, effective, and ethically developed while also protecting the innovations behind them [33].

28.5.1 Quality control in tissue engineering

Quality control (QC) plays a crucial role in guaranteeing the safety and effectiveness of TEPs (Table 28.3).

28.5.2 Ethical considerations in tissue engineering

Ethical issues in tissue engineering are multifaceted and must be addressed at various stages of product development (Table 28.4).

28.5.3 Intellectual property considerations

Intellectual property protection is vital for fostering innovation in tissue engineering (Table 28.5).

28.6 Discussion

Tissue engineering, a cornerstone of regenerative medicine, focuses on creating functional biological substitutes to restore, repair, or enhance damaged tissues or organs. The process integrates scaffold manufacturing, cell culture expansion, and implantation, aiming to replicate the functionality of natural tissues. Despite breakthroughs such as FDA-approved artificial skin and cartilage, regulatory challenges limit broader applications. Global frameworks like those of the FDA, EMA, and India's ICMR emphasize safety, efficacy, and ethical considerations, yet regional disparities and evolving technologies create significant hurdles. Harmonization of regulatory standards, robust quality control through GMP, and

TABLE 28.3 Key aspects for quality control in tissue engineering.

Interpretation	Aspect	References
Good Manufacturing Practices (GMP)	Compliance with GMP guidelines is critical in the production of tissue-engineered products. This includes establishing a well-designed quality assurance and quality control program that adheres to standard operating procedures (SOPs) to ensure consistent manufacturing processes.	[34]
Analytical Techniques	Various analytical methods are employed to assess the quality of engineered tissues at different stages of development. These techniques can be categorized into destructive methods (which destroy samples) and nondestructive methods, allowing for ongoing assessment without compromising the product.	[16]

TABLE 28.4 Ethical consideration in tissue engineering.

Aspects	Interpretation	References
Informed Consent	Obtaining informed consent from donors is paramount, especially when human tissues or cells are involved. This ensures that donors understand the implications of their contributions.	[34]
Animal Welfare	Ethical treatment of animals used in research is a significant concern. The use of animal models must be justified, and alternatives should be considered whenever possible.	[16]
Societal Impact	The broader societal implications of tissue engineering technologies must be considered, including public perceptions and potential disparities in access to these therapies. Engaging with the public and stakeholders can help address these concerns.	[35]
Regulatory Compliance	Adhering to regulatory frameworks not only ensures safety but also reflects ethical responsibility in the development and clinical application of tissue-engineered products.	[36]

TABLE 28.5 Key aspects for intellectual property considerations in tissue engineering.

Aspects	Interpretation	References
Patent Strategies	Effective patenting strategies are essential for protecting novel inventions related to tissue engineering. This includes understanding the types of patents available (e.g., utility patents and design patents) and ensuring that innovations meet criteria such as novelty and industrial applicability.	[37]
Commercialization Challenges	Translating research into marketable products often involves navigating complex IP landscapes. Companies must consider existing patents and potential infringement issues while seeking to secure their own IP rights.	[37]
Market Trends	The landscape of patents in Tissue Eng. has evolved significantly over the past decade, indicating growing commercial interest and investment in this field. However, the number of scientific publications continues to outpace patent filings, suggesting a gap between research output and commercial application.	[38]

adherence to ethical principles like informed consent are critical to advancing TEPs. Intellectual property protection and strategic collaboration between researchers, regulators, and industry stakeholders are essential to overcoming commercialization challenges, fostering innovation, and ensuring equitable access to transformative therapies.

28.7 Conclusion

Tissue engineering holds transformative potential for regenerative medicine by enabling the creation of biological substitutes that can restore or enhance the functions of damaged organs and tissues. This field encompasses complex processes involving scaffolding, cell culture, and growth stimulation to produce tissues that closely replicate natural structures. However, the journey from development to clinical application is fraught with challenges, particularly in regulatory compliance, quality control, ethical considerations, and intellectual property protection.

Regulatory frameworks in India, the United States, and the European Union vary significantly, each with distinct guidelines for ensuring the safety, quality, and efficacy of TEPs. The lack of global harmonization adds complexity for developers, who must navigate differing requirements and approval processes across regions. QC and adherence to GMP are essential to ensure product consistency and safety, while ethical considerations, such as informed consent and animal welfare, are fundamental to maintaining public trust. Intellectual property protection further supports innovation by safeguarding novel inventions, although the IP landscape in tissue engineering remains challenging due to potential patent conflicts and complex commercialization pathways.

In summary, while the field of tissue engineering is posed to significantly impact healthcare by providing novel therapeutic options, ongoing collaboration among researchers, regulators, and industry stakeholders is crucial to address regulatory, ethical, and commercial challenges. Harmonizing global regulatory standards, strengthening ethical oversight, and fostering innovation through IP protection will be essential to fully realize the potential of tissue engineering in clinical practices [1,39–44].

AI disclosure

During the preparation of this work, the author(s) used Grammarly and ChatGPT to check grammar and spelling. After using this tool/service, the author(s) reviewed and edited the content as needed and take(s) full responsibility for the content of the publication.

References

[1] A. Ostadfar, Tissue Engineering of Cardiovascular System, Elsevier BV, (2016), pp. 323–339. Available from http://doi.org/10.1016/b978-0-12-802408-9.00009-0.

[2] M.M. Farag, Recent trends on biomaterials for tissue regeneration applications: review, Egypt Journal of Materials Science 58 (2) (2023) 527–558. Available from https://doi.org/10.1007/s10853-022-08102-x, https://www.springer.com/journal/10853.

[3] D. Williams, Benefit and risk in tissue engineering, Materials Today 7 (5) (2004) 24–29. Available from https://doi.org/10.1016/S1369-7021(04)00232-9, http://www.journals.elsevier.com/materials-today/.

[4] A. Eltom, G. Zhong, A. Muhammad, Scaffold techniques and designs in tissue engineering functions and purposes: a review, Advances in Materials Science and Engineering 2019 (2019) 1–13. Available from https://doi.org/10.1155/2019/3429527.

[5] X. Zhang, X. Chen, H. Hong, R. Hu, J. Liu, C. Liu, Decellularized extracellular matrix scaffolds: recent trends and emerging strategies in tissue engineering, KeAi Communications Co., China Bioactive Materials 10 (2022) 15–31. Available from https://doi.org/10.1016/j.bioactmat.2021.09.014, http://www.keaipublishing.com/en/journals/bioactive-materials/.

[6] S. Liu, J.M. Yu, Y.C. Gan, X.Z. Qiu, Z.C. Gao, H. Wang, et al., Biomimetic natural biomaterials for tissue engineering and regenerative medicine: new biosynthesis methods, recent advances, and emerging applications, Military Medical Research 10 (1) (2023). Available from https://doi.org/10.1186/s40779-023-00448-w, https://mmrjournal.biomedcentral.com/.

[7] A. Sharma, V. Gamta, G. Luthra, Ensuring patient safety and trust: the critical importance of regulatory compliance in healthcare, Journal of Pharmaceutical Research International 35 (18) (2023) 1–15. Available from https://doi.org/10.9734/jpri/2023/v35i187390.

[8] C. Iglesias-Lopez, A. Agustí, M. Obach, A. Vallano, Regulatory framework for advanced therapy medicinal products in Europe and United States, Frontiers in Pharmacology 10 (JULY) (2019). Available from https://doi.org/10.3389/fphar.2019.00921.

[9] S. Gupta, P.S. Kushwaha, U. Badhera, P. Chatterjee, E.D.R.S. Gonzalez, Identification of benefits, challenges, and pathways in E-commerce industries: an integrated two-phase decision-making model, Sustainable Operations and Computers 4 (2023) 200–218. Available from https://doi.org/10.1016/j.susoc.2023.08.005, https://www.sciencedirect.com/science/journal/26664127.

[10] D. Dinçkol, P. Ozcan, M. Zachariadis, Regulatory standards and consequences for industry architecture: the case of UK Open Banking, Research Policy 52 (6) (2023). Available from https://doi.org/10.1016/j.respol.2023.104760, https://www.journals.elsevier.com/research-policy.

[11] P.C. Johnson, T.A. Bertram, B. Tawil, K.B. Hellman, Hurdles in tissue engineering/regenerative medicine product commercialization: a survey of North American Academia and industry, Tissue Engineering - Part A 17 (1-2) (2011) 5–15. Available from https://doi.org/10.1089/ten.tea.2010.0411, http://www.liebertonline.com/tea.

[12] A.E. Lottes, K.J. Cavanaugh, Y.Y.F. Chan, V.J. Devlin, C.J. Goergen, R. Jean, et al., Navigating the regulatory pathway for medical devices—a conversation with the FDA, clinicians, researchers, and industry experts, Journal of Cardiovascular Translational Research 15 (5) (2022) 927–943. Available from https://doi.org/10.1007/s12265-022-10232-1, https://www.springer.com/journal/12265.

[13] D.B. Kramer, Y.T. Tan, C. Sato, A.S. Kesselheim, Ensuring medical device effectiveness and safety: a cross-national comparison of approaches to regulation, Food and Drug Law Journal 69 (1) (2014) 1–23.

[14] S. Yusuf, E. Hay, E. Sansoucy, The commercialization of tissue engineered medical products: challenges and recommendations, World Journal of Advanced Research and Reviews 23 (2) (2024) 1203–1212. Available from https://doi.org/10.30574/wjarr.2024.23.2.2458.

[15] L. Chen, C. Wang, T. Xi, Regulation challenge of tissue engineering and regenerative medicine in China, Burns and Trauma 1 (2) (2013) 56–62. Available from https://doi.org/10.4103/2321-3868.118927, https://academic.oup.com/burnstrauma.

[16] C.V. Oberweis, J.A. Marchal, E. López-Ruiz, P. Gálvez-Martín, A worldwide overview of regulatory frameworks for tissue-based products, Tissue Engineering - Part B: Reviews 26 (2) (2020) 181–196. Available from https://doi.org/10.1089/ten.teb.2019.0315, http://www.liebertonline.com/teb.

[17] K. Belsky, J. Smiell, Navigating the regulatory pathways and requirements for tissue-engineered products in the treatment of burns in the United States, Journal of Burn Care and Research 42 (4) (2021) 774–784. Available from https://doi.org/10.1093/jbcr/iraa210, http://journals.lww.com/burncareresearch.

[18] Tissue & Tissue Products | FDA, 2024.
[19] Establishment of the Office of Therapeutic Products | FDA, 2024.
[20] M. Lloyd-Evans, Regulating tissue engineering, Materials Today 7 (5) (2004) 48–55. Available from https://doi.org/10.1016/S1369-7021(04)00235-4, http://www.journals.elsevier.com/materials-today/.
[21] Tissues and cells - European Commission. 2024.
[22] Guidance on good manufacturing practice and good distribution practice: Questions and answers | European Medicines Agency (EMA), 2024.
[23] Home | Indian Council of Medical Research | Government of India, 2024.
[24] National Certification System for Tissue Culture Raised Plants (NCS-TCP) | Department of Biotechnology, 2024.
[25] CDSCO, 2024.
[26] National Apex Committee for Stem Cell Research and Therapy (NAC-SCRT) | Department of Health Research | MoHFW | Government of India, 2024.
[27] Press Information Bureau Government of India Ministry of Environment, Forest and Climate Change, 2022.
[28] E.R.A. Begum, K. Subachitra, C. Kirthika, S.P. Angayarkanni, K. Bhavani, R. Shenbagarathai, Production and characterization of bead-free β-chitosan/poly(vinyl alcohol) nanofibers by electrospinning for tissue engineering, Regenerative Engineering and Translational Medicine 10 (1) (2024) 46–55. Available from https://doi.org/10.1007/s40883-023-00302-4, https://www.springer.com/journal/40883.
[29] V. Ahuja, Regulation of emerging gene technologies in India, BMC Proceedings 12 (2018). Available from https://doi.org/10.1186/s12919-018-0106-0, http://www.biomedcentral.com/bmcproc/.
[30] N. Cuende, A. Vilarrodona, E. Vuelta, R. Marazuela, C. Herrera, S. Querol, et al., Addressing risks derived from the commodification of substances of human origin: a european proposal applicable worldwide, Transplantation 107 (4) (2023) 867–877. Available from https://doi.org/10.1097/TP.0000000000004527, journals.lww.com/transplantjournal/pages/default.aspx.
[31] CE marking - obtaining the certificate, EU requirements - Your Europe, 2024.
[32] L. Brévignon-Dodin, F. Livesey, Regulation of tissue-engineered products in the European Union: where are we heading? Regenerative Medicine 1 (5) (2006) 709–714. Available from https://doi.org/10.2217/17460751.1.5.709.
[33] J.P. Pirnay, A. Vanderkelen, D. De Vos, J.P. Draye, T. Rose, C. Ceulemans, et al., Business oriented EU human cell and tissue product legislation will adversely impact Member States' health care systems, Cell and Tissue Banking 14 (4) (2013) 525–560. Available from https://doi.org/10.1007/s10561-013-9397-6.
[34] A.F.J. De Kanter, K.R. Jongsma, M.C. Verhaar, A.L. Bredenoord, The ethical implications of tissue engineering for regenerative purposes: a systematic review, Tissue Engineering - Part B: Reviews 29 (2) (2023) 167–187. Available from https://doi.org/10.1089/ten.teb.2022.0033, http://www.liebertonline.com/teb.
[35] E.D. Ferrill, K. Rajan, 3D bioprinting, nanotechnology, and intellectual property, 3D Bioprinting and Nanotechnology in Tissue Engineering and Regenerative Medicine, Elsevier, United States, (2022), pp. 505–522. Available from http://doi.org/10.1016/B978-0-12-824552-1.00006-2, https://www.sciencedirect.com/book/9780128245521.
[36] D. Hernández-Melchor, E. López-Bayghen, A. Padilla-Viveros, The patent landscape in the field of stem cell therapy: closing the gap between research and clinic, F1000Research 11 (2022) 997. Available from https://doi.org/10.12688/f1000research.123799.4.
[37] P. Walton, A. Pérez-Blanco, S. Beed, A. Glazier, D. Ferreira Salomao Pontes, J. Kingdon, et al., Organ and tissue donation consent model and intent to donate registries: recommendations from an international consensus forum, Transplantation Direct 9 (5) (2023) E1416. Available from https://doi.org/10.1097/TXD.0000000000001416, journals.lww.com/transplantationdirect/pages/default.aspx.
[38] A.K. Kiani, D. Pheby, G. Henehan, R. Brown, P. Sieving, P. Sykora, et al., Ethical considerations regarding animal experimentation, Journal of Preventive Medicine and Hygiene 63 (2) (2022) E255. Available from https://doi.org/10.15167/2421-4248/jpmh2022.63.2S3.2768.
[39] Regenerative Medicine Advanced Therapy Designation | FDA, 2024.
[40] Questions and Answers Regarding the End of the Compliance and Enforcement Policy for Certain Human Cells, Tissues, or Cellular or Tissue-based Products (HCT/Ps) | FDA, 2024.
[41] Premarket Notification 510(k) | FDA, 2024.
[42] K.G.M. Brockbank, Tissue Engineering Constructs and Commercialization (2013).
[43] W.T. Seet, M.A. Mat Afandi, M.F. Ishak, M.N.F. Hassan, N. Ahmat, M.H. Ng, et al., Quality management overview for the production of a tissue-engineered human skin substitute in Malaysia, Stem Cell Research and Therapy 14 (1) (2023). Available from https://doi.org/10.1186/s13287-023-03536-9, https://stemcellres.biomedcentral.com.
[44] M. Islam, S. Bijjahalli, T. Fahey, A. Gardi, R. Sabatini, D.W. Lamb, Destructive and non-destructive measurement approaches and the application of AI models in precision agriculture: a review, Precision Agriculture 25 (3) (2024) 1127–1180. Available from https://doi.org/10.1007/s11119-024-10112-5, https://www.springer.com/journal/11119.

Chapter 29

Regulatory pathways for stem cell and tissue engineering in cancer therapy

Gauri Gaur, and Dhruv Kumar
Department of Biotechnology, School of Health Sciences and Technology, UPES, Dehradun, Uttarakhand, India

29.1 Introduction

All over the world, cancer is becoming one of the main causes of premature death [1]. The World Health Organization reports that, with 9.6 million deaths, cancer was the second leading cause of death worldwide in 2018. An estimated USD 25.2 trillion will be spent on cancer worldwide between 2020 and 2050 [2]. The leading cause of cancer-related deaths is still metastasis [3]. According to a 2022 study, 623,405 Americans were estimated to have metastatic malignancies or metastatic melanoma; by 2025, that figure is predicted to rise to 693,452 [4]. According to a 2007 study, people with metastatic bone disorders bear an economic burden of USD 12.6 billion. According to the National Institutes of Health, this figure amounts to 17% of the overall medical burden [5]. According to a more recent study from 2022, the burden of bone metastases has risen to 20% of total oncology expenses [6]. All survivors of metastatic cancer have a concerning life expectancy of less than five years, with the exception of melanoma [4]. Often, palliative care is the sole option for metastases. For instance, the treatment for bone metastases can frequently be palliative and involves medications like denosumab and bisphosphonates that stabilize skeletal abnormalities [7,8].

Tumor heterogeneity, therapeutic resistance, and severe off-target effects are only a few of the many challenges of cancer therapy. Tumor heterogeneity is characterized by genetic, epigenetic, and phenotypic diversity, which makes it harder to treat since it promotes subpopulations of cancer cells that prevent their destruction and lead to recurrence and a poor prognosis [9–11]. Metastatic cancer remains one of the main causes of cancer fatalities. Its resistance to such conventional treatments as chemotherapy and radiation makes it even more difficult to realize a long-term cure [12,13]. These challenges underscore the urgency for innovative and personalized therapy strategies that can cope with the heterogeneity of primary tumors and their microenvironment.

In oncology, stem cell therapy is an emerging approach that has the potential to overcome these challenges. In this context, MSCs, which possess intrinsic tumor-tropic properties, can be used to deliver various therapeutic substances such as cytokines, nanoparticles, and oncolytic viruses precisely to the site of the tumor [14]. By allowing the creation of tumor models and immunological treatments customized to the individual patient, induced pluripotent stem cells (iPSCs) extend the limits of personalized treatment even further [15]. Targeting cancer stem cells (CSCs), which are significant contributors to the formation, metastasis, and relapse of tumors, has also been highly focused. The disruption of CSC habitats has been found to have promising therapeutic potential [16].

Tissue engineering enhances these developments by offering three-dimensional (3D) tumor models and biomimetic scaffolds that replicate the structural and biochemical complexity of the tumor microenvironment (TME). These designed systems enhance medication delivery methods, aid in drug development, and track the course of tumors. For example, bioengineered scaffolds for chemotherapeutic controlled release have shown improved accuracy and decreased systemic toxicity [17–19]. Organoids derived from patients, which mirror the particular tumor biology of the patient, enable the testing of individualized treatment strategies and associate preclinical models with clinical deployment [20].

Deep insight into regulatory mechanisms underscores the efficiency of such innovative treatments. Such important pathways include Wnt/β-catenin, Notch, Hedgehog, and hypoxia-inducible Factors (HIFs), which are important for the control of stem cell function, tumor progression, and evasion of immunity. These pathways also provide an opportunity for therapeutic modulation, particularly in conjunction with immune checkpoint blockade to enhance antitumor immune

Fundamentals of Translational Tissue Engineering. DOI: https://doi.org/10.1016/B978-0-443-33455-9.00028-6

responses [21]. All these new methods provide an in-depth outline that can mitigate all the side effects of the old treatments and maximize the response rate in patients affected with cancer.

This chapter delves into the challenges of current cancer therapies, the emerging roles of stem cells and tissue engineering, and the regulatory pathways that underpin these approaches. By exploring these revolutionary technologies, the chapter attempts to provide an encompassing understanding of their potential in rectifying conventional therapies' pitfalls and redrawing the cancer therapy map.

29.2 Stem cell applications in cancer therapy

29.2.1 Types of stem cells

Stem cells are different from other sources and are used in antitumor therapy according to their capacity to proliferate, migrate, and differentiate.

29.2.1.1 Pluripotent stem cells

Embryonic stem cells (ESCs) obtained from the undifferentiated inner cell mass of the embryo can develop into all cell types, excluding those of the placenta. Nonetheless, the use of ESCs in clinical trials is limited because of ethical concerns. The discovery of Yamanaka factors in 2006, which allows for the induction of pluripotent stem cells (iPSCs) from cultured somatic cells, signified a significant advancement in cellular biology [22]. While avoiding the ethical issues associated with embryo death, these iPSCs have the same traits as ESCs. As of right now, hESCs and iPSCs are both significant sources for producing anticancer vaccines [23,24] and inducing effector T- and NK cells [25–29].

29.2.1.2 Adult stem cells

Adult stem cells (ASCs) have the potential to develop into various specialized cell types of tissues and organs. Within this category, hematopoietic stem cells (HSCs), mesenchymal stem cells (MSCs), and neural stem cells (NSCs) are frequently employed in the treatment of cancer. All adult blood cells in the body can be formed by HSCs, which are found in bone marrow. So far, the administration of HSCs obtained from cord blood is the sole stem cell procedure that has received FDA approval for treating multiple myeloma, leukemia, and certain types of blood disorders [30]. MSCs are found in many different tissues and organs and play important roles in tissue regeneration and repair. They have the ability to proliferate quickly in vitro and generate a range of specialized cell types, such as osteocytes, adipocytes, and chondrocytes. Due to their distinct biological characteristics, MSCs have been used extensively to treat a variety of cancers by delivering therapeutic medications or assisting other therapies [31,32]. Originally discovered in the central nervous system, NSCs have the ability to self-renew and generate new neurons and glial cells. In mouse models, they have undergone extensive testing to treat primary and metastatic prostate, lung, and breast malignancies [33–35].

29.2.1.3 Cancer stem cells

Stem-like cells, immature progenitors of tumor cells, or tumor-initiating cells are other names for CSCs, which are produced by normal stem cells or precursor/progenitor cells with epigenetic modifications. CSCs are present in tumor tissues and are crucial for the development, spread, and recurrence of cancer [36]. Consequently, there may be hope for treating different kinds of solid tumors by focusing on CSCs.

29.2.2 Mechanisms of action

29.2.2.1 Tumor homing

Targeting tumors enables MSCs to complexly integrate into the tumor edges by migrating to the desired area, which is fueled by biochemical and molecular interdependence between targets. Multipotent stromal cells, which are capable of migrating to injured locations during an inflammatory response, are zapped to faulty tissues in the same way that they relocate to tumors. Chemokine gradients are of vital significance in this. Chemoattractants, including CXCL12 (SDF-1), IL-6, and CCL-25, are secreted by tumor cells and immune constituents and bind to MSCs' CXCR4 and CCR4 receptors, bringing them to the location of the tumor. Moreover, the conditions of hypoxia existing in the tumor's microenvironment increase the expression of these chemokines, which strengthen the previously established homing signal [37,38].

Tumor vasculature adhesion molecules also participate in ensuring that MSCs firmly attach to endothelial cells. Integrins or VLA-4 and CD44 found on MSCs bind to their comparatives found on the cancerous tumor's blood vessels to facilitate VCAM-1 molecules and hyaluronic acid, which enables the MSCs to tether, roll, and then transmigrate into the stroma of the tumor [37,38]. Matrix MMP-2 and 9, which are parts of the matrix metalloproteinases family, are involved in breaking down the extracellular matrix, and then MSCs infiltrate the cancerous tumor to enable deeper penetration.

After penetrating the tumor tissue, MMPs, especially MMP-2 and MMP-9, break down components of the extracellular matrix and allow deeper invasion into the tissue. These proteolytic enzymes further remodel the extracellular matrix and make the TME more permissive for cell migration [37].

The TME is an active site of inflammation, hypoxia, and oxidative stress, which synergistically attracts MSCs. Tumor-associated fibroblasts, immune cells, and cancer cells release pro-inflammatory cytokines such as TNF-α and IL-1β, which increase the recruitment of MSCs. HIFs increase the expression of vascular endothelial growth factor (VEGF), thus promoting neovascularization and MSC homing [36,39].

Despite the fact that MSCs can target tumors, which offers therapeutic potential for targeted drug delivery, there are risks. MSCs may enhance tumor growth through immunosuppressive effects or by augmenting angiogenesis. However, knowledge of the exact molecular mechanisms behind tumor homing provides avenues to engineer MSCs as efficient delivery vehicles for anticancer therapies with minimal adverse effects [38,39].

29.2.2.2 Signaling in cancer stem cell

Tumor development and the complicated microenvironment. CSCs are produced when normal stem cells or precursor/progenitor cells have epigenetic alterations. These cells are essential for the stemness, initiation, maintenance, and metastasis of tumors. Cancer-associated immune cells (T-cells, macrophages (MΦ), dendritic cells (DC), cancer-associated fibroblasts (CAF), and other cells) are present in the tumor niche in addition to tumor cells and CSCs. Furthermore, all of those cells secrete chemoattractant factors (CXCL16, SDF-1, CCL-25, IL-6, TNF-α, and IL-1β) that attract MSCs to tumors and play important roles in the growth of tumors (Fig. 29.1).

29.2.2.3 Immunomodulation

MSCs are becoming the most significant immune modulators in conditions like autoimmune diseases, sepsis, and transplant rejection. MSC-based immune modulation is achieved through a complex interplay of cytokine actions, as well as direct cell-to-cell interactions.

Among the most crucial immunomodulatory functions that MSCs can fulfill is through bioactive molecule secretion, that is, cytokines. Among them, IL-10 and TGF-β are central in the establishment of the antiinflammatory environment. IL-10 drives monocytes and macrophages to a regulatory type 2 phenotype. This will have an effect of decreasing pro-inflammatory cytokines production and promoting Tregs, both effects favoring an immune-tolerant state [40,41].

Furthermore, MSCs further suppress the migration and maturation of DCs, thus decreasing the antigen-presenting and T-cell-stimulating activity of dendritic cells. Such a maneuver is critical because overactivation of the immune system can

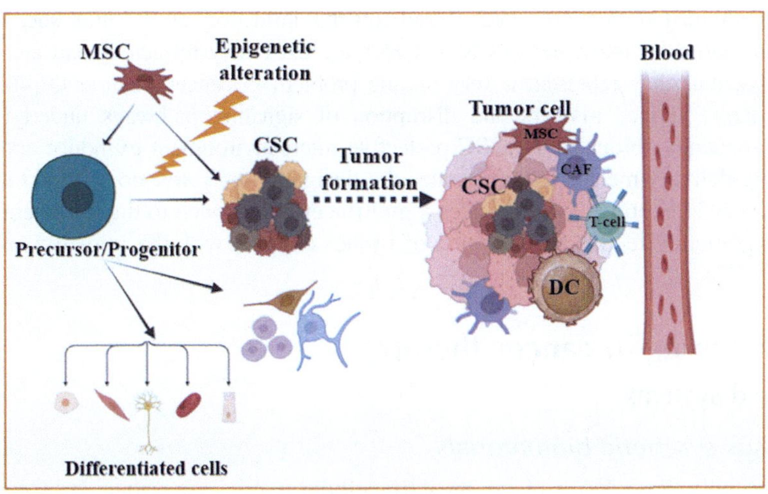

FIGURE 29.1 Mechanism of CSC signaling (T-cells, MΦ, DC, CAF, and other cells). *CAF*, Cancer-associated fibroblasts; *CSC*, cancer stem cell; *DC*, dendritic cells; *MΦ*, macrophages.

lead to damage to tissue [42,43]. In particular, the inhibition of DC function by MSCs is usually mediated through mechanisms involving IL-6 and other inhibitory factors produced by MSCs [43].

The immunomodulatory potential of MSCs has also been found to depend on the activation status. For example, when exposed to inflammatory signals such as IFN-γ and TNF-α, MSCs can upregulate their expression of immunomodulatory factors, including prostaglandin E2 (PGE2) and indoleamine 2,3-dioxygenase (IDO), therefore, augmenting their immune-suppressive effects [41].

A new finding is that both viable and apoptotic MSCs are immunomodulatory, and new therapeutic approaches based on both reduced risks of these effects of MSCs are introduced [40]. When macrophages ingest MSCs, it triggers a change in the macrophage's phenotype, promoting an immunosuppressive environment and facilitating the differentiation of regulatory T cells (Tregs). This interaction also supports the antiinflammatory status needed to translate into therapeutic benefits [40].

Even though the body of evidence for the immunomodulatory potential of MSCs has been gaining momentum, there is still a challenge posed to their clinical use. Evidence-based treatment of MSCs is demonstrated through studies shown to be associated with small numbers of patients, highlighting the necessity of larger and well-designed outcome trials to buttress these observations [41]. Further, variations in the sources of MSCs, preparation methods, and delivery techniques complicate the evaluation of clinical outcomes, making it necessary to establish standardized protocols to improve consistency and facilitate cross-study comparisons [40,41]. MSCs exert a multifactorial therapeutic effect in the regulation of immune responses. Therefore it is extremely important to clarify the cellular and molecular interactions involved in controlling MSC immunomodulation to maximize the clinical translation of MSCs, enhance their efficacy, and minimize their possible risks [40,41].

29.2.2.4 Delivery of therapeutic agents

The delivery system of therapeutic agents is based on the development of biomaterials that mimic the ECM. This imitation is of extreme importance as it provides the cell with the structural and biochemical foundation for growth, differentiation, and migration.

Scaffolds are designed not only to provide mechanical support but also to allow the controlled release of bioactive factors, with a particular emphasis on GFs. MSCs and adipose-derived stem cells (ASCs) can repair and regenerate tissues through their paracrine signaling and their ability to develop into numerous kinds of cells [44]. Moreover, by embedding GFs into these scaffolds, researchers can control cell behavior, which might result in improving treatments involving MSCs and pluripotent stem cells (PSCs) with an unmatched ability to self-renew and differentiate into a wide range of tissues [45].

Drug-delivery systems integrated into these scaffolds, which are accessible by continuous, on-off, and triggered release modes, may be used to control the timing and location of the therapeutic agents. This is especially true for biodegradable scaffolds, which slowly release their constituents as they create new tissue, therefore preserving the maximum degree of bioactivity in a dynamic healing microenvironment [45,46].

The compartmentalized administration of GFs creates gradients that influence cell activities such as adhesion, migration, and differentiation, which is of particular relevance to CSCs. CSCs are extremely refractory and are mainly involved in cancer (recurrence and metastasis); thus they represent a very unique problem in cancer treatments [46]. CSC targeting strategies include the isolation of target surface markers and disruption of signaling pathways underlying CSC self-renewal and differentiation, which consequently abrogated the CSC protective microenvironment to induce apoptosis [46].

Tissue engineering drug-delivery mechanism integrates the design of scaffolds, controlled bioactive factor release, and targeted therapies such as CSC targeting. This provides a multifaceted approach to the therapeutic enhancement of MSCs, PSCs, and ASCs, bringing about effective regeneration of tissues coupled with the impediment of issues associated with CSCs [44–46].

29.3 Tissue engineering in cancer therapy

29.3.1 Scaffold-based systems

29.3.1.1 Natural versus synthetic biomaterials

Natural biomaterials, especially those derived from the extracellular matrix, are critical for the development of an in vivo environment that supports improved cellular performance. These materials strongly affect the transduction of major processes such as cell movement, proliferation, and/or differentiation through the provision of diverse bioactive signals. Amongst these biomaterials, collagen, elastin, and hyaluronic acid are particularly noteworthy, as well as their ability to not only provide structural support but to facilitate biological interactions required for tissue regeneration [47]. Since natural biomaterials are biocompatible and the microstructure is very similar to physiologically normal conditions, their use is desirable for regenerative medicine. Nevertheless, their mechanical stiffness may be insufficient for some applications. In contrast, synthetic biomaterials, such as poly-L-lactide and polyglycolic acid, are designed to offer improved mechanical stability and can be customized for specific degradation rates and precise control over their physical properties [48].

Synthetic materials show good mechanical strength, but they cannot provide a faithful representation of the intricate biological interactions of native ECM, and a potential difficulty lies in the ability to faithfully recapitulate the TME. Indeed, a recent technology that incorporates both natural and artificial biomaterials in one scaffold has proven to be a facilitator for the engineering of scaffolds to better support tissue regeneration and for efficient cancer treatment [49,50].

29.3.1.2 Smart biomaterials for targeted therapy

Smart biomaterials are a new type of material developed exclusively for third-generation advanced cancer therapeutics. These materials change in response to biological cues, for example, pH changes, fluctuations in temperature, or target-specific biomolecules, thereby facilitating drug release control and increased specificity in drug delivery [49]. From that perspective, the electrospun release system provides a sustained release of chemotherapeutic drugs distributes drugs at the tumor action site, and minimizes systemic toxicity. Moreover, bioengineered matrices that mimic ECM components are effective drug-delivery vehicles that overcome the difficulties associated with drug resistance that commonly occurs with CSCs [50]. The characterization of interactions of these smart biomaterials with CSC niches may provide insights into fundamental signaling pathways responsible for tumorigenesis and drug responsiveness. Optimization of such material functionalities that mediate in dynamic TME, scientists move towards creating cancer treatments—personalized and maximally effective, especially tailored to each individuality in a given type of tumor along with its type of CSC that might be affecting it [47,48] (Fig. 29.2).

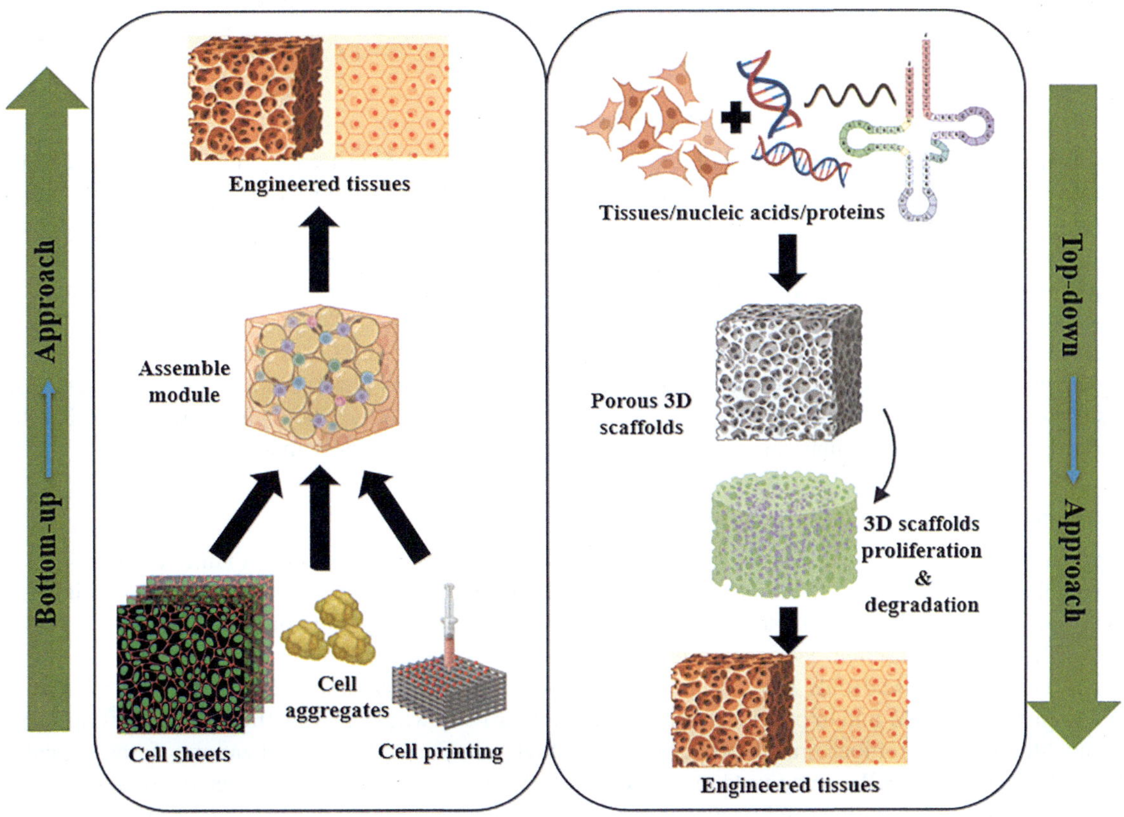

FIGURE 29.2 Process of tissue engineering in CSC therapy using 3D scaffold system. *CSC*, Cancer stem cells.

29.3.2 Organoids and tumor models

29.3.2.1 Organoids for personalized medicine

Organoids refer to a three-dimensional construct consisting of patient-derived tumor cells; they resemble authentic tumor morphology along with its behavior. It renders them important tools for personalized medicines. Organoids develop from a biopsy sample obtained from patients that are further processed to isolate CSCs or epithelial cells. These cells are grown in media with GFs that promote self-organization and differentiation into organoid structures that nearly match the cellular composition of the native tumor. This feature allows for the evaluation of individual responses to different treatment regimens, helping to pinpoint effective strategies tailored to each patient's specific tumor biology. With the help of more sophisticated computational approaches, organoids can also predict tumor dynamics and therapeutic responses based on mathematical modeling, which includes agent-based modeling to change the treatment plans [51–53]. The fast development of patient-specific organoids is particularly useful for the translational development of treatments, thereby leading to better clinical outcomes in oncology.

29.3.2.2 3D bioprinting for tumor modeling

3D bioprinting is a novel technology that advances the modeling of tumors by allowing the controlled stacking and joining of different cell types and biomaterials to form complex patient-specific tumor constructs. The bioprinting procedure begins with computer-aided design (CAD) for defining the shape of a tumor model and is then followed by the application of bioinks consisting of living cells and scaffolding materials. This method allows researchers to replicate the intricate microarchitecture of tumors. In addition to cancer cells, this also includes stromal cells and extracellular matrix components that enhance cell-to-cell interactions along with their environment [51,53]. Several printing techniques, like inkjet and micro-extrusion, are applied to ensure that the viability and functionality of structures are not affected. The resulting printed models are closer to the actual TME than existing two-dimensional cultures, which improve the drug response testing and estimation of combination, which in turn offers an appropriate treatment based on the individual characteristics of patients, hence improving its role in precision oncology [53] (Table 29.1).

29.3.3 Engineering vascularized tissues

29.3.3.1 Importance of vascular networks in tissue viability

Vascular networks are critical for the survival of engineered tissues not only because they deliver oxygen and nutrients but also for their ability to efflux metabolites. Constructs larger than 400 μm in depth are dominated by vascularization to prevent necrosis, as diffusion alone is inadequate for providing enough nutrition to the deeper cells. Efficient vascularization is very important for both transplantation and integration, especially since cells that are more than 100–200 μm beyond capillaries are exposed to great survival difficulties [54]. This problem is even more so for cancer treatment, where the metabolic requirements of tumor tissues are high [55].

The engineered vascular network that could later be interconnected with the host's circulation has turned out to be a stumbling block. To address this, researchers are investigating strategies that stimulate robust vascularization and capillaries close to cells to achieve metabolic requirements. This is critical not only to maintain tissue survival but also for improving drug delivery systems that can respond to the evolving needs of cancer therapy [56].

29.3.3.2 Techniques for vascularization

Several approaches have been created to improve vascularization in engineered tissues. Plan(s) emphasizing the role of endothelial cells underline their importance in neoangiogenesis. One such approach, prevascularization, consists of seeding scaffolds with endothelial cells to form capillary networks before implantation. This methodology allows better integration with the host's vascular bed, consequently increasing tissue survival [55].

Biological and synthetic scaffolds have vital roles in promoting vascularization. Biologically derived scaffolds are engineered to mimic the ECM and to provide a substrate for cell growth, whereas synthetic scaffolds afford the possibility of designing structural and biochemical properties with high precision. Functionalization of these scaffolds with vasculotropic GFs such as VEGF can be used to promote vascular growth, allowing for targeted vascularization of tissues [57]. Novel techniques, such as microfabrication and modular assembly, enable the generation of vascular networks as close to the natural vasculature as possible [56].

Despite these advances, issues are still prevalent in grafting the engineered vascular networks into the host circulatory system. Limitations include the limitation of oxygen diffusion and the requirement of functional anastomosis with host

TABLE 29.1 Comprehensive overview of 3D scaffold-based models in cancer therapy: cancer types, cell lines, and FDA-approved drugs.

Cancer type	3D model	Cell lines used	FDA-approved drugs	References
Lung cancer	Hydrogel-based scaffolds, bioprinted models	A549, H1975, H460, NCI-H23	Gefitinib, Erlotinib, Crizotinib, Pembrolizumab	
Breast cancer	Collagen-based scaffolds, alginate hydrogels	MCF-7, MDA-MB-231, BT-474	Trastuzumab, Paclitaxel, Tamoxifen, Everolimus	[51,52]
Prostate cancer	Decellularized tissue scaffolds, 3D bioprinting	LNCaP, DU145, PC3	Enzalutamide, Abiraterone, Docetaxel	[51]
Colon cancer	Matrigel-based models, polymeric scaffolds	HT-29, Caco-2, SW480, HCT116	Cetuximab, Bevacizumab, Irinotecan, 5-FU	
Ovarian cancer	Hydrogel scaffolds, ECM-inspired models	OVCAR-3, SKOV-3, A2780	Paclitaxel, Carboplatin, Bevacizumab	
Pancreatic cancer	Gelatin-based scaffolds, hydrogel models	PANC-1, MiaPaCa-2, AsPC-1	Gemcitabine, Nab-paclitaxel, Erlotinib	[51,52]
Melanoma	Alginate scaffolds, silk fibroin models	A375, SK-MEL-28, B16-F10	Dabrafenib, Trametinib, Pembrolizumab, Ipilimumab	[51–53]
Hepatocellular carcinoma	Collagen and fibrin scaffolds, 3D spheroids	HepG2, Huh7, PLC/PRF/5	Sorafenib, Lenvatinib, Regorafenib	[51–53]
Glioblastoma	Alginate-based scaffolds, bioprinted brain models	U87, U251, T98G	Temozolomide, Bevacizumab	
Leukemia	Gel-based scaffolds, gelatin methacrylate (GelMA)	K562, HL60, Jurkat	Imatinib, Dasatinib, Nilotinib	[51–53]

vessels that need to be considered. Novelty, including mechanical preconditioning and exposure to pulsatile flow, is being investigated to improve graft integration and functional graft performance [57].

In summary, vascular networks are essential for tissue engineering, especially in the context of cancer therapy. Integration of advanced vascularization technologies into engineered constructs is of paramount importance for successful drug delivery and improved therapeutic results.

29.4 Regulatory pathways in stem cell and tissue engineering

Tissue engineering and stem cells have become key elements of contemporary regenerative medicine, providing novel approaches to the treatment of a range of illnesses, including cancer [58]. Key regulatory pathways are crucial for controlling the development, function, and carcinogenesis of stem cells. Developing therapeutic applications relies on our comprehension of these pathways, particularly stem cell behavior, tissue regeneration, and treatments for cancer [21]. The following review focuses on key regulatory pathways for stem cell biology and tissue engineering concerning their application for tissue regeneration and cancer treatment [59]. The signaling pathways Notch, Wnt/β-Catenin, Hedgehog, and HIF1α are crucial in regulating stem cell behavior, tissue engineering, and cancer progression.

29.4.1 Notch signaling

Transmembrane proteins consist of notch receptors (Notch1–4) and ligands (DLL1, DLL3, DLL4, JAG1, and JAG2). When a ligand expressed on one cell attaches itself to a receptor on a neighboring cell, the pathway is triggered [60]. As a result of this interaction, the cytoplasmic domain of the receptor is initially broken down by disintegrin and metalloproteinases (ADAMs) and subsequently by γ-secretase. The Notch intracellular domain (NICD) is released into the cytoplasm by this dual cleavage, where it moves into the nucleus and triggers target gene transcription through the transcription factor CBF1, Suppressor of Hairless, LAG-1/recombination signal binding protein for immunoglobulin κ J region (CSL/RBPJ) [60,61].

Similar to numerous other signaling pathways, the Notch pathway is extremely conserved and plays a crucial role in controlling the differentiation and cell fate specification of stem and progenitor cells during the development of the embryonic heart, eyes, pancreas, intestine, vasculature, central nervous and hematopoietic systems, and other organs [60–62]. It has been demonstrated that in adults, Notch signaling affects the roles of stem and progenitor cells in the gut, skeletal muscle system, hematological system, and skin [62].

In a variety of malignancies, such as leukemia, glioblastoma, and those of the breast, colon, pancreas, and lung, among others, it has been shown that the Notch pathway controls the characteristics of tumor cells [63]. However, our understanding of the role of Notch signaling in the control of cancer's tumor-initiating cells is still developing [63]. Activated Notch signaling can function as a tumor promoter or suppressor in various tissues, depending on the situation, which contributes to the misunderstandings around its possible role in CSCs [63]. For instance, activating mutations in NOTCH1 are responsible for the development of T-cell acute lymphoblastic leukemia, similar to the APC mutations in colorectal cancer [64].

It has also been shown that notch signaling regulates cell fate in intestinal crypts. Transit-amplifying cells had a complete transformation into differentiated goblet cells under the inhibition of Notch signaling. When Notch signaling was inhibited, transit-amplifying cells were completely transformed into differentiated goblet cells. Both Notch and Wnt activation may be required to preserve the undifferentiated CSC state, as evidenced by the fact that suppression of Notch in adenomas resulting from APC mutations forced the adenoma cells to develop into goblet cells [65].

The Notch signaling genes Notch1, Notch3, Jag1, Jag2, and the Notch target gene Hes1 were indicated to be expressed at higher levels in patient-derived pancreatic CSCs in a different investigation examining the function of Notch in maintaining CSC stemness [66]. The CSC subpopulation and tumorsphere formation frequency were reduced when the CSCs were treated with a gamma-secretase (γ-sec) inhibitor [66]. Similarly, CSC tumorsphere formation was reduced and increased by Hes1 knockdown and CSC treatment with delta/Serrate/Lag-2 peptide, a Notch receptor agonist, respectively, indicating that Notch signaling activity is necessary for CSC stemness [66]. Treatment with γ-secretase inhibitors resulted in increased apoptosis and hindered cell cycle progression, according to cell death assays. This suggests that Notch signaling may support pancreatic CSC stemness by promoting cell survival [66].

Another example is that, in comparison to normal esophageal mucosa, the Notch pathway may be activated in almost three-quarters of primary esophageal adenocarcinoma samples; surprisingly, tumors that were partially or completely resistant to treatment showed increased Notch signaling activity [67]. When esophageal cancer cells were used in xenograft models of carcinogenesis, initial tumor growth was significantly decreased by blocking the Notch system using a γ-secretase inhibitor. The subsequent inability of secondary tumor growth after retransplantation indicated that the cancer-initiating cell population was depleted as a result of Notch loss [67]. Last but not least, an examination of CSC growth, survival, and gene expression revealed that Notch signaling was concentrated in this cell subpopulation and that inhibiting it caused the population to decrease by half [67].

In contrast, Notch may decrease tumors in the skin of mice [68]. BCC-like skin tumors developed spontaneously when Notch1 was deleted in the epidermis. Furthermore, these mice were more likely to develop tumors brought on by exposure to carcinogenic substances because of the lack of Notch1 in their epidermis. Four distinct tumor kinds—papillomas, benign dysplastic lesions, BCC-like, and squamous cell carcinoma—developed in this situation, indicating the possible involvement of extra mutational hits in the development of diverse tumor types. Gli2 and free β-catenin were found to be abnormally upregulated in the epidermis of Notch1-null mice when potential mechanisms for the resulting tumorigenesis were further investigated. This suggests that activated Hedgehog and Wnt pathways were significant mediators of tumorigenesis in the absence of Notch signaling [68].

29.4.2 Hedgehog pathway

The key components in the Hedgehog pathway consist of three secreted Hedgehog ligands—Sonic, Desert, and Indian—their corresponding receptor Patched, the transmembrane protein Smoothened, and three Gli transcription factors (Glis1–3; Gli was named due to the discovery and extraction of Gli1 from a glioma cell line) that influence the activation or inhibition of the pathway. The Patched receptor acts as a constant inhibitor of Smoothened when not bound by a ligand. In this condition, the transcription of the target gene is inhibited by Gli3 and Gli2-R (Gli2 in the form of a repressor). When a ligand binds to Patched, the inhibition on Smoothened is lifted, enabling the transcriptional activators Gli1 and Gli2-A (the activator form of Gli2) to promote the transcription of target genes [69,70].

By regulating cellular proliferation, differentiation, and migration, the Hedgehog pathway plays a crucial role in the development and appropriate patterning of numerous organs throughout embryogenesis, including the nervous system, skeleton, limbs, lung, heart, and gut [18,33]. Except for the adult central nervous system, skin, hair, and teeth, the Hedgehog pathway is mainly dormant in the majority of postnatal tissues, in contrast to many other pathways discussed here. It has recently been demonstrated that resident stem and/or progenitor cell populations are regulated by hedgehog activity [69,71].

Mouse experiments and in vitro studies of cancer cell lines and patient samples have validated the occurrence of abnormal Hedgehog signaling in over a dozen cancer types [69]. The involvement of Hedgehog signaling in CSC function has been recorded in several cancers, such as basal cell carcinoma (BCC), multiple myeloma, glioblastoma, chronic myeloid leukemia (CML), and colon cancer [69,72].

Individuals with alterations in the patched 1 (PTCH1) gene are at a higher risk of developing medulloblastomas. Furthermore, examination of the related mouse gene, Ptch1, suggested that mice with a heterozygous mutation in this gene exhibited a comparable susceptibility [72].

Moreover, alterations in the PTCH1 gene result in Gorlin syndrome, a condition that makes individuals more susceptible to advanced BCC [36]. A recent study that removed Ptch1 from various cellular regions in mouse skin revealed that stem cells situated in different parts of the hair follicle can develop BCC following the loss of Ptch1. The absence of Ptch1 in the interfollicular epidermis stem cells, however, showed no impact. Interestingly, the capacity of abnormal Hedgehog signaling to trigger BCC relied on the existence of sensory nerves in the stem cell niche [73].

These results reinforce earlier research suggesting that BCCs could develop from hyperactive Hedgehog signaling (through Gli2) in limited groups of residual, long-lasting cancer-initiating cells found in skin and hair follicles [74].

The Hedgehog pathway has been demonstrated to influence the characteristics of CSCs in multiple myeloma, glioma, and CML [75–77]. In a study of progenitor cells derived from a human multiple myeloma cell line, the SMO gene, which codes for the Smoothened protein, was found to be overexpressed and associated with elevated Gli1 transcriptional activity when compared to nonstem cancer cells [75]. The chemical inhibition of the Smoothened protein reduced proliferation, maintenance of stemness, and self-renewal in CSCs, strongly indicating that Hedgehog signaling supports these CSC functions in multiple myeloma [76]. In human glioma, CSCs were found to have high levels of the Hedgehog signaling genes Gli1, SHH, and PATCHED1 [75]. When glioma CSCs were treated with a Hedgehog signaling inhibitor, there was a reduction in proliferation, survival, self-renewal, and clonogenicity, along with decreased expression of the stemness genes NANOG, SOX2, and OCT4 [75]. Additionally, it was shown that Hedgehog signaling promotes tumor growth in nude mice using glioma sphere cells that express a conditionally suppressible version of SMO [75]. This evidence suggests that glioma CSCs might depend on Hedgehog signaling for functional support [75].

In CML, a mouse model of BCR-Abl1–induced CML was utilized to show that removing SMO greatly diminished the occurrence of CML CSCs [77]. SMO overexpression in an SMO-lacking mouse model of CML raised CSC frequency four times and notably accelerated CML progression [77]. In vitro expression studies on Numb, a differentiation gene, revealed an inverse correlation between SMO and Numb, indicating that SMO-deficient CML CSCs exhibited elevated Numb levels. These findings were consistent with experiments showing that increasing Numb levels in SMO-expressing CML CSCs leads to a decrease in colony formation [77]. Additionally, the chemical inhibition of SMO in CSCs was shown to hinder the spread of CML in CSC transplantation tests [77]. Together, these results indicate that Hedgehog signaling regulates the frequency and preservation of stemness in CSCs, and could play a crucial role in CSC-driven CML propagation [77]. Research has shown a comparable function for CSCs in colon cancer, where genetic knockdown of SMO in CSCs led to the complete elimination of malignant stem cells after just one in vivo serial tumor passage [78]. Furthermore, the genetic knockdown of PTCH1 in CSCs caused an increase in their population during serial tumor passaging [78]. These findings imply that Hedgehog signaling might be essential for the stemness and survival of colon CSCs, potentially aiding in tumor maintenance and proliferation [78]. Additionally, CSCs derived from liver metastases of colon cancer were observed to express elevated levels of Hedgehog signaling genes Gli1, Gli2, and HIP compared to nonmetastatic controls [78]. Notably, Gli1 was exclusively found in CSCs. Alongside evidence showing that the percentage of CSCs rises with disease progression [78], this information suggests that Hedgehog signaling activity may intensify as colon cancer advances [78]. CSCs from liver metastases demonstrated higher expression levels of SNAIL1, a target of Gli1, compared to nonmetastatic controls [78].

Since SNAIL1 aids in epithelial-mesenchymal transition, a mechanism closely associated with enhancing metastasis [79], the gene transcription data indicate that Hedgehog signaling plays a significant role in the pro-metastatic function of CSCs, as suggested by this study [78].

29.4.3 Wnt/β-catenin pathway

The Wnt/β-catenin pathway, which includes around 15 receptors and 19 Wnt ligands, is extensively genetically preserved [80]. The Wnt ligands are known to control CSC pluripotency and signal through β-catenin for their biological roles during embryonic development and tissue homeostasis [81–83]. Since this system is essential for CSC self-renewal, dedifferentiation, apoptosis inhibition, and metastasis [84], it is a desirable target for therapeutic treatment options meant to interfere with CSC progression and maintenance.

The production of Wnt proteins and inhibitors such as bone morphogenetic protein (BMP) and Delta in the TME activates the capacity of CSCs to self-renew [85]. It has been documented that in colorectal cancer, proto-oncogenes like c-Myc and downstream overexpression of Cyclin D1 cause dormant CSCs to become active CSCs by activating the Wnt pathway [82,86,87]. It has also been demonstrated that the CD44+/CD133+ isolated colorectal CSCs are strongly linked to the Wnt pathway [88]. Other forms of CSCs are also maintained by canonical Wnt signaling, in addition to colorectal cancer.

29.4.4 Immune checkpoints

By focusing on pathways that tumors use to elude immune surveillance, such as PD-1/PD-L1 and CTLA-4, immune checkpoint inhibitors (ICIs) have transformed cancer immunotherapy. By preventing immune checkpoint proteins and their ligands from interacting, these inhibitors improve the immune system's capacity to identify and eliminate tumor cells. However, many patients continue to develop primary or acquired resistance to ICIs, which restricts their therapeutic efficacy even though they have been successful in treating a variety of malignancies [89] (Table 29.2).

This problem has prompted researchers to investigate the combination of ICIs with various treatment techniques, one of the most promising of which is stem cell therapy [118].

Stem cells, specifically CSCs, are a type of tumor cell that has self-renewal ability and can initiate and sustain tumor growth. Conventional treatments like radiation and chemotherapy, which frequently target quickly dividing cells but are unable to eliminate the slower dividing CSCs, are historically ineffective against CSCs. CSCs live in a special TME that contains immunosuppressive elements that aid in their ability to avoid immune identification, which contributes to their resistance. By boosting the immune response against these resistant CSCs, the combination of ICIs and stem cell therapy may be able to resolve this problem. CSCs can be considered more vulnerable to immune-mediated death by altering the immunological microenvironment through immune checkpoint inhibition [119].

Furthermore, it has been demonstrated that stem cells, particularly MSCs, can alter/modulate the immune system by releasing substances that have the dual ability to destroy tumor cells directly and attract immune cells to the tumor site, which may increase the effectiveness of ICIs [120–122].

TABLE 29.2 Cancer stem cell (CSC) surface markers: expression in specific cancers, associated signaling pathways, and therapeutic strategies.

CSC surface marker	Cancer type	Expression in specific cancer	Signaling pathway	Therapies	References
CD133	Glioblastoma, colon, liver, pancreatic	High expression in glioblastoma (tumor-initiation), colon (chemoresistance), and liver cancers	Wnt/β-catenin, Notch, Hedgehog	Anti-CD133 monoclonal antibodies, siRNA targeting CD133, inhibitors of Wnt/Notch pathways	[90,91]
CD44	Breast, prostate, colon, ovarian	Overexpressed in breast cancer (metastasis), prostate cancer (invasion), and colorectal cancer	Wnt/β-catenin, EGFR	Anti-CD44 antibodies, hyaluronic acid-based drug-delivery systems, EGFR inhibitors	[92]
CD24	Breast, pancreatic	Overexpressed in pancreatic cancer (immune evasion) and breast cancer (stemness properties)	Akt, STAT3	Monoclonal antibodies targeting CD24, STAT3 inhibitors	[93,94]
ALDH1	Breast, ovarian, lung, prostate	Highly expressed in breast cancer (poor prognosis) and ovarian cancer (therapy resistance)	Notch, PI3K/Akt	ALDH1 inhibitors (e.g., disulfiram), Notch inhibitors	[95,96]
EpCAM	Breast, colorectal, hepatocellular	Overexpressed in hepatocellular carcinoma and colorectal cancer (self-renewal and metastasis)	Wnt/β-catenin, MAPK	Anti-EpCAM monoclonal antibodies, EpCAM-targeted CAR-T-cell therapies	[97,98]
CD90 (Thy-1)	Liver, leukemia, glioblastoma	Highly expressed in liver cancer (tumor-initiating cells) and glioblastoma (invasion potential)	TGF-β, MAPK	Anti-CD90 antibodies, TGF-β inhibitors	[99,100]
CD10	Colorectal, breast	Overexpressed in colorectal cancer (stem-like properties) and breast cancer (metastatic potential)	Wnt/β-catenin, MAPK	Anti-CD10 antibodies, MAPK pathway inhibitors	[101]
CD166 (ALCAM)	Colorectal, breast, prostate	Overexpressed in colorectal and breast cancer (metastasis and progression)	Wnt/β-catenin, PI3K/Akt	Anti-CD166 antibodies, Wnt inhibitors	[102,103]
CXCR4	Breast, pancreatic, lung	Highly expressed in pancreatic cancer (metastasis) and lung cancer (poor prognosis)	SDF-1/CXCR4 axis, Akt, ERK	CXCR4 antagonists (e.g., AMD3100), inhibitors of SDF-1/CXCR4 signaling	[104,105]
CD47	Multiple cancer types	Overexpressed in various cancers (immune evasion and metastasis)	Immune checkpoint signaling	Anti-CD47 antibodies (e.g., magrolimab), immune checkpoint blockade therapies	[106,107]
CD271 (NGFR)	Melanoma, head and neck	Overexpressed in melanoma (stemness and metastasis)	MAPK, NF-κB	Anti-CD271 antibodies, NF-κB pathway inhibitors	[108,109]
LGR5	Colorectal, gastric, liver	Overexpressed in colorectal cancer (self-renewal) and gastric cancer (tumor-initiation)	Wnt/β-catenin	Anti-LGR5 antibodies, Wnt pathway inhibitors	[110,111]

(*Continued*)

TABLE 29.2 (Continued)

CSC surface marker	Cancer type	Expression in specific cancer	Signaling pathway	Therapies	References
CD117 (c-Kit)	GIST, melanoma	Highly expressed in GIST (proliferation and survival)	KIT/SCF, PI3K/Akt	Tyrosine kinase inhibitors (e.g., imatinib), c-Kit inhibitors	[112,113]
CD49f (Integrin α6)	Breast, prostate, lung	Overexpressed in breast cancer (stem-like traits) and prostate cancer (metastatic potential)	FAK/Src, Wnt/β-catenin	Antiintegrin antibodies, FAK inhibitors	[114,115]
CD25 (IL-2Rα)	Leukemia, lymphoma	Overexpressed in leukemia (immune modulation and stemness)	JAK/STAT	Anti-CD25 monoclonal antibodies (e.g., daclizumab), JAK inhibitors	[116,117]

In preclinical and clinical settings, a number of research investigations have looked into the possibility of combining ICIs with stem cell therapy. For example, when paired with ICIs, MSCs have been utilized to release immunomodulatory cytokines or stimulate immune responses aimed at tumors. One study used MSCs that had undergone genetic modification to express cytokines such as interleukin-12 (IL-12), which is known to stimulate T-cells and increase immunity against tumors. This strategy improved tumor regression in mouse models when combined with PD-1 inhibitors [89].

To increase the action of ICIs in a more focused way, research is also being done on immune system engineering and stem cell-based vaccinations [123]. Furthermore, combining stem cells with ICIs may minimize immune-related adverse events (irAEs), a major side effect of ICI monotherapy, by leveraging stem cells' immunomodulatory characteristics. With this approach, systemic inflammation is reduced, and tumor cells are specifically targeted by a more balanced immune response [89].

29.5 Challenges and future directions

29.5.1 Challenges

29.5.1.1 Tumorigenicity risks

Stem cell therapies have enormous promise for treating cancer, but before they can be widely used, several issues need to be resolved. One of the biggest concerns is still tumorigenicity. Stem cells, especially pluripotent and induced pluripotent stem cells (iPSCs), can multiply unchecked and develop into tumors if they are not carefully controlled. If these cells do not effectively develop or integrate into the host tissue, they may accumulate mutations that could result in oncogenesis [124]. There is still a significant obstacle in the way of efforts to genetically alter stem cells to regulate their proliferation or guarantee that they are completely differentiated before transplantation.

29.5.1.2 Immune rejection of stem cell therapies

Immune rejection is another important obstacle. The recipient's immune system may react negatively to stem cells, particularly those obtained from allogeneic (nonself) sources, which could lead to the transplanted cells being rejected. Because the immune system recognizes stem cell-based therapies as alien, this is especially problematic. Although autologous stem cells—which are generated from the same person—reduce this problem, they have drawbacks of their own, like limited supply or the requirement for intricate processes to collect and grow the cells. To combat this, methods such as immune modulation, immunosuppressive medication, and gene editing to lower immunogenicity are being investigated [125,126].

29.5.1.3 Ethical and regulatory concerns

There are additional difficulties because of *ethical and legal issues* with stem cell treatments. Because using ESCs frequently necessitates the killing of embryos, there are ethical questions surrounding the source of these cells. Furthermore, because different stem cell types (such as adult stem cells and iPSCs) have different regulatory requirements, the regulatory environment around stem cell therapies is complicated. Successfully integrating stem cell-based therapies into clinical practice requires striking a balance between patient safety, ethical issues, and the need for innovation [127].

29.5.2 Future directions

With developments in personalized medicine, hybrid medicines, and regenerative oncology frameworks opening the door for novel methods, the use of stem cell therapies in cancer treatment has the potential to revolutionize the field in the future. By addressing therapy resistance and overcoming existing limitations, these strategies seek to enhance patient outcomes.

29.5.2.1 Personalized medicine applications

By using molecular and genetic profiling to customize treatments to each patient's unique tumor features, personalized medicine is transforming the way that cancer is treated. When paired with cutting-edge diagnostic technologies such as next-generation sequencing (NGS), stem cell-based strategies enable accurate targeting of pathways unique to cancer. For example, Patient-specific induced pluripotent stem cells (iPSCs) can be derived from a tumor or healthy tissues, providing models to identify weaknesses in cancer cells and design tailored treatment [128].

To deliver therapeutic substances like cytokines or pro-apoptotic signals directly to the tumor site, iPSCs can be genetically modified to carry certain alterations. These developments reduce systemic toxicity and off-target effects, which are frequent in traditional cancer. treatments, while also improving therapeutic efficacy [129].

Another integration, personalized medicine with AI, could lead to more sophisticated. Treatment planning. Machine learning algorithms are capable of analyzing large data sets, such as genetic and clinical information, to predict patient responses to stem cell therapies. This may be an avenue towards extremely tailored treatment plans that evolve based on the tumor changes and the patient responses.

29.5.2.2 Hybrid therapies (e.g., combining nanotechnology with stem cells)

Stem cell combined therapies that include nanotechnology provide one novel way of cancer treatment. Nanoparticles provide the potential for carrier application to enhance targeted delivery of stem cells as well as therapeutic agents. For instance, MSCs modified to produce anticancer agents may be covered with nanoparticles, guaranteeing accurate targeting of the TME. These nanoparticles can additionally carry chemotherapy medications, immunomodulators, or gene-editing tools such as CRISPR-Cas9, allowing for multifunctional therapeutic impacts [130–132].

Nanomaterials can additionally improve the tracking and observation of stem cell therapies. Fluorescent or magnetic nanoparticles can be attached to stem cells, enabling real-time imaging to evaluate their movement, distribution, and therapeutic results. This method not only increases safety but also boosts treatment effectiveness by making certain that stem cells arrive at their designated targets. Furthermore, hybrid therapies might aid in the elimination of CSCs, which frequently resist standard treatments and contribute to tumor recurrence and metastasis [133].

29.5.2.3 Development of regenerative oncology frameworks

To treat cancer more comprehensively, regenerative oncology aims to combine the concepts of oncology with regenerative medicine. Regenerative oncology is concerned with healing and regenerating tissues that have been harmed by cancer treatments such as chemotherapy, radiation, or surgery, in addition to eliminating cancer cells. MSCs in particular can be employed to improve immunological function, promote tissue regeneration, and diminish the adverse outcomes of toxic therapies. MSCs are perfect for resolving treatment-induced toxicity in organs like the liver, heart, and bone marrow since they have been demonstrated to support angiogenesis and tissue repair [120,122,134].

Stem cells may also be used to develop more precise preclinical cancer models. Organoids made from stem cells unique to each patient offer a three-dimensional representation of tumors that closely resembles the in vivo environment of tumor.

The discovery of more potent cancer treatments can be pushed up by using these models to test new medications, investigate tumor biology, and uncover resistance mechanisms [135,136].

29.5.2.4 Integration of multidisciplinary approaches

In the future, oncology, immunology, bioengineering, and computational biology will probably work together in multidisciplinary collaborations to develop stem cell therapeutics for cancer. For example, developments in bioprinting may make it possible for researchers to develop bioengineered tissues that contain therapeutic stem cells. These tissues may then be implanted to replace or repair tissues destroyed by cancer [137]. In the same way, adoptive T-cell treatments or immune checkpoint inhibitors combined with stem cell therapies may have potent synergistic effects that improve tissue regeneration and antitumor immunity [118].

29.6 Conclusion

In conclusion, research into the regulatory pathways for tissue and stem cell engineering in cancer treatment shows promise for revolutionizing the field of oncology. Hedgehog, Wnt/β-catenin, and Notch are only a few of the signaling pathways that interact intricately to highlight the intricacy of stem cell behavior and its significance in tumor growth. These pathways offer new therapeutic options for intervention in addition to shedding light on the mechanisms behind carcinogenesis and CSC maintenance.

The potential for individualized, focused, and regenerative approaches to cancer treatment is what makes stem cell therapies so promising. Researchers are developing novel approaches to deal with the problems caused by tumor heterogeneity, therapeutic resistance, and the shortcomings of traditional treatments by utilizing the special qualities of different stem cell types, such as MSCs and iPSCs. Furthermore, more efficient drug-delivery systems and individualized treatment plans are being made possible by developments in tissue engineering, such as the creation of smart biomaterials and organoid models.

However, there are still a lot of obstacles to overcome before these treatments can be used in healthcare facilities. For stem cell-based therapies to be safe and effective, issues with immunological rejection, tumorigenicity, and ethics must be resolved. These obstacles could be removed and the therapeutic landscape improved by integrating multidisciplinary methods, such as developments in computational biology, nanotechnology, and regenerative oncology frameworks.

To successfully traverse the intricacies of stem cell therapies, cooperation between researchers, clinicians, and regulatory agencies is encouraged. The future of cancer treatment may move towards a more accurate, comprehensive, and patient-centered paradigm by tackling the current issues and utilizing the knowledge gathered via regulatory pathways, thereby improving outcomes for individuals dealing with this difficult illness.

References

[1] F. Bray, M. Laversanne, E. Weiderpass, I. Soerjomataram, The ever-increasing importance of cancer as a leading cause of premature death worldwide, Cancer 127 (16) (2021) 3029–3030. Available from https://doi.org/10.1002/cncr.33587.

[2] S. Chen, Z. Cao, K. Prettner, M. Kuhn, J. Yang, L. Jiao, et al., Estimates and projections of the global economic cost of 29 cancers in 204 countries and territories from 2020 to 2050, JAMA Oncology 9 (4) (2023) 465–472. Available from https://doi.org/10.1001/jamaoncol.2022.7826, http://oncology.jamanetwork.com/journal.aspx.

[3] J. Fares, M.Y. Fares, H.H. Khachfe, H.A. Salhab, Y. Fares, Molecular principles of metastasis: a hallmark of cancer revisited, Signal Transduction and Targeted Therapy 5 (1) (2020). Available from https://doi.org/10.1038/s41392-020-0134-x, www.nature.com/sigtrans.

[4] L. Gallicchio, T.P. Devasia, E. Tonorezos, M.A. Mollica, A. Mariotto, Estimation of the number of individuals living with metastatic cancer in the United States, Journal of the National Cancer Institute 114 (11) (2022) 1476–1483. Available from https://doi.org/10.1093/jnci/djac158, http://jnci.oxfordjournals.org/.

[5] K.L. Schulman, J. Kohles, Economic burden of metastatic bone disease in the U.S, Cancer 109 (11) (2007) 2334–2342. Available from https://doi.org/10.1002/cncr.22678.

[6] M.R. DiCaprio, H. Murtaza, B. Palmer, M. Evangelist, Narrative review of the epidemiology, economic burden, and societal impact of metastatic bone disease, Annals of Joint 7 (2022). Available from https://doi.org/10.21037/aoj-20-97, https://aoj.amegroups.com/article/view/6807/html.

[7] J.E. Brown, H. Neville-Webbe, R.E. Coleman, The role of bisphosphonates in breast and prostate cancers, Endocrine-Related Cancer 11 (2) (2004) 207–224. Available from https://doi.org/10.1677/erc.0.0110207.

[8] P.J. Kostenuik, H.Q. Nguyen, J. McCabe, K.S. Warmington, C. Kurahara, N. Sun, et al., Denosumab, a fully human monoclonal antibody to RANKL, inhibits bone resorption and increases BMD in knock-in mice that express chimeric (murine/human) RANKL, Journal of Bone and Mineral Research 24 (2009) 182–195. Available from https://doi.org/10.1359/jbmr.081112.

[9] Z.-F. Lim, P.C. Ma, Emerging insights of tumor heterogeneity and drug resistance mechanisms in lung cancer targeted therapy, Journal of Hematology & Oncology 12 (1) (2019). Available from https://doi.org/10.1186/s13045-019-0818-2.

[10] M. Roszkowska, Multilevel mechanisms of cancer drug resistance, International Journal of Molecular Sciences 25 (22) (2024) 12402. Available from https://doi.org/10.3390/ijms252212402.

[11] E. Anastasiadou, L.S. Jacob, F.J. Slack, Non-coding RNA networks in cancer, Nature Reviews Cancer 18 (1) (2018) 5–18. Available from https://doi.org/10.1038/nrc.2017.99.

[12] M. Eslami, O. Memarsadeghi, A. Davarpanah, A. Arti, K. Nayernia, B. Behnam, Overcoming chemotherapy resistance in metastatic cancer: a comprehensive review, Biomedicines 12 (1) (2024) 183. Available from https://doi.org/10.3390/biomedicines12010183.

[13] B.A. Mengistu, T. Tsegaw, Y. Demessie, K. Getnet, A.B. Bitew, M.Z. Kinde, et al., Comprehensive review of drug resistance in mammalian cancer stem cells: implications for cancer therapy, Cancer Cell International 24 (1) (2024). Available from https://doi.org/10.1186/s12935-024-03558-0.

[14] T. Lan, M. Luo, X. Wei, Mesenchymal stem/stromal cells in cancer therapy, Journal of Hematology & Oncology 14 (1) (2021). Available from https://doi.org/10.1186/s13045-021-01208-w.

[15] Y.S. Chun, K. Byun, B. Lee, Induced pluripotent stem cells and personalized medicine: current progress and future perspectives, Anatomy & Cell Biology 44 (4) (2011) 245. Available from https://doi.org/10.5115/acb.2011.44.4.245.

[16] E. Batlle, H. Clevers, Cancer stem cells revisited, Nature Medicine 23 (10) (2017) 1124–1134. Available from https://doi.org/10.1038/nm.4409.

[17] A. Angelopoulou, Nanostructured biomaterials in 3D tumor tissue engineering scaffolds: regenerative medicine and immunotherapies, International Journal of Molecular Sciences 25 (10) (2024) 5414. Available from https://doi.org/10.3390/ijms25105414.

[18] X.-D. Chen, X.-Y. Zhang, H.-Q. Zhu, H.H. Lu, M. Wang, Three-dimensional printing of hydrogel blend tissue engineering scaffolds with in situ delivery of anticancer drug for treating melanoma resection-induced tissue defects, Journal of Functional Biomaterials 15 (12) (2024) 381. Available from https://doi.org/10.3390/jfb15120381.

[19] W.H. Abuwatfa, W.G. Pitt, G.A. Husseini, Scaffold-based 3D cell culture models in cancer research, Journal of Biomedical Science 31 (1) (2024). Available from https://doi.org/10.1186/s12929-024-00994-y.

[20] B. Swain, S. Maddi, Patient-derived tumor organoids: generation and applications in disease modeling and personalized therapy, Nature Cell and Science (2024). Available from https://doi.org/10.61474/ncs.2024.00008.

[21] Z. Zhang, Y. Zhang, Transcriptional regulation of cancer stem cell: regulatory factors elucidation and cancer treatment strategies, Journal of Experimental & Clinical Cancer Research 43 (1) (2024). Available from https://doi.org/10.1186/s13046-024-03021-y.

[22] K. Takahashi, S. Yamanaka, Induction of pluripotent stem cells from mouse embryonic and adult fibroblast cultures by defined factors, Cell 126 (4) (2006) 663–676. Available from https://doi.org/10.1016/j.cell.2006.07.024.

[23] N.G. Kooreman, Y. Kim, P.E. de Almeida, V. Termglinchan, S. Diecke, N.-Y. Shao, et al., Autologous iPSC-based vaccines elicit anti-tumor responses in vivo, Cell Stem Cell 22 (4) (2018) 501. Available from https://doi.org/10.1016/j.stem.2018.01.016.

[24] X. Ouyang, M.L. Telli, J.C. Wu, Induced pluripotent stem cell-based cancer vaccines, Frontiers in Immunology 10 (JULY) (2019). Available from https://doi.org/10.3389/fimmu.2019.01510.

[25] D.A. Knorr, Z. Ni, D. Hermanson, M.K. Hexum, L. Bendzick, L.J.N. Cooper, et al., Clinical-scale derivation of natural killer cells from human pluripotent stem cells for cancer therapy, Stem Cells Translational Medicine 2 (4) (2013) 274–283. Available from https://doi.org/10.5966/sctm.2012-0084.

[26] Y. Li, Human iPSC-derived natural killer cells engineered with chimeric antigen receptors enhance anti-tumor activity, Cell stem cell 23 (2) (2018) 181–192.

[27] M. Ruella, S.S. Kenderian, Next-generation chimeric antigen receptor T-cell therapy: going off the shelf, BioDrugs 31 (6) (2017) 473–481. Available from https://doi.org/10.1007/s40259-017-0247-0.

[28] F. Timmermans, I. Velghe, L. Vanwalleghem, M. De Smedt, S. Van Coppernolle, T. Taghon, et al., Generation of T cells from human embryonic stem cell-derived hematopoietic zones, The Journal of Immunology 182 (11) (2009) 6879–6888. Available from https://doi.org/10.4049/jimmunol.0803670.

[29] F. Cichocki, S.J.C. van der Stegen, J.S. Miller, Engineered and banked iPSCs for advanced NK- and T-cell immunotherapies, Blood 141 (8) (2023) 846–855. Available from https://doi.org/10.1182/blood.2022016205.

[30] E.A. Copelan, Hematopoietic stem-cell transplantation, New England Journal of Medicine 354 (17) (2006) 1813–1826. Available from https://doi.org/10.1056/nejmra052638.

[31] I. Christodoulou, M. Goulielmaki, M. Devetzi, M. Panagiotidis, G. Koliakos, V. Zoumpourlis, Mesenchymal stem cells in preclinical cancer cytotherapy: a systematic review, Stem Cell Research & Therapy 9 (1) (2018). Available from https://doi.org/10.1186/s13287-018-1078-8.

[32] W. Lin, L. Huang, Y. Li, B. Fang, G. Li, L. Chen, et al., Mesenchymal stem cells and cancer: clinical challenges and opportunities, BioMed Research International 2019 (2019). Available from https://doi.org/10.1155/2019/2820853, http://www.hindawi.com/journals/biomed/.

[33] D. Kanojia, I.V. Balyasnikova, R.A. Morshed, R.T. Frank, D. Yu, L. Zhang, et al., Neural stem cells secreting anti-HER2 Antibody improve survival in a preclinical model of HER2 overexpressing breast cancer brain metastases, Stem Cells 33 (10) (2015) 2985–2994. Available from https://doi.org/10.1002/stem.2109.

[34] H.J. Lee, S.W. Doo, D.H. Kim, Y.J. Cha, J.H. Kim, Y.S. Song, et al., Cytosine deaminase-expressing human neural stem cells inhibit tumor growth in prostate cancer-bearing mice, Cancer Letters 335 (1) (2013) 58–65. Available from https://doi.org/10.1016/j.canlet.2013.01.048.

[35] B.-R. Yi, S.U. Kim, K.-C. Choi, Co-treatment with therapeutic neural stem cells expressing carboxyl esterase and CPT-11 inhibit growth of primary and metastatic lung cancers in mice, Oncotarget 5 (24) (2014) 12835–12848. Available from https://doi.org/10.18632/oncotarget.2547.

[36] J.C. Chang, Cancer stem cells: role in tumor growth, recurrence, metastasis, and treatment resistance, Medicine (United States) 95 (1) (2016) S20. Available from https://doi.org/10.1097/MD.0000000000004766, http://journals.lww.com/md-journal.

[37] I.V. Kholodenko, A.A. Konieva, R.V. Kholodenko, K.N. Yarygin, Molecular mechanisms of migration and homing of intravenously transplanted mesenchymal stem cells, Journal of Regenerative Medicine and Tissue Engineering 2 (1) (2013) 2. Available from https://doi.org/10.7243/2050-1218-2-4.

[38] A.D. Becker, I.V. Riet, Homing and migration of mesenchymal stromal cells: how to improve the efficacy of cell therapy? World Journal of Stem Cells 8 (3) (2016) 73–87. Available from https://doi.org/10.4252/wjsc.v8.i3.73, https://www.wjgnet.com/1948-0210/e-journal.htm.

[39] D.-T. Chu, T. Tien Nguyen, Nguyen, L.B. Tien, D.-K. Tran, J.-H. Jeong, P.G. Anh, et al., Recent progress of stem cell therapy in cancer treatment: molecular mechanisms and potential applications, Cells 9 (3) (2020) 563. Available from https://doi.org/10.3390/cells9030563.

[40] A.R.R. Weiss, M.H. Dahlke, Immunomodulation by mesenchymal stem cells (MSCs): mechanisms of action of living, apoptotic, and dead MSCs, Frontiers in Immunology 10 (2019). Available from https://doi.org/10.3389/fimmu.2019.01191, https://www.frontiersin.org/journals/immunology#.

[41] F. Gao, s Chiu, Mesenchymal stem cells and immunomodulation, Cell Death and Disease (2010).

[42] K. English, F.P. Barry, B.P. Mahon, Murine mesenchymal stem cells suppress dendritic cell migration, maturation and antigen presentation, Immunology Letters 115 (1) (2008) 50–58. Available from https://doi.org/10.1016/j.imlet.2007.10.002.

[43] F. Djouad, L.M. Charbonnier, C. Bouffi, P. Louis-Plence, C. Bony, F. Apparailly, et al., Mesenchymal stem cells inhibit the differentiation of dendritic cells through an interleukin-6-dependent mechanism, Stem Cells 25 (8) (2007) 2025–2032. Available from https://doi.org/10.1634/stemcells.2006-0548France, http://stemcells.alphamedpress.org/cgi/reprint/25/8/2025.pdf.

[44] A. Nourian Dehkordi, F. Mirahmadi Babaheydari, M. Chehelgerdi, S. Raeisi Dehkordi, Skin tissue engineering: wound healing based on stem-cell-based therapeutic strategies, Stem Cell Research & Therapy 10 (1) (2019). Available from https://doi.org/10.1186/s13287-019-1212-2.

[45] M. Biondi, F. Ungaro, F. Quaglia, P.A. Netti, Controlled drug delivery in tissue engineering, Advanced Drug Delivery Reviews 60 (2) (2008) 229–242. Available from https://doi.org/10.1016/j.addr.2007.08.038.

[46] D.L. Dragu, Therapies targeting cancer stem cells: current trends and future challenges, World Journal of Stem Cells 7 (9) (2015) 1185. Available from https://doi.org/10.4252/wjsc.v7.i9.1185.

[47] S. Hinderer, S.L. Layland, K. Schenke-Layland, ECM and ECM-like materials—biomaterials for applications in regenerative medicine and cancer therapy, Advanced Drug Delivery Reviews (2016).

[48] J. Yuan, R. Xie, X. Xu, X. Chen, J. Gao, Tailoring biomaterials through controlled synthesis and self-assembly for achieving enhanced drug delivery, Journal of Materials Chemistry B (2020).

[49] Y. Zhang, Y. Hu, Y. Wei, Y. Yu, Applications of natural and synthetic biomaterials in tissue engineering and regenerative medicine, Materials Science & Engineering C (2021).

[50] G. Hassan, S.M. Afify, S. Kitano, A. Seno, H. Ishii, Y. Shang, et al., Cancer stem cell microenvironment models with biomaterial scaffolds in vitro, Processes 9 (1) (2021) 1–19. Available from https://doi.org/10.3390/pr9010045, https://www.mdpi.com/2227-9717/9/1/45.

[51] R. Augustine, S.N. Kalva, R. Ahmad, A.A. Zahid, S. Hasan, A. Nayeem, et al., 3D Bioprinted cancer models: revolutionizing personalized cancer therapy, Translational Oncology 14 (4) (2021) Available from https://doi.org/10.1016/j.tranon.2021.101015, https://www.journals.elsevier.com/translational-oncology/.

[52] Z. Gilazieva, A. Ponomarev, C. Rutland, A. Rizvanov, V. Solovyeva, Promising applications of tumor spheroids and organoids for personalized medicine, Cancers 12 (10) (2020) 2727. Available from https://doi.org/10.3390/cancers12102727.

[53] A. Karolak, D.A. Markov, L.J. McCawley, K.A. Rejniak, Towards personalized computational oncology: from spatial models of tumour spheroids, to organoids, to tissues, Journal of the Royal Society Interface 15 (138) (2018). Available from https://doi.org/10.1098/rsif.2017.0703, http://rsif.royalsocietypublishing.org/content/royinterface/15/138/20170703.full.pdf.

[54] F.A. Auger, L. Gibot, D. Lacroix, The pivotal role of vascularization in tissue engineering, Annual Review of Biomedical Engineering 15 (2013) 177–200. Available from https://doi.org/10.1146/annurev-bioeng-071812-152428.

[55] M. Lovett, K. Lee, A. Edwards, D.L. Kaplan, Vascularization strategies for tissue engineering, Tissue Engineering - Part B: Reviews 15 (3) (2009) 353–370. Available from https://doi.org/10.1089/ten.teb.2009.0085.

[56] E.C. Novosel, C. Kleinhans, P.J. Kluger, Vascularization is the key challenge in tissue engineering, Advanced Drug Delivery Reviews 63 (4) (2011) 300–311. Available from https://doi.org/10.1016/j.addr.2011.03.004.

[57] H.-H.G. Song, R.T. Rumma, C.K. Ozaki, E.R. Edelman, C.S. Chen, Vascular tissue engineering: progress, challenges, and clinical promise, Cell Stem Cell 22 (3) (2018) 340–354. Available from https://doi.org/10.1016/j.stem.2018.02.009.

[58] A. Soltani Khaboushan, A. Neishabouri, N. Rezaei, Tissue engineering and regenerative medicine in cancer therapy, Springer Science and Business Media LLC, (2023) 1–28, https://doi.org/10.1007/978-3-030-80962-1_293-1.

[59] The Kwon, et al., Journal of Electroceramics. (2018), 1–8.

[60] C. Karamboulas, L. Ailles, Developmental signaling pathways in cancer stem cells of solid tumors, Biochimica et Biophysica Acta (BBA) - General Subjects 1830 (2) (2013) 2481–2495. Available from https://doi.org/10.1016/j.bbagen.2012.11.008.

[61] N. Takebe, L. Miele, P.J. Harris, W. Jeong, H. Bando, M. Kahn, et al., Targeting notch, hedgehog, and wnt pathways in cancer stem cells: clinical update, Nature Reviews Clinical Oncology 12 (8) (2015) 445–464. Available from https://doi.org/10.1038/nrclinonc.2015.61.

[62] S. Chiba, Concise review: notch signaling in stem cell systems, STEM CELLS 24 (11) (2006) 2437–2447. Available from https://doi.org/10.1634/stemcells.2005-0661.

[63] P. Ranganathan, K.L. Weaver, A.J. Capobianco, Notch signalling in solid tumours: a little bit of everything but not all the time, Nature Reviews Cancer 11 (5) (2011) 338–351. Available from https://doi.org/10.1038/nrc3035.

[64] A.P. Weng, A.A. Ferrando, W. Lee, J.P. Morris, L.B. Silverman, C. Sanchez-Irizarry, et al., Activating mutations of NOTCH1 in human T cell acute lymphoblastic leukemia, Science 306 (5694) (2004) 269–271. Available from https://doi.org/10.1126/science.1102160.

[65] J.H. van Es, M.E. van Gijn, O. Riccio, M. van den Born, M. Vooijs, H. Begthel, et al., Notch/γ-secretase inhibition turns proliferative cells in intestinal crypts and adenomas into goblet cells, Nature 435 (7044) (2005) 959–963. Available from https://doi.org/10.1038/nature03659.

[66] E.V. Abel, E.J. Kim, J. Wu, M. Hynes, F. Bednar, E. Proctor, et al., The notch pathway is important in maintaining the cancer stem cell population in pancreatic cancer, PLoS ONE 9 (3) (2014) e91983. Available from https://doi.org/10.1371/journal.pone.0091983.

[67] Z. Wang, T.G. Da Silva, K. Jin, X. Han, P. Ranganathan, X. Zhu, et al., Notch signaling drives stemness and tumorigenicity of esophageal adenocarcinoma, Cancer Research 74 (21) (2014) 6364–6374. Available from https://doi.org/10.1158/0008-5472.can-14-2051.

[68] M. Nicolas, A. Wolfer, K. Raj, J.A. Kummer, P. Mill, M. van Noort, et al., Notch1 functions as a tumor suppressor in mouse skin, Nature Genetics 33 (3) (2003) 416–421. Available from https://doi.org/10.1038/ng1099.

[69] A.A. Merchant, W. Matsui, Targeting hedgehog—a cancer stem cell pathway, Clinical Cancer Research 16 (12) (2010) 3130–3140. Available from https://doi.org/10.1158/1078-0432.ccr-09-2846.

[70] K.W. Kinzler, S.H. Bigner, D.D. Bigner, J.M. Trent, M.L. Law, S.J. O'Brien, et al., Identification of an amplified, highly expressed gene in a human glioma, Science 236 (4797) (1987) 70–73. Available from https://doi.org/10.1126/science.3563490.

[71] R. Petrova, A.L. Joyner, Roles for hedgehog signaling in adult organ homeostasis and repair, Development 141 (18) (2014) 3445–3457. Available from https://doi.org/10.1242/dev.083691.

[72] J.Y.H. Kim, A.L. Nelson, S.A. Algon, O. Graves, L.M. Sturla, L.C. Goumnerova, et al., Medulloblastoma tumorigenesis diverges from cerebellar granule cell differentiation in patched heterozygous mice, Developmental Biology 263 (1) (2003) 50–66. Available from https://doi.org/10.1016/s0012-1606(03)00434-2.

[73] S.C. Peterson, M. Eberl, A.N. Vagnozzi, A. Belkadi, N.A. Veniaminova, M.E. Verhaegen, et al., Basal cell carcinoma preferentially arises from stem cells within hair follicle and mechanosensory niches, Cell Stem Cell 16 (4) (2015) 400–412. Available from https://doi.org/10.1016/j.stem.2015.02.006, http://www.journals.elsevier.com/cell-stem-cell/.

[74] M.E. Hutchin, Sustained hedgehog signaling is required for basal cell carcinoma proliferation and survival: conditional skin tumorigenesis recapitulates the hair growth cycle, Genes & Development 19 (2) (2005) 214–223. Available from https://doi.org/10.1101/gad.1258705.

[75] V. Clement, P. Sanchez, N. de Tribolet, I. Radovanovic, A. Ruiz i Altaba, HEDGEHOG-GLI1 signaling regulates human glioma growth, cancer stem cell self-renewal, and tumorigenicity, Current Biology 17 (2) (2007) 165–172. Available from https://doi.org/10.1016/j.cub.2006.11.033.
[76] C.D. Peacock, Q. Wang, G.S. Gesell, I.M. Corcoran-Schwartz, E. Jones, J. Kim, et al., Hedgehog signaling maintains a tumor stem cell compartment in multiple myeloma, Proceedings of the National Academy of Sciences 104 (10) (2007) 4048–4053. Available from https://doi.org/10.1073/pnas.0611682104.
[77] C. Zhao, A. Chen, C.H. Jamieson, M. Fereshteh, A. Abrahamsson, J. Blum, et al., Hedgehog signalling is essential for maintenance of cancer stem cells in myeloid leukaemia, Nature 458 (7239) (2009) 776–779. Available from https://doi.org/10.1038/nature07737.
[78] F. Varnat, A. Duquet, M. Malerba, M. Zbinden, C. Mas, P. Gervaz, et al., Human colon cancer epithelial cells harbour active HEDGEHOG-GLI signalling that is essential for tumour growth, recurrence, metastasis and stem cell survival and expansion, EMBO Molecular Medicine 1 (6-7) (2009) 338–351. Available from https://doi.org/10.1002/emmm.200900039.
[79] J.H. Tsai, J. Yang, Epithelial–mesenchymal plasticity in carcinoma metastasis, Genes & Development 27 (20) (2013) 2192–2206. Available from https://doi.org/10.1101/gad.225334.113.
[80] M. Kahn, Can we safely target the WNT pathway? Nature Reviews Drug Discovery 13 (7) (2014) 513–532. Available from https://doi.org/10.1038/nrd4233.
[81] Y. Komiya, R. Habas, Wnt signal transduction pathways, Organogenesis 4 (2) (2008) 68–75. Available from https://doi.org/10.4161/org.4.2.5851, http://www.landesbioscience.com/journals/organogenesis/article/KomiyaORG4-2.pdf.
[82] B.T. MacDonald, K. Tamai, X. He, Wnt/β-catenin signaling: components, mechanisms, and diseases, Developmental Cell 17 (1) (2009) 9–26. Available from https://doi.org/10.1016/j.devcel.2009.06.016.
[83] B. Bao, A. Ahmad, A.S. Azmi, S. Ali, F.H. Sarkar, Overview of cancer stem cells (CSCS) and mechanisms of their regulation: implications for cancer therapy, Current Protocols in Pharmacology 61 (2013). Available from https://doi.org/10.1002/0471141755.ph1425s61.
[84] Yuan, Tao, Zhang, Zhang, Sun, Wu, Role of Wnt/β-catenin signaling in the chemoresistance modulation of colorectal cancer, BioMed research international 2020 (2020).
[85] V. Ingangi, M. Minopoli, C. Ragone, M.L. Motti, M.V. Carriero, Role of microenvironment on the fate of disseminating cancer stem cells, Frontiers in Oncology 9 (2019). Available from https://doi.org/10.3389/fonc.2019.00082, http://www.frontiersin.org/Oncology/about.
[86] F.G. Giancotti, Mechanisms governing metastatic dormancy and reactivation, Cell 155 (4) (2013) 750–764. Available from https://doi.org/10.1016/j.cell.2013.10.029.
[87] J.P. Medema, L. Vermeulen, Microenvironmental regulation of stem cells in intestinal homeostasis and cancer, Nature 474 (7351) (2011) 318–326. Available from https://doi.org/10.1038/nature10212.
[88] S.S. Zhang, Z.W. Huang, L.X. Li, J.J. Fu, B. Xiao, Identification of CD200+ colorectal cancer stem cells and their gene expression profile, Oncology Reports 36 (4) (2016) 2252–2260. Available from https://doi.org/10.3892/or.2016.5039, https://www.spandidos-publications.com/or/36/4/2252/download.
[89] S. Vafaei, A.O. Zekiy, R.A. Khanamir, B.A. Zaman, A. Ghayourvahdat, H. Azimizonuzi, et al., Combination therapy with immune checkpoint inhibitors (ICIs); a new frontier, Cancer Cell International 22 (1) (2022). Available from https://doi.org/10.1186/s12935-021-02407-8.
[90] H. Long, R. Xie, T. Xiang, Z. Zhao, S. Lin, Z. Liang, et al., Autocrine CCL5 signaling promotes invasion and migration of CD133 + ovarian cancer stem-like cells via NF-κB-mediated MMP-9 Upregulation, STEM CELLS 30 (10) (2012) 2309–2319. Available from https://doi.org/10.1002/stem.1194.
[91] A. Barzegar Behrooz, A. Syahir, S. Ahmad, CD133: beyond a cancer stem cell biomarker, Journal of Drug Targeting 27 (3) (2019) 257–269. Available from https://doi.org/10.1080/1061186x.2018.1479756.
[92] M.H. Mesrati, S.E. Syafruddin, M.A. Mohtar, A. Syahir, CD44: A multifunctional mediator of cancer progression, Biomolecules 11 (12) (2021) 1850. Available from https://doi.org/10.3390/biom11121850.
[93] A. Jaggupilli, E. Elkord, Significance of CD44 and CD24 as cancer stem cell markers: an enduring ambiguity, Clinical and Developmental Immunology 2012 (2012) 1–11. Available from https://doi.org/10.1155/2012/708036.
[94] G. Mudduluru, P. Ceppi, R. Kumarswamy, G.V. Scagliotti, M. Papotti, H. Allgayer, Regulation of Axl receptor tyrosine kinase expression by miR-34a and miR-199a/b in solid cancer, Oncogene 30 (25) (2011) 2888–2899. Available from https://doi.org/10.1038/onc.2011.13.
[95] N.S. Holah, H.A.E.S. Aiad, N.Y. Asaad, E.A. Elkhouly, A.G. Lasheen, Evaluation of the role of ALDH1 as cancer stem cell marker in colorectal carcinoma: an immunohistochemical study, Journal of Clinical and Diagnostic Research 11 (1) (2017) EC17. Available from https://doi.org/10.7860/JCDR/2017/22671.9291, http://www.jcdr.net/articles/PDF/9291/22671_CE[Ra1]_F(GH)_PF1(PI_RK)_PFA(AK)_PF2(PSH)%20(1).pdf.
[96] S. Deng, X. Yang, H. Lassus, S. Liang, S. Kaur, Q. Ye, et al., Distinct expression levels and patterns of stem cell marker, aldehyde dehydrogenase isoform 1 (ALDH1), in human epithelial cancers, PLoS ONE 5 (4) (2010) e10277. Available from https://doi.org/10.1371/journal.pone.0010277.
[97] T. Wang, M.P. Gantier, D. Xiang, A.G. Bean, M. Bruce, S.-F. Zhou, et al., EpCAM Aptamer-mediated survivin silencing sensitized cancer stem cells to doxorubicin in a breast cancer model, Theranostics 5 (12) (2015) 1456–1472. Available from https://doi.org/10.7150/thno.11692.
[98] O. Gires, M. Pan, H. Schinke, M. Canis, P.A. Baeuerle, Expression and function of epithelial cell adhesion molecule EpCAM: where are we after 40 years? Cancer and Metastasis Reviews 39 (3) (2020) 969–987. Available from https://doi.org/10.1007/s10555-020-09898-3.
[99] K. Ho Tang, Y.D. Dai, M. Tong, Y.P. Chan, P.S. Kwan, L. Fu, et al., A CD90+ tumor-initiating cell population with an aggressive signature and metastatic capacity in esophageal cancer, Cancer Research 73 (7) (2013) 2322–2332. Available from https://doi.org/10.1158/0008-5472.can-12-2991.

[100] G. Rezaei, M. Habibi-Anbouhi, M. Mahmoudi, K. Azadmanesh, S. Moradi-Kalbolandi, M. Behdani, et al., Development of anti-CD47 single-chain variable fragment targeted magnetic nanoparticles for treatment of human bladder cancer, Nanomedicine 12 (6) (2017) 597–613. Available from https://doi.org/10.2217/nnm-2016-0302.

[101] A.N. Dhande, S.G. Sinai Khandeparkar, A.R. Joshi, M.M. Kulkarni, N. Pandya, N. Mohanapure, et al., Stromal expression of CD10 in breast carcinoma and its correlation with clinicopathological parameters, South Asian Journal of Cancer 08 (01) (2020) 18–21. Available from https://doi.org/10.4103/sajc.sajc_56_18.

[102] M. Tachezy, H. Zander, F. Gebauer, A. Marx, J.T. Kaifi, J.R. Izbicki, et al., Activated leukocyte cell adhesion molecule (CD166)—Its prognostic power for colorectal cancer patients, Journal of Surgical Research 177 (1) (2012) e15. Available from https://doi.org/10.1016/j.jss.2012.02.013.

[103] A.G. Hansen, S.A. Arnold, M. Jiang, T.D. Palmer, T. Ketova, A. Merkel, et al., ALCAM/CD166 Is a TGF-β–responsive marker and functional regulator of prostate cancer metastasis to bone, Cancer Research 74 (5) (2014) 1404–1415. Available from https://doi.org/10.1158/0008-5472.can-13-1296.

[104] Z. Liang, J. Brooks, M. Willard, K. Liang, Y. Yoon, S. Kang, et al., CXCR4/CXCL12 axis promotes VEGF-mediated tumor angiogenesis through Akt signaling pathway, Biochemical and Biophysical Research Communications 359 (3) (2007) 716–722. Available from https://doi.org/10.1016/j.bbrc.2007.05.182.

[105] C.V. Hinton, S. Avraham, H.K. Avraham, Role of the CXCR4/CXCL12 signaling axis in breast cancer metastasis to the brain, Clinical & Experimental Metastasis 27 (2) (2010) 97–105. Available from https://doi.org/10.1007/s10585-008-9210-2.

[106] N. Gajendran, M.G. Hernandez, A. Yende, Z. Alahmadi, X. Li, Z. Munoz, et al., Abstract 666: targeting the CD47/SIRPα "do not eat me" phagocytic pathway in macrophages to improve anti-CD47 immune therapy, Cancer Research 83 (7_Supplement) (2023) 666. Available from https://doi.org/10.1158/1538-7445.am2023-666.

[107] P.P. Manna, W.A. Frazier, CD47 Mediates killing of breast tumor cells via Gi-dependent inhibition of protein kinase a, Cancer Research 64 (3) (2004) 1026–1036. Available from https://doi.org/10.1158/0008-5472.can-03-1708.

[108] A.D. Boiko, O.V. Razorenova, M. van de Rijn, S.M. Swetter, D.L. Johnson, D.P. Ly, et al., Human melanoma-initiating cells express neural crest nerve growth factor receptor CD271, Nature 466 (7302) (2010) 133–137. Available from https://doi.org/10.1038/nature09161.

[109] T. Redmer, Y. Welte, D. Behrens, I. Fichtner, D. Przybilla, W. Wruck, et al., Chen, the nerve growth factor receptor CD271 Is crucial to maintain tumorigenicity and stem-like properties of melanoma cells, PLoS ONE 9 (5) (2014) e92596. Available from https://doi.org/10.1371/journal.pone.0092596.

[110] J.C. Snyder, L.K. Rochelle, H.K. Lyerly, M.G. Caron, L.S. Barak, Constitutive internalization of the leucine-rich G protein-coupled receptor-5 (LGR5) to the trans-golgi network, Journal of Biological Chemistry 288 (15) (2013) 10286–10297. Available from https://doi.org/10.1074/jbc.m112.447540.

[111] M. Huch, C. Dorrell, S.F. Boj, J.H. van Es, V.S.W. Li, M. van de Wetering, et al., In vitro expansion of single Lgr5+ liver stem cells induced by Wnt-driven regeneration, Nature 494 (7436) (2013) 247–250. Available from https://doi.org/10.1038/nature11826.

[112] A. Ouban, P. Muraca, T. Yeatman, D. Coppola, Expression and distribution of insulin-like growth factor-1 receptor in human carcinomas, Human Pathology 34 (8) (2003) 803–808. Available from https://doi.org/10.1016/s0046-8177(03)00291-0.

[113] M. Youssif, E. Abdelzaher, Immunohistochemical study of C-KIT (CD117) expression in renal cell carcinoma, Journal of Urology 183 (2010) e245.

[114] L.G. Villa-Diaz, J.K. Kim, A. Laperle, S.P. Palecek, P.H. Krebsbach, Inhibition of focal adhesion kinase signaling by integrin α6β1 supports human pluripotent stem cell self-renewal, Stem Cells 34 (7) (2016) 1753–1764. Available from https://doi.org/10.1002/stem.2349.

[115] G.D. Bigoni-Ordóñez, D. Czarnowski, T. Parsons, G.J. Madlambayan, L.G. Villa-Diaz, Integrin α6 (CD49f), The microenvironment and cancer stem cells, Current Stem Cell Research & Therapy 14 (5) (2019) 428–436. Available from https://doi.org/10.2174/1574888x13666181002151330.

[116] B. Bielekova, M. Catalfamo, S. Reichert-Scrivner, A. Packer, M. Cerna, T.A. Waldmann, et al., Regulatory CD56 bright natural killer cells mediate immunomodulatory effects of IL-2Rα-targeted therapy (daclizumab) in multiple sclerosis, Proceedings of the National Academy of Sciences 103 (15) (2006) 5941–5946. Available from https://doi.org/10.1073/pnas.0601335103.

[117] K.H. Toomer, J.B. Lui, N.H. Altman, Y. Ban, X. Chen, T.R. Malek, Essential and non-overlapping IL-2Rα-dependent processes for thymic development and peripheral homeostasis of regulatory T cells, Nature Communications 10 (1) (2019). Available from https://doi.org/10.1038/s41467-019-08960-1.

[118] M.V. Shaikh, S. Custers, A. Anand, P. Miletic, C. Venugopal, S.K. Singh, Cancer stem cells: current challenges and future perspectives, Springer Science and Business Media LLC 2777 (2024) 1–18. https://doi.org/10.1007/978-1-0716-3730-2_1.

[119] S.N. dos Santos, T.H. Witney, Molecular imaging of cancer stem cells and their role in therapy resistance, Journal of Nuclear Medicine 66 (1) (2025) 14–19. Available from https://doi.org/10.2967/jnumed.124.267657.

[120] R. Margiana, A. Markov, A.O. Zekiy, M.U. Hamza, K.A. Al-Dabbagh, S.H. Al-Zubaidi, et al., Clinical application of mesenchymal stem cell in regenerative medicine: a narrative review, Stem Cell Research & Therapy 13 (1) (2022). Available from https://doi.org/10.1186/s13287-022-03054-0.

[121] Y. Peng, W. Li, Q. Zhang, Editorial: immunomodulation of MSCs in tissue repairing and regeneration, Frontiers in Immunology 14 (2023). Available from https://doi.org/10.3389/fimmu.2023.1150106, https://www.frontiersin.org/journals/immunology#.

[122] V.V. Maldonado, N.H. Patel, E.E. Smith, C.L. Barnes, M.P. Gustafson, R.R. Rao, et al., Clinical utility of mesenchymal stem/stromal cells in regenerative medicine and cellular therapy, Journal of Biological Engineering 17 (1) (2023). Available from https://doi.org/10.1186/s13036-023-00361-9.

[123] R. Chen, G. Monaco, B. Stefanini, M. Marseglia, S. De Lorenzo, F. Tovoli, Combination makes strength: a narrative review of transarterial radioembolization plus immune checkpoint inhibitors for hepatocellular carcinoma, Chinese Clinical Oncology 13 (5) (2024) 71. Available from https://doi.org/10.21037/cco-24-2771.

[124] Z. Wang, Assessing tumorigenicity in stem cell-derived therapeutic products: a critical step in safeguarding regenerative medicine, Bioengineering 10 (7) (2023) 857. Available from https://doi.org/10.3390/bioengineering10070857.

[125] R. Otsuka, H. Wada, T. Murata, K.-ichiro Seino, Immune reaction and regulation in transplantation based on pluripotent stem cell technology, Inflammation and Regeneration 40 (1) (2020). Available from https://doi.org/10.1186/s41232-020-00125-8.

[126] T.B. Meissner, H.S. Schulze, S.M. Dale, Immune editing: overcoming immune barriers in stem cell transplantation, Current Stem Cell Reports 8 (4) (2022) 206–218. Available from https://doi.org/10.1007/s40778-022-00221-0.

[127] N.M.P. King, J. Perrin, Ethical issues in stem cell research and therapy, Stem Cell Research & Therapy 5 (4) (2014). Available from https://doi.org/10.1186/scrt474.

[128] S.C. Selvakumar, K.A. Preethi, K. Ross, D. Tusubira, M.W.A. Khan, P. Mani, et al., CRISPR/Cas9 and next generation sequencing in the personalized treatment of cancer, Molecular Cancer 21 (1) (2022). Available from https://doi.org/10.1186/s12943-022-01565-1.

[129] Z. Wang, H. Chen, P. Wang, M. Zhou, G. Li, Z. Hu, et al., Site-specific integration of TRAIL in iPSC-derived mesenchymal stem cells for targeted cancer therapy, Stem Cells Translational Medicine 11 (3) (2022) 297–309. Available from https://doi.org/10.1093/stcltm/szab031.

[130] R. Abbasi, R.M. Mesgin, F. Nazari-Khanamiri, N. Abdyazdani, Z. Imani, S.P. Talatapeh, et al., Mesenchymal stem cells-derived exosomes: novel carriers for nanoparticle to combat cancer, European Journal of Medical Research 28 (1) (2023). Available from https://doi.org/10.1186/s40001-023-01556-y.

[131] P. Samuel, S. Sundarraj, D.N.P. Sudarmani, Nanotechnology-based stem cell therapy: current status and perspectives, IntechOpen, (2023). Available from https://doi.org/10.5772/intechopen.109275.

[132] M. Chehelgerdi, M. Chehelgerdi, O.Q.B. Allela, R.D.C. Pecho, N. Jayasankar, D.P. Rao, et al., Progressing nanotechnology to improve targeted cancer treatment: overcoming hurdles in its clinical implementation, Molecular Cancer 22 (1) (2023). Available from https://doi.org/10.1186/s12943-023-01865-0.

[133] S. Vinogradov, X. Wei, Cancer stem cells and drug resistance: the potential of nanomedicine, Nanomedicine 7 (4) (2012) 597–615. Available from https://doi.org/10.2217/nnm.12.22.

[134] Y. Peng, W. Li, Q. Zhang, Immunomodulation of MSCs in tissue repairing and regeneration, Frontiers in Immunology 14 (2023) 1150106.

[135] A. Golebiewska, R.C. Fields, Advancing preclinical cancer models to assess clinically relevant outcomes, BMC Cancer 23 (1) (2023). Available from https://doi.org/10.1186/s12885-023-10715-7.

[136] B. Dave, J. Tailor, Human stem cell models to unravel brain cancer, BMC Cancer 24 (1) (2024). Available from https://doi.org/10.1186/s12885-024-13187-5.

[137] W. Zakrzewski, M. Dobrzyński, M. Szymonowicz, Z. Rybak, Stem cells: past, present, and future, Stem Cell Research and Therapy 10 (1) (2019). Available from https://doi.org/10.1186/s13287-019-1165-5, https://stemcellres.biomedcentral.com.

Chapter 30

Social issues regarding organ donation

Devanshi Kapoor[1], and Nishant Tyagi[2]
[1]*School of Biotechnology, Gautam Buddha University, Greater Noida, Uttar Pradesh, India,* [2]*Voxvess Biotech Lab Private Limited, New Delhi, India*

30.1 Introduction

Organ donation, the very pinnacle of meditative practice, provides hope to all those patients who are suffering from organ failure, and it has the added virtue of attacking the global shortage of transplantable organs. But even such a well-justified practice is still buried in the complexities of ethical, legal, social, and cultural considerations. Religious beliefs, gender disparities, and public misapprehensions are the big determinants of organ donation rates, while enforceable regulatory frameworks work to guarantee fairness and ethical integrity in the organ transplantation domain. This chapter makes an effort to address the supply issues involved in organ donation by analyzing its history, ethical dilemmas, religious and cultural influences, and gender discrepancies. It also investigates the global and national legal mechanisms that control organ donation and the efforts made in this regard to reflect the truth about it and bury the so-called taboos. It is anticipated that the comprehensive discussions covered in this chapter will provide a highly enlightened account of the issues in, and the potential opportunities provided, for organ donation, calling for an equally just and very ethical transplant process. The evolution of organ transplantation shows ingenious human intelligence, hard work, and continual medical development. From ancient myths and crude attempts to 3D print organs, as it is known today, this organ transplantation journey knew a tremendous progressive change administered through scientific revolutions, ethical debates, and the genius contributions of a few figures. This section pinpoints the important milestones, technological revolutions, and the figures whose contributions defined organ transplantation migration through history.

30.2 Historical background

Transplantation of organs, from ancient myths through crude attempts at achieving 3D printing today, was a truly ingenious journey led by human intellect, effort, and continuity in medicine. This fantastic metamorphosis of organ transplantation saw its dawn of consciousness, with scientific revolutions that continued to heighten the discourse, ethical discussions, and genius contributions from a very minute number of individuals. This section reveals the milestones, technological revolutions, and the individual contributions along the evolution path of organ transplantation in history.

30.2.1 Evolution of organ transplantation

In examining the transition from ancient myths to modern-day marvels, it is quite easy to see the influence of certain remarkable milestones, of special personalities that left their marks in their respective fields. The origin of organ transplantation had moved in line with the realm of myths and legends created by some of the ancient civilizations. In these stories, people believe there are divine seals and healers and mythical beings that perform miraculous transplants of body organs from the bodies of the dead or from that of animals. Here are some of these magnificent legends: Hindu mythology imbues Lord Ganesha's decapitated head-a symbolic meaning of organ transplantation- with its elephant head to be resurrected [1] and the "Sushruta Samhita," around 600 BCE, is one of the great Indian texts on medicine that mentions some advanced surgical procedures for reconstruction of the nose from skin flaps taken from the forehead, a clear precursor to modern reconstructive surgery [2]. The basic idea of a chimera, which is a mix of human and animal beings, was introduced by Greek mythology around 500 BCE, revealing early notions of biological modification [3]. These legends, while mythical, mirror in essence the fundamental notions of transplanting or altering body parts for functional restoration. Around the same time, advancements in organ transplantation surfaced in other ancient

Fundamentals of Translational Tissue Engineering. DOI: https://doi.org/10.1016/B978-0-443-33455-9.00027-4

civilizations. In China (c. 108–208 AD), Hua Tuo was the pioneer of many surgical techniques, reportedly performing abdominal operations and even early organ transplants with the use of anesthesia [4]. In 348 AD, a Christian tradition recounted the Miracle of the Black Leg, in which Saints Cosmas and Damian replaced a cancerous leg with one taken from the corpse of an Ethiopian man, showing how literal limb transplantation was being interpreted from the very beginning [5]. These stories, coupled with broad discoveries in surgical skills, depict the age-long human interest in surgical repair and transplant, wherein germination was laid for modern transplant medicine.

30.2.2 Early beginnings and theoretical foundations

Before the 20th century, transplant primarily concerned with technical issues like attaching severed limbs, nose, teeth, and so forth. It was not until the 20th century that graft failure rates were mentioned. John Hunter, the great experimentalist of the 18th century who transplanted human teeth and autotransplanted cocks' spurs into their combs, did not seem to have a clue that homograft would fail [6]. Only in the late half of the 20th century was there agreement that the fates of homografts and autografts were different. The first experimental attempts at transplantation were with skin grafts, standing as the first thing done by Swiss surgeon Jacques-Louis Reverdin: He successfully performed autografts in 1869 (Fig. 30.1) [7]. These early works proved that tissue transfer was feasible but also introduced the difficulty of immune rejection: a challenge that would dominate transplantation research for decades to come. By the 1930s and '40 s, surgeons began experimenting with kidney transplants in animals, steadily improving the techniques for their surgery and the

FIGURE 30.1 Progression of organ transplantation through pivotal achievements by medical pioneers, reflecting the historical evolution. The image traces landmark innovations that transformed transplant medicine from experimental procedures to life-saving operations.

understanding of immunological barriers. 1954 marked a turning point when Dr. Joseph E. Murray wrote: The first successful human organ transplant was that of a kidney transplant between identical twins. Because of him, the operation raised the potential of organ transplantation as a therapeutic modality, and he would earn the Nobel Prize for Physiology or Medicine in 1990 [8,9]. The success of that procedure relied on the genetic identity of the donor and the recipient, thus skipping past the block of immune rejection.

30.3 Ethical concerns

In many nations, India included, the opt-in system of organ donation relies on the express consent from individuals for donation, whether donor registration or family authorization. Though studies have shown family members frequently override the deceased person's wishes as they suffer emotional distress or do not fully comprehend the donation process [10]. Here, serious questions are posed on individual autonomy and decision-making by family members, especially within cultures that prioritize collective decision-making in normal circumstances. On the contrary, the opt-out system presumes consent unless individuals specifically refuse to donate an organ; this is a possible explanation for the boost in donation rates in countries like Spain and Belgium [11]. However, such systems have been faulted on the grounds that they undermine the autonomy of the individual and further perpetuate systemic inequities, particularly in low-resource settings where access and awareness of healthcare are limited. The issue of equity and access adds a more stuffed challenge to the area of organ donation. The gap in organ availability is truly very huge; low- and middle-income countries (LMICs) mainly suffer the brunt of chunk shortage in number of organs. In India alone, 500,000 people per year require organ transplants, while only a small fraction gets them [12]. Due to poverty, for instance, the situation becomes more complicated industrially and through lack of healthcare. The commercialized illegal organ trafficking places various crimes against humanity, while the shortened profit volumes in some places forced this project into creation. Even if many WHO opinions were stated against the trade of organs for benefit, unfortunately, it is still moving lividly in many areas due to compulsion put forth with poverty [13]. This adds to the exploitation framework and allows for easier work within periods piled with inequalities.

30.3.1 Informed consent and autonomy

Informed consent is one of the cornerstones of medical ethics, illustrating the inherent value of autonomy and the right to self-determination about one's own body. Informed consent specifically for organ donation requires that prospective donors be provided with relevant and sufficient information regarding the nature of their procedure, the risks and benefits that come along with it, alternatives that exist, and a prediction of what to expect. The reason for this is to ensure that a potential donor has voluntarily made an informed decision on their choice: to donate their organs or to live without one or some of them. The conundrum is opaque because of the multidimensionality of such a procedure. Organ transplantation, as opposed to any other biomedical procedure, is applied discreetly to a donor on one side and to the recipient at the opposite side, thereby placing a number of medical, legal, and ethical issues in the whole process. One may, for instance, begin speaking of explaining to the whole person the possible risks that he or she would undergo as a donor: risks of surgical complications, long-term effects, and psychological impact. Equally, the person is also reminded of what such a decision implies; that is, it explains the consequences, including organ rejection, and the lifelong medical follow-up the recipients have to undergo. Lately, research has started to identify challenges to truly informed consent in organ transplantation. For example, Raza and Neuberger [14] propose a multifactor approach to informed consent that includes understanding risk, effective communication, and robust review processes. The main thrust is the responsibility of the healthcare professionals to use their judgment and discretion to decide that consent is genuinely informed rather than simply a legal requirement [14].

30.3.2 Ethical dilemmas in living and deceased organ donation

The living donor organ donation raises many ethical problems. Should living donation really be allowed to incur risks for the donor in terms of surgical complications, long-term illnesses, and psychological effects, among many others, with the aim of helping others? This raises a grave ethical issue on how justifiable it is to expose a healthy person to any risk for the sake of another individual [15]. In deceased organ donation, the ethical questions largely revolve around the determination of death and obtaining permission. Brain death, according to the medical definition of brain death, is an irreversible cessation of brain function, including the brain stem, which is accepted widely as a valid concept in medical practices. Though this belief has some ground, it remains controversial for the fact that different cultures and religions view brain death differently. Some ethical debates question whether brain death equals true death, challenging the moral background for the procurement of organs from brain-dead donors [16]. Further complicating this problematic question is the "dead-donor rule"-an ethical principle that denotes that organ donors should be declared dead prior to the organs' removal. Some people criticize the dead-donor rule and claim that not all cultures or individuals agree with its view of what constitutes death, thus causing certain ethical and legal challenges in organ transplant practices [17]. The matter is further complicated by challenging issues on consent for organ donation. In many jurisdictions, the Family often has the upper hand, since any one representative of the family may override the presumed consent shown in a will, and that may mean not only for the wishes of the deceased, but this situation further constricts suitable transplant filtering the available organs [18]. Or, conversely, where consent was never documented by the deceased, the family now makes a very painful decision in a time of grief whether to allow a donation. The beliefs of families regarding organ or tissue donation are often do so inform by their understanding of their departed loved ones' wishes, website, or cultural religious beliefs, along with quality communication with healthcare providers. Some studies indicate that families well informed about favorable views held by the deceased in relation to donation are more compliant with giving consent, hence stress on public awareness and debates on organ donation preferences [19].

The allocation of organs also presents significant ethical challenges. Given a limited supply of organs and ever-increasing demand, allocation practices need to consider fairness, equity, and medical urgency. The principles of ethics of justice and equity demand that organs be allocated according to need and prognosis; for example, it should not be dependent on one's socioeconomic status or where he or she lives. Balancing these considerations is extremely difficult and beset by ethical concerns [20].

30.3.3 Emotional coercion and exploitation risks

Emotional coercion and exploitation are two major ethical concerns in relation to organ donation, especially wherein living donation is concerned. Emotional coercion occurs when prospective donors feel they are expected to donate due to familial or social obligations, instead of making a decision that is entirely free and made based on informed consent. This pressure can be applied by family members, health professionals, or by societal expectations of the donor and is able thus to infringe upon the donor's autonomy. This exploitation risk is accentuated in vulnerable populations wherein organ donation becomes a purse-line between life and death. This arcana of selling of organs gives rise to extreme ethical considerations as the very values of justice, fairness while upholding the dignity of a human being get fouled since poor people are often exploited for doing something they have little knowledge of, that being organ transplantation and care received after this process [21]. WHO's "Guiding Principles on Human Organ Transplantation" emphasize the need to prevent commercialization and exploitation regarding organ donation and transplantation.

30.3.4 WHO guiding principles on human organ transplantation

WHO has always been an important body for the establishment of ethical guidelines for organ transplantation. Formulated in 1991 and continually updated during the last few decades, the WHO Guiding Principles on Human Organ Transplantation established a further ethical framework for organ transplantation [22]. Among the said principles, one of the most emphasized is voluntary and unpaid donation, outlawing organ trafficking, and guaranteeing equitable access to transplantation services.

The WHO Guiding Principles hinge on many views:

- **Voluntary and unpaid donation:** Organ donation should be on a voluntary basis and, most importantly, not applied for financially, as there should be no coercion or financial inducement for the donor to donate his organs.
- **Prohibition of the organ trafficking:** Strictly prohibits organ trafficking and any commercial dealings that involve human organs.
- **Equitable access:** Access to transplantation services should be equitable; that is, each patient must be afforded the right to organ transplantation irrespective of standing in the community.
- **Transparency and accountability:** This requires transparency in the donation process and accountability for organ procurement and allocation [22,23].

30.3.5 Declaration of Istanbul on organ trafficking and transplant tourism

The Declaration of Istanbul on Organ Trafficking and Transplant Tourism, originally adopted in 2008 and revised in 2018, is yet another major international instrument. This declaration was formulated by professional organizations such as The Transplantation Society (TTS) and the International Society of Nephrology (ISN) to combat organ trafficking and transplant tourism by promoting ethical practices that respect human dignity and rights in organ transplantation.

The Declaration of Istanbul outlines several salient principles:

- **Prohibition of organ trafficking and transplant tourism:** The declaration campuses the impermissibility of organ trafficking and transplant tourism. The exploitation of vulnerable populations and the undermining of the integrity of organ transplantation are highly penalized.
- **Promotion of ethical practices:** The declaration demands promotion and encouragement of ethical practices in disorgan procurement and organ allocation with the establishment of transparent and accountable systems.
- **Protection of donors and recipients:** The declaration champions that organs and tissues donors and transplant recipients be protected and that they have their rights and their welfare assured all the way through the transplantation pipeline.
- **International cooperation:** The Declaration calls for international cooperation on the matter of the traffic of organs and transplant tourism. This coordinated manner represents very well the already stated need for global responses to these problems [24].

The Declaration of Istanbul laid the basis for an increasing recognition of contemporary organ transplantation practices. It helped raise awareness on the moral issues surrounding transplant tourism and trafficking whilst encouraging various nations to alleviate them. The Declaration itself induced the formation of national policies and a body of legislation aiming at preventing organ trafficking and promoting ethical practices in organ transplantation.

30.3.6 Global challenges and ethical considerations

Global organ donation patterns exhibit glaring inequalities between high-income countries and LMICs. In high-income countries, organ donation has been facilitated by technological advancements, public awareness campaigns, and supportive legal frameworks regarding donation. In 2020 Spain recorded one of the highest organ donation rates at 49.6 per million population (pmp), all with the help of coordinated national transplant programs and opt-out systems [25]. By contrast, LSMS includes little dots that registered in India at a lonely 0.8pmp, which stresses the need for systemic reforms and capacity-building initiatives [26]. This huge gap between these regions reinforces the understanding that the organ donation paradigm of both those realities is intertwined with socioeconomic development, healthcare system, and public policy (Fig. 30.2). In line with these concepts is a panoply of other influencing elements in Indian context that amp

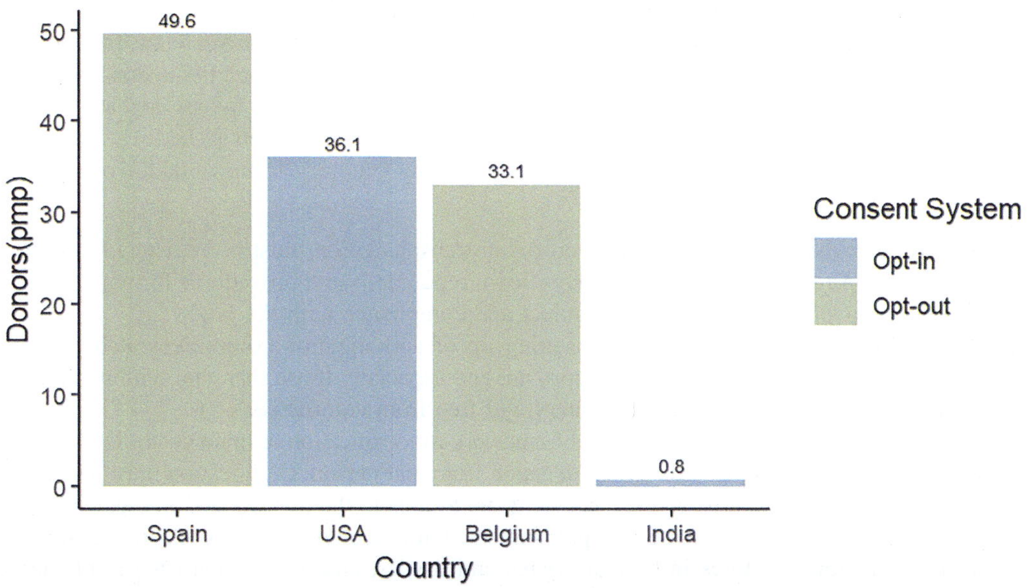

FIGURE 30.2 Comparison of organ donation rates across high-income and LMICs, underscoring stark global disparities in donation infrastructure and accessibility. Notably, Spain stands out with a rate of 49.6 donors per million population, attributed to its opt-out legal framework and highly coordinated national transplant program. In contrast, India reports significantly lower rates, at around 0.8 pmp, reflecting systemic challenges such as inadequate healthcare infrastructure, poor awareness, and sociocultural resistance to organ donation. The visual reinforces the need for equitable global health policy reforms and enhanced local awareness campaigns to bridge the divide and promote ethical organ donation practices worldwide. *LMICs*, Low- and middle-income countries.

the organ donation challenge. Urban centers such as Delhi and Chennai have achieved much more in organ donation promotion, yet rural areas lag far behind, with atrocious accessibility to health services and below-par levels of awareness [27]. Next, rigid societal perception about brain death and lack of trust on the part of people in the healthcare system have all too often been identified as considerations that block the path of donations [28]. With issues like financial inducements, coercion in donation, and organ trafficking remaining prevalent in LMICs, these unethical practices take center stage in the opinion of many as deriving from the global shortage of organs. An area of major concern for policymakers and researchers remains the interface among international ethics principles and domestic regulatory frameworks [29].

- One of the main challenges faced in organ transplantation involves being assured that donations are genuinely voluntary and free of coercion. Financially driven issues exert undue pressures which obfuscate the very ethics of organ donation, with the potential of the practice to descend to exploitation, and that's truly hazardous to the trapped donors. The WHO Guiding Principles and the Istanbul Declaration highlight the requirement for voluntary and unpaid donation, but the practical realization of these principles remains ever elusive [23,24].
- The other crucial issue is equitable access to the transplant service. Some socioeconomic disparities may imply a lack of ready access to organ transplantation, with relatively better facilities for the wealthy. Although equitable access is an emphasis of the WHO Guiding Principles, the attainment of this noble goal works against larger social and economic inequalities [30].
- The posttransplant care would demand some ethical obligations as well. The long-term welfare of the different individuals involved in the transplant process will require adequate medical follow-up and support for both donors and recipients. Having access to posttransplant care, especially in LMICs, remains a major challenge [31].

This should be tackled through a combination of initiatives embracing community engagement, capacity-building, and policy reform to provide enabling conditions for organ donation.

30.4 Legal and regulatory framework

The main legal framework of organ transplantation in India is provided by the Transplantation of Human Organs Act (THOA) of 1994 and other amendments made to enhance its provisions. The THOA was passed in 1994 with the goal of regulating the removal, storage, and transplantation of human organs and preventing all forms of commerce in human organs. The Act is applicable across India, including the states of Goa, Himachal Pradesh, and Maharashtra, along with the Union territories [32]. After its being brought into operation, the THOA of 1994 has been revised from time to time in consonance with emerging challenges and to give it an improved regulatory frame. The most significant of these is the amendment of 2011, which extended the Act's scope to include tissues alongside the organs, thus the renaming of the Act to the Transplantation of Human Organs and Tissues Act, 1994. It states requirements that a national registry for organ and tissue donation and transplantation has been established, stricter penalties for commercial dealings in human organs and/or tissues, and authorization committees to monitor organ transplantation should be put in place at both national and state levels [33].

The THOTA, 1994, and its amendments constitute a wide-ranging statutory framework for organ transplantation in India. Some of these include

- **Hospitals having to be registered:** Hospitals that do organ transplantation must be registered, and the authorization must be done by the appropriate authority. The Act lays down regulations and procedures for registration, as well as the responsibilities of the hospitals to ensure ethical conduct and compliance with the law.
- **Authorization committee:** The Act mandates the setting up of authorization committees at both state and national levels for cases of organ transplantation that need approval, and especially those that deal with unrelated donors. These committees would ensure that the donation is voluntary and free from commerce.
- **Brain death certification:** The Act lays down the legal framework for certification of brain death. This involves doctors who must ascertain brain death as a prerequisite for organ donation from that person. Certification must take place by a medical expert committee which includes a neurologist or neurosurgeon, to uphold the honesty and transparency of the process.
- **Punishments:** The Act further prescribes severe punishment, which could include both imprisonment and fines for violations, to avert commercial dealings in human organs and tissues, and to safeguard the rights and dignity of the donors and recipients [32,33].

30.4.1 Role of regulatory bodies

The Apex body for overseeing organ and tissue donation and transplantation in India is the National Organ and Tissue Transplant Organization. NOTTO is the apex body under the Directorate General of Health Services, under the Ministry of Health and Family Welfare, to implement the National Organ Transplant Program and ensure compliance with the THOTA, 1994 [34].

Key functions of NOTTO include

- **Coordination and networking:** NOTTO serves as the central coordinating agency for organ and tissue procurement and distribution across the country. It maintains a national registry of donors and recipients and facilitates the matching and allocation of organs and tissues based on medical criteria and urgency.
- **Policy formulation and implementation:** NOTTO is responsible for formulating policies, guidelines, and standard operating procedures for organ and tissue donation and transplantation. It works closely with state and regional organizations to ensure uniformity and adherence to ethical standards.
- **Awareness and advocacy**: NOTTO conducts awareness campaigns and educational programs to promote organ donation and address misconceptions and cultural barriers. It collaborates with nongovernmental organizations, healthcare institutions, and media to reach a wider audience.
- **Training and capacity building**: NOTTO organizes training programs for healthcare professionals, including transplant coordinators, surgeons, and nurses, to enhance their skills and knowledge in organ and tissue transplantation. It also provides technical support and consultancy to hospitals and transplant centers.

Monitoring and evaluation: NOTTO monitors transplant centers and hospitals with respect to conformity with legal and ethical standards laid down by THOTA, 1994, and collects and analyzes data for organ and tissue transplantation, aiming to identify trends, challenges, and areas for improvement [35].

30.4.2 Comparison with international laws and guidelines

The regulation of organ donation and transplantation varies significantly across different countries, reflecting diverse legal, ethical, and cultural contexts.

30.4.2.1 United States

In the United States, organ donation is regulated by the National Organ Transplant Act (NOTA), 1984. Among other things, the act prohibits the sale of organs and tissues for transplantation and establishes the OPTN that coordinates the organ donation and transplantation process. The OPTN is responsible for providing comprehensive oversight through a national registry of organ donors and recipients, ensuring that organs are allocated fairly on the basis of medical criteria, and raising public awareness for organ donation. One of the major differences between Indian and American legal frameworks is that the United States has OPTN that has a far more centralized format of coordinating organ donation and transplantation at a national level, while India has regulatory bodies such as the NOTTO. Moreover, the American legal system emphasizes equity in organ allocation; this means organs are allocated according to medical need and the likelihood of successful outcome [36].

30.4.2.2 European Union

The European Union (EU), via Directive 2010/53/EU on the Quality and Safety of Human Organs Intended for Transplantation, has a wider initiative in creating a legal framework for organ donation and transplantation. This directive establishes common standards for the quality and safety of organs intended for transplantation and bans organ sales for transplantation purposes. It also mandates EU member states to set up national competent authorities to watch over the directive's implementation and its compliance with its provisions. One of the key features of difference between Indian and EU legal systems is that in the EU, there is a strong emphasis on enforcing quality and safety standards for organ donation and transplantation. While THOTA in India is mainly focused on the regulation of organ donation and transplantation and prevention of commercialization, the EU directive further stresses ensuring the quality and safety standards concerning the organs through enhanced standards and mechanisms of oversight [37].

30.4.2.3 United Kingdom

The laws governing organ donation in the United Kingdom are based on the Human Tissue Act, 2004, which regulates the removal, storage, and use of human tissue, including organs for transplantation. The act establishes the Human Tissue Authority (HTA) as its overseer, which must ensure the provisions of the act are being followed. HTA provides licensing powers and inspects organizations involved in organ donation and transplantation; additionally, it offers professional guidance and public information on organ donation. A salient difference between the legal frameworks of India and the

United Kingdom is that, while in India, committees for authorization exist at both the state and national levels to oversee the organ donation and transplantation process, the HTA in the United Kingdom has explicit regulation and supervision of the entire process. Besides that, the informed consent principle warrants serious attention in the UK legal framework; that is, donors and family members must have knowledge of the whole process of donation and their rights [38].

30.4.2.4 Spain

Spain has been hailed many times as a role model country looking into organ donation and transplantation because of its much higher donation rates and effective legal framework. The process of organ donation in Spain is overseen by the Spanish Transplantation Law, which is based on the theory of presumed consent. This system presumes that all individuals are organ donors except those who expressly opt out. The coordination of the organ donation and transplantation in Spain is done at the national level under the authority of the Spanish National Transplant Organization (ONT), which also distributes organs equitably across the country and makes efforts to create public awareness on organ donation. Besides, Spain offers a significant contrast from the Indian legal framework, mainly due to the methodology of donation: presumed consent versus opt-in. While in India, the individual must agree to organ donation, Spain gives its system a capital credit for grossly enhancing the rate of organ donations because of the presumed consent system. Further, ONT has assumed a snoozing role in arbiter plus coordinator activities in organ donation and transplantation, while in India, there is no such central authority [39].

30.5 Social and cultural barriers

One of the most important social issues surrounding organ donation is an alarming lack of awareness and education regarding it. Various studies suggest that certain serious misconceptions about organ donation, like the fears of medical neglect or religious prohibitions, have acted as the most serious barriers to organ donor registration [40]. Since organ donation takes place in the Indian context, it has strong religious and cultural underpinnings, with both constructive and destructive contributions to this practice. Being a diverse country, India has multiple religions, with each teaching their own set of beliefs and practices about attitudes towards some issues. Therefore it is pertinent to take cognizance of these religious and cultural perspectives if barriers to organ donation are to be addressed and a culture for organ donation promoted in India. The myriad religions, including Hinduism, Islam, Christianity, Sikhism, and Buddhism, often have a bearing on people's decisions concerning organ donation.

Beliefs about organ donation are variously held within Hinduism, but there are central issues. In Hindu philosophy, karma refers to laws of action and their consequences, while samsara infers the cycle of rebirth. Some Hindus worry that organ donation could thwart the soul's journey into the afterlife. However, much of Hindu teaching emphasizes selfless giving (daan) and righteous living (dharma), which could support organ donation. The Hindu Temple Society of North America notes that Religious law does not prohibit organ donation, and that such a choice is an individual's decision [41].

Different scholars and communities offer different perspectives on organ donation from the Islamic view, weighing out among the competing factors of religion and ethics. Various issues are regarded as key to both the discussions of the broader subject of organ donation and Islamic decisions about organ donation: sanctity of the human body, obligation to save lives, and conditions under which donation becomes permissible. Islam places tremendous emphasis on the careful treatment of the human body, both in life and after death. This respect raises concerns about interventions that would appear to violate this respect, such as the removal of an organ. While some scholars argue that organ donation violates this deference and is therefore not to be sanctioned under any circumstances, a majority of Islamic scholars see saving a life as a virtue; in fact, they will cite a Quranic verse to the effect that saving one life is equivalent to saving all of humanity [42]. This reasoning supports organ donation, since it is viewed as a charitable act (sadaqat) or a show of compassion. Considering organ donation as part of charity, an act of humility, and based on compassion and mercy, the Fiqh Council of North America has recently ruled that organ donation is licit in Islamic law, if done under a set of certain conditions. Islamic Fiqh Academy of India has also issued a fatwa supporting organ donation, restricting it to those cases where consent is duly informed and where the trade in organs in any form has been expressly ruled out [43].

Christianity, especially in the catholic part, traditionally endorses organ donation as an act of love and compassion. The Catholic Church views organ donation as a noble act of charity. Pope John Paul II termed it "a generous act of charity" and an example of "self-giving love." Pope John Paul II stated in the 18th International Congress of the Transplantation Society, 2000, that the donation of organs is a "genuine act of love," wherein he also stressed upon the importance of informed consent and the ethical concerns surrounding the determination of death [44]. "The Catechism of the Catholic Church" supports organ donation, affirming it as a noble and meritorious act, even an expression of generous solidarity worthy to be encouraged [45]. However, the morally acceptable way the act must be performed is also stressed. Ethically acceptable organ donation should be done with consent and without commercialization. In all cases, the dignity of the human person ought to be respected; organ donation should not bring harm to the donor [46].

Buddhism views organ donation differently between individuals: that difference relies upon the core teachings in the concepts of compassion (karuṇā) and rebirth (saṃsāra). A few Buddhists may raise concern that organ donation can impact rebirth, whereas others adopt the viewpoint that organ donation is an act of compassion and generosity [47]. Within Buddhism, the decision of organ donation is, by and large, a matter of personal conscience. This faith highly regards acts of compassion and relieving suffering, which tend to favor organ donation. Reverend Gyomay Masao, president and founder of the Buddhist Temple of Chicago, expressed this with the statement: "We honor those people who donate their bodies and organs to the advancement of medical science and to saving lives" [41].

30.5.1 Gender disparities in organ donation and transplantation

Apart from organ donation and transplantation, gender inequalities are an important social issue in India. Evidence from the National Organ and Tissue Transplant Organization (NOTTO) indicates that the number of organ donors is skewed towards women, while a majority of organ recipients are male. The gender imbalance highlights the wider inequalities based on social stratification and patriarchal norms within Indian society. Manvitha B S [48] conducted a study that discovered that 80% of organ recipients in India are men, while female donors constitute 75%. The sociocultural aspects that drive men as the principal breadwinners and the expectation from women to prioritize family needs in their decisions influence this uneven ratio. Financial dependence and the stronger mentality of self-sacrifice among women also lead to this imbalance. Women are perhaps likely to donate organs to relatives of their close circle, especially in contexts involving living donation. Mothers are, therefore, 73% of the total donation from parents, while wives contribute 91% of donations in married couples. Such trends strongly reflect the stereotypes and societal comportment aligned towards women to catering familial needs [48]. The gender gap in organ transplantation further extends into the domain of healthcare access. Women are in an inferior position to receive an organ transplant than men, even when women appear more willing to donate. Such inequalities stem from prejudices inherent in medical institutions, entrusting men the priority for life-saving treatments from the perspective of their value in economic terms [49].

30.5.2 Public awareness and misconceptions

Public awareness and misconceptions surrounding organ donation are among the most formidable barriers obstructing an increase in donation rates across the Indian Subcontinent. While a few efforts have been made over the years to drum up support for organ donation, many still cast a wary eye, mainly because of a gross lack of information and rampant assumptions. In a fascinating study carried out by Neha Joe et al. [50], they found that, while 95% of their respondents supported organ donation, only 51% expressed willingness to donate. Such a great divide between support and willingness portrays multiple hurdles built on misunderstandings about the process of donation, the fear of its misuse, or/imposed social and cultural beliefs [50]. One of the more common misconceptions is that organ donation is somehow against religious principles. While most of the major religions in India endorse organ donation as an act of charity or compassion, this very endorsement is something the majority is, however, completely unaware of. So the factors that play into the overall hesitance in agreeing to become a donor. Then there is this fear surrounding the very concept of brain death, that is, brain death is thought to be reversible in time. Brain death, according to most medical and legal definitions worldwide, is accepted as death based purely on the fact that there is no longer any function in the brain. Ensuing misunderstandings regarding brain death only add to the hesitancy to go in for organ donation [51]. Public awareness campaigns and educational interventions are the best way to dispel myths and stereotypes surrounding the whole concept of organ donations. As Dr. P Vikranth Reddy (2023) puts it, it is imperative to educate the general public about the benefits of donating organs, and such awareness programs should rightly snuff out the fire of myths surrounding the concept [52]. Education interventions done by Anjum A K Sayyed et al. [53] have shown that improving general knowledge and awareness about organ donation contributed remarkably to an increased willingness to donate [53].

30.6 Conclusion

Organ donation is an immensely important practice through medical, social, and ethical lenses and serves as a beacon of hope for individuals afflicted by end-stage organ failure, standing tall on the horizon of a crisis of a shortage of viable organs for the purpose of transplantation. The process is bound to improve the quality of life of recipients; besides, its nature is founded on the principles of altruism and solidarity, instilling consciousness of community and interdependence among members. Cultural and religious beliefs in India play a significant role in shaping views towards organ donation; the diversity of religions dictates the decision of their followers. The positive features of these beliefs should be promoted, gender disparities must be addressed, and effective public awareness campaigns should be started in India to boost acceptance and willingness for organ donation further. All countries have put up structures and laws, international and national parenthesis, to develop organ donation itself: one compares the THOTA with international laws and, thus, highlights the specific challenges and opportunities within the Indian context. Various ways that are used in an effort to popularize organ donation in India are public awareness campaigns, educational interventions, and initiatives run by organizations such as the National Organ and Tissue Transplant Organization.

There is a need to promote the social, psychological, and economic support that should underlie the principle of donation as a value of the family. Once this is done, a large-scale allocation of other elements could work for organ donation in alliance with the support of pubic awareness campaigns, resolution of myths, and anchoring of cultural values and religious principles, resulting in many lives saved with more opportunities for transplantation and a society based on compassion, solidarity, and altruism. The book chapter thus far has described different social issues concerning organ donation: ethical issues, legal and regulatory frameworks, as well as the overall influence of cultural and religious beliefs. The principles of informed consent and autonomy hinge on the assurance that donors and family members are fully informed, aware, and knowledgeable of product implications and consequences following such action. However, one needs to unscramble the intricacies of related ethical dilemmas such as emotional coercions and risks of exploitation into clear digestible bits of raw meat if one cares to maintain integrity in the donation process. In a nutshell, organ donation is a very profound practice, not only for the individual and family involved, but also for society at large. Addressing these social issues influencing organ donation helps advance towards a world where organ transplantation would be accessible and equitable, where organs would be donated based on ethical guidelines that would invariably work to the benefit of others while creating a world of giving and compassion.

AI disclosure

During the preparation of this work, the author(s) used Grammarly and ChatGPT to check grammar and spelling. After using this tool/service, the author(s) reviewed and edited the content as needed and take(s) full responsibility for the content of the publication.

References

[1] P. Harikrishnan, Head transplant: surprising and interesting facts from the Hindu mythology, Journal of Craniofacial Surgery 30 (5) (2019) 1329–1330.

[2] M.J. Niaz, W. Zakaria, The elaborate history of skin grafting, Journal of the American Academy of Dermatology 72 (5) (2015). Available from https://doi.org/10.1016/j.jaad.2015.02.117. AB271.

[3] E. Bazopoulou-Kyrkanidou, Chimeric creatures in Greek mythology and reflections in science, American Journal of Medical Genetics 100 (1) (2001) 66–80.

[4] R.S. Tubbs, S. Riech, K. Verma, J. Chern, M. Mortazavi, A.A. Cohen-Gadol, China's first surgeon: Hua Tuo (c. 108-208 AD), Child's nervous system: ChNS: official journal of the International Society for Pediatric Neurosurgery 27 (9) (2011) 1357–1360. Available from https://doi.org/10.1007/s00381-011-1423-z.

[5] P. Hernigou, Bone transplantation and tissue engineering, part I. Mythology, miracles and fantasy: from chimera to the miracle of the black leg of saints cosmas and damian and the cock of john hunter, International Orthopaedics 38 (2014) 2631–2638.

[6] C.E. Martin, John Hunter and tissue transplantation, Surgery, Gynecology & Obstetrics 131 (2) (1970) 306–310.

[7] J.L. Reverdin, De la greffe epidermique, P. Asselin, successeur de Béchet jeune et Labé, (1872).

[8] S.Y. Tan, J. Merchant, Joseph Murray (1919-2012): first transplant surgeon, Singapore Medical Journal 60 (4) (2019) 162–163. Available from https://doi.org/10.11622/smedj.2019032.

[9] J.E. Murray, D.E. Reid, J.H. Harrison, J.P. Merrill, Successful pregnancies after human renal transplantation, The New England Journal of Medicine 269 (1963) 341–343. Available from https://doi.org/10.1056/NEJM196308152690704.

[10] M.T. Li, G.C. Hillyer, S.A. Husain, S. Mohan, Cultural barriers to organ donation among Chinese and Korean individuals in the united states: a systematic review, Transplant international: official journal of the European Society for Organ Transplantation 32 (10) (2019) 1001–1018. Available from https://doi.org/10.1111/tri.13439.

[11] B.H. Willis, M. Quigley, Opt-out organ donation: on evidence and public policy, Journal of the Royal Society of Medicine 107 (2) (2014) 56–60. Available from https://doi.org/10.1177/0141076813507707.

[12] National Organ and Tissue Transplant Organization, Ministry of Health and Family Welfare, Government of India, NOTTO annual report 2023–2024, Government of India. Available from: https://notto.mohfw.gov.in/WriteReadData/Portal/News/858_1_Updated_NOTTO_ANNUAL_REPORT__08-05-24_pdf, 2024.

[13] A. Adair, S.J. Wigmore, Paid organ donation: the case against, Annals of the Royal College of Surgeons of England 93 (3) (2011) 191–192. Available from https://doi.org/10.1308/147870811X565061a.

[14] F. Raza, J. Neuberger, Consent in organ transplantation: putting legal obligations and guidelines into practice, BMC Med Ethics 23 (2022) 69. Available from https://doi.org/10.1186/s12910-022-00791-y.

[15] Institute of Medicine, Ethical considerations in living donation, Organ Donation: Opportunities for Action, The National Academies Press, (2006), pp. 263–299. Available from https://doi.org/10.17226/11643.

[16] S.J. Schweikart, Reexamining the flawed legal basis of the "dead donor rule" as a foundation for organ donation policy, AMA Journal of Ethics 22 (12) (2020) E1019–E1024. Available from https://doi.org/10.1001/amajethics.2020.1019.

[17] S. Dutchen, A fine line, Harvard Medicine Magazine. Available from: https://magazine.hms.harvard.edu/articles/fine-line, 2016.

[18] M. Toews, T. Caulfield, Evaluating the "family veto" of consent for organ donation, CMAJ: Canadian Medical Association journal=journal de l'Association medicale canadienne 188 (17-18) (2016) E436–E437. Available from https://doi.org/10.1503/cmaj.160752.

[19] C.V.R. Brown, K.H. Foulkrod, S. Dworaczyk, et al., Barriers to obtaining family consent for potential organ donors, The Journal of Trauma 68 (2) (2010) 447–451. Available from https://doi.org/10.1097/TA.0b013e3181caab8f.

[20] B.-J. Crigger, AMA Code of medical ethics' opinions relevant to organ transplantation and procurement, AMA Journal of Ethics 18 (2) (2016) 122–125. Available from https://doi.org/10.1001/journalofethics.2016.18.2.coet1-1602.

[21] L. Semrau, Exploitation, coercion, and other problems with kidney donation, Think 22 (65) (2023) 1–10. Available from https://doi.org/10.1017/S1477175623000349.

[22] Guiding principles on human organ transplantation, The Lancet 337 (8755) (1991) 1470–1471. Available from https://doi.org/10.1016/0140-6736(91)93150-8.

[23] WHO, WHO guiding principles on human cell, tissue and organ transplantation1, Transplantation 90 (3) (2010) 229–233. Available from https://doi.org/10.1097/TP.0b013e3181ec29f0.

[24] International summit on transplant tourism and organ trafficking, The declaration of istanbul on organ trafficking and transplant tourism, Clinical journal of the American Society of Nephrology: CJASN 3 (5) (2008) 1227–1231. Available from https://doi.org/10.2215/CJN.03320708.

[25] La Moncloa, Spain reaffirmed its global leadership in organ donation during 2019 and now accounts for 20% of EU donors and 6% worldwide, La Moncloa. Available from: https://www.lamoncloa.gob.es/lang/en/gobierno/news/Paginas/2020/20200907organ-donation.aspx, 2020.

[26] K.G. Zirpe, P. Suryawanshi, S. Gurav, A. Deshmukh, P. Pote, A. Tungenwar, et al., Increase in cadaver organ donation rate at a tertiary care hospital: 23 years of experience, Indian journal of critical care medicine: peer-reviewed, official publication of Indian Society of Critical Care Medicine 24 (9) (2020) 804–808. Available from https://doi.org/10.5005/jp-journals-10071-23578.

[27] K. Guleria, A. Singh, B. Kumar, et al., Trends of organ donation and awareness in Ernakulam, Kerala, BMC Proceedings 6 (Suppl 4) (2012) P48. Available from https://doi.org/10.1186/1753-6561-6-S4-P48.

[28] P. Abbasi, J. Yoosefi Lebni, P. Nouri, A. Ziapour, A. Jalali, The obstacles to organ donation following brain death in Iran: a qualitative study, BMC Medical Ethics 21 (1) (2020) 83. Available from https://doi.org/10.1186/s12910-020-00529-8.

[29] B. Aguilera, D. DeGrazia, A. Rid, Regulating international clinical research: an ethical framework for policy-makers, BMJ Global Health 5 (2020) e002287.

[30] K.W. Kizer, R.A. English, M. Hackmann, National Academies of Sciences, Engineering, and Medicine, Committee on a fairer and more equitable, cost-effective, and transparent system of donor organ procurement, allocation, and distribution, National Academies of Sciences, Engineering, and Medicine; Health and Medicine Division, (2022).

[31] R. Beyar, Challenges in organ transplantation, Rambam Maimonides medical journal 2 (2) (2011) e0049. Available from https://doi.org/10.5041/RMMJ.10049.

[32] Government of India, The Transplantation of Human Organs and Tissues Act, 1994, Government of India. Available from: https://www.indiacode.nic.in/bitstream/123456789/19260/1/thoa_act_1994.pdf, 1994.

[33] Government of India, The Transplantation of Human Organs (Amendment) Act, 2011, Government of India. Available from: https://www.india.gov.in/transplantation-human-organs-amendment-act-2011, 2011.

[34] Ministry of Health and Family Welfare, Directorate General of Health Services, Transplantation of Human Organs and Tissues Act (THOTA), National Organ Transplant Program (NOTP) including NOTTO/ROTTOs/SOTTOs, Government of India. Available from: https://dghs.mohfw.gov.in/mainsitedghs/uploads/assets/5dNgS3HryhmaL3GL47bkDozwl5GjkWW2ApJcz8Z1.pdf, 2020.

[35] Ministry of Health and Family Welfare, National Organ and Tissue Transplant Organization, Guidelines for implementation of National Organ Transplant Program (Period: 2021–22 to 2025–26), *Government of India*. Available from: https://notto.mohfw.gov.in/WriteReadData/Portal/News/791_1_GUIDELINES_04_oct_2021__Approved_.pdf, 2021.

[36] Organ Procurement and Transplantation Network, U.S. Department of Health and Human Services, Health Resources and Services Administration, Ethical principles in the allocation of human organs, Organ Procurement and Transplantation Network. Available from: https://optn.transplant.hrsa.gov/professionals/by-topic/ethical-considerations/ethical-principles-in-the-allocation-of-human-organs/, 2025.

[37] European Commission, Proposal for a Directive of the European Parliament and of the Council on the Union code relating to medicinal products for human use, and repealing Directive 2001/83/EC and Directive 2009/35/EC (COM(2023) 192 final), EUR-Lex. Available from: https://eur-lex.europa.eu/legal-content/EN/TXT/?uri=CELEX%3A52023PC0192, 2023.

[38] Department of Health, Explanatory notes to Human Tissue Act 2004: Chapter 30, Legislation.gov.uk, (2004). Available from: https://www.legislation.gov.uk/ukpga/2004/30/notes/division/7

[39] R. Matesanz, B. Miranda, Organ donation: the "Spanish model", Transplantation Proceedings 28 (1) (1996) 11.

[40] V. Shrivastav, Y. Murugan, R. Gandhi, J. Nagda, Knowledge, attitudes, and practices regarding organ donation among medical students in India: a mixed methods study, Cureus 16 (3) (2024) e56136. Available from https://doi.org/10.7759/cureus.56136.

[41] United Network for Organ Sharing, Theological perspective on organ and tissue donation, UNOS. Available from: https://unos.org/transplant/facts/theological-perspective-on-organ-and-tissue-donation, 2025.

[42] NHS Blood and Transplant, Islam and organ donation, NHS. Available from: https://nhsbtmediaservices.blob.core.windows.net/organ-donation-assets/pdfs/islam_and_organ_donation.pdf, 2025.

[43] A.I. Padela, J. Auda, The moral status of organ donation and transplantation within Islamic Law: the *Fiqh* Council of North America's position, Transplantation direct 6 (3) (2020) e536. Available from https://doi.org/10.1097/TXD.0000000000000980.

[44] John Paul II, P., Address of the Holy Father John Paul II to the 18th International Congress of the Transplantation Society, The Holy See. Available from: https://www.vatican.va/content/john-paul-ii/en/speeches/2000/jul-sep/documents/hf_jp-ii_spe_20000829_transplants.html, 2000, August 29.

[45] Archdiocese of Milwaukee, Catholic teaching on organ donation, Archdiocese of Milwaukee. Available from: https://www.archmil.org/offices/social-justice/Catholic-Social-Teaching/Catholic-Teaching-on-Organ-Donation.htm, 2025.

[46] S. Trasancos, When is organ donation moral?, Catholic Answers Magazine. Available from: http://catholic.com/magazine/online-edition/when-is-organ-donation-moral, 2018, June 13.

[47] M. Oliver, A. Woywodt, A. Ahmed, I. Saif, Organ donation, transplantation and religion, Nephrology Dialysis Transplantation 26 (2) (2011) 437–444. Available from https://doi.org/10.1093/ndt/gfq628.

[48] B.S. Manvitha, Gender disparities in organ donation and transplantation in India: a call for equality, Centre for Human Rights and Legal Pluralism, McGill University. Available from: https://www.mcgill.ca/humanrights/article/gender-disparities-organ-donation-and-transplantation-india-call-equality, 2024, May 14.

[49] E. Yee, S.M. Hosseini, B. Duarte, S.M. Knapp, M. Carnes, B. Young, et al., Sex disparities in organ donation: finding an equitable donor pool, Journal of the American Heart Association 10 (19) (2021) e020820.

[50] N. Joe, S.T. Chakala, F. Fathima, N. Shankar, Awareness & perceptions about organ donation among patient attendants in a tertiary-care hospital in South India: an observational study, The Indian Journal of Medical Research 160 (6) (2025) 552.

[51] OnlyMyHealth, Dispelling misconceptions about organ donation in India, OnlyMyHealth. Available from: https://www.onlymyhealth.com/dispelling-misconceptions-about-organ-donation-in-india-1694019330, 2023.

[52] CARE Hospitals, Doctor dispels misconceptions about organ donation in India, CARE Hospitals. Available from: https://www.carehospitals.com/news-media-detail/doctor-dispels-misconceptions-about-organ-donation-in-india, 2025.

[53] A.A.K. Sayyed, S.A. Pratinidhi, D.S.J. D'Souza, S.T. Dhanmane, A study of religious beliefs affecting organ donation and influence of educational intervention in western Maharashtra, International Journal of Clinical Biochemistry and Research 8 (2) (2021) 123–128. Available from https://doi.org/10.18231/j.ijcbr.2021.024.

Abbreviations

2D	Two-Dimensional
3D	Three-Dimensional
ADAMTSs	A Disintegrin and Metalloproteinase with Thrombospondin Motifs
AFM	Atomic Force Microscopy
AI	Artificial Intelligence
AIIMS	All India Institute of Medical Sciences
ALP	Alkaline Phosphatase
AM	Additive Manufacturing
ATP	Adenosine Triphosphate
AU	Author Query
BCE	Before Common Era
BIO300	Nanoparticle Formulation of Genistein
BMJ	British Medical Journal
BMP/BMPs	Bone Morphogenetic Protein(s)
BMSC	Bone Marrow-Derived Mesenchymal Stem Cell
BrdU	5-Bromo-2′-Deoxyuridine (Bromodeoxyuridine)
CAD	Computer-Aided Design
CAF	Cancer-Associated Fibroblast
CCK-8	Cell Counting Kit-8
CD	Cluster of Differentiation
C_{60}	Fullerene (Carbon-60 Buckminsterfullerene)
cm	Centimeter
CMAJ	Canadian Medical Association Journal
CNTs	Carbon Nanotubes
CO_2	Carbon Dioxide
COVID-19	Coronavirus Disease 2019
CRISPR	Clustered Regularly Interspaced Short Palindromic Repeats
CRP	C-Reactive Protein
CS	Chitosan
CT	Computed Tomography
DBT	Department of Biotechnology
DDS	Drug Delivery Systems
DNA	Deoxyribonucleic Acid
DOI	Digital Object Identifier
DRDO	Defence Research and Development Organisation
DMA	Dynamic Mechanical Analysis
ECM	Extracellular Matrix
ECO-COMPASS	Ecological and Cost-Effective Composites for Future Aircraft Applications
EdU	5-Ethynyl-2′-Deoxyuridine
ED	Editor Query
EASA	Electrochemically Active Surface Area
ELISA	Enzyme-Linked Immunosorbent Assay
ESC/hESC	Embryonic Stem Cell/Human Embryonic Stem Cell
EthD-1	Ethidium Homodimer-1
EU	European Union
F/g	Farads Per Gram

FDA	Food and Drug Administration
FDM	Fused Deposition Modeling
FGF/FGFs	Fibroblast Growth Factor(s)
FN	Fibronectin
Fmoc-EF-NH_2	Fluorenylmethoxycarbonyl-Glutamic Acid-Phenylalanine Amide
FRET	Förster/Fluorescence Resonance Energy Transfer
FTTE	Fundamentals of Translational Tissue Engineering
GAGAGA	Glycine-Alanine-Glycine-Alanine-Glycine-Alanine
GAGAGS	Glycine-Alanine-Glycine-Alanine-Glycine-Serine
GAGAGY	Glycine-Alanine-Glycine-Alanine-Glycine-Tyrosine
GBR	Guided Bone Regeneration
GF	Gas Foaming
GFP	Green Fluorescent Protein
g	Gram/Gravitational Acceleration
h	Hour(s)
HA	Hyaluronic Acid
HAp	Hydroxyapatite
HAVIC	Human Aortic Valve Interstitial Cell
HCl	Hydrochloric Acid
HFF	Human Foreskin Fibroblast
HGS	High-Grade Serous
HIF	Hypoxia-Inducible Factor
HNF-4	Hepatocyte Nuclear Factor 4
HTA	Human Tissue Authority
iPSC/iPSCs	Induced Pluripotent Stem Cell(s)
IL	Interleukin
ISS	International Student Series
kDa	Kilodalton
LiBr	Lithium Bromide
LFP	Latent Fingerprint/Luminescent Fingerprint Powder
LMIC/LMICs	Low- and Middle-Income Country/Countries
LOX	Lysyl Oxidase
M	Molar
MA	Massachusetts
mg	Milligram
min	Minute(s)
mL	Milliliter
ML	Machine Learning
mm	Millimeter
MMPs	Matrix Metalloproteinases
MN	Minnesota
MPS	Macmillan Publishing Solutions
MSC	Mesenchymal Stem Cell/Mesenchymal Stromal Cell
mW/m^2	Milliwatts Per Square Meter
Na_2CO_3	Sodium Carbonate
NaCl	Sodium Chloride
NaOH	Sodium Hydroxide
NASA	National Aeronautics and Space Administration
NCI	National Cancer Institute
NCSC	Neural Crest Stem Cell
NEJM	New England Journal of Medicine
NGO	Non-Governmental Organization
nm	Nanometer
NOx	Nitrogen Oxides
NOTA	National Organ Transplant Act
NOTTO	National Organ and Tissue Transplant Organization
OoC	Organ-on-a-Chip
ONT	Organización Nacional de Trasplantes

OPTN	Organ Procurement and Transplantation Network
P25	Glycoprotein P25
PAHs	Polycyclic Aromatic Hydrocarbons
PCL	Poly(ε-Caprolactone)/Polycaprolactone
PDGF	Platelet-Derived Growth Factor
PDLLA	Poly(D,L-lactide)
PEG	Polyethylene Glycol
PGA	Polyglycolic Acid
pH	Potential of Hydrogen
PLA	Polylactic Acid
PLGA	Poly(Lactic-co-Glycolic Acid)
PLLA	Poly(L-Lactic Acid)
PLOD2	Procollagen-Lysine,2-Oxoglutarate 5-Dioxygenase 2
PVA	Polyvinyl Alcohol
PVP	Polyvinylpyrrolidone
qPCR	Quantitative Polymerase Chain Reaction
RGD	Arginine-Glycine-Aspartate Acid
RhoA	Ras Homolog Family Member A
RNA	Ribonucleic Acid
rpm	Revolutions Per Minute
RUNX2	Runt-Related Transcription Factor 2
S.No.	Serial Number
Ser	Serine
SEM	Scanning Electron Microscopy
SLA	Stereolithography
SLS	Selective Laser Sintering
SMC	Smooth Muscle Cell
SNAP-25	Synaptosomal-Associated Protein 25
SOX9	SRY-Box Transcription Factor 9
SOx	Sulfur Oxides
SOTTO	State Organ and Tissue Transplant Organization
TAMs	Tumor-Associated Macrophages
TE	Tissue Engineering
TEER	Transepithelial/Transendothelial Electrical Resistance
TGF-β	Transforming Growth Factor-Beta
THOA	Transplantation of Human Organs Act
THOTA	Transplantation of Human Organs and Tissues Act
TIPS	Thermal-Induced Phase Separation
TME	Tumor Microenvironment
TNF-α	Tumor Necrosis Factor-Alpha
Tregs	Regulatory T Cells
TGRGDSPA	Threonine-Glycine-Arginine-Glycine-Aspartate-Serine-Proline-Alanine
TS/TY	Typesetter Query
TTS	The Transplantation Society
TWIST1	Twist Family BHLH Transcription Factor 1
UK	United Kingdom
UN	United Nations
US/USA	United States/United States of America
UV	Ultraviolet
VEGF	Vascular Endothelial Growth Factor
VIC	Valve Interstitial Cell
VOCs	Volatile Organic Compounds
WST-1	Water-Soluble Tetrazolium-1
wt.%	Weight Percent
XTT	2,3-Bis(2-Methoxy-4-Nitro-5-sulfophenyl)-2H-Tetrazolium-5-Carboxanilide
μm	Micrometer (Micron)
VML	Volumetric Muscle Loss
ECM	Extracellular Matrix

PCL	Polycaprolactone
PU	Polyurethane
PEEK	Polyether Ether Ketone
PMMA	Polymethyl Methacrylate
ESC	Embryonic Stem Cells
BMP-2	Bone Morphogenetic Protein 2
iPSCs	Pluripotent Stem Cells
EPCs	Endothelial Progenitor Cells
QC	Quality Control
GMP	Good Manufacturing Practices
PNIPAAm	Poly(N-Isopropylacrylamide)
ABMs	Agent-Based Models
PDEs	Partial Differential Equations
FEA	Finite Element Analysis
MD	Molecular Dynamics
GEO	Gene Expression Omnibus
PDB	Protein Data Bank
FDA	Food and Drug Administration
RMAT	Regenerative Medicine Advanced Therapy
CBER	Centre for Biologics Evaluation and Research
EMA	European Medicines Agency
ATMPs	Advanced Therapy Medicinal Products
CAT	Committee for Advanced Therapies
PMDA	Pharmaceuticals and Medical Devices Agency
NSCs	Neural Stem Cells
GLP	Good Laboratory Practice
OECD	Organization for Economic Cooperation and Development
GCP	Good Clinical Practice
IB	Investigational Brochure
CTP	Clinical Trial Protocol
CRFs	Case Report Forms
SDS	Sodium Dodecyl Sulfate
MTT	3-(4,5-Dimethylthiazol-2-yl)-2,5-Diphenyltetrazolium Bromide

Index

Note: Page numbers followed by "*f*" and "*t*" refer to figures and tables, respectively.

D

E

F

G

T

9780443334559